AF884971

Sonderheft 6 zu Band 24, September 2001

Offizielles Organ der Deutschen Gesellschaft für Hämatologie und Onkologie, der Österreichischen Gesellschaft für Hämatologie und Onkologie, der Österreichischen Krebsgesellschaft – Krebsliga unter Fortführung der «Österreichischen Zeitung für Onkologie»

ONKOLOGIE

• International Journal for Cancer Research and Treatment

Schriftleitung

W. Queißer, Mannheim
W. Scheithauer, Wien

Fachschriftleitung

C. Bokemeyer, Tübingen
Kommentare
Th. Cerny, St. Gallen
Internistische Onkologie
R. Hartenstein, München
Kasuistik
R. Kreienberg, Ulm
Gynäkologische Onkologie
K. Mross, Freiburg i.Br.
Klinisch-experimentelle Onkologie
H. Rübben, Essen
Urologische Onkologie
P.M. Schlag, Berlin
Chirurgische Onkologie
Th. Wendt, Jena
Radioonkologie

Wissenschaftlicher Beirat

Internistische Onkologie
L. Bergmann, Frankfurt/M.
V. Diehl, Köln
P. Drings, Heidelberg
A. Engert, Köln
H. H. Fiebig, Freiburg i.Br.
D. Hoelzer, Frankfurt/M.
H. Huber, Wien
H. J. Illiger, Oldenburg
U. R. Kleeberg, Hamburg
G. Kornek, Wien
R. Morant, St. Gallen
M. Pfreundschuh, Homburg/Saar
K. Possinger, Berlin
H. Samonigg, Graz
H.-J. Schmoll, Halle

Chirurgische Onkologie
D. Henne-Bruns, Kiel
A. H. Hölscher, Köln
K. H. Link, Ulm
J. Roder, Wasserburg/Inn

Neurochirurgie
G. Schackert, Dresden

Mund-Kiefer-Gesichts-Chirurgie
J. Zöller, Köln

Gynäkologische Onkologie
Th. Bauknecht, Bonn
G. Emons, Göttingen
F. Jänicke, Hamburg
D. G. Kieback, Freiburg i.Br.
M. Neises, Hannover

Urologische Onkologie
J. Breul, München
R. Heicappell, Berlin
G. Mickisch, Rotterdam
T. Otto, Essen
M. Wirth, Dresden

Pädiatrische Onkologie
J. Boos, Münster
M. Brandis, Freiburg i.Br.
H. Gadner, Wien

Dermatoonkologie
D. Schadendorf, Mannheim
W. Tilgen, Homburg/Saar

HNO-Onkologie
F. Bootz, Leipzig
M. Burian, Wien

Radioonkologie
W. Budach, Tübingen
A. Schratter-Sehn, Wien
F. Wenz, Mannheim
T. Wiegel, Berlin

Diagnostische Radiologie in der Onkologie
W. Golder, Berlin

Psychoonkologie
F. A. Muthny, Münster
R. Schwarz, Leipzig

Pathologie
A. Chott, Wien
R. Parwaresch, Kiel

Klinisch-experimentelle Onkologie
A. Hochhaus, Mannheim
Ch. Mannhalter, Wien
R. Schulte-Hermann, Wien

Biometrie
L. Edler, Heidelberg
M. Schumacher, Freiburg i.Br.

Onkologische Rehabilitation
H. H. Bartsch, Freiburg i.Br.

Onkologische Palliativmedizin
I. Strohscheer, Berlin

KARGER

Basel · Freiburg · Paris · London · New York · New Delhi · Bangkok · Singapore · Tokyo · Sydney cell pharm

Erfahrung & Innovation

Die Abstracts auf CD-ROM - exklusiv bei cell pharm

Auf dem Stand Nr. 211 der cell pharm GmbH erhalten Sie auch in diesem Jahr **Ihre persönliche CD-ROM** mit den Abstracts des Kongresses zum Nachschlagen und Recherchieren - völlig kostenlos!

Bibliographische Dienste

Biological Abstracts
Current Contents/Clinical Medicine
Excerpta Medica/EMBASE

Index Medicus/MEDLINE

Medical Documentation Service
Reference Update
Research Alert
Science Citation Index
SCISEARCH Database

Abstracts und Autorenverzeichnis im vorliegenden Sonderheft der Zeitschrift ONKOLOGIE wurden aus den gelieferten Daten im Verlag programm- und drucktechnisch verarbeitet. Auf den Inhalt der Abstracts und der Verzeichnisse hatte der Verlag keinen Einfluss und nahm keine redaktionelle Bearbeitung bzw. Lektorierung vor. Für den Inhalt von Abstracts und Verzeichnissen sind Autoren und Kongressorganisation verantwortlich.

S. Karger · Medical and Scientific Publishers · Basel · Freiburg · Paris · London · New York · New Delhi · Bangkok · Singapore · Tokyo · Sydney

Die Zeitschrift erscheint zweimonatlich; pro Jahr erscheint 1 Band zu je 6 Heften. Mitglieder der Deutschen Gesellschaft für Hämatologie und Onkologie sowie der Österreichischen Gesellschaft für Hämatologie und Onkologie erhalten die Zeitschrift im Rahmen ihrer Mitgliedschaft. Der Bezug ist im Rahmen des Mitgliedsbeitrags abgegolten. Bezugspreis für Jahrgang 24, 2001, DEM 238,– / CHF 179,–, einschließlich MWSt., zuzüglich Postgebühren. Der Abonnementpreis ist im voraus zahlbar. Das Abonnement der Zeitschrift läuft weiter, wenn es nicht spätestens 4 Wochen vor Abschluß eines Bandes abbestellt wird. Abonnementbestellungen können bei jeder Buchhandlung oder direkt beim Verlag aufgegeben werden:

Deutschland:
S. Karger GmbH
Lörracher Str. 16a
D-79115 Freiburg
Tel. +49 761 45 20 70
Fax +49 761 4 52 07 14
E-mail Information@Karger.de

Übrige Länder:
S. Karger AG
Allschwilerstr. 10
Postfach
CH-4009 Basel
Tel. +41 61 3 06 11 11
Fax +41 61 3 06 12 34
E-mail Karger@Karger.ch

Anzeigen:
S. Karger
Verlag für Medizin und Naturwissenschaften GmbH,
Lörracher Str. 16a, D-79115 Freiburg,
Tel. +49 761 45 20 70
Gültig ist die Preisliste Nr. 11 vom 1. Januar 2001.

Für den Inhalt außerhalb des redaktionellen Teiles (insbesondere Anzeigen, Industrieinformationen, Pressezitate und Kongreßinformationen) übernehmen Schriftleitung, Beirat und Verlag keine Gewähr.
Eine Markenbezeichnung kann warenzeichenrechtlich geschützt sein, auch wenn bei ihrer Verwendung in dieser Zeitschrift das Zeichen ® oder ein anderer Hinweis auf etwa bestehende Schutzrechte fehlen sollte. Für Satz- und Übertragungsfehler, insbesondere bei Dosierungsangaben, wird keine Gewähr übernommen.
Die Zeitschrift sowie alle in ihr enthaltenen einzelnen Beiträge und Abbildungen sind urheberrechtlich geschützt. Jede Verwertung, die nicht ausdrücklich vom Urheberrechtsgesetz zugelassen ist, bedarf der vorherigen Zustimmung des Verlags. Das gilt insbesondere für Vervielfältigungen, Bearbeitungen, Übersetzungen, Mikroverfilmungen und die Einspeicherung und Verarbeitung in elektronischen Systemen.

Sonderheft 6 zu Band 24, September 2001
ISBN 3-8055-7332-4

© Copyright 2001 by S. Karger
Verlag für Medizin und Naturwissenschaften GmbH,
Lörracher Str. 16a, D-79115 Freiburg

Verlagsleitung und presserechtlich verantwortlich:
Sibylle Hopf
Anzeigenleitung: Susanne Meister
Produktionsleitung: Georg Brunner

Gesamtherstellung: KONKORDIA GmbH, Bühl
Das Medienunternehmen

KARGER

Fax +49 761 4 52 07 14
E-mail Information@Karger.de
www.karger.com

© 2001 S. Karger GmbH, Freiburg

Home Page der Zeitschrift:
www.karger.com/journals/onk

ONKOLOGIE

Sonderheft 6 zu Band 24,
September 2001

Gemeinsame Jahrestagung der Deutschen und Österreichischen Gesellschaften für Hämatologie und Onkologie

DGHO
Deutsche Gesellschaft für Hämatologie und Onkologie e.V.

Mannheim, 30.9. bis 3.10.2001

ABSTRACTS

Herausgeber:

R. Hehlmann,
W. Queißer, Mannheim

KARGER

Basel · Freiburg · Paris · London · New York ·
New Delhi · Bangkok · Singapore · Tokyo · Sydney

Kongress-Präsident: Prof. Dr. Rüdiger Hehlmann

Ko-Präsident: Prof. Dr. Wolfgang Queißer

Redaktion: Dr. Ute Berger (Kongress-Sekretärin)
PD Dr. Andreas Hochhaus (Kongress-Sekretär)

Dr. Georg Engelich
Dr. Ole Maywald
Dr. Alexander Schreiber

III. Medizinische Universitätsklinik
Fakultät für Klinische Medizin Mannheim der Universität Heidelberg
Wiesbadener Strasse 7-11
68305 Mannheim, Deutschland

Geschützte Warennamen (Warenzeichen) werden nicht besonders kenntlich gemacht. Aus dem Fehlen eines solchen Hinweises kann also nicht geschlossen werden, dass es sich um einen freien Warennamen handele.

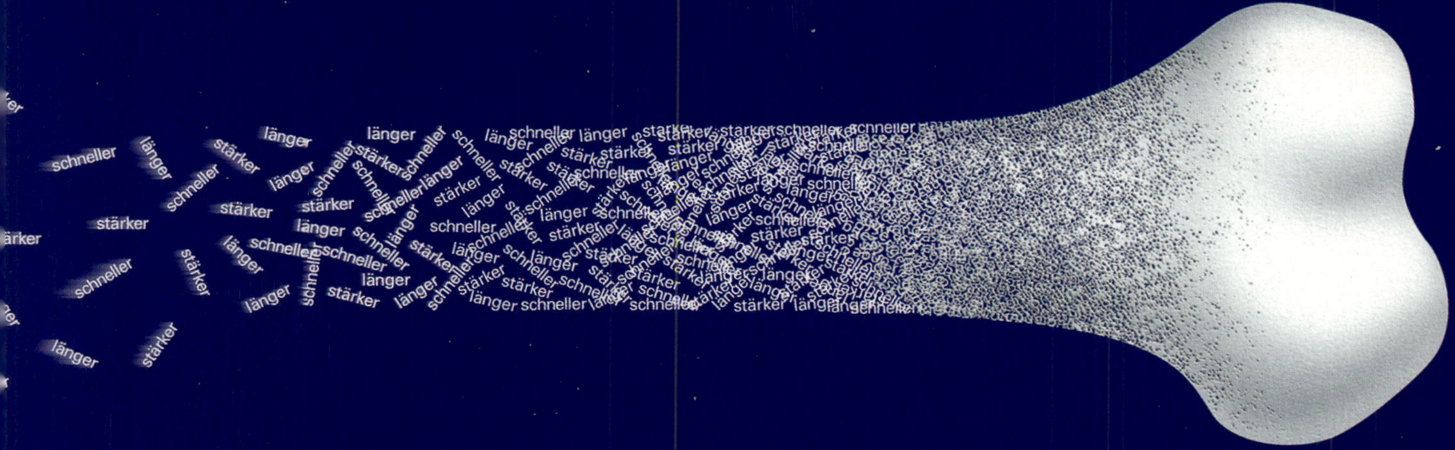

neu ZOMETA®

Schnellerer Wirkungseintritt*
Stärkere klinische Wirksamkeit*
Längere Wirkdauer*

Eine Klasse für sich.

ZOMETA® 4 mg Pulver und Lösungsmittel zur Herstellung einer Infusionslösung Wirkstoff: Zoledronsäure **Zusammensetzung:** 1 Durchstechflasche mit Pulver enthält 4 mg Zoledronsäure (wasserfrei), entsprechend 4,264 mg Zoledronsäuremonohydrat, sowie Mannitol und Natriumcitrat. 1 Ampulle mit Lösungsmittel enthält 5 ml Wasser für Injektionszwecke. **Anwendungsgebiete:** Behandlung der Tumor-induzierten Hyperkalzämie (TIH). **Gegenanzeigen:** Schwangerschaft und Stillzeit sowie Patienten mit klinisch bedeutsamer Überempfindlichkeit gegenüber Zoledronsäure, anderen Bisphosphonaten oder einen der sonstigen Bestandteile von ZOMETA®. Keine Therapieempfehlung bei Patienten mit schwerer Nieren- oder Leberfunktionsstörung, es sei denn die Vorteile einer Behandlung überwiegen die potenziellen Risiken. Keine Erfahrungen mit ZOMETA® bei Kindern. **Nebenwirkungen:** Die Nebenwirkungen von ZOMETA® sind vergleichbar mit denjenigen anderer Bisphosphonate. Am häufigsten kommt es zu einem Anstieg der Körpertemperatur. Auch über grippeähnliche Symptome mit Fieber und Knochenschmerzen wurde berichtet. Gelegentlich traten Arthralgie, Müdigkeit, Verwirrung, Durst, Panzytopenie und Bradykardie auf. Es gibt keine Informationen darüber, ob diese Nebenwirkungen reversibel sind. Die verminderte Kalziumausscheidung im Urin geht häufig mit einer nicht behandlungsbedürftigen Abnahme des Serumphosphatspiegels einher. Der Serumkalziumspiegel kann auf asymptomatische, hypokalzämische Werte abfallen. Gastrointestinale Beschwerden, wie Übelkeit und Erbrechen, sowie lokale Reaktionen an der Einstichstelle, wie Rötung und Schwellung, traten ebenso auf wie einige Fälle von Hautausschlag, Juckreiz, Schmerzen in der Brust, Thrombozytopenie, Kopfschmerzen und Geschmacksstörungen. Wie bei anderen Bisphosphonaten wurden Fälle von Konjunktivitis und Hypomagnesiämie beschrieben. Es gibt einige Berichte über das Auftreten von Nierenfunktionsstörungen bis hin zum akuten Nierenversagen (sehr selten), zu denen jedoch auch andere Risikofaktoren in dieser schwerkranken Patientenpopulation beigetragen haben können. **Wechselwirkungen:** Es sind keine Wechselwirkungen mit häufig angewendeten antitumorösen Arzneimitteln sowie mit Diuretika, Antibiotika und Analgetika bekannt. Zoledronsäure wird nur unwesentlich an Plasmaproteine gebunden und hemmt *in vitro* keine humanen P450-Enzyme. Spezielle klinische Studien zu Wechselwirkungen wurden jedoch nicht durchgeführt. Vorsicht bei gleichzeitiger Gabe von Aminoglycosiden, weil beide Substanzklassen einen additiven Effekt zeigen können, der zu einem niedrigeren Serumkalziumspiegel über einen länger als erforderlichen Zeitraum führen kann. Es ist auch auf eine möglicherweise während der Behandlung auftretende Hypomagnesiämie zu achten. **Dosierung, Art und Dauer der Anwendung:** *Erwachsene und ältere Patienten:* Die Gesamtdosis eines Behandlungsganges bei Hyperkalzämie (Albumin-korrigierter Serumkalziumspiegel (12 mg/dl oder 3,0 mmol/l) beträgt 4 mg. Die rekonstituierte ZOMETA®-Infusionslösung wird mit 50 ml 0,9 % w/v Natriumchlorid- oder 5 % w/v Glukoselösung weiterverdünnt und in einer einzigen 15-minütigen intravenösen Infusion verabreicht. Vor und im Anschluss an die Anwendung von ZOMETA® muss bei den Patienten eine ausreichende Hydratation gewährleistet sein. Weitere Angaben siehe Fachinformation. Verschreibungspflichtig. **Darreichungsformen und Packungsgrößen:** Packung mit 1 Durchstechflasche + 1 Ampulle mit 5 ml Lösungsmittel (N1), Packung mit 4 Durchstechflaschen + 4 Ampullen mit je 5 ml Lösungsmittel (N2). Klinikpackung mit 10 Durchstechflaschen + 10 Ampullen mit je 5 ml Lösungsmittel Stand: März 2001. Novartis Pharma GmbH, 90327 Nürnberg. www.pharma.de.novartis.com

neu ZOMETA®
Zoledronsäure

NOVARTIS
ONCOLOGY

Schneller, stärker, länger

*Major P, Lortholary A, Hon J et al. (2001) Zoledronic Acid is Superior to Pamidronate in the Treatment of Hypercalcemia of Malignancy: A Pooled Analysis of Two Randomized, Controlled Clinical Trials. *Journal of Clinical Oncology* 19, 558-567

Contents · Inhalt

ONKOLOGIE

Vol. 24, Sonderheft 6, September 2001

EPI-cell® 10 mg / 20 mg / 50 mg / 200 mg
Wirkstoff: Epirubicinhydrochlorid, verschreibungspflichtig.
Zusammensetzung: arzneilich wirksame Bestandteile: 1 Durchstechfl. EPI-cell® 10 mg/20 mg/50 mg/200 mg mit 5/10/25/100 ml Inj.-Lösung enth.: 10/20/50/200 mg Epirubicinhydrochlorid (2 mg/ml). Sonst. Bestandteile: Natriumchlorid, Natriumlactat und Salzsäure zur Pufferung, Wasser für Inj.-Zwecke. **Anwendungsgebiete:** Mamma-CA, fortgeschrittenes Ovarial-CA, fortgeschrittenes Magen-CA, kleinzell. Bronchial-CA, fortgeschrittenes Weichteilsarkom. **Gegenanzeigen:** ausgepr. Knochenmarkdepression; ausgepr. Mukositis/Stomatitis; akute Infektionen; ausgepr. Beeintr. der Leberfunktion; muskuläre Herzinsuffizienz Grad III/IV (z. B. nach Infarkt); akuter Myokardinfarkt; akute entzündl. Herzerkr.; Rhythmusstör. mit gravierenden hämodynam. Auswirkungen, auch in der Vorgeschichte; vorausgegangene Behand. mit Epirubicin oder and. Anthrazyklinen bis zur max. kumulativen Dosis; Überempfindlichk. gg. Epirubicin oder and. Anthrazyklinen; Schwangerschaft, Stillzeit. Pat. mit and. Stadien der angeführten Herzerkr. u./o. einer Vorbehandl. mit and. Anthrazyklinen bedürfen einer individuellen therapeut. Entscheidung, einschl. Verlaufskontrolle. Hinweise: besond. Vorsicht bei Pat. mit vorangegang., gleichzeit. o. geplanter Radiother., da erhöhtes Risiko von Lokalreakt. im Bestrahlungsfeld. Vorangegang. Bestrahlung des Mediastinums erhöht Kardiotox. Kardiales Monitoring ab Epirubicin-Dosen >450 mg/m² KOF; max. kumulative Dosis sollte (900)-1000 mg/m² KOF nicht überschreiten. Keine Impfung mit Lebendvakzinen im zeitl. Zusammenhang. Kontakt des Pat. mit Polio-Impflingen vermeiden. EPI-cell® nicht oral, s.c., i.m., intrathekal verabreichen. Epirubicin kann erbgut-, fruchtschädigend, krebsauslösend wirken. Schwangerschaft vor Behandl. ausschließen bzw. während/bis zu 6 Monate danach vermeiden, auch bei Behandl. des männl. Partners. Tritt Schwangerschaft während der Behandl. ein, auf mögl. Schädigungen des Kindes/genet. Beratung hinweisen. Beratung über Spermakonservierung, da Infertilität mögl. **Nebenwirkungen:** blutbildendes System: häufig (>5%): Myelosuppression, Leuko-, Granulozyto-, Neutro-, Thrombozytopenie, Anämie, Fieber, Infektionen. Weniger häufig (<5%): Sepsis, sept. Schock, Hämorrhagien, Gewebshypoxie. Blutbildkontrollen! Einzelfälle von sek. Leukämie bei gleichzeit. Therapie mit DNA verändernden Zytostatika. Kardiotoxizität: dosisunabhängiger Soforttyp (im allgem. reversibel): Arrhythmien (supra-/ventrikuläre Extrasystolen, ventrikuläre Tachy-, Sinustachy-, Bradykardie, AV-Block, Schenkelblöcke) u./o. unspezif. EKG-Veränd. (ST-Streckenveränd., Niedervoltage, verläng. QT-Zeit). Dosisabhängiger Spättyp: Kardiomyopathie/dekomp. Herzinsuffizienz (Dyspnoe, Ödeme, Lebervergrößerung, Aszites, Lungenödeme, Pleuraergüsse, Galopprhythmus); Risiko hierfür stark ansteigend bei kumulativer Dosis von 900-1000 mg/m². Bei kardialen Risikofaktoren niedrigere kumulative Dosis kardiotox. Manifestation meist während/2-3 Monate nach Behandl., auch mehrere Monate/Jahre danach mögl. Herzfunktion kontrollieren! Perikardergüsse möglich. GI-Trakt: häufig (>5%): Übelkeit, Erbrechen, Diarrhöe, Dehydratation, Mukositis (v. a. Stomatitis) mit Schmerzen, brennendem Gefühl, Erythemen, Erosionen, Ulzerationen, Blutungen. Weniger häufig (<5%): Ösophagitis, Appetitlosigk., Hyperpigmentier. der Mundschleimhaut. Haut, allerg. Reaktionen: häufig (>5%): Alopezie (meist reversibel). Weniger häufig (<5%): Rötungen, Hyperpigmentier. der Haut/Nägel, Licht- oder Überempfindlichk. bei Bestrahlung, Urtikaria, anaphylakt. Schock. Injektionsstelle: Rötungen häufig, lokale Phlebitis/Thrombophlebitis, Phlebosklerose möglich. Paravenöse Fehlinjektion: lokale Schmerzen, schwere Fettgewebsentzünd., Gewebsnekrosen. Andere NW: Schwäche, Unwohlsein, Kopfschmerzen, Schwindel, Hyperurikämie, Amenorrhöe, Azoospermie, bedeutungslose Rotfärbung des Urins. **Wechselwirkungen:** Zytostatika verstärkten Gesamttox. (insbes. hämatolog./gastrointest.). Kardiotox. AM bzw. Bestrahlung des Mediastinums verstärkten Kardiotox. Knochenmarkfunktion-beeinflussende AM verstärken Hämatotox. Potenziell hepatotox. AM, über Leber metabolisierte AM: verstärkte Tox. u./o. veränderte Wirksamkeit. Verstärkte Hyperurikämie mögl. durch AM, die Harnsäureausscheid. verzögern. Verapamil: erhöhte Clearance von Epirubicin. Epirubicin bindet an Heparin, Wirkungsverluste mögl. **Dosierungs- und Anwendungshinweise:** Anwendung nur durch erfahrene Onkologen. Applikation strikt i.v., cave: Paravasate! Weitere Hinweise s. Fach- und Gebrauchsinformation. **Stand der Information:** März 2001 (EPI-cell® 10/20/50 mg) bzw. August 2001 (EPI-cell® 200 mg)

cell pharm GmbH ○ Feodor-Lynen-Straße 23 ○ 30625 Hannover ○ www.cellpharm.com

Dienstag, 2.10.2001

Mittwoch, 3.10.2001

Secondary Neoplasias following Chemotherapy, Radiotherapy, and Immunosuppression

Editors
U. Rüther
C. Nunnensiek
H.-J. Schmoll

Oncology; Transplantation, Radiology

Complete remissions and survival times of tumor patients are increasing steadily – and so is the risk of developing therapy-associated secondary malignancy. At the same time, we face the growing diagnostic and therapeutic problem of tumorigenesis during post-transplantation immunosuppressive therapy.

The *theoretical section* delves into the various causes of tumorigenesis, with special chapters on secondary chemo- and/or radiotherapy-induced tumorigenesis and neoplasia occurring during posttransplantation immunosuppressive therapy. The risk of individual cytostatic agents causing secondary malignoma is covered, as are the functions and dysfunctions of immunosurveillance. The immune system's methods of recognizing tumors are also examined, and specific information on the role of herpesviruses in organ transplantation is included. The *clinical part* of the book portrays in depth major hematologic and oncologic malignancies that can arise after chemo- and/or radiotherapy administered after organ transplantation.

With this book, clinical hemato-oncologists and those working in transplantation will gain greater insight into the risks, genesis and complications of secondary malignancy – but also into its treatment today and in the future. Last but most certainly not least, the importance of international collaboration for the advancement of diagnosis and therapy is underscored – as is reflected by the international team of contributing authors.

Contents

Preface

Basic Concepts and Pathogenesis

Basic Concepts of Tumor Biology: ***Nordheim, A.; Lüscher, B.***
Herpesviruses in Secondary Tumorigenesis: ***Schwarzmann, F.; Wolf, H.***
Carcinogenicity of Cytostatic Agents – Secondary Malignancies after Chemotherapy: ***Haussmann, W.***
Mechanisms of Action of Immunosuppressants and Immunoinductors: ***Haussmann, W.***
Immune Surveillance and Tumor Recognition: ***Eichhorst, S.T.; Elser, B.***
Malignant Lymphomas and Lymphoproliferative Disorders following Organ Transplantation and Immunomodulatory Therapy: ***Müller-Hermelink, H.K.; Rüdiger, T.***
Chromosomal Aberrations and Molecular Changes in Therapy-Related Leukemias: ***Staratschek-Jox, A.; Bullerdiek, J.***
Secondary Tumorigenesis after Radiotherapy: ***Ammon, J.***
Secondary Primary Cancers after Cancer and Immunosuppressive Chemotherapy and Radiotherapy: ***Soedirman, J.R.***
Secondary Malignancies after Allogeneic or Autologous Marrow Transplantation: ***Witherspoon, R.P.; Deeg, H.J.; Fischer, L.; Schoch, G.; Storb, R.***
Malignancy following Kidney Transplantation: ***Olbricht, C.J.***
Malignancy after Heart and Lung Transplantation: ***Wendler, O.; Schäfers, H.-J.***
Secondary Neoplasms following Liver and Pancreas Transplantation: ***Nashan, B.***

Emerging Therapies and Preventive Measures

Therapy of Secondary Acute Myeloid Leukemia and Secondary Myelodysplastic Syndrome: ***Rüther, U.; Rothe, B.; Nunnensiek, C.***
Immunological Aspects of the Pathogenesis of Tumors Associated with Epstein-Barr Virus: Vaccines for Prevention and Therapy: ***Benninger-Döring, G.; Wolf, H.***
Live Attenuated Varicella Vaccine: ***Gershon, A.A.***
Posttransplant Lymphoproliferative Disorder – New Approaches to Monitoring and Management: ***Leitch, H.A.; Cantarovich, M.; Mangel, J.E.; Laneuville, P.***
Cytoprotection as a New Supportive Strategy in the Treatment of Solid Tumors: ***Büntzel, J.; Glatzel, M.; Fröhlich, D.; Küttner, K.***

Subject Index

KI 01225

www.karger.com/bookseries/coonc

Contributions to Oncology / Beiträge zur Onkologie, Vol. 55
Series Editors: Queißer, W. (Mannheim); Scheithauer, W. (Wien)
ISSN 0250–3220

Secondary Neoplasias following Chemotherapy, Radiotherapy, and Immunosuppression
Editors: Rüther, U. (Ludwigsburg); Nunnensiek, C. (Reutlingen); Schmoll, H.-J. (Halle/Saale)
VIII + 376 p., 38 fig., 8 in color, 22 tab., hard cover, 2000
CHF 198.– / DEM 257.– / USD 172.25
Prices subject to change
DEM price for Germany, USD price for USA only
ISBN 3–8055–7116–X

KARGER

Order Form

Please send: ____ copy/ies
Postage and handling free with prepayment

Payment:
☐ Check enclosed ☐ Please bill me
Please charge to my credit card
☐ American Express ☐ Diners ☐ Eurocard
☐ MasterCard ☐ Visa

Card No.: ____________

Exp. date: ____________

Orders may be placed with any bookshop, subscription agency, directly with the publisher or through a Karger distributor.

Fax: +41 61 306 12 34
S. Karger AG, P.O. Box, CH–4009 Basel (Switzerland)
E-Mail karger@karger.ch, **www.karger.com**

Name/Address: ____________

Date: ____________

Signature: ____________

ONKOLOGIE

Abstracts

Onkologie 2001:24(suppl6):1–235

Sonntag, 30. 9. 2001

Eröffnungssymposium

1

Cancer-risk in persons exposed to radiation from the Chernobyl nuclear accident

R.P. Gale (London, UK, Los Angeles, USA)

The 1986 Chernobyl accident exposed many persons to potentially adverse long-term effects of ionizing radiations. Populations at risk include: (1) workers involved in accident mitigation (emergency [N=600]; recovery [N=6x10^5]); and (2) the general public (evacuees [N=100x10^5] and residents living in contaminated areas [5x10^6]) defined as average 137Cs.ground deposition density >37kBq m^{-2} (1 Ci km^{-2}). These areas are in Belarus, Russian Federation and Ukraine. Acute radiation disease occurred in 134 emergency workers including 93 who received >2.1 Gy whole body external radiation. There were 28 radiation-related deaths. Recovery workers received an average dose of about 100 mSv. There are no convincing data of serious adverse health effects in this population. Power to detect an increased cancer-risk is small. Doses received by the general public came predominantly from 131I (1,760 PBq released) and 137Cs (85 PBq released). 131I was the main contributor to internal thyroid dose; average dose about 0.2 Gy and was about 10-fold higher for children than adults. 137Cs was the main contributor to external and internal whole body dose; average effective dose was about 10 mSv from 1986-96. About 1500 excess cases of thyroid cancer are reported in children exposed to 131I. Different age groups at time of exposure have different risk estimators. In contrast, there are no convincing data of an increased risk of any other cancer in exposed persons. This is not surprising since most persons were exposed to radiation levels comparable to or a few times higher than natural background radiation levels worldwide. These data indicate an increased risk of thyroid cancer but no detectable increase in any other cancer in persons exposed to ionizing radiations from the Chernobyl nuclear accident up to 2000.

2

Twenty years later: HIV and AIDS in 2001

R.C. Gallo (Baltimore, USA)

Human retroviruses are among viruses which play a role in cancer causation. HTLV-1, the first known human retrovirus (1980), is a cause of adult T cell leukemia, a rare malignancy in most parts of the world but relatively common in Japan, certain other areas of Asia, in Africa, and in some black populations of the Americas. Some practical outcomes of HTLV research may be in eventually learning exactly how an environmental factor can cause leukemia and neurological disease. Others include: (a) giving us insights in the early years (1981-83) as to the causative agent of AIDS, another retrovirus, HIV and (b) the capacity of HTLV to immortalize human T cells so they can be continuously grown and studied in the laboratory, a result that sometimes also brought important advances in HIV/AIDS research. In this report, I will focus on the current views of my colleagues and myself of the major problems and needs in HIV/AIDS research, and our approaches to try to solve them.
A broad outline of how HIV infects cells and ultimately results in immune deficiency is now in our hands, and we have also learned that HIV can also act as a co-carcinogen, significantly increasing the incidence of several kinds of cancer. Two important practical advances have also been made in AIDS research: these are the development of the HIV blood test in 1984 and the development of drugs (1990's) which have dramatically inhibited disease progression when judiciously used. These drugs target two HIV encoded enzymes, reverse transcriptase or protease. Nonetheless, several problems are now evident, such as:

1) the toxicity and/or resistance occurring with these drugs,
2) the unavailability of these agents for the bulk of the world,
3) the slowly emerging multi-drug resistant HIV variants, and
4) the horrendous continuous spread of HIV in parts of the world, most notably in Africa.

For these reasons, improved and more practically applicable therapeutic approaches are needed and obviously, so is a preventive vaccine. I believe that novel, practical, less toxic, less prone to viral resistance, and more affordable therapies, as well as better vaccine candidates, will come from: (a) studies related to HIV entry and development of methods to block it; (b) studies of HIV proteins which enter the extracellular space and impair the immune response of uninfected cells, while also sometimes promoting the susceptibility of some of these cells to infection, and finding means to block their effects; and (c) studies of naturally occurring HIV suppressive factors.
As examples of (a) are the new class of inhibitors, already in clinical trials, that target the fusion of the HIV envelope with the cell membrane and those agents which target CCR5, (the HIV co-receptor), mimicking the effects of some of the (?-chemokines) which will soon be in clinical trials. However, these agents will likely also be difficult to bring to the third world. Ultimately, the best approaches may be immunological (therapeutic vaccines) to block entry, and a „spin-off" of the development of such therapies is that they will likely be directly applicable to preventive vaccines. I will highlight progress in our group (particularly by A. DeVico, T. Fouts and G. Lewis) in the development of just a therapeutic & preventive vaccine, which in animal tests induces antibodies which neutralize infection of cells by primary isolates from different clades of HIV, which to my knowledge is the first time this has been achieved.
As an example of (b) I will review our studies in collaboration with D. Zagury (Paris), A. Gringeri (Milan) and A. Burny (Bursseis) and their co-workers on the role of the HIV Tat protein in AIDS pathogenesis, not only as an essential viral protein involved in HIV replication in infected cells, but also as a factor released by these cells which has detrimental effects on nearby uninfected cells, and I will include attempts to block these effects with vaccines against this protein. Again, this would be a therapeutic vaccine but like the above mentioned vaccine which induces neutralizing antibodies, the vaccine against Tat should also be helpful as a component of a preventive vaccine.
As examples of (c) I will summarize our discoveries of several naturally occurring HIV inhibitors, the mechanisms involved, and how after we learn the mechanism of inhibition we can apply the information to the development of novel approaches to therapy.

3

The European response to Mad Cow Disease

R. Kurth (Berlin, D)

Mad Cow Disease or, more scientifically, Bovine Spongiform Encephalopathy (BSE) first appeared in the United Kingdom in the early eighties of last century. It is estimated that approx. one million infected cows entered the food chain in UK until the end of the century. BSE has spread, albeit on a much lower scale, to most countries in Western Europe, including Germany, where until the end of June 2001 more than 80 animals have been diagnosed as BSE positive. BSE is now thought to have most likely originated from a spontaneously occurring TSE index cow some time in the seventies (and not from scrapie of sheep) whose highly infectious offals entered the feed chain as they were recycled as meat and bone meal (MBM). MBM infected additional animals and infectiousness was thus amplified.
There is virtually no doubt that the new variant form of Creutzfeld-Jacob Disease (vCJD) represents BSE in humans, caused by the consumption of contagious beef. The extent and future development of the vCJD epidemic in the UK and elsewhere is still very difficult to quantitate, depends on cofactors and will be discussed. The European Response to BSE/vCJD encompasses a wide variety of measures to ensure an interruption of BSE and vCJD transmission via food, blood and blood products or by medical products and procedures. These precautions will be presented and discussed.

Wissenschaftliches Symposium: Chronische myeloische Leukämie

4

STI571 alone or in combination - perspectives for a cure in CML?

B.J. Druker, M. Mauro, M. O'Dwyer (Portland, USA)

The BCR-ABL tyrosine kinase inhibitor, STI571, has demonstrated significant activity in all phases of CML. Particularly noteworthy are the data in chronic phase patients who have failed interferon therapy. In a non-randomized Phase II study, a 91% complete hematologic response rate, 55% major cytogenetic rate and a 36% complete cytogenetic response rate has been reported. These data compare favorably to the best data reported for interferon in newly diagnosed CML patients. Response rates to STI571 in chronic phase patients are superior to those with more advanced disease, suggesting that as with most agents, the earlier in the disease that one treats with STI571, the better the results. Although data for STI571 in newly diagnosed patients has not yet been reported, higher response rates in patients with even earlier disease would seem likely. As yet, the durability of responses to STI571 are not known, nor it is not known if cytogenetic responses to STI571 will correlate with improved survival. Although it is possible that STI571, through its targeted mechanism of action could delay disease progression from the chronic phase, possibly indefinitely, it seems more likely that if residual BCR-ABL is present, that resistance, as seen in more advanced phase patients will eventually be observed. Currently, the percent of STI571-treated patients who achieve molecular remissions or BCR-ABL levels that are known to correlate with low risk of relapse in interferon-treated patients is not known. Similarly, it is not known if achieving these low levels of molecular positivity will correlate with prolonged survival in STI571-treated patients. However, if one accepts that molecular remissions or low levels of positivity are the goals of therapy if a cure is to be possible, it would seem unlikely that STI571 monotherapy would achieve this in the majority of CML patients. In vitro work has shown that interferon or Ara-C additively to synergistically enhance the anti-proliferative effects of STI571 and clinical trials with these combinations has begun in chronic phase CML patients. The current results with STI571 monotherapy, STI571 plus interferon and STI571 plus Ara-C will be reported and the prospects for improved results with these combinations will be discussed.

5

Will the use of oral Ara-C facilitate CML treatment?

F. Guilhot (Poitiers, F)

Interferon, cytarabine, homoharringtonine and the recent anti-tyrosine kinase STI571, are selected drugs which suppress the Philadelphia positive hematopoiesis. Preliminary clinical experience suggested that a combination of Interferon-alpha and cytarabine improved the results of IFN-alpha treatment alone. An update of our large French trial showed after a median follow-up of 46 months (range 5-82) a 5 years probability of survival of 58% (95%CI : 51-65) for IFN-alpha treated patients and 70% (95%CI : 64-76) for patients receiving IFN-alpha and cytarabine. Significant side effects partially related to the use of subcutaneous injection of ARA-C were responsible for the discontinuation of the treatment. Cytarabine ocfosfate (YNK01) is a prodrug analogue of cytarabine which is resistant to systemic deamination after oral application. Previous small trials suggested an antitumor activity. Therefore we initiated a phase II trial in order to assess the tolerability and the effectiveness of a combination of IFN-alpha2b with YNK01 in recently diagnosed CML chronic phase patients (n=98). The treatment was subdivided into cycles consisting of 4 weeks of continuous administration of IFN ($3x10^6$ IU/m^2/day 1st week and then $5x10^6$ IU/m^2/d) and 14 days of oral YNK01 (600 mg/day 1st cycle). At the end of each cycle the dose of YNK01 was adjusted according to the blood count observed during the previous 4 weeks. The median time from diagnosis to inclusion in the trial was 2 months (range 6 days to 7.5 months). Pre treatment with hydroxyurea was administered to 95 patients for a median period of time of 1 month (range 4 days to 7 months). At 12 weeks, out of the 98 patients, 62 (63%; 95%CI, 54 - 73) achieved a complete haematological response with the study treatment. Cytogenetic evaluation was obtained for 79 patients at 24 weeks. At that time 2 patients achieved a complete cytogenetic response, 14 a partial response (16% of major responses; 95%CI; 9-24) and 34 a minor response. During the trial, 20 patients progressed to accelerated (6) or blastic phases (14). The median time to progression was 15 months (range 2-38 months). At 3 years the overall survival was 79% (95%CI, 70%-88%). When allogeneic patients are censored at the time of allogeneic transplantation the survival at 3 years is 88% (95%CI, 81%-95%). Although the complete haematological response rate compared favourably with the response rate previously obtained with the subcutaneous formulation of Ara-C, the cytogenetic response rate was less than expected. Most of the patients experienced side effects and all permanently stopped YNK01. Diarrhea, vomiting, and thrombocytemia were the most frequent side effects attributed to YNK01. Although the combination seems attractive the initial dose of 600mg per day is probably too high and should be reconsidered in further trials. Other studies using YNK01 alone or in combination with IFN suggested that a fixed dose of 600 mg produced significant antileukemic effect with acceptable toxicity.The efficacy of a continuous administration of YNK01 in combination with IFNalpha2b was evaluated in patients who failed previous IFN-based regimens. Dose escalations up to 900 mg were permitted for those patients who failed to response to the initial 300 mg/day dosage. Of the 10 patients enrolled in the study, 4 achieved complete haematological responses and one a complete cytogenetic response. In this study, 7 patients went off protocol because of side effects. Toxic effects attributed to YNK01 were fatigue, diarrhea and thrombocytopenia. The question of the relationship between dose of YNK01 was discuss. Although the number of patients was small, it was suggested that treatment-associated side effects worsened with dose escalation of YNK01. While YNK01 therapy was stopped in 3 patients (33.3%) at the low-dose level of 300 mg, 2 patients at 600 mg (40%) and 2 patients at 900 mg (100%) discontinued the study treatment due to YNK01-related toxicity. Moreover, dose levels beyond 300 mg were associated with a higher incidence of WHO grade 2 and 3 toxicities. It was concluded that the combination of YNK01 at the dose of 300 mg in combination with IFN could be an effective regimen in pre-treated late chronic phase patients. The preliminary results of the Australia and New Zealand YNK01 CML trial were recently reported. The study design was exactly the same as the French. Forty patients with a recent diagnosis of CML chronic phase were enrolled. Patients were required to complete two cycles of combination therapy to be evaluable for efficacy. Nine patients were considered non-evaluable due to IFN intolerance. On 21 evaluable patients at 9 months, 43% achieved a major cytogenetic response, the median time to achieve such result being 20 weeks; moreover 5 have achieved a complete cytogenetic response. These results are superior to ours and similar to those achieved previously with the subcutaneous formulation of cytarabine. The question of the modality of administration of cytarabine has been addressed in several trials. It has been suggested that the combination of IFN plus continuous administration of cytarabine could be superior as compared to the intermittent administration of cytarabine in patient with early chronic phase CML. The optimal modality of administration of YNK01 is also currently studied. The Austria CML study group initiated a phase III trial to compare the efficacy of IFN plus continuous oral YNK01 versus intermittent administration of YNK01 in newly Ph+ diagnosed CML patients, the dose of YNK01 being 300 mg total dose. The French study group also initiated a dose escalation study comparing intermittent versus continuous administration of YNK01, 100 mg being the initial daily dose. These studies may provide indication for further use of this oral formulation of cytarabine in combination with other drugs. Preliminary in vitro studies have suggested that an additive effect of STI571 and cytarabine could be demonstrated. Thus in vivo administration of a combination of YNK01 and STI571 would be of some interest.

6

Interferon treatment in the elderly with chronic myeloid leukemia

U. Berger, O. Maywald, G. Engelich, A. Hochhaus, A. Reiter, M. Pfirrmann, J. Hasford, B. Heinze, H. Heimpel, D.K. Hossfeld, H.-J. Kolb, H. Pralle, W. Queißer, R. Hehlmann, and the German CML Study Group (Mannheim, D)

Elderly patients with CML have not been well studied so far and interferon-alpha (IFN) treatment in the elderly is controversial. To assess the long-term outcome and the clinical features of older patients with CML in chronic phase we analyzed 199 patients aged 60 years and older, which represent a subgroup (23%) out of 856 patients enrolled in the consecutive randomized German CML Studies I (IFN vs. hydroxyurea (HU) vs. busulfan (Bu)) and II (IFN+HU vs. HU alone). Median age in the subgroup at diagnosis was 66 years (range 60-85). Spleen size, performance status and white blood count were lower and fever and organomegaly-related symptoms less frequent in the elderly patient whereas an excessive thrombocytosis was seen more often. 45 patients were treated with Bu, 63 with HU, and 91 with IFN-based therapy. Risk distribution according to the New CML score was similar concerning the different treatment arms in the elderly. Five year survival revealed superior outcome with IFN (46%) as compared to HU (33%) and Bu (29%). Overall survival of patients ≥60 years was significantly lower (p=0.0003) than of patients <60 years with five year survival rates of 38% vs. 47%, respectively. Chemotherapy-treated patients aged ≥60 years had a significantly inferior survival (p=0.0058 for HU, p=0.042 for Bu) than chemotherapy-treated younger patients. In contrast, no significant difference in survival (p=0.08) could be verified in IFN-treated patients ≥60 vs. <60 years with five year survival rates of 46% vs. 53%, respectively. Administered IFN dose was significantly lower in the elderly (mean 2.9 vs. 3.3 x 10^6 IU/m^2/day; p=0.017) whereas median duration of IFN-treatment was similar in both age groups (17 vs. 18 months). Hematologic remission was achieved in 82% of IFN-treated patients aged ≥60 years, which is similar to the younger patients (87%). In 15% a major cytogenetic response was observed, 6% being partial and 9% being complete, vs. 21% in younger patients. The overall frequency of adverse effects and hematotoxicity of IFN-based therapy did not differ between both groups. Older patients developed more often neuro-psychiatric (14% vs. 8%) symptoms. Since causes of death in patients ≥60 years were less frequently CML-related (84% vs. 91%), calculation of relative survival was carried out indicating no difference in IFN-treated patients ≥60 vs. <60 years in contrast to HU- or Bu-treatment. Also in a subgroup analysis of patients ≥65 years (n=117, median age 70 yrs) relative survival of IFN-treated patients (n=50) was not lower than in patients <65 years (50% vs. 54%) and the survival advantage was sustained when compared to patients treated with Bu (30% vs. 39%) or HU (28% vs. 47%). We conclude that also elderly CML patients benefit from IFN-therapy and should be offered IFN as first-line treatment option even if dosage has to be reduced because of adverse effects.

7

Randomized comparison of allogeneic stem cell transplantation vs. IFN-based therapy in chronic myeloid leukemia (CML): First interim analysis

A. Reiter, R. Hehlmann, U. Berger, A. Hochhaus, M. Pfirrmann, J. Hasford, H. Heimpel, D. K. Hossfeld, C. Huber, H.-J. Kolb, H. Löffler, H. Pralle, W. Queißer, A. Gratwohl, A. Tobler, and the German CML Study Group (Mannheim, D)

In chronic myeloid leukemia (CML) the algorithm of treatment options in newly diagnosed patients has changed considerably in the recent past. This is mainly due to the fact that favourable long-term survival results after therapy with Interferon-alpha (IFN) have evolved whereas transplant-related early mortality (TRM) has remained high at 25 – 30% after allogeneic stem cell transplanation (SCT). Since no randomized comparison of the two treatment strategies has been published, the German CML Study Group initiated a randomized trial to compare early SCT and the best available drug therapy. Patients were randomized according to availability of a related donor. Evaluation was by intention to treat. Drug treatment was based on IFN, intensive or high dose chemotherapy with autografting was recommended in cytogenetic non-responders. Since 1/95, 676 Ph-positive patients have been recruited of which 376 (56%) patients were eligible for SCT. A related donor was found in 121 (38%) of 319 evaluable patients of which 110 (91%) were transplanted. An unrelated donor was available for 98 patients of which 86 (88%) were transplanted. Within 12 and 24 months, 62% and 96%, respectively, of the related (median 10 months) and 28% and 80%, respectively, of the unrelated transplants (median 16 months) were performed. Peripheral blood stem cells were used in 27% of the related and 9% of the unrelated transplants, respectively. TRM was 27% and 25%, respectively. Median observation time is 23 months after related and 17 months after unrelated transplants, respectively. Median observation time of the 100 patients who received drug treatment is 38 months. The 3-year survival of patients transplanted with a related donor is projected at 64% in all patients and in low risk patients. The 3-year survival of all drug treated patients is projected at 82% and of low risk patients at 91%. The results of this first interim analysis of CML study III indicate that, with regard to survival, drug therapy is superior to allogeneic SCT during the first 3 years due to TRM. The advantage of drug treatment is more pronounced in low risk patients. High risk patients will benefit from early transplantation. The study plan provides a second analysis seven years after the start of the study and the final analysis 2 years later. It is expected that the results of the study will provide a reliable basis for the estimation of survival after allogeneic SCT vs. drug treatment.

8

Reduced intensity conditioning for allografting in high-risk patients with chronic myeloid leukemia: a multicenter analysis of 44 patients

M. Bornhäuser, M. Kiehl, W. Siegert, J. Schetelig, B. Hertenstein, H. Martin, R. Schwerdtfeger, H. Sayer, V. Runde, N. Kröger, C. Theuser, G. Ehninger (Dresden, Idar-Oberstein, Berlin, Hannover, Frankfurt, Wiesbaden, Jena, Essen, Hamburg, D)

This retrospective study describes the outcome of patients with chronic myeloid leukemia following allografting using dose-reduced conditioning with fludarabine and busulphan. Forty-four Philadelphia chromosome (Ph) positive patients were transplanted in nine German centres; 26 patients were in chronic phase, 11 in accelerated phase and seven in blast crisis. Thirty-four patients achieved complete remission, with 18 alive and disease-free at a median follow-up of 562 days (range 244–922 days). Grade II–IV acute graft-versus-host disease (GVHD) incidence was 43%. Twenty patients died, 15 of causes unrelated to relapse. Risk factors predisposing to graft failure by univariate analysis were an unrelated donor (8/23 compared with a related donor (2/21), P=0.07) and interferon therapy within 90 days of transplant (4/6 versus 3/17, P=0.025). At last follow-up, of 31 patients for whom molecular or cytogenetic data were available, 16 (52%) were PCR-negative, and seven (23%) were Ph-negative by fluorescent in situ hybridisation. These findings demonstrate that dose-reduced conditioning with fludarabine and busulphan provides durable engraftment and a low rate of relapse. However, in this population, many of many of whom were not eligible for high-dose conditioning due to age, reduced performance status, prior complications, or extensive pre-treatment, these data highlight the need for effective anti-infectious and GVHD prophylaxis. In addition, this study supports the discontinuation of interferon therapy at least 90 days before transplant.

Wissenschaftliches Symposium:
Neue Substanzen in der Onkologie (Phase I und II) unter Einbeziehung von CESAR (Central European Society for Anticancer Drug Research)

9

Central European Society for Anticancer Drug Research-EWIV (CESAR-EWIV): Phase I- and II-studies

C. Dittrich, J. Beck, D. Behringer, C. Bokemeyer, M. Borner, M. De Santis, H. Fiebig, G. Gastl, R. Hofheinz, D. Hossfeld, H. Illiger, R. Morant, K. Mross, W. Queißer, D. Reichert, M. Scheulen, J. Schüller, J. von Pawel (Wien, A)

Since the cristallization of its mental creation in June 1999 until the finalization of all logistic and legal processes resulting in its foundation, CESAR–EWIV has undertaken great efforts in pursuing its mission of new drug development by performing phase I- and II-studies or participating therein, respectively. Two derivatives of the anthracene-9,10-diones with preclinically high antitumor efficacy and less cardiotoxicity than the parent compound were investigated on the base of an accelerated dose escalation scheme (Novuspharma, Hesperion). Both drugs were started at a dosage of 1 mg/m^2, q. 3 wks. After 10 mos of recruitement of 15 pts over 8 dose steps, the MTD of BBR 3438 was found at 64 mg/m^2, q. 3 wks; the recommended dose for further phase II-studies in gastric and prostate cancer, respectively, was defined as 50 mg/m^2, q. 3 wks. Within 10 mos, BBR 3576 was escalated in 10 dose steps in 27 pts until 150 mg/m^2, q. 3 wks, the dose for further phase II-testing in gastric and prostate cancer, respectively. Neutropenia was the DLT in both trials. CESAR–EWIV has been participating in a phase II-programme of Klinge Pharma–Fujisawa testing iv administered methotrexate-albumin (MTX-HSA) in pts with metastatic renal cancer w/wo prior cytokine therapy, with hormone-refractory prostate cancer and mesothelioma. Albumin-binding of MTX increases its terminal half-life up to 2-3 wks. MTD of previous phase I-studies ranged between 50–80 mg/m^2/wk. For all 3 studies, the recommended dose was 50 mg/m^2/wk. Only in mesothelioma, the only still on-going study, mixed responses (3/22) were observed. CESAR–EWIV has been participating in a randomized phase II-study investigating CCI-779, an analogue of the immunesuppressive sirolimus, at different dosages (75 mg/wk and 250 mg/wk) in pts with advanced/metastatic breast cancer. The binding of CCI-779 to an intracellular cytoplasmic protein results in a complex blocking a kinase, the mammalian target of rapamycin (mTOR), and subsequently key signalling pathways resulting in a block of the G1-phase of the cell cycle. In addition, the protein translation initiation and elongation factor eIF-4E(4E) was reported to be overexpressed in fresh breast cancer samples. Activation of 4E protein is inhibited by sirolimus and potentially also by CCI-779. Dramatic response in a far advanced heavily pretreated breast cancer pt after 2 cycles of submaximal dosage was seen in a phase I-study. Furthermore, CESAR–EWIV has been participating in a phase II-study with vinflunine (Pierre Fabre) as second line therapy in pts with advanced transitional cell carcinoma of the bladder. Vinflunine, a semi-synthetic vinca alkaloid, disposes on a favourable safety profile, especially on reduced neurotoxicity in comparison to known vinca alkaloids. Actually, CESAR–EWIV is planning a prospective phase II-study of the combination of gemcitabine and cisplatin alone or in combination with ZD1839 (Iressa®) in advanced or metastatic transitional cell carcinoma of the urothelium.

10

Modern target oriented high throughput procedures for the discovery of new anticancer agents

H. Fiebig, A. Burger (Freiburg, D)

The present strategy for the discovery of new anticancer agents is based on broad in vitro testings either against isolated molecular targets in a high throughput setting or against a panel of human cell lines. The pharmaceutical industry is focusing today on high throughput screens against isolated molecular targets. With modern processing elements including robots complete compound libraries including several 100.000 of compounds can be screened within weeks. The drawback of this concept is that it is not clear if a new compound can cross cell and nuclear membranes, if the target has any relevance for growth inhibition and if there is any selectivity towards cancer compared to normal cells. Subsequent test systems are crucial for the further selection of lead compounds. At Oncotest Institute for Experimental Oncology we are pursuing 2 strategies. Target expression is verified using tissue microarrays composed of ~ 100 human tumor models and normal tissue by immunohistochemistry if antibodies are available against a specific target. For testing large numbers of compounds, we have developed a cellular high throughput screen of 12 moderately and well differentiated cell lines growing in a monolayer assay using 4 to 6 days drug incubation time. The slowly growing permanent cell lines were developed from xenografts whose characteristics are quite similar to patient tumors. Hits from the tissue microarray (tumors overexpressing a given target) and from the cellular screen (differential active compounds) are then tested in human tumor xenografts from nude mice in the clonogenic assay in vitro. This assay has demonstrated a high predictivity of drug response in the clinic. For follow-up studies in vivo, investigations in preselected models are of crucial importance to demonstrate a selectivity against tumors in comparison to the most sensitive normal tissue. In this respect, human tumor xenografts are considered as the most relevant models since the patients tumors are growing as a solid tumor, they develop a stroma, vasculature, a central necrosis and show more or less differentiation. The response to standard agents of nude mouse grown tumors in comparison to patient tumors was very similar, 90 % (19/21) in sensitive and 97 % (57/59) in resistant tumors respectively. This high correct predictivity validates the xenograft system for drug development. Using these combined in vitro and in vivo testing procedures, novel compounds were identified as active. Agents which are in clinical trial include the methotrexate coupled chemically to human serum albumin, NOAC, a liposomal fatty acide derivative of Ara-C, flavopiridol, 17-AAG, spicamycin, ecteinascidine, and recombinat mistletoe lectin. Compounds under development include salicylicacid derivatives, and cyanocyclin A. These compounds were obtained from German Academia, The US-NCI and the EORTC.

11

Effect of food intake on the pharmacokinetics of the oral camptothecin analogue rubitecan (Rfs-2000): Phase II- and pharmacologic evaluation in non-pretreated patients with metastatic colorectal cancer

P. Schöffski, A. Herr, J. van den Brande, J. Vermorken, J. Beinjen, H. Rosing, F. Reinke, A. Ganser, S. Adank, J. Wanders (Hannover, D, Antwerpen, B)

Rubitecan (9-nitro-20(S)-Camptothecin) has considerable cytotoxic activity in a wide array of preclinical and clinical models with various schedules and routes of administration. Clinical responses with this oral agent were achieved in breast, ovarian, and pancreatic cancer patients. The present study evaluates the antitumor activity of Rubitecan in chemotherapy-naive patients with metastatic colorectal cancer and characterises the pharmacokinetic (PK) profile in relation to the food intake by intra-patient comparison. For PK assessment, oral Rubitecan 1.5 mg/m^2 was initially administered twice with a one week interval either under fasting conditions (last food intake 10 h prior to treatment) or immediately after a 30 min high-calorie breakfast (1122 kcal, fat 58%, carbohydrates 27%, protein 15%). Patients continued with weekly Rubitecan 1.5 mg/m^2/d (days 1-5) under fasted conditions, to evaluate the safety profile (Common Toxicity Criteria) and antitumor activity (WHO criteria). 19 eligible pts. entered the trial (male 10, female 9, median age 60 yrs, range 40-78, previous surgery 19, previous adjuvant chemotherapy 4). The PK results clearly indicate that Rubitecan should be given under fasting conditions to achieve optimal drug exposure. 27 four-week treatment cycles were given (median per pt. 2, range 1-3). Toxicity consisted of diarrhea grade 1-2 (10 cycles), gr. 3-4 (1), vomiting gr. 1-2 (10), fatigue gr. 1-2 (9), gr. 3 (1), nausea gr. 1-2 (10), anemia gr. 1-2 (9), lymphopenia gr. 2-3 (3), alopecia gr. 1-2 (6), anorexia gr.1 (5), and leukopenia gr. 1-4 (4). At this dose and schedule the topoisomerase I inhibitor Rubitecan did not show evidence of activity in non-preteated pts. with colorectal cancer.

12

Development of anticancer drugs targeting cellular signaltransduction: Phase I study of the Raf kinase inhibitor Bay 43-9006

D. Strumberg, R. Bauer, J. Möller, R. Hilger, D. Diaz-Carballo, H. Richly, S. Kredtke, C. Beling, M. Faghih, R. Heinig, D. Hedley, M. Scheulen, S. Seeber (Essen, Wuppertal, D, Toronto, CDN)

Recent advances in cancer biology have led to the identification of signaling proteins that may be rational therapeutic targets. Raf is a protein kinase that acts as a downstream effector of Ras in the mitogen-activated protein kinase pathway, and is thus a significant contributor to the malignant phenotype driven by activated Ras signaling. Raf is also activated in tumor cells containing enhanced growth factor signaling pathways, such as those induced by the EGF receptor family members. Therefore, the collective evidence suggests that Raf is an interesting anticancer drug target. Preclinical experiments have demonstrated that eliminating Raf kinase function reverses the Ras transformed phenotype. BAY 43-9006 is a novel potent and orally active inhibitor of Raf kinase and the first compound in this class to enter clinical trials. The primary objectives of the present study are to: define the dose limiting toxicities (DLTs), determine the maximum tolerated dose (MTD), characterize the pharmacokinetic profile and describe preliminary evidence of antitumor activity.
Administration of BAY 43-9006 was started at weekly doses and developed into continuous daily treatment at doses of 100 and 200 mg bid. To date, 36 advanced stage cancer patients, median age 56, PS 0-2 with refractory malignancies (13 colorectal, 6 hepatocellular, 3 breast, 2 non-small cell lung, 2 gastric,10 others) received BAY 43-9006. The patients were considered fully eligible for consideration of dose-escalation if they had completed at least 4 weeks of treatment. Clinical toxicities observed so far included diarrhea (CTC 1-3), pancreatitis (CTC 3) and fatigue (CTC 1), but these were not dose limiting. Prolonged stabilization > 3 months of previous progressive disease was observed in 33% of the patients. The range of time to progression was 4 - 20(+) weeks in 13 patients with colorectal cancer and 12 - 44(+) weeks in 6 patients with hepatocellular cancer. PK profiles (0-12h) were obtained at the start of treatment and at steady state (after day 7), which showed an AUC0-12h,ss = 47 mg*h/L, Cmax = 6.7 mg/L, and tmax = 2 h at the 200 mg bid level. Biologic correlative studies in peripheral blood lymphocytes of treated patients by flow cytometry revealed inhibition of ERK1/2 phosphorylation at the 200 mg bid dose level. These preliminary results provide evidence that BAY 43-9006 is a promising new Raf protein kinase inhibitor that is generally well-tolerated using a continuous oral dosing regimen. Patient accrual continues at higher doses to define the MTD.

13

Growth inhibition of vascular endothelial growth factor (VEGF) producing acute myeloid leukemia (AML) cell lines by the VEGF-receptor inhibitor SU5614

F. Faber, M. Fiegl, W. Hiddemann, K. Spiekermann (München, D)

Emerging data suggest that vascular endothelial growth factor (VEGF) is an important mediator of tumor angiogenesis, cell survival and proliferation not only in solid tumors but also in leukemic cells. The VEGF binding to VEGFR2 (KDR / FLK-1) induces the proliferation of endothelial cells and may support leukemic cell survival and proliferation. Furthermore VEGFR2 is required for initiation of hematolymphopoiesis and may be a major functional marker for postnatal hematopoietic stem cells (HSC). In contrast VEGF binding to VEGFR1 (FLT-1) elicits endothelial cell-cell interactions and capillary tube formation. In hematopoietic cells VEGFR1 expression increases during monocytic differentiation. In order to evaluate the role of VEGFR signal transduction in AML we quantified the VEGF levels in cell culture supernatants of ten different AML cell lines using a specific hVEGF165 ELISA. In addition, we performed also Western blot analyses of crude lysates to confirm our findings. The expression of different VEGF isoforms, VEGFR1 and VEGFR2 was assessed by RT-PCR with specific primer for each mRNA. VEGF protein expression was detected in all cell lines by Western blot and levels of expression were correlated with VEGF secretion measured by ELISA. RT-PCR analysis showed strong expression of all three VEGF isoforms. Seven of ten cell lines expressed VEGFR1 and two of ten cell lines show VEGFR2 expression. Furthermore we analyzed the effects of SU5614, a tyrosine kinase inhibitor targeting the VEGFR2 and the PDGFR in a proliferation assay using inhibitor concentrations from 0.1μM to 10μM. Three cell lines (Mono-Mac-6, Mono-Mac-1 and Kasumi-1) showed a significantly reduced proliferative activity at concentration of 1μM SU5614. However, response of AML cell lines to SU5614 was not strictly related to VEGFR expression. The strong level of VEGF expression in almost all leukemic cell lines supports the hypothesis that AML blast cells can induce angiogenesis in the bone marrow environment. In addition, a direct effect of VEGF, e.g. by autocrine or paracrine stimulation might directly stimulate leukemic cell growth. Targeting of the VEGFR pathway might represent a new therapeutical approach for the treatment of AML.

14

Efficacy of BBR 2778, a new aza-anthracenedione, in patients with relapsed aggressive non Hodgkin's lymphomas, results of a phase-II study

P. Borchmann, R. Schnell, F. Morschhauser, K. Kutz, D. Leynardier, B. Gurglio, E. Verdi, G. Camboni, A. Engert (Köln, D)

Purpose: BBR 2778 is a new aza-anthracenedione, that had shown superior activity compared to doxorubicine and mitoxantrone in disseminated Y8-lymphoma and murine leukemia. In contrast to its parent compounds, no cardiotoxicity could be observed in the animal model. A phase-I study in pts with non-Hodgkin's lymphoma (NHL) showed promising anti-tumor activity. Therefore, a Phase II study, in patients (pts) with relapsed aggressive NHL according to the REAL-classification has been performed. Methods: Primary objective of this phase-II study was to determine the efficacy of 85 mg/m^2 BBR 2278 for a q1w x 3 treatment schedule (day 0, 7 and 14, repeat on day 28). Secondary objectives included the evaluation of the safety. The study was an open-label, non-randomized, international, multicenter trial. Pts with potential sensitivity to anthracyclines/anthracenediones could be included, i.e.: 1) response to prior treatment with these compounds, and 2) relapsing after more than six months, or 3) anthracycline-naive pts in case of a contra-indication for other anthracyclines.
Results: 12 centres in France and Germany participated in the trial. Recruitment started 01/00 and was finished by 3/01. A total of 33 pts were included, and there are preliminary efficacy and toxicity results on 27 pts available at the time this report is written. A median of two prior regimens had been applicated, and the median prior cumulative exposure to doxorubicine was 300 mg/m^2. The median age was 65 years (range 41-80 y; 15/27 aged > 65 y). 16/27 pts had stage IV disease, 5 stage III, 5 stage II, and 1 pt stage I. The majority of patients had diffuse large B-cell lymphoma (17), or mantle-cell lymphoma (6). Tumor response included 3 complete and 3 partial remissions. Overall, the drug was well tolerated. There were no treatment related deaths. The main acute toxicity was neutropenia, warranting dose reduction in 4 cases and delay or ommision of one dose in 14 cases. There were no episodes of neutropenic fever. Conclusions: These results indicate that BBR 2778 using the q1w x 3 treatment schedule with 85mg/m^2 is effective as a single agent in patients with relapsed aggressive NHL. Even in this group of elderly patients with advanced stage disease, 6/27 pts achieved a CR or PR. In the phase-I study, 3/6 pts with relapsed aggressive NHL responded to this new compound at the corresponding dosage. Although treatment with BBR 2278 was generally well tolerated, neutropenia leading to dose reduction or treatment delay occured frequently. Cardiac safety is under evalution. Further assessment of this new compound in polychemotherapy-regimens is warranted.

Wissenschaftliches Symposium: Konzepte und Strukturen der Palliativmedizin

15

Goals of a palliative care unit and implementation

C. Bausewein (München, D)

In Germany palliative care units are independent units integrated in a hospital setting. The aim of a palliative care unit is to care for patients with far-advanced, progressive disease where the prognosis is limited and the focus of care is to improve quality of life. Symptom control, psycho-social and spiritual support are offered to patients and families by a multiprofessional, interdisciplinary team. Under the leadership of a specially trained palliative care physician, doctors and nurses work closely with social workers, chaplains, physiotherapists and others. Each team member is required to have a profound knowledge of patients' needs in the situation of a life-threatening disease, as well as an understanding of symptom management, communication, team work and ethical issues. Aim of the treatment is to discharge the patients with stable symptom control, so that they can be looked after in the home setting. The average length of stay in a palliative care unit is around 13 days. To guarantee good quality of care at home close co-operation with a palliative home care service is necessary. As of 2000, 65 palliative care units were operating in Germany. With an estimated need of 40–50 beds/million inhabitants the actual situation with 12 beds/million inhabitants (including hospice beds) shows that the implementation of palliative care in Germany is still in its initial phase. It is neither possible nor efficient to create palliative care beds for all patients who are dying of a far-advanced disease. Rather, palliative care units should be mainly open to patients with complicated symptoms (e.g. intractable pain, unrelieved dyspnea, inoperable bowel obstruction), communication problems or difficult psycho-social issues. Palliative care is not a cheap option of care for dying patients. There is less need for expensive diagnostic or therapeutic procedures but because of the high patient-staff ratio (e.g. nurses:patients = 1.4:1) the average cost for an in-patient day is approximately the same or higher as in an oncology ward. On the other hand, good palliative care has been shown to reduce the actual numbers of days spent in hospital during the terminal phase. Since palliative care beds are hospital beds, re-allocation of beds through official bodies, e.g. the ministry of health, is necessary. Extra costs have to be negotiated with health insurances, which may lead to shortcomings due to cost restrictions. Most palliative care units are therefore dependent on charities and/or donations.

16

The role of the hospital support team in palliative care

S. Collins, R. Hartenstein (München, D)

Palliative medicine in Germany is a relatively new medical discipline concerned with the care of patients who are terminally ill. It focusses on quality of life rather than on quantity of life. One way of making palliative medicine accessible to hospital patients who are not looked after in a palliative care ward is the hospital support team (Konsiliardienst). So far, only about three such teams have been established in Germany. The function of the hospital support team is primarily an advisory one. The team often consists of a doctor experienced in palliative medicine, a nurse, and possibly a social worker or secretary. The main function of the team is to advise primary carers within the hospital setting on questions of pain control, symptom control, nursing management and psychosocial support for patients who are terminally ill. The team will support not only the patient and his/her relatives, but also the primary carers. Another vital function of the team is continuing education of health care professionals involved in caring for the dying. The team will also liaise with community organisations to ensure that the patient receives appropriate palliative care outside the hospital setting. It is particularly important that members of the hospital support team have substantial experience in the care of the terminally ill. The concept of the hospital support team in palliative medicine is not yet widely established, but is proving extremely valuable. It will undoubtedly be an important direction for the future of palliative care to ensure that the needs of terminally ill patients are met with competence, compassion and dignity. The hospital support team at the 4. Medizinische Abteilung, Städtisches Krankenhaus München-Harlaching is discussed as an example.

18

Hospice home care services with specialist palliative care support in the state of Rhineland-Palatinate/Germany

M. Weber (Mainz, D)

In recent years the number of hospice home care services in Germany has increased steadily. 582 services were counted in the year 1999. There were considerable differences in structure and quality however. Actually four types of home care services for the terminally ill can be discerned in Germany. Most of these services consist of volunteers giving psychosocial support to patients and families. In some services the work of the volunteers is assisted by half- oder fulltime coordinators. Rather few home care teams are provided with specialist palliative care nurses. These nurses usually give specialist advice in palliative care working alongside with the patient's own doctor and community nursing team. In some services with extended home nursing the specialist palliative care nurses provide nursing together with specialist advice in palliative care. In the state of Rhineland-Palatinate (four million inhabitants) a major effort has been made since 1997 to establish home care services for the terminally ill. Up to now 15 of 43 services are able to provide specialist advice in palliative care by trained nurses. It can be estimated that a total number of about 30 services providing specialist palliative care advice is needed in the state of Rhineland-Palatinate. This would correspond to a ratio of one service for about 130.000 inhabitants (range 60.000 inhabitants in rural regions to 300.000 in urbanized regions). A solid financial fundament for hospice home care services is urgently needed. As the care for the dying is a task of the whole society financing of hospice home care services should rely on health insurance companies as well as on public and private funding.

19

Implementation of a psychooncological support team in a tertiary care hospital

A. von Kries, S. Kesting, M. Kohlmann, J. Petri, N. Frickhofen (Wiesbaden, D)

Supported by the Deutsche Krebshilfe e.V., a psychooncological support team was established in an acute care hospital, treating about 5000 cancer patients per year. Primary goals were (1) to support patients during the diagnostic workup and treatment, (2) to build a multidisciplinary network of related services and (3) to document the activities as a basis for financing and research purposes. It was immediately apparent, that there are several key factors for a psychooncological support team to succeed in an acute care hospital: It is essential that the head of a major Department spreads the idea and convinces the hospital administration of the need of this service. The team should aim at establishing itself as a distinct organizational unit. It must quickly adjust to the fast pace of medical interventions and concurrent rapid changes in the needs of patients. In addition, rotation of the hospital staff requires constant renewal of trust and working relationships. Starting the work in a single Department, supporting the concept of psychooncology, provides a „protective environment" to build the team and develop its program. Spreading the service to other Departments too early may create unrealistic expectations and frustration of the team and its clients. Building a network of partners facilitates patient accrual and improves patient satisfaction. Documentation of the work is important for negotiation of budgets. Patient and caretaker needs can be described as follows: On a personal level, patients have to find a new balance living with cancer. On an interpersonal level, patients, relatives and friends have to find appropriate ways to interact and care for each other. The psychooncological team should be instrumental in supporting communication and decision making between patients and physicians. Finally, the team should use its position to promote a multidisciplinary approach of caregivers. Within the first year, the psychooncological support team carried out 2167 interventions in 493 patients. 58% of the patients were women. The median age was 67 (28 – 86). 61% of the contacts lasted 20 minutes or less and 76% were at the bedside. Relatives were counseled with or without the patients in 20% of the conversations. Physicians and nurses arranged 82% of the contacts. A series of talks for patients and relatives and the website of the team (www.hsk-wiesbaden.de/psyonk²start.htm) help to promote the service.

Wissenschaftliches Symposium: Hodentumoren und Nierenkarzinome

20

New options in immunotherapy of renal cancer: State of the art

P. Brossart (Tübingen, D)

Recent advamces in tumor immunology and the identification of tumor associated antigens (TAA) resulted in the development of vaccination strategies to treat patients with metastatic malignancies including renal cell carcinoma. Dendritic cell (DC) based vaccines are gaining increasing popularity as DC are the most potent antigen presenting cells (APC) with the unique capacity to initiate primary immune responses. For clinical use DC can be generated in sufficient numbers in vitro from CD34+ peripheral blood progenitor cells or monocytes. They can be pulsed with tumor antigens (peptides or proteins), transduced with recombinant viruses, fused to tumor cells or loaded with tumor-RNA and dead cells to induce a protective tumor specific CTL response. Several in vitro studies demonstrated that these strategies can induce zytotoxic T-cell responses that specifically lyse RCC derived tumor cells. Clinical applications of DC in cancer patients was shown to be effective in immunotherapies of malignant diseases including RCC. We analyzed the clinical and immunological responses in patients with malignant diseases using autologous mature DC derived from peripheral blood monocytes. DC were pulsed with HLA-A2 binding peptides deduced from Her-2/neu and/or MUC1. For the recruiting and activation of CD4+ T-helper lymphocytes DC were furher incubated with the PAN-DR binding peptide PADRE. DC vaccinations were performed sc every two weeks four times and repeated afterwards monthly until tumor progression. After the four DC injections patients additionally received 3 injections/week of low dose IL-2 (1Mio IU/m^2) sc. The vaccine-induced enhancement of T cell precursor was monitored using intracellular ELISPOT assay. Furthermore, we tested the ability of postvaccination PBMC to produce cytokines in response to challenge with vaccine-related epitopes by quantitative real-time PCR. In one patient with metastatic renal cell carcinoma (RCC) a partial remission of the metastatic sites was induced after the first 4 vaccinations with MUC1 peptides pulsed DC, that was ongoing after the next cycles. Vaccine-induced peptide specific T-cell responses in vivo were detected in the PBMNC of this patient and in peptide specific DTH reactions. Another promissing approach to treat metastatic RCC is based on the assumption that RCC may be susceptible to agraft-versus-tumor T-lymphocyte response promoted by allogeneic stem cell transplantation. Several phase I/II studies demonstrated that nonmyeloablative allogeneic stem-cell transplantation is feasible and can induce sustained regression of metastatic RCC in 30-50% of patients. However, this approach is limited by its toxic effects, mostly graft-versus host disease. Current reserch is directed at the identification of new TAA that may be target of a more focused cellular therapy and in optimizing of the protocols for the nonmyeloblative allogeneic transplantations.

21

Status of therapy recommendations in testis cancer: results of the consensus conference and new developments

H.-J. Schmoll, S. Krege, R. Souchon (Halle/Saale, Essen, Hagen, D)

The treatment of testicular cancer is under constant evaluation with the aim to reduce toxicity in low stages and increase efficacy in far-advanced stages.
• Seminoma stage I:
If adjuvant radiation is chosen, 20 Gy is a new standard according to the recent MRC-study.
The alternative is „surveillance strategy". A planned study of the „German Testis Tumor Study Group" is prospectively evaluating the role of rete testis involvement and a size > 4 cm to discriminate between patients who are good candidates for surveillance and those patients who have high risk for relapse and therefore are candidates for adjuvant irradiation.
• Seminoma stage II A, B:
The study with single agent carboplatin (N = 100) has shown unsatisfying results, therefore definitive radiation remains the standard.
• Non-seminoma stage I:
The ESMO guidelines favor surveillance strategy for good risk patients and adjuvant chemotherapy (2 x PEB) for high risk patients according to vascular invasion; the factor has also been demonstrated by recent results of the German Testis Cancer Study.
• Nonseminoma stage II A, B:
The ESMO guidelines favor primary chemotherapy instead of surgery.
• Advanced disease:
Good prognosis: PEB can be given for 5 days or 3 days with equivalent results according to the results of the recent EORTC study. Three cycles are sufficient.
Intermediate prognosis: The ongoing EORTC study compares PEB x 4 vs PEB plus Taxol x 4 to improve survival beyond 80%.
Poor prognosis: The standard is PEB x 4; the current EORTC/European Intergroup Study investigates 1 cycle PEB followed by 3 cycles high-dose PEI plus stem cell support.
• Salvage chemotherapy:
The role of high-dose chemotherapy has been investigated in the past EBMT study; preliminary results will be available at the meeting. The current standard is 4 cycles PEI or PVbI.
• Failure after 1 line chemotherapy:
High dose chemotherapy and/ or 2-drug or triple-drug combination including gemcitabine, oxaliplatin, taxol; further agents are potential activity include temozolomide.

22

Phase I/II study of sequential high dose VIP plus paclitaxel supported by PBSC in patients with 'poor prognosis' germ cell tumor (GCT)

J.T. Hartmann, N. Schleucher, B. Metzner, C.-H. Köhne, A. Gerl, O. Rick, J. Schleicher, G. Derigs, J. Beyer, L. Kanz, C. Bokemeyer (Tübingen, Essen, Oldenburg, Rostock, München, Berlin, Stuttgart, Mainz, Marburg, D)

Three consecutive cycles of HD-VIP at dose level 6 (etoposide 250 mg/sqm, ifosfamide 2 g/sqm, cisplatin 20 mg/sqm, days 1-5, q22 days) with PBST had been identified as the recommended dose level during a previous esca-

Nicht kleinzelliges Bronchialkarzinom

Metastasiertes Mammakarzinom

Besuchen Sie unsere Homepage:
www.oncosite.com
www.oncosite.de

Wirksamkeit

Verträglichkeit

Kosteneffizienz

NAVELBINE®

vinorelbin

NAVELBINE® **Zusammensetzung:** NAVELBINE® 10mg/50mg: 1 Inj. Fl. mit 1ml/5ml Infusionslösungskonzentrat enthält: Vinorelbin bis [(R, R,) tartrat] 13,85/69,25mg entsprechend Vinorelbinbase 10mg/50mg. Das Infusionslösungskonzentrat ist nur verdünnt anzuwenden. **Anwendungsgebiete:** Zur Anwendung als Monotherapie oder in Kombination mit Cisplatin zur Behandlung des fortgeschrittenen nicht kleinzelligen Bronchialkarzinoms (Stadium III oder IV) bei Patienten mit gutem Allgemeinzustand. Zur Behandlung des fortgeschrittenen anthrazyklinresistenten Mammakarzinoms bei Patientinnen mit gutem Allgemeinzustand. **Gegenanzeigen:** NAVELBINE® kann erbgutschädigend wirken. Männer sollen während der Behandlung und bis zu 6 Monaten danach kein Kind mehr zeugen. Schwangerschaft: Bei Eintritt der Schwangerschaft ist die Möglichkeit einer genetischen Beratung zu nutzen. NAVELBINE® darf nicht angewendet werden: in der Stillzeit, bei schwerer Leberinsuffizienz, bei bekannter Überempfindlichkeit auf Vinorelbin oder andere Vincaalkaloide, bei akuten Infektionen, bei Neutropenie (<2000/µl) oder Thrombozytopenie (<100.000/µl). **Nebenwirkungen:** nicht kumulative Neutropenie mit stärkster Ausprägung zwischen dem 5. und 7.Tag nach Anwendung, schnell reversibel (5-7 Tage). Anämie, Thrombozytopenie, Ausfall der tiefen Sehnenreflexe, Parästhesien, intestinale Parese, Obstipation, paralytischer Ileus, Diarrhoe, Dyspnoe und Bronchospasmen, lokalisierte Phlebitis, Haarausfall. Paravasale Injektionen können schwere Gewebeschäden bis hin zu lokalen Nekrosen verursachen. Selten: Stomatitis, Anorexie, lokalisierte oder generalisierte Hauterscheinungen, Muskelschwäche oder Schmerzen im Kiefer, Angina pectoris und Myokardinfarkt. Einzelfall eines Guillain-Barré-Syndroms und einer akuten respiratorischen Insuffizienz mit Todesfolge bei vorausgegangener Bestrahlungstherapie. Laborkontrollen: engmaschige hämatologische Kontrollen notwendig (Hämoglobin, Leukozyten, Granulozyten und Thrombozyten), routinemäßige Kontrolle der hepatischen und renalen Laborparameter und Serumelektrolyte. Bei einer klinisch relevanten Verschlechterung der Leberfunktion ist die Dosis zu reduzieren. Bei Überdosierungen werden besondere Maßnahmen empfohlen, siehe Gebrauchsinformation. **Darreichungsformen und Packungsgrößen:** NAVELBINE®: Infusionslösungskonzentrat zur intravenösen Applikation nach Verdünnung Packungen mit 1/10 Durchstechflaschen zu je 1ml (10mg) Infusionslösungskonzentrat. Packungen mit 1/10 Durchstechflaschen zu je 5ml (50mg) Infusionslösungskonzentrat. Stand Juli 2000.

Pierre Fabre Pharma GmbH • Jechtinger Straße 13 • D-79111 Freiburg • Tel 0761/45261-0 Fax 45261-55

Pierre Fabre

ZU HAUSE!

- einfach praktisch – für zu Hause
- einfach verträglich – für Alterspatienten
- einfach stark – bei Tumorprogression

Palladon®

In der Schmerztherapie zu Hause.

Palladon® 4 mg / 8 mg / 16 mg / 24 mg Wirkstoff: Hydromorphonhydrochlorid. Verschreibungspflichtig. Zusammensetzung: Arzneilich wirksamer Bestandteil: 1 Retardkapsel enthält 4 mg / 8 mg / 16 mg / 24 mg Hydromorphonhydrochlorid (entsprechend 3,56 mg / 7,12 mg / 14,24 mg / 21,36 mg Hydromorphon) Sonstige Bestandteile: (alle Wirkstärken): Methylhydroxypropylcellulose, mikrokristalline Cellulose, Ethylcellulose, gereinigtes Wasser, hochdisperses Siliciumdioxid, Dibutyldecandioat, Gelatine, Natriumdodecylsulfat, Schellack, Phospholipide aus Sojabohnen, Dimeticon. Zusätzlich: 4 mg: Erythrosin (E127), Indigocarmin (E132), Titandioxid (E171), Eisen(II, III)-oxid (E172). 8 mg: Erythrosin (E127), Titandioxid (E171), Eisen(II, III)-oxid (E172). 16 mg: Eisen(III)-oxid (E172), Eisenoxidhydrat (E172), Eisen(II, III)-oxid (E172), Titandioxid (E171). 24 mg: Indigocarmin (E132), Titandioxid (E171), Eisen(II, III)-oxid (E172). Anwendungsgebiete: Starke und stärkste Schmerzen. Gegenanzeigen: Nicht einnehmen bei bekannter Überempfindlichkeit gegenüber Hydromorphon, bestehendem Ileus, gleichzeitiger Anwendung von oder innerhalb von 14 Tagen nach Absetzen von MAO-Hemmern, in Schwangerschaft und Stillzeit. Die Wirkstärken 8 mg, 16 mg und 24 mg sind nicht für eine initiale Therapie geeignet. Nur unter besonderer Vorsicht anwenden bei Abhängigkeit von Opioiden, Bewusstseinsstörungen, Krankheitszuständen, bei denen eine Störung des Atemzentrums und der Atemfunktion vorliegt bzw. vermieden werden muss, Zuständen mit erhöhtem Hirndruck, Hypotension bei Hypovolämie, Gallenwegserkrankungen, obstruktive und entzündliche Darmerkrankungen, Prostatahypertrophie mit Restharnbildung, Phäochromozytom, Myxödem, Epilepsie, Alkoholabhängigkeit, Kinder unter 12 Jahre. Abhängigkeitsentwicklung möglich; Entzugssyndrom. Reaktionsvermögen! Nebenwirkungen: Gelegentlich Erbrechen, Verstopfung, Störungen bei der Blasenentleerung, Krämpfe der Bronchialmuskeln, Druckanstieg in den Gallengängen, Verwirrtheitszustände, Schwindel, Kopfschmerzen. Hydromorphon führt dosisabhängig zu einer Atemdepression und Sedierung in unterschiedlichem Ausmaß (von leichter Müdigkeit bis zur Benommenheit) sowie zu Übelkeit. In Einzelfällen Urtikaria. Als charakteristische Begleiterscheinungen gelten Euphorie und Miosis. Gelegentlich Stimmungsveränderungen (missmutige Stimmung, Dysphorie). Selten Exitationszustände. Toleranz und Abhängigkeit können sich entwickeln. Stand der Information: Februar 2000. Mundipharma GmbH, 65549 Limburg (Lahn).

mundipharma

Info-Line: Tel. 08 00 - 8 55 1111. Internet: http://www.mundipharma.de

Aktuell von Mundipharma

lation study for the first-line treatment of patients (pts) with poor prognosis GCT according to IGCCCG classification (Bokemeyer, Sem Oncol 1998). Paclitaxel (T) has shown significant activity in cisplatin-refractory GCT. In order to identify the maximal tolerated dose, toxicity and feasibility, the design of the present study included three different dose steps of paclitaxel (135-175-225 mg/sqm) added to high dose VIP. One cycle of standard-VIP plus G-CSF to collect autologous PBSC was given prior to HD-VIP/T. Pts received filgrastim 5 µg/kg s.c. from day +1 after treatment to leukocyte recovery >2000/µl and reinfusion of PBSC at day +2 after each cycle of HD-VIP/T. T was applied as a 3-h infusion on day 1 prior to HD-VIP. Dose-limiting toxicity was defined as °III non-hematological toxicity or complicated neutropenia. To date, 50 pts were entered onto this multicenter trial. Median age was 34 years (range, 18-55). Forty-three pts (137 cycles) are currently evaluable for toxicity. Overall, the dose intensity of the 4-drug regimen was 98 % at level 1, 99 % at level 2 and 95 % at level 3. All patients experienced grade III/IV granulocyto- and thrombocytopenia with a median time to recovery at days 15 and 15 of the treatment cycles, respectively. Severe stomatitis was seen in 1/3 of patients regardless of the paclitaxel dose applied. No severe neurosensory or neuromotor toxicity were observed. Bowel side effects ranged from 6 to 12% with increasing severity at higher paclitaxel dosages. Severe infections occurred in 9, 0 and 6 % of cycles at dose-levels 1, 2 and 3. Dose reductions have been performed due to paraesthesia, encephalopathia, allergic reactions related to paclitaxel and due to GI toxicity. Of 37 evaluable pts 29 (78%) achieved CR/PRm-/NED after HD-VIP/T ± surgery. One early death occurred due to tumor progression. In conclusion, the combination of HD-VIP with T up to the maximal planned dose of 225 mg/sqm as first-line therapy in pts with 'poor prognosis' GCT is feasible without a significant increase in hematological or neurological toxicity compared to HD-VIP alone. Response data appear to be at least as favorable as for HD-VIP alone. HD-VIP/T is currently investigated for its therapeutic potential in a expanded cohort of 'poor prognosis' germ cell tumor pts.

23

Prospective comparison of [18F]-FDG PET versus conventional assessment by CT scans and serum tumor markers in the evaluation of residual masses in patients (pts) with non-seminomatous germ cell cancer

K. Oechsle, C. Kollmannsberger, B.M. Dohmen, A. Pfannenberg, C.D. Claussen, L. Kanz, R. Bares, C. Bokemeyer (Tübingen, D)

The present study investigates the ability of [18F]fluorodeoxyglucose (FDG) positron emission tomography (PET) to predict the histology of residual masses after chemotherapy in pts with metastatic non-seminomatous germ cell tumors (GCT). The role of PET was compared in a blinded analysis to CT scan and serum tumor marker changes (TUM) as established methods of assessment. Patients and methods: Eighty-five residual lesions in 45 pts were evaluated by PET and corresponding CT scans/TUM by independent reviewers blinded to the results of each other. All pts had been treated within prospective clinical trials receiving primary/salvage high-dose chemotherapy with autologous blood stem cell support for primary poor prognosis disease/relapsed disease. PET was assessed visually and by quantifying glucose uptake (SUV). Results were validated either by histologic examination of a resected mass/biopsy (n=28 lesions) or by a clinical follow-up over 6 months after evaluation (n=57 lesions). Results: In 32 of 85 residual lesions FDG-PET showed an increased tracer uptake. Twenty-nine of these were true positive (TP) and 3 false positive (FP). Fifty-three lesions were classified by PET as being negative (no viable GCT), 33 true negative (TN) and 20 FN. In the blinded reading of the corresponding CT scans/TUM, 39 residual lesions were correctly considered to contain viable carcinoma/teratoma. Forty-six lesions were classified as non suspicious by CT scan/TUM (32 TN, 14 falsely classified). PET correctly predicted the presence of viable carcinoma in 5 and the absence of viable carcinoma in 3 of these 14 lesions. This results in the following sensitivities/specificities for the prediction of residual mass histology: PET 59% / 92%; radiological monitoring 55% / 86%; TUM 42% / 100%. The positive and negative predictive values of PET were 91% and 62%, respectively. The diagnostic efficacy of PET was not improved when pts with teratomatous elements in the primary tumor were excluded from the analysis. In pts with multiple residual masses, a uniformly increased residual FDG uptake in all lesions was a strong predictor for the presence of viable carcinoma. Conclusion: [18F]-FDG PET imaging performed in conjunction with conventional staging methods offers additional information for the prediction of residual mass histology in pts with non-seminomatous GCT. A positive PET is highly predictive for the presence of viable carcinoma. Other useful indications for a PET examination include pts with multiple residual masses and pts with marker-negative disease.

24

13cis-retinoic acid, IFN-alpha2a, IL-2 and chemotherapy in advanced renal cell carcinoma: Results of a prospectively randomised trial of the German Cooperative Renal Carcinoma Chemoimmunotherapy Group (DGCIN)

J. Atzpodien, H. Kirchner, U. Jonas, L. Bergmann, H. Oesterwitz, H. Heynemann, S. Loening, M. Ziegler, S. Müller, H. Westerhausen, H. Bodenstein, S. Pomer, H. Illiger, U. Rebmann, A. Hofstetter, R. Obernader, I. Wandert, T. Patzelt, M. Reitz (Hannover, Bonn, D)

Background. We conducted a prospectively randomised clinical trial to compare the efficacy of three outpatient therapy regimens: (A) subcutaneous interferon-alpha2a (sc INF-alpha2a), subcutaneous interleukin-2 (sc IL-2), intravenous 5-fluorouracil (iv 5-FU), (B) sc INF-alpha2a, sc IL-2, iv 5-FU, peroral 13-cis-retinoic acid (po 13cRA), and (C) sc INF-alpha2a, iv vinblastin in 347 patients with progressive metastatic renal cell carcinoma. Methods. Patients were stratified according to known clinical predictors and were subsequently randomised. Treatment courses consisted of (A) IFN-alpha (5 x 10^6 IU/m^2, day 1, week 1 + 4; days 1, 3, 5, week 2 + 3; 10 x 10^6 IU/m^2, day 1, 3, 5, weeks 5 - 8), IL-2 (10 x 10^6 IU/m^2, days 3 - 5, week 1 + 4; 5 x 10^6 IU/m^2, days 1, 3, 5, weeks 2 + 3) and 5-FU (1000 mg/m^2, day 1 weeks 5 – 8); (B) treatment arm A combined with 13cRA (20 mg 3 x daily); (C) IFN-alpha (5 x 10^6 IU/m^2, day 1, week 1; 10 x 10^6 IU/m^2, days 1, 3, 5, weeks 2 - 8) and vinblastin (6 mg/m^2, day 1 weeks 2, 5 and 8). Findings. Among 135 patients treated with INF-alpha2a, IL-2 and 5-FU there were 7 complete (5%) and 34 partial responders (25%), with an overall objective response rate of 30% (95%CI 22, 38). In 149 patients receiving IFN-alpha2a, IL-2, 5-FU, 13cRA an overall objective response of 26% (95%CI 19, 33) occurred with 12 complete (8%) and 27 partial responders (18%). 63 Patients treated with combined IFN-alpha2a and vinblastin yielded 4 complete (6%) and 9 partial responders (14%), with an overall objective response rate of 20% (95%CI 10, 31). While there were no significant differences in median overall survival comparing all three treatment regimens (A: 25 months, B: 27 months; C: 18 months), the combination of IFN-alpha2a, IL-2, 5-FU and 13cRA yielded a significantly improved progression-free survival when compared against the IFN-alpha2a, vinblastin therapy (all patients, p< 0.0453; low risk patients, p< 0.0011) and against the IFN-alpha2a, IL-2, 5-FU combination without 13cRA (low risk patients, only, p< 0.0195), respectively.

Wissenschaftliches Symposium: Chancen und Risiken der Stammzellforschung

26

Plasticity potential of adult stem cells - an alternative to embryonic stem cells?

A.D. Ho (Heidelberg, D)

There is increasing evidence that stem cell populations isolated from adult tissues can show remarkable plasticity upon transplantation. For example stem cells derived from the bone marrow have been shown to give rise to cartilage, bone, muscle, liver, and neurons. Neural stem cells have been demonstrated to give rise to blood cells. Other tissue specific stem cells such as those from skeletal muscle, oval cells from the liver, etc. have also been shown to give rise to cells of other tissues. Currently, we can only speculate about the mechanisms involved in such dramatic changes in cell fate. Experiments in in vivo models showed that after transplantation, pluripotent stem cells may be able to undergo transdifferentiation or reprogramming if they find themselves in a new stem cell niche. Several factors might enhance this probability. For example the ability of the cell to respond to differentiation may be stimulated by neighboring cells that are differentiating at the same time. Removal of a stem cell from the usual neighborhood may lead to the upregulation or activation of surface receptors for a wide range of circulating factors, some of which are present in the new niche. Our own experiments have demonstrated that this plasticity potential is inversely related to ontogeneic age, i.e. fetal tissues have a far higher potential than stem cells from the umbilical cord, which in turn is higher than those derived from adult bone marrow. The reprogramming of tissue specific stem cells to other, hierarchically higher phases of development is dependent on ontogeneic age. However, the scarcity of tissue-specific stem cells in adults as well as their extremely limited self-renewal capacity, may render them impractical for clinical application. To enhance our knowledge on differentiation and self-renewal for regenerative medicine, there is a need for research in embryonic stem cell lines.

27

Chimerism in embryos and adults following the injection of neural and hematopoietic stem cells into murine blastocysts

A.M. Müller, S. Kreutzfeldt, S. Petrovic, N. Kirchhof, C. Schmittwolf, B. Mühl, A. Merkel, F. Harder (Würzburg, D)

Until recently, it was generally believed that the differentiation potential of a mammalian somatic stem cell is restricted to one tissue only. Recent reports, however, show that mammalian somatic stem cells can, in the course of regeneration, repopulate heterologous cell systems and therefore possess an unexpected plasticity. In order to analyze and compare the developmental potential of different somatic stem cell types, we have isolated murine adult bone marrow-derived hematopoietic stem cells (HSCs; LIN-, ckit+, SCA1+ cells) and murine fetal brain-derived neural stem cells (NSCs). NSCs are grown in vitro in media containing EGF and bFGF. Somatic stem cells were derived from transgenic mouse lines (i.e. eGFP, lacZ, bcl2). About 40 somatic stem cells were injected into pre-implantation murine blastocysts and the repopulation pattern in developing chimeric animals was analyzed.
We detect progeny of the stem cells in more than 40% of the developing animals. Analysis of the tissue distribution of the injected donor cells revealed that progeny of the HSCs are preferentially detected in hematopoietic and at a lower frequency in non-hematopoietic tissues. Following the in vitro propagation of NSCs and their injection into blastocysts, NSCs are also capable of generating chimeric embryos and adults. Analysis of the donor contribution showed evidence of NSC-derived erythroid repopulation, but overall, progeny of NSCs are mostly detected in neural tissues.
Our results indicate, that even if somatic stem cells from different tissue origins are injected into an environment which is permissive for the development of all cell lineages of the developing embryo, they show plasticity but mostly retain their tissue-specific identity.

29

Telomere elongation in hematopoietic stem cells

S. Zimmermann, S. Glaser, R. Ketteler, U. Klingmüller, C. Waller, U. Martens (Freiburg, D)

Loss of telomere repeats has been causally linked to replicative senescence and aging in human cells. In contrast to normal somatic cells, stem cells have low activity of telomerase, which can be transiently up-regulated upon cytokine activation. Despite the activity of telomerase, proliferation of progenitor cells is accompanied by telomere shortening possibly leading to a reduction of their replicative potential. In fact, short telomeres have been associated with reduced bone marrow reserve in patients with aplastic anemia and in mice genetically deficient for telomerase activity. To test whether ectopic expression of telomerase can overcome telomere erosion in hematopoietic stem cells, we transduced hTERT, the catalytic component of telomerase, into CD34+ cord blood (CB) cells using a retroviral pBABE vector containing an insert with hTERT- IRES-GFP (n=5). CD34+ cells were cultured for 3-4 weeks in serum-free medium in the presence of SCF, Flt-3, IL-3, G-CSF, and IL-6. The hTERT transduced progenitor cells exhibited significantly elevated telomerase activity compared to untransduced or empty vector transduced CD34+ cells. Although a continous decline of telomerase activity was also detectable upon proliferation of hTERT transduced CD34+ cells, telomerase activity was still 6-fold higher compared to negative control groups that showed low to absent telomerase activity after 22 days in culture. Using the Flow-FISH technique mean telomere length was found to be significantly elevated in hTERT transduced CB cells, although telomere attrition was not prevented. The maximum lengthening of 570 bp was observed by day 14 which translates to a „rejuvenation" of 19 years assuming a telomere loss of 30 bp per year. Overall expansion potential was found to be similar in hTERT transduced cells relative to control cells. However, we observed a significant increase in BFU-E but not in CFU-C suggesting a growth advantage for erythroid progenitors (132±47 vs. 72±32, p= 0.005 and 46±19 vs. 54±18, p=0.4, respectively). These results indicate that telomerase appears to be strictly regulated in the hematopoietic stem cell compartment. Although overexpression of this enzyme can lead to telomere elongation, telomerase activity is readily down-regulated as a consequence of cell differentiation. Furthermore, the overall expansion potential in CB cells with very long telomeres does not seem to be affected by high levels of telomerase, but telomerase activation may force differentiation of erythroid progenitors. In the future, it would be interesting to investigate, if telomere elongation in progenitor cells of patients with short telomeres (e.g. bone marrow failure syndromes) may improve proliferative capacity of the hematopoietic system.

30

Gene expression profile of primary human CD34+ hematopoietic stem cells

U. Steidl, U.-P. Rohr, R. Fenk, O. Modlich, C. Maercker, P. Neubert, S. Kliszewski, J. Koch, S. Martin, M. Aivado, P. Schneider, U. Germing, H. Bojar, N. Gattermann, R. Haas, R. Kronenwett (Düsseldorf, Heidelberg, Berlin, D)

The molecular phenotype of CD34+ hematopoietic stem cells and the molecular signals steering mobilization, differentiation and self-renewal are only partially understood. In this study, we measured gene expression profiles of primary human CD34+ cells by means of cDNA array technology. CD34+ cells of 20 volunteers were selected immunomagnetically (purity >90%) from steady state bone marrow (BM-CD34+ cells) or peripheral blood (PB-CD34+ cells) after G-CSF-mobilization to compare these two sources of stem cells. Following RNA-extraction, reverse transcription and radioactive labelling with 32P, samples were hybridized with cDNA arrays covering 1185 genes. Expression of 11 genes involved in cell cycle progression and DNA synthesis during S phase (PLK, UBCH10, CDC20, CDC25A, CDC28, B-MYB, MCM2, MCM5, MCM7, PCNA and Lig1) was 2- to 5-fold higher in BM-CD34+ cells compared to PB-CD34+ cells, reflecting the greater cycling activity of BM-CD34+ cells. Expression of caspases 3, 4, and 8, DAPK1, and DNase II was 2-fold higher in PB-CD34+ cells suggesting a greater apoptotic rate in circulating CD34+ cells. Examining transcription factors, we found GATA2, which is known to inhibit differentiation of hematopoietic cells, to be more highly expressed by PB-CD34+ cells. Beside GATA2, expression of other transcription-related genes (GATA3, CLK, cyclin H, BRF1, TFDP2) was greater in PB-CD34+ cells, suggesting a differentiation inhibiting function of those genes, too. Proteinase 3, cathepsin C and D, and tripeptidyl peptidase I, which are able to cleave connective tissue protein, as well as oncostatin M, which activates transcription of collagen alpha 2, showed a 2-fold higher expression in BM-CD34+ cells. Expression of those genes might be essential for BM-CD34+ cells to organize and maintain a suitable microenvironment for hematopoiesis. Thrombin receptor, known to play a role in monocyte chemotaxis, was 3.6-fold more highly expressed by circulating CD34+ cells in comparison to BM-CD34+ cells, suggesting a chemotactic function not only for monocytes but also for CD34+ cells. Released thrombin may guide CD34+ hematopoietic stem cells to sites of cellular damage or increased consumption to facilitate regeneration processes. Interestingly, we found on the one hand a 3-fold higher expression of M-CSF by circulating CD34+ cells and on the other hand a 3.6-fold higher expression of the corresponding M-CSF-receptor by BM-CD34+ cells, suggesting that PB-CD34+ cells and BM-CD34+ cells might communicate via M-CSF. In summary, our data provide a gene expression profile of CD34+ hematopoietic stem cells. The comparison of gene expression of BM-CD34+ and PB-CD34+ cells molecularly confirms the model that BM-CD34+ cells are more rapidly cycling, differentiating hematopoietic progenitors, whereas PB-CD34+ cells consist of a higher number of quiescent, non-differentiating stem cells. Moreover, our data give novel insights into molecular physiology of hematopoiesis.

Wissenschaftliches Symposium:
Tumoren des oberen Gastrointestinaltraktes

31

New treatment options for patients with pancreatic carcinoma

G. Kornek (Wien, A)

Despite certain advances in diagnosis and surgical procedures, the prognosis of pancreatic cancer patients remains dismal. The large majority of patients presents with disease that is beyond the scope of surgical cure, their clinical course is characterized by debilitating symptoms, and the median survival duration in case of distant metastases is generally < 3 months. In 1996 a randomized trial of 5-FU-based chemotherapy compared to best supportive care in advanced pancreatic carcinoma (Glimelius et al.) showed a significant better overall survival (2.5 vs. 5 mos) and improved quality of life. Since 1997, when Burris published a randomized trial, which showed that the novel cytidine analogue gemcitabine was more effective than 5-FU-monotherapy, the former drug has become the new standard regimen for advanced pancreatic carcinoma. Randomized trials comparing gemcitabine-monotherapy versus gemcitabine plus 5-FU ± leucovorin showed no benefit for the combination regimen. The administration of higher doses of gemcitabine, however, yielded a promising overall survival duration of 8.8 months and a clinical benefit response rate in 45% of patients. More recently, several new anticancer drugs (capecitabine, irinotecan, oxaliplatin, raltitrexed and docetaxel) and their combination with gemcitabine have been investigated. In a recently published phase II trial the combination of gemcitabine, epirubicin, cisplatin and infusional 5-FU showed a particularly encouraging objective response rate of 58% and an overall survival duration of 9.4 months, a result that warrants further investigation. During the past five years a number of noncytotoxic approaches have also undergone clinical evaluation. Recognition that K-ras mutations are present in more than 90% of pancreatic carcinomas has resulted in an effort to target the ras-signaling pathways. Farnesyltransferase inhibitiors (FTIs) interfere with posttranslational modification and activation of ras. Currently, two FTIs are under investigation (R115777 plus 5-FU/LV and SCH 66336 orally plus gemcitabine) for the treatment of metastatic carcinoma of the pancreas. Matrix metalloproteinases (MMPs), that have a high expression in pancreatic cancer, play an important role in the process of invasion and metastasis, by degradation of basement membrane and stromal tissue. MMPIs (BAY12-9566) have shown not to affect tumor cells, but tumor growth delay and reduction in the number of metastases. Several monoclonal antibodies have been investigated for the treatment of metastatic carcinoma of the pancreas including the anti-epidermal growth factor receptor antibody (cetuximab), the anti-HER-2/neu (herceptin), both in combination with gemcitabine, and the anti-gastrin antibodies. Early results have been presented at the ASCO meeting 2001. Another type of new anticancer agent is ONYX-015, which is an attenuated adenovirus that replicates and lyses cells lacking functional p53. In a phase I/II trial the intratumoral application of ONYX-015 in combination with i.v. gemcitabine was investigated and resulted in minor and partial responses in 3 and 2 out of 18 patients. In conclusion, gemcitabine monotherapy is still the standard regimen for the treatment of patients with advanced pancreatic carcinoma, though its therapeutic potential might be enhanced by combination with other cytotoxic agents. Other approaches currently under investigation include immunotherapeutic antigen targeting and gene therapy.

32

Conserving treatment strategies for primary lymphoma of the stomach

M. Raderer (Wien, A)

Extranodal lymphomas account for between 25 – 40% of all lymphomas, with a pronounced geographic variation. While being encountered in almost every organ of the human body, the large majority arise within the stomach. Following the definition of lymphoma of the mucosa associated lymphoid tissue (MALT) by Isaacson and Wright in 1983, this type of lymphoma has become a focus of interest in recent years due to its unique pathological and clinical properties. While surgical intervention has been the preferred form of treatment in the past, organ conserving strategies are increasingly being applied. The link between helicobacter pylori (HP) and the development of gastric MALT-type lymphoma has revolutionized treatment options, as elimination of the antigen by means of antibiotic eradication has been demonstrated to result in complete remission of the lymphoma. To the current knowledge, up to 80% of patients with early gastric MALT-type lymphoma restricted to mucosa and submucosa achieve long-term complete remission of the tumor following eradication of HP. In addition, absence of a t(11;18) translocation as well as absence of an underlying autoimmune disease appear to be predictive of response to eradication. Nevertheless, in about 50% of patients gastric biopsies show persistent evidence of a monoclonal B-cell population as judged by PCR in spite of normal findings upon conventional histology and immunohistochemistry. The persistence of monoclonality in post-treatment biopsies without evidence of lymphoma is consistent with the notion that eradication of HP might suppress but does not necessarily eradicate the neoplastic clone in all cases. In case of unresponsiveness to HP-eradication, radiation has given promising results with the potential for stomach conservation. The potential of chemotherapy, however, has not widely been tested. Preliminary results, however, with application of alkylating agents or nucleoside analgues have demonstrated a high rate of CR, approaching 100% in some series. In case of transformation to DLCL, chemotherapy appears to be the management of choice irrespective of clinical stage. Both retrospective analyses as well as emerging prospective data have shown a rate of complete remission approaching 100% using CHOP. In order to minimize the risk of bleeding and perforation, the intake of a proton-pump inhibitor for the whole duration of chemotherapy is advisable. At the moment, it is unclear whether chemotherapy in analogy to nodal disease comprising of 6 – 8 cycles is really superior to shorter courses of treatment. In addition, the role of combining radiation with short-term chemotherapy has to be prospectively assessed in this highly chemosensitive disease. In patients with evidence of HP-infection, however, additional antibiotic therapy should be given in order to treat an underlying MALT-type lymphoma, and recent data have also indicated that DLCL might regress following eradication of HP in selected patients.

34

Biweekly high dose gemcitabine alone or in combination with capecitabine in patients with advanced pancreatic adenocarcinoma: A randomized multicenter phase II trial

H. Ulrich-Pur, G. Kornek, B. Schüll, M. Raderer, W. Fiebiger, B. Schneeweiss, F. Lang, K. Haider, W. Kwasny, D. Depisch, W. Scheithauer (Wien, Kirchdorf, Neunkirchen, Wr. Neustadt, A)

Gemcitabine is an active antitumour agent in the treatment of advanced pancreatic cancer (APC), and has shown synergistic activity with the oral 5-FU prodrug capecitabine in a preliminary phase I/II trial. Based on this background, and in order to determine the antitumour potential and tolerance of this combination more precisely, a multicenter phase II trial was initiated in previously untreated patients with APC. Eligible patients were randomly assigned to arm A (biweekly gemcitabine 2200 mg/m^2 given as a 30 min. infusion) versus arm B (biweekly gemcitabine 2200 mg/m^2 plus oral capecitabine 2500 mg/m^2 given from days 1 to 7). Patients were evaluated for objective response rate, progression-free survival (PFS) and overall survival (OS), as well as for clinical benefit response (defined as >50% reduction in pain intensity, >50% reduction in daily analgesic consumption, and/or >20% improvement in Karnofsky performance status that was sustained for >4 consecutive weeks).

75 patients have entered the study, of whom 59 are presently evaluable. Their pretreatment characteristics were well balanced between the two arms. 5/29 partial responses (17.3%) were observed in arm A, and 6/31 (19.4%) in arm B. Median PFS was 4 months in A (range, 1.3-14) and 6.5 months in B (range, 1.5-13) (P=0.0699). Median OS was 10 and 12 months, respectively (P=0.78). Among patients with tumour-related symptoms who were considered evaluable for clinical benefit response, 9/23 (39.1%) and 14/24 (58.3%) experienced significant palliation in arm A and B, respectively. Chemotherapy was well tolerated in both arms with only 4 vs. 3 patients experiencing WHO grade 3 toxicities. Apart from occurrence of a hand-foot syndrome in 8 patients, no increase in incidence and/or degree of adverse reactions was noted in the combination arm. Overall incidence rates (%) were as follows (A/B): granulocytopenia 55/48, nausea/emesis 48/42, diarrhoea 14/19, fatigue 45/48, and transient elevation in liver functional parameters 48/39. Early results of this trial suggest a promising therapeutic index for the combination of gemcitabine and capecitabine for the treatment of APC, though until today no significant advantage over single agent gemcitabine was noted.

35

MUC4 mRNA in peripheral blood mononuclear cells as a potential tumor marker for pancreatic cancer

G. Faulmann, J. Ringel*, N. Moniaux, R. Brand, S. Tarantolo S. Joshi, S.K. Batra, M. Löhr (Mannheim, D; Omaha NE, USA)*

Background: Despite improvements in diagnostic techniques and increasing knowledge regarding genetic alterations and serum-based tumor markers, the diagnosis of pancreatic cancer remains a pressing clinical problem. New specific diagnostic tools, therefore, are necessary. Pancreatic tumor cells show an altered mucin pattern with an aberrant MUC4 expression, and they interact, especially, with immune cells. We investigated whether mucins expressed in peripheral blood mononuclear cells (PBMCs) can be used as a tumor cell-independent marker for pancreatic cancer. Methods: We analyzed the expression of MUC1, MUC11 and MUC4 mRNA in PBMC samples from 97 patients: 27 with pancreatic cancer, 13 with pancreatitis, 47 with various types of epithelial and haematopoietic malignancies, 10 with non-malignant diseases, and 8 healthy volunteers using RT-PCR. The presence of circulating pancreatic cancer cells was investigated by using a specific nested PCR for cytokeratin-20 (CK-20) and chymotrypsinogen. Results: MUC1 was found in all PBMC samples. In contrast, MUC4 mRNA was detectable in the PBMC from 18 (67%) pancreatic cancer patients, but was absent in all other PBMC samples investigated (100% specificity). Two pancreatic cancer PBMC samples were CK-20 and chymotrypsinogen positive. Cell sorting analyses revealed that MUC4 was expressed in CD3+ lymphocytes. In addition, first data suggest a downregulation of MUC11 in PBMC of pancreatic cancer patients (49%), whereas it was detectable in most of the other samples. Conclusion: Our results show an altered mucin expression in PBMCs of pancreatic cancer patients. The expression of MUC4 mRNA is specific for pancreatic cancer in and has the potential use as a novel cancer cell-independent tumor marker. The reasons for the specific expression of MUC4 are under investigation. (supported by the NIH grant CA 78590 and Boehringer Ingelheim Foundation)

* contributed equally.

37

Multimodal treatment of liver metastases

M. Lorenz, E. Staib-Sebler (Frankfurt/M., D)

Up to 50% of all patients with colorectal cancer present themselves with synchronous metastases or develop metachronous liver metastases. Only a minority is resectable and even in these patients recurrence occurs in up to 80%. Therefore modalities to increase resectability and to reduce relapse rate are needed to improve prognosis. To date, postoperative chemotherapy after resection remains still controversial. Concerning systemic application only retrospective studies are available. Of all postoperative local treatments, only hepatic artery infusion was tested prospectively. Whereas hepatic arterial infusion (HAI) of FUdR plus systemic Mayo reduced the incidence of hepatic recurrence, the overall survival was not improved. The missing impact on overall prognosis was demonstrated already by our own study using adjuvant HAI with 5-Fluorouracil (5-FU)/Folinic acid (FA). Recently significant improvement of the results of chemotherapy were demonstrated. Effective regimen with 5-FU/FA in combination with oxaliplatin were potent enough to reduce size of liver metastases by at least 50 % in half of all patients and made a resection possible in 20-50% of all cases. After resection of these initially unresectable patients a 5 year survival rate up to 50 % was reported. Therefore this preoperative modality is evaluated by a prospective EORTC-study in resectable patients. Rates of resectability may further increase by different local ablative treatments such as radiofrequency and HAI. Despite all progress of surgery and additive treatments, prospective trials are essential before the introduction of these multimodal approaches into clinical practice.

39

Phase II trial of gemcitabine (GEM) with 24-hour infusion of 5-fluorouracil (FU) and folinic acid (FA) in patients with advanced pancreatic cancer

H. Oettle, B. Peters, U. Pelzer, P. Reitzig, C. Ziske, C. Keitel-Wittig, D. Hennesser, R. Hoffmann, A. Kretzschmar, M. Arning, H. Riess (Berlin, D)

Objective: To access activity of Gem/5-FU/FA in patients with locally advanced and metastatic pancreatic cancer. Methods: Following our phase I study (Anticancer-Drugs, 1999, 10:699-704) we started a multicenter phase II trial with GEM (1000mg/m^2; 30min) in combination with FA (200mg/m^2; 2h) and FU (750mg/m^2; 24h) continuous infusion via portable battery driven pump (Walkmed™ 350). Therapy was given on days 1, 8, 15 and 22 with start of the next cycle on day 43. Patients remained on study until disease progression (MRT scans every 6 to 8 weeks). Treatments were given on an outpatient basis. Results: 78 consecutive chemonaive patients (stage III/IV, 17/61) have been enrolled so far. The median age was 61 years [30-80], 67% of the participants were male. 361 cycles were administrated. The median number of cycles administered was 4 [1-12]. 31 patients (40%) developed grade III 48/361 (13,3%) or IV 5/361 (1,4%) side effects. Out of 361 cycles grade III/IV hematological toxicities were: granulocytopenia 8/-, anemia 5/1 and thrombocytopenia 9/1. Main Grade III/IV non-hematological toxicities were: nausea/vomiting 4/1, pain 6/-, elevation of alkaline phosphatase 5/- and hepatic toxicities 7/1. We observed 10 PR (12,8%), 60 pts. (77%) achieved SD, and 6 pts. (7,7%) had progressive disease. There were two early deaths in the first cycle of therapy due to brain metastasis and heart failure. 30 pts. (38,5%) had SD for more than 6 months. Median progression free survival was 6,6 [1 to 18] months [95% CI: 4.8-8.4]. The median survival of the entire group was 9.1 months [95% CI: 7.2-10.8]. The 1-year OS was 36,2%. 12 of 78 pts. are still alive. In a multivariate Cox-analysis stage (p=0.61) and sex (p=0.45) were not prognostic factors for OS. Conclusion: Our results indicate that this combination is feasible, active and of low toxicity in advanced pancreatic cancer. It is currently compared against single agent Gem in a randomized phase III setting.

41

Allografting with immunosuppressive conditioning of low-dose fludarabine-cyclophosphamide after autografting.

A.M. Carella, M.T. Corsetti, G. Beltrami, M. Cavaliere, E. Lerma, D. Occhini (San Giovanni Rotondo, Genova, I)

AlloSCT induces a potent graft-vs-tumor (GVT) response but conditioning regimen toxicity and/or GVHD are still a major cause of post-transplant mortality. Non-myeloablative conditioning regimens are recently used in an effort to minimize conditioning regimen toxicity while maintaining GVT effects. Since it is likely that immunotherapy will be more effective in a minimal disease status, we designed a strategy to follow an AutoSCT with AlloSCT to take advantage of both the antineoplastic and immunosuppressive effects of AutoSCT to allow successful allogeneic cell therapy. Eligibility required a controindication to the use of conventional allograft and/or refractory disease. Sixty-two patients with a median age of 43 years (range, 16-69), were treated. Follow-up is at a median of 374 days (range, 15-1353). Diagnoses were Lymphomas n=34: (HD=19 NHL=15), Solid Tumors n=12: (Breast=10 Kidney=1 Germ cell=1), Leukemias n=7: (CML=6 AML=1), MDS (n=5) and Multiple Myeloma (n=4). All but ten received the combined procedure Autograft/Mini-Allograft. The patients were conditioned with an immunosuppressive regimen consisting of fludarabine 30 mg/m^2/d x 3 days with cyclophosphamide 300 mg/m^2/d x 3 days (Flu-Cy protocol). PBSC grafts from HLA-identical sibling donors mobilized with G-CSF were given on day 0. CSP/MTX was given post-transplant to control graft rejection and GVHD. Donor lymphocyte were given for persistent mixed chimerism and/or progression of malignancy. No myelosuppression, no mucositits, no alopecia were observed after Flu-Cy protocol. Major disease responses were observed in 25 (50%) of 50 patients who had measurable disease pretransplant and have had sustained engraftment. These included 16/28 lymphoma, 2/6 CML (both patients achieved and maintain molecular remission 29 and 10 months after), 2/5 RAEB and 5/10 breast cancer patients. Acute GVHD occurred in 20 (32%) out of 62 patients and in 9 (15%) was of grade >II. Two extremely advanced patients developed and died of diffuse cGVHD. Seventeen other patients died of progressive disease between 15 and 1071 days after allografting (Breast: n=4; HD: n=4; NHL: n=3; blastic phase-CML: n=2; MM: =1; Kidney: =1; MDS: =1; Germ cell: n=1). No fatal rejection occurred in 16 patients (RAEB/RAEB-T: 4; CML: 4; AML: 1; MM: 4; NHL: 2, kidney: 1). These data demonstrate that this novel combined approach dramatically reduced the acute toxicities of conventional allografting even in highly pretreated patients.

43

In vivo T-cell depletion with ATG generates excellent results in allogeneic bone marrow transplantation from unrelated donors in patients with early chronic phase CML

M. Schleuning, O. Stötzer, R. Pihusch, H.J. Kolb (München, D)

Antithymocyte globulin (ATG) is widely used as additional immunosuppression to allow engraftment and prevent graft-versus-host disease (GVHD) in the setting of unrelated donor bone marrow transplantation (URD-BMT). However, there is dispute about the adequate dosing of this drug. In order to find the optimal dosing we conducted a dose escalation study in patients receiving grafts from unrelated donors. In this retrospective analysis the results of this study were evaluated for patients with chronic phase CML. Eighty-nine adult patients, median age 37 years (range 16-52 years), with chronic myeloid leukemia in first chronic phase (CP1) received grafts from unrelated donors which have not been fully matched in 9 cases. Thirty-six patients were transplanted in early disease within 2 years after diagnosis and 53 in late disease more than 2 years after diagnosis. ATG-S (Fresenius) 10-120 mg/kg b.w. was given prior to transplantation. Thirty-one patients received a total dose of less than 60 mg/kg b.w. and 58 patients received 60 mg/kg b.w. or more. All patients received bone marrow as stem cell source and GVHD prophylaxis consisted of cyclosporin A and short-course methotrexate. Graft failure did not occur and neutrophil engraftment >0.5 G/l was reached at a median of day +19 in the group given 60 mg or more of ATG and at day +23 in the group receiving less than 60 mg of ATG (p=0.001). The incidence of severe acute GVHD grade III-IV was 21.3% and of extended chronic GVHD 12.4%. The overall relapse rate was 11% with no significant differences according to ATG doses. With a median fol-

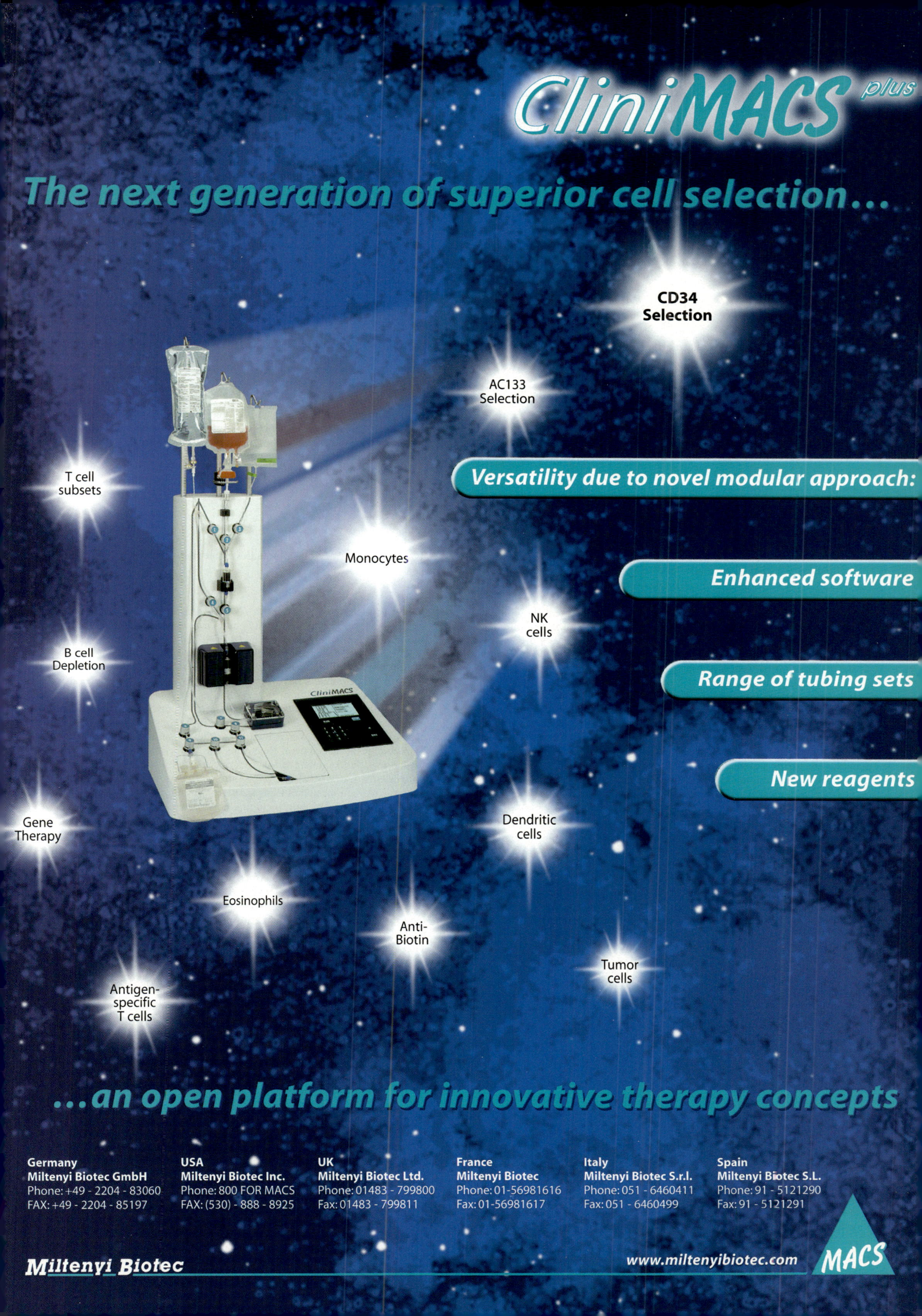
CliniMACS plus
The next generation of superior cell selection...
CD34
Selection
AC133
Selection
T cell
subsets
Monocytes
Versatility due to novel modular approach:
Enhanced software
NK
cells
B cell
Depletion
CliniMACS
Range of tubing sets
New reagents
Gene
Therapy
Dendritic
cells
Eosinophils
Anti-
Biotin
Tumor
cells
Antigen-
specific
T cells
...an open platform for innovative therapy concepts
Germany
Miltenyi Biotec GmbH
Phone: +49 - 2204 - 83060
FAX: +49 - 2204 - 85197
USA
Miltenyi Biotec Inc.
Phone: 800 FOR MACS
FAX: (530) - 888 - 8925
UK
Miltenyi Biotec Ltd.
Phone: 01483 - 799800
Fax: 01483 - 799811
France
Miltenyi Biotec
Phone: 01-56981616
Fax: 01-56981617
Italy
Miltenyi Biotec S.r.l.
Phone: 051 - 6460411
Fax: 051 - 6460499
Spain
Miltenyi Biotec S.L.
Phone: 91 - 5121290
Fax: 91 - 5121291
Miltenyi Biotec
www.miltenyibiotec.com
MACS

low-up of 4.4 years (range 0.5-10.4 years), 5-year disease-free and overall survival for all patients was 53% and 65%, respectively, and for patients transplanted within 2 years of diagnosis it was 74% and 83% respectively. Within this cohort of patients there was a highly significant difference for eventfree and overall survival in respect to the ATG dose used. Patients receiving 60 mg or more of ATG had a 5-year disease-free and overall survival of 93% and 93%, respectively, and for patients receiving less than 60 mg of ATG 5-year disease-free and overall survival was only 33% and 56% respectively. This difference could not be observed in those patients transplanted later than 2 years after diagnosis was established. Our observations indicate that URD-BMT is an excellent treatment option for patients with chronic phase CML, if it is performed early during disease and if a sufficient amount of ATG is included in the preparative regimen.

44

Chimeric status of gut epithelium after human bone marrow transplantation

A. Spyridonidis, A. Dwenger, T. Tomann, M. Follo, D. Behringer, J. Finke (Freiburg, D)

Background: Recently, a number of groups have reported that stem cells derived from adults may be capable of surprising plasticity or versatility. In humans, bone marrow derived cells have been found to give rise to cells with hepatic cell like features. The gastrointestinal epithelium of humans undergoes continuous and rapid renewal throughout life. We have investigated whether bone marrow cells contribute to repopulation of the adult gut epithelium. Methods: We studied biopsies of the large bowel of 6 female patients who underwent a sex-mismatch allogeneic hematopoietic stem cell transplantation (BMT: 4 pts; PBSCT: 1 pt; BMT followed by PBSCT 5 months later: 1 pt) for the presence of epithelial cells of donor origin. Frozen biopsies were taken at different time points varying from 3 weeks to 18 months after transplantation. Samples were examined with a combination of fluorescence in situ hybridization (FISH) in order to detect the Y chromosome, fluorescence immunohistochemistry to detect the epithelial specific marker cytokeratin and TOTO staining to visualize the cell nuclei. Findings: Nontransplanted male and female gut samples were used as controls and showed that the Y chromosome was restricted to the male intestine, demonstrating hybridization specifity. On the other hand, Y chromosome positive, cytokeratin negative cells were found in the gut interstitium of female samples, representing donor derived hematopoietic cells.
Y chromosome positive cells were also found within the epithelium of 5/6 patients (3 BMT; 1 PBSCT; 1 BMT + PBSCT). Using triple staining and laser scanning confocal microscopy we confirmed that there were individual cells expressing cytokeratin which contained a Y chromosome within their nuclei. By using parameters that resulted in a calculated optical section thickness of 0.5 µm, this rigorous analytic method ensured that the pattern observed with these markers represented a localization of the markers within the same cell. Y chromosome staining was localized to nuclei of cytokeratin positive cells and was not associated with any other neighboring nuclei in the x, y, or z planes. 3-dimensional animations of individual cytokeratin positive / Y chromosome positive cells reconstructed from serial 2-dimensional images will be presented. Semiquantitative analysis showed a higher frequency of donor-derived epithelial cells in the sections taken a longer time after transplantation (14 months and 18 months). Interpretation: Our data show that after hematopoietic stem cell transplantation in humans donor-derived hematopoietic cells yielded cells that enter the gut and expressed epithelial specific features. The role of these cells in epithelial renewal has to be confirmed by clonal methods.

Wissenschaftliches Symposium: Schmerz – Diagnostik und Therapie

45

Barriers to effective cancer pain control

I. Strohscheer (Berlin, D)

The WHO published their guidelines for the treatment of cancer pain in 1986. These practice guidelines are validated and well established. It is estimated that in Germany, there are about 220,000 patients in need of tumor pain related therapy each day. Estimates indicate that only 10% of those patients get adequate therapy. A survey from 1996 revealed that 20% of cancer patients received strong opioids with 14.5% of these patients getting morphine for more than eight weeks. Despite the options available for successful treatment by following those simple guidelines, in only a minority of cancer patients adequate pain control is achieved. The intention of this report is to analyse the barriers to an effective cancer pain management. In addition to inadequate social consciousness for the importance of tumor pain control, there are several additional reasons of concern regarding patients and medical personell. Principal causes of this within the medical professions are: deficiencies in pain treatment education, lack of experience, insufficient sensitiveness for the significance of pain, inadequate communication with patients and between the medical professionals, fatalism, fear of addiction and side effects, inadequate knowledge and misconceptions of medication and insecurity concerning regulations about prescribing opioids. The patient's part on insufficient pain treatment is: fatalism, fear of addiction, anxiety of side effects, fear 'that the doctor is concentrating on pain therapy but not on treatment of his cancer', the wish 'to be an uncomplicated patient', anxiety of injections, stigma of receiving opioids only under restricted prescription regulations and fear of developing an opioid-tolerance with the consequence of untreatable pain in the terminal situation of the illness.
According to these problems, conceivable possibilities of solution result in the improvement of communication between both patients and medical staff and within the medical profession, while elevating the sensitiveness for the importance of pain therapy, training programs for education in pain control, elimination of the 'morphine mythos' and negative misconceptions about cancer pain by explaining the effectiveness of opioids and loosening the opioid prescribing restrictions.

46

The importance of non-opioids in cancer pain treatment

M. Kloke (Essen, D)

The effectiveness of non-opioids (NOA) in the treatment of mild cancer pain is well established. The WHO guidelines recommend the their continuation even when opioids are needed (steps II and III of the analgesic ladder). Nevertheless, neither their effectiveness in addition to different dosages of opioids nor their long-term toxicity with regard to the specific risk-benefit ratio in cancer patients have been evaluated by randomized clinical trials (e.g. level A studies). Improved knowledge in the working mechanisms of non-opioids - today mainly conceptionalized as spinally and centrally acting analgesics - and the development of new drugs (e.g. selective COX-II inhibitors) enforce a renewed asignment of their position in the pharmacologic treatment of cancer pain. Regarding the validation studies of the WHO guidelines , the documented treatment periods with NOA are really limited (maximum a few weeks) and informations concerning acute and long-term toxicity, interactions and reasons for discontinuing are rare. NOA have been applied in all types of pain including neuropathic pain. The fact that these drugs are not able to induce physical dependance or addiction and that the initially perceived side effects are less burdening for the patient than those of opioids, NOA are regarded as drugs of first choice. Yet, their use is limited because these drugs exhibit a ceiling dosage. The meaning of the opioid sparing effect of NOA may be questioned with regard to their additional specific toxicity and the subjective burden for the patient enlarging the number of pills to be taken. NSAID are the most frequently used non-opioids. Up to now, the hypothezied superiority of NSAID in somatic pain has not been verified regardless their proven peripheral antiinflammatory potency. They exhibit significant acute and long-term toxicity especially in exsiccated, elderly, cachectic and / or multimorbide cancer patients. NSAID interact with many relevant anticancer drugs (e.g. methotrexat). Their gastrointestinal toxicity increases dramatically with the concomittant use of corticosteroids. The decreased gastrointestinal toxicity of selective COX II inhibitors may therefore be favourable in these patients. Due to the poor evidence for the spasmolytic effect of metamizole (dipyrone), a specific indication for colicky visceral pain could not be delineated. In cases of severe chemotherapy induced bone morrow suppression the distinction between delayed recovery and toxic reaction towards metamizole is difficult. Paracetamol acts only as a weak analgesic. The antipyretic effect of every NOA may contribute to a delay in the recognition of serious infections in the immunocompromized patients.
In conclusion, NOA are worthful drugs for treating all types of cancer pain. Nevertheless, their specific toxicity and possible interactions require a careful indication and close surveillance of side effects

47

New opioids in cancer pain treatment

F. Nauck, C. Ostgathe, E. Klaschik (Bonn, D)

Introduction: It has been estimated by the WHO that cancer pain affects at least 4 million people world-wide. Fifty to eighty per cent of these patients do not obtain satisfactory pain relief. With increasing acceptance of the WHO-guidelines for cancer pain management pain relief may be achieved for the majority of patients. Oral morphine, by international consensus, is the opioid of choice in the treatment of cancer pain, but not all patients obtain sufficient pain relief under such treatment. The new strong opioids available can be the answer to difficult/complex pain syndromes. Patients and Methods: In our palliative care unit we treat cancer patients with difficult pain syndromes. In a computer engineered database and on a daily basis, we collect data focussing on pain and other symptoms as well as on the administration of drugs for pain treatment and symptom control. This includes the assessment of pain on a numeric or verbal rating scale. Our analysis refers to data collected over a period of 20 months. Results: Between 10/99 and 05/2001 263 patients were admitted to our palliative care unit. 76 % of these patients suffered from pain. In about 90 % of our patients we were able to achieve satisfactory pain relief (pain reduction >50 %) or even freedom from pain. Most of these patients were treated with morphine (51 %). But in some patients we could not achieve adequate pain relief, or side effects occurred under the morphine treatment so that we were forced to change to an alternative drug. At discharge, from the 26 % of patients under strong opioids other than morphine 42.3 % received hydromorphone, 30.7 % fentanyl, 19,2% l-methadone and 8,7 % oxycodone. According to the WHO – guidelines, patients were also treated with non-opioids, supplemented with co-analgesics where appropriate. Comment: For a sufficient treatment of cancer pain physicians should be able to identify causes and types of pain, to discuss the side-effects of opioids, and to differentiate between myths and facts as well as the different classes of drugs such as opioids, adjuvants and non-opioid analgetics. A successful cancer pain therapy requires accurate assessment and precise diagnosis of the underlying pain pathomechanism. The majority of patients responds to oral analgesics. Conclusion: Morphine still is the goldstandard in cancer pain relief but the wide range of inter-and intraindividual effects and side effects of opioids in different patients make it necessary to look for alternatives in order to achieve a sufficient treatment. Our survey shows that the new opioids are able to relief strong pain in cancer patients , but until now there are no differential indications for the selection of the right opioid at the right time. The lecture will focus on treatment with different opioids and the distinction of indications for an appropriate individual choice.

48

Invasive therapy in patients with cancer pain

J. Beck, K. van Ackern (Mannheim, D)

One of the most prevalent symptoms in cancer diseases is pain. Therapies aimed at reducing pain are therefore one of the main modalities in palliative care. A detailed diagnosis of how pain is experienced in such patients is of crucial importance for subsequential intervention. In addition, the stadium of disease, the prognosis and the psycho-social status of the patient should be considered in treatment strategies. The treatment strategy should be dynamic in its nature in order to fulfill a high degree of flexibility in the case of changes in pain symptoms or localization and therefore different techniques may be used during the period of therapy. Although there exist a variety of other therapeutic options, to date, oral pharmacotherapy according to WHO is considered to be the golden standard for the treatment of pain in patients with solid tumors. Through transdermal, continuous subcutane or intravenous application of drugs, almost all individual needs can be achieved. Like in other treatment strategies, effectivity is limited by the occurrence of side effects. It has been demonstrated in studies on oral pharmacotherapy that in 5 to 25 % of patients insufficient pain reduction was obtained. Following the WHO guidelines however, only in 5 to 7 % of patients intolerable side effects could be seen. Invasive pain therapy, i.e. neurale blockade, local anaesthesia or neurolysis is therefore an important additional tool that may be implemented in the treatment strategy. As today's knowledge on pathological alterations in the central nerve system, due to peripheral induced pain has markedly been improved, it is now generally believed that for neurolysis only the sympathetic neurolysis, neurolysis of the coeliac plexus and intrathecal neurolysis should be considered. There are basically two ways with which neurale blockade can be accomplished, i.e. by intrathecal or epidural catheters. Due to a low degree of side effects, epidural catheters, which are fixed by suture to the skin after 5 cm of subcutaneous tuneling, are preferentially used in our center. Depending on the localisation of the most intensive pains, the site of punction can vary from lumbal up to cervical. Although its known that the analgetic effects of morphine can be achieved by intrathecal apllication, one should distinguish between lipophilic and hydrophilic morphines. Futhermore, for invasive pain therapy local anesthetics or drugs like clonidin or ketamin can be used. Its of great importance for the effectivity of invasive medicine that a balance between advantages and risks for each patient individually should be judged by the practitioner and patient in a personal interview.

49

Efficacy and toxicity of COX-2-inhibitors in the treatment of tumour pain - a pilot study

A. Heider, T. Reddemann, N. Niederle (Leverkusen, D)

Introduction: Besides opioids nonsteroidal antirheumatica (NSAR) proved to be most effective in the treatment of skeletal or soft tissue pain due to malignant disease. The permanent application of NSAR is limited by gastrointestinal toxicity. The new COX-2-inhibitors have shown significantly fewer gastric ulcers and erosions during treatment of rheumatic illness. In this pilot study, therefore, the efficacy and toxicity of COX-2-inhibitors in tumour pain was investigated. Methods: So far, 36 patients (20 male/16 female, median age 66 years (range, 35 – 84)) with tumour-related pain were included. Fourteen patients (pts) showed solid tumours without bone metastases, 14 pts solid tumours with bone metastasis , 6 pts multiple myeloma with diffuse skeletal involvement, another two patients high-grade lymphoma. In 28 pts, there was no or insufficient pain control prior to treatment entry. In the other 8 pts a sufficient treatment with NSAR had to be stopped due to gastrointestinal toxicities. Treatment consisted in all 36 pts of either rofecoxib 25 – 50 mg (n = 22) or celecoxib 200 – 400 mg (n = 14) daily, while concomitant specific drugs (no other pain treatment n = 14, opioids alone or in combinations with antipyretic drugs n = 17, antipyretic drugs alone n = 5) were not changed. All pts received corticosteroids at individual doses. Protons-pump-inhibitors were given to 20 pts. So far, median follow-up time is 4 months (range, 1 – 10). Efficacy and toxicity were evaluated after 1 and 4 weeks.

Results: In 36 pts , a median pain reduction consisted of 2 steps (range, 0 – 8) according to NAS could be demonstrated. In five pts, however, there was no positive influence on pain control (rofecoxib n = 2, celecoxib n = 3) by the COX-2-inhibitors. Two pts treated with rofecoxib demonstrated gastrointestinal toxicities, both showed no endoscopic findings. Until now, no other side-effects can be documented.

Conclusions: COX-2-inhibitors are effective in the treatment of tumour pain. So far, it remains unclear if they are as effective as NSAR, since these substances are commonly used though not systematically investigated in tumour pain. The side–effects of the COX-2-inhibitors are low, especially gastrointestinal toxicities. Therefore it might be possible to dispense with protons-pump-inhibitors in patients who are not treated with corticosteroids regular. Further prospective randomised studies are warranted.

Wissenschaftliches Symposium: Hochmaligne Lymphome

50

Clinical impact of novel molecular factors in aggressive lymphomas

M. Bentz (Ulm, D)

Molecular findings have provided important clues to the pathogenesis of Non-Hodgkin´s Lymphomas (NHL). Moreover, they can contribute to the development of treatment strategies in aggressive NHL at the following steps: (i) refinement of lymphoma classification, (ii) use of clonal molecular markers for follow-up analyses during lymphoma therapy, (iii) identification of novel (biological) risk factors more relevant than currently used clinical parameters and (iv) specific molecular factors may be used as targets for novel therapeutic agents. An increasing number of genomic aberrations, like e.g. the classical IgH-BCL2 or the NPM-ALK fusion or more recently the IgH-BCL11A or the API1-MALT1 fusion have been shown to be closely associated with specific lymphoma entities. Such gene fusions as well as clonal markers derived from clonal T-cell-receptor- or immunoglobulin-gene rearrangements can be used for polymerase chain reaction (PCR) analyses for the detection of minimal residual disease (MRD) during follow-up under lymphoma therapy. This approach may become more important, as quantitative PCR techniques have opened new avenues for the assessment of MRD. For several molecular markers, a prognostic impact in aggressive lymphomas has been suggested. Mutations of the TP53 gene as well as the presence of a IgH-BCL-2 fusion or expression status of the p27 protein have been associated with an inferior outcome in aggressive lymphomas. Also rearrangement of the BCL6 gene has been suggested as prognostic marker. For all these analyses, however, studies in homogeneous patient populations treated within the same clinical trial are pending. The novel approach of DNA chip technology may greatly enhance our knowledge about molecular factors in NHL. With this technique, comprehensive gene expression analyses have been performed. The data strongly indicated a subdivision of diffuse large cell lymphomas into a „germinal center cell type" with a better prognosis and an „activated B-cell type" with an inferior prognosis. Currently genomic DNA chip protocols are developed that will allow a comprehensive detection of genomic aberrations in the setting of clinical trials. It can be foreseen, that the arrival of DNA microarray technology together with the sequencing of the human genome will revolutionize the biological evaluation of NHL. This may lead to a rapid increase of clinically relevant molecular markers in the near future.

51

Dose escalation and antibody treatment for aggressive lymphomas: new standards?

M. Pfreundschuh (Homburg, D)

To improve treatment results in patients with aggressive NHL by shortening treatment intervals of and/or adding etoposide to CHOP, 890 untreated young low-risk (18 to 60 years) and 807 patients >60 years of all IPI risk groups were randomized to 6 cycles of CHOP-21, CHOP-14, CHOEP-21 (CHOP plus etoposide 100 mg/m^2 d1-d3), or CHOEP-14. A planned interim analysis of 50% of the expected events was performed and included 456 elderly patients (61- to 75 years) who had either an event or a follow-up ^{2}3 months after therapy. CR rates were 56.8% for CHOP-21 (n=118), 69.1% for CHOEP-21 (n=112), 74.3% for CHOP-14 (n=110) and 70.5% for CHOEP-14 (n=116; p = 0.028). In patients with elevated LDH, CR rates rose from 39% with CHOP-21 to 67.3% with CHOP-14 (p = 0.015). At the median time of observation of 34 months, TTF (50% vs. 38%; p=0.03) and overall survival (62% vs. 48%; p=0.04) are significantly better for CHOP-14 and CHOEP-21 than for CHOP-21. Leukocytopenia (WBC <2000/mm^3) occurred significantly less after CHOP-14, while CHOEP-21 and CHOEP-14 had more cycles with platelets <50 000/mm^3. We conclude that in patients >60 years with aggressive NHL, time intensification (CHOP-14) or addition of etoposide (CHOEP-21) improve results significantly over standard CHOP-21 to an extent which appears at least as pronounced as the improvement recently reported by the addition of rituximab to CHOP-21, whereas no improvement was achieved with double-intensive CHOEP-14, most likely because of dose erosion due to toxicity. CHOP-14 is the preferred regimen for elderly patients due to its excellent tumor control and toxicity profile.

A similar improvement has recently been demonstrated in elderly patients for the addition of the monoclonal anti-CD20 antibody rituximab to classical CHOP-21. In 328 evaluable patients, the addition of the antibody resulted in an increase of CR rates 15%, and a dropt of primary progressions by 17%. After a median time of observation of only 12 months, event-free and overall survival are better for the rituximab arm. Pending longer follow-up, CHOP-21 + rituximab appears to be the second breakthrough after CHOP-14 in the treatment of aggressive lymphomas in the last 25 year. Rituximab should now be evaluated in a randomized fashion in the combination with CHOP-14 to determine whether combining time interval shortening and rituximab will result in complimentary, additive or even synergistic efficacy. Moreover, since the recent advantages have only been demonstrated in elderly patients with aggressive NHL, their superiority in the treatment of other patient populations must be confirmed in prospective trials in both young low-risk and young high-risk patients. Supported by Deutsche Krebshilfe.

52

The prognostic significance of molecular staging in diffuse large B-cell lymphoma (DLBCL)

G. Mitterbauer, C. Mannhalter, C. Skrabs, M. Mitterbauer, K. Winkler, A. Chott, K. Lechner, U. Jäger (Wien, A)

Polymerase chain reaction (PCR) analysis of clonal lymphoid cells is of diagnostic value in many lymphoproliferative diseases. Immunoglobulin heavy chain (IgH), immunoglobulin kappa light chain (Igkappa), and T-cell receptor gamma (TCRgamma) gene analysis by PCR amplification are sensitive and specific methods for assessment of peripheral blood (PB) and bone marrow (BM) involvement. However, the diagnostic and prognostic value of the detection of clonal rearrangements in PB and/or BM of patients with diffuse large B-cell lymphomas (DLBCL) at diagnosis has not been evaluated in larger series. We studied clonal rearrangments of Ig and TCRgamma genes in the PB (n=16), BM (n=15), or both (n=55) in 86 patients with DLBCL before the initiation of treatment. Corresponding tumor specimen were analyzed in 53 cases. Circulating tumor cells were detected in 20/86 (23%) patients with DLBCL. The detection of Ig gene rearrangements in PB/BM was related to the clinical stage (CS I-III, 6/58 (10%) vs. CS IV, 14/28 (50%); P=.00005), the histopathological detection of BM involvement (P=.00001), and the International Prognostic Index (P=.001). Ig gene rearrangement positive cases had a significantly lower CR rate (7 of 20, 35%) than Ig gene rearrangement negative patients (45 of 66, 68%; P=.008). The CR rate of the patients with negative BM histology but positive PCR results was 56% (5 of 9). The estimated overall survival (OAS) at 5-years in the group with Ig gene rearrangements (15%) was significantly inferior to that in the group without Ig gene rearrangement (45%, P=.002). There was a significant difference between patients with negative PCR results and negative BM histology, patients with positive PCR but negative histology (28% at 5-years), and patients with positive PCR and histopathological BM involvement (0% at 2-years, P=.001). Our results indicate that molecular methods improve the accuracy of staging in DLBCL. Moreover, we demonstrate a significant prognostic impact of detection of clonal Ig gene rearrangements in PB/BM on CR rate and survival.

53

Immunohistomorphology and genomic analysis of endemic and sporadic Burkitt´s lymphoma with proven MYC-disruptions

T.F.E. Barth, P. Möller, R. Siebert, S. Müller, N. Henn, G. Mechtersheimer, M. Pawlita, M. Bentz (Ulm, Kiel, Heidelberg, D)

Histopathological differentiation between classical Burkitt´s lymphoma (BL) and Burkitt-like lymphoma is difficult. In order to better understand the characteristics of classical BL, we performed a comprehensive immunohistological and molecular cytogenetic study. Seven cases of endemic BL (eBL) were obtained from the Ocean Road Hospital in Dar-es-Salaam, Tanzania. German sporadic cases of BL (sBL) were selected according to the following criteria: typical Burkitt morphology, expression of CD20, proliferation indices > 90 % and the presence of a t(8;14) typical of BL or variant translocations. Nine sBL fulfilled these criteria, seven with a t(8;14) and two with a t(8;22) detected by fluorescence in-situ hybridization using suitable DNA probes. Southern blot analyses were performed for EBV-screening. Immunohistology was performed with a broad panel of monoclonal antibodies and genotypes were analyzed using comparative genomic hybridization (CGH). Results for eBL: 6/7 cases were analyzed for EBV genome, which was present in all these cases. Immunophenotype was very homogenous. All cases expressed MHC class I and class II molecules as well as CD10, CD20, CD38 and CD77. 6/7 cases expressed IgM and 7/7 cases lambda light chains. 6/6 analyzed cases exhibited a t(8;14). By CGH, 3/6 cases had no aberrations. In the remaining three cases, 2, 2 and 4 aberrations were detected, respectively. Results for sBL: EBV was negative in 8/8 cases tested. MHC Class I molecules were expressed in 8/9 cases. MHC class II molecules and IgM were expressed in 7/9 cases. Lambda light chains were expressed in 3/9 cases, while 6 cases had no light chain expression at all. CD10 was detected in 4/9 cases only. CGH analysis revealed a normal pattern in 2/9 cases. In the other cases, 1 - 6 (median, 4) chromosomal imbalances were identified. Two cases each had gains on 1q and 17q. These data show that eBL are very homogeneous both immunologically and genetically. sBL are still quite homogeneous, although less than eBL, provided that the diagnosis is restricted to tumors with proven MYC disruption.

54

High-dose chemotherapy with autologous PBSCT and hyperfractionated radiotherapy as first-line treatment for primary CNS Lymphoma (PCNSL) - Update of a multicenter phase II study

G. Illerhaus, R. Marks, G. Derigs, N. Frickhofen, R. Guttenberger, F. Hirsch, S. Högerle, U. Kaiser, C.B. Ostertag, J. Finke (Freiburg, Mainz, Wiesbaden, Offenburg, Marburg, D)

Primary NHL of the CNS (PCNSL) carries a poor prognosis despite initial response to steroids and radiotherapy. Addition of methotrexate to radiotherapy has improved the prognosis of patients with PCNSL, but a significant proportion of patients are still not cured. To improve relapse free survival in patients with PCNSL and to reduce neuro-toxicity we initiated a multicenter phase II study with early dose intensified chemo-therapy and

PBSCT followed by hyper-fractio-nated whole-brain radio-therapy for patients aged <65 years. The use of high-dose lipophilic blood-brain-barrier-penetrating agents (BCNU, Thiothepa) in addition to maximum doses of water-soluble agents (MTX, AraC) is an novel approach in the treatment of PCNSL. To reduce the risk of delayed neurotoxicity, intrathecal chemotherapy was completely avoided in this protocol and chemotherapy was administered before hyperfractionated radiotherapy. Induction treatment included three repetitive cycles of high-dose MTX (8g/m^2). AraC (2x 3g/m^2) and Thio-tepa (40mg/m^2) followed by rG-CSF were used for stem-cell-mobilisation. The conditioning regimen included BCNU (400mg/m^2) and Thiotepa (2x5mg/kgBW) prior PBSCT. Additional hyper-fractionated whole-brain radio-therapy (45 Gy, 2x1Gy/day) was administered as consoli-dation. Twenty Patients (age < 65 y) have been enrolled in the study so far (median age 55, range 30-65y). 13 of 20 evaluable pts. responded to high-dose MTX (3 CR, 10 PR), and 15/20 received high dose chemotherapy and autologous PBSCT according to the protocol until now. Four patients refractory to MTX proceeded to radiotherapy directly and 1 patient died from PD subsequently. Two Patients in PR are presently prior high-dose chemotherapy. Beside cytopenia induced by high-dose BCNU and Thiothepa treatment, no severe toxicities [WHO Grade 3 or 4] were observed. One Patient (62y, Karnofsky-Index initial 30%) developed leucencephalopathy one year after treatment. Until now all 14 evaluable patients having completed the therapy including hyperfractionated radiotherapy responded to the therapy (14/14 CR) and are alive, one patient experienced a systemic relapse after one year. With a median follow-up of 17 months (range 2 - 34 mo) 19/20 patients are alive. In conclusion sequential systemic application of high-dose differential acting cytostatic agents with consolidating hyperfractionated radiotherapy is very well tolerable with little acute toxicity. This novel approach results in a high rate of complete remissions in patients with PCNSL.

55

Mega-CHOEP: A Phase I/II study of the German High Grade NHL Study Group For Primary Treatment of Aggressive Non-Hodgkin´s Lymphoma: Efficacy of dose level 1 + 2 and feasibility of dose level 3

B. Glass, M. Kloess, W. Berdel, A. Engert, M. Bentz, L. Trümper, M. Loeffler, M. Pfreundschuh, N. Schmitz (Göttingen, Leipzig, Münster, Köln, Ulm, Homburg, Kiel, D)

To optimize high dose therapy for aggressive NHL, we evaluated a new treatment strategy comprising of four to six cycles of dose escalated CHOP + Etoposide chemotherapy requiring stem cell transplantation after three of these cycles. Patients with newly diagnosed aggressive NHL, 18-60 years of age, and LDH above normal were included. Dosage at respective dose levels (DL) were as follows: DL1, Cycle 1: cyclophosphamide (CY) 1500 mg/m^2, adriamycin (ADR) 70 mg/m^2, vincristine 2 mg, Etoposide (ETO) 450 mg/m^2, and prednisone 500 mg/m^2: Cycle II and III : CY 4500 mg/m^2 and ETO 600 mg/m^2, cycle IV: CY 6000 mg/m^2 and ETO 1000 mg/m^2. At DL2 ETO was further intensified: 600, 960, 960 and 1480 mg/m^2 at cycles 1-4, respectively. At DL3 patients were randomized to a four or six cycle variant of Mega-CHOEP with the following doses of CY: 1500, 6000, 6000, 6000 mg/m^2 ETO 600, 1480, 1480, 1480 mg/m^2(arm A) vs. CY 1600, 1600, 1600, 4500, 4500, 6000 mg/m^2, ETO 600, 600, 600, 960, 960, 1480 mg/m^2(arm B). From February 97 to May 01, 208 pts were enrolled in the study, 47 at dose level 1, 77 at dose level 2 and 84 at dose level 3. Five cases (2 %) of treatment related mortality did occur. So far, no case of secondary leukemia was reported. 101 / 110 eligible patients at dose level 1 and 2 were evaluable. In 100 / 101 pts, sufficient numbers (> 2 x 10^6 CD34+ cells / kg BW) of PBPC could be harvested for all scheduled cycles of stem cell transplantation. Recovery of leukocytes (> 1/ nl) was achieved after 12, 14, 15 and 16 days, recovery of the platelets (> 80/nl) occurred at day 14, 16, 17, 20 (medians) after courses 1-4 of MEGA-CHOEP. 75 % of the patients achieved CR or CRu 3 months after last cycle of therapy. At two years, overall survival was 69.6% and freedom from treatment failure was 66%. Actuarial median observation time is 21 months. At DL 3, results were different in arm A and B: 3 / 33 patients in arm B showed progress under therapy and 3 patient did not mobilize sufficient numbers of stem cells. This was not observed in 51 patients treated in arm A. Therefore Mega-CHOEP DL3A (four cycles) is the basis of the current protocol combining Rituximab with Mega-CHOEP therapy. Mega-CHOEP is feasible (DL1-3) and effective (DL 1+2) as primary treatment of aggressive NHL. Mega-CHOEP will be included in future Phase III studies of the German High Grade NHL study group for pts. with high risk features.

56

Usefulness of a registry for planning and conducting clinical trials

M. Horowitz, C. Bredeson, J.P. Klein, J.D. Rizzo, M.J. Zhang (Milwaukee, USA)

Large, representative observational databases (registries) are valuable but underutilized tools for conducting clinical trials. Studies of observational data can focus clinical trials efforts on areas most likely to be productive and aid in clinical trial planning, implementation and interpretation. Precise estimates of outcomes and accrual patterns can aid sample size calculations, monitoring strategies and implementation plans. Identification of appropriate centers for specific trials can increase efficiency. Information about most commonly used supportive care measures can increase acceptability of protocols. Failure to accrue can be systematically addressed by studying the treatment of patients eligible but not enrolled. Comparison of clinical trials outcomes with observational outcomes can give insight about generalizability and patient selection practices. An example of the use of the large clinical database maintained by the IBMTR/ABMTR to design a trial comparing four graft-versus-host disease regimens for allogeneic bone marrow transplantation will be presented, illustrating the use of registry data to identify appropriate endpoints, eligibility criteria and conditioning regimens and to help in statistical design.

57

Adaptive multistage designs in clinical trials

P. Bauer (Wien, A)

A new type of flexible designs is explained, which allows adaptive interim analyses with substantial mid trial design modifications without compromising on type I error probability. The basic principle of using a pre-fixed combination of the test results calculated separately for the disjoint samples at the different stages will be explained for the two stage test. Types of design modifications considered are sample size reassessment change of doses, dropping of treatments, selecting of endpoints, adapting the testing criterion, and others. The recursive application of this principle allows to generalize flexibility to the choice of the number of interim analyses performed: In each interim analysis the experimenter has to decide if he wants to end the trial with the next stage or if he plans for a further interim analysis (where the same decision will come up again). The implications of flexibility for the practice of clinical trials will be discussed for a simple example of a two stage design in a large scale clinical dose finding study. Such methods need a short term availability of validated data in order to allow important mid trial decisions based on information from the ongoing trial. They also will ascribe an increasingly important role to the statistician who will have to support the decisions by sufficient data and appropriate methodology.

59

Metaanalysis of the time-dependent parameter cytogenetic response of ten international studies: prognostic significance despite considerable amount of missing values

M. Pfirrmann, J. Hasford (München, D)

At diagnosis of chronic myeloid leukemia, the Philadelphia chromosome (Ph) is observed in metaphases of bone marrow cells in over 90% of the patients. In a data pool of 804 patients from 10 studies out of 9 countries, all treated with interferon-alpha, the reduction of such Ph-positive metaphases to 35% and less (major cytogenetic response (MCR)) and its prognostic influence on survival time was investigated. Survival time was measured from the start of interferon-alpha therapy, when it was presumed that all patients had more than 35% Ph-positive metaphases (no MCR). Investigators were asked to report for each patient any cytogenetic evaluation performed: the date, the percentage of Ph-positive metaphases, and the total number of metaphases analysed. For analyses of prognostic influence, only cytogenetic results based on at least 20 metaphases were considered, since, with binomial distribution assumption, the estimated percentage p of Ph-positive metaphases depends on the sample size n and p itself. With p around 35%, the lesser n, the larger the part of the confidence interval reaching into the other of the two response categories. It was examined whether in relation to all recorded results, the percentage of MCR based on at least 20 metaphases was representative within each study and, taking heterogeneity into account, whether it was in accordance with published proportions. Depending on patient and study, protocol violation with regard to predetermined examination times varied considerably. However, even if examinations were performed at the assigned 3 – 6 months, in almost every case a time lag between real and observed cytogenetic result was to be expected. To further reduce dependence on data quality, neither the duration of MCR, nor relapse or recurrence of MCR were considered. Starting from category „no MCR" at the date of first interferon taking, a patient changed to category „MCR" as soon as the namely was observed during the course of therapy. Simon-Makuch curves and a Mantel-Byar test displayed a statistically significant survival advantage for patients with MCR ($P < 0.0001$). Of 230 patients with MCR, 192 MCR (83%) were observed within 21 months. Landmark analyses of different points in time suggested 21 months as a good compromise between waiting for an MCR and an essential change in therapy in case of no MCR. Despite the obvious lack in protocol compliance, the time-dependent variable cytogenetic response was able to also refine two of three risk groups of the New CML score (JNCI 1998;90:850). Both within the intermediate risk group and the low risk group the patients with MCR and no MCR were successfully separated by clearly differing Kaplan-Meier curves from month 21 and corresponding statistically significant logrank tests. The results underline the prognostic significance of the time-dependent variable cytogenetic response and should also be understood as a motivation to aspire after higher accuracy of measurement and protocol compliance. For more recent and future studies an appropriate decision time earlier than 21 months would be expected. At current, additional statistical approaches allowing for further improved prognosis at various points in time are investigated.

60

A database for tumor cytogenetic data management

H. Rieder, H. Balz, H. Döhner, B. Heinze, A. Jauch, B. Mohr, C. Schoch, J. Bradtke (Marburg, Ulm, Heidelberg, Dresden, München, D)

Background: Detailed chromosome banding (CBA) and molecular cytogenetic analyses including fluorescence in situ hybridization on metaphase chromosomes (C-FISH) and interphase nuclei (I-FISH) and also reverse cytogenetics, e.g. comparative genomic hybridization (CGH) and FISH with microdissected chromosomes (D-FISH), contribute to a growing amount of data in cytogenetic investigations in leukemias. The International Standard for Human Cytogenetics Nomenclature (ISCN) allows for a complete description of the cytogenetic findings but is not sufficient to search for distinct chromosome breakpoints or partial gains and losses of chromosomal material. The presence of a complexly aberrant karyotype is part of the stratification in therapy protocols but several criteria are being used to define complex chromosome aberrations. Quality assessment and the interpretation of minimal changes in FISH analyses on interphase nuclei requires detailed records of the respective findings. Aims: A system for flexible management and detailed analyses of complex cytogenetic data should be developped. Methods: A minimal cytogenetic dataset was defined for findings in each cytogenetic method. Data from cytogenetic analyses in leukemias and a commercially available database program (Microsoft Access) were used. Results: A case orientated database was developped which allows for detailed management and analyses of tumor cytogenetic findings. For the documentation of findings of CBA, C-FISH, I-FISH, CGH and D-FISH analyses, the data were recorded in separate tables which are relationally connected. For complex search strategies, e.g. for chromosomal breakpoints and partial chromosomal gains and losses, the cytogenetic findings were recorded in separate fields using a specially defined data formate. Complex karyotypes are defined by scoring ploidy, numerical, and balanced as well as unbalanced chromosomal aberrations. For quality assessment and the correct interpretation of rare signal patterns in I-FISH findings, the signal patterns were recorded in detail for each probe set. Conclusions: A tumor cytogenetic database was developed which allows for detailed analyses of complex cytogenetic data and also quality assessment in I-FISH investigations. Supported by the Bundesministerium für Bildung und Forschung (BMBF), Kompetenznetz in der Medizin „Akute und chronische Leukämien".

Wissenschaftliches Symposium:
Akute lymphatische Leukämie

61

Minimal residual disease in childhood acute lymphoblastic leukemia

C. Bartram, T. Flohr, R. Beier, M. Schrappe (Heidelberg, Hannover, D)

The analysis of minimal residual disease (MRD) provides important information on the in vivo efficacy of chemotherapy and contributes to the individualization of treatment. The MRD status in childhood ALL appears to be superior to classical stratification criteria. The International BFM Study Group used clone-specific T-cell receptor and immunoglobuline gene rearrangements as PCR targets with a sensitivity of smaller than 10-4 on two follow-up time points to define three risk groups: a low-risk group (LR) of 43 percent of patients with MRD negativity at the end of induction therapy (day 33) and before the start of consolidation treatment (week 12) and a 5-year RFS of 98 percent; a high-risk group (HR) of 15 percent of patients with MRD levels larger than 10^{-3} at timepoint 2 and a 5-year RFS of 16 percent; an intermediate-risk group (IR) containing the 43 percent remaining patients with a 5-year RFS of 76 percent. Based on this experience the BFM study group introduced MRD-based treatment stratification in the current trial ALL-BMF 2000. Complex logistics had to be developed between the participating hospitals, the study center and MRD laboratory to guarantee on-time MRD results. 273 eligible patients were enrolled in a pilot phase (8/99-8/2000). MRD-based stratification was possible in 78% of patients. Taking into account additional risk group definitions (prednisone response, translocations 9;22 and 4;11) patients were classified as LR (32 percent), IR (51 percent) and HR (18 percent), respectively. Until 6/2001 the remission status of more than 500 children with ALL has been analysed. On-time MRD-based stratification is feasible in a multicenter trial if very reliable logistics and high laboratory standards are available.

63

Distinction of STI571 (Glivec) sensitivity or resistance of Ph+ ALL cells by gene expression profiling: Implication of diverse mechanisms of resistance

W.K. Hofmann, S. de Vos, D. Elashoff, H. Gschaidmeier, H.P. Koeffler, D. Hoelzer, O.G. Ottmann (Frankfurt/Main, D; Los Angeles, USA; Basel, CH)

The ABL tyrosine kinase inhibitor STI571 is a promising agent for the treatment of advanced Philadelphia chromosome positive (Ph+) acute lymphoblastic leukemia (ALL), but resistance develops within a few months in most patients. Utilizing gene expression profile analysis of 25 bone marrow samples from patients with Ph+ ALL undergoing treatment with STI571, we attempted to predict resistance to STI571. Ten of these samples were from individuals who were classified as good responders to STI571 (sensitive), 7 samples were collected from individuals who were found not to respond to STI571 (primary resistant). Eight samples from patients who initially responded to STI571 but relapsed after 6-12 weeks were obtained at the time of relapse while they were on treatment with STI571 (secondary resistant). Oligonucleotide microarray analysis was performed utilizing high-density microarrays (HuGeneFL, Affymetrix, Santa Clara, CA). The expression data were analyzed using the Genespring software (Silicon Genetics, San Carlos, CA). Class membership prediction analysis as well as hierarchical clustering based on expression data of 95 selected genes enabled us to discriminate all sensitive samples from the resistant samples exclusively according to their gene expression profile. Fifty-six highly differentially expressed genes were identified in the STI571 resistant leukemic cells, 20 were overexpressed and 36 had reduced or absent expression. These results demonstrate that gene expression profiles obtained prior to specific antileukemic therapy is clinically relevant in terms of predicting response and generating hypotheses about mechanisms of resistance. This may have profound implications for clinical studies of tyrosine kinase inhibitors in general.

CAMPTO® 40 mg / 2 ml, CAMPTO® 100 mg / 5ml. Wirkstoff: Irinotecan – Verschreibungspflichtig. **Zusammensetzung:** CAMPTO®-Durchstechflaschen 40 mg/ 2 ml bzw. 100 mg / 5 ml enthalten 40 mg bzw. 100 mg Irinotecanhydrochlorid $3H_2O$. **Sonstige Bestandteile:** Sorbitol, Milchsäure Natriumhydroxid, Wasser für Injektionszwecke. **Anwendungsgebiete:** Behandlung von erwachsenen Patienten mit fortgeschrittenem Dickdarm/Mastdarmkrebs: ohne Vortherapie in Kombination mit 5-Fluorouracil und Folinsäure, nach erfolgloser Vorbehandlung mit 5-Fluorouracil als Monotherapie. **Gegenanzeigen:** Schwangerschaft und Stillzeit; chronisch entzündliche Darmerkrankungen und/oder Darmverschluss. Bekannte schwere Überempfindlichkeitsreaktionen auf Irinotecanhydrochlorid $3H_2O$ oder einen der sonstigen Bestandteile von CAMPTO®. Bilirubinwerte > 1,5fache des oberen Normalwertes. Schwere Störung der Knochenmarkfunktion. WHO Performance Status > 2. Hinweis: Da dieses Arzneimittel Sorbitol enthält, ist es bei erblicher Fructoseintoleranz nicht geeignet. Während der Behandlung und mindestens drei Monate danach Kontrazeption notwendig. **Nebenwirkungen:** Verzögert einsetzende Diarrhoe (dosisbegrenzend), Übelkeit und Erbrechen, Dehydration, Obstipation. Selten Darmverschluss, intestinale Perforationen, gastro-intestinale Blutungen oder pseudomembranöse Kolitis. Anorexie, Bauchschmerzen, Mucositis. Neutropenie (dosisbegrenzend), reversibel und nicht kumulativ; Fieber, Infektionen, Sepsis; Anämie, Thrombozytopenien, 1 Fall peripherer Thrombozytopenie mit Thrombozyten-Antikörpern. Passageres schweres akutes cholinerges Syndrom (frühzeitige Diarrhoe, Bauchschmerzen, Konjunktivitis, Rhinitis, Hypotension, Vasodilatation, Schwitzen, Schüttelfrost, Unwohlsein, Schwindel, Sehstörung, Pupillenenge, Tränenfluss, übermäßiger Speichelfluss). Dyspnoe, Fieber ohne Infektion, Muskelkontraktionen oder -krämpfe, Parästhesien, Asthenie, Alopezie, milde Hautreaktionen. Selten: allergische Hautreaktionen und Reaktionen an der Einstichstelle, Niereninsuffizienz, Hypotension, Kreislaufversagen. Stomatitis. Transitorischer Anstieg der Transaminasen, der alk. Phosphatase, des Bilirubins, des Serum-Kreatinins. Die Befähigung zum Fahren eines Fahrzeug oder Bedienen einer Maschine kann nach Applikation eingeschränkt sein. **Stand:** Dezember 2000. AV 706 01 156. **Pharmazeutischer Unternehmer:** Rhône-Poulenc Rorer GmbH, D-50829 Köln. **Gekürzte Angaben,** vollständige Information siehe Fach- bzw. Gebrauchsinformation, die wir Ihnen auf Wunsch gern zur Verfügung stellen. Aventis Pharma Deutschland GmbH, Frankfurt (Main), Postanschrift: Postfach 1109, D-65796 Bad Soden/Ts.

Aventis

Gemeinsame Jahrestagung

der

Deutschen und Österreichischen Gesellschaften für Hämatologie und Onkologie

vom
25. bis 30. Oktober 2002
in
München

www.dgho.de

<u>Kongreßpräsident:</u>

Prof. Dr. W. Hiddemann
Medizinische Klinik III
Klinikum Großhadern der
Ludwig-Maximilians-Universität München
Marchioninistr. 15
D–81377 München
Tel.: +49-(0)89/7095-2551
Fax: +49-(0)89/7095-8875

Links: Text zur Anzeige auf der 4. Umschlagseite

64

STI571 (Glivec) to enable allogeneic stem cell transplantation (SCT) in relapsed or refractory Philadelphia-chromosome positive acute lymphoblastic leukemia (Ph+ALL)

B. Waßmann, J. Atta, H. Pfeifer, U. Scheuring, H. Martin, P. Brück, H. Gschaidmeier, D. Hoelzer, O.G. Ottmann (Frankfurt, Nürnberg, D)

Ph+ALL refractory to or relapsing after conventional chemotherapy has shown high initial response rates upon treatment with the ABL-tyrosine kinase inhibitor Glivec (formerly STI571, Novartis, Basel) in two multicenter phase II-trials (CSTI 109 and 114,). Responses were not sustained however, and allogeneic SCT presently remains the only potentially curative treatment of Ph+ALL. We investigated whether salvage treatment with Glivec could facilitate alloSCT. 33 of 52 consecutive Ph+ALL patients who were enrolled in phase II studies of Glivec at our center had not previously received an alloSCT and were evaluated for allogeneic transplantation. 23 pts. were considered eligible for allogeneic SCT based on performance status and age (median 35 yrs., range 18-58) versus 64 yrs. (range 44-73) in the group considered ineligible. At the time Glivec was started, the disease status of the 23 pts. considered for transplantation was refractory (n=13) or relapsed (n=10)(1st relapse n=6; 2nd relapse n=4) Ph+ALL. A suitable donor was available in 20 patients. 14 pts. underwent allogeneic SCT. Glivec was discontinued prior to conditioning therapy. Reasons for not performing a transplant in 9 of the 23 patients were primary resistance to Glivec (n=4), relapse during treatment before a donor was identified (n=3) or before SCT could be performed (n=1), and patient refusal (n=1). Outcome in SCT-patients:

Status at allo SCT	n	response after alloSCT				
		CR	TRM	n.a.	relapse	CCR
Resistant / relapse	8	5	2	1	1	4
CR	5	4	1	-	1	3
PR	1	-	1	-	-	-

Despite the high risk status of the pts. undergoing alloSCT, 7/14 pts. remain in continuing hematologic remission with a median follow-up of 156 (range 31-398) days. Severe posttransplant toxicity was in the expected range, TRM was due to respiratory failure (n=1, d=+32), multi-organ failure (n=2, d +5 and +52) and secondary graft failure (n=1, d +60). Median duration of STI therapy in the transplant group was 48 (22-70) days, the median interval from start of STI to SCT was 66 (41-152) days. In conclusion, Glivec is promising as salvage therapy for patients with Ph+ALL in whom an alloSCT is planned, although the time interval between starting Glivec and allogeneic SCT has to be short due to the development of resistance. A definite assessment of the efficacy and safety of Glivec in this setting requires future studies with larger numbers of patients and longer follow-up.

65

Long-term survival after autologous BMT/PBSCT in BCR-ABL+ ALL

H. Martin, J. Atta, S.A. Klein, F. Fauth, D. Bunjes, F. Graf Finckenstein, W. Knauf, K. Kolbe, R. Kuse, W. Langer, B. Wassmann, A. Weiss, D. Hoelzer (Frankfurt, Ulm, Hamburg, Berlin, Mainz, Essen-Werden, Mannheim, D)

Between 1992 and 1999 a total of 40 patients with BCR-ABL-positive acute lymphoblastic leukemia and a median age of 45 (range 6 - 64) years were autografted in a prospective multicenter GMALL study. Twentyseven patients received single bone marrow (n=19) or peripheral blood stem cell (PBSC) transplants (n=8), while 13 patients received sequential double PBSCT. At transplant all patients were either in first (n=34) or second (n=6) complete remission (CR) by morphology, however, all except one had evidence of residual BCR-ABL+ cells by semiquantitative RT-PCR. The autografts were purged with CD19+CD10 ± AB4 immunomagnetic beads achieving a 2 - 4 log in vitro depletion of residual BCR-ABL+ cells. Ten patients received posttransplant maintenance therapy with alpha-interferon (IFN) and interleukin-2 (rIL-2) (n=9) or 6-mercaptopurin (6-MP)/ methotrexate (MTX) (n=1). Twentyseven patients including all 6 autografted in CR2 relapsed between 55 and 413 (median 161) days posttransplant, another 5 died from causes other than leukemia. All patients autografted in CR 1 and receiving moderate pretransplant chemotherapy without posttransplant maintenance therapy relapsed (n=10), while 8 of 24 patients receiving either more intensive pretransplant chemotherapy or double autologous transplants and/or posttransplant maintenance therapy remain in continuing CR between 15 and 100 (median 52) months after transplant (p=0.02). Probabilities of leukemia-free survival for all patients, after single and after double autologous transplantation in CR1 are 22% ± 7%, 19% ± 8% and 27% ± 13%, respectively. We conclude that autologous transplantation in CR1 cures some patients and that overall treatment intensity including pre- and posttransplant therapy is of critical importance.

Wissenschaftliches Symposium: Akute myeloische Leukämie I

66

Functional proteomics of AML1-ETO in t(8;21)-AML

G. Behre, R. Vangala, V. Reddy, M. Aneja, T. Pabst, S. Schnittger, C. Schoch, T. Haferlach, M. Ueffing, D.G. Tenen, W. Hiddemann (München, D; Boston, USA)

Acute myeloid leukemia (AML) is characterized by a block in myeloid differentiation and altered proliferation and cell death. Genetic approaches have already identified many cytogenetic abnormalities with corresponding fusion genes or mutations in AML. However, little is known how those genetic aberrations encoding for fusion proteins lead to the biology of AML. Thus, functional genomics approaches aim to identify the target genes of fusion proteins in AML. One of those approaches is functional proteomics which aims to identify the targets of a fusion protein on protein rather than mRNA level. AML1-ETO is a fusion protein encoded by the translocation t(8;21), which is found in 7-15% of AML patients. Our recent results indicate that interactions of transcription factors with other nuclear proteins play not only an important role in lineage commitment and differentiation in the hematopoietic system, but also in the pathogenesis of myeloid leukemias. The transcription factor C/EBPalpha is crucial for the differentiation of granulocytes. Dominant negative mutations of C/EBPalpha have been found in some patients with AML, but not in AML patients with the t(8;21) translocation. However, AML1-ETO-positive AML blasts had undetectable C/EBPalpha protein levels. AML1-ETO can bind to C/EBPalpha, suppress its transcriptional activity, and thereby interfere with the autoregulation of the C/EBPalpha promoter, leading to suppressed C/EBPalpha expression. We suggest an application of proteomic methods (2D gel electrophoresis and mass spectrometry) towards identifying further regulated protein targets of AML1-ETO (interacting proteins of AML1-ETO and proteins altered in expression by AML1-ETO) and demonstrate the utility of this discovery-based global strategy in elucidating novel AML1-ETO pathway effectors.

67

Minimal residual disease assessed by fusion transcripts

U. Jäger, G. Mitterbauer, K. Lechner, C. Mannhalter (Wien, A)

The majority of reciprocal chromosomal translocations in acute myeloid leukemia (AML) results in the production of fusion transcripts, which can be assessed by reverse-transcriptase polymerase chain reaction (RT-PCR). AML-specific transcripts can currently be detected in approximately 50% of patients. During recent years the quantification of fusion transcripts has been made possible by competitive and real-time PCR technologies. This has greatly improved our ability to monitor minimal residual disease (MRD). The major translocations (PML/RARalpha, AML1/ETO, CBFß/ MYH11, MLL & partners) have been quantified during the course of disease. We have gained insight into the expected time-course of MRD in patients reaching complete remissions after various treatment modalities, including autologous or allogeneic stem cell transplantation. Assessment of MRD has been shown to be of decisive prognostic value and improved outcome of patients treated for molecular relapse. Reviewing the data currently available the following issues will be discussed in this presentation: When and for how long should MRD in bone marrow and peripheral blood be assessed for molecular remission? In which instances should quantitative PCR be preferred over qualitative two-step PCR? Which level of MRD is associated with long term clinical remission? How does the detection of molecular relapse influence clinical decisions?

68

The Core Binding Factor (CBF) leukemias: Recent advances in treatment and prognosis

G. Heil, J. Krauter, A. Ganser (Hannover, D)

In about 60% of the acute myeloid leukemias (AML) nonrandom numerical and structural chromosome aberrations can be identified by routine cytogenetic analysis. Two of the most common chromosomal lesions in AML, the translocation t(8;21) and the inversion of chromosome 16 are detectable in 15-20% of adult AML patients. Both aberrations target the genes for the CBFα/CBFβ transcription factor complex essential for normal hematopiesis: In the t(8;21) the 5' region of CBFα (AML1) is fused to the MTG8 gene resulting in a chimeric AML1/MTG8 fusion transcript. The AML1/MTG8 fusion protein retains its DNA binding domain and recruits transcriptional repressors which impair the expression of genes involved in myeloid differentiation. In the inv(16) the CBFβ-gene is fused to the MYH11 gene. The resulting fusion protein most probably acts as a dominant negative repressor of AML1-dependent gene expression. Recently the t(8;21) and the inv16 positive leukemias have been adressed as the „core binding factor (CBF)- associated leukemias. In almost all studies dealing with adult de novo AML patients a CR rate up to 90% is achieved in the CBF leukemias as a strong evidence of their high initial chemosensitivity. There are, however, contradictory results with respect to the longterm outcome of these patients. The CALGB 8525 trial with a high-dose Ara-C based postremission therapy has produced encouraging cure rates of over 80%. As a consequence these leukemias were grouped together as the „favorable risk" AMLs. This superior outcome, however, could not be confirmed by other study groups. The reasons for these discrepancies are not known for the present. One has to take into consideration that these data are based on relatively small patient numbers and different treatment strategies. In addition, other prognostic factors such as additional chromosomal lesions and/or the white blood cell count might have adversely affected the outcome. To further elucidate this issue intergroup analyses should be performed to get a more precise information about the longterm prognosis of these patients as a prerequisite for an optimized treatment strategy. In this context the quantification of minimal residual disease (MRD) by quantitative RT-PCR techniques is of interest. Despite the fact that most CBF patients retain their molecular markers even during longtrem remission quantification of MRD during and after chemotherapy can provide further information about the chemosensitivity of the leukemic clone, the prognostic relevance of MRD and finally may help to identify the optimal procedure to eliminate the residual leukemic blasts. In conclusion CBF leukemias have a well defined molecular defect in common, respond favorably to initial chemotherapy while the optimal postremission strategy remains to be defined.

69

High-risk acute myeloid leukemia

H. Döhner, D. Bunjes, R. Schlenk (Ulm, D)

In acute myeloid leukemia (AML), the karyotype and the response to first induction therapy are one of the most important predicting factors for disease-free (DFS) and overall survival (OS). Cytogenetic high-risk features include abnormalities of e.g. the long arms of chromosomes 3, 5, 7, and of the short arms of chromosomes 12 and 17. CR rates in these subgroups of patients range between 0% and 40% only. Furthermore, there is data from prospective multicenter treatment trials indicating that these patients have dismal DFS even when treated with intensive consolidation therapies such as high-dose cytarabine or high-dose chemotherapy followed by autologous transplantation. The large cooperative study groups are now embarking on treatment protocols in which these subgroups of patients are stratified for experimental induction or postremission therapy. In our ongoing multicenter treatment trial AML HD98A from the AML Study Group Ulm (AMLSG ULM) patients (16 to 60 yrs) with cytogenetic high-risk leukemia and/or patients with persistent disease (>25% bone marrow blasts) after the first induction cycle with idarubicine, cytarabine and etoposide (ICE) are assigned to allogeneic hematopoietic stem cell transplantation with grafts either from matched related donors, matched unrelated donors or haploidentical family donors. In an interim analysis from March 2001, 87 of 381 (23%) eligible patients fulfilled high-risk criteria: 22 patients had an unfavourable karyotype and achieved PR/CR after induction therapy with ICE; the remaining 65 patients had persistent disease after ICE 1 and either had a normal or an unfavourable karyotype. So far, 50 of the 87 patients were transplanted, 23 from a matched related donor, 24 from a matched unrelated donor, and 3 from a haploidentical family donor. In order to improve the results after allogeneic transplantation experimental conditioning regimens using radioimmunoconjugates are currently being evaluated. This approach has been pioneered in both animal models and patients by investigators from Seattle and New York. In the AMLSG ULM, we are currently investigating the use of different strategies to intensify the conditioning regimen: 188Re or 90Y-labeled anti-CD66 (a,b,c,e,) antibody in combination with busulfan/cyclophosphamid for patients in CR or PR at the time of tranplantation; and bisphosphonate targeted radionuclide therapy with 186Re HEDP in combination with busulfan/cyclophosphamid in patients with persistent leukemia. Results from cytogenetics/molecular genetics, from feasibility and early results of allogeneic transplantation, and from intensification of conditioning regimen in this subgroup of AML patients will be presented.

70

Combinations of all-trans retinoic acid and chemotherapy in acute promyelocytic leukemia

E. Lengfelder, R. Hehlmann, T. Büchner (Mannheim, Münster, D)

In patients with APL, complete remission rates over 90% can be achieved with all-trans retinoic acid (ATRA) alone. The addition of intensive chemotherapy during or after ATRA treatment is essential in order to stabilize the remission. Whereas the beneficial impact of anthracyclines seems certain, the effectiveness of ara-C in standard dose or in high dose (HD) is still discussed. In the treatment protocol of the AMLCG for newly diagnosed APL, an intensified chemotherapy regimen is combined with ATRA. The treatment regimen consists of ATRA and double induction therapy (TAD/HAM) including HD ara-C, followed by TAD consolidation and three years maintenance. In patients over 60 years, the second cycle (HAM with dose reduced ara-C) is only given, if no complete remission is reached after the first cycle. Since November 1994, 99 patients with newly diagnosed APL, median age 45 (17-82) years, were treated according to the protocol. Confirmation of the morphologic diagnosis APL (M3 69%, M3v 31%) by the translocation t(15;17) (n=89) or PML/RARα (n=86) was mandatory for elegibility. The initial white blood cell count (WBC) was <5000/µl in 66% and >5000/µl in 34% of the patients. The median observation time of all patients was 29 months (1 day to 6.0 years).

	n	CR (%)	PR (%)	ED (%)	OS (%)*	EFS (%)*	RR (%)*
all	90	89	1	10	84	84	2
<60 y	67	93	0	7	88	88	0
>60 y	17	71	5	24	69	71	12

y: years, CR,PR: complete, partial remission, ED: early death, OS, EFS: overall, event free survival, RR: relapse rate, *after 2 years.

These results indicate a particularly high antileukemic effectiveness of the intensified double induction therapy including high dose ara-C in combination with ATRA. The relapse rate at two years is by far the lowest as com pared to all other treatment regimens in APL published by now.

Wissenschaftliches Symposium:
Therapie kolorektaler Karzinome

71

Adjuvant therapy of colon cancer: State of the art

W. Queißer (Mannheim, D)

The primary therapy of colon cancer is surgical resection suggesting a definite cure rate for all tumor stages of about 50%. However, within the last decade several prospectively randomized clinical trials have convincingly demonstrated that postoperative adjuvant chemotherapy is of benefit to all patients with high-risk disease, especially for those with positive lymph nodes (stage III, Dukes C) and possibly also for those with negative lymph nodes but large primary tumor mass (stage II, Dukes B 2-3). Four pivotal and several confirmatory studies have demonstrated the highly statistical significant survival advantage of 5-Fluorouracil/Levamisole – and 5-Fluorouracil/Folinic Acid – based treatments in Dukes´ C colon cancer patients suggesting a median survival benefit of 6% (4% to 16%). The postoperative use of 5-Fluorouracil/Folinic Acid for approximately six months represents standard care in view of most investigators. According to the high spontaneous postoperative 5-years survival of 70 to 80% in Dukes B colon cancer patients the efficacy of adjuvant treatment is more difficult to prove by statistical analysis. Actual subgroup analyses of two large international studies suggest improved prognosis for Dukes B under adjuvant chemotherapy though it reached statistically significance only in one of them. For this reason at present adjuvant treatment should be limited to high-risk Dukes´ B colon cancer patients such as tumor perforation, tumor adherence, invasion of adjacent organs or early venous, lymphatic or perineural invasion. Current clinical trials are evaluating the use of activated chemotherapy. High-dose 5-Fluorouracil via an intravenous port and Folinic Acid is compared to standard 5-days-schedule of both compounds. 5-Fluorouracil/Folinic Acid standard schedule is being studied with and without addition of Irinotecan and Oxaliplatin, respectively, since recent studies in advanced colorectal cancer have demonstrated that combinations with Irinotecan or Oxaliplatin are superior to 5-Fluorouracil/Folinic Acid alone. Edrecolomab (Panorex[R]), a murine IgG2a monoclonal antibody, was investigated in an initial German trial of 189 patients with resected Dukes C colorectal cancer suggesting a reduced relative risk of mortality by 32% compared with observation alone. Therefore a large European study was initiated to compare Edrecolomab versus 5-Fluorouracil based chemotherapy versus Edrecolomab plus chemotherapy (900 patients in each arm). Recent interim analysis showed that Edrecolomab is not able to improve prognosis compared to chemotherapy alone and in combination with chemotherapy. The aim of recent investigations is to develop new methods that could identify those patients at risk of recurrence in early postoperative stage. Such biological markers may be MIB-1, tumor cell ploidy, microsatelite instability, p 53, K-ras mutation and the DCC gene.

73

Randomized multicenter phase II trial of oxaliplatin plus irinotecan versus raltitrexed as first-line treatment in advanced colorectal cancer

W. Scheithauer, B. Schüll, G. Kornek, M. Raderer, W. Fiebiger, C. Gedlicka, H. Ulrich-Pur, S. Brugger, B. Schneeweiss, F. Lang, A. Lenauer, D. Depisch (Wien, St.Pölten, Kirchdorf, Neunkirchen, Wr.Neustadt, A)

Irinotecan and oxaliplatin are two new agents with promising activity in advanced colorectal cancer (ACC). Based on preclinical and clinical evidence that both drugs act synergistically, a randomized phase II study has been initiated to investigate the therapeutic potential and tolerance of this combination in the front-line setting. 92 patients with measurable ACC previously untreated for metastatic disease were randomly allocated to receive oxaliplatin 85 mg/m^2 + irinotecan 175 mg/m^2 both administered as a 2h-infusion on days 1 + 15 every 4 weeks or to raltitrexed 3 mg/m^2 given as a 15-min infusion on day 1 every 3 weeks. In both treatment arms, chemotherapy was continued for a total of 6 months unless prior evidence of PD. According to a cross-over design, in patients progressing as well as in those relapsing after withholding therapy, second-line treatment with the opposite arm was effected. Patients allocated to oxaliplatin/irinotecan had a significantly better radiological confirmed response rate (43.5% vs. 19.6%; P = 0.0025) and longer progression-free survival (median, 7.1 vs. 5.0 months; P = 0.0033). The improvement in overall survival, however, did not reach the level of significance (median, 16.0 vs. 16.5 months; P = 0.3943). Overall, 52 patients were crossed-over or received other second-line treatment upon the development of PD. The response rate was 8/24 (33.3%) with oxaliplatin/irinotecan compared to 3/21 (14.2%) evaluable patients treated with second-line raltitrexed. Oxaliplatin/irinotecan caused more haematological and gastrointestinal toxicities, necessitating dose reductions in 10 of the first 20 patients. After subsequent adjustment of the irinotecan starting dose from 175 mg/m^2 to 150 mg/m^2, the tolerance of treatment was acceptable; the most commonly encountered events (all grades) included neutropenia (81%), alopecia (65%), nausea/emesis (62%), peripheral sensory neuropathy (62%), and diarrhea (46%). In the raltitrexed control arm, the most frequently reported events were neutropenia (52%), transient elevation of liver functional parameters (52%), nausea/emesis (43%), asthenia (30%), and diarrhea (30%). The oxaliplatin/irinotecan combination seems beneficial as first-line therapy in advanced colorectal cancer, demonstrating a superior response rate and a prolonged progression-free survival with an acceptable toxicity profile at the reduced irinotecan dose level. The impression of a promising therapeutic potential of the experimental combination is being supported by the high response activity (33.3%) noted in the raltitrexed control arm after cross-over to oxaliplatin/irinotecan, which is also likely to explain the lack of a difference in overall survival. Concerning the latter aspect, the best therapeutic sequence in patients with metastatic colorectal cancer remains to be determined.

74

CPT-11 combined or alternated with 5-fluorouracil / folinic acid (5-FU / FA) vs. MAYO clinical regimen in first line therapy of advanced colorectal cancer (ACRC)

U. Graeven, K. Ridwelski, M. Manns, J. Schölmerich, P. Espana, A. Murias, G. Carlsson, M. Borner, D. Oulid-Aissa, B. Boussard, W. Schmiegel (Bochum, Magdeburg, Hannover, Regensburg, D; Madrid, Gran Canaria, E; Göteborg, S; Bern, CH; Paris, F)

In first line chemotherapy of ACRC, CPT-11 combined with 5 FU/FA has shown significant higher antitumor activity than 5 FU/FA alone, leading to a significant increase in survival. A randomised multicenter Phase II study was initiated to assess the efficacy and safety of different combinations of CPT-11 and 5 FU/FA vs the Mayo clinic regimen. Arm A = CPT-11 125mg/m^2 plus 5 FU bolus (b) / FA 20mg/m^2, weekly for 4 weeks (wk) every 6 wk. Arm B = CPT-11 350 mg/m^2, day (d) 1 alternating with 5 FU 425mg/m^2 b/ FA 20mg/m^2, d1 to d5, q6wk. Arm C = 5 FU 425mg/m^2 b/ FA 20 mg/m^2, d1 to d5, q4wk. Pts had measurable ACRC, performancs status (PS) 0-2, and normal marrow, hepatic and renal function. Results (Table 1):

	Arm A	Arm B	Arm C
Evaluable patients	(n = 46)	(n = 50)	(n = 50)
CR	2 (4%)	2 (4%)	4 (8%)
PR	10 (22%)	10 (19%)	9 (18%)
Overall PR + CR	12 (26%)	12 (23%)	13 (26%)
MR	8 (18%)	8 (15%)	4 (8%)
SD	10 (22%)	8 (15%)	11 (22%)
PD	9 (20%)	13 (25%)	16 (32%)
Tumor growth control CR + PR + MR + NC	66%	53%	56%
G3/4 diarrhea/pts (%)	7	21	18
G3/4 stomatitis/pts (%)	2	9	10
G3/4 neutropenia/cy (%)	2	14	2
Median TTP(months)	8.3	6.9	5.9
Median Overall survival	19.1	14.8	17.7

In conclusion the two CPT-11, 5 FU/FA schedules can be given at full dose. These results show that the combined vs the alternated schedule does not compromise the good tolerance and has a very good efficacy.

75

Second-line treatment of advanced colorectal cancer with irinotecan, 24-hour high-dose 5-fluorouracil (HD-5-FU) and folinic acid (FA) after failure for HD-5-FU / FA - first results of a phase II study

R. Hofheinz, G. Hartung, W. Queißer (Mannheim, D)

Background: Irinotecan (IRI) has shown to prolong survival and improve quality of life compared to best supportive care in colorectal cancer (CRC)-patients with pretreatment of bolus 5-FU. After first-line 24-hour HD-5-FU/FA an objective response rate of 12% with 3-weekly IRI 350mg/m^2 has been reported (Schöffski, ASCO 2000;1155). Toxicity (NCI-CTC 3-4) in this study was as follows: neutropenia 16%; diarrhea 23%, nausea 12%. The present study investigated weekly IRI in combination with 24-hour HD-5-FU and FA as a second-line treatment after prior exposure to 24-hour HD-5-FU 2,600 mg/m^2 and FA 500 mg/m^2. Clinically known synergy between 5-FU and IRI was the rationale to combine both drugs for second-line therapy in order to overcome resistance to 5-FU. Weekly IRI was chosen to reduce toxicity compared to a 3-weekly schedule. Patients and methods: Since 11/99 35 patients (m/f 22/13; median WHO performance status I) were recruited in a single institution to receive 6x weekly IRI 80 mg/m^2, FA 200 mg/m^2 and 24-hour HD-5-FU 2,000 mg/m^2. Treatment was repeated on day 57. Results: So far 25 patients are evaluable for response. 4 patients (16%) achieved a partial remission. In 10 patients (40%) no change was observed, while in 11 patients (44%) disease progressed.12 patients have died after a median of 7.3 months (3.5-12 months). Median time to progression and overall survival are pending. 68 treatment cycles (28 pts with at least 1 cycle) are evaluable for toxicity (NCI CTC grade 3-4): Neutropenia 4%, thrombocytopenia 0%, vomitus/nausea 7%, diarrhea 7%. No further grade 3-4 toxicity was observed. Conclusion: Weekly IRI/HD-5-FU/FA is an active combination chemotherapy with tumor control rates in the range of other IRI-based second-line schedules. The toxicity profile in our study appears to be favourable compared to 3-weekly IRI.

76

Extended phase I study of capecitabine in combination with weekly irinotecan as a first-line chemotherapy in metastatic colorectal cancer

M. Tewes, N. Schleucher, W. Achterrath, S. Cao, S. Frings, J. Wilke, A. Harstrick, Y.M. Rustum, S. Seeber, U. Vanhoefer (Essen, Bad Soden, Grenzach-Wyhlen, D; Buffalo, USA)

Capecitabine (CAP) demonstrated efficacy in metastatic colorectal cancer (CRC). Preclinical data on nude mice bearing human colon tumor xenografts demonstrated significant synergistic antitumor activity for the combination of CAP and irinotecan (CPT). Based on these data a phase I study of CAP combined with weekly CPT was conducted in patients (pts) with measurable metastatic CRC as first-line chemotherapy. Methods: CAP (bid) d 1-14 and d 22-35, CPT 30 min. inf. weekly x 6 followed by a one week rest (one cycle). Dose level [DL]1: CAP 1000 mg/m^2 bid and CPT 70 mg/m^2; DL2: CAP 1250 mg/m^2 bid and CPT 70 mg/m^2; DL3: CAP 1250 mg/m^2 bid and CPT 80 mg/m^2. Results: 37 patients (pts) were entered on 3 DL and 89 cycles have been administered. Pts characteristics: Male/female 26/11 pts; PS 0 (0-2); median age 60 years (32-71); prior adjuvant CTx or/and RTx 14 pts. In the first 17 pts treated at DL1-3 the MTD has been reached at DL3 with diarrhea and neutropenia being dose-limiting. In order to confirm the recommended dose (RD), this DL2 was extended to 15 pts, demonstrating an incidence of DLTs in 5 (33%) out of 15 pts (diarrhea, neutropenia, one toxic death). Because the incidence of DLTs was considered to be too high, DL1 was extended to a total of 16 pts. DLTs were observed in 3 (19%) out of 16 pts. The main toxicity observed was diarrhea. So far, 13 out of 29 response evaluable pts (DL 1-3) (45%; 95%CI: 26-63%) and 6 out of 13 pts (46%; 95% CI 19-75%) at the RD (DL1) achieved an objective response. Discussion: CAP 1000 mg/m^2 bid and CPT 70 mg/m^2 is the recommended dose for further studies. The combination of CAP and CPT is feasible and showed promising efficacy as first-line chemotherapy in advanced colorectal cancer.

Wissenschaftliches Symposium: Diagnostik und Therapie in der Palliativmedizin

77

Gastroenterological diagnosis and therapy in palliative medicine

H. Weiss, A. Weiss (Ludwigshafen, Mannheim, D)

Despite a difference in the meaning of „palliative medicine „ and „palliative therapy",
gastroenterolocical interventions in palliative medicine are mostly used for palliative treatment.
The tools of gastroenterological diagnosis are used freely in the staging of the disease and disease progression, especially in emergencies (bleeding, diarrhea and ileus, vomiting, jaundice, ascites)
Therapy comprises treatment of the tumor itself as well as of the concomitant diseases: such as the control of gastrointestinal bleeding (both from varices and from ulcers), tumor-resection or-reduction,dilatation of tumorstenosis with or without laser-therapy, implantation of stents throughout the whole gastrointestinal tract, management of gastroesophageal reflux, percutaneous endoscopic gastrostomy and jejunostomy. Therefore gastroenterological diagnosis and therapy are essential parts of disease management in palliative medicine.

78

Pulmonary symptoms: Diagnostic and therapy in the terminal phase

H. Kaiser (Gütersloh, D)

Pulmonary symptoms, e.g. dyspnoe, cough are very common complaints in patients with advanced illness and often have a devasting impact on patients, family and caregivers. The pathophysiology of these symptoms and the appropriate investigations are discussed. The correlation between subjective assessments and objective parameters is very poor and palliative care assessment therefore should focus on dyspnoe as a symptom. The treatment is directed at the underlying cause when ever possible. Although evidence on the use of many treatments for these common and frightening symptoms is lacking, some symptom management strategies employed to relieve respiratory symptoms and anxiety can be described. When specific therapies no longer exist, palliative interventions based on medical treatment, psychosocial support and breathing control can help patients to cope with the symptom of breathlessness and reduce physical and emotional distress.

80

The role of psychooncology in todays oncology

J.U. Rüffer (Köln, D)

Besides being an illness that seriously harms the body cancer puts a heavy psychological burden on the patient. More than any other disease the diagnosis cancer confronts a patient with the possibility of early dying. In this life threatening situation there is a great need for supportive care. In a modern understanding of cancer treatment psychooncology support plays an important role. In general, psychooncology deals with the ressources a patient has and tries to support the patients coping process. Furthermore, psychooncology makes strong efforts to evaluate the impact of cancer on the psychological functions of the patient, the patient´s family, and staff. On the other hand the role of psychological and behavioral variables in respect of cancer risk and survival is of major interest.
In modern treatment strategies it is mandatory that psychooncology interventions should start directly at the time of diagnosis. Psychooncologic interventions may be helpful from cancer diagnosis until the end of live. This time span is characterized by different phases of coping depending on several variables like the patients age, type of tumor and prognosis. All interventions should aim on encouraging the individual being affected by cancer to regain self-control and self-conciousness. Every patient must be looked at individually and an individual strategy for support has to be developed. Besides developing these support strategies psychooncology research gives new insights in the underlying mechanisms of the relationship and communication between patients and their doctors. For example the cancer fatigue syndrome of which for a long time the importance for the patients has not been recognized by the doctors has become a field of major interest through psychooncology research. With the rapid development in cancer diagnosis and treatment also psychooncology is faced with new challenges. One of these future challenges will be the investigation of the psychosocial burden of genetic testing for the predisposition of cancer. Whereas improvement of Quality of Life by psychooncology interventions is proven in many studies, the impact of these strategies on the disease progression or overall survival is not finally evaluated. Investigations leading to a better understanding of the psychoneuroimmunological effects being observed in several studies are still pending. Thus, there are three main challenges in the future: establishing psychoncology as a supportive tool that should be offered routinously to cancer patients, clarifying the role of psychooncology in respect of disease free and overall survival, and investigating the new developments in cancer diagnostics and treatment focussing on the psychooncology consequences.

81

Evaluation of an ultrasound-guided technique for central venous access by the internal jugular vein in 493 patients

U. Mey, C. Hahn, M. Gorschlüter, C. Ziske, B. Schöttker, M. Mergelsberg, T. Sauerbruch, A. Glasmacher, I.G.H. Schmidt-Wolf (Bonn, D)

Background: Ultrasound guided cannulation of the internal jugular vein clearly offers advantages over the classic external guided landmark technique, particularly in complicated cases (e.g. thrombocytopenia, obesity, dyspnoe). In this prospective examination we evaluated an ultrasound-guided method for central venous access in 493 hematological patients and analyzed patient and operator dependent variables with regard to success and complication rates. Methods: 493 percutaneous cannulations of the internal jugular vein in patients treated for hematological diseases were prospectively recorded using the same two-operator catheterization technique. Success rate, frequency and type of complications were recorded, as well as patient-dependent risk factors (e.g. obesity, thrombocytopenia, vessel diameter, patient compliance) and experience of both operators (performing the catheterization and ultrasound guidance respectively), which were then analyzed with regard to their impact on success and complication rate. Results: Internal jugular vein cannulation was successful in 95.1% of all patients. The jugular vein was accessed on the first attempt in 87.6%. Complication rates for carotid puncture and hematoma were 1.4% and 4.3% respectively. Regarding patient dependent risk factors only low patient compliance during the procedure and a vessel diameter < 7 mm negatively impacted outcome. Experience of the operator performing the puncture did not impact complication and success rate. In contrast, the degree of the experience of the operator in performing ultrasound guidance significantly impacted both success and complication rates. Conclusion: Ultrasound-guided cannulation of the internal jugular vein provides safe central venous access with high success rates and low complication rates. Difficult cases due to patient-dependent risk factors (e.g. thrombocytopenia, obesity, dyspnoe etc.) can be managed well using ultrasonographic guidance. Success and complication rates are not impacted by the experience of the operator performing the puncture, but are impacted by the experience of the operator performing ultrasound guidance. Therefore, we would recommend that, particularly in difficult cases, the more experienced physician should perform ultrasonographic guidance to ensure as high a degree of quality and safety as possible.

Wissenschaftliches Symposium: Infektiologie

82

Treatment of fungal infections – Guidelines of the Infectious Diseases Working Party (AGIHO) of the German Society of Hematology and Oncology (DGHO)

A. Böhme, K. Blumenstengel, D. Buchheidt, H. Einsele, S. Guth, G. Heußel, C. Heußel, C. Junghanß, M. Karthaus, W. Kern, T. Kubin, H. Ostermann, J. Ritter, G. Silling, M. Ruhnke, O. Sezer, T. Südhoff, H. Szelényi, A. Ullmann (Frankfurt, D)

Fungal infections are a major cause of morbidity and mortality in patients with hemato-oncologic malignancies. Whereas the treatment of oropharyngeal and esophageal candidosis with topical or systemic antimycotic therapy is mostly unproblematical, the lethality of invasive fungal infections is furthermore high. The standard treatment of systemic candidoses is amphotericin B (AmB) with/without 5-flucytosine (5-FC) or fluconazole, whereby most experts prefer AmB in clinically unstable patients. The optimal dose of fluconazole (400mg, 800mg or more) is controversely discussed. The combination therapy with AmB plus 5-FC represents the gold standard in cryptococcal infections, but its benefit remains unclear for the most cases of candidoses or aspergilloses. For the treatment of aspergilloses AmB and itraconazole are available. Considering the high lethality of mould infections an adequate dosage of AmB is mandatory. Due to the well-known toxicity of conventional AmB a switch to the better tolerable AmB-lipid-formulations may be sometimes unavoidable. But definite criteria for that switch do not exist and may depend on financial considerations and physicians experience in supportive therapy options. New antimycotics and formulations are under investigation. Intravenous itraconazole, caspofungin and voriconazole will be licensed in Germany soon. Clinical trials with posaconazole, ravuconazole, liposomal nystatin, new echinocandins and various combination therapies are ungoing. A clear evidence of a benefit by additional treatment with hematopoietic growth factors (G-CSF or GM-CSF) is furthermore lacking. Recently, granulocyte transfusions are re-evaluated in severe infections, particularly invasive fungal infections. In mould infections with organ lesions invasive strategies such as surgical intervention may be successful in the management of pulmonary bleeding and CNS or sinus manifestations, but is also discussed for secondary prophylaxis in patients with further intensive chemotherapy regimen. For many types of fungal infections large studies in the special group of neutropenic patients are lacking and procedures analogue to non-neutropenic, intensive care or HIV-infected patients or individual treatment are necessary. The Infectious Diseases Working Party of the DGHO presents evidence-based guidelines for the treatment of proven and probable fungal infections in patients with hemato-oncologic malignancies in consideration of published study results, clinical experiences and expert-group recommendations.

83

Practice guidelines for infection prevention in neutropenic cancer patients

W.V. Kern, DGHO Arbeitsgruppe Infektionen (Ulm, D)

Research on prophylaxis of infections in neutropenic patients has been fraught with controversy. There have been numerous recommendations based on observational studies or on findings made in critically ill nonneutropenic patients, and there are very few metaanalyses of the limited controlled clinical trials of chemoprophylaxis that were published during the last two decades. The DGHO Working Group Infections in Hematology-Oncology recently has published practice guidelines for infection prevention in neutropenic cancer patients summarizing the consensus achieved among German infectious disease experts and hematologists and oncologists working in this field. New studies not yet available at the time of submission of the current guidelines have now been published and will be reviewed in terms of their influence on updated recommendations. Relevant new studies have primarily addressed antifungal chemoprophlaxis and airborne fungal spore loads in different patient environments. The issue of general systemic antifungal chemoprophylaxis versus preemptive therapy in selected risk groups still awaits thorough clinical research initiatives. There is no new data on the issue of antibacterial chemoprophylaxis or antiviral chemoprophylaxis that would alter previous recommendations. Studies, however, are planned or under way in different countries that hopefully will better answer the question of whether antibacterial prophylaxis with a fluoroquinolone provides benefit to the patients and health care system or has only minor effects on such important outcome measures as fever length, need for hospitalization, total consumption of antimicrobial drugs.

84

Three-armed multicenter randomized study on the empiric treatment of neutropenic fever in a high risk patient population (PEG Study III)

O.A. Cornely, D. Reichert, D. Buchheidt, G. Maschmeyer, M. Wilhelm, X. Schiel, B. Glaß, G. Fätkenheuer, H. Link, M. Helmerking, D. Adam, W. Hiddemann (Köln, Oldenburg, Mannheim, Berlin, Würzburg, München, Göttingen, Kaiserslautern, D)

Background: The safety and efficacy of monotherapy in high risk neutropenic fever is a controversial issue Methods: A prospective, randomized, multicentric trial was performed to evaluate the efficacy and safety of meropenem (1g q8h), cefepime (2g q8h), and pip/tazo (4.5g q8h) plus aminoglycoside qd in the treatment of neutropenic fever. In case of fever (oral temperature >38,5°C on once or >38.0°C twice within 24 hours) and neutropenia <500/µl adult patients with hematological malignancies were randomized. The primary endpoint was defervescence after 72h followed by 7 afebrile days. Success was defined as <38°C at 72h without treatment modification. Secondary endpoints were overall response 7 days after end of treatment and toxicity. Results: 562 patients were included, 454 (81%) were evaluable. Acute leukemia accounted for 64%, high grade lymphoma for 36%. Episodes were due to fever of unknown origin (319/70%), pneumonia (36/8%) and other clinically or microbiologically documented infection (99/22%). Success without modification at 72h was achieved with meropenem, cefepime and pip/tazo + aminoglycoside for fever of unknown origin in 62%, 66%, 65%, for pneumonia in 28%, 38%, 40%, for documented infections in 49%, 42%, 39% (statistically not significant). At end of study success over all types of infection was met in 91% equally for all three treatment groups. In acute leukemia 87% and in lymphoma 97% responded ($p<0.01$). Conclusion: Monotherapy is still a viable alternative in the treatment of neutropenic fever.

85

Surveillance of the incidence of respiratory viral, adenovirus and CMV-infections in the first 6 months following allogeneic or autologous stem cell transplantation - a prospective study.

E. Holler, A. Wodzynski, J. Hahn, H.J. Kolb, G. Jäger, H. Wandt, W. Ebell, B. Hertenstein, K. Schilling, A. Ullmann, T. Klingebiehl, U. Duffner, B. Schaaf, R. Andreesen (Regensburg, München, Nürnberg, Berlin, Hannover, Jena, Mainz, Tübingen, Freiburg, D)

In order to collect reliable data on the incidence and relevance of infections and diseases caused by respiratory viruses such as respiratory syncytial virus (RSV), Influenza and Adenovirus (ADV) in patients (pts) receiving allogeneic or autologous stem cell transplantation (SCT), a 1 year surveillance study observing 495 consecutive transplants in 480 pts for the first 6 months following SCT was performed in 11 German SCT centers: Virus surveillance was performed by weekly screening of throat swabs/nasal washes for RSV and ADV and urine for ADV using sensitive antigen-ELISAs. In symptomatic pts, additional viral tests addressing further viruses such as Influenza and confirmatory tests such as viral cultures and PCR were initated. CMV infections and diseases as monitored by local strategies were recorded as well. In 365 pts analyzed so far, the incidence of ADV-,CMV- and Influenza infections was increased in recipients of allogeneic SCT (ADV 31.5%, CMV 33.3%, Influenza 19.9%) as compared to recipients of autologous SCT (ADV 7.8%, CMV 15.8%, Influenza 1%), whereas RSV occured to a similar extent in both groups of pts (6,4-9,8%). Symptomatic diseases affecting predominantly the respiratory and gastrointestinal tract were observed in 30/103 ADV infections, in 5/25 RSV infections, in

MSD

Bald kommt...

C

das erste
Echinocandin-Antimykotikum, MSD

Mein YAHOO! Für Ärzte
www.meinearztseite.de/Hotline:0800/6730673

MSD Infocenter
Tel.: 0800/673 673 673
Fax: 0800/673 673 329
e-mail: infocenter@msd.de

CAN-01-D-2758-J

14/41 Influenza infections and in 13/107 CMV infections. In contrast to CMV, RSV and Influenza, graft-versus-host disease (GvHD) had a major impact on ADV infections: The incidence increased from 20% in pts without GvHD to 55.2% in pts with GvHD grade III/IV; vice versa, ADV infected pts had a significantly increased transplant related mortality (53% vs 31.6% in pts without ADV infections, p<0.001). In summary, our data demonstrate for a large cohort of consecutive pts the relevance of ADV- and further respiratory viral infections which occur with a similar incidence as CMV-infections. Multivariate analyses will be used ot define high risk pts which might be candidates for preemptive treatment strategies in the future.

86

Empiric antimicrobial monotherapy in patients after high-dose chemotherapy and autologous stem cell transplantation. Analysis of a randomized trial

G. Reich, T. Wörtz, O. Cornely, M. Sandherr, T. Kubin, G. Hildebrandt, H. Einsele, H. Szelenyi, B. Dörken, G. Maschmeyer (Berlin, Köln, München, Karlsruhe, Regensburg, Tübingen, D)

No prospective, randomized study on the efficacy of empiric antimicrobial intervention in patients (pts) with infectious complications after high-dose chemotherapy and autologous stem cell transplantation (ASCT) has been published by now. We present the first outcome analysis of 200 pts undergoing ASCT entered from 7 transplant centers into a multicenter, randomized trial comparing the efficacy and tolerability of meropenem (MPM) versus piperacillin/tazobactam (P/T) as empiric antimicrobial first-line therapy for febrile neutropenia. Of 215 documented pts (45.4% with lymphoma, 19.0% with breast cancer, 11.6% with germ-cell cancer, 11.6% with multiple myeloma, 4.6% with sarcoma, 1.9% with ovarian cancer and 6.0% with other solid tumors), 200 (93%) were evaluable. Age, sex and Karnofsky index were equally distributed. Complete response rate to first-line antimicrobial therapy was 65.4% in the MPM group and 54.2% in the P/T group (p=0.11). In 23.1% of pts in the MPM group and 18.8% in the P/T group, therapy was initially supplemented with a glycopetide in case of venous catheter infection or with an aminoglycoside in case of documented Pseudomonas aeruginosa infection. Overall complete response rate after modification was 94.2% in the MPM group and 94.8% in the P/T group (n.s.). Median to defervescence was 3.5 days in the MPM group and 4.0 days in the P/T group (n.s.). The most frequently isolated pathogens were Gram-positive cocci. Tolerability of both drugs was excellent in the MPM group and good in the P/T group. One patient (o.5%) died from infection. Empiric first-line therapy with MPM as well as with P/T is safe and effective in febrile episodes emerging after ASCT.

Kasuistisches Forum 1

87

Extreme eosinophilia and pulmonary infiltrates: A case report

Th. Lipp, K. Namberger, K. Rettler, R. Schulz, T. Castorph, F. Woitinas, C. Nerl (München, D)

A 43–year old woman with a history of allergic asthma bronchiale presented with worsening dyspnea despite inhalative corticosteroids , persistent cough, wheezes and fatigue. She complained about a weight-loss of 12 kg within 6 months. She denied any other medication. Chest X-Ray and CT scan of the thorax showed multiple bilateral pulmonary infiltrates. Blood smear showed an elevated white blood count of 20/nl with a remarkable eosinophilia of 70 to 90 percent. IgE was also elevated (300 IU/ml). ANA, ANCA, ACE and rheumatoid factor in the serum were negative. Bronchoscopy showed signs of chronic bronchitis with significant eosinophilia and lavage was negative for all kinds of specimen including tuberculosis, fungal, viral and parasitic disease. Extreme eosinophilia and Charcot Leyden christals were also found in the bone marrow biopsy. Flow cytometry was unremarkable. Diagnostic thorascopy and open lung biopsy showed chronic eosinophilic pneumonia demonstrating infiltration of interstitial tissue and alveolar spaces. Granulomatous inflammation and vasculitis could be excluded. After treatment with oral prednisone (100 mg per day) for two days the patient was asymptomatic and blood smear was normal. There were no infiltrations at the chest X-Ray any more .We found the same results 8 weeks later. Chronic eosinophilic pneumonia (CEP) , a rare eosinophilic lung disease of unknown cause, is characterized by peripheral blood eosinophilia, bilateral pulmonary infiltrates, absence of extrapulmonary involvement and prompt response to corticosteroid therapy. Most cases occur in middle aged women with a history of asthma bronchiale.
We present this patient as an unusual case of chronic eosinophilic pneumonia with an extreme eosinophilia like it is usually to be seen with hypereosinophilic syndrome. The main differential diagnoses include eosinophilic infections, drug toxicity, Löfflers pulmonary infiltrates, bronchiolitis obliterans organizing pneumonia and Churg Strauss vasculitis. Diagnosis ist best verified by open lung biopsy because transbronchial biopsy might not get enough vessels to exclude vasculitis.

88

Secondary acute myeloid leukemia after treatment of follicular lymphoma with cladribine and mitoxantrone

U. Fabry, O. Galm, R. Osieka (Aachen, D)

Secondary leukemia is a well known long term adverse effect of intensive treatment with antineoplastic chemotherapy compromising curative intent. To our knowledge, there are no reports about leukemia after treatment with modern nucleoside analogues in patients with follicular lymphoma. The female patient (borne 1933) was first diagnosed in 1996 with follicular lymphoma stage IV (abdominal bulky disease, splenomegaly, atypical lymphocytes in blood, and bone marrow involvement). Because of her symptomatic bulky disease the patient received several combination treatment regimens including the CD20 antibody rituximab from 1996 to 1999. Partial remission of lymphadenopathy and complete clearance of bone marrow of lymphomatous infiltration was achieved. In November 1999, however, chemotherapy had to be resumed because of abdominal progression of disease. 15 cycles with cladribine and mitoxantrone were applied to a total amount of 100 mg cladribine and 158 mg mitoxantrone until october 2000. The treatment was then stopped because of prolonged thrombocytopenia. Bone marrow biopsy at that time revealed some myelodysplastic stigmata but no infiltration by the follicular lymphoma. In april 2001 there was a further decrease of thrombocyte levels. The peripheral blood smear revealed 6% myeloblasts without differentiation. The bone marrow was infiltrated by peroxidase positive myeloblasts in excess of 60%. FACS analysis demonstrated a strong expression of CD34 in combination with other markers of myelocytic lineage in these blastic cells. Cytogenetics from the bone marrow aspirate showed multiple chromosomal alterations including monosomy 7 and deletion of 5q, typical for secondary acute myeloid leukemia. The presence of cells positive for t(14;18) was excluded by FISH-analysis. Treatment with a course of idarubicin and cytosine-arabinoside so far did not change the the clinical course of disease. Although our preliminary report does not allow definitive conclusions about the risk of secondary leukemia after treatment with modern nucleoside analogues, such adverse events should be recognized as an early complication of an otherwise well tolerated treatment.

89

Persistent aplasia after chemotherapy for acute myeloid leukemia successfully treated by microtransplant from a matched unrelated donor

L. Hambach, K. Oevermann, J. Schubert, E. Dammann, A. Ganser, B. Hertenstein (Hannover, D)

Persistent aplasia is a rare complication after intensive chemotherapy for acute leukemias leading to a high rate of morbidity and mortality. We report the case of a 60 year old man with acute myeloid leukemia, FAB M2 secondary to a myelodysplastic syndrome who developed persistent aplasia after polychemotherapy with idarubicine, VP16 and cytarabine (IVA). The patient had reached a complete remission after the first cycle of chemotherapy. After the second cycle of chemotherapy the patient remained aplastic and after >2 month of persistent aplasia a search for a matched unrelated stem cell donor was started. The patient received supportive care with transfusion of red blood cells and platelets and antibiotic treatment for one episode of E.coli septicemia. A donor was identified and after 7 months of persistent aplasia an allogeneic peripheral stem cell transplantation (PBSCT) from a matched unrelated donor (MUD) was performed. Prior to PBPCT the bone marrow of the patient showed severe hypoplasia without a significant number of blasts. Due to the age of the patient and the bone marrow findings we chose a dose reduced conditioning regimen with TBI 2 Gy (d0) and fludarabine 30 mg/m^2 (d-5 to d-1) according to Seattle/Leipzig protocol. Immunosuppressive treatment consisted of cyclosporin A and mycophenolate mofetil. A total of 5.8x10^6 CD34+ PBPC was transplanted. The patient reached white cell counts >1 G/l on d+13 and platelet counts >50

G/l on d+15. 6 weeks after transplantation the patient was independent from transfusions, has stable peripheral blood counts with full donor chimerism and thus far no signs of GvHD. In conclusion we have shown that allogeneic MUD PBSCT after a dose reduced conditioning might be successful to treat chemotherapy-induced persistent aplasia even in older patients.

90

Multiple myeloma and lung cancer: coexistence or coincidence? A case report

K. Horvat-Karajz, G. Muzes, L. Sreter (Budapest, H)

A 55-year-old male patient is presented with fatigue, weakness and intermittent claudication. His medical history included hypertension and type 2 diabetes started in 1997. In 1998 he was treated due to angina. Upon clinical admission routine blood tests indicated extremely elevated ESR, moderate anemia and hyperproteinemia. Serum electrophoresis showed monoclonal gammopathy. Multiple myeloma was diagnosed by bone marrow examination and immunelectrophoresis (IgG kappa monoclonal protein). Chemotherapy was given. Five month later he started to complain about caughing. Routine chest x-ray revealed a solide left pulmonary tumor. Chest CT scan also proved the solid tumor formation. The result of the transbronchial needle biopsy was squamous cell carcinoma. Since the chemotherapeutic period of the multiple myeloma was relatively short the associated lung cancer should be a coexistence rather than a secondary tumor.

91

Indications for allogeneic stem cell transplantation (SCT) in two adolescent patients with paroxysmal nocturnal hemoglobinuria (PNH)

B. Strahm, H.U. Kontny, U. Duffner, C.M. Niemeyer (Freiburg, D)

Paroxysmal nocturnal hemoglobinuria (PNH) is a rare acquired clonal disorder of hematopoietic stem cells. Currently allogeneic stem cell transplantation (SCT) is the only intervention that offers cure. Because of the high variability of the clinical course the decision for SCT must be evaluated carefully. We present two indications for SCT in patients with PNH.
Case 1: A seven year old girl presented with pancytopenia. 4 years later PNH was diagnosed (positive HAM's test and deficiency of GPI linked surface proteins). Therapy with cyclosporine and G-CSF was initiated. One month later she developed a Budd-Chiari-Syndrome due to thrombosis of the hepatic veins. Implantation of a intrahepatic portosytemic shunt led to the resolution of the Budd-Chiari syndrome. Despite oral anticoagulation therapy and thrombocytopenia thrombosis of the intrahepatic portosystemic shunt and the lienal vein occured. At the age of 12 years the girl received an allogeneic SCT from a matched unrelated donor. 2 years after SCT the girl is in good clinical condition without symptoms of PNH.
Case 2: A 12 year old boy presented with pancytopenia. The diagnosis of very severe aplastic anaemia was made and sucessfully treated with cyclosporine. Two years later he again developed thrombocytopenia, and PNH was diagnosed (positive HAM's test and deficiency of GPI linked surface proteins). Hemolytic episodes were treated with G-CSF in addition to cyclosprine. After two years with stable disease pancytopenia occurred and a myelodysplastic syndrome was diagnosed. At the age of 20, the patient received an allogeneic SCT from a matched unrelated donor because of a myelodysplastic syndrome related acute myoloid leukemia (85% blasts). 14 months after SCT he is in a good clinical condition without disease. Most patients with PNH have a long stable course. However, severe complications of PNH, such as severe thrombosis or development of a myelodysplastic syndrome are indications for SCT. The two cases presented here demonstrate that an unrelated allogeneic SCT can be performed successfully even in high risk patients.

92

Auer-rod-like inclusions in a patient with multiple myeloma

G. Metzgeroth, W. Back, A. Willer, Y. Dencausse, R. Hehlmann, J. Hastka (Mannheim, D)

A 59-year-old male patient was admitted to the hospital due to severe bone pain in the lumbar region. In serum electrophoresis a monoclonal IgA-kappa protein was demonstrable. Urine immunoelectrophoresis detected the presence of a kappa light chain protein. Bone marrow was occupied by 40 % of atypical plasma cells. Some of these contained numerous bundles of pink-stained Auer-rod-like inclusions within their cytoplasm. These inclusions were strongly alpha-N-esterase positive. In immunocytochemical studies the cells were brightly reactive with IgA-antibodies. The staining was, however, totally absent in the region of the intracytoplasmic inclusions. The plasma cells were stained positive with CD38 and showed no expression of CD11c, CD14, CD19, lysozyme and amyloid. Ultrastructural studies showed mostly rod-like but sometimes also smaller crystalline inclusions with a smooth surface in the cytoplasm of plasmacytoma cells. On cross sections the rod like inclusions showed no tubular substructure. The periodicity of the filamentous striations in these inclusions was about 10 nm (the periodicity of typical Auer rods in AML). Some smaller crystalline inclusions showed a more irregular filamentous substructure (not resembling fibrillary deposition disease, i.e. amyloidosis). The patient was treated with high dose dexamethasone due to hypercalcaemia and decreased creatinine clearance without response to treatment.

Wissenschaftliches Symposium:
Chronische lymphatische Leukämie

93

Molecular genetics of CLL: genomic aberration and IgV mutation status in pathogenesis and clinical course

S. Stilgenbauer, L. Bullinger, A. Kröber, A. Benner, P. Lichter, H. Döhner (Ulm, D)

Fundamental aspects of the cellular origin, transforming events and risk assessment have recently been elucidated by genetic analyses in chronic lymphocytic leukemia (CLL). Two major parameters, the mutation status of the immunoglobuline variable genes (IgV) and acquired genomic aberrations have been used to define subgroups of CLL which are derived from different stages of B-cell development and are characterized by distinct clinical outcome. We have studied the most frequent genomic aberrations in CLL by fluorescence in-situ hybridization (FISH) with a comprehensive probe set. Among 325 CLL cases from a single institution the most frequent genomic aberrations were deletion 13q14 (55%), deletion 11q23 (18%), trisomy 12q13 (16%), deletion 6q21 (7%) and deletion 17p13 (7%). The molecular characterization of these genomic aberrations has allowed the identification of genes involved in the transformation of CLL cells such as p53 in cases with 17p13 deletion and ATM in a subset of cases with 11q23 deletion. Furthermore, genomic aberrations allowed the identification of prognostic subgroups of CLL patients and five genetic risk groups were defined in a hierarchical survival model: 17p13 deletion, 11q23 deletion, 12q13 trisomy, normal karyotype, and 13q14 deletion as sole abnormality; with median survival times (OS) of 32, 79, 114, 111, and 133 months, respectively. In multivariate analysis, 17p13 deletion, 11q23 deletion, age, stage, LDH, and WBC gave significant prognostic information. In addition to genomic aberrations the IgV mutation status has recently been identified as another important genetic parameter in CLL. Therefore, we have complemented the analysis of the genomic aberrations as described above with the study of the IgVH mutation status by PCR and direct sequencing using family specific VH- and JH-consensus primers (n=300) and the CD38 expression level (n=156) by triple color FACS. All CLL cases showed at least one clonal IgVH rearrangement. Applying the 98% germline homology cut-off level IgVH was mutated in 45% and unmutated in 55% cases. Comparison of the two groups showed the following characteristics (IgVH mutated vs. unmutated): 1.) Similar distribution of stage, age and sex; 2.) VH gene usage: VH1 (8% vs. 39%), VH3 (61% vs. 43%), VH4 (28% vs. 14%), VH2,5,6 (3% vs. 4%); 3.) CD38+ >30% (25% vs. 43%) (p=.0267); 4.) Similar overall incidence of chromosomal aberrations (79% vs. 83%); 5.) Different incidence of high-risk aberrations: p53 abnormality (3% vs. 12%), 11q- (4% vs. 26%). 6.) Significant difference in median OS (not reached vs. 77m) (p<.0001). The median OS of the groups with or without CD38+ > 30% was 85m vs. 112m (p=.1315). Within the IgVH unmutated group two subgroups with significantly different median OS (58m vs. 98m, p<.0001) were defined according to the presence or absence of high-risk aberrations (p53 abnormality and/or 11q-). In multivariate analysis p53 aberration and IgVH mutation status were identified as independent adverse prognostic factors. In conclusion, these data indicate that genetic parameters such as genomic aberrations, p53 abnormalities and IgVH mutation status may allow the definition of prognostic subgroups of CLL patients and that there may be a complementary impact of these factors. Therefore, genetic parameters are currently part of the diagnostic work-up within the prospective multicenter trials of the German CLL Study Group to evaluate their impact in relation to other clinical and biological markers.

94

Current concepts in stem cell transplantation for chronic lymphocytic leukemia

P. Dreger (Kiel, D)

Allogeneic and autologous stem cell transplantation (SCT) are increasingly considered for treatment of patients with chronic lymphocytic leukemia (CLL). In order to assess the potential therapeutic value of SCT for CLL, this presentation aims at answering the following crucial questions: 1. Is SCT a curative treatment? 2. Does SCT improve the dismal prognosis of poor-risk CLL? 3. Do risk factors exist which are useful for defining prognostic groups in terms of feasibility and post transplant outcome? The efficacy of auto-SCT relies exclusively on the cytotoxic therapy administered. To date, there is only limited hope that autotransplantation can cure the disease even in this favorable subgroup. Nevertheless, the results of the published series suggest that auto-SCT is capable of improving the prognosis of CLL with defined poor-risk features. Favorable conditions for successful autografting are less advanced disease, a sufficient remission status at SCT and use of a TBI-containing high-dose regimen. The crucial anti-leukemic principle of allo-SCT consists in the immune-mediated GVL effects conferred with the graft. The GVL activity should be responsible for the fact allografting seems to be a curative treatment for at least a subset of poor-risk patients. However, as long as allo-SCT in CLL is still associated with an excessively high treatment-related mortality, only selected patients with advanced poor-risk disease should be considered for allografting. The development of conditioning regimens with reduced intensity may allow to extend the indications of allogeneic SCT for CLL in the near future.

96

B-CLL stage Binet-A: Incidence and correlation of genetic aberrations with clinical and biological risk factors within the CLL1 trial of the GCLLSG.

L. Bullinger, C. Kräutle, B. Schmitt, B. Emmerich, M. Hallek, P. Lichter, S. Stilgenbauer, H. Döhner (Ulm, München, Heidelberg, D)

In early stage B-cell chronic lymphocytic leukemia (B-CLL) treatment has not been shown to improve overall or progression free survival irrespective of risk factors or disease progression. Therefore, the CLL1 trial of the German CLL Study Group (GCLLSG) tests if treatment with fludarabine can improve the outcome of stage Binet-A B-CLL patients at high risk for rapid disease progression as determined by clinical (bone marrow infiltration pattern or lymphocyte doubling time) and biological (elevated TK or ß-2MG) risk factors. Other important biological markers such as genetic abnormalities of the tumor cells are assessed in addition. The genomic regions most frequently affected by deletions (del), trisomies (+), and translocations (t) in B-CLL have been analyzed by interphase FISH among 230 patients enrolled in CLL1 so far. In 178 cases (77%) an identification of aberrations was possible. The incidences of individual chromosome abnormalities were: del(13q14) (62%), +(12q13) (11%), del(11q23) (9%), t(14q32) (6%), del(17p13) (6%), del(6q21) (3%), +(3q27) (1%), and +(8q24) (2%). Correlation of genetic aberrations with the other risk factors showed an association between del(13q14) as single aberration and biological/clinical low risk parameters. Trisomy +(12q13) was shown to correlate with the biological/clinical high risk group (p=0.010) and there was a correlation of chromosomal aberrations del(11q23) and/or del(17p13) with parameters indicating a high risk of rapid disease progression (p=0.016). The conclusion of these results is that specific genetic aberrations are associated with clinical/biological risk factors for disease progression in Binet-A B-CLL. Therefore, the ongoing CLL1 trial may allow to develop a hierarchical model of clinical and biological risk factors identifying early stage B-CLL patients who are at high risk for rapid disease progression.

97

Identification of novel tumor antigens as potential targets for immunotherapy in chronic lymphocytic leukemia

A.M. Krackhardt, M. Witzens, J. Zauls, P. Barrett, J.G. Gribben (Boston, USA)

Chronic lymphocytic leukemia (CLL) is incurable using standard therapeutic approaches and clearly new therapies are required. Although patients with CLL have a number of immunological disturbances, CLL is an attractive candidate to examine immunotherapeutic approaches. We have previously shown that cytotoxic T cell responses can be generated against naturally presented and immunoglobulin derived peptides, however, these responses are weak and we therefore sought to identify additional candidate tumor antigens in this disease using SEREX (Serological identification by recombinant expression cloning) and subtractive hybridization approaches. As hypogammaglobulinemia and autoimmune phenomena are common in CLL, SEREX might be of limited use in this disease setting. However, we have identified six novel antigens using this methodology. Only one of these has been previously described as an antigen whereas the others represent novel antigens. Several antigens have been repeatedly selected and are represented by multiple independent clones. Four antigens represent genes with known protein function whereas the other two genes have not been previously described and have been newly cloned. Investigation of antibody responses from allogeneic sera from CLL patients and healthy donors against these antigens revealed that four antigens were detected from serum of several CLL patients but not from serum of most healthy donors. Of note, patients with hypogammaglobulinemia showed less humoral responses against these antigens. One antigen was detected by the serum of 11 out of 16 CLL patients but only weakly by two sera of 12 healthy donors. This antigen was identified selecting three independent clones from two different patients' libraries and represents an RNA-binding protein with unknown function. Interestingly most of the other antigens represent nuclear factors. Moreover, one antigen is represented by two different clones showing splice variants of a Krüppel-like zinc finger protein which have not been previously described. One splice variant lacks the transcription suppressing Krüppel-associated box domain and is expressed only in hematopoietic cells as assessed by northern blot and polymerase chain reaction. The gene is located on chromosome 12q and of note the cDNA library was obtained from a patient with trisomy 12. Another clone represents an alternative open reading frame (ORF) of a transcription factor. It has previously been reported that tumor-infiltrating CD8-positive cells also recognized an alternative ORF of the Granulocyte-Macrophage Colony Stimulating Factor gene. Additional 36 genes were identified by suppression subtractive hybridization to be overexpressed in CLL compared to normal B cells. Ongoing studies are underway to determine the potential of these tumor antigen candidates to induce humoral and cellular anti-tumor responses as well as the significance of these genes in the pathogenesis of CLL.

98

Prognostic relevance of increased bone marrow angiogenesis in patients with chronic lymphocytic leukemia

K. Niemöller, M. Schweigert, J. Eucker, C. Jakob, E. Fietze, I. Zavrski, U. Heider, I. Blau, K. Possinger, O. Sezer (Berlin, D)

The impact of angiogenesis is well known for the prognosis of solid tumors, while only a few such studies have been performed in hematological malignancies. We have recently published the prognostic relevance of bone marrow angiogenesis for overall survival in multiple myeloma patients. In B-cell chronic lymphocytic leukemia (B-CLL) the prognostic implications of bone marrow angiogenesis have not been published. Thus, the aim of our study was to evaluate the bone marrow microvessel density (MVD) in patients with B-CLL and to investigate the prognostic relevance. Immunohistochemical CD34 stained bone marrow specimens of 90 patients with B-CLL were studied. The median bone marrow MVD was 22.5 vessels/mm^2 (range 0-81). There was no difference between the MVD in Binet stage A patients (n = 50, range 2-77 vessels/mm^2) and in stage B & C patients (n = 40, range 0-81 vessels/mm^2, P = 0.4). In the cohort of patients with Binet stage B & C the bone marrow MVD was found to be of prognostic importance. Using the median MVD as cut-off (< versus > 18.5 vessels/mm^2) in those 40 patients, in whom bone marrow biopsy was performed at the time of diagnosis (n = 33) or in relapsed patients before the initiation of a new treatment (n = 7), significant differences in the progression-free survival (PFS) could be demonstrated. In patients with a high MVD the PFS was significantly shorter than in patients with a low MVD (median PFS 11 months versus median PFS not reached, P = 0.03). Figure 1 demonstrates the Kaplan-Meier curves of PFS in stage Binet B-C patients with a high MVD (dashed line) and low MVD (solid line). Furthermore in Binet stage B & C, patients with a diffuse lymphocytic bone marrow infiltration (n = 18) showed a significantly increased angiogenesis in comparison to patients with a nodular pattern of infiltration (n = 22). The median bone marrow MVD was 33 and 9 vessels/mm^2 in patients with a diffuse and nodular infiltration of the bone marrow, respectively (P = 0.006). In contrast to the advanced stage patients no significant difference in the PFS was found in stage A patients with a high and a low bone marrow MVD (P = 0.66). These results underscore the prognostic impact of bone marrow MVD in advanced stages in CLL, while in the early stage no impact of bone marrow MVD was found. As a potential treatment strategy chemotherapeutic drugs combined with antiangiogenic agents might be particularly beneficial in CLL patients in advanced disease.

MABCAMPATH
Alemtuzumab

MabCampath, der erste CD52-spezifische monoklonale Antikörper zur Behandlung der CLL. Hohe Wirksamkeit bei Fludarabin-refraktären Patienten.

Die zielgerichtete Behandlung der CLL

Bezeichnung: MabCampath 10 mg/ml Konzentrat zur Herst. einer Inf.-Lsg. **Zusammensetzung:** Jede Ampulle enth. 30 mg Alemtuzumab. Hilfsstoffe: Dinatriumedetat, Polysorbat 80, phosphatgepufferte Kochsalzlsg. best. aus: Kaliumchlorid, Kaliumdihydrogenphosphat, Natriumchlorid, Dinatriumphosphat, Wasser f. Inj.-Zwecke. **Anwendungsgebiete:** Zur Behandlung von Patienten m. chronischer lymphatischer Leukämie (CLL), die m. Alkylanzien behandelt wurden u. bei denen nach d. Fludarabinphosphat-Therapie keine komplette od. partielle Remission od. nur eine kurzzeitige Remission (kürzer als 6 Mon.) aufgetreten ist. **Gegenanzeigen:** Überempfindlichkeit gegenüber Alemtuzumab, murinen Proteinen od. einem d. Hilfsstoffe; Pat. m. floriden systemischen Infekt., HIV-Infekt. od. Zweittumor, Schwangerschaft u. Stillzeit. **Nebenwirkungen:** Sehr häufig: Akute infusionsbed., i.d.R. leichte bis mittelschwere Reaktionen wie Fieber, Rigor, Übelkeit, Erbrechen, Hypotonie, Durchfall, Müdigkeit, Ausschlag, Urtikaria, Juckreiz, vermehrtes Schwitzen, Dyspnoe, Kopfschmerzen, deren Häufigkeit nach der 1. Therapiewoche deutlich abnimmt. Die Symptome können durch eine Prämedikation u. Dosiseskalation gelindert od. vermieden werden. Anorexie u. Infekt. (Sepsis, Pneumonie, Herpes simplex) 3. oder 4. Grades. Häufig: Reaktionen an d. Inj.-Stelle, opportunist. Infekt. einschl. *Pneumocystis carinii-/Aspergillus*-Pneumonie, Zytomegalovirus-Infekt., Herpes zoster, Pilzinfekt., Moniliasis, Abszesse, Bronchitis, Sinusitis, Pharyngitis, Harnwegsinfekt., kardiovask. Nebenwirk. (Hypertonie, Tachykardie, Gefäßspasmen, Hitzewallungen, Palpitationen), Tremor, Hypästhesie, Schwindel, Benommenheit, Hyperkinesie, Parästhesie, Verwirrung, Ängstlichkeit, Depression, Schläfrigkeit, Schlaflosigkeit, Abdominalschmerzen, gastrointest. Blutung, Stomatitis (auch ulzerierend), Schleimhautentz., gestörte Leberfunktion, Obstipation, Dyspepsie, Flatulenz, schwere Blutungsreaktionen, Panzytopenie, Anämie, Leukopenie (Granulozytopenie, Lymphopenie), Thrombozytopenie, Purpura, Pneumonitis, Bronchospasmus, Husten, Hämoptyse, Hypoxie, neutropen. Fieber, Schmerzen (Rücken, Brust, Skelett), Mundödeme, Asthenie, Unwohlsein, grippeart. Sympt., Ödeme, veränd. Temperaturempf., Geschmacksverlust, Bindehautentz., Gewichtsabnahme, Dehydrierung, Durst, Hyponatriämie, Hypokalziämie, Arthralgie, Myalgie, erythematöser Ausschlag, bullöse Eruptionen. Gelegentlich: Hämatom, Dermatitis od. Schmerzen an d. Inj.-Stelle, Synkope, Herzstillstand, Myokardinfarkt, Vorhofflimmern, supraventrikul. Tachykardie, anormales EKG, Arrhythmie, Bradykardie, periphere Ischämie, Dystonie, muskuläre Hypertonie, anormaler Gang, Hyperästhesie, Neuropathie, Nervosität, Denkstörungen, Depersonalisation, Persönlichkeitsstörung, Impotenz, Gastroenteritis, Zahnfleischentz., Aufstoßen, Schluckauf, Mundtrockenheit, Schleimhautulzeration, Zungenulkus, Knochenmarksdepression bis -aplasie, Verr. d. Haptoglobins, dissem. intravasale Gerinnung, hämolyt. Anämie, Nasenbluten, Zahnfleischbluten, Laryngitis, Rhinitis, Lungenödem, Stridor, Lungeninfiltrat, Atembeschwerden, verr. Atemgeräusche, Rachenenge, Pleuraerguss, aggravierter Diabetes mellitus, Hypokaliämie, lymphomart. Erkrank., periphere u. periorbitale Ödeme, Schmerzen i. d. Beinen, allerg. Reaktionen, Endophthalmitis, Taubheit, Tinnitus, Geschmacksstörungen, Pilzdermatitis, Onychomykose, makulopap. Ausschlag, gestörte Nierenfunktion, Polyurie, Hämaturie, Harninkontinenz, verringerter Harnfluss. Einzelfälle: Eine letal verlauf. idiopath. thrombozytopen. Purpura, eine rhinozerebrale Mucormykose. **Dosierung, Art u. Dauer d. Anwendung:** Die Verabr. v. MabCampath muss unter Aufsicht eines i. d. Krebstherapie erfahrenen Arztes erfolgen. Die zuber. MabCampath-Lsg. wird durch i.v. Inf. über ca. 2 h verabreicht. Vor d. 1. Dosis, jeder Eskalation u. vor jeder weiteren Inf. sollte, soweit klin. indiziert, eine Prämedikation m. einem geeigneten Antihistaminikum u. Analgetikum erfolgen. Alle Pat. sollten während u. nach d. Therapie routinemäßig m. Antibiotika u. antiviralen Substanzen behandelt werden. In d. 1. Therapiewoche wird MabCampath in eskalierenden Dosen von 3 mg am 1.Tag, 10 mg am 2. Tag u. 30 mg am 3. Tag verabreicht, vorausgesetzt, dass die vorangeg. Dosis gut vertragen wurde. Danach sollten 3x wöchentl. m. jeweils eintägiger Unterbrechung 30 mg verabreicht werden. Die Behandl. kann bis zu 12 Wochen lang fortgesetzt werden. Im Falle einer schweren Infekt. od. schwerwieg. Hämotoxizität ist MabCampath abzusetzen, bis das Ereignis verschwindet. **Wechselwirkungen:** MabCampath sollte nicht innerh. v. 3 Wochen vor u. nach d. Behandl. m. and. Chemotherapeutika verabreicht werden. Pat. mind. 12 Mon. nach d. MabCampath-Therapie nicht m. Lebendimpfstoffen behandeln. **Besondere Hinweise:** Während u. bis 6 Mon. nach der Therapie ist eine wirksame Kontrazeption erforderlich. Vorsicht b. ischämischer Herzerkrankung, gleichzeitiger antihypertensiver Therapie u. b. Pat. m. eingeschränkter Nieren- od. Leberfunktion. Regelm. Kontrolle d. Blutbildes u. d. Thrombozytenzahl während d. Therapie erforderlich. Vorsicht beim Autofahren u. Bedienen von Maschinen, da Verwirrung u. Somnolenz auftreten können. Arzneimittel für Kinder unzugänglich aufbewahren! Bei 2–8°C lagern. Nicht einfrieren. Vor Licht schützen. And. Arzneimittel dürfen nicht d. MabCampath-Inf.-Lsg. zugesetzt od. gleichz. durch dieselbe Inf.-Leitung infundiert werden. Die rekonstituierte Lsg. b. Aufbewahr. b. 15–30°C innerhalb von max. 8 h verbrauchen. Verschüttetes od. überschüssiges Material durch Verbrennung entsorgen. **Verschreibungspflichtig.** **Pharmazeutischer Unternehmer:** Millennium & ILEX UK Ltd, 1&3 Frederick Sanger Road, The Surrey Research Park, Guildford, Surrey GU2 7YD, Vereinigtes Königreich. Stand: 07.2001. **Lokale Vertretung:** MedacSchering Onkologie GmbH, Nördliche Auffahrtsallee 44, D-80638 München.

MedacSchering
Onkologie

Wissen teilen. Wissen mehren.

99

Endocrine treatment of metastatic breast cancer: Standards and new developments

P. Schmid, K. Possinger (Berlin, D)

Approximately one-third of human breast cancers are estrogen-dependant and exhibit regression after estrogen deprivation. Most endocrine agents act either by inhibiting the production of estrogens or the action of estrogen at the cellular level. Antiestrogens have recently been named selective estrogen receptor modulators (SERM) since most of them exhibit both agonist and antagonist activity. For many years, the first-generation SERM tamoxifen has been the drug of choice for endocrine treatment of advanced hormone-responsive breast cancer in postmenopausal women. In premenopausal women, the combination of ovarian suppression and tamoxifen has shown superior results compared to monotherapy. The benefits of tamoxifen are thought to derive primarily from its blockade of the estrogen receptor, thus removing the stimulus to continued proliferation and resulting in regression of the tumor. However, since tamoxifen is also a partial estrogen agonist, its does not induce maximal suppression of the effects of estrogen. More complete estrogen receptor blockade by "pure" antiestrogens might result in longer lasting responses, activity against tumors with acquired tamoxifen resistance, and absence of endometrial stimulation and associated cancers. These "pure" antiestrogens can induce down regulation of the estrogen receptor. Faslodex, the currently best established "pure" antiestrogen has shown activity in both tamoxifen-resistant and tamoxifen-responsive tumors. Two large phase III studies comparing faslodex to anastrozole as first-line endocrine agent showed that faslodex is at least as effective as anastrozole with respect to response, duration of response and overall survival. In postmenopausal women, effective inhibition of the enzyme aromatase can result in nearly complete blockade of estrogen production. Until recently, third generation aromatase inhibitors have been the agents of choice for 2nd-line hormonal therapy. However, two randomized trials, identical in design and objectives, indicate that in postmenopausal patients with advanced breast cancer anastrozole is at least as effective as tamoxifen as first-line endocrine therapy. Furthermore, a randomized, cross-over trial showed that first-line treatment with letrozole is superior to tamoxifen in terms of response and duration of response. Thus, third generation aromatase inhibitors are the first group of hormonal agents that have been shown to be more efficacious than tamoxifen for the first-line therapy of metastatic breast cancer. These results will have a major impact on future standard practice for postmenopausal women.

100

First line single agent mitoxantrone in the treatment of high risk metastatic breast cancer patients is as effective as combination chemotherapy, but less toxic: No difference in survival, but higher quality of life shown in a multicenter randomized trial

R. Mück, M. Woehr, E. Heidemann, H. Stoeger, R. Souchon, W. Hirschmann, H. Bodenstein, C. Oberhoff, M. Schulze, M. Clemens, R. Andreesen, M. Kaufmann (Stuttgart, D)

In metastatic breast cancer, palliation is an important goal and has to be outweighed against survival and toxicity. The dogma that pts. fulfilling high risk criteria have to be treated with combination chemotherapy up front therefore had to be challenged. From July 1992 until April 2001 527 pts. after risk stratification were randomized to receive either single agent mitoxantrone (N) (n=260) or combination treatment as follows: 125 pts. fulfilling high risk criteria received fluorouracil, epirubicin, cyclophosphamid (FEC) from 1992 til June 1997. Thereafter high risk pts. in the combination arm received mitoxantrone plus docetaxel (NDoc) (n=82 as of April 2001). Pats. stratified as low risk received cyclophosphamide, methotrexate and fluorouracil (CMF). While the N vs. NDoc study is still going on (n=160 as of April 2001), the results of the N vs. FEC and the N vs. CMF trials after additional follow-up of 3 years can now be reported. There was no statistically significant difference between N and CMF or N and FEC in terms of response, objective remission rate, remission duration, time to progression or overall survival. There was however a significant difference in quality of life using a modified „Brunners Score". Whereas the low risk group appears too small to draw valid conclusions, the high risk group (n=260) is large enough to justify the abolition of the dogma of the necessity of combination chemotherapy for these patients. To enlarge the basis of this message, the study has been continued from July 1997 using an even more aggressive combination (NDoc).

101

Status quo of antibody therapies in breast cancer

D. Lüftner, K. Possinger (Berlin, D)

The introduction of monoclonal antibodies into the clinic has relevantly enlarged the therapeutic panel. In metastatic breast cancer, the humanized anti-HER-2/neu antibody Herceptin® is in the focus of attention having shown an improvement of survival, response rate, duration of response and quality of life in combination with chemotherapy in HER-2/neu positive breast cancer. In Germany, Herceptin was approved in September 2000 for therapy of HER-2/neu positive, stage IV breast cancer in combination with paclitaxel or as single agent. The established dosing regimen is a loading dose of 4 mg/kg for the first infusion, followed by weekly therapy with 2 mg/kg. Recently published pharmacokinetic studies also recommend an alternative regimen with a loading dose of 8 mg/kg, followed by a 3-weekly therapy with a dose of 6 mg/kg (Gelmon et al., Proc ASCO 2001). The most frequent, but self-limiting adverse events are fever and chills in 40% of the first infusions which generally disappear with longer treatment. The limiting side effect is congestive heart failure in 8% of patients in combination with anthracycline-based chemotherapy, while this event is rare in the combination with paclitaxel. The first German adjuvant Herceptin trial (HERA trial) is just about to begin evaluating the benefit from the addition of a 1-year or 2-year, 3-weekly adjuvant Herceptin therapy after the termination of adjuvant chemotherapy of choice. The main focus of this adjuvant trial must be on the question whether the potential benefit in terms of decreased or postponed relapses exceeds the side effects of the medication, before all cardiac toxicity. Except from Herceptin, there are other antibody therapies coming up which could also change clinical practice. The EGF receptor is an interesting target for the chimeric, human-to-murine EGFR antibody cetuximab (C225) or the humanized monocloncal antibody ABX-EGF. First clincial trials on anti-EGFR antibody therapy included very low numbers of breast cancer patients, of whom 30-50% present EGFR overexpressing tumors. Another antibody-based approach could be the blockade of vascular endothelial growth factor (VEGF) action, either by neutralizing VEGF in the microcirculation or by preventing VEGF binding and signal transduction. VEGF has been shown to be significantly associated with decreased relapse-free and overall survival in breast cancer. While Herceptin and the anti-(V)EGF receptor antibodies are new tools in the oncologist's hands, the anti-17-1A or Ep-CAM antibody Panorex® was initially available for the adjuvant therapy of colorectal cancer. In breast cancer with relevant, Ep-CAM positive minimal residual disease in bone marrow aspirates, therapy with Panorex led to a significant reduction of cytokeratin positive cells (Braun et al., 1999). In contrast to Panorex, the new humanized antibodies don't induce the production of HAMAs making a prolonged systemic therapy possible. Based in the reduction of circulating tumor cells by monoclonal antibodies in breast cancer, adjuvant antibody therapies could translate into a reduction of relapses. By the time this has been proven in clinical trials, Herceptin remains restricted to the therapy of metastatic breast cancer where it has become a standard medication for HER-2/neu overexpressing tumors.

102

Prediction of survival in women with early breast cancer

P. Schmid, M. Wischnewsky, I. Schönborn, W. Lichtenegger, K. Possinger (Berlin, Bremen, D)

Introduction: The most common prognostic parameters in early breast cancer are tumour size (T), nodal status (N), grading (G), age and receptor status (RS). However, the number of additional prognostic parameters is steadily increasing. The purpose of this paper was to analyse the quality of individual prediction of overall survival (OS) using conventional prognostic factors and additional parameters like HER2, PCNA, p53, and EGF-R, respectively. Conventional statistics (COX and logistic regression) were compared to techniques from knowledge discovery in databases (KDD; e.g. self organising maps (SOM), entropy-based decision-tree-algorithms). Patients and Methods: 471 patients with primary breast cancer [median age 43.7 years, maximum age 54 years; median follow-up: 62 mths; G_1 12.8%, G_2 39.7%, G_3 32.1%; T_1 28.7%, T_2 58.0%, T_3 8.7%, T_4 4.7%; N_0: 45.9%; ER (0-19%) 28.0%, ER (20-50%) 53.3%, ER (>50%) 18.7%; HER2-overexpression 21%; EGF-R positive: 44.4%; p53 negative 50.1%] were analysed. Results: Logistic regression (backward conditional) revealed 5 significant

ACNU® • Azulon® • Endoxan® • Exitop® • Fareston® • Holoxan®
Honvan® • Ixoten® • Miltex® • Onkotrone® • Tamoxasta® • Uromitexan®

www.astamedica-oncology.com
Passwort abfragen unter 069/4001-2798

DEM LEBEN DIE HAND REICHEN.

ASTA MEDICA

parameters for OS: T, N, G, histology (H) and PgR. Using these parameters prediction of OS was possible in 77.6% of pts. COX regression (backward conditional) delivered 4 significant parameters for OS: T, N, G, and H. Using conventional statistics, additional parameters like HER2 or EGF-R could not improve the quality of prediction of OS. For N-positive pts with $T_{3/4}$-tumors, no further prognostic factor had any additional significance for OS (e.g. ER: p=0.83; HER2: p=0.30; H: p=0.42). In contrast, in N-negative pts with $T_{1/2}$-tumors, various prognostic factors were additionally significant for OS (e.g. G: p=0.02; ER: p=0.04). For N-positive pts with T1/2-tumors, p53 was the most significant additional parameter (p=0.01). Prediction of OS based on KDD was correct in >90% of patients using T, N, G, H, EGF, p53, HER2, PCNA and age. Conclusion: The classical parameters T, N, G, H and RS remain the most important factors for OS. With conventional statistics prediction of OS is possible in about 75% of pts on the basis of classical prognostic parameters, whereas additional parameters have only significance in subgroups. In contrast, KDD can predict OS in 90% and more, but require in addition to classical parameters various biomolecular parameters.

103

Detection of her-2/neu in breast cancer - a comparison of PCR, RT-PCR, IHC and FISH

N. Burdinski, N. Max, K. Evers, G. Schaller, I. Anagnostopoulos, E. Thiel, U. Keilholz (Berlin, D)

Her-2/neu has become an important prognostic and therapeutic factor in breast cancer because it has been found to be amplified and/or overexpressed in about 20 to 30% of tumors. her-2 routinely is detected by IHC or FISH, but so far it still is not clear which method is more reliable, as discordant results occur in a certain percentage of cases. Also, detection by PCR and RT-PCR has been investigated but has not entered routine use so far. In our experiments, we established a real-time PCR and RT-PCR for all four members of the EGFR-family (her-1, her-2, her-3 and her-4), using the LightCycler with fluorescent TaqMan and Hybridization probes. 39 breast cancer and 2 healthy breast tissue samples, 5 ovarian cancers and one healthy ovarian tissue sample were studied. Of these, 6 showed gene amplification for her-2 and overexpression in 4 cases when defining a gene copy number of >4 as a cutpoint. Gene amplification and overexpression occurred simultaneously in 3 cases only. For her-1, 9 samples were amplified and 2 overexpressed with this definition, for her3 0 and 23, and for her-4 4 and 0, respectively, although, based on the distribution, the latter three genes had different cutpoints or did not show a clear distinction into two groups. The results for her-2 were then compared to IHC and FISH analysis for available samples. Concordance of FISH and IHC was 79.4% for all samples. Of FISH and PCR concordance was 71.0%, of FISH and RT-PCR 59.3%, of IHC and PCR 79.5% and of IHC and RT-PCR 70,0%. Concordance of IHC, FISH and PCR was 71.0%. Remarkably, IHC and PCR have an equal correlation to IHC and FISH, whereas RT-PCR failed to detect most her-2-postive samples. Also, not all cases showing gene amplification had protein overexpression as well, neither in the comparison of IHC and FISH nor of IHC and PCR. In the 2+ cases of IHC, FISH had a gene copy number of either <4 or 5-10, while PCR was negative in 3 of 4 cases. On the other hand, 5-10 gene copy numbers by FISH were either negative or 2+ by IHC, yet PCR detected them all as negative. Therefore these cases show a tendency towards negative interpretation, taking PCR into account. PCR should further be evaluated in clinical studies as an alternative for IHC and FISH, being an automated, fast, cheap and equally reliable method.

Kompetenznetz „Maligne Lymphome":
State of the Art und neue Entwicklungen

104

Innovative treatment options in indolent lymphomas and mantle cell lymphomas – recent results of the German Low Grade Lymphoma Study Group and the European MCL Working Party

M. Dreyling, M. Unterhalt (München, D)

The clinical course of indolent lymphoma is characterized by frequent relapses and a steadily declining survival curve. However, even in relapse, the disease is sensitive to conventional chemotherapy in a significant subset of patients. In contrast, despite an indolent morphology, mantle cell lymphoma (MCL) is characterized by an aggressive clinical course with a median survival of less than 3 years and virtually no long term survivors. In a number of prospectively randomized multicenter trials, innovative treatment options like high dose chemotherapy with autologous stem cell transplantation, interferon alpha (IFN) maintenance treatment and anti-CD20 antibody-based immunotherapy were evaluated to define their role in indolent lymphomas and MCL. Patients <60-65 years with advanced lymphomas were randomly assigned to receive either intensive consolidation (fractionated total body irradiation with 12 Gray plus cyclophosphamide 2 x 60 mg/kg followed by transfusion of unpurged autologous stem cells) or IFN maintenance after initial CHOP-like chemotherapy. In recurrent disease, patients were randomly assigned to receive either an fludarabine-containing chemotherapy (FCM: fludarabine 25 mg/m^2 d 1-3, cyclophosphamide 200 mg/m^2 d 1-3, mitoxantrone 8 mg/m^2 d 1) alone or in combination with an anti-CD20 antibody (375 mg/m^2; R-FCM). Currently, 231 patients with indolent lymphomas are evaluable for interim analysis with a maximum follow-up of 4 years. A recurrence of lymphoma was observed in 22 of 103 (21%) patients of the high dose therapy arm, whereas the lymphoma progressed in 45 of 128 patients (35%) receiving IFN maintenance, corresponding to an estimated risk reduction of relapse of 53% in favor of the high dose arm (p=0.00253). Similarly, in 76 MCL evaluable, high dose therapy led to an estimated risk reduction of relapse of 63% (p=0.0107). Both treatment options were well tolerated with <2% treatment related deaths in the high dose therapy arm. In 70 cases with relapsed disease, the combined immuno-chemotherapy (R-FCM) achieved a significantly superior remission rate (89% OR, 36% CR, 53% PR) compared to chemotherapy alone (53% OR, 15% CR, 38% PR; p=0.000715). So far, no improvement of overall survival has been detected in the high dose therapy arm. However, although the benefits of this treatment option may be diminished by late toxic side effects (secondary malignancies), our study represents the first randomized trial confirming the superiority of high dose chemotherapy in MCL. In addition, to the best of our knowledge, our results for relapsed lymphomas represent the first prospectively randomized study which confirms the synergistic benefits of a combined immuno-chemotherapy in this disease.

105

State of the art and new developments in aggressive lymphomas

M. Pfreundschuh (Homburg, D)

For 25 years the chemotherapy combination CHOP (cyclophosphamide, doxorubicin, vincristine and prednisone) has been the standard regimen for the treatment of aggressive NHL and the Intergroup Trial (Fisher et al., 1993) clearly demonstrated that newer 2nd or 3r generation regimens were only more toxic, but no more efficacious than CHOP. The German High-Grade Non-Hodgkin´s Lymphoma Study Group (DSHNHL) therefore tested modifications of the classical 3-week CHOP (CHOP-21) in a prospective randomized trial. 890 untreated young low-risk (normal LDH; 18 to 60 years) and 807 patients >60 years of all IPI risk groups were randomized to 6 cycles of CHOP-21, CHOP-14, CHOEP-21 (CHOP plus etoposide 100 mg/m^2 d1-d3), or CHOEP-14. Patients in the 2-weekly regimens received G-CSF starting from day 4. All patients received radiotherapy (36 Gy) to sites of initial bulky disease (>7.5cm). A planned interim analysis of 50% of the expected events was performed and included 456 elderly patients (61- to 75 years) who had either an event or a follow-up ^{2}3 months after therapy. Median age was 67 years, and adverse prognostic factors were well balanced between the treatment arms. CR rates were 56.8% for CHOP-21, 69.1% for CHOEP-21, 74.3% for CHOP-14 and 70.5% for CHOEP-14 (p = 0.028). In patients with elevated LDH, CR rates rose from 39% with CHOP-21 to 67.3% with CHOP-14 (p = 0.015). At the median time of observation of 34 months, TTF (50% vs. 38%; p=0.03) and overall survival (62% vs. 48%; p=0.04) are significantly better for CHOP-14 and CHOEP-21 than for CHOP-21. In patients >60 years with aggressive NHL, time intensification (CHOP-14) or addition of etoposide (CHOEP-21) improve results significantly over standard CHOP-21 to an extent which appears at least as pronounced as the improvement recently reported by the addition of rituximab to CHOP-21, whereas no improvement was achieved with double-intensive CHOEP-14, most likely because of dose erosion due to toxicity. CHOP-14 is the preferred regimen due to its excellent tumor control and toxicity profile and the new reference standard for elderly patients. In contrast to elderly patients with any pretreatment LDH, shortening of time intervals did not improve results in young patients with normal pretreatment LDH, i. e. supposedly slowly growing tumors. However, the addition of etoposide (CHOEP-21; CHOEP-14) was well tolerated in the younger population and led to significant increases in CR rates and a trend in favor of CHOEP-14/-21 with respect to FFTF. In young high-risk patients the reference control therapy is open to debate and can only discussed in light of recent progress in the elderly. Randomized controlled trials with high-dose regimens using cytotoxic drugs with prime efficacy in aggressive NHL (e.g. Mega-CHOEP) appear to be promising. Since the additional efficacy of rituximab has only been demonstrated in elderly patients treated with CHOP-21, well-controlled trials must evaluate the role of the antibody in other subpopulations and/or in combination with more effective chemotherapy regimens such as CHOP-14 to determine whether combining these regimens with the antibody will result in complimentary, additive or even synergistic efficacy. Supported by Deutsche Krebshilfe.

106

Treatment of Hodgkin's lymphoma

A. Engert, V. Diehl (Köln, D)

The German Hodgkin Study Group (GHSG) was founded more than 20 years ago by Professor Volker Diehl. Since then, both, the number of centers contributing and the number of patients enrolled into the GHSG studies have steadily increased. We currently recruit more than 1400 patients with Hodgkin's disease per year. The database contains more than 8000 patients enrolled between 1983 and now. The GHSG discriminates between early, intermediate and advanced stages depending on clinical stage and additional risk factors. In early stages, the addition of two cycles of ABVD has decreased the number of relapses and allowed a reduction in field size (EF-IF). In intermediate stages, four cycles of chemotherapy followed by EF radiotherapy had been the treatment of choice. We demonstrated that also in this group of patients the extended field is unnecessary so that patients are currently being irradiated in IF-strategy. Thus, the aim in patients with early and intermediate stage HD is to reduce toxicity rather than to improve survival. In contrast, the prognosis of patients in advanced stage was unsatisfactory with no major improvement since the introduction of MOPP or ABVD. In order to improve on these results, we developed a condensed schema containing most of the effective drugs used in patients with Hodgkin's disease (BEACOPP). This regimen was developed in standard doses given in 21 days intervals and also in an escalated version which used GCSF (Neutrogen)in order to maintain the 21 day rhythm. In the HD9 study of the GHSG, the old standard arm COPP ABVD (8 cycles) was randomised against BEACOPP baseline (arm B) and BEACOPP escalated (arm C). A total of 1212 patients was randomised. Arm A was closed primaturely at the second interim analysis due to a significant inferiority compared with the two BEACOPP arms. The latest analysis confirms these data showing significant differences between arms A, B and C in terms of FFTF and also significant differences between arm A and the two BEACOPP arms in terms of survival. Thus, BEACOPP escalated has been selected standard in the ongoing HD12 study.

107

Treatment options for elderly patients with non-Hodgkin's lymphoma

L. Trümper, W. Jung, B. Glass (Göttingen, D)

The incidence of non-Hodgkin's lymphoma (NHL) has been steadily increasing during the last two decades. This rise in incidence is only partly due to the increase in the elderly population . The increase is mainly seen in follicular and diffuse large cell B-cell lymphoma (DLCL) (WHO classification). In addition, the majority of pts with NHL are abover 60 years of age. A high percentage of pts has at least one co-morbidity that frequently leads to an increase in potentially dangerous complications during therapy. Therefore, elderly pts are often underrepresented in reports from clinical trials and are often treated less intensively than younger pts. However, several lymphoma subgroups can be treated and possibly cured, and since the normal life expectancy of 60-year old people is 15-20 years, these curative options should be used in elderly pts. DLCL especially can be cured by systemic therapy, and several trials groups have shown that full-dose chemotherapy with growth factor support can lead to durable remissions. Monoclonal antibodies in combination with chemotherapy may increase the cure rate even further. In order to determine before treatment initiation whether a patient is suitable to standard chemotherapy or not, assessment scores similar to the ones established in geriatric practice have to be deloped. These should supplement clinical prognostic score systems and allow a stratification of lymphoma pts into three groups: those that may and should be treated according to standard protocols with curative intent, those that may receive dose- reduces palliative chemotherapy with the intent of improving the quality of life and those that may not be treated with currently available modalities due to the severe risk of life-threatening complications and the low chance of achieving remissions or improvements in the quality of life. In addition, clinical trials should be designed to specifically include elderly pts and record the complications and chances for success.

109

Post transplant lymphoproliferative disorders (PT-LPD): State of the art and new developments

S.H. Oertel, J. Grefer, B. Dörken, H.B. Riess (Berlin, D)

Post Transplant Lymphoproliferative Disorder (PT-LPD) is a new entity of lymphatic diseases which arose due to the progress in transplantational medicine. The highly effective rejection prophylaxis causes a distinct lost of T-cell mediated immunosurveillance. Primary infection or reactivation of Epstein Barr virus (EBV) induces viral triggered B-cell lymphoproliferation. PT-LPD are in up to 90% of cases associated with the EBV. The histologic spectrum ranges from plasmacytic hyperplasias (5% of cases), polymorphic lymphoproliferations (30%) to diffuse large cell lymphomas (50%). Some rare histologies are multiple myeloma, Hodgkin's disease and Burkitt's lymphoma. Circa 50% of cases occur in the first year after transplantation but PT-LPD can occur even 10 years after transplantation. In EBV-negative children PT-LPD occurs in up to 50% of cases. The incidence in adults ranges from 0.5% after kidney transplantation to 25% after small bowel transplantation. The clinical presentation and course is different from that in normal individuals with lymphomas. Extranodal sites are involved in up to 70% of cases. Preferred sites are gastrointestinal tract, liver, central nervous system, kidney and the transplant itself. The clinical course may be aggressive and rapid lethal but also delayed courses are observed. The clinical management depends on the interval from transplantation, kind of transplant, kidney-, liver- and bone marrow functions, infectious complications, rejection periods and performance state itself. A commonly accepted treatment strategy does not exist. The reduction of immunosuppression may result in disappearance of PT-LPD and should be performed in co-operation with the transplantational team. An antiviral treatment with ganciclovir+arginine butyrate or foscarnet should be performed in case of EBV-association and acceptable performance state. First experiences demonstrate that the well tolerated monoclonal anti-CD20 antibody rituximab induced complete responses in up to 60% of cases with PT-LPD. The effectiveness and tolerability of rituximab is currently investigated in a french and a german multicenter trial. In up to 50% of cases with PT-LPD the patients do require treatment with chemotherapy because the above described treatment approaches are ineffective or due to an initially aggressive course of PT-LPD. Most centres apply anthracyklin-based chemotherapies like CHOP as first line chemotherapy. The haematologic toxicity is the main complication of cytotoxic chemotherapy in organ transplant recipients. Strict dose adaptions and use of granulocyte-colony stimulating factor (G-CSF) are necessary. Carboplatin and etoposide are a tolerable and effective second-line combination in relapsed or progressive patients with PT-LPD. In conclusion PT-LPD is a new entity of lymphatic disease with a different clinical presentation and histologic spectrum than lymphomas in normal individuals. Successful treatment is possible. Various treatment approaches like antiviral treatment, monoclonal antibodies, adoptive T-cell strategies or cytotoxic chemotherapy are currently under investigation. All patients should be treated in multicentre studies.

Wissenschaftliches Symposium:
Ethik in der Palliativmedizin

110

Limits of therapy in palliative care

T. Haferlach, D. Lange, W. Hiddemann, C. Bausewein (München, D)

One has to clearly separate two therapeutic scenarios in oncological patients in advanced stages of their disease: i) a palliative therapy approach, where cure is not achievable but disease specific treatment to prolong life is the main focus, or ii) a setting of palliative care, where reduction of suffering is the main goal using supportive interventions. The ones who treat have to define and discuss such alternatives for every individual patient open-minded. We ask the question, who decides about further treatment approaches: the patient alone, his family, the nurses and/or the physicians. Is there an algorithm for decision making based on distinct rules that could be helpful for physicians and nurses, for patients as well as families to define their interests at the end of life? We also have to discuss the problems of limiting the application of intravenous antibiotics in case of infection, the support with intravenous nutrition and saline infusions, the reasonable time point to apply or to interrupt drugs such as dopamine or adrenaline. Although all these points of interest have to be discussed with respect to every single patient every day again, we are far away from a nihilism in palliative care medicine. Quit the reverse, we think that even when the patient, the family or the palliative care team accept therapy with limitations, this does not mean that we are at any limit of therapy in a palliative care situation.

111

Informed consent in palliative medicine

G. Maio (Freiburg, D)

Since the 1960s there has been a considerable change in the ideal of patient-physician relationship. There has been an important shift from a physician-based commitment of promoting well-being to a patient-based right of information arising from individual autonomy instead of a Hippocratic paternalistism. Consent requirement is ethically derived from the duty to treat persons as ends, not merely as means. So the physician has a special moral obligation to assure the patient's moral agency by reducing the inequality in information between patient and physicain. Therefore a morally valid consent has to be a continuing process rather than a singular event. In this paper there will be first outlined some philosophical thoughts about informed consent and in a second step results from an empirical study will be presented which show the attitudes of physicians and nurses towards informed consent especially in the context of palliative care.

112

Clinical decision making between advance directives and medical conscience

A. van de Loo (Freiburg, D)

Patient self determination in clinical situations of critical terminal illness where communication with the physician in charge is impossible has become a major issue recently. The law offers three ways to represent the patients will in these cases : 1. Vorsorgevollmacht, 2. Betreuungsverfügung and 3. Patientenverfügung (P). The 1. and 2. form require a formal procedure with a judicial order signed by a judge whereas for P until now no formal requirements have been established. However a recently dated signature, detailed knowledge of the actual illness and its complications and precise directives for the physician are criteria that increase the impact on clinical practice. „Advance directives". (German: Patientenverfügung) today are frequently presented by relatives on intensive care units or palliative care wards. Physicians have to consider P as a probable living will of the patient and must treat him/her accordingly. It is under discussion wether non compliance with a P requires law prosecution as grievous bodily harm. In the case of end of live decisions many conflicts between a physicians view and clinical experience and the written interest of the patient may occur. With the increasing dominance of patients self determination expressed by P on clinical decision making and medical jurisdiction physicians tend to reduce their work to neutral information and treatment according to the patients expressed will. However an essential dimension of patient-doctor interaction particularly in the field of end of life medicine is being neglected in this development: Patients and their relatives want to trust in their doctor competently leading them a way in this extremely difficult phase of their lives. Without this dimension of confidence and trust patients and relatives are left essentially alone in a critical situation. Furthermore a fascinating aspect of a doctors clinical work is given up for routine execution of medical guidelines only. Conclusion: P should be considered as an important information to find out the presumed patients will in a clinical end of life situation. However, it is essential for future patient-doctor interaction that the dimension of trust is not eliminated in favour of a neutral customer-client relation.

113

Palliative medicine as an alternative to active euthanasia

H. Csef (Würzburg, D)

Public opinon polls have consistently shown that between 60 and 80 per cent of the European and American population favor the legalization of active euthanasia as a last option to end the suffering of terminally ill patients. The controversial debate about physician-assisted suicide and euthanasia is important for oncology because cancer patients are the largest group of patients for whom active euthanasia is considered. It is striking that in many discussions palliative care ist not taken into account as an option which could euthanasia make unnecessary. Remarkably, those countries which neglect palliative medicine show the highest acceptance of active euthanasia. For instance, depressive cancer patients may request physician-assisted suicide. However, sufficient psychotherapeutic or psychopharmacological treatment is able to reduce the wish for dying. German psychiatrists give suicide prevention a high priority. In contrast, Dutch psychiatrists tend to be more active in suicide assistance. Those physicians who provide good palliative care guarantee an ethical standard. Thus, palliative care is a good and humane alternative to active euthanasia.

Young-Masters-Turnier

114

Ex vivo inhibition of proliferation enhances the transplantability of expanded human cord blood stem cells in NOD/SCID mice

H. Glimm, P. Tang, P. Tang, C. von Kalle, C. Eaves (Freiburg, D; Vancouver, CDN)

Proliferation of hematopoietic stem cells (HSCs) ex vivo is important for numerous applications in cellular and gene therapy. We have previously shown that ex vivo stimulated cord blood HSC in S/G2/M do not engraft (Blood 2000; 96: 4185-4193). Recently, we observed that exposure of human hematopoietic cells to TGF-ß1 or SDF-1 in vitro (in the presence of stroma) or in vivo (in engrafted NOD/SCID mice) can block the S-phase entry of primitive human hematopoietic progenitor cells defined either by HPP-CFC or 6 week LTC-IC assays. We therefore reasoned that appropriate ex vivo exposure to these agents might similarly arrest the proliferation of NOD/SCID repopulating stem cells in cord blood expansion cultures and thereby increase their engrafting activity. Lin- cord blood cells were stimulated for 96 hr in serum-free medium containing Flt-3-ligand, Steel Factor (SF), IL-3, IL-6 and G-CSF and then transferred to cultures containing, SF and TPO with or without TGF-ß1 and /or SDF-1 for 48 hr. TGF-ß1 (but not SDF-1) increased the proportion both of viable G0/G1 cells and of apoptotic cells. TGF-ß1 (but not SDF-1) also reduced the yield of clonogenic progenitors in the secondary cultures, due primarily to a decrease in erythroid progenitors. LTC-IC numbers were not affected by exposure to either TGF-ß1 or SDF-1. Limiting dilution analyses showed that SDF-1 but not TGF-ß1 significantly increased the number of NOD/SCID engrafting human HSCs detectable (2.3 fold, by comparison to control cultures to which these agents were not added). The ability to manipulate human stem cell engraftment activity with specific cytokines will facilitate future investigation of the molecular basis of these changes and will also be of potential importance for numerous applications in cellular and gene therapy.

115

Prognostic variables for response and outcome in patients with Extragonadal germ cell tumors - A International multivariate analysis

J.T. Hartmann, C. Nichols, J.-P. Droz, A. Horwich, A. Horwich, A. Gerl, S. Fossa, H.-J. Schmoll, J. Beyer, J. Pont, L. Kanz, L. Einhorn, C. Bokemeyer (Tübingen, Halle, Marburg, München, D; Portland, Indianapolis, USA; Lyon, F; Sutton, UK; Oslo, N; Wien, A)

This investigation evaluates prognostic variables in patients with seminomatous and nonseminomatous extragonadal germ cell tumors (EGCT) in order to identify relevant factors for long-term outcome following cisplatin-based chemotherapy. Patients from six countries treated in at eleven centers in Europe and United States between 1975 and 1996 were retrospectively evaluated. Multivariate analyses of prognostic variables for survival and for response to chemotherapy were performed. Data were available from 635 EGCT patients, 104 seminomas and 524 nonseminomatous EGCT (n=7 not specified). For nonseminomatous EGCT the following independent adverse factors were identified: presence of either liver [hazard ratio (HR): 1.7], lung (HR: 1.4) or central nervous system (CNS) metastases (HR: 2.5), primary mediastinal tumor site (HR: 2.3), elevation of pretreatment ß-HCG (HR: 1.5). For extragonadal seminoma no adverse feature was identified. Integration of this variables produced the following prognostic grouping: 'excellent prognosis', all seminomatous EGCT, comprising 17% with a 89% 5-year survival rate; 'intermediate high', 'intermediate low' and 'poor', all nonseminomatous EGCT with score values of 0/1, 2/3 or >3, respectively, comprising 20%, 52% and 11% of EGCT with a 69%, 55% and 17% 5-year survival rate. The decreased survival among the different groups was due to a lower rate of favourable objective remissions and a higher rate of relapses.

In addition CART modelling confirmed histology and location of primary as the major prognostic division points. Multivariate testing for the probability of response to chemotherapy has revealed nonseminomatous histology, primary mediastinal tumor site, presence of liver, lung and CNS metastases as independent adverse factors. In EGCT prognostic variables for the outcome and for response to chemotherapy could be identified which in part differ from gonadal GCT. The proposed models might help to better understand the specific prognosis of EGCT and to tailor risk-adapted treatment strategies.

116

Interactions between breast cancer cells and macrophages influence invasiveness and matrix metalloprotease expression

Th. Hagemann, M. Schulz, G. Jahns-Streubel, L. Trümper, C. Binder (Göttingen, Hannover, D)

Interactions between tumor cells and the surrounding stromal compartment have been implicated in tumor progression and metastasis. Degradation of the extracellular matrix as a prerequisite for malig-nant invasion is facilitated by matrix metalloproteases (MMPs), a family of zinc-endopeptidases. Tumor cells, stromal fibroblasts and macrophages show different profiles and intensity of MMP expres-sion. It is still largely unknown, how expression and activity of these enzymes are influenced by inter-actions between the various tumor components. To investigate one aspect of this problem, the human breast cancer cell lines MCF-7 and SK-BR3 were co-cultivated on Engelbrecht-Holm-Schwarm matrix with human macrophages, derived from peripheral blood monocytes. Cells were grown on special membranes without direct cell-to-cell con-tact. The effect of the co-culture on in vitro-invasiveness was evaluated using a microinvasion model (Boyden chamber). MMP expression was determined by quantitative RT-PCR on the Light Cycler System (R) (Roche Diagnostics, Mannheim) as well as by zymography on gelatine- and casein-containing gels. After 96h of co-incubation, invasiveness of MCF-7 cells was increased from 1.75 ± 0.5 to 121.25 ± 7.63 cells per high power field. SK-BR3 cells showed a similar increase. Analysis of MMP protein production in single culture yielded only barely detectable intracellular levels of MMP-2/-9 and MMP-3/-7 in MCF-7 and SK-BR3 cells, while slightly higher amounts of these MMPs were secreted into the supernatant. Macrophages contained and secreted medium levels of all 4 MMPs. Upon co-incubation, MMP-2/-9 as well as MMP-3/-7 concentrations in the joint supernatant were significantly increased. On the mRNA level, all measured MMPs were strongly expressed in control macrophages and only weakly in tumor cells. Co-cultivation lead to significant upregulation of MMP-2 and –9 mRNA in macrophages. The respective mRNA concentrations in the tumor cells remained unchanged, suggesting macrophages as the primary source of increased MMP secretion. MMP-3 and –7 mRNA levels in both macrophages and tumor cells were not influenced by co-cultivation, pointing to post-transcriptional mechanisms of upregulation. MMPs -1, -10, -11, -12, -13, –14 and tissue inhibitors of metalloproteases-1, -2, -3, -4 did not play any obvious role in this context, as mRNA concentrations were negligible under both conditions. Taken together, co-cultivation with macrophages leads to enhanced migration of breast cancer cells through a model basement membrane, presumably due to upregulation of MMP expression in the macrophages. Further studies are necessary to clarify how the interaction between tumor cells and macrophages induces MMP upregulation and whether these mechanisms play a role for invasiveness of tumor cells also in vivo. (Supported by grant nr 10-1270-Bi I, Deutsche Krebshilfe)

117

Evaluation of the incidence of myeloma-specific genetic abnormalities - An interphase cytogenetic approach using 10 different DNA probes for FISH

P. Liebisch, C. Wendl, G. Schilling, K. Roth, N. Baßermann, A. Viardot, H. Goldschmidt, H. Einsele, M. Bentz, S. Stilgenbauer, H. Döhner (Ulm, Hamburg, Heidelberg, Tübingen, D)

Introduction: Specific genetic abnormalities are of prognostic significance in various hematological malignancies. In multiple myeloma (MM), deletions of chromosome 13q (13q-) have been shown to be a marker of poor prognosis despite intensive treatment. So far, no other myeloma-associated chromosomal aberration has been proven to be of prognostic value when tested in a multivariate analysis. We used comparative genomic hybridization (CGH) and tri-color fluorescence in-situ hybridization (FISH) to study genetic changes in patients (pts.) with Salmon & Durie stage II/III MM. Aims are: 1. identification of new genomic aberrations in MM applying CGH analysis to a subgroup of pts., 2. evaluation of the incidence of myeloma-associated aberrations by FISH, and 3. determination of the prognostic impact of these genetic changes within multicenter treatment trials. Results: CGH detected gains or losses of chromosomal material in 28/46 pts. (61%). Commonly gained chromosomal segments were found involving chromosome bands 6p21-p22, 9q31-qter, and 11q23-qter. Chromosome bands 1p22, 6q21-q22, 8p11-p12, and 13q14-q21 were identified as minimally deleted regions. On the basis of our CGH data as well as of data from published studies, a DNA probe set for interphase FISH studies was generated. It consists of 10 BAC and PAC probes that map to chromosome bands 1p22, 1q43, 6q21, 8p11, 9q34, 11p15, 11q22.3-q23.1, 13q14, 17p13, and 22q11. Screening a large series of pts. with these probes, FISH revealed deletions of chromosomes 13q14 in 76/171 pts. (44.4%), 6q21 in 14/105 pts. (13.3%), 22q11 in 13/108 pts. (12.0%), 1p22 in 11/105 pts. (10.4%), 17p13 in 9/161 pts. (5.6%), and 11q23 in 5/126 pts. (3.9%). Trisomies of chromosomes 11q23 were shown in 46/126 pts. (36.5%), 1p22 in 14/105 pts. (13.3%), 6q21 in 10/105 pts. (9.5%), 17p13 in 15/161 pts. (9.3%), and 22q11 in 5/108 pts. (4.6%). No trisomy of chromosome 13q14 was detectable in 171 pts.. Conclusions: CGH allows the detection of genomic alterations in a high proportion of pts. with stage II/III MM and facilitates the delineation of various commonly gained and deleted chromosomal regions. Recurrent loss of chromosome 8p has not been described yet as a myeloma-associated genetic change. Using FISH, 13q- is the most frequent deletion in MM, followed by deletion of chromosome 6q21. Deletion of chromosome 17p13, the locus of the p53 tumor suppressor gene, is rarely observed. Deletion of chromosome 11q22.3-q23.1, frequently seen in B-cell chronic lymphocytic leukemia, is infrequent in MM, but our FISH and CGH data suggest partial or complete trisomy 11q as a common genetic feature. In contrast to our CGH data, FISH shows a much higher proportion of trisomies than deletions involving chromosome 1p. Deletion of chromosome 22q, recently reported to be detectable in more than a third of newly diagnosed myeloma pts. and introduced as a novel independent prognostic parameter, is much less frequently found in our series.

118

Cloning and Characterization of ARAP: A novel nuclear protein interacting with NPM-ALK in large cell anaplastic lymphoma (LCAL)

F. Bassermann, T. Ouyang, R-Y. Bai, C. Peschel, J. Duyster (München, D)

The NPM-ALK fusion gene, formed by the t(2;5) translocation in a subset of large cell anaplastic lymphomas, encodes a 75 kDa hybrid protein that contains the N-terminal portion of the nucleolar phosphoprotein nucleophosmin (NPM) joined to the entire portion of the anaplastic lymphoma kinase (ALK). This translocation event results in the constitutive activation of the NPM-ALK tyrosine kinase.

We report here the cloning and identification of ARAP (ALK regulated anti-apoptotic protein), a novel nuclear protein which interacts with NPM-ALK in a tyrosine-kinase dependent manner in a yeast two hybrid system. Northern analysis with a human multiple tissue blot identified the m-RNA of this protein as a 2.1 kb mRNA species showing widespread expression except in lung and peripheral leukocytes. The 5'-end of ARAP, missing in the initially found clone Q3, was identified by 5'-RACE and revealed the complete c-DNA sequence of 1.7 kb. A raised rabbit antibody identifies a 66 kDa protein. A GST-ARAP fusion protein was shown to interact with NPM-ALK in lysates of KARPAS-299 cells, a cell line derived from a patient with LCAL. When coexpressed in 293 cells, ARAP and NPM-ALK showed complex formation in-vivo. Moreover, coexpression of NPM-ALK resulted in strong phosphorylation of ARAP which was not detectable upon treatment with phosphatase. Neither co-immunoprecipitation nor phosphorylation of ARAP was observed when coexpressed with kinase-dead NPM-ALK. To determine intra-cellular localization, we fused ARAP to E-GFP. Under fluorescence microscopy exclusive nuclear expression of ARAP was observed. Co-expression of NPM-ALK did not affect the intracellular localization. We identified the nuclear localisation site (NLS) of ARAP as a seven aminoacid long stretch in the C-terminal segment of ARAP. Point mutations within this site abolished the nuclear restriction of ARAP and both cytoplasmatic and nuclear lokalisation was observed. We further assessed the role of ARAP in the control of apoptotic cell death. In Ba/F3 cells cultured under interleukin-3 withdrawal, expression of wildtype ARAP significantly reduced apoptotic cell death when compared to wildtype Ba/F3 cells by the methods of TUNEL and PI-staining. Expression of an ARAP construct with a mutation in the NLS did not exhibit this anti-apototic effect emphasizing the essential role for the nuclear restriction of ARAP. In addition, we analyzed induction of apoptosis in NPM-ALK expressing Ba/F3 cells treated with the PI3K-inhibitor wortmannin. We observed a significant increase of apoptotic cells upon co-expression of the NLS-mutant of ARAP compared to coexpression of wildtype ARAP indicating a vital role of ARAP in NPM-ALK mediated mitogenicity. The identification of ARAP as a novel nuclear protein which interacts with NPM-ALK in a kinase dependent manner and displays a significant anti-apoptotic function may help to unravel the complex mechanisms by which NPM-ALK mediates transformation of hematopoietic cells.

119

Tsg101 is a novel modulator of the p53 / mdm2 feedback loop

J. Ruland, L. Li, C. Sirard, A. Elia, D. MacPherson, A. Wakeham, J. de la Pompa, S. N. Cohen, T. W. Mak (Toronto, CDN; Stanford, USA)

The p53 tumor suppressor gene is the most commonly mutated gene in human cancer. p53 transcriptionally activates several targetgenes involved in cell cycle control and apoptosis. These include the gene encoding the cyclin-dependent kinase inhibitor p21WAF-1/CIP-1 and the oncogene mdm2. As a ubiquitin ligase mdm2 in turn regulates p53 protein levels in a negative feedback loop by promoting ubiquitin-mediated degradation of p53. This control loop itself is regulated, to allow rapid p53 accumulation upon p53 activating stimuli. However, the molecular details how the p53 / mdm2 loop is regulated are not well understood. The tsg101 tumor susceptibility gene was initially identified by the transformation caused by it's antisense expression in mouse 3T3 cells and their consequent ability to form metastatic tumors in mice. It encodes for a protein with a Ubc domain, characteristic of the catalytic domain of ubiquitin conjugating enzymes but it lacks the active site cystein crucial for enzymatic activity. To investigate the physiological function of tsg101, we generated a null mutation of the mouse gene by targeted disruption in embryonic stem cells. Homozygous tsg101-/- embryos fail to develop past day 6.5 of embryogenesis (E6.5). They are reduced in size, and do not form mesoderm. Mutant embryos show a decrease in cellular proliferation in vivo but no increase in apoptosis. In addition, the inner cell mass of isolated mutant blastocysts does not proliferate in vitro. These results demonstrate that tsg101 is essential for the proliferative burst prior to the onset of gastrulation. Although levels of p53 transcripts are not affected in tsg101-/- embryos, p53 protein accumulates dramatically, implying altered posttranscriptional control of p53. In addition, transcription of the p53 effector p21WAF-1/CIP-1, is increased 5-10 fold. Introduction of a p53 null mutation into tsg101-/- embryos rescues the gastrulation defect and prolonges survival until E8.5 demonstrating that the gastrulation block is p53 mediated. Together these results imply a functional connection between tsg101 and the p53 pathway in vivo. In vitro we found that tsg101 binds to mdm2 as well as to p53. The Ubc domain of tsg101 interferes with ubiquitination of mdm2, and upregulation of tsg101 inhibits mdm2 decay. Upregulated tsg101 therefore elevates mdm2 steady-state levels and leads to down regulation of p53 protein. These in vitro data are consistent with and complementary to our in vivo analysis and provide a molecular basis for the upregulation of p53 in tsg101 -/- embryos. Our genetic and biochemical results identify tsg101 as a novel essential regulator of the p53 / mdm2 feedback loop.

Kasuistisches Forum 2

120

Rapid tumour lysis in a patient with mantle cell lymphoma treated with an anti-CD20 monoclonal antibody (IDEC-C2B8, Rituximab)

H. Simon, G. Brittinger, L. Trümper (Göttingen, D)

Background: The chimeric human monoclonal anti-CD20 antibody IDEC-C2B8 (rituximab) is in the meantime well established in the treatment of non-Hodgkin´s lymphomas, especially follicular lymphoma. Rituximab has been demonstrated to have mild side effects. But there exist also several reports on severe side effects, such as tumour lysis syndrome (TLS). Especially in patients with large numbers of antigene-dense cells such as B-cell chronic lymphocytic leukemia and in case of a high mitotic index such as prolymphocytic leukemia and high grade lymphoma, there is an increased risk of TLS. We report on a 66-year old man with mantle cell lymphoma and growing infiltration of the liver who developed a severe tumour lysis syndrome after a second infusion of rituximab.
Case report: After diagnosis of a mantle cell lymphoma stage IIIA in May 1999, a 66-year old male patient had been treated with 6 cycles of the CHOP-chemotherapy (stable disease) followed by Interferon alpha maintenance. In view of a progress in July 2000 with enlarged lymph nodes cervical, axillary, inguinal and in the area of the porta hepatis two more cycles of CHOP were applied out-patient. Subsequently the patient was scheduled for a non-myeloablative radioimmunotherapy. In October 2000, the patient was admitted in our hospital because of a worsening of his clinical performance status. We diagnosed an advanced infiltration of the liver with liver insufficiency, portal hypertension and ascites. In view of the bad clinical performance status and a progressive pancytopenia we instituted a palliative treatment with the monoclonal antibody rituximab. The first administration of rituximab with a reduced dose rate of 375 mg led to a merely moderate increase of the lactate dehydrogenase without any sign of organ failure. Therefor we continued the treatment in standard dose of 375 mg/m^2 one week later. Despite the common prophylactic measures the patient developed a severe acute TLS with renal failure. In spite of intensive therapeutic measures inclusive haemodialysis the patient died because of multiple systems organ failure. Conclusions: Also in case of mantle cell lymphoma and high tumour mass physicians must be aware of the risk of acute tumour lysis after treatment with the monoclonal antibody rituximab. Moreover this case shows that rituximab treatment could be effective even in chemotherapy refractory mantle cell lymphoma.

121

Clear cell carcinoma of the lung - Report on two advanced cases responding to tamoxifen treatment

F.W. Hirsch, B. Weber, I. Dresel, E. Stöckl, M. Heiland, M. Hofstetter, A. Jakob (Offenburg, D)

Introduction: Clear cell carcinoma is the most common histological subtype in renal cell cancer. These tumors often metastasize to the lung. However, there are also clear cell carcinomas primarely confined to the lung. We report on two cases with multilocular lung disease without primary tumor in the kidneys. Both responded well to hormonal treatment with tamoxifen.
Case 1: A 77-y-old woman was found to have two lung tumors 10 y ago. 3/92 these tumors were resected. The histology was: clear cell carcinoma. No other primary was found. In 7/96 multiple lung metastases and a nodule in the vagina were detected; the latter was removed. Since 7/96 the patient is treated with tamoxifen starting with a dose of 60 mg daily; since 1/98 the dose is reduced to 30 mg daily. An impressive durable partial response was achieved. To date, the patient is well and without symptoms.
Case 2: A 68-y-old woman presented with dyspnoe in 2/2000. X-ray of the chest showed numerous nodules in both lungs. No primary was found. The needle biopsy yealded the diagnosis clear cell corcinoma. Hormonal treatment with 30 mg tamoxifen daily was started. 15 month later dyspnoe and radiological findings are slightly improved. X-rays of both patients will be presented. Conclusion: Clear cell carcinomas confined to the lung may have a different clinical course as compared to the clear cell renal carcinomas with pulmonary metastases. The two cases presented here show a surprisingly good response to hormonal treatment with tamoxifen.

122

MRI of the spine - Incidental diagnosis of indolent non-Hodgkin's lymphoma in two patients

G. Geibel-Werle, M. Hoffmann, M. Uppenkamp (Ludwigshafen, D)

Indolent non-Hodgkin's lymphoma are either diagnosed on clinical symptoms, lymphadenopathy or because of a high blood cell count. The patients are usually of older age and may present with splenomegaly or signs of an immunodeficiency syndrome. First line treatment varies depending on stage from irradiation therapy to „wait and watch" or chemotherapy in progressive systemic disease. We report about two patients (pts.) who were presented with the question of a hematological disorder after having had a MRI of the spine for lumbal disc prolaps.In both cases MR-imaging revealed a diffuse decrease of signals in the spine and was interpreted as an activated blood-forming marrow. In the case of a 49-year old women no clinical symptoms of a hematological disease were noticed. Laboratory investigations were regular, there were no splenomegaly or lymphadenopathy. The trephine biopsy showed a 60-70% infiltration with malignant lymphocytic cells, which could be phenotyped by flow cytometry as a CD103-positive population without the typical morphology of a hairy cell leukemia. The same CD103-positive cell population was found in the blood. IL-2-receptors were strongly expressed on the cells. The second pt. was a 58-year-old man with a very discrete blood lymphocytosis of 4800/µl. No lymphadenopathy or splenomegaly was diagnosed. The trephine biopsy revealed an indolent lymphoma, which was phenotyped as a chronic lymphocytic leukemia of B-cell type. In both pts. the lymphoma of low grade malignancy was diagnosed by MRI of the spine in an early stage, lacking any typical symptoms and without the necessity of treatment. Lower back pain is a common disorder in the general population and MRI has become a standard diagnostic tool. It would not be surprising, if the MRI bone marrow signal is examined carefully that hematological disorders could be diagnosed early in the course of disease. This information would give a better understanding of the latency period from very early to final manifestation of the disease. A decrease of the bone marrow signal in the MRT scan of the spine should be regarded as a possible sign of a hematological disorder. Further investigations should be performed.

123

Dermatitis and diabetes mellitus - symptoms of a rare oncological disease. A case of glucagonoma

Th. Lipp, K. Knochel, C. Günther, C. Nerl (München, D)

A 33-year-old woman was admitted to the hospital with a newly diagnosed insulin-dependent diabetes mellitus and a two year history of recurrent dermatitis which was treated repeatedly with local and systemic antibiotic therapy. Further she reported on weight loss of 8kg during the last year and of nearly 30kg during the last 5 years. The clinical examination showed a necrolytic erythema on legs, perineum and groins and a marked hepatomegaly. Hyperglycemia and normochromic normocytic anemia were shown on laboratory findings. The ultrasonographic exam demonstrated a pancreatic tumor and multiple liver metastases. In conjunction with the pathognomic necrolytic migratory erythema and the diabetes this strongly suggests an endocrine tumor of the gastrointestinal tract. The glucagon serum level which was initially over 800 U/l confirmed the diagnosis of a glucagon producing tumor, i.e. a glucagonoma. Surgical tumorreduction was performed with partial pancreatic resection, splenectomy and resection of several large liver metastases. The patient received monthly subcutaneously octreotide and eight cycles of a streptozotocin and doxorubicin containing chemotherapy. The clinical symptoms, especially the dermatitis improved rapidly and the liver metastases showed a progressive reduction in size. We will report on the clinical course of this very rare malignant glucagone producing tumor and the therapeutic strategies.

124

Primary cardiac non-Hodgkin's lymphoma in a 62-year old HIV-negative Caucasian male patient

C. Gerlach, C. Hoppe, B. Jung, F.-J. Beck, W.-P. Klövekorn, N. Frickhofen (Wiesbaden, Bad Nauheim, D)

A 62-year old male patient presented with a 6 week history of declining vigor, angina on exertion, fainting on bending forward, and venous congestion of the head, neck and upper thorax. A 7 x 8 x 10 cm mass filling the right cardiac atrium and obstructing the upper V. cava was revealed by echocardiography. NMR scanning demonstrated a solid mass without necrosis originating from the atrial septum. Staging did not reveal additional physical abnormalities, in particular no evidence for extranodal or nodal lymphoma. Laboratory studies were unremarkable except for an LDH of 950 U/l without evidence for hemolysis. Cardiac surgery confirmed the right atrium to be almost completely filled by a tumor infiltrating the atrial septum and extending to the left atrial wall, the upper vena cava and the right ventricle. The mass could be completely removed with resection of parts of the upper vena cava and atrial wall. The defects were remodeled by a vascutek-prosthesis and patches for the atrial wall. Histology demonstrated a diffuse large cell lymphoma of B cell type. The patient was treated with 6 cycles of cyclophosphamide, adriamycin, vincristin and prednisone, given at 14 day intervals and supported by filgrastim (CHOP-14). He is alive and well 17 months later. Primary cardiac lymphoma represents less than 1% of aggressive B cell lymphomas and about 2% of primary neoplasms of the heart. It is extremely uncommon in patients without known immune compromise. There is no evidence for a causal link to myocarditis or other risk factors, which might explain the origin of the lymphoma in cardiac tissue. Case reports suggest lasting remissions following standard chemotherapy with or without additional radiotherapy. Resection of the lymphoma was considered necessary in our patient to quickly alleviate clinically significant obstruction of the cardiac blood flow.

125

Richter's syndrome: Complete molecular remission after autologous stem cell transplantation

U. Apfelbeck, M. Eibl, R. Brezinschek, G. Höfler, W. Linkesch (Graz, A)

Approximately 3% of patients with B-cell chronic lymphocytic leukemia develop during their course of disease a high-grade large-cell lymphoma consistent with Richter`syndrome, which is associated with an unfavorable prognosis and a median survival of six months. We report the case of a 53-year-old male who presented with cervical lymphadenopathy in November 1997. Diffuse large-cell lymphoma was diagnosed by biopsy. Staging revealed a CS II. The patient was treated with six courses of combination chemotherapy (CHOP). A complete remission was seen after the second course. Sixteen months later the patient presented again with lymphadenopathy. Bone marrow biopsy and excisional biopsy of the lymphnodes revealed infiltration by CLL. Polymerase chain reaction (PCR) on the heavy chain immunoglobulins showed a band of the same size as in the initial lymphnode specimen and bone marrow biopsy. The initial diagnosis of diffuse large cell lymphoma was changed to Richter`s syndrome and the patient treated with three courses of CHOP combined with hyperthermy After four courses of Rituximab (375 mg/m^2 once weekly) a complete remission was achieved. G-CSF-mobilization and peripheral stem cell apheresis with CD 34 positive and CD 19 and CD 20 negative selection was performed. In October 2000 the patient underwent autologous stem-cell transplantation. Conditioning consisted of TBI and Cyclophosphamide. Seven months after stem-cell transplantation and 42 months after the initial diagnosis the patient remains in complete hematological and molecular remission.

Wissenschaftliches Symposium:

Die Bedeutung neuer bildgebender Verfahren für die Diagnostik maligner Tumoren

126

Somatostatin-receptor scintigraphy for oncologic applications

I. Virgolini (Wien, A)

The high level expression of somatostatin receptors (SSTR) on various tumor cells has provided the molecular basis for successful use of radiolabeled octreotide / lanreotide analogs as tumor tracers in nuclear medicine. The vast majority of human tumors seem to over-express the one or the other of five distinct hSSTR subtype receptors. Whereas neuroendocrine tumors frequently overexpress hSSTR2, intestinal adenocarcinomas seem to overexpress more often hSSTR3 or hSSTR4, or both of these hSSTR. In contrast to 111In-DTPA-DPhe1-octreotide (OctreoScan®) which binds to hSSTR2 and 5 with high affinity (Kd 0.1-5 nM), to hSSTR3 with moderate affinity (Kd 10-100 nM) and does not bind to hSSTR1 and hSSTR4, 111In / 90Y-DOTA-lanreotide was found to bind to hSSTR2, 3, 4, and 5 with high affinity, and to hSSTR1 with lower affinity (Kd 200 nM). Based on its unique hSSTR binding profile, 111In-DOTA-lanreotide was suggested to be a potential radioligand for tumor diagnosis, and 90Y-DOTA-lanreotide suitable for receptor-mediated radionuclide therapy. As opposed to 111In-DTPA-DPhe1-octreotide and 111In-DOTA-DPhe1-Tyr3-octreotide, discrepancies in the scintigraphic results were seen in about one third of (neuroendocrine) tumor patients concerning both the tumor uptake as well as detection of tumor lesions. On a molecular level, these discrepancies seem to be based on a „higher" high-affinity binding of 111In-DOTA-DPhe1-Tyr3-octreotide to hSSTR2 (Kd 0.1-1 nM). Other somatostatin analogs with divergent affinity to the five known hSSTR subtype receptors have also found their way into the clinics, such as 99mTc-depreotide (NeoSpect®; NeoTect®). Most of the imaging results are reported for neuroendocrine tumors (octreotide analogs) or nonsmall cell lung cancer (99mTc-depreotide), indicating high diagnostic cabability of this type of receptor tracers. Consequently to their use as receptor imaging agents, hSSTR recognizing radioligands have also been implemented for experimental receptor-targeted radionuclide therapy. Beneficial results were reported for high-dose treatment with 111In-DTPA-DPhe1-octreotide, based on the emission of Auger electrons. The Phase IIa study „MAURITIUS" (Multicenter Analysis of a Universal Receptor Imaging and Treatment Initiative, a eUropean Study) showed in progressive cancer patients (therapy entry criteria) with a calculated tumor dose > 10 Gy / GBq 90Y-DOTA-lanreotide, the proof-of-principle for treating tumor patients with peptide receptor imaging agents. In the „MAURITIUS" study, cummulative treatment doses up to 200 mCi 90Y-DOTA-lanreotide were given as short-term infusion. Overall treatment re-

sults in 70 patients indicated stable tumor disease in 35% of patients and regressive tumor disease in 10% of tumor patients with different tumor entities expressing hSSTR. No acute or chronic severe hematological toxicity, change in renal or liver function parameters due to 90Y-DOTA-lanreotide treatment, were reported. 90Y-DOTA-DPhe1-Tyr3-octreotide may show a higher tumor uptake in neuroendocrine tumor lesions and may therefore be superior for treatment in patients with neuroendocrine tumors. However, there is only limited excess to long-term and survival data at present. Potential indications for 90Y-DOTA-lanreotide are radioiodine-negative thyroid cancer, hepatocellular cancer and lung cancer. Besides newer approaches and recent developments of 188Re-labeled radioligands, no clinical results on the treatment response are yet available.
In conclusion, several radioligands have been implemented on the basis of peptide receptor recognition throughout the last decade. A plentitude of preclinical data and clinical studies confirm their potential use in diagnosis as well as „proof-of-principle" for therapy of cancer patients. However, an optimal radiopeptide formulation does not yet exist for receptor-targeted radionuclide therapy. Ongoing developments may result in peptides more suitable for this kind of receptor-targeted radionuclide therapy.

127

MRI and multislice helical CT of tumors of the liver, biliary tree and pancreas

J. Gaa, Ch. Düber (Mannheim, D)

The dramatic success that magnetic resonance imaging (MRI) has achieved in the diagnosis of diseases of the central nervous system has not been mirrored in abdominal applications. Problems associated with MRI of the abdomen arise almost entirely because of artifacts secondary to gross physiologic motion (e.g. respiration, bowel peristalsis and cardiac pulsation). Therefore, helical CT has been the workhorse in most radiology departments for the evaluation of abdominal pathologies.
Recent advances in abdominal MRI, including faster pulse sequences, breath-hold imaging, use of intravenous contrast agents, and surface coils have improved image quality and examination times. In addition, the development of novel MR techniques such as Magnetic Resonance Cholangiopancreatography (MRCP) and 3D-MR-Angiography (MRA) have greatly enhanced the diagnostic utility of abdominal MRI. For these reasons, MRI has evolved from a problem solving study to become the primary imaging examination in many patients with hepatopancreatico-biliary diseases.

128

FLT-PET for the detection of tumor manifestations in patients with malignant lymphoma

M. Bommer, I. Buchmann, F. Leithäuser, A. Viardot, H. Döhner, M. Bentz (Ulm, D)

Positron emission tomography (PET) has advanced rapidly in the past few years and has become a valuable diagnostic procedure for patients with Non-Hodgkin Lymphomas (NHL). The most frequently used tracer is the glucose analogue F-18 fluorodeoxyglucose (FDG), which provides a marker of the metabolic activity in different tissues. We studied the thymidin analogue [F-18]3'-deoxy-3'-fluorothymidine (FLT) as a tracer to detect tumor proliferation in lymphomas. Aim of this study was the evaluation of side effects and whole-body biodistribution of [F-18]FLT. In addition, in subsets of patients FLT data were compared with data obtained by FDG-PET and with immunological proliferation markers in tumor samples. Fourteen patients with lymphoproliferative diseases (6 patients with aggressive lymphomas or lymphomas in transformation, and 5 patients with indolent lymphomas), Hodgkin's disease (1) and multiple myeloma (2, one had extramedullary myeloma) were examined. Tumor diagnosis was established by biopsy and histological and/or cytological examination. Proliferative activity was evaluated by immunohistostaining of Ki-67 antigen by MIB-1 staining (3 patients) or flowcytometry using FITC-labelled MIB-1 antibody (Clone MIB-1, IgG1; Immunotech, Coulter) (3 patients). PET studies were performed using a full ring-scanner (ECAT Exact, Siemens, Erlangen, Germany). Mean doses of 279 MBq FLT were administered. Patients who received both FLT- and FDG-PET (n=8) received 571 MBq FDG in addition to the FLT on different days. PET scans were performed 50-70 minutes after application of the tracer. No side effects due to FLT injection were observed. In 13/14 patients, lesions were detected by FLT-PET (median 2; range, 0 – 13 lesions). Standard uptake values (SUV) ranged from 1.53 to 13.3 (median 5.44). In patients with aggressive or transformed Lymphoma (6 patients) the average SUV was 6.35, in patients with indolent lymphoma (5 patients) the average SUV was 4.67 and in the myeloma patients the average SUV was 4.75. The only patient with Hodgkin`s disease had an SUV of 2.70. In the nine patients who had both FLT- and FDG-PET, 47 lesions were detected by FLT-PET, and 43 lesions were detected by FDG-PET. Using routine diagnostic procedures (physical examination, CT-scan, bone marrow biopsy and ultrasonic examination) 39 lesions were identified in these patients. In six patients tissue samples for Ki-67 measurement were available. So far, there were no clear-cut correlations between proliferation indices and mean SUV values. More detailed analyses are currently being performed. These results indicate, that FLT is a suitable tracer to detect manifestations of malignant lymphoma.

129

The value of positron-emission-tomography (PET) in the assessment of recurrences after radiotherapy in an experimental tumour system

M. de Wit, A. Raabe, H. Beck-Bornholdt, B. Seegers, R. Buchert, D. Hossfeld (Hamburg, D)

PURPOSE: The study was performed in order to implement an experimental system for using 18FDG-PET in the diagnosis of experimental tumours and to determine a) the influence of time and tumour volume after therapy on the sensitivity and specificity of 18FDG-PET, and b) the benefit of 18FDG-PET concerning early recognition of recurrences after fractionated radiotherapy. METHODS: Subcutaneously growing Rhabdomyosarcoma R1H of the rat were irradiated applying total doses of 80 or 85 Gy (30 fractions in 6 weeks). Tumour volume was determined twice a week. PET was performed weekly before, during and for 6 months after therapy using a conventional full-ring whole body PET-Scanner (Siemens CTI Ecat/Exact). For PET measurements animals received 440 MBq 18FDG s.c. 3 h prior to the PET procedure. The PET results were evaluated qualitatively using a 6 score scale. PET results and actual tumour volume were compared. Sensitivity and specificity were calculated and the quality of diagnosis of recurrences by 18FDG-PET was determined using the Receiver Operating Characteristics (ROC). RESULTS: After irradiation 8/20 tumours (40%) were locally controlled. Recurrent tumours were detectable by volume measurements when they reached a volume of 0.1 ccm. The intra-observer concordance of tumour diagnosis by PET was reproducible and a kappa coefficient of 0.78 (95% C.I.: 0.71-0.85) was determined. Sensitivity and specificity were dependent on the score used for evidence of tumours. The sensitivity increased with time, reflecting the increase in volume of recurring tumours. For recurrences of 0.2 ccm neither false negative nor false positive PET results were obtained, whereas for tumours of 0.1 ccm sensitivities of 91% or 70% and specificities of 67% or 100% were obtained, according to the score used. On average tumours were recognised by PET 14, 57, 68, and 68 days (median) before approaching volumes of 0.2 ccm, 0.5 ccm, 0.8 ccm, or 1.0 ccm, respectively. CONCLUSION: The experimental system allows reproducible detection of R1H-tumours using 18FDG-PET. The quality of PET as diagnostic test for R1H-tumours is very good for tumours of more than 0.1 ccm, or when PET is performed later than 12 weeks after start of irradiation treatment.

130

Ultrasound guided fine needle aspiration cytology for early identification of melanoma metastases (update)

T. Mayer, C. Voit, M. Kron, L. Weber, G. Moser, U. Zeelen, A. Schoengen (Ulm, Berlin, D)

Within the spread of melanoma cells lymph node recurrences as well as subcutaneous metastases may occur. Ultrasound (US) is a non invasive tool for the detection of these lesions and is evolving to become the diagnostic standard, since palpation alone has also known to be an inaccurate technique for assessment of metastases. To verify these lesions as malignant, FNAC may be helpful. To test the feasibility and effectiveness of presurgical free-hand style FNAC in melanoma metastases - originally without and since introduction of US with ultrasound guidance -, we performed a retrospective review of melanoma patients, who were treated between 1984 and 2001. Clinical examination and high resolution US was done during follow-up of melanoma patients. Validation of findings was established by FNAC as non-invasive technique. Cytodiagnostic evaluation was performed morphologically. The cytologic findings were confirmed either by histopathologic diagnosis or prolonged clinical follow-up. The study included 1053 FNACs in 469 melanoma patients. Our study achieved a sensitivity (Se) of 97,7% (95%CI: 96.3%;98.6), a specificity (Sp) of 99,7% (95%CI: 98.1%;100%), a positive predictive value (ppV) of 99,9% (95%CI: 99.2%;100%) and a negative predictive value (npV) of 94,5% (95%CI: 91.3%;96,7%). In 214 FNACs derived from lesions with a diameter of less than 1 cm Se was 95,3% (95%CI: 90.5%;98.1%) with a Sp of 100% (95%CI: 93.4%;100%) and a ppV and npV of 100% (95%CI: 97.4%;100%) and 88,5% (95%CI: 77.8%;95.3%). Diagnosis has been established within one day of sampling in an outpatient setting without side effects. Conclusion: FNAC is highly predictive in early detection of melanoma metastases, is a rapid, cost-effective method free of complications and helps to avoid unnecessary invasive surgery in non-malignant lesions. FNAC is highly useful in combination with US in melanoma follow-up and is able to establish the diagnosis of lesions smaller than 10 mm without impairment of Se or Sp.

Postersitzung:
Akute Leukämien - Klinik

131

FLAG (Fludarabine, cytosine arabioside, G-CSF) in the treatment of refractory and relapsed high risk hematological malignancies (AML, ALL, NHL). A single center experience

T. Nösslinger, E. Koller, H. Tüchler, K. Jadrna, M. Bernhart, E. Pittermann-Höcker, M. Pfeilstöcker (Wien, A)

Background: The objectives of this analysis were to evaluate the efficacy and toxicity profiles of the combination of fludarabine and high-dose cytosine arabinoside (Ara-C) in refractory or relapsed high risk hematological malignancies. Patients and Methods: Thirty-six patients (28 AML, 5 ALL, 3 NHL) were treated with a total of 60 cycles of FLAG (30 mg/m^2 fludarabine days 1-5, 2 g/m^2 Ara-C days 1-5 and 5µg/kg G-CSF day 0 to polymorph-nuclear recovery, in 5 cycles 10mg/m^2 idarubicine days 1-3 was added). The median age was 50 years (range 18-65) and the sex ratio was 1:1. Before administration of FLAG the patients had received in median five cycles of conventional chemotherapy (range 1-12): five patients (14%) were refractory to these regimens, twenty-four (67%) were in first and seven (19%) were in second or subsequent relapse, with a median response duration of twelve months to induction therapy. Results: Eleven patients (31%) achieved a complete (CR) and six (17%) a partial remission (PR), twelve patients (33%) had resistant disease and seven (19%) died during remission induction, four due to progressive disease and three due to pneumonia. The median CR duration was 8.5 months and the median survival duration was 6 months for all patients and 10 months for those who had response to treatment. Eleven patients (31%) are still alive with six of them in continous CR. Hematological toxicity was quite severe, more pronounced in patients receiving additional idarubicine. Median time to neutrophil (>1,000/µl) and to platelet (>20,000/µl) recovery were 22 days (ranges 14-31 and 13-36 respectively). A median of 6 red cell units (range 0-25) and of 7 platelet concentrates (range 2-28) had to be given. Median duration of febrile neutropenia was 5 days (range 0-35) and i.v. antibitoics were administered in all patients for a median of 19 days (range 8-50). There were 49 documented infections with 29 septical events and 20 pneumonias. Nonhematological toxicity was low. The median period of hospitalization was 29 days (range 17-60). Conclusion: FLAG is an efficient regimen with tolerable toxicitiy in relapsed AML. Subsequent courses are recommended, if CR or PR is achieved after the first cycle. Addition of idarubicine seems to lead to increased side effects, without improving the remission rate.

132

High dose cytarabine in consolidation therapy for acute promyelocytic leukemia

R. Schlenk, S. Fröhling, B Füllgraf, P. Dreger, J. Fischer, A. Glasmacher, K. Götze, W. Grimminger, U. Germing, F. Hartmann, E. Koller, H. Mergenthaler, H. Salwender, C. Waterhouse, W. Weber, F. del Valle, H. Pralle, K. Döhner, H. Döhner for the AMLSG ULM (Ulm, Kiel, Karlsruhe, Bonn, München, Stuttgart, Düsseldorf, Homburg, Hamburg, Trier, Oldenburg, Gießen, D; Wien, A)

All-transretinoic acid (ATRA) differentiates abnormal promyelocytes into mature granulocytes in APL and induces complete remission (CR) in 80 to 90% of patients (pts). However, duration of CR with ATRA alone is usually brief and postremission chemotherapy is required. This study was designed to evaluate the influence of high-dose cytarabine (HiDAC) in consolidation therapy of APL on disease free survival. In two consecutive clinical trials (AML HD93 (1993-1998) and AML HD98A (1998-ongoing) patients with APL (age 16-60 yrs) were treated for first induction therapy according to the AIDA protocol (ATRA 45mg/m^2, day 1 to 28, idarubicin 12mg/m^2 continuously day 2,4,6,8). Second induction therapy was given according to the ICE protocol (idarubicin, cytarabine, etoposide) and both first and second consolidation therapies were given according to the HAM protocol (cytarabine 3g/m^2 12 h day 1 to 3, mitoxantrone 10mg/m^2 day 2,3). From 1993 to 1998 21 and since 1998 from 35 pts data were available. Of 56 pts with cytogenetically proven t(15;17) 49 were diagnosed morphologically. Seven pts were initially misdiagnosed and treated with ICE as first induction therapy, of whom two received AIDA as second induction therapy, three maintenance therapy with ATRA 45mg/m^2 for 6 months after completion of consolidation therapy and one patient relapsed after consolidation therapy before start of maintenance therapy with ATRA. Six pts died due to uncontrolled coagulopathy before start of treatment or during the first two days of first induction therapy and 2 pts died during first induction because of severe infection leading to an early death rate of 14%. All other pts achieved first CR (86%). Of the 43 pts receiving AIDA as first induction therapy 3 developed an ATRA syndrome necessitating premature discontinuation of ATRA. One pt developed severe keratitis during ATRA therapy. This pt did not receive consolidation therapy, relapsed, achieved second CR and is in continuous CR after autologous transplantation. Three pts died during second induction therapy in 1.CR because of severe infection. Forty-two pts completed the whole treatment with two cycles of HAM and none of them relapsed. The two consolidation cycles were well tolerated. The median follow up duration is 2 years. Relapse probability, disease free and overall survival are 4,3%, 90% and 82% at two years, respectively. These results suggest that HiDAC is effective in consolidation therapy for preventing relapse in APL.

133

Monitoring of minimal residual disease in patients with acute lymphoblastic leukemia in clinical remission using multicolor flow cytometry

H. Engel, H. Diedrich, E. Richter, C. Elser, T. Weihkopf, R. Kirstein, A. Ganser (Hannover, D)

Morphological techniques, used in clinical practice for monitoring acute leukemia patients, are able to detect only 1-5% of malignant cells. Sensitive methods to detect minimal residual disease (MRD) may allow better estimates of the leukemia burden and help the selection of appropriate therapeutic strategies. Flow cytometric detection of MRD is based on the identification of immunophenotypic combinations expressed on leukemic and not on normal hematopoietic cells. In this study we describe the presence and frequency of residual leukemic cells detected by multicolor flow cytometry in patients with acute lymphoblastic leukemia (ALL) in clinical complete remission (CCR). Sixty-three bone marrow samples of 27 patients (25 B-lineage B-ALL, 2 T-ALL) were analyzed at diagnosis, the time of entering complete remission and subsequently during CCR and relapse for the coexpression of atypical antigens (CD2/CD24/CD34, CD3/CD22/CD34, CD5/CD19/CD33, CD7/CD15/CD34, CD10/CD19/CD34, CD19/CD 7.1/CD34, CD20/CD33/CD34). 4/27 patients relapsed after a median remission duration of 5.5 months (range; 2-32). 22/27 patients are still in CCR after a median follow-up of 5 months (range; 1-44) and 1 patient died after 19 months in CR. MRD levels were significantly higher (p<0.0001) in patients who relapsed (mean ± SEM: 2.21% ± 0.95%) compared to those who are still in clinical remission (mean ± SEM: 0.45% ± 0.05%). We conclude, that the coexpression of atypical antigens in CCR provides additional information that is useful in the identification of patients with high and low likelihood of relapse.

134

Two delayed granulocyte colony-stimulating factor administration schedules after high-dose cytarabine therapy in adult patients with acute lymphoblastic leukemia: Results from a prospective randomized trial

W.K. Hofmann, G. Seipelt, S. Langenhan, R. Reutzel, D. Schott, O. Schöffski, H.J. Illiger, F. Hartmann, L. Balleisen, A. Franke, F. Fiedler, C. Huber, H. Rasche, L. Bergmann, A. Ganser, C. Pott, R. Pasold, C. Rudolph, O.G. Ottmann, N. Gökbuget, D. Hoelzer (Frankfurt/Main, Miltenberg, Bad Soden, Erlangen-Nürnberg, Oldenburg, Homburg/Saar, Hamm, Magdeburg, Chemnitz, Mainz, Bremen, Hannover, Kiel, Potsdam, Cottbus, D)

In acute lymphoblastic leukemia (ALL), treatment with granulocyte colony-stimulating factor (G-CSF) during remission induction shortens granulocytopenia and may decrease morbidity due to infections. However, the optimal timing of G-CSF administration after chemotherapy is not known. In a prospective randomized multicentre study adult ALL patients were treated with high-dose ARA-C (HDAC, 3 g/m^2/bid (1 g/m^2/bid for T-ALL) days 1-4) and mitoxantrone (MI 10 mg/m^2 days 3-5). They were randomized to receive recombinant human G-CSF (Lenograstim) 263 µg/d SC starting either from day 12 (Group 1) or day 17 (Group 2). Fifty-five patients (41 male, 14 female) with a median age of 34 years (range 18-55) were enrolled into the study. 50 patients were evaluable. The median duration of neutropenia <500 /µL after HDAC/MI was 12 days (range 7-22) in the early G-CSF group 1 and equally with 12 days (range 4-22) in the late G-CSF group 2, but it was shorter than in the historical control group (15 days, range 4-43, n=46) where the patients received identical cytotoxic treatment without G-CSF. Twenty-seven patients in group 1 (90 %) had fever >38 °C compared with 23 patients in group 2 (92 %). Seventeen infections were observed in 14 patients in group 1 (47 %) and 13 infections in 10 patients in group 2 (40 %) compared to 27 infections in 49 patients of the historical control (54 %). In group 1 the patients received a median of 11 injections with G-CSF (range 7-22) compared to 7 injections (range 4-19) in group 2. The total administered dose of G-CSF in group 2 was significantly reduced by 40 % (p<0.0001). The delayed start of G-CSF after HDAC/MI in ALL achieves the same clinical benefit compared to the earlier initiation of G-CSF. The reduction of treatment costs by reducing the total G-CSF dose may be important in future treatment with this hematopoietic growth factor.

135

T-cell acute lymphoblastic leukemia occuring in a patient with acute myeloid leukemia after three years of remission

D. Adorf, T. Lipp, C. Nerl (München, D)

Secondary acute lymphoblastic leukemia following acute myeloid leukemia is a rare event. Only few cases have been reported. In literature, according to Tsukasaki (Cancer 1998, Reb 1;82(3):488-94) only 9 patients with the two entities of acute leukemias were described.
We report a patient with acute myeloid leukemia in hematological remission, who developed a T-lymphoblastic leukemia three years after diagnosis of AML. The 49 year old patient was admitted first to our hospital with a leukocytosis of 23,6/nl, a thrombocytopenia of 59/nl and anemia (hemoglobin of 9,9 g/dl). In the differential blood count about 78% blasts with positive myeloid-peroxidase staining were found. According to the cytomorphological analysis and cytochemistry of the bone marrow smears an acute myeloid leukemia of FAB- M2-subtype was diagnosed. In immunphenotypic analysis the blast population expressed the myeloid antigens CD33, MPOcy, and CD117, partly HLA-DR. The blast-population showed no expression of T-cell-antigens (CD3, CD4, CD5, CD7, CD8) and no minor pathological T-cell-subclone could be identified. Cytogenetics showed a completely normal karyotype (46, XY). The patient received two cycles of polychemotherapy according to the ICE-scheme (cytarabinosid, idarubicin, etoposid) with complete remission after the first treatment. After two cycles of consolidation chemotherapy according to HAM-scheme (high dose cytarabinosid and mitoxantrone) he remained in complete remission for about 2 1/2 years. He then again presented with a leukocytosis of 35,8/nl with 10% blasts in the peripheral blood and thrombocytopenia of 40.8/nl without anemia.
According to the morphological, cytochemical (myeloidperoxydase staining negative) and phenotypic analysis of the bone marrow now a T-lymphoblastic leukemia (cyCD3, CD2, CD5, CD7, partlyCD8, CD4 positive; CD1a negative), subtype of a Pre-T-ALL, with a mediastinal bulky disease could be diagnosed. There was no expression of myeloid antigens on the blasts and no pathologic myeloid subclone. Cytogenetics showed again a completely normal karyotype (46, XY). After induction chemotherapy according to the GMALL-scheme the patient again achieved complete remission. This case illustrates the very rare event of developement of two morphologically, cytochemical and phenotypical completely different entities of acute leukemias in one patient with normal karyotype within relatively short time. Furthermore, the usefulness of immunophenotyping and multiparametric flow cytometry (even in the routine diagnostic of AML) is shown; otherwise the exclusion of minor leukemic populations at the different times of diagnosis could not have been made.

136

Unusual chloromas in a patient with acute myeloid leukemia - successful pallitive treatment with low dose 6-mercaptopurine

B. Weber, A. Jakob, I. Dresel, A. Peuckert, D. Messerschmidt, E. Stöckl, F. Hirsch (Offenburg, D)

A 77-year old female patient presented with a 2 month history of submandibular lymphadenopathy and progressive facial skin infiltrates (chloromas). On CT-scan she had enlarged cervical and mediastinale lymph nodes. Complete blood count revealed a pancytopenia (WBC 3,2/nl, Haemoglobin 9,0 g/dl, platelets 129/µl) with 25% of immature blasts. Bone marrow biopsy and skin biopsy were performed, both showing an infiltration with immature cells. With help of cytochemistry and FACS-analysis the diagnosis of acute myeloid leukemia FAB M 4 could be established. Due to patient age and comorbididy a palliative treatment with oral idarubicine and low dose subcutaneous Ara-C was begun. 2 month later the patient was admitted again with fever due to a pneumonia. The skin infiltrates had slightly improved but the lymphadenopathy had worsened and the blood count showed severe pancytopenia. Treatment was changed to low dose 6-mercaptopurine (150 mg per week).
Under this treatment the skin infiltrates and lymphadenopathy improved markedly.The platelet count rose from 39/µl to 194/µl while the patient remained leucopenic and was still dependant on red cell transfusions. The haematological improvement was paralleled by an improve in the patient´s general condition. This case represents an unusual presentation of acute myeloid leukemia in an elderly patient with a good partial response on a treatment with low dose 6-mercaptopurine.

137

CMA 676 in combination with ara-C and liposomal daunorubicin in the treatment of relapsed AML

S. Günzelmann, M. Koldehoff, N. Basara, A.A. Fauser, M.G. Kiehl (Idar-Oberstein, D)

Treatment of patients with high-risk acute myelogenous leukemia (AML) (e.g. complex chromosomal abnormalities), primarily refractory AML or relapse of AML is associated with a poor outcome. Herein we report on the first results of our ongoing study to evaluate the efficacy of CMA 676 (Mylotarg)(10 mg/m^2 on day 1) in combination with ARA-C (500 mg/m^2 twice daily on days 2,3 and 4) and liposomal daunorubicin (daunoxome)(100 mg/m^2 on days 3 and 4). All patients demonstrate a CD33 surface expression of > 70% on blasts. Grade 4 mucositis was observed in 2 patients and 1 patient (No. 4) died due to septic shock in aplasia. Median duration of neutrophils < 500/µl was 13 days (range 9 – 23 days) and platelets < 20.000/µl was 18 days (12 – 35 days). Three patients received allogeneic stem cell transplantation and out of these one patient is still alive in complete remission. One patient relapse 4 months later and received a second course. From these preliminary data we conclude that a chemotherapy protocol using an antibody-targeted agent in combination with arabinosid cytosine and liposomal daunorubicin is feasible in high-risk AML patients.

138

Reliability and clinical impact of prospective BCR-ABL analysis by polymerase chain reaction (RT-PCR) in adult acute B-lineage lymphoblastic leukemia (ALL)

B. Gleissner, C. Bartram, H. Rieder, N. Goekbuget, L. Janssen, C. Fonatsch, J. Janssen, B. Heinze, C. Schoch, D. Hoelzer, E. Thiel, The GMALL Study Group (Berlin, Heidelberg, Marburg, Frankfurt, München, Ulm, D; Wien, A)

Background: Evaluation of t(9;22) provides important prognostic information in acute lymphoblastic leukemia (ALL). However, the reliability of routine BCR-ABL RT-PCR has not adequately been determined and comparisons to cytogenetic data have been done mainly on small patient cohort. Patients and Methods: This prospective study analyzed the reliability of routine BCR-ABL RT-nested-PCR in 1453 B-lineage ALL or hybrid leukemia at initial diagnosis by RT-nested-PCR. All BCR-ABL-positive (n = 640) and 176 BCR-ABL-negative samples underwent a second RT-PCR. In 519 patients, karyotyping and/or FISH was compared to the BCR-ABL status. 478 c-ALL and pre-B ALL patients were stratified according to the RT-PCR result and treated in two consecutive GMALL multicenter trials (04/89; 05/93). In these patients, the prognostic impact of BCR-ABL was analyzed in comparison to other clinical risk features. Results: Confirmed

RT-PCR detected BCR-ABL fusion transcripts consistently in 487 out of 1453 ALL samples (c-ALL: <3%, pre-B ALL: 34%, pro-B ALL: 5%, B-ALL: 0%, hybrid leukemia: 5/11). The two main BCR-ABL subtypes were similarly distributed in different immunophenotypes with a predominance of p190. Concerning reliability of RT-nested-PCR, the second RT-PCR revealed in 155/640 initially positive samples a divergent result (153 BCR-ABL-negative, two other transcripts) that in most cases turned out to be caused by contaminations in the first RT-nested-PCR. Confirmatory RT-PCR detected 2/176 false negative first RT-nested-PCR results. Thirty-nine specimens remained ambiguous despite multiple RT-PCR assays. As far as cytogenetic evaluation and FISH is available (n = 23), the majority but not all patients with an ambiguous RT-PCR result were Ph-negative (n = 19). RT-nested-PCR and cytogenetics yielded in 346 of 383 evaluable samples a concordant result. Differing results are given and account in part to the lower sensitivity of karyotyping. In c-ALL and pre-B ALL patients, the presence of BCR-ABL predicts a lower chance of initial treatment response (68.4% vs 84.6%; p = .001) and a substantially shortened probability of disease free survival at 3 years (0.13 vs 0.47; p = .001). No significant different clinical outcome was evident for p190-positive (n = 34) or p210-positive (n = 135) patients (survival at 3 years: 0.19 vs 0.03). In comparison to clinical risk criteria, univariate and multivariate analysis revealed BCR-ABL as the leading poor prognostic factor (RR = 2.766; p = 0.0001).
Conclusion: This large prospective study clarifies the incidence of the BCR-ABL recombination in B-lineage immunophenotypes. Since false-positive initial RT-nested-PCR data were frequent, either confirmatory second RT-PCR or FISH analysis is warranted to guarantee sensitive and reliable BCR-ABL RT-nested-PCR results of utmost clinical relevance. p190 as well as p210 are unequivocally demonstrated as the most important predictors of outcome with a poor long term survival.

139

Cooperative analysis of core binding factor acute myeloid leukemia treated in the German multicenter treatment trials AML 2/95 (SHG Hannover) and AML HD93 (AMLSG Ulm)

R. Schlenk, A. Ganser, J. Krauter, K. Döhner, A. Müllek, N. Papendorf, H. Döhner, G. Heil (Ulm, Hannover, D)

The gene transcriptional products of both t(8;21) and abn(16q22) lead to structural alterations of different components of the core binding factor (CBF), which is involved in the regulation of hematopoiesis. Therefore t(8;21) and abn(16q22) karyotypes are referred to as CBF acute myeloid leukemia (AML), which are characterized by high first complete remission (CR) rates and by superior disease free survival (DFS) following high dose cytarabine (HD Ara-C) for consolidation therapy. Between 1993 to 1999 forty-two and forty-eight adults with CBF AML from the AML SG Ulm and the AML SHG were included in two prospective multicenter treatment trials AML HD93 and AML2/95 in which CBF AML was assigned to consolidation therapy with HD Ara-C. After double induction therapy patients were assigned in both studies to HD Ara-C in first and second consolidation with a cumulative dose of 36g/m^2 and 44g/m^2 for the AML HD93 and AML 2/95 studies, respectively. The patient characteristics at diagnosis differed between the two studies concerning median age (AML HD93: 46 years; AML 2/95: 39 years, p=0.007) and karyotype distribution (AML HD93: inv(16) n=25, t(8;21) n=17; AML 2/95: inv(16): n=17, t(8;21) n=31; p=0.04) and between the karyotype with significantly higher WBC in the inv(16) group (median WBC inv(16): 35/µl ; t(8;21): 11/µl; p=0.0004). The median follow-up was 3.3 years. The CR rate was 90%, 4.5% of the patients died and 5.5% had persistent disease. The overall survival at 3 years was 76% (95% confidence interval: 67 - 86). Seventy-seven patients were eligible for consolidation therapy in first CR, n=6 received allogeneic transplantation, n=6 autologous transplantation and n=65 HD Ara-C. The DFS at 3 years was 62% (95% confidence interval: 50 - 74). In the group of patients receiving HD Ara-C for consolidation 24 relapses occurred between 4 and 30 months and 20 patients achieved a second CR. The DFS of these 65 patients at 3 years was 59% (95% confidence interval: 47 - 72). The white blood count (WBC) was the only factor predicting relapse. The cut-off point was 28/nl with an risk ratio of 5.04 for WBC above 28/nl. Neither karyotype, blast count in the bone marrow or peripheral blood, nor age were informative variables. In conclusion CBF-AML showed good overall survival. However, about one third of the patients treated with HD Ara-C relapsed.

140

Efficacy and safety of an antibody targeted chemotherapy with gemtuzumab ozogamicin in acute myeloid leukemia in first relapse: An updated report on 188 patients over and under 60 years of age

D. Voliotis, M. Theobald, O. Staak, E. Stadtmauer, R. Larson, E. Sievers, E. Estey, B. Löwenberg, M. Berger, F. Appelbaum (Köln, Mainz, D; Philadelphia, Chicago, Seattle, Houston, Radnor, USA; Rotterdam, NL)

Gemtuzumab ozogamicin (MylotargTM, CMA-676) is an antibody-targeted chemotherapy agent that is used for the treatment of CD33-positive acute myeloid leukemia (AML). Gemtuzumab ozogamicin was administered to patients with AML in first relapse at a dose of 9 mg/m^2 IV for 2 doses 14 days apart. Patients were required to have a duration of first complete remission of [greater than or equal to]3 months and could not have an antecedent hematologic disorder or secondary AML. A pooled analysis of 188 patients entered into three phase 2 studies as of 31 March 2000 was performed. Fifty-eight (31%) patients achieved a remission (characterized by [less than or equal to] 5% blasts in the marrow, [greater than or equal to]1500/µL neutrophils, [greater than or equal to] 9 g/dL hemoglobin, and red blood cell and platelet transfusion independence). Remission was obtained in 34% (30/87) of patients <60 years and in 28% (28/101) of patients [greater than or equal to] 60 years. Median overall survival was 5.9 and 5.4 months for patients <60 and [greater than or equal to] 60 years, respectively. Adverse events included an infusion-related symptom complex (self-limited fever, chills, and, infrequently, hypotension and shortness of breath), grade 4 neutropenia and thrombocytopenia, and transient increases in bilirubin and/or hepatic transaminases. The early mortality rate was 14% both for patients <60 years and for patients [greater than or equal to] 60 years. Gemtuzumab ozogamicin administered as a single agent to patients with AML in first relapse has similar efficacy and safety in younger and older patient groups.

141

A multicenter trial with topotecan, mitoxantrone and cytarabine in patients with relapsed or refractory acute myeloid leukemia or blast crisis in chronic myeloid leukemia - interim results

B. Höchsmann, H. Döhner, A. Ganser, G. Heil, C. Denzlinger, H. Einsele, A. Raghavachar, H. Dürk, D. Hoelzer, L. Bergmann (Ulm, Hannover, Tübingen, Wuppertal, Hamm, Frankfurt / Main, D)

Based on in vitro data suggesting a synergystic antitumor-activity of topoisomerase I- and II-inhibitors we designed a new combination therapy to improve the results of previous studies with topotecan in myeloid leukemias. In the ongoing prospective multicentre phase II study up to now 55 patients (pts) were enrolled, 46 pts were analysed for patient characteristics. Diagnosis included 4 refractory de novo AML, 3 refractory secondary AML, 22 relapsed de-novo AML, 10 relapsed secondary AML and 7 CML-BC. Median age was 54 years (18-69). Male/female ratio was 1 (23pts/23pts). Karyotyp was normal in 13 pts, 22 pts showed abnormalities, 6 pts had complex abnormalities and no result was available in 5 pts. 27% of the pts were pretreated with high-dose AraC, 13% got a PBSCT/BMT in pretreatment. As induction therapy pts received MTC I+II consisting of mitoxantrone 12mg/m^2 day 1+2, topotecan 1,5mg/m^2 day 3-7 and cytarabine 1g/m^2 day 3-7 per cycle. Consolidation was done by allo-PBSCT/BMT or a third course MTC. Pts < 60 years without complications in the previous MTC cycles had the option of intensified MTC III by increase of the topotecan dose to 1.75 mg/m^2. For toxicity and response so far 41 enrolled pts were evaluable after MTC I, 16 pts completed induction therapy with MTC II and 10 pts received MTC III. Median duration of neutropenia < 0.5 G/l (thrombopenia < 25 G/l) was 21 (19) days during MTC I, 16 (11) days during MTC II and 15 (16) days during MTC III. Median duration of fever >38.5°C in MTC I, MTC II and MTC III was 2 days. Mucositis grade 3 arised in 3 cases of 67 cycles. 1 pt suffered from a reversible cardiac rythm disturbance by hypokaliemia. During MTC I 2 pts died in aplasia, after hematopoietic reconstitution 1 pt died in sepsis. Post MTC I remission-rate was 79% for relapsed AML with a RFS > 6 months after initial therapy (19 pts), 17% for relapsed AML with a RFS < 6 months after initial therapy (6 pts), 57% for refractory AML (7 pts) and 58% for CML-BC (7 pts) respectively. Consolidation part is finished by 16 pts, 6 pts received regular MTC III, 4 pts dose intensified MTC III, 10 pts got PBSCT/BMT. The results suggest an encouraging treatment response in this poor prognosis patient group combined with tolerable toxicity of the MTC-regimen . The ongoing study has to evaluate the efficiacy in a wider patient group and a longer follow-up period.

142

Treatment of relapsing acute lymphocytic leukemia

M. Freund, R. Arnold, G. Heil, W. Ludwig, T. Lipp, D. Niederwieser, H. Rieder, E. Thiel, F. Fiedler, N. Grobe, A. Grote-Metke, D. Hähling, F. Hartmann, N. Gökbuget, D. Hoelzer (Rostock, Berlin, Hannover, München, Leipzig, Marburg, Chemnitz, Neubrandenburg, Hamm, Schwerin, Homburg/Saar, Frankfurt/M, D)

The German Relapsed Acute Lymphocytic Leukemia Study Group has conducted three subsequent studies on the treatment of relapsed acute lymphocytic leukemia. In the present study treatment ist stratified according to the duration of the preceeding complete remission. In patients with <18 months CR treatment is started with a combination of prednisolone and cyclophosphamide. Patients with late relapses are given a three-weeks-combination of vindesine, DNR, PEG-asparaginase and prednisolone as first induction. Thereafter therapy is continued in all patients with an intensive combination of ara-C, VP16, ifosfamide, high-dose MTX and prednisolone followed by a course of high-dose VP16. To as many patients as possible blood stem cell transplantation is offered. Results: As on Feb. 2001, 32 patients have been recuited. 25 patients are evaluable for toxicity, and 24 for response. 9 patients had early relapses (7 B-lineage, 2 T-lineage). 4 CR have been achieved in these patients. 4 of 9 patients have received an allogeneic, another one an autologous transplant. The median survival has been 10.2 months (transplant results included). 11 of 16 patients with late relapse (13 B-lineage, 2 T-lineage) entered CR. 7 patients have received an allogeneic, one an autologous transplant. The median survival was 7.2 months (course after transplant included). The treatmed was tolerable with significant mucositis but quick hematopoietic regeneration after the high-dose MTX-based regimen. The overall results are not different from the preceeding two studies of the German Relapsing ALL Study Group:. The present treatment regimen seems to be not inferior to high-dose ara-C based combinations studied before. However the patient numbers are limited and further follow-up is necessary.

143

Limited response of patients with acute myeloid leukemia (AML) to treatment with anti-CD33-antibody calicheamicin conjungate

R. Fenk, G. Kobbe, U. Germing, A. Hünerlitürkoglu, K. Schulte, A. Giagounidis, U. Rohr, G. Meckenstock, P. Schneider, R. Haas (Düsseldorf, D)

We report on 7 patients (pts.) with refractory or relapsed AML who recieved calicheamicin-conjugated anti-CD33-antibody (Mylotarg, CMA-676). These were extremely poor-risk pts. as shown by the following characteristics: median age was 47 years (range 20-66), 2 pts. had an unfavourable karyotype, 2 pts. had a prior myelodysplastic phase, 2 patients were primary refractory to conventional chemotherapy, 4 pts. relapsed after autologous or allogeneic peripheral blood stem cell transplantation (PBSCT), 1 patient had an extramedullary relapse, 2 pts. had a second or greater relapse. The median time of prior remission duration was 7,4 month (range: 3-12), and all pts. showed active disease. Pts. were treated with one (4 pts.) or two (3 pts.) doses (9mg/m^2) of CMA-676. A second dose was not given because of progressive disease (2pts.) or severe hepatotoxicity (2pts.). Response to treatment was observed in 2 pts. One of these two pts. achieved a blast-clearance in the peripheral blood for 35 days. This patient received donor lymphocytes as concomitant treatment and suffered from cutaneous graft-versus-host disease grade 4. The other responding patient had a complete clinical remission for 115 days before further progression of AML occurred. All pts. died after CMA-676 administration with a median survival of 89 days (range: 14-261). Five pts. received further therapy (chemotherapy: 4 pts., allogeneic PBSCT: 1 pt.) after CMA-676-treatment. After administration of CMA-676 a post-infusion syndrome with fever and chills was observed in 3 pts. and one patient showed a severe tumour lysis syndrome after application. Other severe adverse events included neutropenia grade 4 (4 pts.), thrombocytopenia grade 3-4 (4 pts.) and hepatotoxicity grade 3-4 (4 pts.) including liver function failure with need for fresh frozen plasma transfusions (3 pts.). One of these pts. with normal liver enzymes and normal function tests before CMA-676 application died because of progressive liver failure 26 days after treatment. One additional patient, who received an allogeneic PBSC transplantation as further salvage therapy, also developed fatal liver failure early after transplantation. In our pts. the efficacy of CMA-676-therapy in patients with advanced AML was low and associated with substantial hepatotoxicity.

144

Combined chromosome banding and immunologic analysis in acute lymphoblastic leukemia using immunomagnetic beads

S. Busch, I. Bachmann, B. Reckzeh, S. Schwartz, E. Thiel, D. Hoelzer, H. Rieder (Marburg, Berlin, Frankfurt, D)

Background: The combined analysis of cytogenetic and immunologic features provides important insights into the relation between genotype and phenotype of leukemias. Chromosome banding analyses (CBA) have successfully been performed on FACS sorted cells of peripheral blood and bone marrow samples of AML and MDS patients. However, a high propensity to undergo apoptosis and lack of adequate in vitro cultivation methods hamper the use of FACS sorted cells for combined immunopenotyping and CBA in acute lymphoblastic leukemia. Aims: Immunomagnetic beads selection for the combined cytogenetic and immunologic analysis in B- and T-lineage neoplasias and its applicability on samples of ALL patients was investigated with respect to the efficacy and accuracy of the cell separation as well as to the compatibility with CBA. Methods: Cell separation was performed using immunomagnetic beads coated with anti-CD19 or anti-CD7 antibodies on cell lines containing defined mixtures of CD19+ and CD7+ cells to determine the efficacy and accuracy of the technique, and on bone marrow or blood samples of 7 ALL patients. The target cells were identified by the presence or absence of respective cytogenetic abnormalities as revealed by FISH or CBA.

Results: In titration experiments with 5 cell lines using positive cell selection, CD19+ or CD7+ target cells were enriched by a factor up to 6.4 or 4.7, and a purity up to 99.6% or 100% was achieved, respectively. In all 7 samples of ALL patients investigated by chromosome analysis on immunomagnetically selected cells metaphase chromosomes with clonal chromosome abnormalities were obtained in the CD19+ positive cell fraction. Immunophenotyping revealed c-ALL in all patients.

Conclusions: Chromosome analysis of immunomagnetically selected cells may efficiently be employed in the cytogenetic analysis of ALL samples and contribute to the correlation of immunologic and cytogenetic features.

Supported by the Deutsche José Carreras Leukämie-Stiftung e.V., grant R24.

145

Treatment of acute myeloid leukemia (AML) in patients under and over the age of 60: Report of the AML96-#033 and AML97-#038 study of the East German Hematology and Oncology Study Group (OSHO)

L. Uharek, R. Krahl, W. Pönisch, C. Becker, L. Mantovani, U. Wedding, M. Herold, M. Freund, A. Franke, F. Fiedler, R. Pasold, M. Peter, R. Subert, J. Meran, E. Faßhauer, K. Dachselt, N. Grobe, M. Aßmann, E. Kettner, G. Dölken, D. Kämpfe, J. Steglich, R. Huhn, M. Schulze, W. Schultze, T. Ittel, C. Klinkenstein, V. Schirmer, A. Florschütz, J. Doepper, C. Winkelmann, W. Knauf, H.J. Schmoll, G. Schott, W. Helbig, D. Niederwieser (Leipzig, Jena, D)

We report the results of 347 patients with de novo (n=254) and secondary (n = 93) AML that were treated within two OSHO protocols between 1996 and 2000. Patients under the age of 60 received one or (in case of PR/NR) two induction cycles (ARA-C 2g/qm day 1,3,5,7; IDA 12 mg/qm day 1-3 or MITOX 10 mg/qm) followed by three consolidation courses. The first was identical to the induction cycle, the second was reduced by ever a day, and the third consisted of either hematopoetic stem cell transplantation (SCT) or chemotherapy identically to second consolidation. All pts < 60 were randomized to receive ARA-C either with a standard infusion rate of 2x1g over 3 hours (=333 mg/m^2/h) or in a dose that was considered to allow optimal intracellular Ara-CTP concentrations (1x2 g/m^2/d over 8 hours = 250 mg/m^2/h). Pts. >60 yrs received an identical induction treatment but only two consolidation courses with lower doses (ARA-C 240 mg/qm day 1-5; MITOX 10 mg/qm day 1+2). Older Patients with contraindications for intensive chemotherapy received either palliative therapy (idarubicin 10 mg po day 1; thioguanin 40 mg po day 1-5 or Ara-C 80 mg sc day 1-5 or etoposide 100 mg po day 1-5) or supportive treatment only. To avoid a general selection bias, all patients with AML were registered and monitored regardless of the final treatment decision (i.e. curative/palliative/ supportive). The overall results for pts. treated with curative intent were as follows:

Group		N	%	CR %	ED %	Med.surv.(mo.)	LFS (2 yrs.)
<60	novo	173	74	70	9	17 mo.	0.41
	sec	60	26	52	18	8	0.25
>60	novo	81	71	70	14	12	0.29
	sec	33	29	55	18	17	0.20

Since 03/98 additional 72 pts >60 yrs. entered the palliative and 22 the supportive arm. Median overall survival (OS) in the palliative arm was 73 days. Of the 79 pts < 60 who terminated the treatment completely, 23 (29%) pts were allografted, 24 (30%) autografted, and 32 (41%) have received a third chemotherapy consolidation. Our results compare well with other studies in patients with AML. In particular, they demonstrate that standard induction therapy is feasible and effective in most elderly patients with AML. The worse prognosis of elderly AML pts can be attributed to the higher proportion of sec. leukemias and, until recently, to the fact that the treatment option of allogeneic transplantation was not available for this group of patients.

146

A case of HIV-associated B-ALL successfully treated by intensive chemotherapy (B-ALL-protocol)

M. Hentrich, R. Munker, H. Pohlmann, N. Brack, R. Hartenstein (München, D)

Acute lymphoblastic leukemia is rare in the setting of HIV-infection and the prognosis is usually poor. We report on a patient with B-ALL successfully treated by intensive chemotherapy. In June 1999, a 55-year-old heterosexual man presented with a 4-week history of fatique, fever, polyarthralgia/myalgia, night sweats and weight loss. He was known to be HIV-seropositive since 1992. AIDS defining events had not been noted (CDC A2). Antiretroviral treatment (ART) with zidovudine was started in October 1996 and subsequently switched to lamivudine/stavudine/indinavir. On admission the patient did not declare to be HIV-infected. CT-scanning demonstrated enlarged cervical and pretracheal lymph nodes of 2 cm in greatest dimension. Within a few days his leucocytes increased to 25.7x10^9/l and the differential showed 29% neutrophils, 8% band cells, 6% myelocytes, 3% promyelocytes and 5% blasts. Hemoglobin was 12.3 g/dl and platelets decreased to 87/nl. LDH was elevated to 3200 U/l. A bone marrow biopsy showed hypercellularity with a diffuse infiltration by lymphoblasts (>70%). After excision of a cervical lymph node a diagnosis of Burkitt's-like lymphoma was made and confirmed by a reference pathologist. The patient was treated according to the GMALL-protocol starting with a pretreatment period (prednisone/vincristine (VCR) and cyclophosphamide/dexamethasone) followed by 3 cycles of methotrexate (MTX), VCR, ifosfamide, teniposide, cytarabine and intrathecal (i.th.) prophylaxis (block A) alternating with MTX, VCR, cyclophosphamide, doxorubicin and i.th. prophylaxis (block B). After the first cycle doses were reduced by 20%. A bone marrow aspirate performed after the completion of block A/1 revealed a complete remission (CR). Apart from one episode of central venous line associated bacteremia, neutropenic fever, stomatitis and mild peripheral neuropathy therapy was well tolerated. However, after the completion of chemotherapy (CT) peripheral neuropathy temporary deteriorated (WHO grade 3) and the patient developed a right pulmonary infiltrate which resolved under antibiotic therapy. Because the CD4-lymphocyte count had declined to 20/ml in August 1999 (as compared with 503/ml in 4/99) highly active antiretroviral therapy (HAART) was reinitiated in September 1999 (while under block B/2). The viral load had increased from 3.400 to 19.000 copies/ml under CT but decreased to 740 copies/ml as a consequence of HAART. There was no evidence of recurrent ALL 17 months after the completion of CT. The patient is currently on a 4-drug-ART and his latest CD4-lymphocyte count was 397/ml. Less than 20 patients with HIV-associated B-ALL have been reported in the available literature, the majority of them died of resistant leukemia or infectious complications. However, durable remissions can be achieved and the risk of opportunistic infections might be reduced by the combination of both, CT and HAART.

147

Apoptosis modulating pathways triggered by IL-7 and dexamethasone are closely associated in childhood T-ALL

L. Karawajew, C. Wuchter, V. Ruppert, B. Dörken, W.-D. Ludwig (Berlin, D)

Glucocorticoids play a crucial role in the therapy of acute lymphoblastic leukemia (ALL), but their signaling pathways remain to be identified. Recently, sensitivity of precursor T-cell ALL (T-ALL) to IL-7 in terms of spontaneous apoptosis inhibition in vitro has been proposed to reflect distinct developmental stages and apoptotic programms in childhood T-ALL (Karawajew, Blood 96:297, 2000). In the current study, we used this parameter as a tool to approach mechanisms underlying sensitivity/resistance to glucocorticoids in T-ALL. To this end, leukemic cells from T-ALL patients were investigated as to their susceptibility to induction of apoptosis by dexamethasone (detected by annexin V-staining; incubation periods of 24 hrs) as well as to IL-7 in terms of modulation of spontaneous and dexamethasone-induced apoptosis. IL-7 was highly effective in the rescue of leukemic cells from dexamethasone-induced apoptosis (inhibition observed in 37 of 46 samples). Furtermore, the susceptibility to dexamethasone-induced apoptosis was highly associated with the extent of IL-7-mediated rescue of T-ALL cells from spontaneous apoptosis (p<0.0001, n=80). In addition, dexamethasone and IL-7 effects disclosed striking similarities regarding maturation dependence including positive correlation with surface CD3 expression and negative correlation with CD34 expression in individual T-ALL cell samples. In addition to primary leukemias, we investigated a series of T-ALL cell lines (n=8) as to their susceptibility to dexamethasone and IL-7 signaling. Cell lines disclosed an overall resistance to dexamethasone. In two cell lines, KE37 and P12/Ichikawa, low levels of dexamethasone-induced apoptosis could be observed after extended periods of incubation with dexamethasone (48 hrs). In contrast to the low susceptibility to apoptosis, these cell lines were highly susceptible to dexamethasone in terms of cell cycle perturbations (G0/G1-phase arrest observed after 24 hrs of incubation with dexamethasone). Interestingly, IL-7 almost fully inhibited dexamethasone-specific apoptosis but did not reverse cell accumulation in G0/G1-phase. Taken together, our data strongly indicate that anti-apoptotic and pro-apoptotic pathways triggered by IL-7 and glucocorticoids, respectively, are closely related in T-ALL cells and might share common checkpoints responsible for susceptibility of T-ALL to glucocorticoid treatment in vivo.

148

Mylotarg therapy in acute myeloid leukemia

R. Weber-Nordt, A. Lenze, S. Steiger, S. Gottschling, M. Rieger, A. Ho (Heidelberg, D)

Specific targeting of cytotoxic therapies may lead to reduced toxicities and elevated treatment efficacies. Mylotarg is a novel antibody based therapy specific for CD33 positive cells. Because the majority of AML cells express CD33 antigen anti-CD33 antibody linked to the calicheamicin cytotoxic drug (Mylotarg) has been developed and proved to be effective in AML patients resistant to conventional treatment options. Here we report a single center experience with Mylotarg therapy (9 mg/m^2, i.v. as a single infusion) in patients who became therapy resistant to conventional regimens (4), showed relapse after conventional or reduced conditioning and allogeneic stem cell transplantation (2) or were chosen for Mylotarg because of reduced clinical condition or old age (3). AML blasts were characterized for their expression of CD33 and MDR protein P-gp190. Peripheral blood derived mononuclear cells were analyzed by flow cytometry for the content of myeloid blasts as well as the numbers of CD3, CD4, CD8 and CD19 positive cells. Furthermore these parameters were measured upon Mylotarg therapy until recovery of normal blood values or until day +60. Of the six patients refractory to conventional therapies or stem cell transplantation only two patients responded with a reduction of AML blasts in their PB. These two patients were negativ for MDR expression. Upon CD33+ cell reduction an absolute and relative rise in CD4 and CD8 cells was observed in the PB of these responders. In contrast, AML blasts of the nonresponders displayed high levels of P-gp190. Furthermore, in patients who received Mylotarg in a first line setting, blast reduction was positively related to the level of CD33 expression and inversely related to the presence of MDR. Thus, these observations implicate that high expression of MDR protein and low expression of CD33 are predictive for poor response to Mylotarg whereas patients with low MDR expression and high expression of the CD33 antigen may be chosen for the anti-CD33 based cytoreductive therapy. The change in the CD4 cell numbers may indicate an optimized effector to target ratio upon Mylotarg infusion leading to a rationale to combine AML specific cytoreduction and immunotherapy.

149

Treatment of elderly patients with AML non-eligible for standard therapy with continous infusion of idarubicin and cytarabine at reduced doses

A. Glasmacher, C. Hahn, G. Dölken, H. Kirchner, T. Südhoff, P. Ehscheidt, K. Rieche, E. Jaeger, F. Gieseler, I. Schmidt-Wolf (Bonn, Greifswald, Hannover, Bochum, Neuwied, Gelsenkirchen, Kiel, D)

More than 1/3 of patients (pts) with acute myeloid leukemia (AML) >60 years cannot be treated with standard therapy protocols. Higher toxicity, altered pharmacokinetics, multiple comorbidity and adverse leukemia risk factors contribute to low response and high early death rates (30-50% each). AIM: (1) To evaluate efficacy and toxicity of a new induction regimen for

AML pts with contraindications to standard therapy. (2) To evaluate the duration of neutropenia after chemotherapy after immediate or delayed application of filgrastim. PATIENTS AND METHODS: Pts who did not fulfil the entry criteria for standard therapy protocols were eligible. They received 1-2 courses (crs) of idarubicin (5 mg/m^2/d d1-5 cont. inf.) and cytarabine (100 mg/m^2/d d1-5 cont. inf.), thereafter filgrastim (5 mg/kg/d) was randomly started either on d6 or d12. RESULTS: 51 pts are registered in the ongoing trial, 37 pts were completely evaluable. The median age was 70 years (range, 40-79), 54% had severe comorbidity (kidney, heart, liver or lung), 42% severely reduced activity scores, 38% had >50 G/l leukocytes at diagnosis. CR/ABR was achieved in 17/37 pts after the 1st crs (46%), 35% no response, 19% death. Neutropenia lasted 19 d (2nd crs 10 d) in pts achieving remission (results of randomisation are not revealed). Toxicity was low (WHO scale, no 4° tox.): Nausea/vomiting: 9% (3°); mucositis: 4% (3°); diarrhoea 35% (3°); kidney dysfunction: 0% (2-3°). The median duration of fever was 1 d (0-5). CONCLUSION: Continuous infusion of idarubicin and cytarabine at reduced doses is an effective and well tolerated treatment for elderly patients with contraindication to standard therapy.

149a

Sequential high-dose cytosine arabinoside and idarubicin (S-HAI) +- fludarabine in patients with refractory and relapsed acute myeloid leukemia: Analysis of prognostic impact of karyotype abnormalities, duration of first remission, and age

W. Kern, A. Kerkhoff, C. Wuchter, C. Kerschgens, C. Huber, P. Staib, T. Haferlach, C. Schoch, T. Büchner, W.D. Ludwig, W. Hiddemann for the German AMLCG (München, D)

To further refine the stratification model for patients with acute myeloid leukemia (AML) undergoing salvage therapy, karyotype abnormalities both at diagnosis and at relapse, duration of first remission, and age were analyzed for their impact on achievement of complete remission and on long-term outcome. Patients being treated within the current protocol of the German AML Cooperative Group for treatment of patients with refractory and relapsed AML with the S-HAI regimen were eligible for this analysis. S-HAI consisted of high-dose cytosine arabinoside (AraC) twice daily on days 1, 2, 8, and 9 as a three-hour infusion and of idarubicin 10 mg/m^2 on days 3, 4, 10, and 11 as a 30-min infusion. AraC was applied at 3 g/m^2 per dose in patients under age 60 with refractory AML and early relapses after a first remission of less than six months duration. In all other cases AraC was applied at 1 g/m^2 per dose. Patients were randomized to receive fludarabine 15 mg/m^2 before each infusion of AraC in addition to S-HAI or S-HAI alone. At current, 200 patients have been entered into the trial, 140 of whom are fully evaluable (median age, 55 years; refractory AML and early relapses, 39%; karyotype favorable/intermediate/unfavorable/n.a. at diagnosis, 7%, 52%, 16%, 25%; karyotype favorable/intermediate/unfavorable/n.a. at relapse, 7%, 46%, 21%, 26%). Response rates were: CR, 45%; PR, 2%; NR, 35%; ED, 18%. Response rates did not depend on age and karyotype at diagnosis. Duration of first remission and karyotype at relapse significantly influenced response rates (p=0.0059 and p=0.0745). Median overall survival and median event-free survival were 6.1 months and 2.4 months, respectively, and were not affected by age. Median overall survival and median event-free survival were shorter in patients with unfavorable karyotype abnormalities at relapse (p=0.0210 and p=0.0023) and in patients with refractory AML and early relapses (p=0.0162 and p=0.0004). These parameters were independent from each other in a multivariate analysis. Future stratification models for patients with AML undergoing salvage therapy should include karyotype abnormalities at relapse in addition to the duration of first remission as stratification parameters.

Postersitzung:
Allogene Stammzelltransplantation - Rezidiv und GvHD

150

Treatment of leukemic relapse after allogeneic stem cell transplantation with cytoreductive chemotherapy and/or immunotherapy or second transplants

E. Prinz, H. T. Greinix, P. Kalhs, K. Lechner, K. Moser, I. Schwarzinger, U. Jäger, N. Worel, C. Mannhalter, W. Rabitsch, A. Schulenburg, M. Mitterbauer, P. Höcker, F. Keil (Wien, A)

We analyzed toxicity and efficacy of chemotherapy (CT) or second stem cell transplantation (SCT) and/ or immunotherapy defined as stop of immunosuppression (IS) or donor leukocyte infusion (DLI) in 50 patients relapsing with acute leukemia. Ten patients received no treatment and 14 patients were treated with CT only. In 14 patients IS was stopped and 3 of them received additional CT. Five patients received DLI after CT as consolidation and two patients as frontline therapy. Five patients received a second SCT. Median overall survival after relapse was 2 months for the untreated patients, 2 months for patients receiving CT only, 3 months in patients after cessation of IS, 16 months in DLI treated patients and 3 months in patients receiving a second SCT. Seventeen patients achieved remission after relapse. Two with CT (2, 2 months), 5 with IS (3,19,3+,7+ 28+ months), 7 with DLI (3,8,9,14,20,36,6+ months) and 3 with second SCT (2,4,6 months). Conventional CT was able do reestablish donor hematopoiesis and patients achieving remission showed a significant better survival than patients with refractory disease. Patients who were brought into remission by (DLI) or cessation of IS had a significant better survival than patients who achieved remission with CT alone or a second SCT. We conclude that a selected group of patients achieving remission with regeneration of donor hematopoiesis following CT might benefit from immunotherapy as consolidation.

151

Fludarabine and Cytosine Arabinoside as salvage therapy in patients with relapsed leukemia after allogeneic stem cell transplantation

A. Lollert, V. Runde, D.W. Beelen, U.W. Schaefer, A.H. Elmaagacli (Essen, D)

Background : Chemotherapy with fludarabine and cytosine arabinoside is an effective treatment option for relapsed acute lymphoblastic leukemia (ALL). The objective of this study was to evaluate its use in relapsed acute and chronic leukemic disease after allogenic stem cell transplantation. Patients and Methods : 14 patients (10 male, 4 female) with relapsed ALL (n=5), AML (n=4) and CML (n=5) after BMT (n=3) or PBSCT (n=11) were included in this study. The median age of patients was 35 years (range 19-57 years). Patients with CML were transplantated in 1st chronic phase (n=2), acceleration (n=2) or blast crisis (n=1) and relapsed in blast crises (n=4) or stable chronic phase (n=1). Patients with acute leukemia underwent transplantation in 1st CR (n=1), 2nd CR (n=1), 1st relapse (n=1) or 2nd relapse (n=5). Relapse after transplant occurred at a median time of 386 days (range 59-1773 days). Chemotherapy consisted of fludarabine 30 mg/m^2 and ara-C 100 mg/m^2 for 5 days. Post chemotherapy 10 of 14 patients received also unfractioned DLI. Results : Thirteen (93 %) patients achieved a complete remission. But 9 of 13 (69%) patients relapsed again. Duration of complete remission ranged from 80 to 781 days with a median at 314 days. Up to date, seven of 14 patients (50%) are still alive. The median survival time after occurrence of relapse post transplant was 362 days. Four of the 7 patients who are alive are without signs of relapsed disease, the median survival time in this group is 473 days. The main side effect was hematotoxicity and worsening of graft-versus-host-disease of skin and liver. Conclusions: Chemotherapy with fludarabine and cytosine arabinoside is an effective antileukemic regimen in patients with advanced relapsed leukemia after stem cell transplantation, which can induce long lasting complete remissions.

152

Leukemic cell vaccination in relapsed patients with CML after allogeneic blood stem cell transplantation

A.H. Elmaagacli, A. Koppelle, H. Ottinger, D.W. Beelen, U.W. Schaefer (Essen, D)

Leukemic cell vaccination were applied three times in 3-week intervals to two patients with relapse of CML after allogeneic stem cell transplantation. One of these two study patients was transplanted for CML in 1st chronic phase and relapsed in a stable phase of CML, whereas the other study patient received a transplantation for CML in lymphatic blast crises, and suffered from an extramedullary relapse of CML after transplant. Leukemic cell vaccination was performed two weeks after applying an unfractioned donor leukocyte infusions (DLI) in the first patient with relapse of CML. In the second study patient, leukemic cell vaccination was applied after achieving a complete remission of CML again using a combination of radiochemotherapy and DLI in order to keep the patient in remission.
The vaccine was generated by culturing monocyte-derived donor-cell derived antigen presenting cells which were pulsed with patient-derived CML cell lysates. We used a quantitative, real-time PCR assay to directly measure the immune reactivity of patients T cells to the vaccine. T cells were tested for immune reactivity after in vitro sensitization of leukemic cell vaccine for Interferon-gamma messenger RNA expression by real-time PCR. A twofold increase in Interferon-gamma transcript accumulation in T cells after vaccination was found and considered to be evidence of specific response. In contrast to in vitro tests, clinically both patients did not profit from the leukemic cell vaccination. Nor did the first patient achieved a remission of CML, neither was the second patient pretended from relapsing again after completing the 3rd vaccination. So, we concluded that leukemic cell vaccination may have clinically marginal anitleukemic effects after allogeneic transplantation.

153

CD34+ donor cell chimerism in bone marrow predicts early relapse of acute myeloid leukemia after allogeneic stem cell transplantation

C. Scheffold (Münster, D)

Detection of imminent relapse in hematologic malignancies after allogeneic stem cell transplantation (SCT) is important in order to employ timely interventions, e.g. donor lymphocyte transfusions (DLT). Quantification of donor cell chimerism (DCC) in peripheral blood (PB) using multiplex STR-PCR with fluorescence detection has recently been shown to be useful in monitoring residual disease after transplantation. We prospectively evaluated, in serial samples of PB and marrow (BM), the kinetics of DCC in different cellular subsets in relation to relapse incidence. 15 patients with high-risk AML underwent PBSC (n=14) or BM (n=1) transplantation after a TBI-based reduced conditioning regimen. Immunophenotyping of the leukemic clone prior to transplantation showed the expression of CD34 in 10 of 13 patients (two excluded for lack of material). Cellular subsets from PB (CD3+, CD14+, CD19+) and BM (CD3+, CD14+, CD19+, CD34+) were separated by fluorescence activated cell sorting (FACS Vantage) following immunomagnetic pre-enrichment. Donor-recipient chimerism of cellular subsets and total cells were compared to marrow cytology, cytogenetics, and immunophenotypic profile.
Results: At engraftment, all patients except one showed complete DCC in PB and BM, both in total cells and each of the above cellular subsets. In contrast, when relapse was cytologically diagnosed in 4/12 patients with initial complete remission, only a small fraction of CD34+ cells in BM (median 5%; range: 1-21%) were of donor origin while the other cellular subsets still showed complete or mixed DCC (median / range in total cells: 65% / 18-83%; CD3+: 96% / 87-96%; CD14+: 24% / 7-81%; CD19+: 95% / 87-99%). Beyond day+30 after transplantation, a CD34+ DCC <80% in BM was highly predictive for relapse (4/6 patients), while the patients with stable CD34+ DCC >80% (n=5) remained in remission after a median follow-up of 359 days ($p<0.05$). Furthermore, all four patients with a decline in marrow CD34+ DCC >10% subsequently relapsed. The decrease in marrow CD34+ DCC preceeded the diagnosis of relapse by cytology or cytogenetics by 1-3 months. In 2 patients with declining CD34+ DCC, early intervention with DLT restored complete DCC with normal BM cytology . We conclude that the analysis of donor-recipient chimerism in CD34+ cells of the BM is a sensitive method for the detection of early or impending relapse in acute myeloid leukemia and provides an additional criterion for further treatment decisions.

154

Treatment of steroid refractory acute GvHD with daclizumab (Anti-CD 25)

M. Weisser, O. Stoetzer, M. Schleuning, H. Kolb (München, D)

Background: Acute GvHD causes major problems after allogenic bone marrow or stem cell transplantation. Especially steroid refractory GvHD remains very difficult to treat and is associated with a high mortality rate. This evaluation presents our experience with a novel blocking antibody directed against the alpha-chain of the interleukin-2 receptor (CD25) - daclizumab (Zenapax). CD25 plays an important role in the activation and proliferation of T-lymphocytes, the effector cells of GvHD.
Forteen patients with steroid refractory acute GvHD post BMT/SCT were treated on day 1, 4, 8, 15, 22 and 28 with 1mg/kg daclizumab. Patients received an average of 4.3 applications. The underlying diseases were myeloma (2), NHL (2), CLL (1), ALL (2), AML (4) and CML (2). The average age of the patients was 38.7 years (range 14-60). Ten of 13 patients had already received OKT or ATG or both antibodys for GvHD treatment. All patients had acute GvHD stage II-IV. The leading organ affected was intestine (7), liver (3) and skin (3). Results: Five of thirteen patients (38%) had a good response to anit-CD25 therapy. Four of those had severe GvHD of the intestine and responded to a combination therapy of daclizumab and sirolimus. These four patients are still alive. The other responder had acute GvHD of the skin and was treated with Daclizumab as second line treatment. Unfortunately this patient died of cardiac failure 2 months after therapy. One patient with GvHD of liver and intestine had only a short partial response. The other 7 patients did not respond and died due to GvHD, infections or relapse. During therapy 1 CMV, 1 HHV6 and 2 EBV reactivations were observed. Daclizumab was well tolerated. Drug reactions related to the infusion of the antibody did not occur. Exemplary FACS-scan analysis showed that CD25 was undetectable on T-lymphocytes the day after daclizumab therapy. Conclusions: Daclizumab is a potent immunosuppressive drug. Response to therapy was seen in 38% despite our patients were heavily pretreated. Combination with sirolimus seems to increase efficacy in the treatment of steroid refractory GvHD. Our data together with the fact that CD25 is an early activation marker may justify an earlier use of anti-CD25. Prospective trials are necessary to clarify the role of daclizumab plus sirolimus in the therapy of steroid refractory acute GvHD.

155

Establishment of complete remission and donor chimerism by cessation of CsA in relapse of AML after allogeneic SCT with dose-reduced conditioning - a case report

E. Prinz, F. Keil, K. Moser, C. Mannhalter, P. Kalhs, H. Loidolt, I. Schwarzinger, K. Lechner, H.T. Greinix (Wien, A)

We report on an 33-year-old woman diagnosed with AML FAB M5 who had experienced a relapse 11 months after an autologous stem cell transplantation (SCT) and underwent allogeneic SCT with dose-reduced conditioning with and HLA-identical unrelated donor in CR2. Methods: For conditioning she was given Fludarabine (30 mg/m^2/d on 3 days) and total body irradiation (TBI) of 2 Gy on day of stem cell infusion. Immunosuppression started on day -3 with Cyclosporine A (CsA) and mycophenolate mofetil (MMF) on day 0. Then, 5.4 x 10^6 CD34+ peripheral blood stem cells/kg body weight obtained from a matched unrelated male donor after G-CSF stimulation were infused. Results: The patient's absolute neutrophil counts (ANC) were below 1 x 10^9/l on 3 days but did not drop below 0.5 x 10^9/l. After 18 days she could be discharged from the hospital. Chimerism was assessed by variable number of tandem repeats (VNTR-PCR) and fluorescence in-situ hybridization (FISH) analysis. On day 27 after SCT 95 % donor chimerism was observed in unseparated periperal blood (PB), bone marrow (BM) and granulocytes, > 90 % donor chimerism in CD3+, CD8+ and CD56+. Only in CD34+ BM cells 28 % of recipient hematopoiesis was seen. BM and PB revealed a continuous complete remission.
On day +59 leukemic relapse was seen with 22 % blasts in bone marrow aspiration. In addition, an increasing number of host-derived cells in FISH- and VNTR-PCR-analysis was observed. Therefore immunosuppression with CsA was immediately stopped. The patient presented acute graft-versus-host disease (GVHD) of skin grade II 2 weeks after cessation of CsA. One month later chimerism analyses showed complete donor chimerism in PB and BM and various cell types. Conclusion: In the investigated patient with relapsed AML discontinuation of CsA and occurence of acute GVHD led to a sustained complete remission with full donor chimerism.

156

Remission induction of severe steroid-refractory acute GvHD after allogeneic PBSCT by the IL-2R-antagonist basiliximab

W.U. Knauf, I.W. Blau, T. Fietz, H. Schrezenmeier, E. Thiel (Berlin, D)

Activated CD25 expressing T-lymphocytes are involved in the development of GvHD after allogeneic hematopoietic stem cell transplantation. Therefore, monoclonal antibodies against the alpha chain of the IL-2-receptor (CD25) should to be an attractive option to treat GvHD. To further test this hypothesis we initiated a phase II study on feasibility and efficacy of basiliximab, an IL-2R-antagonist which is currently used to prevent and treat graft rejection after solid organ transplantation. So far, seven pts (five with AML, one ALL/Ph1+, one CML/CP1) have entered the trial. Three have had a HLA-identical sibling graft, while the others have had a matched unrelated donor (three pts) or a mismatched unrelated donor (one pt). GvHD prophylaxis consisted of CSA (two pts), CSA+MMF (two pts), CSA+MMF+ATG (one pt), and CSA+MTX (two pt), respectively. Nonetheless, six of these pts developed acute cutaneous and/or intestinal aGvHD, that was progressive to grade III/IV disease despite the administration of daily 1-2 mg/kg body weight steroids. After 20 mg basiliximab d0 and d4, respectively, clinical signs of aGvHD disappeared completely or were grade I within one week in five pts. While tapering down steroids, aGvHD reappeared in two pts, but was again succesfully treated with basiliximab. In one pt, however, aGvHD remained refractory leading to death. The seventh pt of our series suffered from extensive sclerodermic GvHD despite treatment with steroids, MMF, and PUVA. He experienced a short-time minor response after basiliximab. In summary, application of basiliximab is safe. Remissions of severe steroid-refractory aGvHD are achievable quite rapidly. However, more pts and longer follow-up are needed to evaluate this antibody-based treatment of GvHD.

157

Dendritic cells (DC) aggravate graft-versus-host (GVH) reaction dependent on the dose of interleukin (IL-12) and peptide pulsed donor DC can elicite leukemia specific T cell responses in vivo

M. Taniguchi, F. Leithäuser, Y. Miyahara, L. Bergmann, H. Döhner, H. Shiku, M. Schmitt (Ulm, Frankfurt, D; Mie, J)

T- lymphocytes and NK cells play an important role in GVH and GVL. IL-12 is known to activate both lymphocyte subpopulations. We performed a major histocompatibility complex (MHC) matched, but minor histocompatibility antigen (mHA) mismatched allogeneic bone marrow transplantation (allo-BMT) from DBA/2 to BALB/c mice. In this murine BMT model, we examined the effect of IL-12 on GVH. When we administered IL-12 at a dose of 200 ng to a recipient BALB/c mouse at the day of BMT and two days later, mice died between day 4 and day 10. The lethality from IL-12 was clearly dose dependent: When 100 ng of IL-12 was administered, 50 % of the animals died, at a dose of 50 ng and less all mice recovered. Blocking experiments in vivo showed that the early lethality is mediated by both FasL and TNFalpha. Early lethality could be abolished by in vivo administration of anti-IFN-g antibody. To investigate the effect of IL-12 at a histological level, we sacrificed three mice per group receiving 200 ng, 25 ng or no IL-12 on day 4 after allo-BMT and IL-12 administration.Hematoxyline-eosin (HE) staining and immunohistochemistry (IHC) were performed. HE staining of the large intestine of IL-12 treated animals showed a severe mucosal injury with edema, epithelial erosions and ulcerations, and a moderate cellular infiltrate of the lamina propria. In IL-12 treated mice as well as untreated control animals, ICH revealed only few CD3+ T cells in the lamina propria. In contrast, numerous CD4+ cells were found in the Peyer's patches. Moderate numbers of CD11c+ DEC205- DC were detectable in the lamina propria. This suggests that DC play an important role in the aggravation of GVH reactions after administration of IL-12 in our model. To assess the graft versus leukemia (GVL)-effect, recipient mice of allo-BMT as described above were challenged with 5x10 E4 cells of the leukemia cell line RLmale1 at the day of BMT,. DC of DBA/2 origin were pulsed with the specific leukemia associated antigen peptide pRL1a and administered on days 1, 7 and 14 after BMT, In vitro analysis of specific cytotoxic T lymphocytes (CTL) by ELISPOT revealed the induction of CD 8+ IFN gamma secreting CTL in the case of vaccination with DC+pRL1a. Vaccination with DC alone resulted only in a weak CD8+ cell response. Without vaccination the CD8+ cell activation was only at the background level. In in vivo experiments such a specific vaccine could significantly prolong the survival of leukemic mice, but not prevent the death of the animals. Experiments to improve the vaccination results by additional administration of CpG-oligodinucleotides are ongoing.

158

Comparison of mycophenolate mofetil with methotrexat for GvHD prophylaxis in allogeneic stem cell transplantation

K. Potthoff, J. Hahn, S. Mielke, R. Wäsch, H. Bertz, E. Holler, J. Finke (Freiburg, Regensburg, D)

Acute and chronic GvHD still remain a major problem in allogeneic stem cell transplantation following reduced intensity conditioning. Several approaches have been pursued to reduce the risks of GvHD. Here we report on promising data from our prospective phase I/II multicenter study using a novel fludarabine-based conditioning regimen in allogeneic stem cell transplantation. Between August 1998 and April 2001 in Freiburg and Regensburg, 84 patients (pts.) median age 55.9 years (range 23.8-70.5) were transplanted for active leukemias/ MDS (39), CML (10), NHL (14), CLL (5), MM (6), MPS (5), melanoma (3) or renal cell carcinoma (2) from matched related (n=61), matched unrelated (n=22) or HLA-different unrelated (n=1) donors. The majority of pts. had advanced disease either primary refractory or relapsing (n=64) or untreated (7). 13 pts. were in CR1 or CP1. All pts. received fludarabine 5x30 mg/m^2. Those patients below age 55 received 2x200 mg/m^2 BCNU and 140 mg/m^2 melphalan (n=39), whereas older pts. received 2x150 mg/m^2 BCNU and 110 mg/m^2 melphalan (n=45). GvHD prophylaxis consisted of cyclosporine A (CSA) and mycophenolate mofetil (MMF, n=52) or „mini-Methothrexat" (mMTX, n=32). In case of unrelated donor grafts (n=23) additional anti-T-cell globulin ATG-S (Fresenius, Bad Homburg) was administered. Results: Engraftment was rapid (median WBC count >$1x10^9$/l d+11 [range 1-24], median platelet count > $20x10^9$/l d+12 [range 7-48]), one secondary graft failure was observed, but this patient engraftet d+9 after second transplantation. Severe acute GvHD °III+°IV occurred in 17/84 pts. (20.2%), in particular in 8/52 pts. (15.4%) in the MMF-group and 9/32 pts. (28.1 %) in the mMTX-group. Overall incidence of extensive chronic GvHD was 22.7 % (15/66). Splitted into MMF- and MTX-subgroups, extensive chronic GvHD was seen in 3/36 (8.3%) MMF-pts. and in 13/30 (43.3%) MTX-pts. (p=0.0013; Fisher's test). With a median follow up of 13 months (31-1014 days) 56 pts. are alive, 47 pts. remain in remission, 28 pts. died (8 due to relapse, 2 with GvHD, 14 from infection, 4 other reasons or not evaluated so far). Conclusion: MMF is a highly effective drug for GvHD prophylaxis. Given the good efficacy and favorable side effect profil the use of MMF for GvHD prophylaxis is a promising strategy and should be investigated in randomized prospective multicenter trials.

159

Determination of an interferon gamma subpopulation quotient (IGSQ) as a predictive marker for acute GvHD

C. Schimmelpfennig, J. Thomas, M. Bornhäuser (Dresden, D)

Introduction: Acute GvHD is mediated by a Th1 polarised cytotoxic T-cell reaction. The rate of gIFN expression in different compartments of PBMNC measured by flow cytometry could indicate the early stage of GvHD and sub-sequent therapy. Methods: Peripheral blood samples of patients with aGvHD not treated (n=12), treated aGvHD (n=19), no GvHD (n=9) and control (n=9) were stimulated 24h with ionomycin and PMA together with Brefeldin A. After permeabilization with saponin intracytoplasmatic expression of gIFN was determined by two colour flowcytometry. Conclusion: In contrast to patients without or with treated GvHD, in patients with acute GvHD the ratio between CD3+gIFN+ and CD3-gIFN+ lymphocytes shifts towards the CD3+gIFN+ population. The gIFN-Quotient might be a predictive marker for the clinical appearance of acute GvHD.

160

Real time in vivo imaging of alloreactive T cells in graft-versus-host disease

M. Edinger, M. Verneris, Y. Cao, M. Bachmann, C. Contag, R. Negrin (Stanford, USA)

Graft-versus host disease (GvHD) after allogeneic bone marrow transplantation (BMT) is initiated by alloreactive donor T cells. We recently developed an in vivo imaging system which allows us to localize primary lymphocytes in vivo and to quantify cell proliferation non-invasively in mice. We employ a mouse Maloney leukemia virus based retroviral transduction system to label lymphocytes with a fusion construct containing the green fluorescence protein (gfp) and the bioluminescent reporter gene luciferase (luc). Using a low light imaging camera we externally detect light emitted from luc+ cells which is transmitted through the tissues of the animals.

Using this method, we can follow the migration of adoptively transfered cells in living animals with high sensitivity. By measuring the photon counts over a 5 min. integration time we are able to quantify the signal intensity which increases as cells proliferate. We used this system to examine the fate of transduced lymphocytes after syngeneic (BALB/c into BALB/c) or allogeneic (C57Bl/6 into BALB/c) BMT. Lymphocytes after syngeneic BMT home preferentially to lymphoid organs (spleen and lymph nodes) and proliferate at these sites late after transfer (d6). In contrast, allogeneic lymphocytes home preferentially to GvHD target tissues (gut) rapidly after injection (1d) and proliferate earlier and more aggressively at these sites. With ongoing disease cells also infiltrate skin, liver, lymph nodes and spleen. This newly developed system permits us to examine the influence of inhibitors of GvHD, such as drugs (i.e.steroids, CyA) or biologic modifiers, (i.e. anti-B7.1 and B7.2 or anti-CD40-L-antibodies) on cell survival, homing and proliferation of pathogenic cell populations in living animals.

161

Pentostatin as salvage therapy for steroid refractory acute intestinal graft versus host disease

S.A. Klein, C.F. Dietrich, G. Bug, S. Hermann, D. Hoelzer, H. Martin (Frankfurt, D)

Steroid refractory acute Graft versus Host Disease (aGvHD) of the bowel is a life threatening complication after allogeneic hematopoietic cell transplantation. No therapy has been proven to increase the survival rate in steroid refractory aGvHD so far. A new strategy is the use of Pentostatin, a purine nucleotide analogue, which is known to decrease function and number of lymphocytes. In a pilot study we treated five patients suffering from steroid refractory acute intestinal GvHD with Pentostatin (two female and three male patients). The mean age was 44 (range 25-60 years). The underlying diseases were CML (1), AML (2 with refractory disease) and Ph+ c-ALL (one in first and one in second CR). Three patients were allografted with HLA-identical sibling donors, one with a matched unrelated donor and one with a related donor with a single HLA A mismatch. All patients were allografted with peripheral blood stem cells. As GvHD prophylaxis Ciclosporine and Mycophenolate mofetil were used. All patients had an acute GvHD stage III. Four patients had a GI-tract involvement grade III, one patient grade II. In all patients the primary treatment with high dose Prednisolone failed. Four patients received ATG (Fresenius) without clinical benefit. After failure of steroid and/or ATG treatment Pentostatin was applied as a salvage therapy (1 mg/m^2 for three consecutive days). One patient received a second Pentostatin course after three weeks. Therapy was well tolerated. No severe neutropenia or impairment of renal function were observed. All five patients achieved a complete remission of aGvHD. After the treatment with Pentostatin it took at least 12-14 days until a clinical improvement was observed. CR was achieved between day 18 and 30 post Pentostatin. Two patients are alive (60 and 334 days post pentostatin). Three patients died (2 patients due to relapse of leukemia 23 and 55 days post Pentostatin, 1 patient at day 50 post pentostatin of idiopathic interstitial pneumonitis). In summary, in this pilot study Pentostatin has been demonstrated to be a highly effective and well tolerated drug for salvage therapy of steroid refractory acute GvHD with intestinal involvement. Further prospective and randomised studies are needed to prove whether Pentostatin can improve the survival rate in steroid refractory aGvHD.

162

Peripheral blood progenitor cell mobilization in related donors: influence of the G-CSF dosing schedule

M.J. Eibl, W. Zinke-Cerwenka, P. Neumeister, U. Apfelbeck, K. Rosskopf, W. Linkesch (Graz, A)

Introduction: Peripheral blood progenitor cells (PBPCs) are increasingly preferred to bone marrow in allogeneic stem cell transplantation for hematologic malignancies. Various rhG-CSF mobilization regimens have been published with different stem cell yields collected, but so far no standard dosing schedule has been established. The aim of our study was to collect a maximum of PBPCs (target CD34+ cell dose: 4x10^6/kg rec. b.w.) despite reducing the total G-CSF dose, the number of apheresis sessions and the costs. Patients and methods: In our institution from October 1998 to January 2001 15 consecutive patients (male : female = 9 : 6; AML: n=7, CML: n=4, NHL: n=2, MM: n=1, MDS: n=1)with a median age of 42 years (28-58) were allografted using PBPCs from their relatives (HLA matched: n=14, 1 antigene mismatch: n=1) The median age of the donors was 44 years (19-71); seven were male and eight were female. They underwent the following G-CSF mobilization schedule: 1x10 µg/kg on day 1, 2 and 4, and 2x10 µg/kg on day 3. The first apheresis session was performed on day 4, on conditon that the CD34+ cell count in peripheral blood was above 10 cells/µl. In case of a harvest of less than 4x10^6 CD34+ cells/kg rec. b.w. the donor additionally received 1x10 microg/kg G-CSF on day 4 and 5, and a second apheresis session was carried out on day 5. Results: More than 2x10^6 CD34+ cells/kg rec. b.w. could be collected in all donors after a single apheresis session. The target CD34+ cell dose of 4x10^6/kg was achieved in 12/15 donors (80%) in one and in 3/15 (20%) individuals in two procedures. The median number of CD34+ cells/kg rec. b.w. was 5.77x10^6 (4.1-18.09) harvested in one (80%) and two (20%) apheresis sessions respectively. The total G-CSF dose was 50 µg/kg in 80% and 70 µg/kg in 20% of the donors. A sustained trilinear hematopoietic engraftment was achieved in all patients. Eight patients suffered from acute or chronic GVHD. Conclusion: In comparison with preexisting G-CSF mobilization regimens we collected a sufficient number of PBPCs with a lower total G-CSF dose and a minimum of apheresis sessions by dose escalation of G-CSF on the day before the first scheduled apheresis.

163

Intracellular cytokine profile of Th1 and Th2 cells in healthy volunteers and in patients with chronic graft-versus-host disease after allogeneic BMT

Th. Lipp, C. Guenther, E. Eberlein, K. Knochel, C. Nerl (München, D)

T helper cells play an important role in immune responses. The Th1 cytokines support the cell-mediated immunity, whereas the Th2 cytokines are supposed to exert a regulatory and immunosuppressive effect. We have established an in vitro assay for detecting intracellular produced cytokines in CD4 cells and elucidate the role of cytokine production in T helper cell subtypes in patients with chronic GvHD after allogeneic BMT. Materials and methods: The frequency of cytokine-producing CD4 cells was measured by FACS-analysis after a short in vitro stimulation of PBMCs with PMA/Ionomycin. Cell activation was determined by CD69 and CD25 expression. Further analyses included the ratio of memory (CD4 CD45RO+) and naive (CD4 CD45RA+) T cells. After permeabilization the cytokine antibodies (IFN-gamma, IL-4, TNF-alpha and IL-10) were added and the cytokine producing cells could be determined quantitativly. This method allows a detection of cytokine production at a single cell level even in low frequency. Another advantage is the simultaneous analysis of different cytokines. Results: In healthy control volunteers (n=10) the CD45RO/RA ratio was 1.07 ± 0.37 and the CD69 expressed in over 90% of cells. After a 4 hour stimulation period the percentage of cytokine producing CD4 cells was low: IFN-gamma 4.2% ± 2.2, TNF-alpha 2.4% ± 1.5, IL-4 1% ± 0.5, IL-10 2% ± 1.4 of CD4 cells. A longer incubation period of 18 hours led to a higher proportion of cytokine-producing T-cells and allowed therefore a better discrimination. We further analyzed patients with chronic GvHD mainly of the skin (n=10). As expected the CD4 cells were decreased and the memory T cells predominated (CD45RO/RA = 6.2 ± 5). CD69 expression was reduced. The patients with chronic GvHD showed a higher frequency of IFN-gamma-positive cells (14.5% ± 5.1) and of IL-4 producing T helper cells (5.4% ± 2.2) compared to controls. Both cytokines were detected in 7 out of 7 tests. In contrast the patients without GvHD showed a predominant IFN-gamma production and a low frequency of IL-4 producing T helper cells. Conclusion: The detection of intracellular cytokines in T helper cells by FACS analysis appears to be applicable to patients with chronic GvHD. The preliminary results of cytokine pattern in healthy controls shows no preferential activation of Th1 and Th2 cells after in vitro stimulation. In patients with chronic GvHD the shift to a high proportion of memory T cells together with a higher frequency of IFN-gamma- and IL-4- producing CD4 cells may reflect the activation of this lymphocyte subset and the imbalance in T helper cells. Further experiments will adress the role of IL-10 and TNF-alpha in chronic GvHD and the functional characterization of the T helper cell subsets.

164

Response and outcome after donor lymphocyte infusions for relapse after conventional and non-myeloablative blood stem cell transplantation

J. Koch, P. Schneider, D. Söhngen, R. Fenk, M. Aivado, U. Rohr, A. Rong, F. Neumann, R. Kronenwett, G. Meckenstock, R. Haas, A. Heyll, G. Kobbe (Düsseldorf, D)

Response and outcome after donor lymphocyte infusions for relapse after conventional and non-myeloablative blood stem cell transplantation. Between 1991 and 2001 nineteen patients with hematologic malignancies

(9 CML, 5 AML, 1 ALL, 2 MDS, 1 MM, 1 NHL) received donor lymphocyte infusions (DLIs) for relapse (14) or persistent (5) malignant disease after conventional myeloablative (6 T-cell depleted, 6 unmodified) or non-ablative (7) blood stem cell transplantation. Of the 9 patients with CML 2 relapsed cytogenetically, 4 in hematologic chronic phase, 1 in blast crisis and two patients had persistent stable (1) or progressive (1) disease after non-ablative stem cell transplantation. Donors were HLA-identical relatives in 14 and matched unrelated donors in 5 cases. Donor lymphocytes were harvested by leukapheresis from steady state hematopoiesis (8) or after G-CSF mobilisation (11). Twelve patients received 1, and 7 patients received 2-3 DLIs. A fraction of patients were treated with chemotherapy (6) before or interferon alpha (7) after DLI. Following DLI, 7 patients (6 CML, 1 NHL, 5 after conventional and 2 after non-ablative blood stem cell transplantation) achieved complete molecular remission and 100% donor chimerism. All of these 7 exerienced ≥II° acute graft versus host disease which was treated successfully. Four (57%) have chronic graft versus host disease (cGvHD, 1 extensive, 3 limited). All responders are long term survivors with a median follow up of 527 days (range 386-3512) after DLI. One additional patient (ALL) achieved blast clearance from the peripheral blood but developed severe pancytopenia and died of aGvHD and infection. One patient (MDS) was not evaluable for response and died in pancytopenia of aGvHD and infection. The remaining ten patients (3 CML, 5 AML, 1 MDS, 1 MM) had progressive disease despite repeated DLIs and aGvHD ≥II° in 5 (50%). Two of the non-responders are alive (1 CML, 1 MDS) with active disease without GvHD 145 and 173 days after DLI and receive further treatment with chemotherapy (1) or interferon alpha (1). The remaining 8 patients died of progressive disease (6) and a combination of progressive disease, aGvHD and infection (2). In our experience patients with CML who relapsed cyogenetically or in chronic phase after allogeneic blood stem cell transplantation had the greatest chance to achieve complete molecular remission and complete donor chimerism (75%), which is in agreement with published data. All responders developed aGvHD. Patients who achieved remission had a high probability to become long term survivors. We did not observe clinical remissions in patients with acute leukemia or CML in blast crisis despite the development of aGvHD in 50%.

165

Basiliximab is well tolerable and effective in the therapy of steroid-refractory graft-versus-host disease (GvHD) after allogeneic stem cell transplantation

G. Massenkeil, I. Genvresse, G. Reich, M. Schweigert (Berlin, D)

Introduction: Basiliximab is a chimeric monoclonal antibody directed against the alpha-chain (CD25) of the interleukin-2-receptor (IL-2-R alpha). IL-2-R alpha is only expressed on activated T-lymphocytes, which are known to play a crucial role in the development of GvHD. We have studied the safety and efficiency of basiliximab in the treatment of GvHD after allogeneic stem cell transplantation (SCT).
Patients: We report on 14 patients (pts.), median age 41 years (19-55) with ALL (n=2), AML (n=3), CML (n=8) and NHL (n=1). Pts. received standard conditioning before allogeneic SCT from an HLA-identical related (n=3) or unrelated (n=11) donor. GvHD-prophylaxis was performed with cyclosporin A (CSA) (n=1), CSA and methotrexate (MTX) (n=4) and CSA, MTX and prednisone (n=9). GvHD was diagnosed 8 to 180 days (median 30 days) after allogeneic SCT. Indications for basiliximab were severe GvHD and failure of first line therapy with CSA and high-dose methylprednisolone in all pts. Basiliximab 20 mg was administered on two consecutive days and weekly thereafter. Results: 38 basiliximab infusions were administered to these 14 patients. None of the pts. showed infusion-associated or cytokine-related symptoms after basiliximab. 12/14 pts. (86%) responded to basiliximab. Time to first response was 1 to 8 days after basiliximab (median 1.5 days). Complete remission of GvHD was achieved in 6/14 pts. (43%). Conclusion: Basiliximab is well tolerable and can be safely administered to pts. with acute GvHD after allogeneic SCT. Our preliminary data suggest that basiliximab is effective as rescue therapy in acute GvHD.

166

Graft-versus-host disease with severe hematochezia: Successful treatment with surgery and infliximab

M. Prumbaum, S. Braeuninger, A. Burchardt, D. Judith, S. Kriener, L. Duschka, H. Baurmann, R. Schwerdtfeger (Wiesbaden, Frankfurt, D)

Introduction: Graft-versus-host disease (GVHD) is a common complication of allogeneic stem cell transplantation and often involves the gastrointestinal system. It might lead to emergencies like bleeding, perforation and peritonitis that require surgery. If gastrointestinal GVHD fails to respond to steroids, mortality rate exceeds 70%. Infliximab, a chimeric antibody to human TNF-alpha, was recently approved for treatment of Crohn`s disease. There are preliminary results about its activity in management of acute and chronic GVHD. Case report: A 36-year-old male underwent allogeneic bone marrow transplantation from a fully matched unrelated female donor for CML in first chronic phase in 10/1999. Conditioning regimen consisted of busulfan 16 mg/kg, cyclophosphamide 200 mg/kg and ATG (Merieux) 10 mg/kg. He received CSA and short-course methotrexate as GVHD prophylaxis. On day + 50 he developed watery diarrhea to a maximum extent of 2300 ml/d. Endoscopy of the upper and lower gastrointestinal tract showed white plaques in the duodenum and severe edema and eythema in the terminal ileum and the ileocecal valve, both consistent with diagnosis of GVHD. Histology confirmed the diagnosis and showed GVHD- typical features also in the colonic mucosa, that had been considered unaffected by endoscopy. Steroid dose was increased to a maximum of 500 mg/d, budenoside and mycophenolate mofetil were added, FK 506 was given instead of CSA. Despite the escalation of the immunosuppressive therapy the situation further deteriorated and the patient developed severe hematochezia. Radiology showed involvement of the terminal ileum with normal appearance of jejunal mucosa. On day + 101 ileocecal resection was performed, the postsurgical course was uneventful. Immunosuppressive therapy was continued as before and infliximab was added in four weekly doses of 10 mg/kg. Gastrointestinal hemorrhage stopped, stool volume and frequency normalized. Resolution of GVHD was confirmed by endoscopy and biopsy, also in the colonic mucosa. Though gastrointestinal GVHD did not relapse, the patient died day + 253 due to fungal septicemia. Conclusions: In otherwise refractory gastrointestinal bleeding due to GVHD surgical resection of a diseased bowel segment is feasible and might be successfully performed. Since histologic findings revealed involvement also in normal-appearing mucosa of the colon, additional systemic immunsuppressive treatment was required. In our case clinical improvement and endoscopical and histological resolution of GVHD can be attributed to the treatment with infliximab.

167

GvH-prophylaxis with antithymocyte globuline (Fresenius) in recipients of matched and mismatched unrelated donor transplants for chronic myelogeneous leukemia: A retrospective analysis

A.R. Zander, M. Schleuning, N. Kröger, F. Finke, T. Zabelina, J. Berger, D.W. Beelen, R. Trenschel, R. Schwerdtfeger, H. Baurmann, M. Bornhäuser, G. Ehninger, A.A. Fauser, M. Kiehl, H.J. Kolb, U.W. Schaefer (Hamburg-Eppendorf, München, Freiburg, Saarland, Essen, Wiesbaden, Dresden, Idar-Oberstein, D)

Matched unrelated donor transplant has an increased risk of severe graft versus host disease and transplant related mortality (TRM). ATG has been introduced to decrease GvHD and to facilitate engraftment. We conducted a retrospective analysis of Fresenius ATG, n = 145, average = 90 mg/kg bw, range 40 – 90 mg/kg bw compared with a no-ATG group, n = 188, age, sex, HLA-matched versus mismatched were comparable. Cell source (bone marrow vs. peripheral blood stem cell) was significantly different in the 2 groups. Conclusion: ATG Fresenius decreases the incidence of acute GvHD and TRM and leads to a significant better overall survival. A prospective randomized study is needed to evaluate the definite role of ATG in HSCT.

168

Matrix metalloproteinase-1 promoter polymorphism in donors and recipients is associated with acute GvHD following hemopoietic stem cell transplantation

A.B. Chukhlovin, T.S. Zabelina, N. Kröger, A.R. Zander, B. Fehse (St. Petersburg, RU; Hamburg-Eppendorf, D)

The acute graft-versus-host disease (aGvHD) is generally regarded as a complex alloimmune disorder caused by HLA differences in donor/recipient pairs. However, some other genetic factors may influence the incidence and severity of aGvHD, e.g., those modulating blood clotting and invasive capacity of immune cells. In this respect, some common mutations in promoter regions of plasminogen activator inhibitor 1 (PAI-1) and matrix metalloproteinase 1 (MMP1, collagenase-1) could be actual, since these polymorphisms are known to affect transcriptional activities and, resp, the production of specific proteins. The aim of our study was to determine the relative occurence of MMP-1 and PAI-1 promoter variants in patients with various hematological disorders and their donors, in order to assess a possible role of their polymorphism in occurence of aGvHD after allogeneic HSCT. Gene promoter polymorphism has been determined in 103 patients with chronic myelogenous leukemia (n=29), AML (n=20), ALL (n=19); multiple myeloma (n=13) and other hematological disorders subjected to dose-intensive chemotherapy followed by allogeneic HSCT. Genomic DNA was isolated from peripheral blood or buccal cells of the pre-transplant patients. 125 normal donors served as a control group. Two gene polymorphisms have been assessed, i.e., 5G/4G variants of PAI-1 gene (4G allele more active), and 1G/2G alleles of MMP-1 gene (2G allele highly active), by means of original allele-specific DNA PCR technique. A near-Mendelian equilibrium state was observed for PAI-1 and MMP-1 alleles among 125 normal subjects. A constellation of PAI/5G+MMP/1G+ genotypes was more often in the patients with aGvHD (p=0.06), thus suggesting both genes to be less functional in this group. Moreover, the MMP-1 alleles were unevenly distributed among the patients and their donors, i.e., 94% of GvHD patients carried the less active MMP/1G allele, as compared to 76% in GvHD-free cases (p=0.02). In the donors of aGvHD patients, a tendency to increased frequency of more active MMP/2G allele was noted (p<0.2). When comparing individual donor/recipient pairs, the constellation of patient's MMP/1G and donor's 2G allele proved to be much more frequent in the cases of clinical aGvHD (this status was found for 80% of GvHD cases vs. 54% in GvHD-free patients; p<0.01). Hence, the genetic variability of MMP-1 gene promoter could be of significant value as a promising predictive marker for acute GvHD and a prospective target for therapeutic efforts.

169

Fate of alloreactive, antigen-specific CD8+ T cells within donor lymphocyte infusions (DLI) administered to mixed hematopoietic chimeras

M. Mapara, Y. Kim, M. Sykes (Berlin, D; Boston, USA)

To investigate the mechanisms of DLI-mediated GVL effects in mixed hematopoietic chimeras, we utilized a recently developed cyclophosphamide-based non-myeloablative conditioning regimen, which results in the establishment of long-term stable mixed hematopoietic chimerism in the same B10.A to B6 strain combination. Administration of DLI on day 35 post-BMT mediated a powerful GVL-effect with complete protection against mortality mortaility induced by the host-type leukemia/lymphoma, EL4. When 500 EL4 tumor cells were administered seven days post-DLI, complete protection from leukemia (100% survival at 164 days) was observed in the DLI-treated group. GVHD did not occur, despite this GVL effect and conversion of mixed chimerism to full donor chimerism. In contrast, the control group receiving tumor without DLI showed 100% mortality with a median survival time [MST] of 68 days, (p<0.001). Rechallenging these surviving DLI recipients with the same tumor dose 17 weeks later, after mixed chimeras had converted to full donor chimerism, led to rapid tumor mortality. Furthermore, when day 35 DLI recipients were challenged for the first time 10 weeks post-DLI with EL4 cells, no significant antitumor effect was detected. Hypothesizing that this lack of long-term GVL reactivity might be attributed to a gradual decline of alloreactive cells from the DLI, we used the 2C TCR transgenic mouse strain which carries an Ld-specific B6 CD8 clone-derived TCR and is alloreactive toward BALB/c mice. We established B6®BALB/c chimeras using the CTX conditioning regimen. Mixed chimeras were then given 5x106 2C spleen cells alone or together with wildtype B6 spleen cells (1x107) on day 35. Control chimeras received wildtype B6 splenocytes alone or no DLI at all. Animals were followed for chimerism and sacrificed 10 days or 10 weeks post-DLI. Maximum expansion of 2C CD8+ T cells was observed after 10 days, and this was followed by a steady decline in 2C CD8 splenocytes until week 10 post-DLI. Thus, our results clearly show that induction of mixed chimerism followed by administration of DLI results in powerful GVL effects. Furthermore, the loss of DLI-mediated GVL effects that occurs following conversion to full chimerism is associated with a significant decrease in the number of donor-derived alloreactive anti-host-specific cells.

170

Role of host-antigen-presenting cells in the induction of graft-versus-leukemia (GvL) effects following donor lymphocyte infusions (DLI)

M. Mapara, Y. Kim, M. Sykes (Berlin, D; Boston, USA)

Rodent studies have shown that donor lymphocyte infusion given to mixed hematopoietic chimeras can mediate a graft-versus-host-reaction which is restricted to the lymphohematopoietic system, as indicated by the conversion from mixed chimerism to full chimerism without the concomitant development of GVHD. The presence of host antigen-presenting cells in mixed chimeras might allow improved presentation of host antigens to the donor lymphocytes, leading to superior alloactivation of the DLI inoculum and hence superior GVL effects in mixed compared to full chimeras. To test this hypothesis, we established mixed or full donor hematopoietic chimeras in the fully MHC-mismatched B10.A to C57BL/6(B6) strain combination. B6 (H2b) mice received 10.25 Gy total body irradiation and were reconstituted with either a mixture of T cell depleted (TCD) B6 plus B10.A (H2a) bone marrow or TCD B10.A bone marrow cells alone. Administration of DLI led to a significant GVL effect in both mixed and full chimeras compared to the control groups receiving EL4 cells alone (p<0.0001 and 0.0005 respectively). However, full donor chimeras which had received DLI displayed only a slight delay in mortality from EL4 leukemia compared to non-DLI controls (MST 91 vs. 86 days post-BMT). In contrast, almost all mixed chimeras receiving DLI were completely protected from leukemic mortality, with 22 of 25 animals still alive at the end of the observation period (i.e. >100 days post-tumor inoculation). The improved survival observed in DLI recipients with mixed hematopoietic chimerism at the time of DLI administration was statistically highly significant (p<0.0001) compared to the survival of full chimeras receiving DLI. To address the role of direct allorecognition of MHC class I on host hematopoietic cells in inducing GVL effects, mixed chimeras were established which were deficient in MHC-class I expression on host hematopoietic cells. This was achieved by reconstituting wild-type B6 mice following lethal irradiation with a mixture of TCD allogeneic BMC plus TCD beta-2-microglobuline deficient (-/-) B6.ß2mB6 BMC. GVL effects were markedly reduced in mixed chimeras in which host-type hematopoieitic cells were deficient in class I MHC expression compared to those with normal class I expression on B6 hematopoietic cells. Furthermore, among animals with class I-deficient B6 hematopoietic cells, GVL effects were only apparent among those with high levels of B6.ß2mB6 and relatively low levels of B10.A donor hematopoiesis. Thus, GVL effects are maximized by the presentation of recipient antigens on normal host-type APC, accounting for the superior GVL effects in mixed vs. full allogeneic chimeras.

171

Increased risk of immunologic organ complications after allogeneic transplantation of peripheral blood stem cells compared to bone marrow

R. Peceny, A.H. Elmaagacli, B. Scheulen, H. Ottinger, U.W. Schaefer, D.W. Beelen (Essen, D)

The occurrence of potentially fatal immunologic organ complications (IOC) after allogeneic blood stem cell transplantation has been studied prospectively in patients with malignant hematologic diseases.

Between October 1994 and December 2000, 182 patients were transplanted with peripheral blood stem cells and 150 patients with bone marrow from HLA identical sibling donors (n=239) or partially matched family donors (n=93). With a median of 622 days post transplant (range 14 days – 2348 days), a total of 28 immunologic organ complications developed in 26 of 332 (8%) patients and manifested as bronchiolitis obliterans (n=7), bronchiolitis obliterans with organizing pneumonia (n=12), acute immune hepatitis (n=3), and acute glomerulonephritis (n=6). The cumulative 6-year-estimate of immunologic organ complications was 6% ± 2.5% after bone marrow transplantation compared to 24% ± 5% after transplantation of peripheral blood stem cells (p < 0.007, log-rank test). Further, this estimate was clearly higher in male recipients of female grafts (36% ± 8%) compared to other gender combinations (8% ± 2.5%) (p < 0.0001, log-rank test) and slightly higher in patients developing chronic GvHD (18% ± 3.5%) compared to those without this condition (6% ± 3%) (p < 0.05, log-rank test). Multivariate analysis confirmed an independently higher relative IOC-risk (95%-confidence limits) of 3.3 (1.3 – 8.1) after transplantation of peripheral blood stem cells (p < 0.02) and of 5.5 (2.6 – 12.6) for male recipients of female grafts (p < 0.0001). All other analyzed factors (chronic GvHD, acute GvHD, donor match, immune prophylaxis of GvHD, underlying disease, donor and patient age) had no significant influence on the IOC-risk in multivariate analysis. In conclusion, a significantly higher rate of potentially fatal immunologic organ complications has to be expected after transplantation of allogeneic peripheral blood stem cells compared to bone marrow with the highest increase in IOC-risk for male patients with female donors.

172

Adeno-associated virus mediated gene transfer of endostatin inhibits angiogenesis and tumor growth in vivo

C. Teschendorf, W. Shi, N. Muzyczka, D. Siemann (Bochum, D; Gainesville, USA)

Inhibition of angiogenesis through the use of angiostatic factors like endostatin has been shown to be an effective strategy in cancer therapy1. However its application has been hampered by difficulties associated with large-scale production needed for the repeated administration of the protein. This hurdle could be overcome through a gene therapy approach aimed at a sustained in vivo release of the therapeutic angiostatic protein from transduced tissues such as skeletal muscle. In the present study, we evaluated the efficacy of endostatin an inhibitor of angiogenesis when expressed through a rAAV vector. In vitro studies indicated that human endostatin (HuEndo) had a direct inhibitory effect on endothelial cells growth and migration Media endostatin levels of 50 ng/ml suppressed both the proliferation (by 48%, p<0.01) and migration (by 27%, p<0.01) of endothelial cells. In vivo, HT29 clones expressing high levels of endostatin were found to induce fewer blood vessels than did parental HT29 cells (32 vs 20, p<0.05). Similarly, HT29 cells injected into mice exhibiting elevated serum endostatin levels induced far fewer vessels than did tumor cells inoculated into normal mice. In order to achieve sustained expression of endostatin in situ, nude mice were injected intramuscularly with various doses of rAAV:HuEndo. Four weeks later elevated serum endostatin levels were detected in a dose dependent manner. High rAAV doses (1 x 10^9 i.u.) resulted in a serum endostatin level of 35-40 ng/mlat 6 weeks. The endostatin levels achieved by these means were sufficient to inhibit tumor angiogenesis in vivo. They also suppressed both the initiation of HT29 xenograft formation and the growth of macroscopic HT29 xenografts once established (time for the tumor to grow from 200 to 1000 mm^3, 25 vs. 16 days). These findings indicate that the delivery of human endostatin via rAAV vectors may provide effective anticancer therapy.

173

New strategies for the therapy of sarcomas and other solid tumors using recombinant adeno-associated virus 2 vectors that contain a suicide gene

M.R. Veldwijk, J. Topaly, S. Laufs, W.J. Zeller, A. Ho, S. Fruehauf (Heidelberg, D)

Adeno-associated virus 2 (AAV-2) vectors are used for clinical gene therapy of hereditary diseases and at this stage for preclinical cancer gene therapy. In previous experiments, we could show a high susceptibility of some solid tumors for a rAAV-2 vector containing the humanized green fluorescent protein (hGFP) gene. Among a series of primary cells and eight solid tumor cell lines infected with these AAV supernatants, we observed the highest infection rates (functional titer: 200 iu/cell) in soft tissue sarcoma cells (HS1; mean: 96% hGFP+ cells) and in breast cancer cells (T47D; mean: 83%, MCF-7; mean: 85%); other cell lines (1 ovarian tumor, 1 germ cell tumor, 1 osteosarcoma and 2 small cell lung cancer) were less permissive (3% - 12%); CD34+ peripheral blood progenitor cells showed the lowest transduction rates (mean 4%). These data suggest that the sarcoma and breast cancer cells are the most suitable candidates for further development of AAV-2 tumor suicide gene therapy. The first requirement for using these vectors in cell kill experiments was the construction of new rAAV-2 vectors containing a suitable suicide gene. The thymidine kinase (TK) gene was chosen in combination with the prodrugs aciclovir and ganciclovir. Three new vectors were cloned containing either the TK gene, the TK and the hGFP gene connected by an IRES-sequence or a TK/eGFP fusion protein, each under control of cytomegalovirus promoter/enhancer sequences. Since some of the new vectors contained no marker gene (GFP) and therefore did not allow fluorescence-based titration, a new titration method based on the quantitative TaqMan real-time polymerase chain reaction was established so that an easy, quick, safe, accurate and high-throughput titering of AAV-2 stocks with and without selectable marker genes became possible. For all the cell lines to be used in further experiments (T47D, MCF-7, HS-1 and an additional sarcoma cell line HT1080) dose-effect curves for both aciclovir and ganciclovir were assessed to determine the cytotoxicity induced by cellular kinases and the optimal therapeutic index. Xenotransplantation tumor models for the human sarcoma and breast cancer lines were established and optimized in our lab for both the intraperitoneal and subcutaneous tumor transplantation routes. In proof of principle experiments that are currently being performed we should now be able to show the effectiveness of our AAV-TK suicide gene therapy approach in the treatment of soft tissue sarcoma and breast cancer. Transferred to a clinical context, suicide gene therapy using the small rAAV-2 particles may be applicable perioperatively, especially in a R1 or R2 situation to eradicate local tumor spread.

174

Thalidomide does not affect the CD4+/CD8+-ratio in patients with haematologic malignancies

C. Strupp, M. Aivado, U. Germing, N. Gattermann, R. Haas (Düsseldorf, D)

Thalidomide is an immunmodulatory drug with a wide spectrum of activity that is not clearly understood. Current data suggest that the action of thalidomide is related to several different mechanisms, including suppression of tumor necrosis factor (TNF-alpha), effects on interleukins and interferons, down-regulation of selected cell surface adhesion molecules, and significant shifts in the lymphocyte-subsets from CD4+ lymphocytes to CD8+ lymphocytes. It is not clearly known if the clinical response to thalidomide in patients with multiple myelomas or myelodysplastic syndromes is related to its ability to inhibit angiogenesis or to its immunmodulatory effects. We therefore examined its effect on lymphocyte subsets in 11 patients with various haematologic malignancies (5 multiple myelomas, 4 myelodysplastic syndromes, 2 non-Hodgkin's lymphomas). All these patients had either a relapse or progressive disease under conventional chemotherapy, so they were started on thalidomide with an initial dose of 100 mg/d p.o. and escalated if tolerated to a maximum of 400 mg/d. T-lymphocyte-subsets (CD4, CD8) were analyzed by fluorescence activated cell scanning (Becton Dickinson FACScan) before starting thalidomide. The median of CD4+ cells was 1196/µl, and the median of CD8+ cells was 757/µl. After a 3-months period of treatment with thalidomide no significant changes in the proportion of CD4+ cells (1106/µl), (p=0.822) or CD8+ cells (567/µl), (p=0.551) were observed. Thalidomide has been shown to cause a decrease in the ratio of CD4+ to CD8+ lymphocytes in the blood of healthy males, also in patients with Erythema Nodosum Leprosum. In our patients with haematologic malignancies, such changes were not observed.

175

Pioglitazone and rofecoxib combined with angiostatic scheduling of chemotherapy in metastatic tumors: preliminary results

A. Reichle, K. Bross, Th. Vogt, B. Coras, R. Dengler, G. Freunek, S. Mayer, M. Voglhuber, Ch. Schicht, M. Zaiss, H. Wagner, H. Thomssen, J. Wilke, A. Wodzynski, R. Andreesen (Regensburg, D)

Thiazolidinediones, like pioglitazone, and selective cyclooxygenase–2-inhibitors, like rofecoxib are modulators of cell cycle and may induce differentiation and apoptosis in tumor cell lines. Therefore, these drugs were combined with chemotherapy (CT) in a phase I/II trial for extensively pretreated patients (pts) with metastatic tumors (mean number of different schedules 1.6 range 1 to 4). Pts received daily 25 mg rofecoxib po and 45 mg pioglitazone po, additionally 2x1 g/m^2 capecitabine po from day 15 to 28 (gastric cancer (n=4) and renal cell carcinoma (n=7)) and 3x50mg trofosfamide po from day 15 to 35 (sarcoma (n=6) and melanoma (n=10)), respectively. Cycles are given every 21 days to a maximum of 6, without interruption of rofecoxib/pioglitazone application. Up to now data on 27 pts (Karnofsky index > 60) with measurable and progressive disease are available for an interim analysis of response after 2 to 6 cycles of capecitabine or trofosfamide (mean 4.2): Up to now 70% of the pts responded (CR+PR 19%) or achieved stable disease (SD 51%). Response and SD are observed in all histological subgroups, are independent of both the number of preceding CTs and of the preceding therapy-free interval, 1.14 months (range 0.4 to 2.77 mo, n=22) vs 16.73 (range 8.87 to 33.27 mo, n=5). Even refractory diseases are responding. Hematologic toxicity was < grade 2, non-hematologic toxicity was grade 3/4 (n=3) due to hand-foot syndrome (n=3). In 8 cases creatinine levels rised maximally up to 132 µmol/l. Therefore, therapy with rofecoxib was interrupted in 3 cases. In summary, tumor response independent of histology, pretreatment and status of disease suggests a biological response to the combined therapy, probably due to its angiostatic activity and/or to the induction of differentiation and apoptosis.

176

In vivo activity of thalidomide and immunomodulatory drugs against multiple myeloma

S. Lentzsch, R. LeBlanc, F. Davies, K. Podar, B. Lin, R. Bargou, B. Dörken, K. Anderson (Berlin, D; Boston, USA)

Introduction: We have previously shown that thalidomide (Thal) and its potent immunomodulatory drug (IMiD) analogs inhibit the growth of Multiple Myeloma (MM) cell lines and patient MM cells that are resistant to conventional therapy by inducing apoptosis or G1 growth arrest. To study the in vivo effect of these drugs we injected $3x10^7$ HS-Sultan cells s.c., together with matrigel, into immunodeficient BNX-mice. After the development of measurable tumors (~150 mm^3), treatment was started with daily i.p. injections of 50 mg/kg Thal, IMiD1 or IMiD3. The endpoint of our study was survival; animals were sacrificed if the tumor was ≥2cm and/or necrotic. Microvessel density (MVD) of tumors was assessed using anti-CD31-staining. Results: No toxicity was observed at a dosage of 50 mg/kg in all treated groups. Tumors occurred within 4-10 days after tumor cell injection. All control animals were sacrificed at median of day 9 due to tumors ≥2cm. Treated animals showed a significantly longer survival and slower tumor progression than untreated animals: IMiD1, IMiD3>>Thal>>control group. Complete eradication of tumor occurred in 4 out of 6 mice treated with IMiD1 at median of day 16 (range 8-23 days) of treatment; the remaining 2 mice showed ≥50% tumor reduction of. Eight of 9 animals treated with IMiD3 are alive on day 27 after tumor cell injection, and showed a significantly slower tumor growth in comparison to Thal and the control group; one animal lost the tumor on day 5 of treatment. All 5 animals treated with Thal had to be sacrificed at median of day 19 of treatment. The MVD of tumors was significantly lower in all treated groups in comparison to the control group. By western blotting we could show that the MAPK and AKT-pathways activated by VEGF in MM1.s cells could be partially blocked by IMiDs. Furthermore, IMiDs and Thal inhibited the VEGF induced migration of MM1.s cells. Conclusions: Our results show that IMiDs have significantly increased in vivo anti-MM-activity compared to Thal, are able to induce sustained complete tumor remission (>8weeks), and prolong survival. There was no evidence of toxicity, and all drugs showed anti-angiogenec effects in vivo. Based upon these results phase 1 trials of IMiDs in MM have been initiated.

177

Thalidomide (Thal) in the treatment of relapsed and refractory multiple myeloma (MM)

K. Neben, T.M. Moehler, A. Kraemer, G. Egerer, B. Kasper, A.D. Ho, H. Goldschmidt (Heidelberg, D)

Thal is a drug with anti-inflammatory, immunomodulatory and anti-angiogenic activity, that was introduced recently in the treatment of MM (N Engl J Med 1999;341:1565-71). This is a prospective analysis of 83 patients (pts.) with refractory and relapsed MM that were enrolled in a phase II study and treated with Thal 400 mg daily in our center. There were 61 males and 22 females with a median age of 59 years (range, 34 - 86). The median time from diagnosis to entry in the study was 37 months (range, 5 - 191). The pts. were pretreated with radiotherapy (n=29) and a median number of 7 (range, 3 - 30) chemotherapy cycles, including one (n=60), two (n=25) and three (n=1) cycles of high-dose chemotherapy and peripheral blood stem cell transplantation. The serum and urine levels of monoclonal protein were reduced by at least 90% in 7 pts. (including 1 complete remission), at least 75% in 10 pts., at least 50% in 17 pts. and at least 25% in 35 pts., resulting in a total rate of response of 42%. Using the criteria of the European Bone Marrow Transplantation Group, 26 of 77 (34%), 30 of 58 (52%), 26 of 41 (63%) and 19 of 28 (68%) pts. showed at least a minimal response after 3, 6, 9 and 12 months of Thal therapy, respectively. Among all pts. the estimated 12 months progression-free and overall survival was 55% and 88%. So far 18 pts. died, 6 were non-responsive to Thal and 12 died in progressive disease after initial response. Major side effects (grade 3 and 4 toxicity) were reported on 15 pts., including infections (n=4), polyneuropathy (n=5) and cardiovascular events (n=6). In conclusion Thal is an active drug against MM that might improve, either alone or in combination with conventional chemotherapy, the prognosis in this presently incurable disease.

178

Enhanced sensitivity of lymphoma cells to immunological effector cells after wild-type p53 gene transfer

P. Buttgereit, F. Schakowski, A. Märten, S. Renoth, C. Ziske, K. Brand, T. Heinicke, I. Schmidt-Wolf (Bonn, Berlin, D)

p53 is a tumor suppressor gene encoding for a nuclear phosphoprotein that functions as a negative regulator of cell cycle progression. Studies examining the cellular response to DNA damage have also demonstrated a role of p53 in triggering apoptosis. After DNA damage, p53 accumulates in the nucleus and stops the cell cycle via induction of cyclin-dependent kinase inhibitor p21WAF/CIP1 and MDM2 genes. Although p53 is not essential for normal development, the loss of wild-type p53 function contributes to cellular acquisition of malignant properties. Several investigators showed that reconstitution of normal p53 expression in tumor cells can lead to suppression of cell growth and to cell death. The aim of this study was to increase the sensitivity of p53 defective lymphoma cells against immunological effector cells. Therefore, we transfected p53 defective lymphoma cells by adenoviral p53 gene transfer and examined the change in sensitivity against cytokine-induced killer cells (CIK; J. Exp. Med. 174:139-149 (1991)). For Raji and Daudi cells an increased sensitivity against CIK cells was found at an effector to target ratio of 20 to 1 by LDH release assay (Buttgereit et al., Cancer Gene Ther.; in press). In further studies we examined if reconstitution of p53 by adenoviral p53 gene transfer in lymphoma cells led to an increase of sensitivity against chemotherapeutic agents (e.g. etoposid). For that, we used the chemotherapeutic concentration which led to a 50% decrease of proliferation and examined the proliferation of p53 transfected, mock transfected and non-transfected cells. In Raji cells we could show that the reconstitution of p53 after adenoviral gene transfer led to a significant increase of sensitivity against chemotherapy.

179

Safety and efficacy of thalidomide in patients with acute myeloid leukemia

M.B. Steins, T. Padró, R. Bieker, S. Ruiz, M. Kropff, J. Kienast, T. Büchner, W.E. Berdel, R.M. Mesters (Münster, D)

Emerging data suggest an involvement of angiogenesis in the pathophysiology of acute myeloid leukemia (AML). Thus, antiangiogenic therapy could constitute a novel strategy for the treatment of AML. To test this hypothesis, we performed a phase I/II dose escalating trial to study the safety and efficacy of thalidomide, a putative inhibitor of angiogenesis, in 20 patients with AML not qualifying for intensive cytotoxic chemotherapy. Thirteen patients were assessable for both toxicity and response tolerating a maximum dose of 200 - 400 mg daily for at least one month. Seven patients had to be withdrawn prematurely from drug administration due to progressive disease and death (3 patients), personal decision (2 patients) or intolerance of thalidomide (2 patients). Overall adverse events of the drug comprised fatigue (12 patients), constipation (9 patients), rash (5 patients) and neuropathy (4 patients). In 4 patients, a partial response defined as reduction in the blast cell infiltration of the bone marrow of at least 50% accompanied by increases in the platelet counts and haemoglobin values was observed. One additional patient showed a haematological improvement without fulfilling the criteria of a partial response. In parallel, microvessel densities significantly decreased in these 5 patients during treatment with thalidomide ($p<0.05$). This decrease was accompanied by declining plasma levels of vascular endothelial growth factor and basic fibroblast growth factor, the most potent angiogenic growth factors. In conclusion, single-agent thalidomide has antiangiogenic and antileukemic activity in AML, although a causal relationship between both effects has still to be proven.

180

Lymphocytopenia and elevated serum immunoglobulin levels are associated with shortened telomere length in patients with systemic lupus erythematosus

C. Amberger, A. Wahl, I. Ersöz, U. Hartmann, H. Wolke, B. Berner, K. Bartolovic, C. Bokemeyer, L. Kanz, I. Kötter, T. Brümmendorf (Tübingen, D)

Telomeres in vertebrates are composed of TTAGGG repeats and associated proteins. In most somatic cells telomere repeats are lost with each cell division. Detailed studies on telomere length dynamics in peripheral blood lymphocytes (including naïve and memory subsets) and granulocytes derived from > 300 samples normal human donors using fluorescence in situ hybridization and flow cytometry (Flow Fish) have been performed previously (J. Exp. Med. 190:157-167). In the current study, we were interested to investigate the relationship between lymphocytopenia and lymphocyte telomere length. Telomere length in lymphocytes and granulocytes of 29 patients with systemic lupus erythematosus (SLE), an autoimmune disease frequently associated with chronic lymphocytopenia were analyzed. The results were compared with age adjusted healthy controls and correlated with lymphocyte count, disease stage and severity. We found significant shortening of age-adjusted telomere length in lymphocytes (mean ± S.E.: -1.7 ± 0.7 molecular equivalents of soluable fluorochrome units (kMESF), n=22) both compared to granulocytes of the same individual (-0.8 ± 0.8 kMESF, n=29, p=0.002, paired t-test) and to lymphocytes from normal controls (p=0.02, 2-sided t-test). Samples from SLE patients with lymphocyte counts < 1000/µl were found to have shorter telomeres that differed significantly from controls (-2.2 ± 1.0 kMESF, n=11, p=0.03) whereas patients with > 1000 lymphocytes/µl did not show significant telomere shortening (-1.2 ± 1.2 kMESF, n=10, p=0.3). No significant correlation of age-adjusted lymphocyte telomere length with duration of the disease, duration of treatment, affected organ site or inflammatory parameters (e.g. CRP, ESR) was detected. Nevertheless, in samples from patients with high plasma immunoglobuline levels (IgG > 1.2 g/l) telomere length was significantly shorter than in controls (-3.1 ± 1.0 kMESF, n=10, p=0.002) whereas in patients with lower IgG levels (IgG < 1,2 g/l) a significant difference was not detectable (-0.7 ± 1.3 kMESF, n=9, p=0.6). In summary, these findings support the hypothesis of an increased lymphocyte turnover in SLE. Future studies will use the combination of cell surface staining with Flow FISH in order to be able to specifically analyze telomere length in phenotypically defined lymphocyte subpopulations in this disease.

181

Centrosome aberrations occur at a higher frequency in diffuse large B-cell lymphoma as compared to follicular lymphoma

S. Schweizer, K. Neben, P. Lichter, U. Krause, C. Giesecke, H.K. Müller-Hermelink, A.D. Ho, G. Ott, A. Krämer (Heidelberg, Würzburg, D)

Genetic instability is a common feature of almost all human cancers including non-Hodgkin's lymphoma (NHL). This condition is frequently characterized by an abnormal number of chromosomes, although little is known about the mechanism that generates this altered genetic state. In NHL, numerical chromosome aberrations have been shown to provide information about the clinical course and the risk of transformation to a high grade malignancy. Recently, centrosome aberrations have been described as a possible cause of aneuploidy in many human tumors (Marx J, Science 292: 426-429, 2001). Centrosomes assure accurate chromosome segregation during mitosis and undergo duplication precisely once before cell division as a function of cyclin A/cyclin-dependent kinase 2 activity at the G1/S boundary. To investigate whether centrosome aberrations do occur in NHL and correlate with the histologic subtype, we examined a set of 16 follicular lymphomas (FL) and 12 diffuse large B-cell lymphomas (DLBL) using antibodies to centrosomal proteins. Evaluation was performed by analyzing at least 200 cells per sample using fluorescence microscopy. Centrosomes were considered abnormal if they were present in more than two copies per cell or if they had diameters > two times the diameter of centrosomes in nontumor control cells. First, we found that all 28 B-cell NHL analyzed displayed numerical as well as structural centrosome aberrations as compared to normal tonsillar tissue. Centrosome abnormalities were detectable in 32.50 ± 9.74% of the cells in NHL but only in 5.25 ± 0.7% of the cells in normal tonsillar tissue (p<0.001). Whereas DLBL contained 42.17 ± 4.14% cells with abnormal centrosomes, the FL harbored centrosome aberrations in only 25.25 ± 5.23% of the cells (p<0.001). These results indicate that centrosome defects are a common feature of NHL and suggest that they may contribute to the aquisition of an increasing karyotypic instability typically seen in high grade NHL.

182

Thalidomide in endstage acute myeloid leukemia (AML) and myeloproliferative disorders - therapeutic benefit versus side effects

S. Wittwer, J. Krauter, A. Kohlenberg, J. Schubert, A. Ganser, G. Heil (Hannover, D)

Thalidomide regained interest due to its immunomodulatory and antiangiogenic properties. Evidence for antitumor activity has been found against a variety of tumors, e.g. multiple myeloma and glioblastoma. We wished to analyze the efficiency and safety of thalidomide treatment in patients with AML or with myeloproliferative disorders. Eleven patients (5 male, 6 female) were treated with thalidomide because of refractory hematologic disorders after written informed consent under a compassionate-use protocol. Medication was kindly provided by Gruenenthal GmbH. Treatment was initiated with 200 mg daily and dose was escalated by 200 mg every other week if the drug was well-tolerated. Diagnosis was acute myeloid leukemia (n=7), advanced polycythemia vera (n=3) and myeloid blast crisis of chronic myeloid leukemia (n=1). Median age was 65 years (range 54-77 years). All patients had received prior chemotherapy and were refractory to standard treatment. Dose reductions were made in 2 pts because of skin rash and sedation. Time of treatment ranged from 16 to 208 days (median 54 days). Median cumulative dose of thalidomide was 7000 mg (range 2450 mg to 99200 mg). During treatment we observed a raise of the platelet count in two patients. One patient had a reduction of trisomy 8 positive blasts with an increase of granulocytes. In one patient with CML blast crisis response to hydroxyurea treatment was restored. Side effects observed were fatigue (n=8), constipation (n=5), paresthesia (n=1), skin rash (n=2), vertigo (n=1) and nausea (n=2). Treatment was stopped because of skin rash (n=2), sedation (n=2), progressive disease (n=3), psychotic episode (n=1) and death (n=3, two progressive disease, one intracranial hemorrhage). During follow-up 5 patients died in median 73 days after the end of treatment (range 6-96 days). Reasons of death were progressive disease (n=3), intracerebral hemorrhage (n=1) and pneumonia (n=1). 3 patiens are alive 133, 367 and 414 days after end of treatment. Thalidomide is applicable in patients with advanced hematologic disorders. Its moderate side effects allow the broader application of the drug, possibly in combination with chemotherapy. Due to its effects observed in this study its application to a larger number of patients who are no longer candidates for standard treatments appears justified to get more information about its efficiency.

183

Centrosome aberrations in hematological malignancies

C. Giesecke, U. Krause, C. Giesecke, K. Neben, P. Lichter, S. Schweizer, M. Hensel, H. Goldschmidt, A. Ho, A. Krämer (Heidelberg, D)

Karyotypic alterations, including whole chromosome loss or gain, ploidy changes, and a variety of chromosome aberrations are common in cancer cells. If proliferating cells fail to coordinate centrosome duplication with DNA replication, this will inevitably lead to a change in ploidy, and the formation of monopolar or multipolar spindles will generally provoke abnormal segregation of chromosomes. It has long been recognized that errors in the centrosome duplication cycle may be an important cause of aneuploidy and thus contribute to cancer formation. This view has recently received fresh impetus with the description of supernumerary centrosomes in almost all solid human tumors (Marx J, Science 292: 426-429, 2001). As the primary microtubule organizing center, the centrosome assures accurate chromosome segregation during mitosis. Centrosomes undergo duplication precisely once before cell division as a function of cyclin-dependent kinase 2 activity that couples centriole duplication to the onset of DNA replication at the G1/S phase transition. Since alterations of G1/S phase regulating proteins like the retinoblastoma protein, cyclins D and E, cdk4 and 6, cdk inhibitors p16INK4A and p15INK4B, and p53 are among the most frequent aberrations observed in human malignancies, we examined a set of five acute lymphoblastic leukemias (ALL), 24 acute myeloid leukemias (AML), 24 chronic lymphocytic leukemias (CLL), and 21 multiple myelomas (MM) for the presence of centrosome aberrations using antibodies to centrosomal proteins. Evaluation was performed by analyzing at least 200 cells per sample for ALL, AML, and CLL and an average of 195 plasma cells (range, 39-400) per multiple myeloma sample using fluorescence microscopy. Centrosomes were considered abnormal if they were present in more than two copies per cell or if they had diameters > two times the diameter of centrosomes in normal peripheral blood mononuclear cells (PBMC) from healthy individuals. First, we found that all 74 samples analyzed displayed numerical as well as structural centrosome aberrations. Centrosome abnormalities were detectable in 35.21 ± 11.67% of the cells from all 74 samples as compared to 6.67 ± 1.89% in normal PBMC (p<0.001). Specifically, abnormal centrosomes were found in 43.9 ± 10.21% of the cells in ALL samples, 45.42 ± 21.25% of the cells in AML samples, 20.77 ± 10.57% of the cells in CLL samples, and 30.73 ± 6.9% of the cells in MM samples. These results indicate that centrosome defects are a common feature of hematological malignancies. Since centrosome defects might lead to karyotypic instability, a correlation to chromosome aberrations of these samples is under way.

184

Induction of cell killing and gene transfer by parvovirus H1 in human tumor cells but not in primary cells of the immune system

M. Möhler, M. Zeidler, J. Scaede, N. Weiskopf, J. Rommelaere, J.J. Cornelis, P.R. Galle, M. Heike (Mainz, Heidelberg, D)

Objective: The autonomous parvovirus H1 preferentially replicates in transformed cells and reduces the incidence of spontaneous and induced animal tumors. So far, only few studies investigated the H1-mediated cell killing. Since H1 virus can induce apoptosis in animal tumors, apoptotic cell death of human tumor cells may enhance cell mediated immune response in vivo. Therefore, susceptibility of human immune cells to H1 infection would be considered as a drawback for the potential use of parvovirus H1 as antitumor agent and as vector for cancer gene therapy.Methods: We investigated infection, cell death and gene transfer of wildtype and recombinant parvovirus H1 in gastric, colon, hepatoma and melanoma tumor cells as well as in primary peripheral blood monocytes, dendritic cells, B- and T-lymphocytes. Cell death was determined by FACScan staining with Annexin V and propidiumjodide. Results: Only human tumor cells but not primary immune cells were susceptible to H1 virus-induced cell killing. Viability of H1-infected gastric tumor, hepatoma and melanoma cells was reduced to less than 10% after 5 days. The H1 virus-induced cell death correlated with H1 DNA replication and the viral nonstructural protein expression. FACScan presented apoptotic cell death in gastric tumor, melanoma and p53-deficient hepatoma cells. Infection with recombinant H1 virus carrying the luciferase gene (luc) under parvoviral promoter control resulted in high transgene activity in these tumor cells. In contrast, H1 virus-incubated immune cells showed neither signs of apoptosis nor presented high transgene activity.Conclusions: Parvovirus H1 infects, replicates and induces apoptosis in human tumor cells but not in primary human immune cells. Consequently, recombinant H1 viruses carrying small transgenes may be suitable in vivo tools for treatment of human tumors.

185

Activity of the Bcl-2/Bcl-Xl bispecific antisense oligonucleotide 4625 on tumor growth in combination with chemotherapy.

M. Posch, E. Elstner, U. Zangmeister-Wittke (Berlin, D; Zürich, CH)

Impaired apoptosis signaling contributes to cancer development and resistance towards chemotherapy, since apoptosis normally eliminates cells with damaged DNA or increased malignant potential. Bcl-2 and Bcl-xL are inhibitors of apoptosis frequently overexpressed in malignant tumor cells. Downregulation of either Bcl-2 or Bcl-xL lowers the apoptotic threshold and tumor cells undergo apoptosis. Antisense oligonucleotides targeting bcl-2 or bcl-xL mRNA revealed anti-tumor potential in different pre-clinical and clinical approaches. The 4625 antisense oligonucleotide targets a region of high homology in the bcl-2/bcl-xL mRNAs and simultaneously downregulates Bcl-2 and Bcl-xL. We investigated the anti-tumor potential of the 4625 oligonucleotide in combination with chemotherapeutic drugs in vitro and in vivo. Methods: 4625 mediated inhibition on Bcl-2 and Bcl-xL protein expression was assessed by immuno-blotting. Growth inhibition and induction of cell death were determined by MTT Assay and FACS analysis on different lung, breast and colorectal tumor cell lines. For in vivo experiments mice were treated systemically with 4625 alone or in combination with chemotherapy over 21 days starting one week after transplantation of cancer xenografts. Paclitaxel and 5-FU were chosen as standard therapy for the treatment of breast and colorectal cancer. Results: 4625 downregulates Bcl-2 and Bcl-xL expression in cancer cell lines of diverse histological origins. It further leads to proliferation inhibition at concentrations ranging from 75–600 nM and cell death induction in a dose-dependent manner. The anti-tumor activity of 4625 in vitro is independent from cellular Bcl-2/Bcl-xL expression patterns and synergistically interacts with various chemotherapeutic agents. The ip. application of 20 mg/kg 4625 with or without Paclitaxel/5-FU led to a growth inhibition of established human carcinomas xenografts in nude mice, relative to those treated with the 4626 control oligonucleotide with or without chemotherapy. Discussion: Bcl-2 and Bcl-xL play nonredundant roles in tumor growth and are often heterogeneously expressed in solid tumor tissues. Most lung carcinomas do overexpress both proteins. Our data suggest that state-of-the-art antisense technology offers a potent approach to inhibit the expression of the two major anti-apoptotic proteins Bcl-2 and Bcl-xL with one single oligonucleotide. The broad activity of 4625 against different tumor cells could make additional patients benefit from a treatment with this antisense compound.

186

Salvage therapy in poor prognosis multiple myeloma with thalidomide plus CED chemotherapy

T.M. Möhler, K. Neben, B. Kasper, G. Egerer, A.D. Ho, H. Goldschmidt (Heidelberg, D)

Thalidomide (Thal) was recently shown to have anti-tumor activity in Multiple Myeloma (MM) (N Engl J Med 1999, 341:1565). We report on savety and efficacy of Thal combined with CED Chemotherapy (TCED) in 56 patients (pts.) with poor prognosis MM (primary refractory, secondary refractory after ABSCT or conventional chemotherapy, plasma cell leukemia): T (400 mg p.o. per day, continuously), cyclophosphamide (C) 400 mg/m^2, etoposide (E) 40 mg/m^2 both as continuous i.v. infusion d1-4 and dexamethasone (D) 40 mg p.o. repeat after 28 days until best response or maximum of 6 cycles. 50 pts. were evaluable for response with 4 % (2) CR, 64 % (32) PR, 18 % (9) MR, 6 % (3) SD, 8 % (4) PD resulting in an objective response rate (bigger thab MR) of 86.6 % (intend-to-treat-analysis: 76.7 %; n=56). Subsequent to successful remission induction 18 pts. received autologous (n=9) or allogeneic (n=9) hematopoietic stem cell transplantation (SCT). The estimated median follow up duration was 14 months. The median progression-free-survival (PFS) was 16 months. The median overall survival was not reached so far (12+ months). Thal associated WHO grade I and II adverse effects were in the same range as reported previously including somnolence (32%), constipation (28%), tingling or numbness (23%), weakness (12%), tremors (11%), dizziness (10%). Severe adverse effects (WHO III/IV) included infectious complications (35.7% of pts.) and cardiovascular events (7.1% of pts.).

Thal in combination with CED chemotherapy is effective in inducing meaningful responses in poor prognosis multiple myeloma. Our data suggest that Thal improves antitumor activity of combination chemotherapies in poor prognosis MM.

187

Detection of aneuploid multiple myeloma clones in bone marrow samples by flow cytometry

H.-D. Kleine, D. Ezrodt, U. Buthmann, M. Freund (Rostock, D)

Multiple myeloma cells are characterized by the expression of CD38 and CD138. Flow cytometric analysis of these two markers combined with DNA-staining is a useful tool in the detection of aneuploid multiple myeloma cells. Flow cyto-metric list-mode-data of 112 samples were analyzed by the new „Continuous-Gating"-Software (Version 2.0.1), The new version is able to detect aneuploid tumorcells with an unknown DNA-index automatically. Bone marrow samples were measured in a FACSort flow cytometer, after staining with CD38-FITC and CD138-FITC and propidiumiodide. When analyzed by using the „Continuous Gating"-software a gate is set in the DNA parameter, defining the normal G0/G1-peak. Afterwards a 200x200 channel-gate (out of 1023x1023 channels) is defined automatically by the program, scanning the whole side-scatter(SSC)/CD38/CD138 dotplot 10,000 times. The cells defined by the 10,000 different dot plot gates were analysed for percentage of cells within the DNA-gate. The DNA-gate is moved by the program to higher channel numbers and the analysis is done repeatedly with increasing DNA-indices. The 112 bone marrow samples were divided into three subgroups after analysis: < 1% (62.5% of all samples), >1% to <10% (26.7%), >10% (10.7%) plasmacell content. In the first group 25% of the samples showed at least one aneuploid clone, in the second group 53.3%, and in the third group 58.3% respectively. The mean DNA-indices measured in samples with aneuploid clones varied from 1.22 to 1.32. 58.8% of all aneuploid plasmocytoma clones are CD56 positive in samples with a plasmacell content <1%, increasing to 68.7% of aneuploid samples >1% to <10% and 85.7% of aneuploid samples containing more than 10% plasma cells. In summary more than half of all samples with more than 1% plasma cells in the bone marrow are showing at least one aneuploid plasmocytoma clone. Samples with less than 1% plasma cells, and therefore bone marrow involvement not proven by standard diagnostics, are showing aneuploid plasma cell clones in 25% of all samples.

188

Apoptosis is a key mechanism in anti-thymocyte globulin-induced immunosuppression

M. Germer, S. Stegmann, D. Rohm, M. Kloft (Dreieich, D)

Anti-thymocyte immunoglobulin preparations (ATGs) serve as potent immunosuppressive agents especially in organ transplantation. Despite of their well established clinical effectiveness their mechanism of action is still debatable. But an ATG induced rapid decline in T cell numbers contributes to immunological silence. However, the active principle of T cell elimination has not been elucidated non-ambiguously so far. Whereas previous data indicate a role both for cellular mechanisms (e.g. ADCC) and for humoral factors (complement-mediated lysis) recent data suggest the involvement of apoptosis via an CD95-L dependent pathway. Aim of this study was to clarify the role of ATG-induced apoptosis. Antibody specificities were determined by competitive flow cytometry. Our studies demonstrate the presence of anti-CD95 antibodies in ATGs which are a mixture of activities against multiple surface antigens. Induction of apoptosis was monitored by loss of membrane integrity, fragmentation of DNA and change of the phospholipid bilayer composition. ATGs induced apoptosis in an anti-CD95 sensitive but not an corresponding anti-CD95 resistent cell line. This finding demonstrates that the anti-CD95 antibodies are functionally active. The rapid onset of apoptosis in cell lines, activated and even resting T cells makes the protein synthesis dependent involvement of additional mediators unlikely but suggests the direct activation of CD95. Although activated T cells exhibit increased ATG sensitivity also resting T cells are susceptible to apoptosis under therapeutic ATG concentrations. Two different flow cytometric assays were developed for the quantization of ATG-induced apoptosis, i.e. the fragmentation of DNA and the specific change of the phospholipid bilayer composition. DNA fragmentation is vastly superior in terms of sensitivity and intraassay precision. The relevance of these assays for the activity assessment of ATGs is highlighted by the inferiority of preparations with low apoptosis-inducing capacity in a primate skin graft model (allogeneic skin grafting on Rhesus macaques). ATGs are potent and direct inducers of apoptosis. Since the anti-apoptotic activity in vitro correlates with graft survival in vivo studies of the immunosuppressive mechanism must take this pathway into account.

189

Role of MDM2 in the survival effects of BCR-ABL and IL-3 in hematopoietic cells

A.W. Götz, H. van der Kuip, M. Oren, W.E. Aulitzky (Stuttgart, D; Rehovot, IL)

The p53/Mdm2-pathway plays an important role in the induction of cell cycle arrest or apoptosis in response to genotoxic stress. Both the oncogene BCR-ABL and physiological growth factors such as IL-3 can modulate the outcome of cellular exposure to DNA damage. To determine whether BCR-ABL and growth factors can affect the p53/Mdm2-pathway, we studied the expression of Mdm2 in the IL-3 dependent pre-B cell line BaF3 and its BCR-ABL-transfected derivative BaF3p185 after IL-3 deprivation or treatment with the c-Abl tyrosine kinase inhibitor STI571. We found that both growth factor withdrawal and inhibition of BCR-ABL kinase lead to a rapid downregulation of Mdm2. The early decrease of Mdm2 protein was not due to transcriptional regulation or to caspase mediated cleavage. On the other hand, it could be completely blocked by the proteasomal inhibitor lactacystin. Targeted downregulation of Mdm2 protein by antisense oligodeoxinucleotides overcame the survival effects of IL-3 and BCR-ABL and resulted in accelerated apoptosis. Taken together, survival signals provided either by physiological growth factors or by oncogenic BCR-ABL can positively regulate Mdm2, while Mdm2 ablation can reduce cell survival. These findings imply that, similarly to physiological growth factors such as IL-3, BCR-ABL can promote cell survival through modulating the p53-Mdm2 pathway.

190

Cyclin D1 locus alterations and Rb deletions but not p16INK4A promoter hypermethylation are associated with an inferior outcome in multiple myeloma

A. Krämer, B. Schultheis, A. Willer, K. Neben, H. Goldschmidt, R. Hehlmann, A.D. Ho (Heidelberg, Mannheim, D)

The molecular mechanisms underlying the pathogenesis of multiple myeloma (MM) are poorly understood. The retinoblastoma protein (pRb), p16INK4A, D-type cyclins, and their partners cyclin-dependent kinase (CDK) 4 and 6 constitute a G1 regulatory pathway commonly targeted in tumorigenesis. Recent studies have described hemizygous deletions of Rb in 50 to 70% of patients with MM. Also, biallelic deletions of Rb with loss of pRb expression have been reported. Chromosomal translocation of cyclin D1 by a t(11;14)(q13;q32) has been found in 5-10% of MM patients. No deletions or mutations of p16INK4A could be detected so far. Data on inactivation of p16INK4A by 5' CpG island hypermethylation in MM have been controversial with frequencies between 0 and 75%. Several malignancies show a reciprocal correlation between genetic alterations of single members of the pRb pathway, namely Rb or p16INK4A inactivation and cyclin D1 or CDK4 overexpression. Therefore, we determined the frequency of Rb deletions and cyclin D1 alterations by fluorescence in situ hybridization as well as 5' CpG island hypermethylation of the p16INK4A gene using methylation-specific polymerase chain reaction in bone marrow mononuclear cells from three individuals with monoclonal gammopathy of undetermined significance, 73 patients with MM and six patients with plasma cell leukemia. Alterations in at least one of the components of the pathway were found in 75%. Cyclin D1 translocations or amplifications were detected in 14/82 (17.1%), Rb deletions at 13q14 in 23/82 (28%) of the cases, including 3 (3.6%) homozygous deletions. p16INK4A was hypermethylated in 33/57 (57.9%) of the samples. Further analysis revealed a highly significant correlation between cyclin D1 gene locus alteration and extramedullar or leukemic myeloma manifestation (P=0.014; Fisher's test). Whereas Rb deletions seemed to occur alternatively to cyclin D1 alterations, no reciprocal correlation was found between p16INK4A hypermethylations and cyclin D1 or Rb locus aberrations. Seven out of eight patients with plasma cell leukemia or MM with extramedullary manifestations harbored either a cyclin D1 alteration or a Rb deletion. Of particular interest, cyclin D1 locus alterations and Rb deletions were associated with a significantly worse prognosis whereas p16INK4A hypermethylation had no impact on survival. We conclude that cyclin D1 and Rb gene locus aberrations seem to occur as alternative events in MM cell cycle deregulation and contribute to clinical course and prognosis of plasma cell malignancies. In contrast, although p16INK4A 5' CpG island hypermethylation is frequent in plasma cell disorders, inactivation of p16INK4A seems not to be involved in the pathogenesis of MM.

191

Cooperative signaling between BCR-ABL and integrin is required for maximum protection against apoptosis in BCR-ABL positive BaF3 cells

H. van der Kuip, C. Miething, J. Duyster, W.E. Aulitzky (Stuttgart, München, D)

The phenotype of BCR-ABL transformed cells is characterized by a growth factor independent survival and a reduced susceptibility to apoptosis. Furthermore BCR-ABL kinase influences adhesion features by phosphorylating cytoskeletal and/or signaling proteins important for integrin function. Integrin mediated adhesion to extracellular matrix (ECM) molecules is critical for the regulation of growth and apoptosis. However, effects of a possible cross-talk between BCR-ABL and integrin-signaling on regulation of apoptosis are largely unknown. We investigated the influence of adhesion on survival and apoptosis in BCR-ABL positive and negative BaF3 cells. BCR-ABL-transfected BaF3 cells (BaF3p185) adherent to immobilized fibronectin had a significant survival advantage and reduced susceptibility to apoptosis following gamma irradiation when compared to the same cells grown on laminin, polylysin, or in suspension. Both inhibition of BCR-ABL kinase by STI571 and inhibition of specific adhesion reversed this fibronectin-mediated antiapoptotic effect in BaF3p185. The DNA damage response of BCR-ABL negative BaF3 cells was not affected by adhesion to fibronectin. In contrast to parental BaF3 cells, BaF3p185 adherent to fibronectin did not release cytochrome c to the cytosol following irradiation. The fibronectin-mediated antiapoptotic mechanism in BCR-ABL active cells was independent on Bcl-XL or Bcl-2 expression, but required an active phosphatidylinositol 3-kinase (PI-3K). BCR-ABL in combination with fibronectin-induced integrin-signaling led to a hyperphosphorylation of AKT.

Thus, cooperative activation of PI-3K/AKT by BCR-ABL and integrins causes synergistic protection of BCR-ABL positive cells from DNA damage induced apoptosis.

192

Inhibitors of PARP influence a delayed cell death in leukemic cells via at least three different mechanisms

A. Pogrebniak, M. Hasmann, I. Schemainda, R. Pelka-Fleischer, H. Schmetzer, V. Nuessler (München, D)

Poly(ADP-Ribose)-polymerase (PARP) is a nuclear enzyme, that catalyzes the breakdown of NAD+, accompanied by transfer of poly-ADP-ribose moieties on PARP itself and on nuclear proteins such as histones. Our aim was to determine, whether pharmacological inhibition of PARP by 3-aminobenzamide or 4-amino-1,8-naphtalimide interferes with cell death induced by alkylating substances such as 1-methyl-3-nitro-1-nitrosoguanidinium (MNNG) and melphalan. WST-1-reduction was markedly decreased already after incubation of HL60 and CCRF-CEM cells with 10 µg/ml (w/v) MNNG for 30min and remained almost undetectable until at least 70% cell death within first 12 hours after commencement of treatment occurred. This event was accompanied with strong depletion of intracellular NAD+, NADH, reduced glutathione and ATP and increased activity of PARP. Alkaline comet assay showed the appearance of single strand breaks in nearly 100% of cells, treated with MNNG in both cell lines. The caspase-3 dependent cleavage of PARP, oligonucleosomal (180 bp) degradation of DNA and fragmentation of nuclei could however be observed in HL60, but not in CCRF-CEM cells. Pretreatment of CCRF-CEM and HL60 cells with 2mM 3-Aminobenzamide (ABA) or 50µM 4-Amino-1,8-naphtalimide (ANI) resulted in significant cell protection, that was surprisingly associated with switch from necrosis to apoptosis in CCRF-CEM cells and enhanced apoptosis in HL60 cells, accompanied by PARP-cleavage and appearance of apoptotic bodies. PARP-inhibitors significantly delayed the drop of WST-1 reduction and of intracellular NAD+, NADH and ATP levels and prevented further DNA damage at least in CCRF-CEM cells, as shown by comet assay.
After 6 hours of treatment with 100 µM melphalan the cells showed typical morphological and biochemical features of apoptosis. In contrast with MNNG-induced cell death, we could not observe an early activation of PARP, decrease of WST-1 reduction and subsequent wasteful consumption of NAD+, NADH and ATP. In addition, melphalan failed to cause DNA strand breaks and the degradation of DNA occurred simultaeously with progression of apoptosis in both cell lines. Incubation of HL60 cells with the same concentrations of ABA or ANI 1h before starting treatment with melphalan resulted in increased cellular survival, associated with decreased formation of apoptotic bodies. The same effect was caused by Cytochalasin B, an inhibitor of actin polymerization. The time course of apoptosis and the grade of apoptotic degradation of DNA remained unaltered. Taken together, protective effects of PARP inhibitors on cell death induced by alkylating agents in CCRF-CEM and HL60 cells, are due to its ability to retain the intracellular NAD+, NADH and ATP levels , to inhibit the endonucleolytic degradation of DNA and to prevent the polymerisation of actin.

193

Activation of caspases is involved in arsenic trioxide (ATO)-induced apoptosis in myeloid and non-myeloid malignant cell lines

M. Rojewski, S. Körper, E. Thiel, H. Schrezenmeier (Berlin, D)

Arsenic trioxide (ATO) has been used successfully for treatment of refractory or relapsed acute promyelocytic leukemia (remission rate >80%). Lately, we demonstrated induction of apoptosis and reduction of proliferation rate by ATO in myeloid and non-myeloid malignant cell lines (including MDR-resistant cell lines). To elucidate the molecular events we analyzed caspase activation, mitochondrial activity and p53 regulation. Caspase activation was measured by monoclonal antibodies binding only to activated caspase3, cleavage site specific antibodies for PARP and D4-GDI and fluorogenic caspase substrates and inhibitors (PhiPhiLuxG2D2, FITC-VAD-fmk and FITC-YVADAPK-Dnp). The percentage of cells that showed caspase activation as measured by different markers after 12 hours incubation in 10µM ATO is shown in Table 1. The activation of caspases is independent from constitutive CD95 and CD90 expression. Staining of cells with the cell-permeable dye MitoTrackerCMX-Ros showed reduced mitochondrial activity after incubation with 10µM ATO. Depending on the cell line, incubation time necessary to obtain mitochondrial effects was 2 –24 hours. No difference between the MDR-expressing cell lines K-562 and KG-1a could be observed in any parameter analyzed compared to non MDR-cell lines. ATO-induced regulation of p53 was heterogeneous. Upregulation of p53 by 10µM ATO could be seen in the ATO-sensitive cell line LOUCY after 2 hours of incubation, whereas 6 hours of incubation were necessary for ATO-sensitive cell lines 697 and CEM. No p53 regulation could be seen for the less ATO sensitive cell lines HL-60 and Jurkat as well as MDR-resistant cell lines CEM/C1, CEM/C2, HL-60/MX1, HL-60/MX2, K-562(0.02), K-562(0.1) and K-562 and KG-1a after up to 24 hours of incubation with 10µM ATO.
We could demonstrate that ATO induces apoptosis in a substantial proportion of cells and leads to activation of caspases, in particular caspase3, and to reduction of mitochondrial activity. This effect could be observed also in MDR-resistant cell lines. In addition, we could demonstrate ATO-induced upregulation of p53 which was absent in ATO-resistant cell lines. These data demonstrate that ATO-induced apoptosis is not restricted to cells with the translocation t(15;17). Since ATO both induces apoptosis and inhibits proliferation also in MDR-resistant cells, clinical trials with ATO in AML refractory to conventional treatment should be undertaken.

Table 1

cell line	Procaspase3 cleavage	PhiPhiLux cleavage	D4-GDI cleavage	PARP cleavage	VAD binding	YVADAPK binding
697	38	74	38	39	N.D.	N.D.
CEM	50	78	82	28	76	48
CEM/C2	38	38	34	26	54	17
HL-60	34	47	55	49	29	2
HL-60/MX2	2	2	34	4	N.D.	N.D.
Jurkat	61	51	74	10	N.D.	N.D.
K-562	1	3	0	1	14	4
KG-1a	1	1	2	2	2	0
LOUCY	45	48	37	22	51	45

194

Effects of antithymocyte globulin in myeloid cells: induction of apoptosis, activation of caspases and inhibition of proliferation rate

H. Schrezenmeier, T. Mönich, M. Rojewski, E. Thiel (Berlin, D)

Recently we could demonstrate induction of apoptosis in myeloid cells by antilymphocyte globulin and antithymocyte globulin (ATG). ATG contains a number of different antibody specificities with variations between different brands. The effective components of ATG for treatment of aplastic anemia (AA) and myelodysplastic syndrome (MDS) are still unknown. The mechanism(s) of ATG-induced apoptosis and its role for response of AA and MDS to ATG remains unclear. We analyzed regulation of apoptosis and proliferation for various brands of ATG (Lymphoglobulin, Thymoglobulin [IMTIX SangStat], ATG-Fresenius S [Fresenius AG], Tecelac [Biotest]). KG-1a cells were incubated with ATG at concentrations from 1 µg/ml to 1000 µg/ml for up to 10 days. Apoptosis was measured by staining with 7-aminoactinomycin D (7-AAD). At concentrations of 10 µg/ml none of the tested ATG brands was able to induce apoptosis. However, at concentrations $\geq$ 100 µg/ml ATG induced apoptosis. Proportion of apoptotic cells after incubation with 100 µg/ml for 7 days was 36%, 25%, 37%, 87% with Lymphoglobulin, Thymogloblin, ATG-Fresenius S, Tecelac, respectively. Significant ATG-induced apoptosis was only seen after $\geq$ 2 days of incubation. Binding of the fluorochrome labelled caspase binding peptide FITC-VAD-fmk could be seen in more than 40%, 35%, 30% of KG-1a cells after incubation for 4 days with 1000 µg/ml Tecelac, ATG-Fresenius S or Thymoglobulin, respectively. However, incubation with 1000 µg/ml Lymphoglobulin did not result in binding of the specific peptide to activated caspases after incubation for up to 7 days. Proliferation assays were performed using the stable membrane dye PKH67. After 10 days incubation, lower decrease in PKH67 intensity could be observed for samples with 100 µg/ml Tecelac and Lymphoglobulin as compared to 10 µg/ml Tecelac and Lymphoglobulin, or control cells. In summary, this study shows that ATG induced apoptosis in the myeloid cell line KG-1a involves activation of caspases. In addition to induction of apoptosis, ATG has an antiproliferative effect.

195

Implication of reactive oxygen species (ROS) in doxorubicin induced apoptosis in acute leukemia

L. Karawajew, C. Wuchter, G. Czerwony, B. Dörken, W.-D. Ludwig (Berlin, D)

Anthracyclines, e.g. doxorubicin (DOXO), are widely used for treatment of acute leukemia, however, molecular mechanisms underlying their cytotoxic activity are incompletely identified. Here we addressed implication of ROS which have been controversially reported in the apoptosis induced by different stimuli including DNA damage. In an acute leukemia cell line MOLT-3, DOXO was found to cause apoptosis which could be completely inhibited by an antioxidant N-acetylcysteine (NAC). Mutational and expressional analyses revealed wild type p53 gene in MOLT-3 cells and an accumulation of the p53 protein after DOXO treatment. Kinetic studies revealed that NAC was only effective when added prior (15 min) to the addition of DOXO. Moreover, NAC did not inhibit apoptosis or has become even cytotoxic when added shortly (15-30 min) after addition of DOXO, thus indicating ROS involvement at the early, initiating phase of the DOXO-induced pro-apoptotic signaling. The accumulation of p53 was observed considerably later (>3 hours after DOXO addition) and was accompanied by the loss of mitochondrial membrane potential, strong increase of intracellular ROS levels as well as externalization of phosphatidylserine, i.e. events characteristic of the late, execution phase of apoptosis. DOXO-induced ROS-dependent apoptosis which could be completely prevented by pre-incubation with NAC has been also observed in primary acute leukemia cells. Interestingly, pre-treatment of MOLT-3 and primary leukemic stem cells with sublethal doses of H2O2 conferred resistance to DOXO-induced apoptosis, thus suggesting that exogenous oxidative stress might reduce chemotherapeutic efficacy of DOXO. These data provide evidence for the existence and clinical relevance of the DOXO-induced ROS-mediated apoptotic pathway in acute leukemia.

196

Reversible upregulation of the cyclin dependent kinase inhibitor p21/Cip1 in a myeloid cell line by BCR-ABL

K. Forster, N. Simon, C. Schuster, T. Schröder, U. Just, M. Hallek, M. Warmuth (München, D)

Chronic myeloid leukemia is a myeloproliferative disorder associated with the clonal expansion of a tranformed hematopoietic stem cell. It is caused in the majority of cases with the Philadelphia chromosome, a translocation between chromosome 9 and 22 that fuses the bcr and abl genes. Compared to the cellular ABL kinase, BCR-ABL shows increased tyrosine kinase activity and it is located in the cytoplasm only, whereas Abl is found also in the nucleus. These alterations enable the BCR-ABL protein to interact with various cellular proteins and signalling pathways. Amongst the major cellular effects of BCR-ABL are for example increased mitogenic activity, reduced sensitivity to apoptosis, and altered adhesion to extracellular matrices. This study was designed to investigate the influence of BCR-ABL on the expression of the cyclin dependent kinase inhibitors p16INK4a, p21 Cip1 and p27 Kip1. As a model system we used the murine myeloid cell line 32D. Western blots of parental 32D cells and 32D cells stably expressing BCR-ABL (32Dp210) showed that these cells express considerable levels of p21 and p27, but no detectable amounts of p16. We examined if BCR-ABL influences p21. Interestingly, while p27 was found to be mildly downregulted in 32Dp210 cells, p21 was upregulated in 32D cells expressing BCR-ABL, when compared to parental 32D cells. Moreover, in 32Dp210 cells p21 expression was not influenced by interleukin-3, whereas p21 was found downregulated by interleukin-3 withdrawal in 32D cells. Addition of kinase inhibitor STI571 to BCR-ABL expressing 32D cells lead to a significant downregulation of p21 to levels found in IL-3 starved parental 32D cells. Downregulation of p21 was associated with upcoming apoptosis. In order to verify that p21 expression is directly influenced by BCR-ABL, we produced a cell line (32DTp210) expressing BCR-ABL under the control of a doxycycline (dox)-inducible promoter. This cells grew factor independent in the absence of dox, with similar growth rates than stably expressing 32Dp210 cells. Addition of dox lead to downregulation of BCR-ABL-expression, complete stop in proliferation and induction of apoptosis in up to 95% of the cells due to the induction of factor dependent growth, showing that the system is appropriately regulated. Using this system we found that downregulation of BCR-ABL expression is closely linked to downregulation of p21 expression. The results suggest that p21 is a direct target gene of BCR-ABL. Future research will focus on identifying the signalling pathways leading to p21 expression as well as the possible functions of p21 in BCR-ABL induced leukemogenesis.

197

Effect of arsenic trioxide (ATO) and glycoprotein-P-antagonists on apoptosis and mitochondrial membrane potential in a cell line without expression of glycoprotein-P (gP)

S. Körper, M. Rojewski, E. Thiel, H. Schrezenmeier (Berlin, D)

ATO was successfully used to treat APL. Recently we have shown that ATO is a potent inductor of apoptosis in a wide variety of cell lines. Ciclosporine A (CsA) and verapamil (VER) are well known to overcome multidrug resistance by blocking glycoprotein-P (gP). Nifedipine (NIF) is a Ca-channel blocker, wich has been shown to have protective effects against apotosis. We tried to modulate ATO effects with VER, NIF and CsA in cell lines. Cell line HL-60 was induced to apoptosis with ATO concentrations from 1 to 10μM for 24h and 96h. CsA (50 – 5000μg/l), VER (10 and 100μM) and NIF (30 and 300μM) were tested alone and in combination with various concentrations of ATO. Apoptosis was measured by flow cytometry using AnnexinV, 7-aminoactinomycin-D (7-AAD) or JC-1 as an indicator of depolarisation of the mitochondrial membrane potential (PSIm). The expression of gP was evaluated by flow cytometry using a FITC-conjugated mouse-anti-gP-IgG. Results are given here for JC-1 (24h) or AnnexinV (96h). After 24h incubation with ATO 13.6 $\pm$ 7.4% (5μM) and 27.5 $\pm$ 16.0% (10μM) (control: 9.2 $\pm$ 6.7%) were induced to apoptosis. 50 – 5000μg/l CsA, 10μM VER and 30μM NIF induced no apoptosis during this time. 300μM NIF alone showed marked apotosis. Coincubation of As203 with 5mg/l CsA and 10μM VER and NIF showed an increase in apoptosis induced by 5μM ATO from 12.1 $\pm$ 5.4 (10μM VER) up to 52.7 $\pm$ 4.0% (5mg/l CsA). 250μg/l CsA enhanced the apoptosis induced by 10μM ATO from 24.8 $\pm$ 10.6% to 32.6 $\pm$ 5.5%. The other agents also enhanced apoptosis induced by 10μM ATO. After 96h incubation, 1μM ATO, 1mg/l CsA and 10μM VER showed no enhanced apoptosis when compared to the control (7.1 $\pm$ 4.3%), whereas 5mg/l CsA induced marked apoptosis (39.5 $\pm$8.4%) itself. Coincuabtion lead to enhanced apotosis: 1mg/l CsA: 28.5 $\pm$5.2%, 5mg/l CsA: 93.6$\pm$ 6.3% and 10μM VER: 12.1 $\pm$ 6.6%. The populations which survived after 24h incubation with ATO showed a significant increase of the mean fluorescence in FL2 (JC-1). This indicates that the surviving cells have a higher PSIm than the control population. We have clearly shown that that the gP antagonists CsA and VER anhance apoptosis-induction by ATO in cells lacking gP. Thus, it is likely that there is another drug resistance mechanisms than gP in these cell lines. NIF can induce apoptosis in a AML cell line at high concentrations and enhances also the effect of ATO. One might speculate that capacitive Ca-currents or internal Ca-storages have a function in the apoptosis induction by ATO. It is thought that one mechanism of ATO is the opening of the transitition pore of the inner mitochondrial membrane. The surviving population after ATO treatment have a higher PSIm. Thus we suggest that cells with a higher PSIm are more resistant against ATO.

198

Induction of apoptosis in malignant lymphoid and myeloid cells by trivalent antimonials

H. Schrezenmeier, S. Lösler, M. Rojewski, S. Körper, E. Thiel (Berlin, D)

Arsenic trioxide (As_2O_3) is an emerging treatment of several malignant hematological disorders. Arsenic and antimony - both elements of group V of the periodic table - are chemically similar. Trivalent antimonials (e.g. Sb_2O_3) might interfere with apoptosis-related processes in a similar manner to As_2O_3. This prompted the analysis of effects of Sb_2O_3 on apoptosis in malignant cell lines. Several cell lines (HL-60, K-562, LOUCY, CEM and chemotherapy-resistant derivatives of HL-60 and CEM) were incubated with As_2O_3 (1 and 5μM) or Sb_2O_3 (1 and 5μM) for up to 35 days. The selected cell lines represent different levels of sensitivity to As_2O_3 -induced apoptosis: low (K-562), intermediate (CEM, HL-60) and high sensitivity (LOUCY). Percentage of apoptotic cells was measured by Annexin-V-FITC and 7-amino-actinomycin D (7-AAD). Sb203 (1 and 5μM) did not induce apoptosis in K-562, which is sensitive to 5 μM, but resistant to < 1μM As_2O_3. In the cell line CEM, which became apoptotic after incubation with As_2O_3 (1 and 5μM), Sb_2O_3 did not induce apoptosis at 1μM, but induced apoptosis in about 50 % of cells after > 4 days incubation with 5μM Sb_2O_3. It took about 2 - 4 days longer to reach the same level of apoptosis with Sb_2O_3 compared to As_2O_3 (both at 5μM). A similar pattern was observed with HL-60 and the highly As_2O_3-sensitive cell line LOUCY. Apoptosis induction was lost in mitoxantrone-resistant HL-60 cells. However, camptothecin-resistant CEM cells were still sensitive to apoptosis induction by Sb_2O_3. Sensitivity to both As_2O_3 and Sb_2O_3 can be increased by buthionine sulfoxid which decreases glutathion levels. In summary, Sb_2O_3 can induce apoptosis in malignant cells in-vitro. Apoptosis-inducing activity of Sb_2O_3 is lower than that of As_2O_3 compared on a molar basis. Sb_2O_3 can be effective in cells resistant to cytostatic substances. Given the rapidly expanding clinical use of As_2O_3, also Sb_2O_3 should be explored further as alternative or additive treatment to As_2O_3.

199

Novel therapeutic approach: The combination of pioglitazone and all-trans-retinoic acid induces specific apoptosis of cancer cells

J. Fritz, M. Wächter, M. Fenner, C. Zange, J. Reichardt, M. Becker, A. von Deimling, K. Possinger, E. Elstner, H.P. Koeffler (Berlin, D; Los Angeles, USA)

A new class of antidiabetic drugs termed thiazolidinedione (TZD) has been developed over the last decade and now is widely used for the treatment of insulin-resistant diabetes mellitus. It reduces both insulin resistance and elevated plasma glucose in these patients. The family of TZDs are ligands for peroxisome proliferator-activated receptor gamma (PPARg). Data from us and other groups have shown that TZD, such as troglitazone (TGZ), can inhibit the proliferation of a variety of cancer cells. Pioglitazone (PGZ) is another member of TZD with no evidence of drug-induced hepatoxicity and is a potent and highly selective agonist for PPARg. We studied the effects of this PPARg ligand on the proliferation and apoptosis of breast cancer cell lines (MCF7 and MDA-MB-231) as well as glioblastoma cell lines (N39 and U373). PGZ (10E-5M) alone inhibited the in vitro clonogenic proliferation of these cancer cell lines (30-47% of control). Pulse exposure (6 hrs.) of the cells to PGZ (10E-5M) had only a minimal effect. The retinoid, all-trans retinoic acid (ATRA, 10E-6M) inhibited the clonal growth of these cell lines by about 30-40% in vitro. Furthermore, the combination of PGZ (10E-5M) and ATRA (10E-3M) had an additive effect on the inhibition of the clonal proliferation of all the cell lines tested. Moreover, incubation for 4 days with the combination of PGZ (10E-5M) and ATRA (10E-6M) induced morphological changes associated with apoptosis in about 60% of MCF7, MDA-MB-231 and N39 cells. Taken together, our data demonstrated that PGZ alone reversibly inhibited the proliferation of breast and glioblastoma cancer cells without significant apoptosis. In contrast, combination of PGZ and ATRA additively inhibited the proliferation as well as induced apoptosis in about 40-60% of the MCF7, MDA-MB-231 and N39 cells. In summary, PGZ and ATRA, when used together, may have an adjunctive role in the therapies of breast cancer and glioblastoma multiforme. Therefore, we propose a phase I/II trial of PGZ (ACTOS) and ATRA (Vesanoid) in the treatment of metastatic breast cancer.

200

Functional role of prostate-apoptosis-response-gene-4 in apoptosis of neoplastic lymphocytes

S. Boehrer, K. Chow, F. Beske, E. Puccetti, M. Ruthardt, V. Rangnekar, D. Hoelzer, P. Mitrou, E. Weidmann (Frankfurt, D; Lexington, USA)

Inhibition of apoptosis is a hallmark of malignancies of the hematopoetic system. Previous studies in non-hematopoetic cells demonstrated that Prostate-aptosis-response-gene-4 (Par-4) is upregulated in cells undergoing cell death. We previously demonstrated the aberrant expressional pattern of Par-4 in neoplastic lymphocytes, as well as inverse expressional patterns of Par-4 and Bcl-2 in malignant cells of patients suffering from acute lymphatic leukemia. We here assessed functional consequences of Par-4 overexpression during apoptosis in Jurkat T-lymphocytes. We thus show that in lymphatic cells Par-4 overexpression decreases the level the level of Bcl-2, whereas Bax, the proapoptotic counterpart of Bcl-2, retains unaltered levels. Moreover, Par-4 overexpression is accompanied by cleavage of Poly-(ADP-ribose)-Polymerase (PARP). Despite these effects overexpression of Par-4 alone is not sufficient to induce apoptosis, but markedly increases the rate of apoptosis upon treatment with different chemotherapeutic agents. Upon chemotherapeutic treatment Par-4 overexpression enhances disruption of mitochondrial membrane potential, PARP-cleaving activity, as well as activation of caspase-3. The hypothesis of caspase-dependency of Par-4-promoted apoptosis is further supported by demonstrating complete abrogation of programmed cell death after pretreatment with a broad spectrum caspase-inhibitor. Upon incubation with a specific caspase-3-inhibitor, PARP-cleaving activity was completely abrogated in Par-4 negative cells, but maintained in Par-4 overexpressing cells. Our study is the first to identify Par-4 as a proapoptotic protein in lymphatic cells, and suggests a model of action evaluating the roll of Bcl-2 and Bax, as well as demonstrating the impact of Par-4 overexpression on PARP-cleavage, disruption of mitochondrial membrane potential and caspase activation.

201

The role of caspase-8 as a death-receptor-independent downstream amplifier in cytostatic drug-induced apoptosis in B lymphoid cells

F. Essmann, T. Wieder, K. Schulze-Osthoff, B. Dörken, P.T. Daniel (Berlin, Münster, D)

Caspase-8 has been shown previously to act as an initiator caspase in the signal transduction of death receptor-triggered apoptosis, i.e. by CD95/Fas, the TRAIl receptors and TNF-R1. We recently observed that Caspase-8 is activated during drug-induced apoptosis upon treatment with etoposide, anthracyclines and taxanes. A careful analysis of the CD95/Fas ligand, receptor and signalling pathway in B demonstrated that cytostatic drug-induced apoptosis in both B lineage ALL cells and B lymphoid cell lines is independent from CD95 Fas ligand/receptor binding and FADD signalling. Caspase-8 was shown to occur downstream of Caspase-3 (Blood 2001,97:1378-1387) and subsequent to mitochondrial activation. In the present work we investigate the functional role of Caspase-8 as a downstream amplificator and executioner caspase in drug-induced apoptosis. Blocking of Caspase-8 activity by zIETD-fmk inhibited drug-induced apoptosis which formally shows that this caspase is functionally relevant in the drug response. The same experiment was performed in cells over-expressing a dominant negative FADD mutant (FADD-dn) with similar results. Thus, the Caspase-8 activation following exposure to cytotoxic drugs is independent from death receptor signaling. This mechanism may not only be relevant for the co-activation of additional executioner caspases or other proteases but also for the degradation of regulatory substrates. One of these, Bid, may be cleaved to tBid by the active Caspase-8 which provides a positive feedback amplification loop for the activation of the mitochondrial apoptosis signalling pathway.

202

Change of histological and apoptotic patterns in the small intestine of rats after CPT-11 treatment

W. Voigt, M.Fittkau, P. Kopinski, H. Simon, H.-J. Holzhausen, H.-J. Schmoll (Halle/Saale, D)

Induction of severe delayed diarrhea still limits the therapeutical usefullness of the DNA topoisomerase I inhibitor irinotecan (CPT-11) in colorectal cancer. Symptoms like dehydration or electrolyte imbalance are treated by the use of loperamide or topical steroids. However, since the underlying mechanism of CPT-11 induced delayed diarrhea is not clearly understood yet, no sufficient preventive therapy is available so far. In order to further clarify the nature of CPT-11 induced delayed diarrhea we analyzed the histological changes and changes in the apoptotic pattern in the small intestine of rats treated with 150–350 mg/m^2 i.p. for 4 days. At the onset of treatment induced diarrhea animals were sacrified. Then, sections of the small intestine were fixed in formalin and embedded in paraffin. After dewaxing in xylene sections were rinsed in graded ethanol. For morphological analysis slides were stained with hematoxylin and the tunel assay was used for detection of apoptotic cells. Complementary, we analyzed tissue sections for overexpression of the active form of caspase 3 as a marker of apoptosis. Overall, we observed regular villi and crypt structure in control sections and a marked loss of villi in the small intestine of animals treated with different doses CPT-11. In addition we demonstrated a high amount of apoptotic cells on the villi surface in control sections and hardly any apoptotic cell in the crypts of control sections. However, in CPT-11 treated sections beside flattening of villi there were nearly no apoptotic mucosa cells left on the villi whereas the amount of apoptotic cells in the crypt was significantly increased as compared to control sections. Thus, in conclusion the onset of CPT-11 treatment induced diarrhea in rats is associated with marked histopathological changes like loss of villi in the small intestine. Additionally, we observed a significant increase of apoptotic cells in the basal crypt region in the small intestine of rats subsequent to the exposure with CPT-11.

203

Human growth hormone delays apoptosis and increases superoxide production in neutrophils

W. Springer, D. Decker, S. Daufeldt, I. Schmidt-Wolf, A. von Ruecker (Bonn, D)

Recombinant human growth hormone (hGH) has been used successfully to reverse the high rate of catabolism observed either after an operation or during serious illness. In spite of this advantage, a recent study has shown that treatment with hGH causes a high mortality rate in critical illness. The reason for this phenomenon is still unclear. In a randomised double-blind placebo-controlled study we show that high dose hGH treatment leads to anti-apoptosis and the enhancement of superoxide production in polymorphonuclear neutrophils (PMN) of patients after elective major vascular surgery. High-dose hGH doubled the superoxide production of PMN induced by opsonised zymosan, which still persisted 24 hours after surgery, and elevated the long-term capacity of PMN to generate superoxide. hGH treatment also reduced spontaneous apoptosis in PMN as shown by flow cytometric analysis and lowered the sensitivity of PMN towards Fas-mediated apoptosis prior to or after surgery. These findings suggest that high-dose hGH distinctly prolongs the life span of PMN and leads to a dysfunctional high release of superoxide. The presence of activated and highly apoptosis-resistant cells during a systemic inflammatory response (i.e. sepsis) may contribute to the injury characteristic of systemic inflammatory response syndrome (SIRS) and predispose to the development of multiple-organ failure.

204

Piceatannol, a hydroxylated analog of the chemopreventive agent resveratrol, is a potent inducer of apoptosis in the lymphoma cell line BJAB and in primary, leukemic lymphoblasts

T. Wieder, A. Prokop, B. Bagci, F. Essmann, D. Bernicke, K. Schulze-Osthoff, B. Dörken, H.-G. Schmalz, P.T. Daniel, G. Henze (Berlin, Köln, Münster, D)

The stilbene phytochemicals resveratrol and piceatannol have been reported to possess substantial antitumorigenic and antileukemic activities, respectively. Although recent experimental data revealed the proapoptotic potency of resveratrol, the molecular mechanisms underlying the antileukemic activity have not been studied in detail yet. In the present study, we show that resveratrol as well as the hydroxylated analog piceatannol are potent inducers of apoptotic cell death in BJAB Burkitt-like lymphoma cells with an ED50 concentration of 25 µM. Further experiments revealed that treatment of BJAB cells with both substances led to a concentration-dependent activation of caspase-3 and mitochondrial permeability transition. Using BJAB cells overexpressing a dominant-negative mutant of the Fas-associated death domain (FADD) adaptor protein to block death receptor-mediated apoptosis, we demonstrate that resveratrol- and piceatannol-induced cell death in these cells is independent of the CD95/Fas signaling pathway. To explore the antileukemic properties of both compounds in more detail, we extended our study to primary, leukemic lymphoblasts. Interestingly, piceatannol but not resveratrol is a very efficient inducer of apoptosis in this ex vivo assay with leukemic lymphoblasts of 21 patients suffering from childhood lymphoblastic leukemia (ALL).

205

Induction of cell death by the BH3-only Bcl-2 homologue Bik/Nbk is mediated by a Bax-dependent mitochondrial pathway

B. Gillissen, F. Essmann, L. Stärck, P.G. Hemmati, C. Stroh, S. Radetzki, K. Schulze-Osthoff, B. Dörken, P.T. Daniel (Berlin, Münster, D)

To investigate the mechanism of action of the BH3-only Bcl-2 homolog Nbk and to explore the use of Nbk for induction of cell death in tumors, we constructed a single-vector adenoviral expression system based on the Tet-off system. The E3 region of the Ad5 vector was replaced with the tTA-transactivator for the regulation of the myc-tagged Nbk cDNA in place of the deleted E1-region to produce the Ad5-mycNbk-tTA adenoviral vector for the conditional expression of Nbk. Transduction of carcinoma cells with this vector showed that Bax-positive (SW48, SW480) but not Bax-mutated and therefore negative cells (DU145 and LoVo) were killed by apoptosis upon adenoviral Nbk expression. Overexpression of wild-type Bax in DU145 restored sensitivity to Nbk. Adenoviral Nbk overexpression induces a conformational switch in the Bax N-terminus coinciding with mitochondrial permeability shift transition. Co-localization studies show that Nbk induces redistribution of Bax from to the mitochondria, and disrupts Bcl-x/Bax co-localization. Immunoprecipitation studies furthermore demonstrate that Nbk interacts with Bcl-xL but not Bax. We therefore propose a model where Nbk acts as an indirect killer by disrupting apoptosis-inhibition by Bcl-xL leading to a conformational change and activation of Bax to promote mitochondrial activation and apoptosis.

206

Inactivation of the INK4 gene cluster in mantle cell lymphomas - standardization and initial analysis of differential methylation patterns of cell cycle regulators

G. Hutter, Y. Zimmermann, E. Göcke, G. Herzog, J. Lu-Hesselmann, W. Hiddemann, M.H. Dreyling (München, D)

Mantle cell lymphoma (MCL) is characterized by the chromosomal translocation t(11;14)(q13;q32), resulting in the overexpression of cyclin D1. However, additional genetic alterations are detectable in the vast majority of cases. Thus, deletions of the INK4A gene cluster (p15INK4B, p16INK4A, and p16ARF) on chromosomal band 9p21 is one of the most frequently detected secondary alteration in MCL. So far, little is known on the differential effect of the p15INK4B/ p16INK4A-CyclinD1 and the p16ARF-MDM2-p53 signal pathway. The hypermethylation of CPG islands of the promoter region represents an alternative mechanism of gene inactivation. So far, the hierarchy of p15INK4B, p16INK4A, and p16ARF promoter methylations as well as the interaction of different methylation sites within a distinct promoter region is poorly understood. We have established a comprehensive method to exactly determine the promoter methylation pattern of the INK4 gene cluster. Tumor DNA was modified with bisulfite according to previously published methods (Herman et al., PNAS 1996) and amplified with methylation-specific primers constructed based on recently published promoter sequences. Consequently, amplification products were sequenced on a semi-automated personal sequencer unit (ABI 310). In an initial series of 5 hematological cell lines, the methylation pattern of p15INK4B, p16INK4A, and p16ARF could be confirmed and characterized in more detail in all samples analysed. Even more interestingly, we could identify a p16INK4A promoter methylation in a hematological cell line so far associated only with genomic p16INK4A deletion or point mutation (HL60). Currently, we have initiated a complete methylation analysis of 52 MCL which had been previously characterized for other INK4A gene cluster (genomic deletion) and p53 (mutations, expression) alterations to more accurately define the differential impact of p15INK4B, p16INK4A, and p16ARF methylation patterns in mantle cell lymphomas.

Postersitzung: Bronchialkarzinome

207

Phase I/II study of weekly docetaxel (D) and carboplatin (C) in NSCLC IV

F. Griesinger, J. Leugering, M. Metz (Göttingen, D)

Objective: Docetaxel (D) is one of the most effective agents in the treatment of NSCLC with response rates of 30%. Taxane combinations with carboplatin (C) have shown response rates of upt to >50% in NSCLC stage IV. Increase of dose density by weekly application of paclitaxel and C has shown to be effective and to reduce hematologic toxicity. No data exist on the combination of weekly D and C. Methods: 38 Patients with NSCLC stage IV have been enrolled since Oct. 98 to receive D 34-40 mg/m^2 q 7 days and C AUC 2,5-3,5 q 7 days for 6 weeks, to be repeated every 8 weeks for a maximal cycle number of 3. Hematologic and non-hematologic toxicity were primary endpoints, response, survival and quality of life were documented as secondary endpoints. Restaging was performed after each 6 week cycle. Results: 34/38 patients are evaluable for toxicity. 8 pts. received D35/C2.5 (dose level 1) (median number of cycles =1, 1-3), 5 pts D35/C3.0 (median 2 cycles, 1-3), 10 D35/C3.5 (median 1 cycle, 1-3) and 15 D40/C3.0 (median 2 cycles, 1-3). On dose level 1, no dose limiting toxicity was observed. On dose levels 2-4, hematologic toxicity WHO grade 3-4 was observed in 1/5, 1/8 and 1/13 evaluable cases, respectively. Non-hematologic toxicities (arrhythmia and empyema in 1 case, respectively) led to end of study in 2/34 cases. 37/38 patients are evaluable for response after 1 cycle. 15/37 (40%) pts had a PR, 7/37 (20%) pts stable disease (SD) and 15/37 (40%) progressive disease. 9/14 pts (64%) had PR or SD on dose level 4. Conclusion: Weekly application of D and C with a dose density equivalent to 3-weekly D 100 mg/m^2 and AUC 7.5 is feasible and safe without G-CSF-support. Final recommended MTD as well as survival data will be presented at the meeting.

208

Outpatient treatment in advanced non small cell lung cancer patients

S. Haase, H. Köppler, J. Heymanns, R. Weide (Koblenz, D)

Objective: Evaluation of feasibility, effectivity and toxicity in patients with advanced NSCLC, who were treated completely on an outpatient basis. Methods: Retrospective analysis of 97 consecutive, unselected NSCLC-patients with advanced disease treated between 6/95-6/2000 in a community based oncology group practice. Results: The median age of the patients was 62 years (37-83) and their performance status ranged from 0-3. 79 were male and 18 were female. Clinical stage distribution was as follows: 5 stage I (functional inoperability), 3 stage II, 41 stage III (8 IIIA, 33 IIIB), 48 stage IV. 82 patients received cytotoxic treatment (7 radiotherapy only, 44 radiotherapy and chemotherapy, 31 chemotherapy only). 15 patients received no cytotoxic treatment (8 best supportive care, 6 observations, 1 refused therapy). Chemotherapy was performed as monotherapy in 10 patients and polychemotherapy in 65 patients. A total of 1008 chemotherapy cycles were performed during the study period (100 applications 3-weekly, 908 applications weekly). The observed toxicity was mild. A reversible grade 3 and 4 haematotoxicity was seen after 48 cycles (4.8%). Mucotoxicity and neurotoxicity occurred in 14 patients (14.4%). Therapy associated hospitalisation was seen in 4 patients. 49 patients died during the study period (22 at home, 27 in hospital). The median survival of stage IIIB/IV-patients is 18 months (range 1-118). The median survival of stage IIIB is 22 months (range 5-66) and stage IV is 11 months (range 1-118). The median survival of the whole cohort from diagnosis is 21 months (range 1-118). The 1, 2, 3 and 5 year survival is 60.7%, 43.3%, 36.8% and 17.9% respectively. Conclusions: Treatment of patients with advanced NSCLC can be performed with minimal toxicity and a high patient acceptance in the outpatient setting. The median survival in this unselected cohort is high and hospitalisation could be avoided in the majority of patients.

209

Cytologic diagnosis of lung cancer

M. Engels, J. Karow (Düren, D)

Introduction: Although the cytologic examination of respiratory specimens has been successfully used as an established method for more than forty years its value for diagnosing lung cancer is not widely appreciated. In our institution, a municipal hospital, as well as by several practising pulmologists in the neighbourhood, cytology is routinely employed in the work-up of patients with pulmonary disease, especially with suspected lung cancer. Method and Material: This study presents a retrospective survey of the cellular specimens from the lower respiratory tract which were submitted to our laboratory during a period of 12 months (from March 2000 to February 2001). In total, 477 cytologic specimens from the lower respiratory tract were evaluated. 329 samples were obtained by bronchoscopy: bronchial brushings and/or bronchial washings. For 93 of these specimens a histopathologic diagnosis was available for correlation, in the remaining 236 cases no biopsy had been obtained. In addition, 143 sputum specimens and 5 fine needle aspirates of the lung were submitted. Results: During the period of the study 338 specimens were interpreted as benign, 19 as suspect and 43 were non-satisfactory for evaluation. 77 cytologic specimens contained malignant cells. Of the 329 samples obtained by bronchoscopic examination, 55 were interpreted as diagnostic of cancer, 21 of the carcinomas were confirmed by tissue examination. In 5 additional cases, carcinoma was diagnosed only by histologic examination, the cytologic specimen yielded no malignant cells. In 16 cases, the cytologic specimen was diagnostic of cancer whereas the biopsy contained no malignant tissue. In 18 cases in which no biopsy had been obtained malignant cells were revealed by cytology. Of the 143 sputum specimens 21 were interpreted as diagnostic of cancer. Only 5 fine needle aspirates of the lung were submitted, 1 of them containing malignant cells and carcinoma was confirmed by tissue examination in this case. In all kinds of specimens - bonchial brushings und washings, sputum and fine needle aspirates - a clear distinction between small cell lung cancer and non-small cell lung cancer was possible in each case. Conclusion: The cytologic examination of specimens from the lower respiratory tract, especially of those obtained by bronchoscopy, is a valuable method for diagnosing lung cancer. Without inflicting additional risk or discomfort to the patient it provides an increase of information. In several microphotographs typical examples of different pulmonary neoplasms are demonstrated as they present in routine cytology.

210

Gemcitabine and platinum combination therapy in the treatment of malignant mesothelioma

E. Ulsperger, L. Sakr, G. Baumgartner (Wien, A)

Malignant pleural mesothelioma is a rare disease, closely associated to asbest exposure. The naturally chemoresistant tumor has a median survival ranging from 4 to 18 month. We treated in our department 46 malignant mesotheliomas mainly outpatient based with immunomodulating therapy and cytostatic treatment. One of these patient relapsed after palliative resection. Interleukin-2 therapy followed by Vinorelbine and Carboplatin chemotherapy was performed causing progressive disease. Mesothelioma manifestations were located in mediastinal lymphnodes infiltrating pericardium (7cm in diameter) and pleura (4cm in diameter). Clinical symptoms were chest pain, dyspnoea and cough. We treated the patient in the 4th line with 1000mg/m^2 Gemcitabine and 80mg/m^2 Cisplatin (day 1 and 8; 21-days-scheme). After 5 cycles we were able to determine a partial remission with significant reduction of symptoms. This remission was improved and stabilized during the following cycles. We stopped treatment at the 12th cycle of Gemcitabine/Cisplatin combinationtherapy. Mediastinal and thoracic radiation followed as consolidation. The remission lasted for 15 months. Following this impressive good result we treated in our department 12 further mesothelioma patients with Gemcitabine and Platin-combinations. The patients characteristics: male patients only, 2 patients in 1st-, 4 in 2nd-, 1 in 3rd-, 3 in 4th- and 2 in 5th-line therapy. In 9 patients we determined occupational disease with asbest exposure. Median age 58 years (range 49 - 62). At onset of Gemcitabine therapy 9 patients had pleural (including mediastinal lymphnodes) and in 3 even peritoneal manifestations. 8 patients had chest pain when therapy started. Results: in this group of pretreated patients 3 had a partial response, 3 stable disease or minimal response and 6 progressive disease. Significantly symptoms improvement was found in 6 patients. The median survival was 15.1 month. Toxicity grade 4 was tolerable and acceptable: nausea/vomiting in 2, thrombopenia in 2, leucopenia in 1 and hairloss in 0 patients. Gemcitabine combined with Carbo- or Cisplatin offers a good chance in the treatment of malignant mesothelioma. Prospective multicenter trials have to verify these results.

211

Bendamustin for treatment of relapsed small cell lung cancer: a pilot study

A. Schmittel, G. Hütter, K. Schulze, E. Thiel, U. Keilholz (Berlin, D)

The single agent activity of bendamustin in small cell lung cancer (SCLC) is high with a response rate of 46% in previously untreated patients (Heider et al. 1999). However, a phase 2 trial in patients refractory to primary chemotherapy failed to show activity of second line bendamustin treatment. Therefore we investigated the activity of bendamustin in patients with relapsing SCLC after response to first line treatment. In this pilot study we treated patients with relapsed SCLC extensive disease with bendamustin 120mg/m^2, day 1+2, every three weeks. In the absence of tumor progression a maximum of six cycles was given. Inclusion criteria were absence of brain metastases and time to relapse > 2 months after first line chemotherapy. Patients refractory to first line therapy were excluded from this study. Six out of seven patients are currently evaluable for toxicity and response to treatment. Toxicity was mild. Only grade 1 and 2 hematologic and gastrointestinal toxicity was observed. One patient had a poststenotic pneumonia during therapy. No alopecia was observed. Three out of six patients had a partial response after two and four cycles of therapy. One patient had disease stabilization, two patients progressed. This ongoing study demonstrates already the activity of bendamustin in patients with relapsed SCLC and confirms the low toxicity found in other trials. The study has been extended to a multi-centre phase II trial in 35 patients.

212

Escalation strategy in small cell lung cancer by adding ifosfamide and navelbine to carboplatin and etoposidphosphate: a phase II study

A. Schmittel, J. Siehl, E. Thiel, U. Keilholz (Berlin, D)

Platinum-based chemotherapy with etoposid is standard first line therapy for small cell lung cancer (SCLC). To analyse, whether an escalation strategy employing four active agents is feasible and increases efficacy, we added ifosfamide and vinorelbine to the combination of carboplatin and etoposidphosphate. The four substances were administered as follows: carboplatin

300mg/m^2 on day 1, etoposidphosphate 120mg/m^2 on day 1+2, ifosfamide 2000mg/m^2 on day 1+2 and vinorelbine 30mg/m^2 on day 1. Cycles were repeated every 3 weeks until tumor progression to a maximum of six cycles. G-CSF was administered as secondary prophylaxis after febrile neutropenia. 23 patients with a median age of 58 years (36 - 72 years) have been entered. Response rate was 74% with four (17%) complete responses and thirteen (57%) partial response. In four patients (17%) stabilization of disease was observed, two patients were refractory to therapy. Median survival was 12.0 months (95% confidence interval 9.0 to 15.0 months). One-year survival was 33.8%. One treatment related death due to neutropenic sepsis after the forth cycle occurred. The major toxicities were neutropenia, anemia and alopecia. This phase II study demonstrates the high efficacy of this four agent regimen in SCLC. The response and 1-year survival rate are promising. Due to the hematologic toxicity this regimen should be restricted to younger patients without comorbidity.

213

Retrospective analysis of docetaxel/carboplatin combination chemotherapy in patients with advanced non small cell lung cancer (NSCLC)

A. Reissig, A. Klippstein, U. Hengst, A. Machnik, C. Kroegel (Jena, D)

Combination chemotherapy with taxanes plus platinum derivatives has been shown to induce high response rates in advanced NSCLC. We studied the activity and toxicity of the combination of docetaxel (D) and carboplatin (C) as first-, second- or fourth-line chemotherapy for advanced NSCLC in daily clinical routine. Patients and method: Between 10/96 and 11/00 38 patients (pts) (median age 65 y; 28 m., 10 f.) with histological confirmed stage IIIB (n=17) and IV (n=21) NSCLC and with an ECOG performance status of PS 0-2 were treated. The first 12 pts got D (100 mg/m^2) plus C (300mg/m^2) and the following pts received D (75mg/m^2) plus C (AUC 5). 32 pts were treated as first line chemotherapy and 6 pts as second (5 pts)- or fourth (1pt) line therapy. Cycles were repeated every 3 weeks to a maximum of 6 cycles. A total of 164 cycles were delivered with a median of 4 cycles per patient. Results: Toxicity: All pts are evaluable for toxicity: Neutropenia grade IV was observed in 7 pts, grade III in 15 pts. 19 episodes of febrile neutropenia occured but without treatment-related death. 5 pts experienced grade III infections. Grade III thrombocytopenia was only seen in 3 pts and anemia in one pt. Grade III non-hematologic toxicities were nausea (6) and neuropathy (3).
Response: So far 36 patients were evaluable for response. Partial response (PR) was observed in 5/36 (14%), stable disease (SD) in 30/36 (83%), progression in one pt 1/36 (3%). 13 pts are still alive. Median survival was 287 days (177-413 days, 95% CI., Kaplan-Meier-analysis). Conclusion: These results demonstrate that the docetaxel/carboplatin combination chemotherapy is an active and feasible regimen as front- and second line treatment of advanced NSCLC outside controlled clinical trials. Docetaxel (75mg/m^2) and carboplatin (AUC 5) is suitable as a regimen in daily clinical routine.

214

LD (Stage I – IIIa) small cell lung cancer (SCLC): multimodal therapy including surgery

M. Serke, E. Allica, D. Kaiser, R. Loddenkemper (Berlin, D)

Small cell lung cancer responds to chemo- and radiotherapy but cures are rare. The role of surgery to improve local control in SCLC stage I-IIIa is still not defined. Patients: We analysed our results of multimodal therapy including chemotherapy, radiotherapy and surgery in 150 consecutive patients with SCLC stage I-IIIa operated on in our hospital between 1983-2000. Results: Median age: 58 years. Patients with proven SCLC had induction chemotherapy prior to surgery. All patients received 3 cycles of adjuvant chemotherapy, some with additional radiotherapy. Perioperative mortality: 2%. Median survival: 22,4 months. 1-year-survival: 79%, 2-years-survival: 47%, 5-years-survival: 32%. Discussion: A median survival of 22.4 months in multimodally treated LD-SCLC most of them stage Iib/IIIa appears promising. The results are comparable or better than LD-SCLC studies with radio-chemotherapy without surgery with median survivals between 14 – 23 months. 2- and 5-years survival is promising and indicates that some patients may be cured.

215

Weekly gemcitabine / paclitaxel: feasibility in 3rd line therapy of NSCLC

G. Jacobs, J. Schimke (Saarbrücken, D)

Both Paclitaxel (T) and Gemcitabine (G) are of proven efficiency in NSCLC. The combination of both is not documented up to now. In patients (Pt) with progressive stage IV disease after primary and 1st relapse therapy (Tx) without either P nor G presenting in a good clinical condition deserving, a third line of Tx with acceptable tolerability and sufficient anti-tumour effect is desirable. In a series of 7 Pt (data of 10 will be shown in September) with histologically proven NSCLC in 2nd relapse / progression in good clinical condition (KPS 70% or better) who had neither T nor G in previous treatments and where radiation therapy was inadequate for tumour control resp. impossible because of earlier application, a weekly combination of both (T: 75 mg/m^2; G: 1000 mg/m^2 q 8 d x 6) was offered as a 'salvage regimen'. Up to now 13 cycles are evaluable in respect to tolerability. The weekly schedule for both substances was selected for best control of (hematological) toxicity. The given doses of T resp. G reached in median .86 (range .5 to 1.0) for T and .82 (.5 to .92) for G. Maximum hematological toxicity reached WHO grade III for neutro/thrombopenia and anaemia. All other toxicities were gr. 0 to II. Tumour control was CR=0/7, PR=2/7, SD=1/7 after the first cycle. The weekly schedule of combined T+G in Pt untreated with both substances before is a tolerable effort to reach tumour control in advanced situations. The hematological toxicities were far less than expected (although dose modification was applied) even in intensely pre-treated Pt. The main non-haematological toxicity was mild nausea and mucositis.

216

Paclitaxel, carboplatin plus vinorelbine in NSCLC stage IV: A phase II trial

A. Schmittel, J. Siehl, G. Hütter, R. Gaub, K. Schulze, E. Thiel, U. Keilholz (Berlin, D)

ECOG 1594 trial demonstrated a favourable toxicity profile for the combination of Carboplatin plus Paclitaxel (Proc. ASCO 2000: 2), when compared to cisplatin-based two agent regimens. Triple chemotherapy regimens like Cisplatin, Gemcitabine and Vinorelbine have higher efficacy when compared to two agent regimens in advanced NSCLC (JCO 2000;18:1451). We added Vinorelbine to the combination of Carboplatin plus Paclitaxel in the present phase II study. Carboplatin (AUC 5) + Paclitaxel (175mg/m^2) + Vinorelbine (30mg/m^2) were administered on day 1 every 3 weeks until tumor progression to a maximum of six cycles. Exclusion criteria were: age > 70 years, presence of brain metastases, ECOG PS > 2. G-CSF was administered as secondary prophylaxis after febrile neutropenia. 30 patients have been entered to date and 26 are evaluable for toxicity and response to treatment. Therapy was well tolerated, taxane-induced alopezia occured regularly. The major grade III and IV toxicity were neutopenia and febrile neutropenia, which was observed in 5 patients. Three patients had pneumonia in the absence of neutropenia. Three patients had grade 2 neuropathy, which required discontinuation of vinorelbine administration. Of 26 stage IV patients 5 had a partial response, 14 had disease stabilization and 7 progressed. The data from this study shows that the triple combination is active in advanced NSCLC. Febrile neutropenia and neuropathy are the major toxicities, which makes this regimen only suitable for patients without significant comorbidity.

217

Non small-cell lung cancer 10 years after successful treatment for small-cell lung cancer

M. Schroeder, N. Dreier, B. Röhne, C. Aul (Duisburg, D)

We report a 61 year old woman who died due to cerebral metastases of NSCLC ten years after cured SCLC. In 1/91 the patient first entered our hospital with the diagnosis of cerebral metastases of a histologic proven SCLC. Karnofsky index was reduced to 20 %. The treatment was started on request of the patient's husband and consisted of cerebral irradiation (ad 38 Gy) and polychemotherapy including ADM, VCR, Ifo (AIO) and Carboplat, Etoposid (CE). Achieving a complete remission (CR) primary tumor irradiation was added ad 50 Gy. CR lasted until 2/2001. Parkinsonism was registrated as a form of late toxicity of cranial irradiation since 8/98. In 2/2001 the patient entered our clinic as an emergency presenting with right hemiparesis and unconciousness. Diagnosis of an intracerebral mass bleeding due to a new cerebral metastasis was shown by a CT-scan. The patient died 6 days later without neuro-surgical intervention. Autopsy revealed a new bronchogenic cancer (classified as NSCLC, measuring 1 cm in diameter) in the contralateral lung and a new intracerebral metastasis on a new localisation. Continued tobacco smooking may be the reason for the second neoplasm.

218

Neoadjuvant Chemotherapy (weekly Carboplatin / Docetaxel) in patients with stage IIIa / IIIb non-small cell lung cancer (NSCLC): Interim analysis of feasibility and toxicity

M. Serke, S. Fassheber, N. Schoenfeld, R. Loddenkemper (Berlin, D)

Introduction: The combination and sequence of chemotherapy, surgery and radiotherapy in stage III NSCLC is still under discussion. We investigated the feasibility and toxicity of neoadjuvant chemotherapy of weekly carboplatin (C)/Docetaxel (D) in stage III NSCLC patients. Patients and method: Between 6/00 and 04/01, 14 patients were enrolled, 3 pts with stage IIIa NSCLC (female) and bulky N2 disease (CT-scan and mediastinoscopy) and 11 pts with stage IIIb NSCLC and N3-lymphnodes or T4-tumor. Median age was 53 years. All patients were planned to receive weekly doses of C (AUC2) and D (35 mg/m^2) for 4 weeks followed by 2 weeks rest for a maximum of 3 consecutive cycles. Results: Toxicity: To date 90 chemotherapy infusions of C and D are administered and evaluable for toxicity. Grade III / IV hematotoxicity and non-haematologic toxicity occured in 4% and 2.7 % of cycles respectively. 4 patients developed grade II leucopenia, 2 grade III leukopenia. In order to increase feasibility dose reductions have been undertaken in 40% of the second cycle (D: 30mg/m^2). One mild allergic skin reaction was seen. Two patients reported of diarrhea grade II and only 1 of grade III diarrhea. One patient experienced PNP grade II during the first cycle of chemotherapy. Other toxicities included mild to moderate nail disorder, interference with taste, hoarseness. Response: In an interim analysis 14 patients were evaluable for response. PR: n = 4, NC/MR: n = 4, PD: n= 6. ORR (PR): 29%. Conclusion: Based on the preliminary results, this ongoing phase II study indicates that the combination of C + D given as neoadjuvant chemotherapy is a very feasible and well tolerated regimen with encouraging results.

219

A randomized phase II study of gemcitabine/docetaxel vs gemcitabine/cisplatin followed by docetaxel in patients with metastatic non-small cell lung cancer: Intermediate results

D. Binder, V. Sailer, H. Schweisfurth, C. Kieslich, F. Petri, M. Schweigert, B. Temmesfeld, M. Fleischhacker, T. Beinert, K. Possinger (Berlin, Cottbus, D)

Background: Over the past years, new anticancer agents have been developed and new treatment schedules have been proven in patients with NSCLC. Despite high response rates with triple chemotherapy regimens, cumulative toxicity might limit patients' benefit. Therefore we conducted a phase II study to compare a sequential drug application with a two-drug combination chemotherapy in metastatic NSCLC patients. Patients: 34 patients with stage IV NSCLC were prospectively randomized to one of the following treatment arms: Schedule A: gemcitabine (900mg/m^2, 30-min-infusion) on days 1 and 8 and docetaxel (75mg/m^2, 1-hour-infusion) on day 1, repeated every three weeks, up to six courses. Arm B: gemcitabine (900mg/m^2, days 1 and 8) and cisplatin (75mg/m^2, 1-hour-infusion, day 1) are given for three courses, followed by three courses of docetaxel (100 mg/m^2, day 1, repeated every three weeks). Results: Of 140 pts planned in total, 34 pts were enrolled and analysed so far. In arm A (18 pts) and arm B (16 pts), 8 pts (44%) and 7 pts (43%) showed partial tumour response after the third course, while in arm B one patient achieved a complete response. In treatment schedule A a RDI of 90% was achieved, and in arm B a RDI of 94%. Mild to moderate leukopenia was the most common side effect: In 5 pts/4 pts (A/B) a grade III-IV leukopenia (WHO) occured. None of the pts developed neutropenic fever. Two pts (arm A) developed moderate cutaneous manifestations of hands and neck. Other side effects (WHO grade II-III) have been alopecia (44%/38%), asthenia (26%/21 %) and diarrhea (0 %/12%). Conclusion: In this phase II study, both regimens showed encouraging activity. Toxicity was manageable in both arms without any clinical complications.

220

Topotecan is effective in second-line treatment of small-cell lung cancer with symptomatic brain metastases also after whole brain irradiation

A. Korfel, C. Oehm, J. von Pawel, U. Keppler, M. Deppermann, L. Güttler, E. Thiel (Berlin,Treuenbriezen, Gauting, Immenhausen, München, D)

Brain metastases occur in up to 80% of the patients with recurrent small-cell lung cancer (SCLC). The benefits of whole brain irradiation (WBI) in these patients are palliative and most of them die from the systemic disease after a median survival of <6 months. Systemic chemotherapy has the advantage of simultaneous treatment of systemic spread. The new topoisomerase-I inhibitor topotecan (Hycamtin) may be an alternative in the therapy of SCLC brain metastases because of its significant activity against SCLC and the ability to cross the intact blood-brain barrier. The purpose of this multicenter phase II study was to evaluate the efficacy and toxicity of topotecan in patients with SCLC who relapsed with symptomatic brain metastases after first-line chemotherapy. Topotecan was administered as a 30-minute intravenous infusion of 1.5 mg/m^2 (in 4 patients 1.25 mg/m^2) for 5 consecutive days every 3 weeks. Thirty patients at 5 institutions entered the study. Seventeen patients were pretreated with at least 2 chemotherapy protocols and 12 patients with WBI. Systemic metastases were recorded in 26 patients. Cerebral metastases responded in 10/30 patients (3 CR and 7 PR) and systemic in 6/26 patients (1 CR, 5 PR). Four of the 12 patients pretreated by WBI responded in the brain. Median response duration was 117 days (range 42-200), median overall survival from study begin was 85 days (range 7-425) with 7 patients still alive at 30, 48, 90, 101, 120, 158 and 425 days. Therapy was generally well tolerated with myelotoxicity being the most common adverse event. These first results of monotherapy with topotecan in heavily pretreated patients with SCLC brain metastases are promising and merit further evaluation.

Postersitzung:
Genexpression

222

PTEN Deficient Prostate Cancer Cells show greater Sensitivity to Inhibitors of the PI3 kinase/Akt pathway

V. Grünwald, L. deGraffenried, R. Ray, M. Hidalgo (San Antonio, TX, Saint Louis, MO, USA)

Genetic abnormalities have been reported for a variety of solid tumors. In prostate cancer, alteration of chromosome 10 at the region 10q23 have been recognized to appear frequently, later being known as the locus of the PTEN/MMAC1 tumor suppressor gene. The PTEN protein is a dual-specific phosphatase with putative tumor suppressing abilities, including inhibition of the PI3kinase/Akt signaling pathway. Alterations of the PI3k/Akt pathway have been implicated in a wide range of human cancers, activating a number of proliferative, metabolic, and antiapoptotic pathways Moreover, negative PTEN status is correlated with chemotherapy resistance and advanced clinical stage in a number of malignancies. In order to assess the contribution of PTEN status and its effect on the PI3k/Akt signaling pathway to drug susceptibility, we evaluated growth response of prostate cancer cell lines with varied PTEN status to different treatment protocols. In preliminary MTT growth assay results, the PTEN-negative prostate cancer cell line PC-3 displayed a much greater proliferation rate compared to the PTEN-positive prostate cancer cell line DU-145, even under serum starved conditions, and showed greater resistance to the chemotherapeutic agent, doxorubicin. The PC-3 cells however displayed much greater sensitivity to inhibitors of the PI3k/Akt pathway, including rapamycin, an inhibitor of mTOR, and LY294002, a specific inhibitor of the PI3 kinase pathway. The DU-145 cells were more susceptible to the inhibitory effects of the MAP kinase inhibitor PD98059, suggesting that this alternate growth pathway may not be as important in PTEN-negative cells. In order to study the role of PTEN status in chemotherapy resistance, and to evaluate potential therapeutic agents that target the PI3k pathway, we have established PC-3 cell lines that stabley express PTEN, driven by a CMV promoter. Our preliminary findings suggest that PTEN status plays a crucial role in chemotherapy sensitivity in prostate cancer cell lines and resistance to conventional cytotoxic therapy may be circumvented by combination of PI3K inhibitors with conventional chemotherapy.

223

Histone deacetylases in the pathogenesis of acute leukemias

U. Mahlknecht, D. Hoelzer (Frankfurt am Main, D)

Chromatin structure is gaining increasing attention as a potential target in the treatment of cancer. Relaxation of the chromatin fiber facilitates transcription and is regulated by two competing enzymatic activities, histone acetyltransferases (HATs) and histone deacetylases (HDACs), which modify the acetylation state of histone proteins and other promoter-bound transcription factors. While HATs, which are frequently part of multisubunit coactivator complexes, lead to the relaxation of chromatin structure and transcriptional activation, HDACs tend to associate with multisubunit corepressor complexes, which result in chromatin condensation and transcriptional repression of specific target genes. HATs and HDACs are known to be involved both in the pathogenesis as well as in the suppression of cancer. Some of the genes encoding these enzymes have been shown to be rearranged in the context of chromosomal translocations in human acute leukemias and solid tumors, where fusions of regulatory and coding regions of a variety of transcription factor genes result in completely new gene products, which may interfere with regulatory cascades that control cell growth and differentiation. On the other hand, some histone acetylation modifying enzymes have been located within chromosomal regions that are particularly prone to chromosomal breaks. In these cases gains and losses of chromosomal material may affect the availability of functionally active HATs and HDACs, which in turn disturbs the tightly controlled equilibrium of histone acetylation. We review herein the recent achievements, which further help to elucidate the biological role of histone acetylation modifying enzymes and their potential impact on our current understanding of the molecular changes involved in the development of solid tumors and leukemias.

224

Transdifferentiating effect of Axl-overexpression in the undifferentiated myelogenous leukemia cell line KG1a

A. Burchert, M. Jänike, C. Brendel, A. Neubauer (Marburg, D)

Axl is a receptor tyrosine kinase, containing extracellular adhesion motives, a transmembrane domain and an intracellular kinase domain. It was cloned from patients with chronic myelogenous leukemia, and subsequently shown to be over-expressed in primary myeloid leukemia cells. In vitro, overexpression of Axl in the murine myeloid mouse cell line 32D caused transformation. This suggested a role for Axl in the pathogenesis of human leukemias. However, the biological effects of Axl in primitive human hematopoietic cells are unknown. We therefore investigated consequences of Axl-expression in KG1a cells in vitro. KG1a is an undifferentiated myeloid CD34+ human progenitor cell line. After confirming successful transduction of these cells with the 1B1-Axl cDNA, a previously described acutely transforming Axl-variant-cDNA, by polymerase chain reaction (PCR) and western blotting, the phosphorylation status, morphology, immunophenotype and growth characteristics of KG1a-Axl cells vs. KG1a cells were analyzed. Interestingly, Axl expressing KG1a cells became adherent and dramatically changed there morphology toward a fibroblastic appearance. Their immunophenotypic analysis revealed CD34 negativity and absence of all hematopoietic surface-antigens that are normally being expressed by naive KG1a-cells. Phosphotyrosine-blotting on whole cell lysates showed constitutive phosphorylation of Axl only in KG1a-Axl cells but not in controls. Paradoxically, despite the evident permanent Axl activity in the KG1a-Axl cells, they grew more slowly than their parental counterparts. We conclude that Axl can employ the signal transduction machinery of immature hematopoietic cells thereby exerting its effects. Axl signaling appears to trans-differentiate hematopoietic cells into a non-hematopoietic cell type. These data imply a novel role for Axl in the induction of differentiation of immature cells and demonstrate for the first time an antimitogenic effect of Axl kinase activity.

225

Mutation analysis of the DNA-damage checkpoint gene CHK2 in hematological malignancies

W.K. Hofmann, S. Takeuchi, S. Tavor, C.W. Miller, D. Hoelzer, H.P. Koeffler (Frankfurt/Main, D; Okohcho, Nankoku, J; Los Angeles, USA)

Checkpoint genes code for a family of proteins which sense DNA damage in eukaryotic cells. They play an important role in the control of the cell cycle. The human CHK2 is a homolog of the yeast G2 checkpoint kinases known as CDS1 and RAD53. The CHK2 may be a tumor suppressor gene because it was found to be mutated in some individuals with the Li-Fraumeni syndrome. These cases had a normal, non-mutated p53 gene. We performed a mutational analysis of the 14 exons of the CHK2 gene using polymerase chain reaction-single strand conformation polymorphism (PCR-SSCP) in genomic DNA extracted from a total of 225 tumor samples including 41 myelodysplastic syndrome (MDS), 41 acute myeloid leukemias (AML), 48 non-Hodgkin's lymphomas (NHL), 41 acute lymphoblastic leukemias (ALL) and 54 acute T-cell leukemias (ATL). DNA from aberrantly migrating bands was direct sequenced. We found a novel G to C transversion resulting in a change from Ala to Gly at codon 507 of CHK2 in a MDS sample. Normal lymphocytes from the peripheral blood of this individual did not have the abnormality of the CHK2 gene. Furthermore, a C to T mutation changing Ser to Phe at codon 428 was found in a NHL sample. In addition, we demonstrated a polymorphism at codon 84 (A to G at nucleotide 252) of exon 1 of CHK2 in 13 tumor samples (6 %). The identification of mutations at a potentially critical region of the CHK2 gene in MDS and NHL leads to a number of questions regarding the functional importance of this mutation as well as the frequency of CHK2 mutations in solid tumors. The presence of CHK2 mutations in hematological malignancies highlights the importance of cell cycle checkpoints in alterations of hemato- and lymphopoiesis.

226

mRNA expression profiles of different relevant tumor-associated antigens in myeloid leukemias

J. Greiner, M. Ringhoffer, A. Szmaragowska, S. Hübsch, M. Taniguchi, H. Döhner, M. Schmitt (Ulm, D)

For specific immunotherapy of tumors the identification of tumor-associated antigens (TAA) eliciting an immune response in the tumor-bearing host is mandatory. Significant mRNA expression of antigens is a prerequisite for immunotherapy in leukemias. Simultaneous expression of different antigens is a basis for therapies with polyvalent vaccines. TAA were identified by SEREX (Serological screening of recombinant expression cloning) or biochemical or genetic methods followed by screening with specific CTL. In this study, TAA previously characterized in solid tumors and antigens newly characterized in myeloid leukemias by our group using the method of SEREX were examinated for their mRNA expression in up to 42 AML and in 8 human and murine leukemia cell lines by RT-PCR. To investigate their specifity for leukemic blasts, the mRNA expression level was also characterized in PBMN of healthy volunteers and in CD 34 positive selected cell samples of healthy volunteers serving as germ line cells. mRNA expression of different antigens were found in AML patients / leukemia cell lines: PRAME (48 % / 80 %), WT-1 (64 % / 60 %), RHAMM (75 % / 80 %), BAGE (60 % / 20 %) and MAZ (41 % / 80 %). No mRNA expression of the antigens PRAME, RHAMM and BAGE was found in PBMN and CD 34 positive separated cell samples of healthy volunteers. WT-1 expression was detectable in CD 34 positive cells (54 %) and MAZ expression in both PBMN (13 %) and CD 34 positive cells (13 %) of healthy volunteers. mRNA expression of the TAA SSX-2, MAGE-1, MAGE-3 and G250 was only found in leukemia cell lines but not in samples of AML patients. The antigens PINCH and syntaxin were expressed in AML patients but also in most controls. The antigens HER2/neu, GAGE and RAGE-1 were not expressed in all samples. Experiments for the mRNA expression of the TAA NY-ESO-1 and SCP-1 are ongoing. Taken together, the immunogenic antigens PRAME, RHAMM, WT-1 and BAGE are potential candidates for immunotherapy in leukemia for mono- or polyvalent vaccines.

227

The implication of LDSF motive of E2a-Pbx1 in target gene expression in 3T3 fibroblasts - cloning of EF9 gene.

M. Kolanczyk, M. Gantert, T. Zemojtel, M.P. Kamps, J. Scheele (Freiburg, D; San Diego, USA)

In Pro-B acute lymphoblastic leukemia ALL, a t(17;19) chromosomal translocation results in the fusion of the E2A gene to a gene encoding a novel bZIP family transcription factor called HLF . In 25% of pre-B ALL's, a t(1;19) translocation fuses E2A to a homeobox gene called Pbx1. Both chromosomal translocations result in the expression of the chimeric protein that is composed of the amino-terminal translocation domains of E2A linked to a heterologous DNA binding domain. It is a current hypothesis that leukemias result from inappropriate expression of target genes normally regulated by Pbx1 or HLF In support of this E2A-Pbx1 but not Pbx1 has been shown to be a potent transactivator. The fusion protein activates transcription of target genes in the tissue specific manner. Exact mechanism of

E2a-Pbx1 driven transcription activation is not known. Here we study the activation of target genes of E2a-Pbx1 in mouse fibroblasts. We introduce the point mutation in evolutionarily conserved LDFS motive in AD1 transactivational domain of E2a-Pbx1 and observe effects of that mutation on the target gene expression in NIH 3T3 fibroblasts. Further we study histon 3 and 4 acetylation status of the 5' flanking regions of target genes in leukemic cell lines and NIH 3T3 fibroblasts. We observe transcription activation dependant differences of acetylation levels of histon 4 in 5' regions of E2a-Pbx1 target genes. In that contekst we also present the RACE PCR derived full cDNA sequence of EF-9 direct target gene from 3T3 fibroblasts. The 5' flanking region of that gene has also been cloned, and it's E2a-Pbx1 responsive elements identified. Further we analyze the histon 3 and 4 acetylation status in 5' regions of several E2a-Pbx1 target genes.

228

Leukemic fusion protein AML1-ETO enhances c-jun proto-oncogene expression

A. Elsässer, M. Franzen, M. Weisser, S. Schnittger, R. Vangala, V. Reddy, J. Rangatia, M. Ueffing, P. Pelicci, D. Tenen, W. Hiddemann, G. Behre (München, D; Milan, I; Boston, USA)

AML1-ETO t(8;21) is one of the most common fusion proteins found in acute myeloid leukemia (AML). AML1-ETO functionally represses important myeloid transcription factors, like AML1B and C/EBPa, which results in a block in normal myeloid differentiation. In AML1-ETO knock-in studies, however, AML1-ETO increases cell proliferation and self-renewal capacities. We therefore investigated potential target genes of AML1-ETO linked to cell proliferation. In fact, we found that AML1-ETO positively regulated c-jun proto-oncogene expression in a myeloblastic U937 cell line with inducible expression of AML1-ETO. Most importantly, a positive correlation of AML1-ETO and c-jun mRNA expression was observed in t(8;21) positive AML patient samples. Investigating the mechanism of AML1-ETO induced c-jun expression, we could show that AML1-ETO transactivates the human c-jun promoter through the proximal AP-1/CRE site. This effect is mediated via an indirect mechanism, as AML1-ETO does not directly bind to the AP-1/CRE site in electrophoretic mobility shift assay. We demonstrated that C/EBPa, a bZIP transcription factor like c-jun, downregulates the c-jun promoter. We show that C/EBPa can bind to the AP-1/ CRE site of the human c-jun promoter as well as down-regulate an unrelated AP-1 dependent promoter. C/EBPa is functionally repressed by AML1-ETO. Therefore, we propose an indirect mechanism by which AML1-ETO upregulates c-jun expression through the release of repression by C/EBPa at the AP-1/CRE site of the c-jun promoter. We suggest, that AML1-ETO induced expression of c-jun might contribute to enhanced proliferation and thus promote leukemogenesis.

229

CD44 targeted differentiation therapy down-regulates c-jun expression in human myeloid cells lines and AML patients

A. Peer Zada, S. Singh, V. Reddy, A. Elsässer, T. Haferlach, D. Tenen, W. Hiddemann, G. Behre (München, D; Boston, USA)

A member of the AP-1 transcription factor family, c-Jun, is important for cell proliferation, differentiation and apoptosis. Accordingly, its expression and function are altered in response to diverse stimuli. On the other hand, ligation of CD44 by anti-CD44 monoclonal antibodies (A3D8 and H90) has been shown to induce differentiation and inhibit proliferation of AML cells. In the present study, we investigated the effect on c-jun and the molecular mechanism of CD44 ligation on the proliferation and maturation of HL60 and U937 cells. Ligation of CD44 by A3D8 led to the down-regulation of c-jun expression in HL60, U937 cells and AML patient samples leading to their decreased proliferation. To further investigate the molecular mechanism of c-jun down-regulation by CD44 ligation, U937 cells were transiently transfected with c-jun promoter\luciferase constructs. We report that c-jun promoter (-1780\+731) activity was down-regulated by 12 fold upon ligation of CD44 by A3D8. This reduction of c-jun promoter activity was abolished after deletion of the region between –1780 to –63, where two AP-1 sites (-180 & -64) are located. Down-regulation of c-jun promoter activity was observed when proximal or distal AP-1 sites were deleted compared to the deletion of both AP-1 sites. These findings demonstrate that anti-proliferative effect of CD44 ligation with A3D8 in myeloid cell lines involves c-Jun\AP-1 function. Taken together our results indicate that down-regulation of c-jun plays a pivotal role in the signal cascade (anti-oncogenic) initiated by CD44 ligation causing anti-proliferative response in target cell lines and AML patient samples.

230

Adenovirus-mediated overexpression of p14ARF is capable of inducing apoptosis in p53-mutated/BAX-negative DU145 prostate cancer cells

P. Hemmati, B. Gillissen, D. Stuckenberger, L. Staerck, J. Wendt, S. Hermann, I. Sturm, B. Dörken, P.T. Daniel (Berlin, D)

The human INK4a locus encodes two structurally unrelated tumor suppressor proteins, p16INK4a and p14ARF, which are frequently inactivated in human cancer. Whereas p16INK4a mediates its biologcal effects mainly through the Rb-cdk4/6-cyclin D pathway, both the pro-apoptotic and cell cycle-regulatory actions of p14ARF were shown to be primarily dependent on a functional p53-mdm2 tumor suppressor pathway. However, recent reports have also implicated p14ARF in a p53-independent mechanism of cell cycle regulation. To further explore the pro-apoptotic function of p14ARF in relation to functional cellular p53, we constructed replication-deficient adenoviral vectors overexpressing p14ARF (AdCMV-p14ARF). As expected, the adenovirus-mediated overexpression of p14ARF was capable of inducing apoptosis in p53/Rb-wt U2OS osteosarcoma cells at a low multiplicities of infection (MOI). Apoptosis was observed as early as 48h after infection as indicated by the occurrence of cells with a characteristic sub-G1 DNA-content. Interestingly, AdCMV-p14ARF also induced cell death in the BAX-mutated/p53-mutated DU145 prostate cancer cells. Restoration of BAX expression by retrovirus-mediated overexpression further enhanced p14ARF triggered apoptosis. This shows that p14ARF involves BAX-dependent mechanisms but may also occur via BAX-independent pathways. Since the induction of cell death was not accompanied by induction of effectors such as caspase-3, changes in the mitochondrial membrane potential (DYm) were investigated in more detail. Taken together, these results clearly indicate the involvement of the pro-apotolic BAX irrespective of the presence of functional cellular p53. Nevertheless, p14ARF also induces apoptotic cell death in cells lacking functional p53 and BAX which strongly indicates the participation of additional signaling pathways.

231

C/EBPalpha is a negative regulator of c-jun : significance in granulopoiesis and AML

J. Rangatia, R. Vangala, N. Treiber, A. Elsässer, D. Tenen, W. Hiddemann, G. Behre (München, D; Boston, USA)

C/EBPa, a prototype of a bZIP transcription factor modulates transcription and differentiation in adipocytes, hepatocytes and hematopoietic cells. C/EBPa knockout mice revealed a specific defect in the hematopoietic system characterized by a selective block in differentiation to mature granulocytes. Conditional expression of C/EBPa is capable to drive bipotential myeloid cells to neutrophilic differentiation. These reports strongly point to the critical role of C/EBPa as a master regulator of the granulocyte lineage. In the C/EBPa -/- mice, the level of c-Jun RNA in the fetal liver is dramatically increased. c-Jun is a component of AP-1 transcription factor complexes. c-Jun mRNA is upregulated upon monocytic differentiation of bipotential myeloid cell lines. Induced expression of c-Jun in myeloid cell lines results in partial macrophage differentiation. This role of c-Jun is attributed to its being a co-activator of PU.1 transcription factor. Based on these findings, we addressed the possibility of C/EBPa downregulating c-Jun expression during granulocytic differentiation. Human c-Jun promoter transcriptional activity is downregulated 8 fold in presence of C/EBPa in transient transfections in myeloid and fibroblast cell lines. U937 cell line with inducible C/EBPa expression shows decrease in c-Jun mRNA on induction of exogenous C/EBPa as seen by Real Time PCR and Northern blot analysis. Transient transfections in F9 cell line lacking endogenous c-Jun were carried out to map the C/EBPa binding site in the c-Jun promoter and the repression domain in C/EBPa. We observed c-Jun promoter transcriptional downregulation to be lost in absence of C/EBP consensus binding site in c-Jun promoter and deletion of N-terminal domain in C/EBPa .The importance of this repression domain is emphasized by findings that about 25% of AML patients have mutations in the N-terminal region of C/EBPa leading to either truncated or non-funcitonal protein. This mutated C/EBPa may no longer be capable of supressing c-Jun expression. Thus, c-jun being a proto-oncogene when expressed in deregulated way will cause hyperproliferation. Hence, we hypothesize that C/EBPa besides upregulating target genes for granulocytic differentiation blocks c-Jun expression by downregulating promoter activity to (a) block its co-activator role in monocytic differentiation, and (b) to prevent hyperproliferation of undifferentiatied myeloid cells.

232

Detection of Bax, Bcl-2, Bcl-xL and Bcl-xs mRNA expression in the T-cell line HSB-2 in response to Camptothecin treatment by real-time PCR

A. Radonic, A. Nitsche, N. Steuer, S. Thulke, W. Siegert (Berlin, D)

Using the TaqMan(tm) chemistry, we developed four high-throughput, sensitive PCR assays for specific quantification of Bax, Bcl-2,Bcl-xs and Bcl-xL mRNA, respectively. transcripts per assay and profits from the general advantages of TaqMan(tm) chemistry: high sensitivity, specificity, reproducibility, real-time determination, speed and minimal risk of carry-over contamination. Bax, Bcl-2, Bcl-xs and Bcl-xL detection are performed simultaneously under the same PCR conditions and only small amounts of starting material are required for mRNA quantification. Apoptosis of HSB-2 cells was determined by Annexin V-FITC/PI staining and detected by Flow Cytometry (FACScan). In 9 independent experiments, HSB-2 cells were treated with Camptothecin in DMSO (final concentration 10μM) or with DMSO alone. Apoptosis rate and expression of Bax, Bcl-2,Bcl-xs and Bcl-xL mRNA was determined after 2, 4 and 6 hours of incubation. As expected, we found a significantly increased apoptosis rate in cells treated with Camptothecin after 2 ($p=0.003$), 4 ($p=0.001$) and 6 hours ($p=0.001$) as detected by flow cytometry. In correlation to the increasing apoptosis rate, we found a shift in the ratio of Bax to Bcl-2 mRNA expression. The ratio was shifted towards the pro-apoptotic Bax mRNA after 2 ($p=0.001$), 4 ($p=0.001$) and 6 hours ($p=0.001$). Incubation with Camptothecin for 6 hours resulted in an apoptosis rate of 37% and a 7.5 fold higher Bax to Bcl-2 ratio than following DMSO treatment. A comparable shift in the Bcl-xs Bcl-xL ratio could not be observed. We conclude, that the ratio of Bax to Bcl-2 is rapidly shifted by inductors of apoptosis like Camptothecin and only through long-term processes like development of cancer. This work was supported by a grant of the Wilhelm Sander-Stiftung (# 98.072.1).

233

Correlation of Hsp90, C-erB2 and Bcl-2a expression in a panel of 98 human tumor xenografts derived from patient explants

K. Schandelmaier, G.L. Wirth, E. Tetling, H.H. Fiebig, A.M. Burger (Freiburg, D)

Heat shock protein 90 (hsp90) has recently emerged as novel and attractive target for cancer therapy. The hsp90 molecular chaperone is required for activation of several families of protein kinases and nuclear hormone receptors, many of which are proto-oncogenes and play a prominent role in cancer e.g. c-erbB2. The ATP-binding site of hsp90 is required for the regulation of its chaperone function. Occupation of the ATP-binding pocket inhibits all hsp90-mediated conformational refolding reactions and results in degradation of substrates. Geldanamycins have been shown to block the active binding site, thus altering hsp90 function and inducing c-erbB2 degradation. As a consequence of these events, induction of apoptosis was observed. Here, we examined whether hsp90 is differentially expressed in human tumors, and whether a correlation exits between protein levels of hsp90, c-erbB2 kinase and anti-apoptotic protein bcl-2a in 98 patient-derived tumor xenografts of 13 different histologies. Furthermore, we investigated the effects of 17-allylaminogeldanamycin (17-AAG) treatment on hsp90 levels in responsive xenograft biopsies. To accelerate and parallel the evaluation of gene expression patterns in multiple tumor types and to examine target modulation, we employed a tissue microarray technology. Microarrays were assembled from 98 well characterized, formalin-fixed, paraffin-embedded xenograft specimens, each in triplicates, by using a tissue arrayer. 4 μm sections of the resulting microarray block were cut and transferred onto glass slides by employing the paraffin-sectioning aid system. The expression of c-erbB2, hsp90 and bcl-2a was studied by immunohistochemistry using a peroxidase/diaminobenzidine staining procedure. 36% of all tumors stained positive for hsp90, 38% for c-erbB2 and 42% expressed bcl-2a to a notable extent. We found differential expression of all 3 antigens examined. Bcl-2a was up-regulated in colon, ovarian and several lung carcinomas, but no correlation was seen to either hsp90 or c-erbB2 expression. Significantly higher c-erbB2 levels were observed in lung adenocarcinomas, renal cell cancers and breast tumors compared to small cell lung cancers or melanomas. Interestingly, hsp90 which activates erbB-family members was co-expressed with c-erbB2 (Pearson coefficient, $R = 0.65$). 17-AAG, which has recently entered clinical trials, inhibits hsp90 function. In this study, we could show that hsp90 is downregulated in biopsies taken from 17-AAG treated tumors growing s.c. in nude mice compared to controls. Moreover, the decrease of hsp90 protein was dependent on duration of 17-AAG administration. These data suggest, that measurement of hsp90 levels could serve as surrogate marker for evaluation of 17-AAG efficiency. Our data further indicate that a relationship exists between c-erbB2 and hsp90 expression in tumor tissues and that more than 30% of human cancers might benefit from treatment with hsp90 or c-erbB2 modulating agents.

234

Tissue microarrays of human tumor xenografts for the characterization and validation of new molecular targets in cancer therapy

G.J. Wirth, K. Schandelmaier, E. Tetling, H.H. Fiebig, A.M. Burger (Freiburg, D)

Recent advances in oncogenomics particularly by using DNA microarray technology have led to the discovery of a vast amount of cancer related genes. Latter might prove valuable therapeutic targets or markers for diagnosis and prognosis of cancer. However, their characterization and validation is still very time-consuming. To accelerate this process, we have developed tissue microarrays from human tumor xenografts. The Freiburg xenograft collection comprises more than 400 human tumors of 20 different organs, which grow subcutaneously in serial passage in nude mice. Patient characteristics such as tumor grade, stage and survival as well as the response to standard therapeutic agents in patient and mouse are known. Microarrays were assembled from >100 formalin-fixed, paraffin-embedded tumor xenografts, each in triplicates, by using a tissue arrayer. The array comprised 13 tumor types. 4μm sections of the resulting microarray block were cut and transferred onto glass slides by using the paraffin-sectioning aid system. For 12 xenografts, target protein expression was compared over several passages. Expression of proteins associated with proliferation (Ki67), angiogenesis (VEGF, integrin avß3) and metastasis (cathepsin B, elastase) were studied using standard immunohistochemistry methods. Ki67 was expressed in most tumors, but at a variable extent, 39% were highly proliferative and showed nuclear staining in 75% of the surveyed area. High tissue VEGF was found in melanomas, prostate cancers and the majority of colon tumors. Tumor integrin avß3 was overexpressed in melanomas, ovarian and small cell lung cancers and was absent in sarcomas. Elastase was found widely distributed, but markedly elevated in the lung cancer panel. A high cathepsin B content was observed in melanomas, colon cancers and adenocarcinomas of the lung, whilst the protease was low in breast and small cell lung tumors. In general, overexpression of integrin avß3 and VEGF was more prominent in tumors derived from metastases than primaries.

No differences in staining intensity of any antigen examined were seen with increasing passages of human explants in nude mice. Furthermore, staining patterns remained largely consistent among 4 and more independent experiments. Many drug-targeting approaches such as polymer therapeutics employ tumor neovascularization for enhanced drug uptake or the expression of tumor-associated proteases for prodrug cleavage and activation. Our survey of proteins associated with invasion and angiogenesis revealed that such agents might best be evaluated in certain melanoma, lung and colon cancers. Tissue microarrays can accelerate the translation of basic science into the clinic by means of high-throughput target validation and parallel analyses of multiple tumor types. Moreover, by using patient specimens xenografted into nude mice, tumors expressing target levels can be selected for rational/target-oriented in vivo testing of tumor inhibitory strategies.

235

Identification and characterization of CATS, a novel protein which interacts with CALM and is expressed predominantly in thymus and spleen

L. Archangelo, J. Gläsner, S.K. Bohlander (München, Göttingen, D)

The recurring t(10;11)(p13;14) chromosomal translocation is found in the leukemic cells of patients with acute myeloid leukemia (AML), acute lymphoblastic leukemia (ALL) and malignant lymphoma. This rearrangement results in the fusion of AF10 on 10p13 and CALM (Clathrin Assembly Lymphoid Myeloid leukemia gene) located on 11q14. Molecular analysis of the CALM/AF10 fusion mRNA in patients with different types of hematological malignancies has shown that there is no obvious correlation between the location of the breakoints in CALM or AF10 and the phenotype of the leukemia. CALM can bind to clathrin heavy chains and has a phosphoinositol binding domain in its N-terminal region. CALM plays an important role in the regulation of clathrin-coated pit formation and in receptor mediated endocytosis. It could be shown that overexpression of CALM will block the formation of endocytotic vesicles. The t(10;11)(p13;q14) translocation is the sole structural chromosomal abnormality in several leukemic samples studied implying that the resulting CALM/AF10 fusion protein plays a pivotal role in leukemogenesis. However, nothing is known about the mechanism through which leukemia is caused by the CALM/AF10 fusion protein. To further elucidate CALM function we used the N-terminal half of CALM (aa 1-335) as a bait in a yeast two hybrid interaction screen to identify novel protein interaction partners of CALM. Among $1x10^6$ clones from a HeLa cDNA library several interacting clones were identified. One of these clones was derived from the C-terminal region of an unknown gene encoding 50 amino acids. Multiple tissue Northern blot analysis using this clone revealed a 1.6 kb transcript which is predominantly expressed in spleen and thymus and to a lesser extent in small and large intestines. The gene was named CATS (CAlm interacting protein expressed in Thymus and Spleen). Sequence analysis of a full length CATS cDNA revealed an open reading frame of 238 amino acids. The amino acid sequence of CATS shows no significant homologies to other proteins in the database. CATS is located on chromosome 17, has at least 5 exons and spans approximately 6 kb. Several processed pseudogenes of CATS could be identified in the human genome. The interaction between CALM and CATS could also be verified using GST-pulldown experiments. Further analysis showed that the CATS interacting region of CALM is between amino acids 294 and 335. To clarify the role of CATS in lymphoid tissue development further expression analysis including whole mount in situ hybridization on mouse embryos are underway. Disruption of CATS function by the CALM/AF10 fusion protein might be involved in leukemogenesis.

236

Overexpression of CXCR4 on B-CLL cells is associated with increased SDF-1 alpha plasma levels

H. Nückel, J. Dürig, K. Halfmeyer (Essen, D)

Chemokine receptors and their ligands have recently been shown to play a regulatory role in the migration, dissemination and survival of malignant hematopoietic cells. Using a multicolor flow cytometry approach we demonstrate that CXCR4 is expressed in immature hematopoeitic progenitor cells and up-regulated as CD34+ cells develop into mature B lymphocytes. CD19+/CD5+ B-CLL cells were found to exhibit an increased CXCR4 surface expression as compared to B cells isolated from the peripheral blood (PB) of normal donors (n=31, p<0.05). The presence of functional CXCR4 in these cells was confirmed employing a standard four hour transwell chemotaxis assay revealing a typical concentration dependent response to SDF-1 alpha. To gain further insight into the potential pathophysiological role of SDF-1 alpha / CXCR4 we compared SDF-1 alpha PB plasma levels (using ELISA) in CLL patients with those of normal donors. Interestingly, SDF-1 alpha plasma concentrations in the CLL patient cohort were significantly higher than in the normal control group (2689 ± 424 vs. 2128 ± 418 pg/ml, mean±SD; p=0.0004). SDF-1 alpha plasma levels were relatively homogenously distributed within the patient cohort and there was no apparent association with disease stage.
Finally, CLL cells were incubated in the presence or absence of varying concentrations of SDF-1 alpha and cell survival was studied at successive time points using trypan blue exclusion. In contrast to a recent report, we did not observe any consistent SDF-1 alpha induced effect on cell survival under these experimental conditions.

237

Expression of the tissue-specific cell cycle regulator cyclin A1 in solid tumors

C. Müller-Tidow, J. Potraz, S. Diederichs, S. Klümpen, M. Möller, B. Steffen, U. Vogt, R. Metzger, P. M. Schneider, M. Schrader, W.E. Berdel, H. Serve (Münster, Ibbenbüren, Köln, Berlin, D)

Cyclin A1 is a tissue specific cell cycle regulator which is highly expressed in meiotic cells in testis. Also, cyclin A1 has been shown to induce leukemia and to be highly expressed in acute myeloid leukemia blasts. In the current study, we analyzed cyclin A1 mRNA expression in a wide variety of human solid cancers (n=200) by using a quantitative real-time RT-PCR methodology. Interestingly, cyclin A1 expression was significantly reduced compared to normal controls in non-small cell lung cancer. This was in contrast to cyclin A2 levels which were found to be increased in the same tumors compared to normal controls Also, cyclin A1 expression was consistently very low in breast cancer. On the other hand high levels of cyclin A1 expression were detected in testicular tumors. A high variability in cyclin A1 expression was observed in colon and gynecological cancers. Interestingly, tumors with p53 mutations showed a trend to lower cyclin A1 expression levels in comparison to tumors with wildtype p53. To investigate whether p53 was associated with cyclin A1 induction, we induced p53 by UV irradiation of several wildtype p53 containing colon cancer cell lines. These experiments demonstrated a dose dependent increase in cyclin A1 mRNA expression following UV irradiation. This effect was restricted to p53 wildtype harboring colon carcinoma cells but did not occur in myeloid leukemia cell lines. Luciferase assays showed that UV irradiation led to cyclin A1 promoter induction in colon cancer cell lines. Again, this phenomenon did not occur in several tested myeloid leukemia cell lines where UV irradiation led to a steep decline in cyclin A1 promoter activity. Taken together, cyclin A1 is expressed at significant levels in a subset of solid tumors. The relationship between cyclin A1 and p53 in solid tumors suggests differential mechanisms of transcriptional cyclin A1 regulation in solid tumors compared to myeloid leukemia.

238

KIAA 0537, a ser/thr kinase that is frequently inactivated in primary human brain tumors

M. Schmid, S. Linortner, H. Ruff, H. Döhner (Ulm, D)

Next to deletion and mutation, hypermethylation of CpG-dinucleotides in the promoter area is another mechanism of gene inactivation. Using the SAGE database (http://www.ncbi.nlm.nih.gov/SAGE), we identified the serine/threonine kinase KIAA0537 as a gene with a CpG-rich promoter that is predominantly expressed in normal but not in malignant brain tissue. The gene consists of 7 exons coding for an 661aa protein with a highly conserved catalytic domain. In addition to normal brain, KIAA0537 is also expressed in normal heart and lung tissue. In four matched RNA glioma specimens (normal vs. tumor tissue from the same patient), KIAA0537 RNA was found exclusively in the normal but not in the tumor tissue. Furthermore, gene expression could be induced by the demethylating agent 5´-aza-2´deoxycytidine in 2/5 glioma cell lines (GMS-10 and GOS-3) and none of the lines had a deletion at the KIAA0537 gene locus. Southern blot analysis using methyl-sensitive restriction enzymes indicated that the reinduction of KIAA0537-mRNA was a direct effect of KIAA0537-promoter demethylation. Inhibition of histone deacetylase by Trichostatin A had no effect on KIAA0537 RNA expression. Phylogenetic analysis revealed a high homology to a C.elegans protein with so far unknown function and to a human cdc25C phosphatase co-factor. The Cdc25c-complex is involved in cell cycle control at the G2/M checkpoint. Interestingly, treatment of the human glioblastoma cell line GOS-3 with 5-aza-2´-deoxycytidine for 5 days arrested more than 50% of the cells in G2-phase of the cell cycle. Apoptosis was not observed. Finally, the full length KIAA0537 cDNA was cloned into a mifepristone-regulated mammalian expression vector and transfected into the human glioma cell line GMS-10. The addition of mifepristone resulted in the expression of a 71 kDa protein, which is very close to the theoretical molecular weight of KIAA0537. These data suggest that KIAA0537 might be a new tumor suppressor gene candidate that is frequently inactivated in human gliomas. In addition, our work puts emphasis on the importance of publicly accessible databases. Further studies are necessary to explore the possible role of 5-aza-2´-deoxycytidine as an anti-neoplastic drug in primary human brain tumors. This work was supported by a grant from the Deutsche Krebshilfe (Schm 1083-2/1)

Postersitzung:
Hämostase / Transfusion / Thrombopoese

239

Pathogen inactivation in platelet concentrates using Helinx technology

K. Janetzko, V. Harris, L. Lin, J. Flament, H. Klüter (Mannheim, D; Concord, USA; Nivelles, B)

Purpose: The INTERCEPT Platelet System using photochemical Helinx™ technology with psoralen S-59 and UVA inactivates bacteria, viruses, protozoa and leukocytes in platelet concentrates (PCs). After collection, PC are mixed with S-59, which reversible intercalates into helical regions of DNA and RNA. Upon UVA illumination, S-59 reacts with nucleic acid, forming covalent monoadducts and crosslinks. While platelets, which contain no nuclei, are unaffected, leukocytes and pathogens are unable to further replicate. Before transfusion, the PCs are treated with an adsorption matrix, which reduces the levels of residual S-59 and unbound photoproducts. The study evaluated the impact of the timing of Helinx treatment on in vitro platelet function. Methods: The study had two arms; each was paired. Double-dose apheresis PCs ($7.0^2 10^{11}$ platelets) resuspended in 35% plasma/65% platelet additive solution (PAS III) split into two equal size PCs. One PC unit was retained as an untreated Control and the other PC unit was photochemically treated as an INTERCEPT PC, with 150 mM S-59 and 3 Joules/cm^2 UVA. In study arm A (n=7) the photochemical treatment (PCT) was carried out one day after collection while in study arm B (n=8) PCT was performed 4 hrs after collection. Platelet function was assessed by analysis of pH, blood gas, glucose, lactate, lactate dehydrogenase (LDH), hypotonic shock response (HSR) and CD62 expression over a storage period of 7 days. Results: In vitro data on day 5 were expressed as mean ± SD. Using the ANOVA with time as repeated measure factor analysis of all data from Day 2 to Day 7 showed no significant differences between test (T) and control (C) units in arm A (p>0.05). In arm B the only significant interaction observed was for pH values at a probability of p=0.026. Conclusion: These results demonstrated that the in vitro quality of Helinx™ pathogen-inactivated split-dose PC did not differ significantly from Control PC over the storage period of 7 days whether the platelets were treated 4 hours post-collection ore one day later. To determine the therapeutic efficacy and safety of INTERCEPT platelets in an in vivo setting a randomised, controlled, clinical trials of hemato-oncological thrombocytopenic patients are currently underway.

240

The functional state of steroid versus G-CSF mobilized granulocytes: considerations for the storage of granulocyte concentrates for neutropenic patients

A. Eil, M. Schmitt, P. Reinhardt, B. Maccari, E. Krug, A. Schulz, B. Kubanek, M. Wiesneth (Ulm, D)

Background: High yields of granulocytes by G-CSF mobilization of the donors sparked new enthusiasm for transfusion in patients with severe neutropenia. The functional state of these granulocytes during storage was evaluated, and the clinical outcome of the transfused patients assessed. Material and Methods: Granulocyte concentrates collected from 10 donors receiving prednisone or G-CSF before leukapheresis were evaluated over 48 hrs of storage for differential blood count, pH, electrolytes, viability, phagocytosis and oxidative burst. Secretion of IL-1beta, IL-6, IL-8 and TNFalpha was measured. The follow-up of 5 transfused patients was monitored by blood counts, CRP and clinical outcome. Results: Over the first 24 hrs of granulocyte storage, no major changes were seen after both mobilization regimens for viability, phagocytosis, oxidative burst and cytokine profile. In contrast, these parameters changed significantly after 48 hrs of storage. Steroid and G-CSF mobilized granulocytes showed no major differences in the first 24 hrs, except a more prominent decrease of pH and increase of IL-8 secretion of the G-CSF mobilized granulocytes. Both, fresh granulocyte transfusions and the G-CSF mobilized granulocytes stored overnight were well tolerated without an additional side effect and resulted mostly in a leukocyte increase and in a CRP decrease. Conclusion: G-CSF increases the yield of leukocytes 3.5 fold without enhancing the cytokine mediated toxicity, but also without improving their functional stability during storage. G-CSF mobilized granulocytes can be stored for 24 hrs without significant loss of in vitro and in vivo function.

241

Contribution of p- and k-receptor chains to c-mpl mediated proliferation and signal transduction

S. Koschmieder, W.K. Hofmann, S. Wagner, D. Hoelzer, O.G. Ottmann, U. Kalina (Frankfurt/Main, D)

Introduction: Functional c-mpl is essential for normal megakaryopoiesis and hematopoietic stem cell renewal. Aberrant c-mpl has been implied in acute myeloid leukemia, congenital amegakaryocytic thrombocytopenia, and myeloproliferative disorders. The c-mpl gene encodes two transmembraneous receptor chains (p, k) but their functional role and interaction is incompletely understood. This study was initiated to further characterize the function of c-mpl-p and -k. Methods: c-mpl receptor p and k chains were cloned from human bone marrow-derived CD34+ cells. Growth factor-dependent murine BA/F3 cells which lack c-mpl were stably transfected with either the p or k chain alone (using a neomycin or zeocin-resistant vector, respectively) or both chains together. The clones were analyzed for cell growth and survival (trypan blue exclusion) in liquid cultures, proliferation using ATP release and 3H thymidine uptake assays, and for signal transduction by Western blotting. Results: RT-PCR verified expression of the respective c-mpl receptor chains in the transfected cells. c-mpl-p clones showed thrombopoietin (TPO)-dependent cell growth and survival in culture for at least 14 days in the presence of neomycin, and this was abrogated by dominant-negative STAT5 (no viable cells after 14 days). Likewise, TPO enhanced proliferation in c-mpl-p clones (1956±836 vs. 371±103 relative units) as detected by ATP release assays. Mock clones and c-mpl-k clones failed to grow in a TPO-dependent manner, and TPO did not increase proliferation. Clones containing both p and k chains (c-mpl-pk) showed high proliferation without (11376±3602 units) and with TPO (13271±780 units). Western blot showed phosphorylation of STAT5, STAT3, and MAPK ERK1/2 in c-mpl-pk clones which was enhanced by TPO. c-mpl-p clones showed very weak activation of these molecules while c-mpl-k and mock clones did not show any activity. Conslusion: The c-mpl p chain is the major functional receptor chain which mediates TPO-responsive growth, survival, proliferation, and signaling. However, full proliferation and signaling requires the presence of both receptor chains p and k.

242

Anti-CD 20 monoclonal antibody Rituximab in patiens with refractory immune thrombocytopenia

A. Lang, B. Hartmann, R. Konzett, R. Greil (Feldkirch, Innsbruck, A)

Background: An unresolved problem is the treatment of patients with a clinical diagnosis of severe immune thrombocytopenia who had failed therapy with corticosteroids, intravenous IgG, intravenous anti-D, chemotherapy, plasmapheresis and/or splenectomy. Objective: We would like to report the treatment of patients with severe, refractory immune thrombocytopenia with the chimeric anti-CD20 monoclonal antibody Rituximab in patients who failed conventional therapy. Patients: Five patients with immune thrombocytopenia who failed one or more conventional therapies and platelet counts < 20000/µl. Intervention: Weekly infusion of Rituximab 375 mg/m^2 for four weeks. Measurement: platelet count. Results: 2 patients with complete normalisation of thrombocytes, 2 pts. with marked increase of platelets >100000/µl, 1 pt. no response.

243

Recanalization of cerebral sinus thrombosis after continuous intravenous administration of dalteparin

H.-H. Wolf, M. Stephan, O. Dorligschaw, A. Heider, A. Teuber, H.-J. Schmoll (Halle, D)

Thrombosis of cerebral sinus is a serious, life threatening complication. Ascending jugular and cerebral sinus thrombosis may occur in patients with thrombophilic diathesis as APC resistance, protein C deficiency, or after asparaginase therapy. Predisposition may be mediastinal tumour, vena caval compression, or central venous access. In most of the patients, thrombotic material cannot be resected by thrombectomy, and systemical lysis cannot be performed as platelet counts may be low and plasmatic disorders may induce hemorrhages. We report two patients with thrombophilic diathesis and diffuse thrombosis of cerebral sinus. The 28 year old man suffering from acute T-lymphoblastic leukemia presented mediastinal mass and thrombosis of the subclavian vein, superior caval vein, and venae jugulares, respectively, following central venous catheter. A 47 year old female patient with protein S deficiency suffered from cerebral sinus thrombosis without any history of injury or surgery. In both patients thrombosis was extended, and craniotomy had to be performed for excessive intracerebral pressure despite anti-edematous medication. After initial bolus injection (5 IU / kg body weight iv) we administered dalteparin as continuous intravenous infusion in a dosage of 5 IU / kg body weight / hour in order to achieve anti-Xa plasma concentrations of 0.4 – 0.6 IU/ ml longer than for 3 weeks. In both patients mostly complete recanalisation of cerebral sinus and jugular veins were seen without any hemorrhagic complications. Neurological symptoms improved significantly. We conclude that continuous intravenous infusion of dalteparin presents a high fibrinolytic activity and may be an adequat therapy of sinus thrombosis even in patients with high risk for cerebral bleeding.

244

Intracellular domains of target antigens influence their capacity to trigger ADCC

B. Stockmeyer, C. Frank, K. Tiroch, M. Gramatzki, T. Valerius (Erlangen, D)

Today, an increasing number of antibodies is used in tumor therapy. Antibody- mediated signalling in tumor cells, and antibody- dependent cellular cytotoxicity (ADCC) were both considered as relevant effector mechanisms for antibodies. To address potential interactions between these two mechanisms, we generated chimeric target antigens, which were then expressed in experimental tumor target cells. For these experiments, HER-2/neu and CD19 served as prototypic target antigens. HER-2/neu- directed antibodies were documented to mediate effective ADCC with mononuclear as well as neutrophil effector cells, while antibodies against CD19 were effective only with mononuclear cells, but not with neutrophils. Using splicing-by-overlap-extension- PCR, we generated cDNAs encoding HER-2/CD19 (extracellular/intracellular), or CD19/HER-2 chimeric fusion proteins. For this purpose, cDNAs for extracellular domains of HER-2/neu or CD19 were fused with cDNAs encoding intracellular domains of CD19 or HER-2/neu, respectively. After transfecting wildtype HER-2/neu, or chimeric HER-2/CD19 into RAJI Burkitt's lymphoma cells, or wildtype CD19, or chimeric CD19/HER-2 into SK-BR-3 breast cancer cells, target cell lines were selected by FACS for high membrane expression of transfected target antigens. We then used Cr-51 release assays to investigate the efficacy of tumor cell lysis by neutrophils and mononuclear effector cells in the presence of CD19- or HER-2/neu- directed antibodies. Mononuclear cells triggered effective ADCC against target cells expressing wildtype or either chimeric target antigen. As expected, neutrophils killed wildtype HER-2/neu-, but not wildtype CD19- transfected target cells. Neutrophils were also effective against chimeric CD19/HER-2, but - interestingly - not against chimeric HER-2/CD19- transfected target cells. In conclusion, these results demonstrate that intracellular domains of target antigens contribute substantially to effective antibody- mediated tumor cell killing by neutrophils.

245

CD8+ T cell responses against candidate immunotherapy antigens WT1 and proteinase 3 in patients with acute myeloid leukemia

C. Scheibenbogen, E. Thiel, A. Schmittel, D. Nagorsen, U. Keilholz (Berlin, D)

The Wilms' tumor (WT1) protein and proteinase 3 are both overexpressed antigens in acute myeloid leukemia (AML) blasts and contribute to leukemogenesis. Recently it was shown that cytotoxic T lymphocytes against these antigens can be elicited in vitro and in murine models. We performed this study to investigate whether WT1- and proteinase 3-specific CD8+ T cells exist in AML patients. In two of 9 HLA-A2-positive AML patients in complete remission T cells recognising HLA-A2.1-binding epitopes from WT1- and proteinase 3 could be detected in repetitive blood samples by IFNg-ELISPOT assay ex vivo. T cells reactive with the WT1(187-196) epitope were further characterized by four-colour-flow cytometry, identifying them as CD3+CD8+IFNg+granzymeB+ T cells, resembling cytotoxic effector T cells. These results provide for the first time evidence for T cell reactivity against defined antigens in AML patients. These data further support that WT1 and proteinase 3 are immunogenic in AML and are attractive candidates for vaccination strategies.

246

Characterization of domains in the human peptide transporter subunit TAP2 required for TAP function

S. Han, U. Ritz, C. Huber, B. Seliger (Mainz, D)

The transporter associated with antigen processing (TAP) belongs to the large family of ABC-transporters and has the typical three domain structure of these transporters consisting of a C-terminus spanning the ER-membrane several times, a substrate binding domain and the NTP-binding site near the N-terminus. TAP is the only ABC transporter with a function in the immune system as it transports antigenic peptides from the cytosol into the ER where they are loaded onto MHC class I molecules. In contrast to other ABC transporters TAP consists of two subunits, TAP1 and TAP2. Recently, regions in the peptide binding domain of the TAP1 subunit have been identified to be essential for peptide transport. In order to define domains in the TAP2 subunit required for proper TAP function, a series of human TAP2 deletion mutants was generated and stably expressed in human TAP2 negative cells. Although TAP2 mRNA expression could be detected in all TAP2 mutants, a heterogeneous TAP2 protein expression pattern was observed in these cells. In contrast to wt TAP2 transfectants, the different TAP2 mutants lack MHC class I surface expression. Further functional analyses in terms of intracellular localization, ATP-binding ability and/or peptide transport rate are in progress. However, the results provide evidence that a large region of the putative peptide binding domain of the TAP2 subunit is absolutely crucial for the proper presentation of cytosolic antigens via MHC class I molecules. These data further implicate that in comparison to the TAP1 subunit, TAP2 appears to be more essential for peptide binding, but this hypothesis will be further discussed.

247

In vivo tracking of adoptively transferred melan-A-specific T cells in melanoma patients using peptide-MHC tetramers

N. Meidenbauer, M. Laumer, S. Vogl, J. Heymann, R. Andreesen, A. Mackensen (Regensburg, D)

Adoptive T cell therapy has been successfully used for treatment of viral and malignant diseases. However, little is known about the fate and the frequencies of transferred T cells in the peripheral blood. Since the tetramer technology has been shown to allow detection and characterization of antigen-specific cytotoxic T lymphocytes (CTL) at very low frequencies, we used this method to assess the number of circulating Melan-A-specific CTL in melanoma patients before and after adoptive T cell therapy. Melan-A-specific CTLs were generated from HLA-A2.1-positive patients with advanced melanoma by 4 rounds of in vitro stimulation of CD8+ T cells with autologous Melan-A-pulsed dendritic cells. Up to date, four patients received at least three i.v. infusions of 120-700x10^6 T cells at 2 week intervals together with low dose IL-2 (3x10^6 IE s.c./d for 6 days). The frequency of Melan-A-specific CTL was monitored by staining peripheral blood mononuclear cells with fluorescent-labelled HLA-A2.1-Melan-A tetramers before and 1h, 3h, 24h, and 48 h after adoptive transfer. To ensure the specificity of the staining, an irrelevant tetramer was used as negative control. The transferred T cell product contained between 10% to 90% Melan-A specific CTL as determined by tetramer analysis. Before therapy, the frequencies of Melan-A-specific CTL in all CD8+ T cells ranged from 0.01% to 0.07%. Depending on the amount of transferred T cells, the number of antigen-specific cells rose to maximal frequencies ranging from 0.04% to 1.77% CD8+ T cells as early as 1h after transfer. At 3-6h after transfer we could observe a slight decrease in the number of circulating Melan-A-positive CTL, followed by an increase in 3 out of 4 patients. This pattern was remarkably stable after each transfer in the respective patients. An elevated frequency of antigen-specific T cells could be demonstrated for up to 14 days after transfer leading to increased baseline levels of Melan-A specific CTL after repeated T cell therapy in 2 out of 4 patients. These studies suggest that the infused Melan-A specific CTL remained physiologically intact for more than 2 weeks after retransfer. Moreover, the tetramer technology provides a method to monitor the frequency of adoptively transferred antigen-specific T cells in the peripheral blood thus facilitating new insights into the fate of ex vivo generated T cells after retransfer into the patient.

248

Quantification of WT1 and proteinase 3 antigen expression in AML patients to assess their potential as targets for immunotherapy on molecular basis

N. Max, E. Thiel, K. Wolf, W. Knauf, U. Keilholz (Berlin, D)

WT1 and proteinase 3 are both markedly overexpressed in leukemia with an inverse correlation between expression and prognosis. Although, they are also expressed in normal counterparts, WT1 is thought to be related to leukemogenesis as indicated by antisense oligonucleotides leading to growth inhibition. Both, WT1 and proteinase 3, might be good targets for immunotherapy as shown in several human and murine settings. We here report on WT1 and proteinase 3 specific real-time RT-PCR assays to absolutely quantify WT1 and proteinase 3 transcript amounts in patient blood or bone marrow samples (different stages of disease) and normal counterparts. We analyzed samples from 15 AML patients at diagnosis, 5 at hematological remission, 5 at relapse as well as 6 sorted CD34+ samples (leukapheresis product of peripheral blood stem cells) and PBMCs from 5 healthy individuals. Total RNA was isolated and reverse transcribed into cDNA which was subsequently quantified by WT1, proteinase 3 and porphobilinogen deaminase specific real-time PCRs (Light Cycler). Regarding WT1 (proteinase 3) expression, samples of patients at diagnosis showed in average a 1,674- (1,627-), 129-, 37-(2-) and 7- fold higher expression than those of healthy volunteers, patients at remission, CD34+ cells and patients at relapse. proteinase 3 expression was 3- and 2-fold higher in patients at remission and at relapse. WT1 (proteinase 3) expression of patients at relapse were in average 250-(3.336-), 19- and 6-(4-) fold higher than those of healthy volunteers, patients at remission and CD34+ cells. In CD34+ samples, WT1 (proteinase 3) was in average 45-(877-) and 3-(5-) fold higher expressed than in healthy persons and in patients at remission. Patients at remission showed a WT1 (proteinase 3) expression which was 13- (877-) fold higher than in healthy blood. WT1 seems to be an ideal target for immunotherapy, because it is significantly overexpressed in patients at diagnosis and at relapse and less in patients at remission compared to the normal counterpart. This is not the case for proteinase 3. Furthermore, several investigations showed that WT1/proteinase 3 derived peptides can stimulate/elicit CTL responses against peptide pulsed target cells without killing physiologically WT1 expressing cells, indicating that triggering of CTLs seems to be a treshold penomenon and making WT1 a good target for vaccine and T cell therapy.

249

A phase-I study with a ricin A-chain immunotoxin (Ki-4.dgA) in patients with refractory CD30 positive Hodgkin's and non-Hodgkin's lymphoma

R. Schnell, O. Staak, P. Borchmann, C. Schwartz, B. Matthey, E. Vitetta, J. Schindler, V. Diehl, A. Engert (Köln, D; Dallas, USA)

Background: Ki-4.dgA is an anti-CD30 immunotoxin (IT) constructed by coupling the monoclonal antibody (Moab) Ki-4 via a sterically hindered disulfide linker to deglycosylated ricin A-chain. This IT demonstrated high efficacy in vitro and in a disseminated human Hodgkin's lymphoma SCID mice model. Objectives of this phase I trial included the determination of the maximum tolerated dose (MTD), the dose limiting toxicities (DLT), pharmacokinetic data and antitumor activity. Patients with relapsed CD30-positive lymphoma were included. Patients and methods: The IT was administered in a dose escalation trial in four bolus infusions over seven days (day 1, 3, 5 and 7). 17 patients were treated at total doses per cycle of 5, 7.5, or 10 mg/m^2. Patients received one to three cycles of treatment. Results: All patients had signs of progressive disease and were heavily pretreated with a mean of 6 (range 2-9) different prior therapies including autologous bone marrow transplantation in 13 of 17 patients. 9 patients had primary progressive disease. Mean age was 35 years (24 to 52 years). 10/17 patients had advanced disease (stage IIIB and IV) with massive tumor burdens. Peak serum concentration of intact IT varied from 0.23 to 1.7 µg/ml. Side effects were related to vascular leak syndrome (VLS), i.e. decreases in serum albumin, edema, weight gain, hypotension, tachycardia, myalgia and weakness. At 7.5 mg/m^2 one patient experienced a grade 4 myalgia. All 2 patients receiving 10 mg/m^2 had WHO grade 3 toxicities (VLS, edema, nausea, dyspnea, weakness or tachycardia). 7/17 patients made human anti-ricin antibodies (HARA) (= 1.0 µg/ml) and one patient developed human anti-mouse antibodies (HAMA) (= 1.0 µg/ml). Clinical response included of 15 evaluable patients 2 minor response (MR) and 2 stable diseases (SD). Conclusions: VLS is the DLT of Ki-4.dgA. The MTD is 5mg/m^2. The IT was only moderately tolerated which might be due to binding of the IT to soluble CD30 antigen and prolonged circulation of IT/sCD30 complexes. Thus, native Ki-4 should be given before infusion of the IT.

250

Growing tumors induce myeloid barrier cells which in the presence of IgG-bound TGF-ß prevent proliferation of activated CD8+ CTL

C. Beck, K. Schreiber, D.A. Rowley (Kiel, D; Chicago, USA)

Progressive growth of immunogenic tumors occurs despite priming of specific cytolytic T cell responses, which are required for efficient tumor rejection. Nevertheless, in many murine tumor models, specific cytolytic T cells are barely demonstrable in the tumor-bearing host. Most tumors also induce the formation of tumor specific antibodies as a result of a sustained B cell response and continuous B cell stimulation is shown to induce CTL inhibition in the presence of a third cell. We here identify this „barrier cell" in murine bone-marrow as a 'late' CFU (most likely CFU-M) based on surface-marker expression. „Barrier cell" activation is rapidly induced in vivo by non-specific inflammatory stimuli such as CFA or by growing tumors. Its function results in reversible inhibition of CTL proliferation, requires continuous presence of IgG-bound TGF-ß, intact Fc receptors on the „barrier cell", is MHC-independent but requires cell-cell contact. We therefore propose a model of CTL regulation mediated by 'late' CFU in response to B cell activation and additional inflammatory stimuli, resulting in inhibition of CTL proliferation. Interference with „barrier cell" activity by eliminating B cells or blocking Fc Receptors may be a rational approach for enhancing effective CTL mediated anti-tumor immunity.

251

HLA-A*0201-transfected K562 cells can present endogenously expressed and exogenously loaded peptide antigens to CD8+ T cells in IFN-gamma ELISPOT assays

C.M. Britten, R.G. Meyer, T. Kreer, I. Drexler, T. Wölfel, W. Herr (Mainz, München, D)

ELISPOT assays are widely used for the monitoring of specific T lymphocyte responses in clinical trials assessing antigen-specific immunizations in patients with cancer or chronic viral infections. One major limitation for the broad clinical implementation of ELISPOT assays is the lack of an inexhaustible source of suitable HLA-matched antigen-presenting cells (APC). Currently available allogeneic or xenogeneic APC (such as human T2 or HLA-transfected insect cells) can either lead to strong background spot production by APC-reactive T lymphocytes or they have a low antigen presentation capability which can prevent the detection of low frequency T cell responses in PBMC. In search of alternative APC, the human erythroleukemia cell line K562 that per se does not express HLA class I and class II molecules on its cell surface was transfected with the HLA-A*0201 gene. Resulting HLA-A*0201-expressing K562 (K562/A*0201) cells were able to process and present endogenously expressed and exogenously loaded melanoma peptide antigens to HLA-A*0201-restricted cytolytic T lymphocyte clones both in cytotoxicity and interferon-gamma (IFN-g) ELISPOT assays. K562/A*0201 and T2 cells were then used as APC in IFN-g spot assays to detect ex vivo CD8+ T cells responsive to HLA-A*0201-binding peptide epitopes derived from Epstein-Barr virus (EBV), influenza and cytomegalovirus (CMV). CD8+ T cells isolated from PBMC of 10 HLA-A2-positive donors recognized unloaded T2 cells with frequencies ranging from 15 to 274 spot-forming cells (SFC) per 10^5 CD8+ T lymphocytes, whereas T cell reactivity induced by unloaded K562/A*0201 cells was comparably low and ranged from 0 to 33 SFC per 10^5 CD8+ lymphocytes. Thus, background T cell responsiveness against unpulsed APC was generally reduced when using K562/A*0201 compared to T2 cells. Of major importance, both APC lines were similarly efficient in their ability to visualize EBV, influenza and CMV peptide-reactive CD8+ T lymphocytes (34 to >500 SFC per 10^5 CD8+ T cells). We conclude that K562 cells transfected with HLA-A*0201 or other HLA genes can serve as suitable APC for monitoring T lymphocyte responses against tumor and viral peptide antigens in IFN-g ELISPOT assays, especially when T lymphocyte samples contain a strong level of anti-T2 reactivity.

252

Generation and expansion of Her2–reactive T cell clones for adoptive immunotherapy.

B. Schmidt, D. Busch, E. Schulz, C. Hermann, H. Wagner, Ch. Peschel, H. Bernhard (München, D)

INTRODUCTION: The human epidermal growth factor receptor 2 (Her2/c-erbB-2/neu) is overexpressed in tumors of various origins, such as breast or stomach. Overexpression in breast cancer is associated with poor prognosis. Several Her2-derived peptides have been defined that bind to HLA-A2 molecules and are able to induce specific T cells capable of lysing Her2-overexpressing tumor cells in vitro. Peptides are derived from the intracellular (ELVSEFSRM), transmembrane (IISAVVGIL) and extracellular (VLRENTSPK) domain of Her2. The goal of our work is to improve the generation and expansion of Her2-specific cytotoxic T cells (CTL) ex vivo as a requirement for adoptive transfer of tumor-reactive CTL in patients with Her2-overexpressing tumors. METHODS: Autologous mature dendritic cells were pulsed with Her2-derived peptides and used for stimulating autologous peripheral blood mononuclear cells (PBMC) from HLA-A2 positive healthy donors and cancer patients. Following several weekly restimulations CD8+ T cells were sorted, based on their binding to HLA-Peptide-Tetramers, and further cloned and expanded. RESULTS: Baseline frequencies of CD8+/Tetramer+ Her2-specific T cells were < 0.01% of total PBMC in healthy donors. Following 2 restimulations the frequencies of Her2-ECD specific T cells increased to 0.07% of total PBMC. These T cells were sorted and cloned in the presence of allogeneic PBMC, EBV-B-cells, anti-CD3 monoclonal antibody and IL2. T cell clones were screened for lysis of peptide-pulsed T2 cells and specific T cell clones were further expanded. The calculated yield after 3 rounds of expansions was $>1x10^{10}$ T cells per T cell clone, a number sufficient for several cycles of adoptive immunotherapy. The expanded CTL efficiently lysed peptide-pulsed T2 cells and the Her2 overexpressing, HLA-A2 transfected ovary carcinoma cell line SKOV3-A2. The generation of CTL clones against other Her2-epitopes from both HLA-A2+ healthy donors and patients is currently underway. CONCLUSIONS: Tetramer guided sorting of in vitro primed T cells allows early cloning of low-frequency Her2-specific CTL. The expansion of the tumorreactive T cell clones yields cell numbers sufficient for adoptive immunotherapy for patients with Her2-overexpressing breast cancer. A program of adoptive transfer of Her2-specific CTL is planned.

253

Expansion of colony forming cells of high proliferation potential and long term culture initiating cells by IL-10, SCF and IL-3 in vitro

F. Keil, F. Elahi, H. T. Greinix, G. Fritsch, N. Louda, A. L. Petzer, E. Prinz, T. Wagner, P. Kalhs, K. Lechner, K. Geissler (Wien, A)

Recently, we have shown that interleukin-10 (IL-10), IL-3 and stem cell factor (SCF) expand committed bone marrow (BM) progenitor cells in vitro. Here, we demonstrate that long term culture initiating cells (LTC-IC) can be expanded in a one week suspension culture of unseparated BM cells supplemented with IL-10, IL-3 and SCF. After one week in suspension culture a two fold increase of LTC-IC was seen in limiting dilution analysis if compared with BM cells pre expansion. Additionally an increased proliferative capacity was achieved with a 7 fold increase of committed colony forming cells (CFC) and a 10 fold proliferation of CFC of high proliferative potential. Suspension cultures supplemented with IL-10 alone showed maintenance of LTC-IC without increased proliferation of committed CFC or nucleated cells whereas unsupplemented cultures showed terminal differentiation of all hematopoietic progenitors. Immunophenotyping revealed a 7 fold increase of CD34 positive RA negative cells in IL-10+ IL-3+ SCF stimulated suspension cultures whereas there was no difference in CD34 positive cells. IL-3+ SCF showed also significant stimulation of CFC and LTC-IC but expansion of LTC-IC and lineage negative CD34-positive cells was significant lower if compared with the IL-3+ SCF+ IL-10 supplemented culture. In contrast to LTC-IC and early multilineage colonies IL-10 had no effect on proliferation of final differentiated cells which indicates that IL-10 inhibits terminal differentiation. To our knowledge expansion of LTC-IC by IL-10+ IL-3 + SCF has not been shown so far. In summary, our in vitro model allows expansion of LTC-IC without extensive stem cell manipulation which should be an attractive model for in vitro purging, gene transfer or expansion of stem cells to allow rapid engraftment after myeloablative chemotherapy.

254

The receptor for hyaluronan acid mediated motility (RHAMM) is a new immunogenic tumor-associated antigen highly expressed in acute and chronic myeloid leukemia (AML/CML)

J. Greiner, M. Ringhoffer, M. Taniguchi, A. Szmaragowska, A. Eil, H. Döhner, M. Schmitt (Ulm, D)

Identification and characterization of tumor-associated antigens (TAA) eliciting an immune response in tumor patients is a prerequisite for specific immunotherapy of tumors. The receptor for hyaluronan acid mediated motility (RHAMM) was identified as a new immunogenic antigen using the method of serological screening of expression libraries (SEREX) prepared from the CML cell line K 562. RHAMM is involved in cell growth and motility. Serological responses to RHAMM were detectable in patients with CML (31 %), AML (42 %), renal cell cancer (RCC, 50 %), melanoma (83 %), breast cancer (67 %) and ovarian cancer (50 %), whereas in healthy volunteers (0/12) and patients with autoimmune diseases (0/7) no serological reactions were observed. mRNA expression of RHAMM was detected at a high level in 35 % (14/40), at a low level in 40 % (16/40) and not detected in 25 % (10/40) of AML patients. No mRNA expression was found in PBMN and CD34 separated cell samples from healthy volunteers. Western blot analysis showed expression of the RHAMM protein in 70 % of AML patients and in all leukemia cell lines, but not in PBMN of healthy volunteers. The mRNA expression of RHAMM was quantified by real time RT-PCR using the light cycler technology. mRNA expression of RHAMM was also found in CML patients (39 %) and patients with renal cell cancer (73 %). In normal tissues, mRNA expression of RHAMM was only found in testis, placenta and thymus tissue. Moreover, we examined the expression of the different splice variants RHAMM-48 and RHAMM-147, and we found a new variant of RHAMM expressed in myeloid leukemias and RCC. Taken together, RHAMM is a newly identified tumor-associated antigen inducing serological immune responses in hematological malignancies and solid tumors and showing high expression in AML, CML and in RCC.

255

Relapsed aleukemic leukemia cutis complicated by pulmonary aspergillosis: successful treatment with interleukin-2

H.-G. Kopp, B. Schönberger, M. Sökler, L. Kanz, C. Denzlinger (Tübingen, D)

Leukemia cutis is an uncommon extramedullary manifestation of leukemia associated with a bad prognosis. All reported cases of aleukemic leukemia cutis progressed into overt AML with bone marrow and peripheral blood involvement within months. Induction chemotherapy analogous to protocols for AML is recommended, but relapse rates are high and not all patients qualify for this strategy. Interleukin-2 (IL-2) has been shown to be of potential benefit in AML and may serve as an alternative approach in patients unfit for chemotherapy. We report on an elderly patient who was succesfully treated with IL-2 in relapsed aleukaemic leukemia cutis complicated by pulmonary aspergillosis. At her initial presentation the 62-year-old woman had aleukemic leukemia cutis diagnosed by skin biopsy. She refused chemotherapy, and systemic AML FAB M5 evolved within one year. The patient was treated with combination induction chemotherapy including mitoxantrone, cytarabine, and etoposide. Complete remission was obtained and consolidated by two cycles of idarubicin, etoposide, and cytarabine. However, the patient acquired invasive pulmonary aspergillosis with a retrosternal aspergilloma persisting despite intensive antimycotic therapy with amphothericin B and itraconazol. Only 2 months after the second consolidation chemotherapy course, the patient presented with a relapse of aleukemic leukemia cutis. With continued itraconazol therapy, fungal infection persisted. Therefore, salvage chemotherapy was rejected.Treatment with IL-2 was initiated according to a regimen adapted from an EORTC protocol. Four cycles of high-dose IL-2 were administered, consisting of 5 consecutive days of continuous intravenous IL-2-infusion (4.5 x 10^6 IU/m^2 to 10 x 10^6 IU/m^2 in the first cycle, 10 x 10^6 IU/m^2 to 18 x 10^6 IU/m^2 in the second and 10 x 10^6 IU/m^2 both in the third and fourth cycle) followed by 72 h rest. The cutaneous infiltrates appeared more pronounced and slightly increased in number during the first cycle, and completely disappeared after the second. Complete remission was maintained with subcutaneous IL-2 (4 x 10^6 IU/m^2) given on 5 days every month on an outpatient basis. During this treatment, pulmonary aspergillosis resolved completely. However, six months later, systemic AML evolved again. As the patient had a matched sibling donor, she was assigned to immediate allogeneic stem cell transplantation. The patient had a timely engraftment and there was no major transplant-related toxicity. This case illustrates that treatment with IL-2 can induce complete remission of relapsed leukemia cutis. Moreover, IL-2 may contribute to resolve infectious complications such as pulmonary aspergillosis.

256

CD4+ T cells engrafted with a recombinant immunoreceptor efficiently lyse target cells in a MHC antigen- and Fas-independent fashion

A. Hombach, C. Heuser, T. Marquardt, A. Wieczarkowiecz, V. Groneck, C. Pohl, H. Abken (Köln, D)

T cells engrafted by a recombinant immunoreceptor with predefined antigen specificity can efficiently lyse antigen-positive target cells in a MHC antigen independent manner. It is yet unresolved how receptor grafted CD4+ T cells contribute to MHC antigen independent target cell lysis. To address this issue, we grafted isolated CD8+ and CD4+ T cells from the peripheral blood with recombinant anti-CEA and anti-CD30 receptors, respectively. Cytotoxicity analyses revealed that grafted CD4+ T cells exert cytolysis of antigen-positive target cells with an efficiency similar to that of grafted CD8+ T cells. Lysis by receptor grafted CD4+ T cells is antigen-specific and is inhibited by blocking the target antigen or the antigen binding site of the recombinant receptor. Both Fas-sensitive and Fas-resistant target cells are lysed with equal efficiency and lysis of Fas-sensitive target cells is not blocked by an anti-Fas ligand antibody indicating that cytolysis by receptor grafted CD4+ T cells is independent of the Fas pathway. We conclude that cytolysis by CD4+ T cells equipped with a recombinant immunoreceptor is MHC antigen- and Fas-independent and likely to be mediated by perforin present in receptor grafted CD4+ T cells.

257

Alterations of the MHC class I phenotype in HNSCC due to deficiencies of the antigen processing pathway

M. Meissner, T. Reichert, T. Whiteside, C. Huber, B. Seliger (Mainz, D; Pittsburgh, USA)

MHC class I surface expression is frequently altered in head and neck squamous cell carcinomas (HNSCC) which represents a strategy for their evasion of immune surveillance. In this study we examined the underlying mechanisms of such downregulations by analysing tumour biopsies and long term established cell-lines from HNSCC for the expression of the proteasome subunits, the TAP peptide transporter, various chaperones and MHC class I antigens. HNSCC lesions displayed a heterogeneous expression pattern of the components of the MHC class I antigen processing machinery (APM). In contrast to corresponding normal tissues, the protein expression of LMP2, LMP7, TAP1 and TAP2 was strongly downregulated or even lost in a high frequency of tumour lesions. These abnormalities were neither associated with tumour grading and staging nor with the HPV phenotype.
The HNSCC cell lines demonstrated impaired expression and/or function of a single or combined APM component(s) which could be paralleled by reduced levels of the MHC class I surface antigens. With the exception of LMP7 IFN-gamma treatment was capable to enhance the expression of the different APM components. Thus, the altered expression of the MHC class I antigen processing pathway is a frequent event in HNSCC affecting different steps of the antigen processing pathway and can result in loss or downregulation of MHC class I surface expression in this disease. These alterations may explain the immune escape phenotype of HNSCC cells, which negatively interferes with the development of tumour-specific T-cell response and have an impact on the design of T-cell based immunotherapeutical strategies.

258

The tumor-associated antigen PRAME is a potential marker for minimal residual disease (MRD) and a candidate for immunotherapy in acute myeloid leukemia (AML)

J. Greiner, M. Ringhoffer, M. Taniguchi, S. Hübsch, A. Eil, H. Döhner, M. Schmitt (Ulm, D)

The first tumor-associated cancer/testis-antigen characterized in leukemias is the preferentially expressed antigen in melanoma (PRAME). Specific cytotoxic T lymphocytes (CTL) clones against PRAME could be generated. In conventional RT-PCR, 48% of samples obtained from 42 patients with acute myeloid leukemia at the time of diagnosis expressed PRAME at a high level, 17 % at a low level, while no expression was found in 35% of these AML patients. mRNA expression was also not detectable in all peripheral blood mononuclear cells (PBMN), bone marrow cells and CD34 positive separated cells from healthy volunteers. In normal tissues, mRNA expression of PRAME was only found in testis tissue. Therefore, it is a potential target for specific immunotherapy of leukemias. To evaluate the eligibility of patients for PRAME directed immunotherapy, to monitor PRAME expression during therapy and to detect MRD, we established an RT-PCR for PRAME using the light cycler in the sensitive SYBR green format. For the quantification of the mRNA expression of PRAME against the housekeeping genes beta-actin and TATA-box binding protein (TBP) amplification curves for serially diluted DNA samples of these genes were generated and set into relation for standardization. 12 AML patients were examinated for mRNA expression of PRAME in the course of chemotherapy. In all patients a reduction of the mRNA expression of PRAME was detectable. Patients remaining in complete remission showed no mRNA expression of PRAME. In three patients an increase of mRNA expression was followed by a clinical relapse. Taken together, PRAME might be a marker for minimal residual disease in acute myeloid leukemia, but also a target for specific cellular immunotherapy in AML. Light cycler real time RT-PCR of the antigen PRAME might be both a diagnostic and a therapeutic tool.

259

Effect of in-vivo T-cell depletion with Anti-Thymocyte Globulin (ATG) on relapse and graft

N. Kröger, T. Zabelina, W. Krüger, H. Renges, S. Sonnenberg, F. Tögel, F. Ayukatan, M. Weber, Rischewsky, A.R. Zander (Hamburg, D)

ATG as part of the conditioning regimen resulted in a low incidence of GvHD in unrelated stem cell
transplantation. In a retrospective multicenter study in CML patients who underwent unrelated stem cell transplantation the incorporation of ATG in the preparative regimen resulted in a better OS. We investigated in a sequential, non-randomized trial whether the addition of ATG was in favour of GvHD incidence, overall and disease-free survival after stem cell transplantation from related do-nors in patients with good risk acute myeloid leukemia 1.CR or chronic myeloid leukemia in chronic phase.
102 patients with a median age of 37 years (range: 11-58) were enrolled. The male/female ratio was 52/50. CML patients were conditioned with Busulfan (14-16 mg/kg) and cyclophosphamide (120 mg/m^2), whereas AML patients received Busulfan (14-16 mg/kg), Etoposide (30-45 mg/kg) and cyclophosphamide (120 mg/kg). GvHD prophylaxis consisted of ciclosporine A, short course of Methotrexate 10 mg/m^2 (day +1,+3,+6) and metronidazol 3 x 500mg until engraftement. ATG was given either 30 mg/kg on day –1 (n=26) or 60-90 mg/kg divided on day –3,-2, and –1(n=19).
Despite the shorter follow-up of the ATG group, this data showed that incorporation of ATG in the preparative regimen lowered the incidence of GvHD without obvious increase of relapse in stem cell transplantation from related donors. However, a prospective randomized trial is necessary to confirm these data.

260

Allogeneic NK cells shielded against target cell induced apoptosis do not show improved cytolytic activity in vitro

J. Hasenkamp, S. Dingeldein, N. Schmitz, B. Glass (Kiel, Göttingen, D)

Tumor cell induced apoptosis in NK cells and T cells has claimed to be a relevant tumor escape mechanism against immunologic tumor surveillance and a potential obstacle to cellular immunotherapy. We questioned, if shielding of allogeneic NK cells against apoptosis is feasible and enhances cytolytic activity against tumor cells. NK cells from peripheral blood were enriched by immunomagnetic CD56+ selection. Freshly isolated NK cells were activated by IL-2 to obtain lymphokine-activated killer cells (LAK). Early phase of apoptosis was assessed by an Anexin V – FITC, necrotic cells were identified by propidium jodid (PI) staining. In co-culture experiments LAK cells and target cells were distinguished by membrane dyes PKH26 and DIO, respectively. In two-color flowcytometry we determined the absolute number of completely intact, of apoptotic and of necrotic LAK and target cells (Jurkat, K562), respectively. In control experiments different caspase-inhibitors were tested to block apoptosis induced by daunorubicin . Both, the blastic B-cell line Jurkat as well as LAK cells were susceptible to apoptosis induced by daunorubicin (1 µM). This could completely be blocked in Jurkat cells by pre-incubation with ICE-inhibitor ZVAD-fmk. In contrast ZVAD-fmk (200 µM) alone was not able to shield LAK cells against apoptosis. A combination of ZVAD-fmk and the caspase 1,3 and 7 inhibitor DEVD-cho (800 µM) was needed to grossly reduce daunorubicin-induced apoptosis in LAK cells. In co-incubation experiments with LAK cells and K562 or Jurkat targets, respectively we determined cytotoxicity of LAK cells and target cell induced apoptosis at various effector-target ratios (1:10, 1:1, 10:1). Cytotoxicity against K562 was more intense than against Jurkat cells (93, 61, 11 vs. 72, 41, 9 %). Also target cell induced apoptosis was more pronounced in the LAK-K562 co-cultures as in LAK-Jurkat co-cultures (14, 25, 44 vs. 10, 12, 22 %) After preincubation of LAK cells with ZVAD-fmk and DEVD-cho in optimal concentrations target cell induced apoptosis was significantly diminished in LAK-K562 co-cultures (6, 13, 12 %) and present but less pronounced in the LAK-Jurkat system (5, 7, 18 %). Cytotoxicity of LAK cells was not influenced in both systems. We conclude that intracellular apoptosis mechanisms may differ between freshly generated LAK cells and malignant cell lines (Jurkat). In this in-vitro system, shielding of LAK cells against tumorcell induced apoptosis is feasible but not followed by enhanced cytotoxicity against the target cell population. The in-vivo relevance of this observation remains to be determined.

261

Adenoviral gene transfer into dendritic cells efficiently amplifies the immune response to the LMP2A-antigen - a potent strategy for EBV+ Hodgkin's lymphoma immunotherapy

B. Gahn, E. Yvon, S. Gottschalk, M.K. Brenner, H.E. Heslop, C.M. Rooney (Houston, USA)

The Epstein-Barr virus (EBV) encoded LMP2A protein is consistently expressed in EBV+ Hodgkin's lymphoma (HL) and can be targeted by cytotoxic T-lymphocytes (CTL). CTL stimulated conventionally by lymphoblastoid cell lines (LCL) have limited activity against LMP2A-positive target cells. In 3 healthy donors of different HLA-type and 1 patient with EBV+ HL we used an alternative approach, based on the in vitro stimulation of CTL with dendritic cells (DC) genetically modified with an E1/E3 deleted recombinant adenovirus encoding LMP2A (AdLMP2A). Transduction of DC with the AdLMP2A at an MOI of 100 resulted in LMP2A expression in 40% of DC. Higher MOI resulted in DC death. CTL-lines activated by transduced DC had a higher frequency of LMP2A-tetramer-specific CTL than CTL-lines activated by LCL. CTL stimulated with transduced DC lysed both autologous fibroblasts infected with vaccinia virus LMP2A (FB-vaccLMP2A) expressing LMP2A with high intensity and autologous LCL, which express LMP2A at lower levels. In contrast, CTL generated from the same donors by stimulation with autologous LCL showed minimal lysis of the FBvaccLMP2A. Moreover one donor who did not respond to LMP2A when CTL were stimulated with LCL became a responder when LMP2A was expressed by the transduced DC. Hence this recombinant adenovirus encoding LMP2A effectively transduces DC and directs the generation of LMP2A-specific CTL. This approach needs to be confirmed in additional patients with EBV+ HL and will be a potent strategy in HL immunotherapy.

262

Leukemic cell vaccination in relapsed patients with CML after allogeneic blood stem cell transplantation

A.H. Elmaagacli, A. Koppelle, H. Ottinger, D.W. Beelen, U.W. Schaefer (Essen, D)

Leukemic cell vaccination were applied three times in 3-week intervals to two patients with relapse of CML after allogeneic stem cell transplantation. One of these two study patients was transplanted for CML in 1st chronic phase and relapsed in a stable phase of CML, whereas the other study patient received a transplantation for CML in lymphatic blast crises, and suffered from an extramedullary relapse of CML after transplant. Leukemic cell vaccination was performed two weeks after applying an unfractioned donor leukocyte infusions (DLI) in the first patient with relapse of CML. In the second study patient, leukemic cell vaccination was applied after achieving a complete remission of CML again using a combination of radiochemotherapy and DLI in order to keep the patient in remission.
The vaccine was generated by culturing monocyte-derived donor-cell derived antigen presenting cells which were pulsed with patient-derived CML cell lysates. We used a quantitative, real-time PCR assay to directly measure the immune reactivity of patients T cells to the vaccine. T cells were tested for immune reactivity after in vitro sensitization of leukemic cell vaccine for Interferon-gamma messenger RNA expression by real-time PCR. A twofold increase in Interferon-gamma transcript accumulation in T cells after vaccination was found and considered to be evidence of specific response. In contrast to in vitro tests, clinically both patients did not profit from the leukemic cell vaccination. Nor did the first patient achieved a remission of CML, neither was the second patient pretended from relapsing again after completing the 3rd vaccination. So, we concluded that leukemic cell vaccination may have clinically marginal anitleukemic effects after allogeneic transplantation.

263

Systemic T-cell stimulation with interleukin-2 in individuals with HIV-infection: Immunological changes and safety of an outpatient treatment approach

D. Voliotis, V. Zech, M. von Bergwelt-Baildon, A. Juette, V. Diehl, G. Faetkenheuer (Köln, D)

Introduction: Imbalances of T-cell immunity, especially CD4+ cells are implicated in the pathogenesis of HIV infection. IL-2 treatment seems to increase the number of CD4+ cells in HIV without affecting viral load. Recent findings indicate that IL-2 can reverse reduction of the TH1/TH2 ratio observed in HIV-infection. Patients and methods: In a pilot study 10 HIV-infected patients with a viral load of <10,000 copies/ml and >200 CD4+ cells/µl receiving HAART were included. Patients received $4.5x10^6$ IU of IL-2 intermittently sc. Total duration of therapy was 6 cycles (1 year). IL-2 was self-administered on an outpatient basis. Clinical, virological and immunological markers were monitored. Lymphocyte activation and memory cell status were monitored by FACS as well as T-cell receptor distribution. Distinction of T-helper cell subsets (TH1/TH2) was performed by analyzing intracytoplasmatic IL-2, IL4 and IFN-g Expression. Results: 9 patients received a total of 54 cycles. IL-2 was well tolerated on an outpatient basis. None of the patients experienced WHO toxicity >Gr 3. There was a median increase of >100 CD4 cells/µl after 1 year. The median HIV-load did not increase during the course of therapy. There was an increased percentage of TH1, reflected by an augmented IFN-g and IL-2 expression after IL-2. The level of T-cell proliferation correlates with the immunological status and the progression of HIV-infection. Cytokine analysis revealed a shift in the TH1/TH2 profile during IL-2 treatment, a fact which supports the assumption that IL-2 treatment could restore -at least in part- the cytokine dysregulation caused by the HI virus. Conclusion: Low-dose IL-2 in HIV-infected indivduals is well tolerated and can be administered safely on an outpatient basis. It leads to an increase in CD4+ T-cells and upregulation of TH1 type T-lymphocytes. IL-2 given continuously could restore some immune functions dysregulated by HIV, depending on the individual immunological status.

Postersitzung:

Indolente Lymphome

264

Autologous stem cell transplantation for patients with follicular low grade non-Hodgkin's lymphoma - Long-term outcome and toxicity

S.M. Probst, A. Krämer, E. Leo, A. Abdallah, G. Egerer, M. Körbling, R. Haas (Heidelberg, Düsseldorf, D; Houston, USA)

High-dose radio-chemotherapy with autologous stem cell support in patients with follicular non-hodgkin's lymphoma (NHL) has produced encouraging results. To investigate the long-term benefits and adverse events of high-dose therapy with stem cell transplantation in patients with follicular NHL we retrospectively analyzed the data of 134 patients who received autologous stem cell transplants from 1990 to 2000 with specific focus on long-term side effects and secondary malignancies. The patients received high-dose therapy if they relapsed or had high-risk factors and were transplantated after response to induction or reinduction chemotherapy. As conditioning regimen most patients (n=113) received high-dose cyclophosphamide (200 mg/kg) preceded by TBI at 14.4 Gy. The others received BEAM. The median age of the patients was 46 years (range 22-61 years). 68 were female and 66 male. 96 (71.6 %) patients had Stage IV, 30 (22.4 %) Stage III and 8 patients (6 %) Stage I/II disease. 63 patients remained in remission and 101 patients were alive at the time of analysis. The median survival was not reached at the time 100 months follow-up. A plateau in survival (66.6 %)seemed to be reached at 68.9 months and in event-free-survival (40.4 %) at 56.0 months. Long-term adverse events included bi- or pancytopenia (n=13; 10 %), myelodysplastic syndrome (n=2; 1.5 %), acute myeloid leukemia (n=1) and lung cancer (n=1). Thus the rate of secondary malignancy was 3.0 % at a median follow-up of 40.0 months (range 1-87 months). Age, sex and stage at the time of transplantation did not seem to have any impact on the outcome and the chance of developing posttransplant malignancies. Despite the risk of partial engraftment failure in 10 % and secondary malignancies in 3.0 %, patients with relapsed follicular lymphoma and with high risk factors probably derived a long-term benefit from high-dose radio-chemotherapy with autologous transplantation.

265

Successful treatment of advanced follicular lymphoma plus malignant histiocytosis with allogeneic peripheral blood stem cell transplantation - A case report

I.W. Blau, C. Baldus, T. Fietz, B. Gleisner, H. Schrezenmeier, H. Stein, E. Thiel, W.U. Knauf (Berlin, D)

A 33-years old male patient presented with B-symptoms and generalized lymphadenopathy and was diagnosed with low-grade follicular B-cell lymphoma IV. Surprisingly, additionally malignant histiocytosis was found as second tumor in the bone marrow, liver, and lymph nodes, respectively. After 6 cycles of CHOEP chemotherapy B-symptoms as well as lymphadenopathy have resolved completely, but reevaluation of bone marrow showed persistant infiltrations of both malignant diseases. While in excellent clinical condition the patient was scheduled for allogeneic stem cell transplantation (PBSCT). Two months after the last CHOEP treatment high-dose conditioning started with TBI 12 Gy followed by CY 2 x 60 mg / kg body weight. The allograft contained 3.8 x 10^6 / kg CD34+ G-CSF mobilized peripheral blood stem cells from his HLA-identical sister. Short course MTX plus cyclosporine were administered for GvHD prophylaxis. Leukocytes engrafted on day +12, and platelet recovery was demonstrated on day +21. Full donor chimerism was achieved on day +30. Acute GvHD I of the skin occurred, whereas no other severe complications were observed. However, bone marrow biopsies revealed residual infiltration of the lymphoma without signs of persistant histiocytosis. The patient was discharged on day +34. While tapering cyclosporine limited chronic skin GvHD and a BOOP developed, but were controlled by prednisolone. Half a year post transplantation bone marrow was evaluated again and was found to be free of lymphoma as well as from histiocytosis. The patient went back to work at day +390, and is in ongoing complete remission of both malignant diseases. This case report demonstrates that allogeneic PBSCT can eradicate low-grade follicular lymphoma and malignant histiocytosis. An additional graft versus tumor effect seemed to have played an important role in the course of this particular patient.

266

Fludarabine in combination with bendamustine: An effective therapy for indolent non-Hodgkin's lymphoma

K. Bross, M. Gnad, A. Wodzynski, R. Andreesen, M. Demandt, R. Dengler, A. Hobelsberger, A. Holstege, U. Vehling-Kaiser, J. Weiß (Regensburg, Straubing, Landshut, Weiden, D)

Background: The nucleoside analog fludarabine (F) alone or in combination has been reported to be very effective in indolent lymphoma in various phase II and III trials. Due to encouraging results with bendamustine (B), an alkylating agent similar to cyclophosphamide, the FB combination was investigated in this trial. Aims: The study was designed as a two-step multicenter phase II trial in order to determine the feasibility, tolerability and efficacy of the combination of F and B in untreated and pretreated lymphoma patients with advanced stages (Ann Arbor III/IV) of immunocytoma (IC), follicular lymphoma (FL), mantle cell lymphoma (MCL) or chronic lymphocytic leukemia (CLL; Binet B or C). Methods: Twenty-eight patients with indolent lymphoma (8 CLL, 7 immunocytoma, 3 MCL, 10 FL) have been recruited so far into the first part of the study. The median age of the patients was 60 years (range 44 to 73 years). 16 patients were untreated, 12 patients were pretreated (median 2 regimens). The untreated patients received treatment with 20 mg/m^2 F and 40 mg/m^2 B on days 1-3 (an escalation to 25 mg/m^2 F and 50 mg/m^2 B was possible when the first cycle was tolerated well), the pretreated patients were given a reduced dose of 15 mg/m^2 F and 30 mg/m^2 B on days 1-3 (possible escalation to 20 mg/m^2 F and 40 mg/m^2 B). Treatment was given every 4 weeks for a maximum of 6 courses. Patients received antibiotic prophylaxis with cotrimoxazol throughout treatment and dexamethason (2x8 mg days 1-5) on the first and second cycle. Results: As of 05/2001 24 of 27 evaluable patients achieved a remission, 14 CR (52%) and 10 PR (37%), with an overall response rate of 89%. As expected, more complete remissions could be achieved in 16 evaluable untreated (11 CR, 4 PR, 1 PD) than in 11 evaluable pretreated patients (3 CR, 6 PR, 1 MR, 1 PD). Due to the short follow-up of the study the median progression free survival is 5,5+ (1-15+) months in the pretreated patients and 5+ (2-14+) months in the untreated group. Concerning side effects, in the pretreated group 7 severe events were observed in 51 cycles: 4 x grade 3, 1 x grade 4 leukopenia and 2 x grade 3 thrombopenia. In untreated patients 80 cyles of FB were applied and in 15 cyles severe side effects occurred: 10 x grade 3 and 1 x grade 4 leukopenia, 1 x grade 4 thrombocytopenia, 1 x grade 3 anemia, 1 x grade 3 infection and 1 x grade 3 vasculitis. Other side effects were mostly hematologic and mild. Conclusions: The combination of fludarabine and bendamustine is feasible and well tolerated if dose is adjusted to the treatment status. Its efficacy in untreated and pretreated lymphoma patients is very promising (> 50% CR) but must be confirmed in the second phase of the study by additional recruitment.

267

IL-4 receptor polymorphisms in early gastric marginalzone B-cell lymphoma of MALT-type

N. Schnitzer, B. Alpen, T. D. Kim, M. Schmidt, E. Bayerdoerffer, M. Stolte, A. Neubauer (Marburg, Berlin, Dresden, Bayreuth, D)

Background: Primary gastric B-cell lymphoma of the mucosa associated lymphoid tissue (MALT) are associated with a chronic Helicobacter pylori (H. pylori) infection. The development depends on nonneoplastic H. pylori - specific T-cells (Hussell et al. J Pathol 1996; 178:122-7). T-helper cells can be functionally subdivided into TH1 and TH2 cells corresponding to their cytokine expression profile. Greiner et al. showed in vitro the influence of TH2 but not TH1-cytokines on the MALT-lymphoma cell proliferation (Am J Pathol 1997; 150:1583-93). The IL-4/IL-4 receptor interaction leads to a STAT6 mediated signaltransduction and supports the developement of TH2 cells (Finkelman et al. 2000; 164:2303-10). An regulative influence of IL-4 receptor polymorphisms on the TH1/Th2 balance was discussed in the literature (Hershey et al. NEJM 1997; 337:1720-5; Hackstein et al. Tissue Antigens 1999; 54: 471-7). Aim: To investigate the frequency of IL-4 receptor polymorphisms in early gastric marginal zone B-cell lymphomas of MALT-type. Methods: Genomic DNA extracted from whole blood samples of 106 patients suffering from low-grade gastric MALT-lymphoma as well as 120 healthy controls were analyzed. Allele specific polymerase chain reaction (PCR) for six previously described polymorphisms of the IL-4 receptor gene was performed. The validity of the allele specific PCR was confirmed by cloning of the PCR-products and sequencing analysis. Results: Comparing the allele and the genotype frequency of each polymorphisms between MALT-lymphoma patients and normal controls no difference was figured out. Conclusions: These results point to a lower significance of IL-4 receptor polymorphisms in the pathogenesis of marginal zone B-cell lymphoma of MALT-type.

268

Bendamustine / Mitoxantrone / Rituximab (BMR): A new effective treatment for refractory or relapsed indolent lymphomas

R. Weide, J. Heymanns, H. Köppler (Koblenz, D)

Background: Bendamustine and mitoxantrone have been shown to be potent cytotoxic drugs for the treatment of relapsed or refractory indolent lymphomas. The anti-CD20 monoclonal antibody rituximab has produced an overall response rate of 50% as a single agent in relapsed or refractory indolent lymphomas. We asked the question whether a combination of bendamustine, mitoxantrone and rituximab (BMR) could improve these results. Objective: Open label, single center pilot study for patients with relapsed or refractory, CD20-positive indolent lymphoma or chronic lymphocytic leukemia. The therapy consisted of bendamustine (80 mg/m^2, day 1-3), mitoxantrone (10 mg/m^2, day1), rituximab (375 mg/m^2, week 2,3,4,5). BM was repeated on day 29 or when the haematological parameters had recovered. The maximum therapy consisted of 1 BMR-application, followed by 5 BM applications. Treatment was stopped when the disease responded with a partial remission (PR) or a complete remission (CR). Results: Between 3/99 and 12/00 20 patients received the BMR-regimen (4 secondary high grade lymphoma, 12 indolent lymphoma, 4 B-CLL). The median age of the patients was 67 years (range 36-82) and their performance score ranged from 0-3. Median number of previous treatment regimens was 2 (1-6). Of the lymphoma patients 14 had stage IV disease, 1 stage III and 1 stage II. B-CLL patients were all Rai stage IV (Binet C). Overall response rate was 95% (19/20) with 7 patients achieving a CR (35%) and 12 patients achieving a PR (60%). Median time to progression is 7 months (1-21) with a median observation time of 7 months (1-21). Response is still durable in 15/20 patients (75%) (1+ to 21+ months after therapy). A symptomatic, reversible grade 3 or 4 hematotoxicity occurred in 4/20 patients (20%). A non-symptomatic grade 3 or 4 haematotoxicity was seen in 9/20 patients (45%). No major nonhaematological toxicity was observed. Conclusion: BMR is a well tolerated, very effective outpatient treatment for relapsed and refractory indolent lymphoid malignancies.

269

Monoclonal antibodytherapy with Campath-1H in patients with non-Hodgkin's lymphoma and CLL - single center experience

M. Reiser, T. Elter, H. Schulz, M. Jensen, A. Engert (Köln, D)

Rationale/Objectives: CAMPATH-1H is a monoclonal humanized IgG1-antibody that binds to a pan-lymphocyte antigen, CD52. The antigen is highly expressed on malignant cells from patients (pts.) with CLL and NHL. Several studies have shown significant activity of CAMPATH-1H in pts. with chronic B-and T-cell leukemia, especially T-prolymphocytic leukemia. Therefore, very recently the antibody has been approved by the FDA for the treatment of pts. with Fludarabine-resistant CLL. Methods: From 6/93-5/01 we have treated 23 pts. with CAMPATH-1H. Pts. either had relapsed or refractory NHL (n=4) or Fludarabine-resistant CLL (n=19). Median age was 61 years (range 46-79 years). Median time since first diagnosis was 5.44 years. Pts. had between 3-6 prior treatments. Treatment consisted, after escalation to target dose, of 30 mg CAMPATH-1H intravenously thrice weekly for 2 h for 4-12 weeks. All pts. received at least paracetamol and antihistamines as premedication; steroids were used only when side effects occurred. Antiinfective and antiviral (exept CMV) prophylaxis was given to all pts. up to two months after the end of treatment consisting of cotrimoxazol and aciclovir. Results: 21/23 pts are evaluable for response. 7/19 pts with CLL and 1/4 pts with NHL responded. Treatment was stopped in 7/19 pts with CLL due to hematological and non-hematological side effects.
Conclusion: CAMPATH-1H is effective in pts. with refractory CLL even when the CLL was resistant to prior fludarabine therapy. Side effects were more severe in pts. with advanced disease leading to discontinuation of treatment.

270

Fludarabin plus Bendamustin for relapsed and refractory low grade lymphoma - dose-finding within a multicenter phase I/II study of the East German Society of Hematology and Oncology (OSHO)

M. Koenigsmann, W.U. Knauf, M. Herold, M. Wiermann, R. Bartsch, A. Franke (Magdeburg, Berlin, Erfurt, D)

Background: Low complete remission rate as well as short remission duration have so far hampered the therapy of recurrent indolent lymphoma. A standard treatment has not yet been defined in this situation. As single agents, the nucleoside analogon fludarabin as well as the bifunctional alkylating agent bendamustin have been proven efficacious, so that a combination of these two drugs with different mechanisms of action appears reasonable. Aims: To evaluate dose-limiting toxicities (DLT), the maximal tolerable dose (MTD) and efficacy, a multicenter phase I/II protocol was initiated by the East German Society of Hematology and Oncology (OSHO) in October 1999. Treatment plan: Fludarabin was given at 30 mg/sqm/d, bendamustin at 30 or 40 or 50 mg/sqm/d, corresponding to dose levels I - III. Both drugs were applied sequentially iv. over 30 minutes from days 1 to 3 every four weeks for six cycles. Results: Fourteen pts were evaluable with a median age of 62 years (range 57 to 74). Five pts had centrocytic lymphoma, 9 pts centrocytic/centroblastic lymphoma. Stage of disease at relapse was III in 3 pts and IV in 11 pts. All pts had been pretreated with chemotherapy with or without additional radiotherapy or immunotherapy. Dose level I was given to 7 of our fourteen pts. DLT consisted in leucocyte nadirs Grade III and IV (CTC criteria)in two and was associated with therapy delay beyond day 35 in one pt. Five pts at dose level I completed all 6 courses. Toxicities other than hematological remained Grade I and II. Out of the 7 pts at dose level I, 1 reached CR, 4 PR, and 2 pts had progressive disease, one of whom died under progression. At dose level II, hematological DLT was observed in 3 pts. One of these 3 pts died due to Grade IV infection with Grade IV thrombocytopenia. Four pts completed all 6 courses. Other toxicities remained Grade I and II. Of the six remaining pts two pts reached CR, 3 pts PR, 1 pt had progressive disease. Conclusions: Dose level I of this phase I/II study was well tolerated, whereas toxicity at level II was too high for a relapse situation with low-grade lymphoma. Our preliminary response data in this pretreated and in part refractory patient population are encouraging. Therefore, the efficiency will be further studied in another 20 pts at dose level I.

271

Anti-CD20 monoclonal antibody (Rituximab) in combination with cytotoxic chemotherapy is effective in inducing clinical and molecular remissions in patients with newly diagnosed or relapsed follicular lymphoma

S. Martin, R. Kronenwett, A. Niederste-Hollenberg, M. Free, I. Mlynarczuk, G. Meckenstock, R. Haas (Düsseldorf, D)

Between 1/2000 and 1/2001, 20 patients (pts) with CD20-positive follicular lymphoma were included into a study using Rituximab (R) in combination with cytotoxic chemotherapy. There were 12 men and 8 women with a median age of 48 years (range, 29 – 73). As assessed by nested PCR, all patients had the bcl2/IgH rearrangement at the major or minor breakpoint region in bone marrow (BM) and/or peripheral blood (PB). 13 pts with stage III or IV disease were included for first-line therapy and received either Rituximab (375 mg/m^2) and fludarabine/mitoxantrone (R/FM) or Rituximab plus cyclophosphamide/adriamycin/vincristine/prednison (R/CHOP) followed by Rituximab and high-dose cytarabine/mitoxantrone (R/HAM) as consolidation therapy. 7 pts were enrolled at relapse after conventional therapy and were treated with R/HAM as salvage regimen, which also served for stem cell mobilization. In both groups, R/HAM was followed by high-dose BCNU/etoposide/cytarabine/melphalan (BEAM) with autologous peripheral blood stem cell transplantation. Rituximab in combination with either of the cytotoxic chemotherapy regimens induced remissions in all pts, as confirmed by the results of clinical and pathological staging examinations. The differences in the efficacy of the various immuno-chemotherapies relay on the ratio of complete (CR) versus partial remissions (PR) and on the ability to induce molecular responses, as assessed by sequential t(14;18)-nested-PCR examinations. After R/CHOP, only one third of the pts were in CR. In 60% of the pts receiving this regimen t(14;18)-positive cells persisted in PB and/or BM. The combination R/FM induced a CR in half of the pts, but 75% of the pts scheduled for this therapy remained PCR-positive. The clinical and molecular responses induced by R/HAM were different according to the disease status. In the group of pts in first remission, following consolidation therapy with R/HAM, 6 of 6 pts achieved a CR and converted to PCR-negativity in samples from blood and BM. In the same line, their leukapheresis products were free of tumor cells, obviating the need for in vitro purging. In contrast, when R/HAM was given as salvage therapy, a lower tumor depletion was obtained, with only 4 out of 7 pts achieving a CR. At the molecular level, one third of the pts treated at relapse still had residual disease in PB or BM without clearance of PCR-positive cells in their autografts. In conclusion, Rituximab combined with cytotoxic chemotherapy induces clinical remissions in newly diagnosed and pretreated pts with follicular lymphoma. The probability of achieving a clinical complete remission and a molecular remission depends on the type of chemotherapy combined with the anti-CD20 antibody and the time point of therapy.

272

Targeting immungene rearrangements by consensus fluorogenic probes for quantitation of minimal residual disease (MRD) in mantle cell lymphoma

C. Pott, M. Brüggemann, N. Derner, M. Ritgen, P. Dreger, M. Kneba (Kiel, D)

Mantle cell lymhomas are characterized by the specific t(11;14) translocation which provides a sensitive and specific molecular marker for MRD detection by PCR. However, t(11;14) translocation is detectable by PCR in only 30-40% of the cases. Clonal immune gene rearrangement (IgH-CDR3) provide reliable molecular markers for monitoring of MRD in 70%-90% of MCL. Therefore we developed a real-time quantitative TaqMan PCR (RQ-PCR) assay targeting clonal immungene rearrangements to monitor MRD. We designed 4 JH hybridization probes and reverse downstream primers that anneal to conserved germline JH gene segments of the four most often rearranged JH families. As a consequence, only the selection of a clone specific forward primer matching the VH-N-D-N region of the individual patient is necessary. In contrast to the use of individually designed primer/probe sets for IgH-RQ PCR this method is less time consuming and expensive and has been established successfully for MRD detection in acute leukemias. To evaluate the prognostic impact of MRD on disease-free survival and long-term remission of patients with MCL undergoing high dose therapy followed by peripheral blood stem cell transplantation (PBSCT) we are currently quantifying follow-up samples in a series of 20 MCL patients with a clonal immungene rearrangement undergoing high dose therapy and PBSCT. The sensitivity of the assay allowed detection of 1-10 MCL cells in the background of up to 100,000 normal cells. PCR positivity post PBSCT and increasing MRD level revealed as an indicator of subsequent relapse.

RQ-PCR of IgH-CDR3 rearrangements allow the quantitation of residual lymphoma cells with high reproducibility and accuracy and may contribute to evaluate the efficacy of immunologic purging methods and the impact of MRD on prognosis of patients with MCL.

273

Treatment of follicular and mantle cell non-Hodgkin's lymphoma with anti-CD20 antibody Rituximab after high-dose chemotherapy and autologous CD34+ selected blood stem cell transplantation

W. Brugger, J. Hirsch, R. Repp, G. Schlimok, F. Grünebach, M. Manz, W. Vogel, H.G. Kopp, W. Aulitzky, A. Ganser, M. Gramatzki, K. Fehnle, L. Kanz (Tübingen, Erlangen, Augsburg, Stuttgart, Hannover, München, D)

Objective: In October 1997, we initiated a multicenter phase II study to investi-gate the safety and efficacy of Rituximab in patients with follicular (FL) and man-tle cell lymphomas (MCL) after high-dose therapy (HDT) and autologous CD34+ enriched peripheral blood progenitor cell (PBPC) transplantation. Methods: Eligibility included newly diagnosed stage III/IV FL or newly diagnosed or relapsed patients with stage III/IV MCL [t(11;14)+]. Initial chemotherapy consisted of VACOP-B (6 weeks) or CHOP chemotherapy, followed by further cytoreduction with G-CSF supported VIP-E chemotherapy. PBPC were harvested, and positive selection of CD34+ cells was performed as a purging strategy. HDT consisted of TBI (6x2 Gy) and cyclophosphamide (120 mg/kg), followed by CD34+ selected PBPCT (5.49×10^6 CD34+/kg; range 2.4-20.3). Starting eight weeks after trans-plantation, Rituximab was given at a dose of 375 mg/m^2 weekly for 4 weeks. Results: Up to now, 29 patients were included. 22 patients [FL n=16; MCL n=6] (49, range 31-59) completed the whole treatment program and were evaluable for safety and efficacy as of November 30, 2000. Rituximab treatment after HDT was generally well tolerated. WHO Grade III-IV toxicities included leukopenia (n=4), lymphocytopenia (n=5), nausea (n=1), and infection (n=2). 4 serious ad-verse events (2 pneumonias, 1 fever, 1 fatigue and nausea) were observed which required rehospitalisation. There was one treatment-related death which occurred after HDT but before initiation of rituximab treatment. At a mean follow up of 13.3 months (3-32 months), all patients are in clinical CR and none of the patients relapsed, including none of the MCL patients. As defined by CT-staging criteria, CR developed over time with 50% at 6 months, and 88.9% at 12 months. Eighteen patients were evaluable for minimal residual disease by PCR at various time points after transplantation. Prior to HDT, 41% of the patients were PCR negative for t(14;18) or t (11;14). These numbers converted to 55% PCR negativity in pe-ripheral blood and bone marrow after HDT, 77% after rituximab, and 100% at 6-9 months post transplantation. None of the patients became PCR-positive during follow-up. Conclusion: Rituximab treatment after HDT and CD34+ PBPCT is safe and effective as demonstrated by the induction of complete molecular remis-sions. These data suggest that this approach is a potentially curative option in MCL and FL patients.

274

Alpha-radioimmunotherapy of B-Cell non-Hodgkin's Lymphoma using ^{213}Bi-labelled anti-CD19- and anti-CD20-CHX-A"-DTPA conjugates

S. Heeger, G. Moldenhauer, G. Egerer, H. Wesch, A. Eisenmenger, T. Nikula, C. Apostolidis, W. Janssens, M. Brechbiel, A. Ho, U. Haberkorn, R. Haas (Heidelberg, Karlsruhe, Düsseldorf, D; Bethesda, USA)

The major advantages of alpha-emitters like 213Bi are 1) the high linear energy transfer (100 keV/µm), 2) the short path length of several cell diameters (50-80 µm) compared to conventionally used beta-emitters such as 131I and 90Y, 3) the short half life of 45.6 min. Due to these physical properties unspecific irradiation of healthy tissue can be avoided and systemic cytotoxicity is minimized. In preclinical cell experiments the strong cytotoxic potency of 213Bi-labeled anti-CD19 and anti-CD20 antibodies targeting lymphoma cells was demonstrated. In addition, we investigated in vivo stability, biodistribution and toxicity in a mouse model. Both antibody conjugates showed a similar stability in vivo. After 6 hours 55-70 % of the injected activity was still detectable in blood circulation and 32-45 % after 24 hours, respectively. Biodistribution studies revealed no unspecific uptake or degradation of the compounds. Dose limiting toxicity was myelosuppression. Further toxic side effects were only seen in histological specimens from the spleen of the animals. Other organs were not affected by this type of radioimmunotherapy. LD10 was estimated to be at 6-6.5 MBq whereas LD50 was reached at 7.5 MBq. We are currently conducting a phase I dose escalation trial in humans to determine the safety and feasibility of this novel approach as well as pharmacology and dosimetry. Three patients suffering from non-Hodgkin's lymphoma are treated in the first dose group with an activity of 15 mCi 213Bismuth coupled to 3 mg of the anti-CD20 antibody Rituximab (MabThera®). Pharmacokinetic measurements are performed by taking blood samples every 30 minutes after injection. The 440 keV gamma emissions of the isotope can be used for continuous gamma scan imaging following therapy. Thus, radiation doses to individual organs can be estimated. So far acute toxicities have not been observed in these patients. In one patient mild leukopenia occurred one week after administration of the radioimmunoconjugate. The next patients will be treated at higher dose levels with planned activities of 30 and 45 mCi per patient in 2-5 fractions. Our data indicate that ^{213}Bi-labelled antibody conjugates represent promising tools for an improved therapy of B-cell malignancies.

275

Proteom analysis of mantle cell lymphomas - standardization and initial analysis of differential expression patterns of cell cycle regulators

M. Weinkauf, J. Lu-Hesselmann, G. Herzog, W. Hiddemann, M.H. Dreyling (München, D)

Mantle cell lymphoma (MCL) is characterized by the translocation t(11;14)(q13;q32), resulting in the overexpression of cyclin D1. However, secondary genetic alterations are detectable in the vast majority of cases. Thus, p53 and the p16 gene cluster (p15INK4B, p16INK4A, and p16ARF) have been shown to be frequently involved in the secondary transformation to aggressive MCL variants. So far, little is known on the differential effect of the p16INK4A-CyclinD1 and p16ARF-MDM2-p53 signal pathway. Only expression profiling expression profiling of proteins allows the overall analysis of molecular alterations on the posttranslational level as well as those which may have occurred on the DNA or RNA level. Due to the technical improvements within the last decade, proteomics represents a sensitive and reasonably reliable method of differential molecular analysis. We have established a two-dimensional protein electrophoresis based on isoelectric focusing (IPGphor) and SDS-page (Protean II XL system) which allows the reproducible analysis of protein patterns applying the computer assisted registration of landmarks (CAROL) program (Kriegel et al., Electrophoresis 2000).

A p53-negative, p16-negative hematological cell line (HL-60) was transfected with either p53 wild type or p53 mutant constructs. Comparative two-dimensional proteom analysis of both derivatives and the wildtype cell line (transfected with an empty vector construct) has been performed and differentially expressed protein spots were identified which will be further characterized by MALDI-TOF. Results will be confirmed by western blot and RNA array analysis. In a second step, primary mantle cell lymphomas will be screened for the identified target proteins to more accurately define their prognostic impact. In addition, the established proteom approach will be adapted for the analysis of leukemic mantle cell lymphomas classified according to their p53 and p16 status to further gain insights into the molecular pathogenesis of mantle cell lymphomas.

276

Myeloablative radiochemotherapy with autologous stem cell transplantation in mantle cell lymphoma - results of a prospective randomized European intergroup study

M.H. Dreyling, W. Hiddemann, M. Pfreundschuh, C. Gisselbrecht, V. Diehl, S.M. Boiron, L. Trümper, H.-J. Illiger, C. Rudolph, M. Boogaerts, H. Kluin-Nelemanns, R. Parwaresch, M. Unterhalt (München, Homburg, Köln, Göttingen, Kiel, Oldenburg, Cottbus, D; Paris, Pessac, F; Leuven, B)

Despite an indolent morphology, mantle cell lymphoma is characterized by an aggressive clinical course with a median survival of less than 3 years and virtually no long term survivors. In a large retrospective survey, these patients did not benefit from conventional chemotherapy including anthracyclin-containing combinations. Initial studies on high dose chemotherapy reported a wide range of clinical efficacy with a 2 year overall survival between 16% and 100%. Thus, in 1996, a European intergroup study was initiated to evaluate the role of high dose chemotherapy in a prospective, randomized trial. After initial CHOP-like chemotherapy, patients < 65 years with partial or complete remission were assigned to receive either intensive consolidation (myeloablative radiochemotherapy followed by autologous peripheral stem cell transplantation) or interferon alpha (IFN) maintenance. In patients of the high dose arm, peripheral stem cells were mobilized with DexaBEAM. Subsequently, patients successfully harvested underwent fractionated total body irradiation with 12 Gray plus cyclophosphamide (60 mg/kg) on days -2 and -3 followed by transfusion of unpurged autologous stem cells. Patients of the IFN maintenance arm received 2 additional courses of the cytoreductive chemotherapy followed by 5x 10^6 units interferon alpha 3x weekly until progression. Since August 1996, 186 patients from Belgium, France, Germany, Great Britain, Italy, Netherlands, Spain and Suisse were recruited. 71% of the documented patients (102/143) responded to the initial chemotherapy. Currently, 76 patients are evaluable for interim analysis with a maximum follow-up of 4 years. A recurrence of lymphoma was observed in 6 of 36 (17%) patients of the high dose therapy arm, whereas the lymphoma progressed in 21 of 40 patients (53%) receiving IFN maintenance, corresponding to an estimated risk reduction of relapse of 63% in favor of the high dose arm (p=0.0107). Both treatment options were rather well tolerated. At present time, no significant differences in overall survival have been observed. Although the benefits of high dose therapy may be diminished by late toxic side effects (secondary malignancies), our study represents the first randomized trial confirming the benefits of high dose chemotherapy in mantle cell lymphoma.

277

Efficacy and toxicity of rituximab plus MCP versus MCP alone in advanced indolent NHL - New interim results of a clinical phase III study of the East German Study Group Hematology/Oncology (OSHO)

M. Herold, F. Fiedler, R. Pasold, S. Srock, W. Knauf, M. Freund, R. Naumann, M. Aßmann, R. Rohrberg, U. von Grünhagen, C. Klinkenstein, U. Gericke, A. Franke, U. Kaiser, D. Niederwieser, J. Steglich, M.-R. Clemens, E. Kettner, U. Mey, P. Richter, F.-A. Hoffmann, K. Bremer, G. Dölken, H. Eschenburg, S. Hahnfeld, U. Haak, V. Lakner, R. Subert (Erfurt, D)

In a prospective randomized trial the efficacy and toxicity of rituximab (375 mg/m^2 day 1) plus MCP (mitoxantrone 8 mg/m^2 days 3 + 4, chlorambucil 3x3 mg/m^2 days 3 – 7, prednisolone 25 mg/m^2 days 3 – 7) q 4 weeks x 8 is compared with MCP alone (days 1-5 q 4 weeks x 8). The study includes patients with advanced follicular lymphoma (grades 1 and 2), lymphoplasmacytic lymphoma and mantle-cell lymphoma stages III and IV. According to the results of phase I and II trials we expect better quality remissions, in particular higher CR rates, resulting in an improved freedom from treatment failure (FFTF) and survival in the experimental arm of the study. In this second interim analysis of 124 randomized patients 106 patients are evaluable on an intent-to-treat basis; 55 in the rituximab plus MCP group and 51 in the MCP group. The response rate (CR plus PR) for all patients is 81% (95%-CI 72 % - 88%); at present there is no statistically significant difference concerning the response rate between the treatment arms, and so the respective numbers are still blinded. CR rate for all patients is 40%, PR rate 41%. Toxicity analysis was made on the basis of 414 treatment cycles in the R + MCP group and 347 cycles in the MCP group (7,5 vs. 6,8 per patient). Counting all adverse events (CTC-grades I – IV) there is a tendency to more events in the R+MCP group (375 vs. 266; 0.9 vs. 0.76 per cycle), however this is not statistically significant (p= 0.612). Myelotoxicity is the main cause for CTC-grade IV toxicity, there is no statiscically significant difference for hemoglobin and platelets, but a significantly higher proportion of CTC-grade IV leukopenia occured in the R+MCP group (13% vs. 6% of cycles, p= 0.008), however this did not result in an increased frequency of infections (p= 0.41) or a reduced chemotherapy dose intensity in the R+MCP group. Our preliminary response data demonstrate an overall response rate (81%) comparable or even better than in other multicenter studies with a remarkable high proportion of complete responses (40%) for the entire group of patients. Regarding toxicity, there is a tendency to more adverse events in the R+MCP arm, esp. concerning leukopenia, however this has no impact on infections or dose intensity of chemotherapy. In the analysed cohort of patients we have had no severe infusion related events following rituximab.

278

Secondary chromosome changes in t(11;14)-positive mantle cell lymphomas detected by cytogenetic banding analysis

L. Harder, B. Schlegelberger, S. Gesk, R. Siebert, W. Grote (Kiel, D)

Mantle cell lymphoma (MCL) represents a distinct clinicopathological subgroup of B-lymphoid neoplasms. The cytogenetic hallmark of MCL is the t(11;14)(q13;q32) which can be detected in virtually all cases. The secondary genetic and molecular events involved in the progression of these tumors are widely unknown. By R-banding analysis we analyzed a series of 49 MCL carrying a t(11;14)(q13;q32) comprising both typical and blastoid variants (tMCL and bMCL). Chromosome analyses were performed on affected lymph node, bone marrow, spleen or blood samples. 19 cases (70%) of tMCL and 16 cases (73%) of bMCL were analyzed at primary diagnosis. Chromosomal imbalances were found in 48 of 49 cases. Loss of 1p21p31 was present in both MCL variants (tMCL - 30%, bMCL - 36%). Other frequent chromosomal imbalances in bMCL were gains of 3q (27%), complete chromosomes 3 (19%) and chromosome 12 (18%) as well as losses in chromosome regions 6q (23%), 9p (23%), 10p (18%), 17 (18%), 8p (14%); regions of loss most prominent in tMCL were chromosomes 9 (26%), Y (22%), 1p36pter (22%), 1q41qter (22%) and 13 (19%). Recurrent breakpoints were localized in the following regions: 3q13~21, 6q14q22, 9p12~13, 10p14~15 in bMCL and 1p36, 1q21, 1q31, 6q26, 3q24, 9q21, 12q24, 17p11 and 18cen in tMCL. Preliminary analyses of clinical features suggest bMCL in our series to present more frequently with advanced stage, B-symptoms, elevated LDH and splenomegaly and to be associated with an unfavourable prognosis as compared to tMCL. Our findings indicate tMCL and bMCL to be genetically distinct. The clinical impact of secondary chromosome changes in t(11;14)-positive MCL is currently tested.

279

Upfront stem cell transplantation for mantle cell lymphoma using the Rituximab/Tbi/Cy high-dose regimen

B. Seyfarth, R. Kuse, R. Sonnen, B. Glass, C. Pott, M. Kneba, N. Schmitz, P. Dreger (Hamburg, D)

The purpose of this study was to investigate if incorporation of 2 375mg/m^2 doses of the CD20 antibody rituximab into a myeloablative regimen consisting of total body irradiation and high-dose cyclophosphamide (R-TBI/CY) might improve the results of autologous stem cell transplantation (SCT) in patients with disseminated mantle cell lymphoma (MCL). 12 patients (median age 56 (42-67) years) with stage III/IV MCL were treated with a sequential dose-escalating therapy comprising 2-6 cycles of CHOP for remission induction, intensive DexaBEAM chemotherapy for mobilization of peripheral blood stem cells, and R-TBI/CY with SCT (= group 1). Results: In all 12 patients a complete or partial remission was achieved after CHOP-DexaBEAM. Stem cell collection was successful in all cases, yielding 10.5x10^6/kg (4.1-34.5) CD34+ cells. However, flow cytometry revealed MCL contamination of the collection products in 10/12 patients. R-TBI/CY with SCT was not associated with unexpected toxicity and resulted in prompt engraftment (median days to neutrophil recovery >0.5x10^9/l: 10 (8-10); median days to platelet recovery >20x10^9/l: 11 (6-13)). Hematopoietic reconstitution was not significantly different from that of 20 historical patients with MCL myeloablated with TBI/CY alone (= group 0). Although B cell restoration was delayed after R-TBI/CY, incidence of infections and day +100 Ig levels were similar to that of control patients (504 vs. 788 mg/dl; p = .17). With a follow-up of 7 (1-15) months post transplant all patients of group 1 are alive with no clinical, immunological (10/10 tested by flow cytometry) and molecular (6/6 tested by CDR3 PCR/genescan) evidence of disease. At present, however, it is unclear if this might translate into an improvement of event-free and overall survival of SCT with standard TBI/CY, which was 56% and 100% at 3 years in group 0 (median follow-up 31 (17-65) months). Conclusions: R-TBI/CY is a safe and effective high-dose regimen for upfront SCT in patients with advanced MCL. Longer follow-up is needed to prove that the addition of rituximab to TBI/CY can improve the outcome of SCT by eliminating residual MCL cells surviving conventional myeloablative therapy.

280

Cytogenetic analysis of follicular lymphomas is important in the delineation of biological and clinical disease subgroups

J. Kalla, A. Lohr, T. Katzenberger, T. Rüdiger, S. Kindelberger, M.M. Ott, H.K. Müller-Hermelink, G. Ott (Würzburg, D)

89 follicular lymphomas (FL) with an exclusive or partial follicular growth pattern were analyzed for their cytomorphological features, immunohistochemical marker expression profile, and cytogenetic constitution. Clinical data were retrieved from 62 untreated cases. The cases were stratified according to the number of centroblasts into 29 FL1, 33 FL2, 11 FL3a (with preserved maturation to centrocytes), and 16 FL3b (exclusively consisting of blasts) with (n=12) or without (n=4) a DLBL component. FL of grades 1, 2 and 3 showed different proliferation indices (PI) reliably distinguishing the groups (p<0.001). According to morphology, cytogenetic data, and CD10 expression pattern, a homogeneous group of FL1-3a with CD10+, t(14;18)+ and maturation to centrocytes could be separated from the FL3b group (with CD10±, t(14;18)-, and absent maturation to centrocytes). Survival analysis showed a borderline statistical significance for the distinction of these two groups (p=0.12). In the biologically homogeneous group of

FL1-3a, neither differences in histological grade, nor different proliferation indices proved to be statistically significant indicators of outcome. FL grade 3a showed lower t(14;18) (70%) as compared to tumors of grades 1 and 2 (85%) and were characterized by frequent deletions in 1p. In univariate analysis, cases with >4 aberrations in banding analysis (p=0.05), 6q deletions (p=0.07) and +18 (p=0.01) pursued a worse clinical course, irrespective of histological grade or PI. In multivariate analysis, the number of numerical chromosome aberrations and presence of +18 retained their statistical significance. These results show that cytogenetic analysis is important both with respect to the delineation of biological disease subgroups as well as to the identification of prognostic markers within these groups.

281

Clinical pitfalls in gastric MALT lymphoma

K. Ranft, K.M. Lohner, G. Meyer, R. Westhaus (Penzberg, D)

PROBLEM: Staging as well as therapy of gastric MALT lymphoma (GML) are widely standardized. However, unexpected developments in clinical course could easily become disadvantageous for the patient. CASE 1: A forty-year-old man was admitted to the hospital on suspicion of low malignant GML. He was suffering from upper abdominal pain and discrete weight loss for three months. Serologic testing had proved a high antibody titer for Helicobacter pylori (Hp) which had been followed by eradication therapy. As his complaints had persisted an esophago-gastro-duodenoscopy (EGD) had been performed and a large gastric ulcer being interspersed with small lymphocytes had been found. Endosonography showed near-circular thickening of gastric wall, control of EGD showed further enlargement of the ulcer. Cytologically and histologically we found additional large blastic lymphocytes so that a high malignant GML with possibly low malignant portions was confirmed. A laparoscopic BII-resection and an adjuvant chemotherapy with 6 cycles of CHOP were performed. Patient recovered quickly and is in continuing complete remission for 1 ^{2} years now.
CASE 2: A seventy-year-old man was admitted to the hospital with polypoid mucosa swellings in his fundus ventriculi and numerous intraperitoneal lymphoma of 1 – 2.5 cm of diameter. Gastric mucosa histology revealed Hp gastritis with small cell lymphocytic infiltrates evidential for low malignant GML. Both, gastric mucosa lesions and intraperitoneal lymphoma completely regressed after Hp eradication therapy and 6 cycles of COP chemotherapy. One year later he was readmitted with bulky abdominal lymphoma but intact gastric mucosa. Diagnostic laparoscopy was not feasible due to excessive intraabdominal adhesions and infiltration of serosa and omentum. Sonographic guided biopsy out of an intraabdominal bulky lymph node revealed both, parts of a high- and a low malignant B-Cell lymphoma (CD20 positive) with marked differences in proliferation index. Thereby transformation to a mixed cellularity lymphoma was confirmed; EGD furthermore inconspicuous. Systemic chemotherapy with FCM and CHOP ± Mabthera induced partial remissions but the tumor relapsed soon and spread into supradiaphragmal lymph nodes. After another 1 ^{2} years we applied palliative radiotherapy which was limited by bone marrow insufficiency. DISCUSSION: 1) „Never treat titers“ and „never treat without endoscopy“ is especially valid in Hp infection. 2) Extensive primary histology is indispensable in order to prove GML in transformation. 3) Large polypoid and ulcerous lesions of gastric mucosa are suspicious of high malignant GML. 4) If enlarged abdominal lymph nodes are found laparoscopy is recommended as part of staging procedure in order to compare gastric mucosa and lymph node histology. As an alternative sonographic guided biopsy should be considered.

Postersitzung:
Lymphopoese

282

Chronic lymphocytic leukemia B cells express C5a receptor and are chemoattracted to C5a

K. Goetze, T. Decker, B. Schmidt (München, D)

The anaphylatoxin C5a is a potent chemoattractant for all cells of the myeloid lineage. The biologic effects of C5a are mediated through binding to the C5a receptor (C5aR), a pertussis-sensitive G-protein coupled receptor. Recent evidence suggests that C5a can exert biological activity on lymphoid cells as well. C5a has been demonstrated to be chemotactic for murine B cells, and C5aR was recently shown to be expressed on a subset of human tonsillar B lymphocytes. We investigated the expression and chemotactic function of C5aR on malignant B cells from patients with CLL (n=12) compared to normal purified B lymphocytes isolated from tonsils (n=6) and peripheral blood (n=4) of healthy donors. Using a mouse monoclonal antibody, we found that CLL B lymphocytes from all patients studied expressed high levels of C5aR. The mean percentage of malignant CD19+CD5+ B cells positive for C5aR as assessed by flow cytometry was 56%. In contrast, C5aR was detectable on only a small fraction of CD19+ normal B cells: 5% of tonsillar B cells and 10% of peripheral blood B cells. The mean flourescence intensity ratio of B-CLL cells was 11.8, compared to only 1.0 for tonsillar B cells and 1.2 for peripheral blood B lymphocytes (p<0.001). The chemotactic activity of C5a on B cells was evaluated in a transwell system. While spontaneous locomotion in the absence of rC5a was low, B-CLL cells exhibited a specific and dose-dependent migration towards rC5a, which peaked at a concentration of 1 nM and showed a biphasic curve characteristic of chemoattractant-induced migration. Normal tonsillar B cells showed a markedly increased spontaneous locomotion in the absence of chemoattractant as compared to B-CLL cells, but no effect towards rC5a was observed. Similarly, purified normal B cells from peripheral blood also failed to migrate in response to rC5a. Chemotaxis of malignant B cells towards rC5a was almost completely inhibited by pretreatment of cells with pertussis toxin. Furthermore, preincubation of input B-CLL cells with mAb against C5aR abolished specific migration while preincubation with an isotype matched control mAb had no significant effect. Taken together, these results demonstrate that C5aR is expressed on malignant B lymphocytes from patients with CLL and represents a fully functional receptor, delineating a novel role for C5a in the trafficking of malignant B cells.

283

CD4 and CD56-positive lymphoma

P. Reimer, V. Kunzmann, D. Krämer, F. Weißinger, J. Müller, H.K. Müller-Hermelink, M. Wilhelm (Würzburg, D)

CD4 and CD56-positive lymphoma first described by Adachi et al. in 1996 represent a rare subtype of non-Hodgkin's lymphoma. The origin of CD4+CD56+ lymphoma cells still remains unclear. Recently Chaperot et al. postulated a relation to plasmacytoid dendritic cells using different phenotypic and functional assays. Beside its unusual phenotype this lymphoma shows characteristic clinical features: (initial) skin involvement, rapid spreading (especially to the bone marrow), good response to initial conventional chemotherapy, and high rate of early fatal relapses. Experience with myeloablative chemoradiotherapy followed by autologous peripheral blood stem cell transplantation (PBSCT), as initial treatment to alter prognosis of CD4+CD56+ lymphoma is very limited. We report the clinical course of five patients with CD4+CD56+ lymphoma which were referred to our clinic between 1998 and 2001 with two of them undergoing myeloablative chemoradiotherapy and PBSCT. Table 1 summarizes the clinical features of the five cases. All patients presented with involvement of skin and bone marrow and were in complete remission after initial non-myeloablative chemotherapy. Both patients treated with myeloablative regimen and PBSCT are still in complete remission whereas two of the other three died of relapsing lymphoma. Table 2 shows the immunophenotypic pattern of the five cases revealing expression of CD4 and CD56 without expression of other T-cell or NK-cell markers.
Our data could be a hint for a curative potential of myeloablative chemoradiotherapy and autologous PBSCT in the treatment of this rare disease.

284

Jurkat cell-reactive anti-thymocyte globulin (ATG) assessed ex-vivo by flow cytometry persists three weeks in circulation.

T.H. Eiermann, S. Freitag, L. Cortes-Dericks, H. Sahm, A.R. Zander (Hamburg, D)

Rabbit anti-thymocyte globulin (ATG-Fresenius) is a polyclonal serum raised against the lymphoblastic T-cell line Jurkat. It is used for in vivo depletion of host and donor T cells in allogeneic stem cell transplantation. After 90 mg/kg prior to transplant rabbit IgG was present up to 4-5 weeks, but it is unknown how long T cell reactive antibodies persist. We measured anti-Jurkat antibodies by flow cytometry. The detection limit for Jurkat-reactive antibodies was at 0.1µg/ml rabbit IgG and half-maximal labelling of Jurkat cells was at 183 µg/ml rabbit ATG. The mean half-life of Jurkat-reactive antibodies in 7 patients was 4 days. They persisted up to 3 weeks with antibody levels equivalent from 0.2 to 4.1 µg/ml rabbit ATG. Jurkat-reactive antibodies were eliminated twofold faster as rabbit IgG measured by ELISA. The results showed that Jurkat-reactive antibodies might modulate recovery of T cells.

285

Cell cycle progression of chronic lymphocytic leukemia cells is controlled by cyclin D2, cyclin D3, cyclin dependent kinase (CDK) 4 and the CDK inhibitor p27

T. Decker, F. Schneller, S. Hipp, T. Jahn, J. Duyster, C. Peschel (München, D)

B-CLL cells are arrested in G0/early G1 phase of the cell cycle and are characterized by a marked hyporesponsiveness towards a variety of polyclonal B cell activators. We have previously demonstrated that costimulation with CpG-ODN and IL-2 can overcome this proliferative defect. Cyclin D3 is the principal D-type cyclin which mediates G1 progression in normal B cells, but in B-CLL cells both, cyclin D2 and cyclin D3, were strongly upregulated upon stimulation. Both cyclins were associated with cdk4 but not with cdk6 which is the catalytic partner of D-type cyclins in normal B cells. Moreover, immunecomplexes consisting of cyclin D2 and cdk4 or cyclin D3 and cdk4 were both functional and phosphorylated the RB protein in vitro. The cell cycle inhibitor p27 plays a pivotal role in cell cycle progression of B lymphocytes and has been shown to be overexpressed in B-CLL cells. P27 was rapidly downregulated in B-CLL cells even when stimulated with a non CpG-ODN or IL-2 alone while only moderate regulation could be observed in normal B cells. Taken together, our findings demonstrate that regulation of early cell cycle progression differs between B-CLL cells and normal B cells. These findings do not only contribute to the understanding of B-CLL pathophysiology but might ultimately lead to the identification of new therapeutic targets.

286

Sensitization of B-CLL cells to recombinant immuntoxin by immunostimulatory phosphorothioate oligodeoxynucleotides

T. Decker, S. Hipp, R. Kreitman, I. Pastan, C. Peschel, T. Licht (München, D; Bethesda, USA)

A recombinant anti-CD25-immunotoxin, LMB-2, has shown clinical efficacy in hairy cell leukemia and T-cell neoplasms. Its activity in B-CLL is inferior but might be improved if B-CLL cells expressed higher numbers of CD25 binding sites. We have recently reported that DSP30, a phosphorothioate CpG-oligodeoxynucleotide (CpG-ODN) induces immunogenicity of B-CLL cells by upregulation of CD25 and other antigens. We now investigated the antitumor activity of LMB-2 in the presence of DSP30. To this end, B-CLL cells from peripheral blood of patients were isolated immunomagnetically to >98% purity. Incubation with DSP30 for 48 hours augmented CD25 expression in 14 of 15 B-CLL samples as assessed by flow cytometry. DSP30 increased LMB-2 cytotoxicity dose-dependently while a control ODN with no CpG motif had no such effect. LMB-2 displayed no antitumor activity in the absence of CpG-ODN as determined colorimetrically with an MTS assay. In contrast, B-CLL growth was inhibited in 12 of 13 samples with 50% inhibition concentrations (IC50) in the range of LMB-2 plasma levels achieved in clinical studies. Two samples were not evaluable because of spontaneous B-CLL cell death in the presence of DSP30. Control experiments with an immunotoxin that does not recognize hematopoietic cells, and an anti-CD22 immunotoxin confirmed that sensitization to LMB-2 was sp ecifically due to upregulation of CD25. LMB-2 was much less toxic to normal B- and T-lymphocytes as compared with B-CLL cells. In summary, immunostimulatory CpG-ODN efficiently sensitize B-CLL cells to a recombinant immunotoxin by modulation of its target. This new treatment strategy deserves further attention.

287

Identification of lymph node homing peptides in vivo

M. Trepel, W. Arap, R. Pasqualini (Houston, USA)

Lymphatic organs such as lymph nodes and spleen are the primary sites of the immune response in the body. Antigens have been targeted by local injection into lymphoid organs in order to strengthen the immune response after vaccination. Other approaches to this end have involved targeting of receptors expressed on dendritic cells and other antigen presenting cells to augment immune response after vaccination. Here, we show that targeting antigens to lymph nodes via the bloodstream enhances the humoral immune response. We used in vivo phage display technology to select peptides that home to the vascular endothelium of lymph nodes after intravenous injection in mice. We isolated two peptides from an X2CX4CX library (C = cystein, X = any residue) that mediated homing of phage to lymph nodes after systemic administration („LNHP1" and „LNHP2" peptide, respectively). In contrast to a phage lacking a peptide insert (fd-tet phage), phage displaying either one the peptides LNHP1 and LNHP2 showed preferential homing to the lymph nodes compared to homing to control organs. To investigate the influence of targeting an antigen to the lymph node vasculature on immune response, we vaccinated mice with phage particles, which are known to be potent immunogens. We compared the immunogenicity of phage displaying lymph node-homing peptides to that of the insertless control phage. Mice immunized with lymph node homing phage had a markedly higher titer of anti-phage antibodies than mice immunized with fd-tet phage. Our results show that systemic targeting of antigens to lymph nodes enhances the humoral immune response after vaccination. This novel way of targeting an antigen may have broad implications for the development of effective vaccines, production of antibodies, and immunotherapy.

288

Expansion and characterization of gamma/delta T-cells

V. Bodemüller, M. v. Lilienfeld-Toal, A. Märten, S. Renoth, F. Schakowski, M. Gorschlüter, I. Schmidt-Wolf (Bonn, D)

Gamma/delta T-cells account for less than 10% of circulating T-lymphocytes. They are known to be involved in immune response in infectious, autoimmune and neoplastic dieseases. They have also been shown to be cytotoxic against lymphoma and gastrointestinal tumors. Recent investigations showed that gamma/delta T-cells could play a role in immunotherapy of multiple myeloma. Here, five different cell cultures were established from Ficoll separated peripheral blood lymphocytes of healthy donors: Culture 1 was stimulated with IFN-gamma on day 0, IL-2, anti-CD3 antibody and IL-1 beta on day 1 followed by restimulation with IL-2 every two to three days (Cytokine Induced Killer (CIK) cell protocol, according to J.Exp.Med. 174:139-149 (1991)). Culture 2 was stimulated according to the CIK protocol mentioned above with addition of pamidronate on day 1. Culture 3 was stimulated with pamidronate on day 0 and low dose IL-2 on day 2 followed by restimulation with IL-2 every four to six days. Culture 4 was depleted of CD8+ cells via magnetic activated cell sorting and further stimulated according to the CIK protocol. Culture 5 was also depleted of CD8+ cells and stimulated according to the CIK protocol with addition of pamidronate on day 1. Immunophenotyping using double color flow cytometry showed that the mean number of gamma/delta T-cells increased in all cultures. For the CD8+ depleted cell populations the mean number was increased more than 50 fold (similar to the results of Verneris et al., Blood 94, Suppl.1, p.706a, 1999). Further immunophenotyping showed that the majority of gamma/delta T-cells express the Vdelta2/Vgamma9 T-cell receptor and have an increased percentage of CD3+/CD56+ cells. Europium cytotoxicity assay showed activity of the CD8+ depleted cells against myeloma cell line U266 with a median lysis of 47.3% at an effector to target ratio of 40:1. Additional stimulation with pamidronate appeared to increase cytotoxicity (mean 51.9%).

Our results suggest that expansion of gamma/delta T-cells by depletion of CD8+ cells and stimulation with pamidronate improves cytotoxicity against multiple myeloma cells.

289

Differential expression of chemokine receptors in B-cell malignancies

J. Dürig, U. Schmücker, U. Dührsen (Essen, D)

Chemokines are a family of 8-10 kDa proteins with a wide range of biological activities including the regulation of leukocyte trafficking, modulation of haemopoietic cell proliferation and adhesion to extracellular matrix molecules. Using a panel of chemokine receptor specific monoclonal antibodies (MoAb) in a multicolour flow cytometry approach we analysed the expression of the lymphocyte-associated chemokine receptors CXCR4, CXCR5, CCR5 and CCR6 in B cell acute lymphoblastic leukemia (precursor B-ALL; 6 cases), B cell chronic lymphocytic leukemia (B-CLL; 31 cases), multiple myeloma (10 cases), mantle cell lymphoma (MCL, 4 cases), follicular lymphoma (FL, 3 cases) and hairy cell leukemia (HCL, 5 cases). We demonstrate that CXCR4, CXCR5 and CCR6 are differentially expressed in these B lymphoproliferative disorders depending on the maturational stage of the malignant B cell population investigated. In particular, we found that CXCR4 is strongly expressed on immature ALL blasts whereas no surface immunoreactivity for CXCR5, CCR5 and CCR6 was observed. By contrast, non-Hodgkin`s lymphomas (NHLs) corresponding to more mature peripheral B cell subsets (i.e. B-CLL and MCL) exhibited high expression levels of CXCR4 and CXCR5. Analysis of terminally differentiated myeloma cells revealed a down-regulation of CXCR4, CXCR5 and CCR6. CCR5, which is not expressed in normal B cells, was also absent from the majority of NHLs. However, CCR5 staining was seen in three of 5 cases of HCL, representing the first example of cross-lineage aberrant chemokine receptor expression in malignant hemopoietic cells.

290

Inhibition of PI 3-kinase in chronic lymphocytic leukemia induces apoptosis in a proteinkinase B/AKT independent manner

I. Ringshausen, F. Schneller, C. Peschel, T. Decker (München, D)

Compounds that inhibit protein kinases are currently undergoing clinical evaluation for the treatment of a variety of malignancies. Here we examined the effect of inhibition of the phosphatidylinositol 3-kinase (PI 3-kinase) by a specific inhibitor, LY 294002, on B-cell chronic lymphocytic leukemia (B-CLL) in vitro. LY 294002 induced apoptosis in 15 of 15 tested samples in a dose dependent manner. This effect was specific for the inhibition of PI 3-kinase, as inhibition of other signaling pathways like p70S6Kinase or MAPKinase did not increase the amount of apoptotic cells. In contrast to leukemic B-cells, LY 294002 did not induce apoptosis in B-cells of normal donors. The induction of apoptosis was associated with a loss of the mitochondrial membran potential and could be rescued by preincubation cells with a pan-caspase inhibitor, zvad.fmk. LY 294002 induced striking decrease of the level of the antiapoptotic protein MCL-1, whereas the level of bcl-2 was not affected. The level of the proapoptotic bax-protein was upregulated by the inhibition of PI 3-kinase and an enhanced bax-cleavage was observed. Because proteinkinase B/AKT is known to be a downstream target of PI 3-kinase and is part of signaling pathways of a broad spectrum of antiapoptotic signals, we investigated, if a constitutive activation of AKT is involved in the defect of programmed cell death in B-CLL. We could demonstrate that AKT is expressed in all investigated B-CLL cases, but we could not detect a phosphorylation of AKT. These findings suggest that activated PI 3-kinase is involved in the general defect of programmed cell death in B-CLL cells in a proteinkinase B/AKT independent manner. Further investigations need to be done to figure out if activation of PI 3-kinase is constitutive or related to exogenous factors. Inhibition of PI 3-kinase by a specific inhibitor may be another approach to treat chronic lymphocytic leukemia.

291

B cell chronic lymphocytic leukemia cells express functional CXCR5 chemokine receptors for the chemokine B cell-activating chemokine-1 (BCA-1)

J.A. Burger, A. Bürkle, D. Behringer (Freiburg, D)

Chronic lymphocytic leukemia (B-CLL) is characterized by the relentless accumulation of mature, functionally incompetent B lymphocytes in the blood, secondary lymphoid tissues and the marrow. The mechanisms that regulate the dissemination of CLL cells within these tissue compartments are largely unknown. Chemotactic gradients established by constitutively expressed chemokines may be critical for regulating the trafficking of CLL cells. Recently, we characterized the chemokine stromal cell-derived factor-1 (SDF-1) as an important mediator for the marrow homing and stromal or „nurse-like" cell-dependent survival of CLL cells (JA Burger et al., Blood 94:3658-67, 1999 and Blood 96:2655-63, 2000). In this study, we examined expression and function of the CXCR5 receptor for the chemokine B cell-activating chemokine-1 (BCA-1) in B-CLL. Disturbances of B cell colonization within specific anatomic compartments of lymph nodes and spleen in CXCR5-deficient mice indicate that chemotactic recruitment by locally produced BCA-1 is important for the development of B cell areas of secondary lymphoid tissues. Normal, mature B cells and a subset of memory T cells express CXCR5 chemokine receptors and migrate in response to BCA-1. By means of flow cytometry, we found CXCR5 to be expressed at high levels in each of 14 B-CLL patients samples, and in 4 out of 5 other leukemic B cell lymphomas. Functional studies including chemotaxis and actin polymerization in response to synthetic BCA-1 demonstrate that CXCR5 receptors are functionlly active in B-CLL. These observations indicate that CXCR5 receptors participate in the trafficking and homing of CLL cells to secondary lymphoid tissues.

292

Selective internalization of monoclonal antibodies by B-CLL cells

T. Sieber, D. Schoeler, S. Srock, F. Ringel, A. Smorodchenko, A. Grützkau, M. Pascu, F. Schriever (Berlin, Berlin, Wartenberg, D)

Aims: B-cell chronic lymphocytic leukemia (B-CLL) is the most common type of adult leukemia in western countries. The clinical enigma that B-CLL can not be cured by conventional chemotherapy is based on the fact that in B-CLL cells cell cycle is arrested and programmed cell death (apoptosis) is defective. Therapeutic monoclonal antibodies (mAb), therefore, offer a valid option for the treatment of B-CLL. In that respect, mAbs that are internalized by the leukemic cells may serve as carriers for cytotoxic agents. The aim of the presented study was to identify among a panel of murine mAbs those antibodies that are internalized by B-CLL cells. Methods: B-CLL cells were isolated from the peripheral blood of 21 patients with B-CLL. We developed a new flow cytometric method to investigate internalization of antibodies into B-CLL cells. To differentiate cell membrane binding from internalization of mAbs, B-CLL cells were incubated with a panel of murine mAbs, followed by treatment with a FITC-labeled goat-anti-mouse (GAM) antibody in vitro for 4 hours at 4°C and at 37°C, respectively. Background staining of the cell membrane was obtained by removal of antibody binding with acid washes. Increase of fluorescence intensity compared to controls was defined as internalization. This was confirmed by a so called „capping phenomenon" followed by detection of intracellular fluorescence using fluorescence microscopy and by endosomal uptake of antibodies by immunelectron microscopy using 10 nm-gold-coupled second mAbs. Results: Antibodies against CD5, CD19, CD40 and CD45 were internalized by up to 90% of the B-CLL cells whereas mAbs against CD20, CD21 and CD23 failed to be taken up by the leukemic cells. Internalization was detectable by FACS analysis after 15 min and reached its maximun after 4 hours. Conclusions: To our knowledge this is the first report that freshly isolated B-CLL cells are able to internalize mAbs against CD5, CD19, CD40 and CD45. We are currently testing whether antibodies against these antigens can be employed as tools to carry cytotoxic drugs into B-CLL cells for the therapy of this hematologic malignancy.

293

The Caspase-3/Caspase-8 expression ratio correlates with sensitivity to apoptosis induced by alkylating agents, anthracyclines and nucleoside analogues, but not corticosteroids in B cell chronic lymphocytic leukemia (B-CLL)

T. Wieder, F. Essmann, D. Head, S. Hermann, P. Hemmati, B. Dörken, P.T. Daniel (Berlin, D; Bath, UK)

Chronic lymphocytic leukemia (B-CLL) is characterised by the accumulation of non-proliferating B cells due to defects in apoptosis. Chemotherapy triggers apoptosis and dysregulation of this stringently controlled pathway has been associated with chemoresistance. We previously demonstrated that inactivation of the Bax/p53 pathway is related to chemoresistance in B-CLL. To further investigate the mechanisms of chemoresistance we investigated the role of caspases. These enzymes comprise a large family of cysteine proteases that play a critical role in the amplification, the regulation and the execution of the final stages of apoptosis. Nevertheless, the significance of the caspases as determinants of chemoresponse in hematological malignancies is presently poorly defined. We therefore measured the in vitro chemoresponse of 39 B-CLL patients to a panel of nine cytotoxic drugs by assessing the ex vivo apoptosis reponse of the B-CLL cells. To investigate the underlying mechanism of resistance, we compared endogenous levels of caspases-3, -8 and -9 with cellular drug responses, and with survival. The ratio of procaspase-3 to procaspase-8 (C3/C8) levels correlated with ex vivo response to alkylating agent, anthracycline, vincristine and nucleoside analogue therapy (p for all £ 0.018). Patients expressing lower levels of C3/C8 were particularly more resistant to anthracyclines and alkylating agents (p all £ 0.021). Interestingly, the C3/C8 ratio was not associated with the ex vivo response to corticosteroids (p for all $\geq$ 0.355). These results suggest that, in addition to p53 and Bax, the endogenous ratio of C3/C8 levels may provide a marker of subsequent chemoresponse to multiple cytotoxic agents in B-CLL patients.

294

Peroxisome proliferator-activated receptor gamma ligands induce growth inhibition and apoptosis of human lymphocytic leukemia

C. Zang, M. Wächter, A. Klügel, O. Doudina, H. Lui, M. Becker, M. Ruthardt, K. Possinger, E. Elstner (Berlin, Frankfurt, D)

Peroxisome proliferator-activated receptor gamma (PPARg) is a ligand activated nuclear hormone receptor which plays a key role in the differentiation and lipid metabolism of adipocytes. Recent data frequently indicated that PPARg ligands are also implicated in the growth inhibition, differentiation and apoptosis induction of several human cancers with diverse tissue origin suggesting that PPARg may serve as a novel target for cancer therapy. However, such evidence is still limited in hematopoietic malignancies, especially in human lymphocytic leukemia. In this study, expression of PPARg and the effect of its ligands on the proliferation and apoptosis of 10 lymphocytic leukemia cell lines were investigated in vitro. All cell lines were found to express PPARg as detected by Western blotting and RT-PCR. A marked growth inhibition was induced in 8/10 leukemia cell lines by the synthetic PPARg ligand Pioglitazone as well as by its natural ligand 15-deoxy-delta 12,14-prostaglandin J2 as measured by MTT. This inhibitory effect was further enhanced in 6/8 cell lines when these PPARg ligands were combined with all-trans retinoic acid (ATRA). Furthermore, in 6/8 cell lines, a remarkable apoptosis induction was observed in the cells treated with these PPARg ligands alone and/or their combinations with ATRA as judged by TUNEL-assay and ELISA. Experiments using caspase inhibitor clearly indicated that such an apoptosis induction was caspase-dependent. These findings strongly suggest that PPARg ligands may offer a new therapeutic approach for the treatment of lymphocytic leukemia.

295

In vitro model of malignant human B lymphopoiesis in multiple myeloma

H. Lannert, V. Eckstein, V. Welterlin, P. Schnitzler, E. Wunder, A. D. Ho, H. Goldschmidt (Heidelberg, D; Mulhouse, F)

Purpose of the study: Peripheral blood (PB) precursor cells of multiple myeloma (MM) patients were collected, sorted and cultered under specific conditions to develop a model of malignant human B lymphopoiesis. For this purpose FACS sorted CD20+ pro-B cells from peripheral blood (PB) of three patients with IgG-MM were cultured on a layer of AFT024 stroma cells in X-VIVO20 media containing 5% human serum. The cytokines IL4, IL6, IL10 and soluble CD40 ligand were added at day 0 of cell culture. Methods: Mononuclear cells were isolated by a standard Ficoll-Hypaque gradient separation method from 32 ml of patients blood. To receive an enriched and unmanipulated B-cell fraction we used the Dynal B-cell depletion kit to remove T-cells and monocytes. The purity of the B-cell fraction received was approximately 60%. Using a FACS Vantage SE cell sorter 10^4 CD20-FITC positive cells were sorted in each well of a 96-well plate containing medium described above. The sorted CD20+ cells showed a high coexpression (>95%) of CD19 and CD40 antigens. Results: After 14 days of culture cellcluster occured selectively. We obtained early plasma cells that did not express syndecan-1. These cells were further cultivated in 24-well plates for another 4 weeks without a stroma layer and stimulated with IL-6 only. Flowcytometric analysis showed a specific cytoplasmic kappa and lamdba restriction in two cases respectively. The cells were characterized through an increase of CD38 and a decrease of CD19 antigen expression. IgG secretion in the media was measured with a sensitive ELISA. Clonality was detected by fluorescent immunglobulin heavy chain gene rearrangement (IgH) PCR. IgH-PCR was performed with primers located in the FR4 JH region in combination with a consensus primer in the FR3 region and 6 family specific primers in the FR1 region. Two monoclonal cell populations corresponding to the patients with IgG kappa and IgG lambda could be identified in our new in vitro system. The third case showed a polyclonal cell population. The three cell lines generated are now cultured since several months. The immortality of these cell lines is probably a result of a shown Epstein-Barr virus (EBV) infection. All patients were EBV positiv before we took the blood samples. This fact is not uncommon because more than 90% of adults are latently infected with EBV. In conclusion, we have demonstrated two cases of in vitro differentiation and proliferation of peripheral CD20+ B cells to mature monoclonal malignant plasmablasts expressing no CD19 and strong CD38 antigen. Our B-cell culture system can be a valuable tool to further investigate B cell biology of MM. The two generated monoclonal cell lines are now further analyzed for aberrant surface protein expression in order to identify tumor antigens for immunotherapy protocols.

296

Bedside RNA stabilization with the PAXgene system: a useful approach for residual disease monitoring in multicenter studies

M.C. Müller, K. Merx, A. Weißer, S. Kreil, T. Lahaye, M. Paumen, R. Hehlmann, A. Hochhaus (Mannheim, Hilden, D)

The quality of blood samples plays a crucial role in performing molecular analyses by reverse transcriptase polymerase chain reaction (RT-PCR) on BCR-ABL transcripts. Transit time of outpatients material to the laboratory ranges in multicenter trials usually from 0 to 3 days. To preserve a sufficient amount of RNA a new blood collection system was introduced which is based on bedside RNA stabilization in a nucleic acid preservative. In a pilot study, 11 BCR-ABL+ CML patients after therapy with STI571 were examined within a two week period. A total volume of 16 ml of peripheral blood was taken and divided into two parts of equal size for comparative analyses. The first aliquot was processed after 2 hours while the second aliquot was processed after 72 hours of storage at room temperature. The standard blood preparation system (RNeasy, Qiagen, Hilden, Germany) using EDTA anticoagulated peripheral blood was compared with the new stabilization and purification method of intracellular RNA (PAXgene blood RNA kit, PreAnalytiX GmbH, Hombrechtikon, Switzerland). cDNA synthesis via reverse transcription was performed simultaneously for all samples. Quantitative real-time RT-PCR for BCR-ABL and ABL transcripts was performed using LightCycler technology. After 72 h sample qualities differed significantly depending on the method. Yield of RNA measured by RT-PCR on ABL expression was significantly higher with the PAXgene system as compared to the RNeasy system (median ratio of the yield after 2 and 72 hours 40.1% vs. 13.1%, p=0.014). Ratios BCR-ABL/ABL with both methods at 2 hours showed good correlation (r=0.95, p<0.0001). Whereas the ratios BCR-ABL/ABL comparing 2 vs. 72 h processing of the PAXgene system correlated well (r=0.99, p<0.0001), correlation was poor with the RNeasy kit with unstabilized EDTA blood (r=0.65, p=0.03). We conclude that the new PAXgene blood preparation kit provides good sample stability even after a room temperature incubation time of three days. The central RT-PCR analysis of blood samples in multicenter studies could be improved with this method.

297

Monitoring of minimal residual disease in acute myeloid leukemia associated with t(9;11) using Real-time RT-PCR technique

C. Scholl, H. Breitinger, S. Fröhling, R.F. Schlenk, H. Döhner, K. Döhner (Ulm, D)

One of the strategies being explored to predict clinical outcome in patients with acute myeloid leukemia (AML) exhibiting specific genomic rearrangements is detection of minimal residual disease (MRD) following achievement of hematologic remission. The Real-time RT-PCR is a novel method that allows the identification of MRD by quantification of the chimeric transcripts. Nevertheless, for most of the AML specific gene fusions the prognostic value of MRD still has to be determined. The objective of this study was to design a Real-time RT-PCR assay for quantification of the four common fusions types resulting from t(9;11) and to evaluate the predictive value of MLL/AF9 fusion transcript quantification for disease relapse. All patients (pts) were treated within the multicenter treatment trials AML HD93 and AML HD98A. In both treatment protocols pts receive double induction therapy with idarubicin, cytarabine (ARA-C) and etoposide (ICE). First consolidation therapy consists of HAM (high-dose ARA-C and mitoxantrone), as second consolidation pts were assigned to allogeneic or if no HLA-identical sibling was available to autologous stem cell transplantation (SCT). 87 samples [bone marrow (BM) n=44; peripheral blood (PB) n=36, leukapherese (LP) n=7] from 14 of 17 pts with t(9;11) associated AML were available for analysis. In 3 of the 14 pts only samples from diagnosis were obtained. The following fusion types were detected by conventional RT-PCR: MLL exon6-AF9 site A n=3; MLL exon6-AF9 site B n=4; MLL exon6-AF9 site D n=1; MLL exon7+8-AF9 site A n=6. For each different fusion type a specific probe set for Real-time RT-PCR was established. The fusion transcript copy number in each sample is reported as the normalized value of MLL/AF9 per 104 transcript copies of a housekeeping gene, porphobilinogen deaminase (PBGD), as a control. No correlation was found between the percentage of BM or PB blasts and the copy number of the fusion product at diagnosis. Furthermore, the copy number at diagnosis was not predictive of subsequent clinical outcome. After double induction therapy the fusion transcript could not be detected anymore in 6 pts. 4/6 pts are still alive and in CR, one died in CR because of GvHD after allogeneic BMT. 1/6 pt relapsed 11 months after diagnosis. This pt received sequential HAM instead of autologous or allogeneic SCT. In contrast, 5 pts still had detectable levels of the specific fusion transcript after induction therapy. 4/5 relapsed within 8 months (range 3,6 to 8,6 months); 1/5 died of VOD (veno occlusive disease) after autologous BMT in CR. Although intriguing, these data are preliminary and larger prospective studies are necessary to establish the role of MLL/AF9 fusion transcript quantification in the clinical management of t(9;11) associated AML.

298

Monitoring of AML1-ETO fusion transcripts during anti leukemic therapy using quantitative real time RT-PCR (LightCycler technology)

M. Weisser, C. Schoch, T. Haferlach, W. Hiddemann, S. Schnittger (München, D)

Background: The significance of minimal residual disease (MRD) is of growing interest in the therapy of acute myeloid leukemia (AML). t(8;21) positive leukemic blasts express a constant fusion mRNA (AML1-ETO) that can be detected by reverse transcriptase polymerase chain reaction (RT-PCR) and thus provide a molecular marker for MRD studies. Quantification of AML1-ETO fusion transcripts may identify specific patients that are at high risk for a haematological relapse via specific disease kinetics.
Methods: We established a real-time RT-PCR approach to specifically detect and quantify AML1-ETO. Detection of the PCR product is based on specific fluorescence hybridization probes. Limiting dilution series revealed a sensitivity of 1 AML1-ETO positive leukaemic blast in a background of 100000 normal cells or 10 cell equivalents. AML1-ETO was quantified relative to G6PDH and cABL housekeeping-genes. The ratio of AML1-ETO/housekeeping-gene is normalized against a known concentration of t(8;21) positive Kasumi-1 cells. Quantitative analysis using standard curve analysis as well as normalisation against a calibrator sample correlated well.
Results: So far 66 patients at primary diagnosis were analyzed. AML1-ETO levels showed 200fold variation (range 13500-3450000/median 343918). Eighteen patients were monitored at different stages after chemotherapy and bone marrow transplantation (BMT). Five patients relapsed. One patient had increasing AML1-ETO levels before detection of cytogenetic relapse, four presented with morphological and molecular relapse simultanuosly. Six patients are in molecular remission after conventional chemotherapy, two after allogenic BMT. In five patients we monitored the decline of AML1-ETO in peripheral blood 2-3 times a week during induction chemotherapy. On day 10 after start of chemotherapy AML1-ETO levels declined to 25% and below 1% on day 14. Three patients remained positive at low expression levels, two achieved a molecular remission after day 14. Conclusions: Realtime RT-PCR allows accurate monitoring of increase and decline AML1-ETO levels. Therefore G6PDH as well as cABL are appropiate housekeeping genes. The prognostic relevance of the variation of AML1-ETO levels at primary diagnosis and the quality of decline during induction chemotherapy remains to be clarified. Frequent monitoring of transcript levels in zytomorphologic remission is necessary to detect patients with molecular relapse before clinical manifastation.

299

Monitoring of VLA-4 expression by quantitative real-time RT-PCR

U. Steidl, T. Gräf, R. Fenk, S. Martin, N. Bienek, U.-P. Rohr, F. Neumann, R. Haas, R. Kronenwett (Düsseldorf, D)

The beta 1 integrin very late antigen-4 (VLA-4) is an adhesion molecule expressed in mononuclear cells including CD34+ hematopoietic stem cells and is involved in cell-to-cell and cell-to-matrix interactions via binding to its ligands vascular cell adhesion molecule-1 (VCAM-1) and fibronectin. Inhibition of VLA-4-mediated adhesion using peptides, monoclonal antibodies, or antisense oligonucleotides is a promising molecular approach to mobilize hematopoietic stem cells for transplantation or to treat patients with therapy-refractory acute leukemia. Here, we developed a quantitative real-time RT-PCR using the LightCycler system to measure mRNA levels of the alpha 4 subunit of VLA-4 in peripheral blood or bone marrow samples. Following RNA extraction, reverse transcription and PCR with primers binding to the alpha 4 subunit of VLA-4, PCR products were detected by means of frequency resonance energy transfer (FRET) using two product-specific hybridization probes carrying fluorescein and LC-Red640 serving as donor and acceptor fluorophores. GAPDH was used as external control. The level of alpha 4 expression was described as ratio alpha4/GAPDH. Linearity of the alpha 4 and GAPDH PCR over 4 logarithmic ranges was shown by a dilution series with cDNA reversely transcribed from RNA extracts of the VLA-4-positive cell line K422. We determined the amplification efficacy of the alpha 4 and GAPDH PCR to be 1.93 and 1.83 per cycle, respectively. For validation, we compared VLA-4 mRNA levels measured by RT-PCR with VLA-4 protein levels determined by flow cytometry, using RNA samples of K422, K562, and DHL-16 cells, immunomagnetically enriched CD34+ cells, as well as peripheral blood and bone marrow samples from a clinical phase I study with VLA-4-directed antisense oligonucleotides in patients with therapy-refractory acute myeloid leukemia. The relative mRNA levels measured by our method reflected the respective protein levels assessed by flow cytometry. In conclusion, our real-time PCR is a fast and sensitive method to measure alpha 4 mRNA levels in cell culture and clinical samples, and is therefore suitable for the molecular monitoring of novel therapeutic approaches targeting VLA-4 with peptides, monoclonal antibodies, or antisense oligonucleotides.

300

The usefulness of molecular monitoring (14;18 translocation, clonal Ig-rearrangement) in treating follicular lymphoma

C. Skrabs, G. Mitterbauer, M. Mitterbauer, C. Mannhalter, H. Greinix, K. Lechner, U. Jäger (Wien, A)

Purpose: There is accumulating evidence for longer relapse- and progression-free survival in Follicular Lymphoma (FL) without t(14;18) detection after therapy. If treatment targets on complete clinical remission (CR) molecular remission (MR) is desirable as well. Established therapies of FL were evaluated with respect to their impact on molecular status. Methods: Clinical and molecular data of 39 patients with FL were screened retrospectivly. The molecular status was determined by BCL2/IgH-PCRand CDR3-PCR in blood and bone marrow. Results: 10 of 39 patients had localized stages CS I/II, 29 patients had CS III/IV. The median observation time is 44 months. 2 Patients died. Initial molecular data were obtained from 36 patients. Group I: 13/36 patients were t(14;18) neg and IgR pos/neg. Group II: 23/36 patients were t(14;18) pos and IgRpos/neg. In group I, 12/13 patients (92%) are presently in CR, in group II 12/23 patients (52%) are in CR. Due to localized stages 9 patients were given radiotherapy alone. 3/9 patients with initial t(14;18) negativity and CS I are still in CR, 1/9 PCR-negativ patient reached CR after CHOP, the remaining 5/9 patients were t(14;18) positiv before or after radiatio and relapsed. 3/13 patients, treated with CD20-Ab, are in CR to date with no further therapy, 2 are t(14;18) negativ. 2/13 patients in MR recieved ric-Allo-transplantation. 11/12 patients, treated with CD20-Ab+CHOP, are still in CR (median observation time 17 months), 9 are still t(14;18) negativ. CD20-Ab failed to induce MR in 4/4 patients with t(14;18)+IgR positivity, CD20-Ab+CHOP failed in 2/6 patients with t(14;18)+IgR positivity. 5/6 of these non responders showed progression or stable disease. time to MR (t14;18neg): 2 months, time to MR (IgRneg): 4 months. Conclusion: There is a good correlation between clinical and molecular remissions following therapy with CD20-Ab±CHOP. Patients with persisting t(14;18) and IgR positivity are at risk for relapse or progression of FL. Patients who benefit from radiotherapy are restricted to t(14;18) and IgR negativity and CS I.

301

Molecular monitoring of patients with CML during STI 571 or interferon alpha treatment as well as after myeloablative or non-myeloablative allogeneic transplantation using a quantitative real time RT-PCR

F. Neumann, G. Kobbe, M. Aivado, A. Rong, R. Rössig, B. Hildebrandt, B. Royer-Pokora, N. Gattermann, R. Haas, R. Kronenwett (Düsseldorf, D)

The quantitative evaluation of BCR-ABL mRNA levels with real-time RT-PCR reflects the efficacy of different treatment modalities on a molecular level. Recently, the tyrosine kinase inhibitor STI571 and non-myeloablative allogeneic stem cell transplantation (PBSCT) have been used as new approaches for the treatment of patients with CML. In this study, we assessed the molecular response of these two novel treatment strategies in comparison with the standard therapies with Interferon alpha (IFN) and myeloablative PBSCT using real-time RT-PCR. 190 samples from peripheral blood of 40 patients with CML in chronic phase (10 during STI571 versus 11 during IFN treatment, and 16 after myeloablative allogeneic PBSCT versus 3 after non-myeloablative allogeneic PBSCT) were examined serially by quantitative RT-PCR as well as by standard two step nested RT-PCR. For detection and quantification of BCR-ABL mRNA the LightCycler technology was used. Primers and hybridization probes were chosen to detect the b2a2, b3a2 and e1a2 variants of BCR-ABL. G6PDH served as internal control gene and the relative level of BCR-ABL mRNA was expressed as the ratio BCR-ABL/G6PDH. All patients treated with STI571 reached complete hematological remissions after a median time of 23 days (range, 7-54). With respect to molecular response we found a continuous reduction of the BCR-ABL mRNA down to a constant level of the BCR-ABL/G6PDH quotient between 0.01 and 0.48 over a median time of 80 days. On the other hand, in the IFN group there were 3 of 14 patients with a complete hematological remission after a median time of 126 days (range, 70-209). In 7 patients the BCR-ABL mRNA level persisted on the level of pre-treatment values ranging from 3.1 and 13.6.

In 3 patients an increase was noted and in one patient a reduction down to 0.07. All except for one patient treated with myeloablative allogeneic bone marrow or PBSC transplantation reached BCR-ABL negativity, a complete cytogenetic remission and 100% donor chimerism after a follow-up time between 140 days and 11 years. Of the three patients treated with non-myeloablative allogeneic PBSCT two showed a continuous decrease of the BCR-ABL mRNA to PCR negativity after 57 and 139 days, respectively. In these patients donor chimerism has developed during this time period.

In conclusion, the real-time RT-PCR is a suitable method for molecular monitoring of the different treatment modalities for patients with CML. With respect to complete molecular remissions, allogeneic transplantation is the most efficient modality with non-myeloablative conditioning being a promising alternative. Comparing STI571 with IFN, the tyrosine kinase inhibitor is significantly more efficious in reducing the BCR-ABL mRNA amount.

302

Expression of BCR-ABL mRNA in individual CML cells as determined by in situ amplification

K. Pachmann, M. Andreeff (Jena, D; Houston, USA)

We here present results of a novel method developed for evaluation of in situ amplification, a molecular genetic method at the cellular level. Reverse transcription polymerase chain reaction (RT-PCR) was used to study BCR-ABL transcript levels in individual cells from patients with chronic myelogenous leukemia (CML). After hybridizing a fluorochrome-labeled probe to the cell-bound RT-PCR product, BCR-ABL mRNA positive cells were determined by image analysis. A dilution series of BCR-ABL positive BV173 into normal cells showed a good correlation between expected and actual values. In 25 CML samples, the percentage of in situ PCR positive cells showed an excellent correlation with cytogenetic results (r=0.94, p<0.0001), interphase FISH (r=0.95, p=0.001), and hypermetaphase FISH (r=0.81, p<0.001). The fluorescence intensity was higher in residual CML cells after IFN treatment than in newly diagnosed patients (p=0.004), it was highest in late stage CML resistant to IFN therapy and lowest in CML blast crisis (p=0.001). Mean fluorescence values correlated with BCR-ABL protein levels as determined by Western blot analysis (r=0.62). Laser scanning cytometry (LSC) allowing automated analysis of large numbers of cells confirmed the results. Thus, fluorescence in-situ PCR provides a novel and quantitative approach to monitor tumor load and BCR-ABL transcript levels in CML.

303

Quantitative analysis of CBFß-MYH11 transcription rates during therapy of AML M4eo

S. Schnittger, C. Schoch, G. Mellert, S Koneberg, W. Hiddemann, T. Haferlach (München, D)

Acute myeloid leukemia (AML) M4eo with inv(16)/t(16;16) is assigned to the prognostically favourable subgroup according to cytogenetics in most study groups. However, a subset of M4eo patients relapse. In cytomorphologic remission many patients express the correlated CBFß-MYH11 fusion transcript on levels detectable with PCR over several months. The prognostic importance of this phenomenon is still unknown. We wanted to evaluate 1) whether the kinetics of the initial CBFß-MYH11 reduction are different between certain therapy regimen 2) whether the initial kinetics are predictive for the individual prognosis, and 3) whether it is possible to predict clinical relapses due to increasing CBFß-MYH11 expression levels. To analyze CBFß-MYH11 expression we developed five different quantitative real time PCR assays (LightCycler) that potentially detect all CBFß-MYH11 fusion variants known so far. The sensitivity of the assays was at least 1/10000 up to 1/100000 cells as determined by limited dilution assays. We have monitored 23 patients at at least 4 time points during the first four months and afterwards we continued the monitoring up to two years. The patients were treated with TAD-HAM-TAD (n=14), TAD-TAD (n=2), HAM-HAM-TAD (n=4), IVA (n=1), AIDA (n=1), allo KMT (n=1). During the first 4 months the CBFß-MYH11 expression was reduced to different levels in the individual patients. In 22 patients the range of CBFß-MYH11 expression reduction was 3 to 6 orders of magnitude after 3-4 months of therapy. There were no differences in the molecular response between the different therapies. Two patients had a low reduction of transcripts of only 1 and 2 orders of magnitude, respectively, and both patients relapsed after 4 and 11 months of remission . In addition, 4 further patients with an initial good response relapsed. In one of these cases an elevation of transcript-rates was detected 4 months before clinical and cytogenetical relapse in bone marrow (BM), in the second case continuously at 5 and 2 months before relapse in BM and in a third case at 3 and 1 months before relapse in the peripheral blood (PB). One relapse could not be detected prospectively because of a long time period (11 months) between control time points. These results suggest that the level of initial molecular response is important for ongoing clinical remission and that frequent follow up controlls at intervals shorter than three months are needed for early prediction of relapses.

304

PCR-based detection of minimal residual disease (MRD) in AML with normal karyotype

S. Schnittger, C. Schoch, G. Mellert, S. Koneberg, W. Hiddemann, T. Haferlach (München, D)

Detection of minimal residual disease is of growing importance in disease management of acute myeloid leukemias (AML). Although the sensitivity of multiparameter immunophenotyping is improving mainly because of recent methodological developments of four color detection, in most cases the highest sensitivity is reached with PCR-based methods.The most common targets used for PCR are the transcripts of fusion genes like AML1-ETO, CBFb-MYH11, and PML-RARa. However, about 45% of AML have a normal karyotype. Thus, new molecular markers are urgently needed that can serve as targets for PCR-based detection of MRD. We have screened 1400 and 1050 consecutive AML for partial tandem duplications in the MLL-gene (MLL-PTD) and FLT3 length mutations (FLT3-LM), respectively. AML being positive for one of these markers were analyzed at two to ten time points during and after therapy using a conventional RT-PCR or genomic PCR. In addition, the individual FLT3-LM were sequenced to allow the design of patient-specific primer/probe sets for the quantitative real time PCR. The FLT3-LM was found in 40% and the MLL-PTD in 10% of all cytogenetically normal AML. Thirty three cases with the FLT3-LM and 10 cases with the MLL-PTD were analyzed at two to ten follow up time points. The conventional PCR with a sensitivity of up to 1 positive cell in 10000 negative cells showed a good correlation to the clinical course. In six cases the PCR was again positive after previous negativity 1-2 months before they ultimately had a morphological relapse. As there is few (for FLT3-LM) or no (for MLL-PTD) overlap with other PCR-detectable markers the FLT3-LM and MLL-PTD we suggest that these markers may represent molecular follow up markers for 25% of all AML and with the common fusion genes adding up to 50% of AML that are now evaluable for prospective MRD by PCR.

305

Assessment of rAAV-2 titers using quantitative real-time PCR

U. Rohr, M. Wulf, U. Streidl, S. Stahn, R. Haas, R. Kronenwett (Düsseldorf, D)

Although rAAV-2 is a promising vector in gene therapy determination of viral vector titers is still a time-consuming procedure. In this study, we developed a quantitative real time PCR (qPCR) to determine genomic rAAV-2 titers using the LightCycler technology. Since the CMV promoter is the most commonly used promoter in gene therapeutic approaches, primers were designed which hybridize with the human CMV promoter sequence. PCR products were detected by the addition of cyber green. qPCR of a 5 log spanning serial dilution of the vector plasmid containing one CMV promoter per plasmid molecule was performed in order to calculate a linear regression curve (mean r=0.993+0.01). Using this new method, we were able to detect as few as 8 plasmid copies with an amplification efficiency of 1.94. For qPCR analysis of rAAV-2 preparations viral DNA was released from protein core of the rAAV-2 particle by proteinase K digestion. To quantify the copy number of viral genome, the crossing points of AAV-2 samples applied to qPCR were related to the standard curve. Viral titers were calculated by the volume subjected to qPCR. For validation of the method, rAAV-2 preparations were analysed by a standard method and qPCR in parallel. As standard method flow cytometry was used for titration of infectious viral particles on HeLa cells using the Enhanced Green Fluorescent Protein as a marker gene under the control of the CMV promoter. We found a significant correlation between the results obtained by flow cytometry and the results from the qPCR over a 4 log range (r=0.91). The mean ratio between infectious rAAV-2 particles titrated via flow cytometry and genomic copies of rAAV-2 measured via qPCR of the same sample was 1:644 (range: 1:182 - 1:1235). These differences might be due to multiple transduction of a single cell using the infectious titration assay via flow cytometry and a significant amount of non-infectious particles generated during rAAV-2 preparation. In conclusion, pPCR is a fast and reliable tool for determination of rAAV-2 titers. Further, this method might be useful in monitoring transduction rates in clinical studies using rAAV-2 vectors without a marker gene.

306

Simultaneous quantification of BCR-ABL transcripts and T-cell receptor excision circles after allogeneic bone marrow transplantation

G.K. Przybylski, K.A. Kreuzer, K. Laurisch, S. Siegert, C.A. Schmidt (Berlin, D)

Reconstitution of donor T-cell repertoire after bone marrow transplantation (BMT) is poorly understood. Quantification of T-cell receptor excision circles (TREC) is a new tool for detecting recent thymic emigrants. In order to measure the thymic output and the minimal residual disease after allogeneic BMT, in 14 CML patients, using a real-time PCR, we quantified the TREC count and the BCR-ABL transcript numbers. To more accurately determine the TREC counts we prepared a duplex vector containing both the target (TREC) and the reference gene (RAG2). Since the same vector dilutions were used to obtain the target and the reference standard curve the absolute numbers of TREC were not influenced by the errors in DNA quantification, preparing of dilution series and variation of vector stability. Prior to BMT all patients showed markedly lower TREC count compared to healthy individuals, most likely due to chemotherapy. After BMT in most patients TRECs became undetectable. In 10 patients TRECs reappeared within 4 months after BMT and in 4 patients TRECs remained negative. All these 4 patients developed an acute graft versus host disease (GvHD).
We next compared TREC numbers and BCR-ABL transcript levels. Three patients who remained BCR-ABL positive after BMT were TREC negative. There seems to be an inverse correlation between TREC and BCR-ABL transcript numbers, however, number of patients analyzed is to small to draw a definite conclusion. We herein describe for the first time a more accurate, duplex vector based, real time PCR for TREC quantification. Our data suggest, that patients developing GvHD as well as patients with persisting BCR-ABL transcripts show an impaired reconstitution of naive T-cells.

307

Quantification of minimal residual disease by quantitative real-time RT-PCR of blood from CML patients – Towards a procedure suitable for routine diagnostic

P. Häusler, W. Schwabe, A. Weißer, K. Ott, D. Behnke (Jena, Mannheim, D)

Chronic myelogenous leukemia (CML) is characterised by the reciprocal chromosomal translocation t(9,22) which fuses a segment of the BCR gene on chromosome 9 to a segment of the C-ABL gene on chromosome 22. The BCR-ABL fusion may occur in different regions of the BCR and the ABL gene, respectively, giving rise to different chimeric BCR-ABL mRNA transcripts. The level of residual BCR-ABL chimeric transcripts can be used as a measure of residual leukemia. Molecular information about residual disease is important for clinical decisions, especially in patients after bone marrow transplantation or with good response to other therapies. Recurrence or increasing levels of BCR-ABL may result in a relapse which can be detected already in a very early stage by RT-PCR. However, very low levels of chimeric BCR-ABL transcripts may be detected continuously in patients without any sign of cytogenetic or haematological relapse and even in healthy donors. Therefore quantitative and serial detection of residual BCR-ABL transcripts will provide better information about the risk of relapse. There have been several attempts to quantify the BCR-ABL fusion transcripts by quantitative real-time RT-PCR in order to gain information concerning changes in BCR-ABL transcript level. However, most published procedures attempted to quantify different species of chimeric transcripts using the same PCR amplicon. We present here results indicating that precision and sensitivity of the results are increased when a unique PCR amplicon for each species of transcripts is used. We thus suggest a biphasic protocol consisting of a genotyping step based on real-time PCR and a subsequent quantification step. Furthermore, we present evidence that there is a considerable variation of the expression of the ABL gene (Var = 33%) in blood of different donors. Thus for better standardisation we suggest to use G6PD as an internal standard for BCR-ABL transcript quantification. Finally we show that quantitative PCR results obtained in a blinded manner by two independent laboratories using two different techniques of real-time PCR (LightCycler and TaqMan) correlate significantly ($r^2 = 0.74$; 32 patients).

308

Molecular assays for the detection of rearrangements of chromosome band 3q21 in myeloid leukemia

R. Wieser, U. Schreiner, H. Pirc-Danoewinata, M. Aytekin, U. Vinatzer, H. Schmidt, H. Rieder, C. Fonatsch (Wien, Graz, A; Marburg, D)

Chromosome aberrations affecting band 3q21, i.e., the inv(3)(q21q26), the t(3;3)(q21;q26), and the t(1;3)(p36;q21) are associated with poor responsiveness to conventional as well as intensified treatment regimens in myeloid leukemia. Their unequivocal diagnosis is therefore of great clinical importance. However, detection of these rearrangements by classical cytogenetics (G-band analysis of metapahases) may be hampered by insufficient numbers of dividing cells and/or poor metaphase preparations. A molecular assay would not only provide an additional means to confirm or exclude the presence of such aberrations at diagnosis, but also offer higher sensitivity than metaphase analysis during disease monitoring. The putative 3q26 oncogene, EVI-1, is transcribed at extremely low levels in normal peripheral blood or bone marrow cells, and is overexpressed in many, but not all cases of inv(3) and t(3;3). On the other hand, EVI-1 transcription was also observed in a substantial proportion of patient samples without such rearrangements, but was not per se indicative of poor patient survival. Quantitative reverse transcriptase polymerase chain reaction (RT-PCR) assessing EVI-1 transcription may thus be useful for sensitive disease monitoring in patients overexpressing this gene, but cannot proof or exclude the presence of the prognostically unfavourable 3q21 rearrangements. We therefore established an interphase FISH assay for the detection of these chromosome aberrations. PACs centromeric of the 3q21 breakpoint region were labelled with a red, and PACs telomeric of it with a green fluorescent dye. Analysis of 12 samples with and 27 samples without 3q21 aberrations demonstrated that separation of the normally colocalizing red and green signals is a reliable indicator for the presence of 3q21 breakpoints. In addition, therapeutic success as well as recurrence of disease could be monitired using this assay. Our data show that the new assay provides a means to detect and monitor the presence of cells carrying 3q21 rearrangements in a rapid, efficient, sensitive, and quantitative manner.

309

Quantification of minimal residual disease in patients with acute promyelocytic leukemia (APL) by real time quantitative RT-PCR with specific fluorescent hybridization probes

A. Mitterbauer, C. Mannhalter, G. Mitterbauer, U. Jäger, P. Kalhs, K. Lechner (Wien, A)

The majority of APL cases are associated with the translocation t(15;17)(q21;q12) which results in fusion of the retinoic acid alpha nuclear hormone receptor RARalpha to the PML gene. We have developed an automated real-time RT-PCR assay based on LightCycler technology for reliable and sensitive detection and quantification of PML-RARalpha fusion transcripts. The method allows the reproducible amplification and quantification of 10 molecules per reaction, and the detection of 1 blast cell among 104 (bcr1-/L-form) to 105 (bcr3-/S-form) normal cells. Intra- and inter-assay variation coefficients were lower than 19% and 10%, respectively. In all samples, ABL transcripts were quantified as internal control to determine the amount of amplifiable mRNA/cDNA. The percentage of PML-RARalpha molecules / ABL molecules was calculated. A total of 572 consecutive bone marrow and peripheral blood samples from 13 patients (median, 44 samples/patient; range, 15-83 samples) collected at diagnosis and at various times after conventional chemotherapy and/or bone marrow transplantation were first analyzed by 2-step RT-PCR. Then, all PML-RARalpha-positive samples were quantified by LightCycler-PCR. In 5 patients studied sequentially, a constant logarithmic increase in the number of PML-RARalpha fusion transcripts was detectable in complete remission (doubling time: median, 5.4 days; range, 3.4-11.6 days). All 5 patients experienced a hematological relapse within 0.6-3.2 months. Our data demonstrate that quantification of PML-RARalpha mRNA allows the assessment of the dynamics of the leukemic clone and the early detection of a continuous increase. Thus, quantification is valuable for the evaluation of minimal residual disease following various therapies and for early detection of relapse.

310

Importance of reverse transcriptase step in uncoupled RT-PCR methods on the detection of minimal residual disease in leukemias

F. Schüler, G. Dölken (Greifswald, D)

RT-PCR is a highly sensitive method to detect minimal residual neoplastic cells in patients with leukemic in clinical remission by in vitro amplification of specific DNA and RNA sequences. Controversial results, i.e. persistence of low numbers of specific chromosomal translocation-carrying cells or no specific cells detectable by very sensitive RT-PCR techniques, have been found in patients in long-term remission after successful treatment of t(8;21)-positive, inv(16)-positive and t(15;17)-positive acute myeloid leukemias (AML). In almost all studies uncoupled RT-PCR protocols are used, in which cDNA synthesis can be of great variability and final PCR amplification might be affected by the quality of the template or even inhibited to varying degrees due to the presence of contaminants. Various factors may decrease the overall sensitivity of RT-PCR methods for the detection of minimal residual disease (MRD). Therefore, we have carried out experiments to control and optimize the RT-step for RT-PCR protocols. Total RNA was isolated from peripheral blood mononuclear cells obtained from patients with different subtypes of acute myeloid leukemia (AML), chronic myelogenous leukemia (CML) and the established BCR-ABL-positive cells line K-562. For comparative analyses cDNA synthesis was carried out with 1 [mu]g aliquots of total RNA obtained from the same cell sample. Two different protocols for cDNA synthesis were primarily used, differing in KCl, DTT, use of anti-sense or nonspecific random hexamer primers. These experiments clearly show the inhibitory influence of increasing KCl concentrations present in buffers for cDNA synthesis on the amplification of specific transcripts by RT-PCR, as detected by quantitative real-time PCR. PCR amplification is progressively inhibited by increasing KCl concentrations above 55 mM and reaches almost 100% at concentrations between 65 and 70 mM. Therefore, quantitative results obtained by real-time RT-PCR or any other RT-PCR technique are unreliable and may be even false-negative if more than 1/20 of a cDNA synthesis mixture containing 75 mM potassium chloride (as provided by several suppliers) is used for PCR amplification in a single assay. High KCl concentrations used during cDNA synthesis limit the amount of cDNA that can be analyzed in single PCR assay thereby reducing the sensitivity as defined by the detection of a single copy in a certain number of normal cells. Quite simple technical problems as the one described could be at least in part responsible for controversial results in research on minimal residual disease.

311

Real-time PCR for quantitative analyses of MRD and chimerism in patients with acute and chronic leukemia after allogeneic transplant

A. Elmaagacli, M. Hlinka, U.W. Schaefer (Essen, D)

We report the use of real-time RT-PCR method to quantify fusion transcripts in patients with t(9;21) positive CML and t(8;21) or inv(16) positive AML after allogeneic stem cell transplantation. In 225 patients with CML and 20 patients with AML specific fusion transcripts were determined and normalized by using the GAPDH housekeeping gene product as endogenous reference. The real-time PCR method was not only useful for monitoring patients in relapse of leukemia, but also helpful in identifying patients who are at high risk for a relapse. All real-time RT-PCR methods were performed using the LightCyclerâ device in a single step RT-PCR. Furthermore we found that real-time PCR for Y-chromosome specific gene sequences or single nucleotide gene polymorphism (SNP) is a suitable new method for quantitative chimerism analyses after allogeneic transplantation. The real-time method is not only faster, but also more sensitive than qualitative PCR for Y-chromosome specific gene sequences or VNTR-PCR. Using real-time PCR for Y-chromosome we were able to identify safely 1 male cells of 10.000 female cells. Further we compared the results of qualitative PCR to those of real-time PCR for SNP or Y-chromosome in more than 100 patients. For patients without a MRD-marker real-time PCR for chimerism analyses may offer a faster and more sensitive analyses of the chimerism status, and thereby allows an earlier detection of relapse.

312

Monitoring of residual PML/RARalpha fusion transcripts in acute promyelocytic leukemia (APL) by real-time RT-PCR

A. Weißer, S. Saußele, A. Reiter, Th. Büchner, A. Hochhaus, R. Hehlmann, E. Lengfelder (Mannheim, Münster, D)

The t(15;17) translocation is the cytogenetic hallmark of APL. At the molecular level it results in two chimeric fusion genes PML/RARalpha and RARalpha/PML. RT-PCR methods have been developed for monitoring residual PML/RARA fusion transcripts after therapy. PCR positivity (RT-PCR sensitivity of 10^{-4}) in hematological remission is associated with a subsequent relapse. However PCR negativity is of uncertain prognostic value, as relapses may also occur in PCR negative patients. To improve the identification of patients at high risk for relapse, we established a rapid and reliable real-time RT-PCR approach for the detection and quantification of residual PML-RARA transcripts using the LightCycler technology. A pair of hybridization probes was designed complementary to exon 3 of RARalpha, enabling the detection of the two major transcript types of PML-RARalpha, bcr1 (L-transcript) and bcr3 (S-transcript). After analyzing different primer pairs for optimal sensitivity and specifity a primer set previously described was selected and PCR conditions adapted for the LightCycler. The sense primer for bcr1 is located on exon 6 of the PML gene and for bcr3 on exon 3, respectively. The antisense primer for both transcripts is located on RARalpha exon 3. Amplification products are 190 bp for bcr1 and 165 bp for bcr3. The housekeeping gene ABL was used for standardization and the level of residual disease was expressed as the ratio of PML/RARalpha to ABL. Conditions were established to amplify less than 5 PML-RARalpha plasmid copies per reaction and to detect one NB4 cell in 10^5 HL60 cells. Intraassay comparison of 20 identical samples resulted in a coefficient of variation of 0.098 for bcr1 and 0.062 for bcr3 samples, respectively. Retrospective analysis of 54 samples from 15 APL patients was performed and compared with the results obtained by qualitative nested PCR as previously described. All samples tested negative by real-time PCR were also negative by nested PCR (15/54). Of 39 samples tested positive by real-time PCR (39/54) only 35 were positive by nested PCR. In the 4 remaining samples real-time PCR was positive at a very low level (PML-RARalpha/ABL ratios 0.0007 to 0.021) whereas nested PCR was negative. This is likely to be due to a higher sensitivity of real-time PCR. In two patients who had hematological relapse both methods were positive. Molecular relapse in one of these patients preceded hematological relapse by 6 weeks. We therefore conclude that the established real-time PCR for PML/RARalpha is a reliable and sensitive method for detection and monitoring of residual disease in APL patients.

313

Early detection of relapse in CML patients after allogeneic stem cell transplantation by real-time quantitative PCR (RQ-PCR)

F. Gärtner, M. Rebelo, K. Potthoff, C. Waller, H. Bertz, J. Finke (Freiburg, D)

Background: For patients with chronic myeloid leukemia (CML), allogeneic stem cell transplantation offers the only established curative option. Post-transplant surveillance in these patients requires sensitive and reliable methods for early indication of relapse to counteract when the tumor load is still low (e.g. by donor lymphocyte infusion (DLI)). Methods: In this study we employed a real-time quantitative RT-PCR assay (RQ-PCR) based on TaqMan technology, which allows detection of up to 1 BCR-ABL positive cell in 10^5 normal cells as determined by dilution experiments with suitable cell lines. Reproducible quantitative results are expressed as the BCR-ABL/GAPDH ratio in per cent, where the housekeeping gene GAPDH is required for normalization, compensating for the varying RNA quality encountered in routine clinical samples. 33 patients with CML in first chronic phase (CP1, n=23) or advanced disease (acceleration, blast crisis, second chronic phase, n=10) received allogeneic bone marrow (n=27) or peripheral blood stem cell (n=6) transplantation from sibling (n=12) or unrelated (n=21) donors after conditioning therapy with a high dose busulfan-based regimen (n=27) or a low intensity protocol comprising fludarabine, BCNU and melphalan (n=6). A total of 541 peripheral blood and bone marrow samples (2164 single assays) taken at various time points after transplantation were investigated by RQ-PCR for their suitability in terms of decision making and therapy control with a median follow up of 749 days (70-1577). Results: 7 out of 23 patients transplanted in CP1 and 5 out of 10 patients transplanted with advanced disease received treatment (e.g. DLI, IFNalpha) for relapse (5 hematologic and 7 molecular only relapses). Typically, RQ-PCR showed increases of BCR-ABL/GAPDH over time, which preceded hematologic relapse up to 630 days. On the other hand, successful treatment could be documented by decreasing BCR-ABL/GAPDH ratios. Peripheral blood samples proved sufficient for routine monitoring. Conclusion: We suggest that serial real-time quantitative PCR (RQ-PCR) of BCR-ABL in peripheral blood should be the method of choice in the surveillance of CML patients after allogeneic transplantation, as (i) being a sensitive PCR-based assay it may lead to detection of presumptive relapse early when intervention is most successful, (ii) it allows decisions based on the amount of BCR-ABL transcripts and in particular on their dynamics (increase, decrease) as opposed to mere qualitative RT-PCR which might not be informative, and (iii) it is a robust, reproducible and easy-to-perform method with built-in control for RNA quality, suitable for routine clinical use (unlike e.g. competitive PCR).

Postersitzung: Morbus Hodgkin

314

Induction of selective tumor vessel thrombosis in a mouse model for Hodgkin's lymphoma

J. Philipp, M. Unruh, C. Gottstein (Köln, D)

Background: Selective occlusion of tumor vessels via STIAs (specific thrombosis inducing agents) has been proposed as a new promising approach for cancer therapy. This strategy has three main advantages: i) Endothelial cells are directly accessible to the intravenously administrated agent. ii) An amplification of the effect, because one capillary provides oxygen and nutrients to many tumor cells. iii) Mutation of the target cells and subsequent drug resistance is not very likely. In this study we describe the generation and evaluation of a STIA targeting VCAM-1 (vascular adhesion molecule 1), which is present on the vasculature of Hodgkin`s lymphoma. Methods: We constructed a recombinant protein consisting of a truncated form of the coagulation inducing protein tissue factor and a scFv recognizing VCAM-1. The protein was expressed in E. coli and extracted from the periplasma. Binding was assessed by FACS (fluorescence activated cell stain) analysis and coagulation function by a two stage coagulation assay. To evaluate the protein in vivo, we injected SCID mice with the metastasizing Hodgkin`s Lymphoma cell line L540rec. Tumor bearing mice were treated with an intravenous injection of fusion protein or control reagents. Treatment effects were assessed by histological analysis of tumors and normal organs. Results: We generated a recombinant fusion protein directed against VCAM-1. The protein was active in vitro, i.e. i) was able to bind to VCAM-1 expressing endothelial cells with a similar affinity as the parental antibody and ii) was able to induce factor Xa activation in a concentration dependent manner. In vivo we observed thrombosis in tumor vessels - but not in normal vessels - followed by destruction of tumor tissue. The average value of necrotic tumor tissue was 75% in the group treated with the fusion protein, 43% treated with tissue factor, 25% with scFv and 0% with PBS. Conclusion: A recombinant fusion protein against VCAM-1 induced selective thrombosis in tumor vasculature of Hodgkin`s Lymphoma bearing mice. The amount of tumor necrosis in control mice points to an additional factor. The latter could consist of traces of endotoxin still present in the purified protein preparation. We are currently investigating this hypothesis.

315

Thymic hyperplasia following nodular sclerosing Hodgkin's disease

K. Wieker, R. Bartsch, K. Jentsch-Ullrich, A. Franke (Magdeburg, D)

A young female patient showed up with mediastinal bulky disease and lymph node swelling in her left supraclavicular region. Clinical and histological investigations proved nodular sclerosing Hodgkin's lymphoma. The patient received combined modality treatment according to the protocols of the German Hodgkin's Disease Study Group (HD-11) and achieved complete remission. 6 months later the Chest X-ray and thoracic CT-scan showed mediastinal tumor masses suggesting relapsed Hodgkin's disease. Surprisingly the histological investigation proved thymic hyperplasia as well as the absence of any signs of Hodgkin lymphoma. Thymic hyperplasia is well known as a potential differential diagnosis of mediastinal space-occupying lesions and also as a long-term complication in patients cured of Hodgkin`s disease. A detailed case report and a complete review of literature are provided.

316

Treatment of relapsed CD20-positive Hodgkin's lymphoma with the monoclonal antibody rituximab is well tolerated and extremely effective

H. Schulz, U. Rehwald, M. Reiser, V. Diehl, A. Engert (Köln, D)

Rationale: Treatment of patients (pts) with relapsed follicular NHL using rituximab at standard dose (4 x 375mg/m^2) has resulted in response rates of more than 50%. While Hodgkin-Reed-Sternberg cells are CD20-positive only in the minority of cases of classical Hodgkin's Disease (HD), almost all pts with lymphocyte predominant Hodgkin's disease (LPHD) express this B-cell marker in high density on their malignant cell population. Objectives: This international multicenter phase II trial investigates safety and efficacy of rituximab in pts with LPHD or other CD20-positive subtypes of HD at first or higher relapse or progressive disease after at least one chemotherapy. All histological slides were reviewed by a reference panel. Results: Eleven pts were treated with intravenous infusions of rituximab 375mg/m^2 weekly for four doses. All pts had at least one prior therapy (median 2). LPHD was diagnosed in eight patients, lymphocyte-rich classical HD in one, and two patients with initial HD had T-cell rich B-cell lymphoma at the time of relapse. Median time from first diagnosis was 9 years. Adverse events, such as rhinitis, fever, chills, and nausea, were usually transient and of mild to moderate toxicity. The overall response rate in eleven assessable pts was 100% with nine complete remissions (CR) and two partial remissions (PR). Ten of eleven pts are still in remission with a median duration of response of 14+ months (5+-19+ months). In this study rituximab single-agent therapy proved to be safe and showed significant clinical activity in pts with LPHD. Since freedom of treatment failure and overall survival of LPHD patients have not been significantly improved yet by using intensified chemo- and/or radiotherapeutic protocols, rituximab might be a non-toxic and efficient treatment alternative for this group of young patients.

317

Promising results with a 14 day variant of the BEACOPP regimen in advanced stage Hodgkin's disease: Results of a pilot study of the German Hodgkin's Lymphoma Study Group (GHSG)

M. Sieber, J. Franklin, H. Bredenfeld, R. Naumann, F. Boissevain, P. Koch, P. Worst, M. Soekler, U. Rueffer, U. Paulus, A. Josting, A. Engert, V. Diehl (Köln, D)

Background: Improvement of treatment results in advanced stage Hodgkin´s disease has been achieved by G-CSF-supported dose escalation of the BEACOPP regimen in a 21-day cycle. An alternative approach to intensify therapy is to shorten cycle duration. With the aim of improving toxicity profile while maintaining treatment results of the BEACOPP escalated-dose regimen (B21esc), the present multicenter pilot-study assessed the feasibility and efficacy of the BEACOPP baseline-dose regimen (B21base) in a 14-day variant (B14). Patients and Methods: From September 1997 until 2000, 94 patients with Hodgkin's disease stage IIB and III, IV were enrolled from 32 centres. Patients were scheduled to receive eight cycles of B14 (bleomycin 10 mg/m^2 day 8, etoposide 100 mg/m^2 day 1-3, adriamycin 25 mg/m^2 day 1, cyclophosphamide 650 mg/m^2 day 1, vincristine 2 mg day 8, procarbazine 100 mg/m^2 day 1-7, and prednisone 80 mg/m^2 day 1-7; G-CSF from day 8 to 13; recycle day 15). Radiotherapy was administered to regions with initial bulky disease or residual tumor after chemotherapy. Results: All patients were evaluable for assessment of toxicity and treatment outcome. Eighty-six patients received the intended 8 cycles of B14 and 66 patients received consolidation radiotherapy. Chemotherapy could generally be administered on schedule, with 90% of cycles given within 21 days. Dose reductions varied between drugs: 10% to 25% of patients received less than 80% of the projected doses. Acute toxicity was moderate: WHO grade 3/4 leucopenia occurred in 48% of cycles, thrombocytopenia in 8%, anemia in 26%, and infection in 2%. Of the 94 patients, 88 (93%) achieved a complete remission, 4 patients had progressive disease and 2 patients died while on therapy (septic deaths). At a median observation time of 24 months, only 2 patients have relapsed and overall 3 deaths were documented. The 2-year survival and freedom from treatment failure rates were estimated as 97% (95%-CI: 93-99%) and 91% (85-97%). Up to now, no second tumor was observed.
Conclusion: Acceleration of the BEACOPP regimen in a 14-day variant is feasible at moderate acute toxicity. With a 2-year FFTF-rate of 91%, treatment results are promising, and the B14 regimen may provide a highly effective and yet generally acceptable regimen for advanced Hodgkin's disease.

318

Thoracic positron emission tomography using 18F-fluorodeoxyglucose for the evaluation of residual mediastinal Hodgkin's disease

M. R. Weihrauch, D. Re, K. Scheidhauer, S. Ansén, M. Dietlein, S. Bischoff, H. Bohlen, J. Wolf, H. Schicha, V. Diehl, H. Tesch (Köln, Villingen-Schwenningen, D)

Residual mediastinal masses are frequently observed in patients with Hodgkin's disease (HD) after completed therapy and the discrimination between active tumor tissue and fibrotic residues remains a clinical challenge. We studied the diagnostic value of metabolic imaging by 18F-Fluorodeoxyglucose (FDG) positron emission tomography (PET) in detecting active mediastinal disease and predicting relapse. Twenty-eight HD patients with a residual mediastinal mass of at least 2 cm after initial therapy or after salvage chemotherapy were prospectively assigned to 29 examinations with FDG PET and were evaluated as 29 'subjects'. Patients were monitored for at least 1 year after examination and observed for signs of relapse. Median follow-up was 28 months (range 16 to 68 months). A PET-negative mediastinal tumor was observed in 19 subjects, of whom 16 stayed in remission and 3 relapsed. Progression or relapse occured in 6 out of 10 subjects with a positive PET, whereas 4 subjects remained in remission. The negative predictive value (negative PET result and remission) at one year was 95% and the positive predictive value (positive PET result and relapse) was 60%. The disease free survival for PET negative and PET positive patients at 1 year were 95% and 40%, respectively. The difference was statistically significant. A negative FDG PET indicates that a HD patient with a residual mediastinal mass is unlikely to relapse before one year, if ever. On the other hand, a positive PET result indicates a significantly higher risk of relapse and demands further diagnostic procedures and a closer follow-up.

319

Mobilization strategy and characteristics in patients with Hodgkin's disease

M. Möstl, G. Kandler, M. Fillitz, E. Schlögl, H. Tüchler, E. Pittermann-Höcker (Wien, A)

Patients with stage 2B-4A,B hodgkin´s disease not being cured with conventional chemo±radiotherapy have good chance with abmt to achieve long time remission. We retrospectively analyzed data from 63 mobilized pat. (male n=30; female n=33) diagnosed and treated at our department. Stem cell sources were bone marrow in 7 pat., peripheral blood in 47 pat. and both in 9 pat.. 21 pat. (male n=10; female n=11) received autologous transplantation. 40 pat. were mobilized during first line therapy, in 2 pat. mobilization was performed before first line therapy, 21 pat.were collected after having one or more relapses. Pat. were mobilized with G-CSF alone (n=7), HD-Cyclophosphamid+G-CSF or standard HD-CT+ G-CSF (n=56). CD34 cell counts varied in patients with the same previous chemotherapeutic regimen. The objectives of our investigation were to define predictive factors for poor mobilizers other than number of previous chemotherapy-cycles. The association between risk factors such as stage, involvment pattern, age, kind of mobilization and yield of CD34 pos. cells was analized. We conclude that there is a subgroup of high risk-hodgkin´s disease patients, where stem cell harvesting as part of first line therapy is justified.

320

Up-regulation of the chemokine receptor CCR7 in classical but not in lymphocyte predominant Hodgkin disease correlates with distinct dissemination of neoplastic cells in lymphoid organs

U.E. Höpken, H.-D. Foss, D. Meyer, M. Hinz, K. Leder, H. Stein, M. Lipp (Berlin, D)

Chemokine and chemokine receptors are key mediators for regulating cell traffic and positioning in both homeostatic and inflammatory conditions. It is also presumed that chemokines and their receptors are likely to play a critical role in the localisation of malignant haematopoietic cells in their target organs. We analyzed chemokine and chemokine receptor expression on several Hodgkin disease (HD) derived cell lines and in HD tumors. All HD derived cell lines express functional CCR7 and CXCR4 receptors. CCR7 up-regulation was mediated by constitutive NF-kB activity. Lymphoid tissues in HD revealed differential expression levels of CCR7, CXCR4 and CXCR5 depending on the distinct subtypes of HD. HD of the classical subtypes (cHD), located in the interfollicular zone, showed strong CCR7 and CXCR4 expression and moderate CXCR5 expression. In contrast, the nodular lymphocyte predominant HD (NLP) subtype, located within follicular structures, exhibited no CCR7 reactivity but abundant CXCR4 staining. Their respective chemokine ligands showed marked expression by reactive cells within the tumors of classical HD and outside of the tumor nodules in NLPHD. Functionally, such differential chemokine receptor expression might contribute to specific localisation and confinement of neoplastic cells within the target organ in the distinct HD entities.

321

Expression of cancer testis antigens in Hodgkin and Reed-Sternberg cells does occur in vitro but not in vivo

W. Jung, B. Stürzenhofecker, F. von Bonin, M. Pfreundschuh (Göttingen, Homburg/Saar, D)

Cancer testis antigens (CTA) represent a distinct class of differentiation antigens with a restricted pattern of expression in normal tissue, but variable degree of expression in a wide range of human tumors. CTAs are now being viewed as attractive candidates for the development of vaccines against human cancer. In Hodgkin's disease expression of the CTAs SSX1 and MAGE 4a has been reported in Hodgkin derived cell lines. The aim of this study was to analyse the expression of nine CTAs (MAGE1, MAGE3, MAGE4a, SSX1, SSX2, SSX4, SCP-1, Hom-Tes-85, NY-Eos1) in native Hodgkin's lymphoma. In 29 cases of Hodgkin's lymphoma expression analysis of these CTAs was performed by global RT-PCR and gene specific RT-PCR of micromanipulated Hodgkin and Reed-Sternberg cells and of native lymph node tissue, respectively. None of the nine CTAs was expressed in these 29 cases. In contrast strong expression of most of the CTAs was found in a panel of five Hodgkin derived cell lines including KM-H2 and L428. In the cell line HD-Myz expression of CTAs was increased after culture in the presence of the DNA demethylating agent 5'-azacytidine. From these data we conclude that CTAs may not be attractive targets for immunotherapeutic approaches against Hodgkin's lymphoma. Expression of CTAs in Hodgkin derived cell lines may be only an in vitro phenomen probably related to a different methylation status of the promoter regions of CTAs in vitro and in vivo.

322

Comparing costs and effectiveness of COPP/ABVD and BEACOPP escalated regimens for advanced stages of Hodgkin's disease

R. Walshe, J.P. Glossmann, D. Waldschmidt, G. Schnell, U. Paulus, J. Franklin, B. Pfistner, A. Josting, O. Cornely, B. Standaert, H. Erder, J. Wolf, A. Engert, V. Diehl (Köln, D; Brussels, B; Thousand Oaks, USA)

Objective: To estimate the incremental cost-effectiveness (CE) ratio between COPP/ABVD and BEACOPP escalated (esc) supported by G-CSF in primary prophylaxis. Method: Retrospective analysis of resource use linked to survival data of the HD9 trial of the German Hodgkin's Lymphoma Study Group (GHSG) with 262 patients in the COPP/ABVD treatment arm and 403 patients in the BEACOPP esc treatment arm modeled in a decision tree. Median age of patients when treated was 32 years (y). Median observation time was 40 months. Only direct medical costs of resource use were considered including outpatient visits, lab tests, blood products (packed red blood cells and platelets), G-CSF use, chemotherapy drugs, other drugs and inpatient visits. Unit costs (1999 Deutsche Mark [DM] values) for the inpatient visits were derived from the University Hospital of Cologne (UHC) accounting department. Frequencies of salvage treatment modalities were estimated using the entire GHSG database. Results: Average direct medical cost estimates for patients treated with COPP/ABVD were DM 55,153 ($25,893) and DM 74,728 ($35,084) for patients treated with BEACOPP esc. Cost differences between the two arms were mainly due to differences in inpatient visits (frequency and duration), higher doses of chemotherapy drugs and G-CSF use. A significant survival benefit of 3.2 months in favor of BEACOPP esc was observed in the first 5 y of follow-up; the estimated incremental CE value was therefore 72,499 DM ($34,037) per live year saved. Life expectancy beyond 5 y was modeled using German life expectancy tables. Assuming that the natural death probabilities applied to both arms were increased by a range of equal percentages of zero to 1% as suggested by other studies, the estimated survival benefit during the first 10 y was 7.7 to 7.9 months. The resulting estimated incremental CE ratios were therefore DM 29,659 to DM 30,586 ($13,924 to $14,360) per life year saved. Conclusion: The model suggests an incremental CE ratio for the dose intensified BEACOPP esc treatment arm that is in the range of CE ratios for treatment of other common medical conditions. Consideration of life expectancy beyond 5 y further improves this ratio.

323

Loss of heterozygosity is frequently detected in Hodgkin and Reed Sternberg cells

D. Re, N. Massoudi, A. Staratschek-Jox, V. Dries, V. Diehl, J. Wolf (Köln, D)

Chromosomal aberrations have been frequently described in Hodgkin and Reed Sternberg (H-RS) cells in classical Hodgkin´s Lymphoma (HL) using conventional cytogenetics or fluorescence in situ hybridization. Nevertheless, no specific genetic aberration was identified in these cells so far. In order to define consistent genetic alterations in H-RS cells, we applied a molecular cytogenetic approach and performed „loss of heterozygosity (LOH)" studies in Hodgkin´s lymphoma. Highly polymorphic microsatellites were PCR amplified from pooled single laser-micromanipulated H-RS cells and from the respective germline tissue, i.e. DNA extracted from peripheral blood mononuclear cells or the HL affected whole lymph node. Heterozygosity of the fluochrome labelled microsatellites was determined by fragment length analysis on an ABI sequencer. We first analyzed 11 cases of classical HL with 21 microsatellite markers scattered over the whole genome. In 9/11 cases LOH was detected in over 40% of informative loci. There was no correlation between clinical or pathological characteristics of these cases and frequency of LOH. More frequently, LOH was present on the long arm of chromosome 6 (6q), which was affected in over 90% of cases. We therefore performed a more detailed deletion mapping of 6q in H-RS cells by analyzing a total of 16 cases with 13 microsatellite markers located on chromosome 6q. Our results led to the identification of a 3.3 Mbp region on 6q that is commonly deleted in the majority of HL cases. This region harbors severel well characterized tumor suppressor genes. Thus, it is tempting to speculate thatdeletion of a specific chromosomal region on 6q might be involved in the inactivation of specific tumor suppressor genes during the pathogenesis of H-RS cells.

Postersitzung:
Multiples Myelom - Therapie

324

Allogeneic transplantation for multiple myeloma relapsing after autologous transplantation

M. Görner, L. Kordelas, T. Luft, M. Thalheimer, S. Pfeiffer, T. Möhler, H. Goldschmidt, A.D. Ho (Heidelberg, D)

Background: Existence of a Graft vs Myeloma effect has been well documented by responses to DLI and long-term survival after allogeneic bone marrow transplantation. But the more widespread use of allogeneic transplantation in MM has been limited by the high rates of non-relapse mortality in these older and heavily pretreated patient group. We hypothesized that a lymphoablative conditioning which is non-myeloablative can be used to allow engraftment of allogeneic cells without excessive regimen toxicity or GVHD even in relapsing MM patients. Patients and Methods: From 8/99 to 4/01 20 patients with MM have received related (n=12) or unrelated (n=8) allogeneic bone marrow (n=1) or blood stem cell transplantations (n=19). For conditioning 2 patients received 2 Gy TBI alone, in the other patients fludarabin was combined with TBI. Post transplant immunosuppression consisted of cyclosporine and mofetil mycophenolat. 5 patients received T-cell depleted and 15 patients received unmanipulated grafts with a mean cell number of 5.8×10^6 CD34+cells/kg. Mean age of patients was 48 years. Median time from diagnosis to transplant was 51 months and all patients were heavily pretreated with all but one patient having had at least one previous high-dose chemotherapy with autologous stem cell support. At the time of transplant 3 patients were in CR, 10 had stable disease or showed PR following conventional salvage chemotherapy and 7 had progressive disease despite conventional therapy.
Results: Treatment was well tolerated with one transplant-related death due to GVHD IV. All but 1 patients engrafted with 9 patients never dropping neutrophil counts below 0.5×10^9/l. All but 1 evaluable patients had mixed chimerism >50% by day 50 and > 75% by day 100 with the majority achieving >95% donor chimerism documented by molecular techniques. 5 patients developed acute GVHD grade grade III to IV, 1 patient died due to steroid-resistant aGVHD of the gut. With a median follow up of 228 days 15 patients are alive, with 5 patients beeing either in PR (n=7) or CR (n=6), 2 patients are alive and progressed, whereas 5 patients died due to progressive disease.Conclusion: Allogeneic transplantation after non-myeloablative conditioning can be safely performed in heavily pretreated patients with MM even with an unrelated donor. Transplant-related mortality and morbidity is very low, for interpretation of response and survival longer follow-up is needed.

325

Hyperfractionated cyclophosphamide in combination with pulsed dexamethasone and thalidomide (HyperCDT) in primary refractory or relapsed multiple myeloma

M. Kropff, N. Lang, R. Fenk, G. Innig, W. Berdel, J. Kienast (Münster, München, Düsseldorf, Osnabrück, D)

Thalidomide is active in 30 % of patients with relapsed or refractory multiple myeloma (MM). More-over, thalidomide has been reported to restore the sensitivity of myeloma cells to dexamethasone-induced apoptosis. This phase II trial was intitiated to study the combination of thalidomide with pulsed dexamethasone and hyperfractionated cyclophosphamide. Fractionated cyclophosphamide administered at the dosing schedule employed in this study and combined with the VAD regimen has previously been shown to induce 40 % responses in VAD-resistant MM. 50 patients with advanced MM were treated with 2 to 6 monthly courses of hyper-fractionated cyclophosphamide (300 mg/m^2 IV over 3 h q 12 h x 6 doses, days 1 - 3, total dose 1800 mg/m^2) combined with pulsed dexamethasone (20 mg/m^2/d PO, days 1 - 4, 9 - 12, 17 - 20) and once daily thalidomide at individually escalating doses ranging from 100 to 400 mg/d. All patients received oral antimicrobial prophylaxis with ciprofloxacin and non-absorbable antifungal agents. G-CSF starting on day 7 was added after a previous grade 4 neutropenia. Responding patients were maintained on daily thalidomide and monthly dexamethasone pulses. 15 patients included in this study had primary refractory disease on regimens containing anthracyclines and pulsed dexamethasone. None of these patients was eligible for immediate high-dose therapy. 35 patients had relapsed after previous therapy with anthracyclines/pulsed dexamethasone (e.g. VAD, ID) followed by high-dose melphalan with autotransplant. Patient characteristics included median age 63 years; age > 60, 65 %; B2M > 2,5 mg/L, 25 %; CRP > 4,0 mg/L, 10 %; creatinine > 2 mg/dL, 5 %; and prior standard therapy > 12 mo, 80 %. Among 40 evaluable patients, 30 (75 %) achieved a partial remission; as yet, complete remissions were not observed. After a median follow-up of 9 months (2 - 24), 25 PR patients are alive and free of disease progression. Five patients experienced disease progression (4 deaths). One PR patient died subsequent to an allogeneic stem cell transplant despite non-myeloablative conditioning. Following chemotherapy, 25/50 patients experienced grade 4 neutropenia during at least one treatment cycle; no grade 4 thrombo-cytopenia. There were 4 grade 3/4 infections: 2 patients died during neutropenic sepsis; another patient developed neutropenic colitis, required intensive care but recovered. Other side effects included grade 2 constipation (35 %), grade 2/3 skin reactions (15 %), 2 patients with deep venous thrombosis, asymptomatic sinus bradycardia in 4 patients and hypothyreosis in 3 patients. In 11 patients thalidomide was stopped due to neurotoxicity (9) or skin reaction (2), and treatment was continued with HyperCD. In addition, 11 patients required a dose limitation of thalidomide to 300 mg (2), 200 mg (2), 100 mg (5) or 50 mg (2) due to neurotoxicity. HyperCDT appears to be a highly active and reasonably well tolerated regimen in advanced/resistant MM.

326

A new conditioninig regimen involving total marrow irradiation, busulfan and cyclophosphamide followed by PBSCT in patients with advanced multiple myeloma

H. Einsele, L. Kanz, W. Budach, C. Meisner, C. Straka, H. Schmidt, H. Schmidberger, H. Hebart, L. Trümper, P. Müller, G. Schlimok, B. Hertenstein, D. Peest, B. Metzner, N. Frickhofen, M. Bamberg (Tübingen, München, Göttingen, Augsburg, Hannover, Oldenburg, Wiesbaden, D)

Purpose: The overall survival of patients with advanced multiple myeloma (MM) undergoing high-dose chemotherapy and autologous stem cell transplantation (SCT) was found to depend mainly on the quality of response. Thus, to improve the response rate, a new intensified high-dose chemoradiotherapy was evaluated in a phase I/II study. Patients and Methods: 89 patients (median age of 51, range 32 - 60 years) received an intensified conditioning regimen consisting of total marrow irradiation (9 Gy applied in 2 fractions over 3 days with shielding of the lung and liver), busulfan 12 mg/kg and cyclophosphamide 120 mg/kg (TMI/Bu/Cy) followed by PBSCT. Fourteen patients received busulfan at a reduced dosage (9 - 11 mg/kg) due to organ dysfunction or significant overweight. Prior to high dose chemoradiotherapy an induction and mobilization chemotherapy (4 g/m^2 cyclophophamide) were administered. Isotypes were IgG in 48, IgA in 23, Light chain disease in 14, IgD in 3 and a non-secretory MM in 1 patient. Stage IIIA MM was diagnosed in 54, stage IIIB in 17, stage IIA in 15, stage IIB in 3 patients. 56 (62%) patients were pretreated at the time of inclusion in the study (35 with 1, 21 with > 1 lines of treatment). 27 patients had received conventional therapy for > 1 year prior to high dose chemotherapy. Regime-related toxicity according to WHO criteria and response rates as defined by the EBMT/IBMTR criteria were analysed. Results: TRM was 1% with 1 patient dying of invasive aspergillosis. Three patients developed reversible hepatic veno-occlusive disease (VOD). The main toxicity observed was mucositis WHO grade III/IV in 76%, and fever WHO grade > I in 74% of patients. Among the patients with de novo and pretreated MM a CR rate of 46% and 34%, respectively, was documented. The median duration of event-free survival (EFS) and overall survival (OS) of the whole patient group after transplant were 32 and 58 months, respectively. Low beta2 m (< 2,5 mg/l), a CRP < 4 mg/l and a normal serum LDH level (<190 U/ml) were the most significant parameters associated with prolonged EFS and OS. Conclusions: Administration of this new intensified conditioning regimen was associated with a tolerable toxicity and a high response rate. Due to these promising data, conditioning therapy with TMI/Bu/Cy is currently compared with tandem high dose melphalan in patients with MM stage II/III in a phase III study.

327

Allografts with dose-reduced conditioning in patients with heavily pretreated multiple myeloma

H. Schaefer, P. Bader, H. Hebart, M. Bamberg, L. Kanz, H. Einsele (Tübingen, D)

A reduced-intensity preparative regimen consisting of a single dose TBI (2 Gy, d -1), cyclophosphamide (2 x 20 mg/kg) (d -3 and -2) and fludarabine (5 x 30 mg/dl) (d -9 to - 5) was evaluated for allogeneic transplantation in 26 patients with heavily pretreated multiple myeloma. The median age was 54 years (range 36-66 years). Disease status at transplantation was relapse after autografting (n=22), with 10 patients relapsing after two autotransplants and 4 patients refractory to at least 2 lines of chemotherapy without an option for an autograft. Seven patients received unmanipulated G-CSF-mobilized peripheral blood stem cells from HLA-identical siblings, and 19 patients received unmanipulated marrow from unrelated donors, 4 of them from unrelated donors mismatched in 1 HLA- class I or II antigen. GvHD prophylaxis consisted of serotherapy with ATG, ciclosporine alone (n = 16) or CSA plus MMF (n = 10). Results of the chimerism analysis on day +30 post-transplant revealed complete donor type in 20/24 evaluable patients (2 patients with an early relapse had residual autologous cells on day +30 detected). The probability of grade 2-4 and 3-4 acute graft-versus-host disease was 0.27 and 0.04, respectively. All four patients without a previous autograft died early following allografting. In these chemotherapy-refractory patients, at least two attempts to mobilize a sufficient number of CD34+ cells for autografts had failed. Two of these four patients died due to non-engraftment without autologous recovery, the two other patients due to residual autologous B-cell proliferation associated with severe immune-mediated cytopenia and early relapse. All patients who had received a previous autograft showed engraftment of the allogeneic transplant. Two patients died due to intracerebral hemorrhagic and infections complications. Even among these heavily treated patients, more than 80% achieved at least a minor response (MR). Twelve out of 20 patients achieved a PR, six a VGPR/CR. Among patients achieving at least a partial response, the progression-free survival at 1 year was 58% and for patients who were either in CR or only positive for the paraprotein by immunofixation (VGPR), PFS was 80%, respectively. In summary, reduced intensity preparative regimen and allogeneic SCT from related and partially matched unrelated donors are feasible in older patients with heavily pretreated multiple myeloma and shows a low incidence of GvHD, but long-term disease control can be achieved only in patients with > PR after allografting.

328

Multiple myeloma: Consolidation therapy with the anti-CD20 antibody IDEC-C2B8 and molecular follow-up

F.W. Cremer, C. Gemmel, E. Ehrbrecht, M. Witzens, G. Moldenhauer, A.D. Ho, H. Goldschmidt, M. Moos (Heidelberg, D)

Background: Circulating clonal B cells are discussed to be the source of relapse after high-dose therapy (HDT) and autologous transplantation in multiple myeloma (MM). We have previously shown that in remission, the absolute numbers of clonotypic cells in peripheral blood (PB) are low, and the majority of these cells are CD20-positive. Therefore, a phase I/II study to evaluate a consolidation therapy with rituximab after HDT was initiated. Patients and Methods: Thirteen patients (pts) in remission after HDT (11 in CR and 2 in PR) were treated with 4 weekly infusions of 375 mg/m^2 rituximab. Sequential flowcytometry analyses were performed with B- and plasma cell markers. In 7 pts, the proportion of clonotypic cells in PB and BM samples was determined by a quantitative PCR assay (qPCR) with IgH specific primers based on limiting dilutions. Results: Treatment with rituximab could be completed in 12 pts, while in one pt it had to be aborted due to an allergic reaction WHO grade II. Otherwise, toxicity did not exceed WHO grade I. Ten pts are still in CR or PR, while 2 pts with CR relapsed: on day (d) 90 after treatment (7 months after PBSCT) and d270 (25 months after PBSCT), respectively. The median follow-up after PBSCT is 39 months (range 32 to 62 months), and after rituximab 26 months (range, 17 to 28). Flowcytometry showed in all pts assessed a prolonged depletion of CD20-positive cells in PB and BM. B cell reconstitution did not occur prior to d270 (median 360, range, 270-480). Prior to rituximab, the proportion of clonotypic cells was higher in CD20+ sorted cell fractions than in CD20-fractions in 3 of 4 pts assessed. In the pt who relapsed on d90, the majority of circulating clonotypic cells had been CD20-. In PB collected immediately prior to treatment, a median proportion of clonotypic cells of 0.0004% (range, below DT to 0.0068%) was found by qPCR. On d30 after treatment, in 5 of 7 pts a reduction of circulating clonotypic cells was observed (median, 0.00013%; range, below DT to 0.0037%). In BM, the median proportion of clonotypic cells was 0.001% (range, below DT to 0.0045%) prior to treatment. A reduction was observed on d30 in 2 of 7 pts, while in 4 an increase occured (median, 0.008%; range, 0.0003 to 1.44%). Only in 2 of 7 pts, a continued reduction of clonotypic cells in PB as well as BM was observed (follow up until d360). Conclusion: Rituximab is a feasible therapy with low toxicity for pts with MM, that causes prolonged B-cell depletions in PB and BM. The increase in BM tumor load observed immediately after rituximab treatment warrants further investigations. However, a subgroup of pts might benefit from this consolidation therapy.

329

Phase II trial evaluating clarithromycin in monoclonal gammopathy of unknown significance and multiple myeloma

N. Frickhofen, H. Gamm, D. Gerhartz, M. Klein, H.-H. Wolf, R. Weide, H. Hebart, M. Just, W.-D. Gassel, T. Fischer (Wiesbaden, Mainz, Halle, Tübingen, Bielefeld, Passau, D)

Clarithromycin has been claimed to induce remissions in about 50% of patients with multiple myeloma (B.G.M. Durie, Blood 90, Suppl. 1, 579a). Killing of unidentified pathogens, inhibition of RNA polymerase III and antiangiogenic properties have been proposed as explanation for this observation. To further evaluate this concept, a phase II study was developed for three groups of patients: Patients with high risk monoclonal gammopathy of unknown significance (MGUS) (Blood 87:912; 1996) and patients with myeloma, not treated for at least 3 months, were treated with clarithromycin alone (monotherapy group). Patients with myeloma and stable or progressive disease during standard chemotherapy were given clarithromycin in addition to the current treatment, which had to be continued without changes in dose or schedule (combination therapy group). Clarithromycin was given at a dose of 2 x 500 mg for one month. If there was no response, it was increased to 2 x 1000 mg during the second month for a total treatment time of 2 months. By this study design we wanted to answer two questions: If clarithromycin is active, (1) could it be even more active in early disease such as MGUS and (2) is there a difference between monotherapy and combination therapy in established myeloma? Thirty-six patients were included in this trial. In one patient, treatment had to be discontinued after 7 days due to significantly increased bone pain, headache and malaise. The other patients took clarithromycin for 1 month (35 patients) or the planned period of 2 months (33 patients). Reasons for early discontinuation of study medication were adverse effects and patient decision in one patient each. Gastrointestinal discomfort and a disturbed taste were the principal adverse effects observed in 32% of the patients. Among 15 patients with MGUS and 13 myeloma patients without current treatment, we did not observe any response as defined by a decrease of the monoclonal protein by at least 25% or a decrease of the ECOG bone pain score by at least 2. Six out of 7 patients, who received clarithromycin in addition to their current chemotherapy regimen, showed no response. In one patient who was progressing during treatment with 7 cycles of vinorelbin and dexamethason, the monoclonal protein rapidly declined after addition of clarithromycin. After 2 months on clarithryomycin (and continued vinorelbin and dexmethasone), his monoclonal protein hat declined from 6 g/dl to 3,6 g/dl. His remission has been maintained with clarithromycin, given for 7 days every month, for almost 2 years. No other factor except for clarithromycin has been found to explain the remission. We conclude, that clarithromycin alone is inactive in MGUS and multiple myeloma. Whether clarithromycin might improve response to standard chemotherapy, as observed in 1 out of 7 patients treated in the combination therapy group, will need to be confirmed. Supported by Abbott, Germany.

330

Low treatment related mortality following allografting in advanced multiple myeloma with an intensified conditioning regimen and anti-thymocyte globulin for GvHD prevention

N. Kröger, G. Derigs, H. Wandt, K. Schäfer-Eckart, G. Wittkowsky, R. Kuse, J. Casper, W. Krüger, T. Zabelina, H. Renges, H. Kabisch, A. Krüll, H. Einsele (Hamburg, Mainz, Nürnberg, Rostock, Tübingen, D)

To reduce the incidence of severe graft versus host disease (GvHD) and lower the treatment related mortality(TRM) in allogeneic stem cell transplantation from HLA-identical siblings patients with multiple myeloma, we incorporated anti-thymocyte globulin (ATG, Fa Fresenius, Bad Homburg, Germany) in the conditioning regimen of 12 patients. The conditioning regimen consisted of modified total body irradiation, busulfan and cyclophoshamide (n=9) or busulfan and cyclophoshamide (n=3). The median age was 44 years (range, 29-53) and the median time from diagnosis to transplant was 12 months (range, 6 to 56). The stem cell source was bone marrow in 10 , and peripheral blood stem cells in 2 patients. Grade II – IV acute GvHD occurred in 3 patients (27%percent). Severe grade III and IV GvHD developed in only 1 patient. Major toxicity was mucositis. Grad II according the Bearman score was noted in 10 patients, whereas 2 patients experienced grade III , requiring prophylactic intubation. One patient died of severe GvHD grade IV and one patient developed multi-organ failure on day +13, resulting in a TRM of 17%. A complete response in surviving patients after allogeneic transplantation was seen in 4 (40%), and PR in 6 (60%) patients. Two of the patients with PR received a donor lymphocyte infusion (DLI) for further tumor reduction 8 and 14 months after stem cell transplantation, and converted to CR, which increased the rate of CR up to 60%. After a median follow-up of 25 months (range, 5 -62), no patient with CR after allogeneic transplantation relapsed during follow-up, while 3 out of 4 patients with P.R. experienced progress within 3 years (p=0.07). The estimated overall survival at three years for all patients is 83 percent (CI 95%: 68%-98%). The estimated progression-free survival at three years is 61 percent (CI 95%: 32%-90%). These results suggest that the incorporation of anti-thymocyte globulin may prevent severe GvHD without obvious increase of relapse. DLI should be administered for patients with incomplete response after transplantation to enhance the rate of complete remission, which results in long term freedom of disease.

331

A randomized study of zoledronate versus control in patients with multiple myeloma stage I

M. Freund, O. Sezer, A. Ammon, A. Aldaoud, W. Aulitzky, O. Brudler, H. Derigs, H. Dürk, H. Einsele, B. Emmerich, J. Fischer, R. Fuchs, U. Haus, M. Herold, R. Hoffmann, U. Kaiser, U. Kleeberg, C. Maintz, J. Mittermüller, D. Niederwieser, M. Nowrousian, R. Pasold, D. Peest, K. Pflüger, R. Rohrberg, I. Schmidt-Wolf, G. Seipelt, B. Weinkauf, H. Wolf, H.H. Wolf, H. Wörmann (Rostock, Berlin, Göttingen, Leipzig, Stuttgart, Augsburg, Mainz, Hamm, Tübingen, München, Karlsruhe, Eschweiler, Nürnberg, Erfurt, Norderstedt, Marburg, Hamburg, Würselen, Germering, Essen, Potsdam, Hannover, Bremen, Halle, Bonn, Frankfurt/M, Dresden, Braunschweig, D)

Zoledronate is a new potent bisphosphonate with anti-tumor potential in vitro. In several preceeding randomized studies in advanced Multple Myeloma the question whether bishosphonates can exert an anti-tumor effect in vivo is still controversial. We have therefore initiatied a study in patients with multple myeloma stage I, in which zoledronate 4 mg every 4 weeks i.v. plus calcium 500 mg and vitamin D 400-500 I.U. daily orally is compared to observation. The total study period will be 48 months. Progression free survival and time to the development of sceletal events will be evalutated as primary end points. Secondary parameters are quality of life, over all survival and bone resorbtion parameters. MRI scans, bone marrow histology and cytogenetics will be evaluated in separate protocols. A total of 220 patients are planned to be included. As on end of May 2001 19 centers have been initiated, 11 additional are in the process of initiation. 12 patients have been randomized.
Further data on the study concept and the current status will be presented. This study in Multiple Myeloma Stage I will give insights into the biology of early multple myeloma and into the effect of the bisphophonate zoledronate on the clinical course of the disease

Postersitzung: Myelodysplasien

332

Clinical and differentiating effects of ethyol on hematopoiesis in MDS

M. Pfeilstöcker, M. Möstl, H. Karlic, H. Grüner, R. Reisner, T. Nösslinger, J. Hirsch, H. Tüchler, R. Waldner, E. Pittermann (Wien, A)

Aim of the study: In MDS proliferation and maturation processes are affected in the bone marrow. Ethyol is currently studied for improving maturation in MDS patients. We studied clinical response during ethyol treatment and correlated results with bone marrow morphology, cytogenetics and expression of the hematopoiesis specific G protein alpha subunit G alpha 16 - a marker of proliferation - and 5 lipoxigenase (5-LO) which is expressed after initiation of differentiation. Patients and Methods: 10 MDS patients (3 RA, 4 AISA, 1 RAEB, 1RAEBt, 1 CMML according to FAB; median age 75 years - range 54-87; sex distribution 5:5) received ethyol at a dose of 200mg/m^2 3 times weekly for 3 weeks for a maximum of 3 cycles, nonresponders received an escalated dose of 400mg/m^2. Before and after treatment bone marrow was analyzed for morphology, cytogenetics (according to standard techniques) and molecular markers (using RT-PCR from isolated bone marrow MNC). Results: Clinical results: 1 major response and 5 minor responses in terms of improvement of cytopenias and decrease of transfusion needs were observed at a dose level of 200mg/m^2 (total 13 cycles), administration of 400mg/m^2 (11 cycles) resulted in 1 major and 2 minor responses. Side effects were generally mild, mainly hypotension independently of dose and nausea with a pronounced need for antiemetics at a dose level of 400mg/m^2. Bone marrow examinations before and after treatment showed no changes in FAB type, cellularity, degree of dysplasia and cytogenetics. In two patients bonemarrow blast counts increased slightly. There was no significant change in serum ferritin levels. 7 Patients are available for analysis of genetic markers of proliferation and differentiation: Before treatment with ethyol 3/7 patients expressed G alpha16, 5/7 expressed 5-LO. In one responder positive G alpha 16 and negative 5-LO expressions were reversed. In one minor responder 5-LO expression was initiated by treatment. Conclusion: Single agent ethyol has a minor effect in MDS with tolerable side effects. Differential expression of G alpha 16 and 5-LO in MDS suggests simultaneously activated pathways of proliferation and differentiation in these diseases. Determination of these genes may provide a tool for observing regulation of growth and maturation of hematopoietic cells especially under treatment conditions.

333

Use of erythropoietin (EPO) in combination with granulocyte-(G-CSF) or granulocyte-macrophage colony-stimulating factor (GM-CSF) in patients with myelodysplastic syndromes (MDS)

C. Kasper, J. Zahner (Jena, Düsseldorf, D)

MDS are a heterogeneous group of hemopoietic progenitor cell disorders, and patients (pts) with MDS regularly develop anemia and frequently become transfusion dependent. Treatment with EPO has been tried to correct anemia with only limited success (response rates 16 - 25%). However, it is becoming evident that the generally rather low response rate of EPO in patients with MDS will be dramatically improved by the combination of EPO with either G-CSF, or GM-CSF. Here, we present the results from the literature (6 papers and 1 abstract using EPO + G-CSF, and 3 papers using GM-CSF). Among all trials, the response criteria were not uniform, and cytokine dose and application varied. The average response rate for improving anemia was 40% in 245 pts treated with EPO + G-CSF, and 24% in 145 pts treated with EPO + GM-CSF. There was a higher response rate for RA (41%), RARS (43%), and RAEB (37%) compared with RAEBT (18%) for the treatment with EPO + G-CSF. The response rates of this treatment increased with the duration of cytokine application from 4% (8 weeks), 10% (12 weeks), 50% (24 weeks), to 80% (36 weeks). In conclusion, the combination of EPO with G-CSF is probably superior to EPO + GM-CSF with an positive correlation to the duration of cytokine treatment and negative correlation to aggressive disease. However, controlled studies are mandatory to evaluate the role of the combined cytokine treatment in pts with MDS.

334

Thiazolidinedione-type peroxisome proliferator activated receptor-gamma agonists exert antiproliferative effects in chronic myelomonocytic leukemias

C. Denzlinger, K. Dittmann, A. Möhle, F. Machicao, R. Möhle, P. Brossart, H. Häring, L. Kanz (Tübingen, D)

Current treatment options for patients with chronic myelomonocytic leukemia (CMML) are not satisfactory. Peroxisome proliferator activated receptor g (PPARg) is a nuclear receptor controlling the expression of several metabolic enzymes and regulating the differentiation of a number of cell types. Thiazolidinedione agonist of PPARg (troglitazone, rosiglitazone and related drugs) have recently been introduced in the treatment of type 2 diabetes mellitus. Since myelomonocytic differentiation is sensitive to PPARg-agonists, we investigated whether troglitazone or rosiglitazone have an effect on proliferation of mononuclear cells from healthy individuals and from CMML patients in vitro. At a concentration of 50µM, troglitazone reduced [^{3}H]thymidine incorporation in mononuclear cells from healthy individuals (n=5) or from CMML patients (n=10) during 24 h of incubation to 57±18% (mean±SD) or to 9±7% of control. In monocytic cells isolated by immunomagnetic beads the difference between healthy donors and CMML patients was similarly pronounced. [^{3}H]thymidine incorporation was reduced to 68±10% of control in the former and to 9±7% in the latter. In parallel, troglitazone enhanced PPARg expression in mononuclear cells. Rosiglitazone had similar but less potent antiproliferative effects in CMML cells in vitro whereas [^{3}H]thymidine incorporation in cells from healthy individuals was not significantly affected. Our in vitro data prompted us to offer troglitazone 600 mg/day to three CMML patients. After troglitazone had been withdrawn from market, we have treated another three patients with rosiglitazone, 12 to 40 mg/day. Treatment with either drug was very well tolerated without any evidence of a drug-related side effect. In four of the six patients modest beneficial effects were observed: 54 or 65% reduction of monocyte counts in two patients, reduction of monocytic skin infiltrates, or 20% increase in platelet counts, in one patient each. Our results suggest that proliferation of normal and CMML cells is differentially affected by PPARg agonists and that these drugs may have therapeutic potential in CMML justifying further study of this approach.

335

Treatment of refractory myelodysplastic syndrome (MDS) with a combination of fludarabine and cytosine arabinoside (FA): A report of three cases

P. A. Maubach (Ingolstadt, D)

Rationale: The combination of fludarabine and cytosine arabinoside (ara-C) has shown promising activity in acute myeloid leukemia and also myelodysplastic syndromes (e.g. FLAG-protocol). Fludarabine effectively modulates the metabolism of ara-C leading to higher intracellular levels of ara-CTP, the cytotoxic metabolite. Case Reports: We have treated three consecutive patients with a combination of 100-150 mg Ara-C and 25-50 mg fludarabine one to max. three times per week who were refractory to a previous monotherapy with low dose cytosine arabinoside: The first male patient (64 years) developed a secondary MDS (REAB-T) 3 years after initial diagnosis and successful treatment of a large cell anaplastic Non-Hodgkin´s lymphoma. After 11 months therapy with low- and intermediate dose Ara-C he progressed and was then treated with the FA-regimen. His disease could be controlled for additional 8 months before he died of infectious complications. Secondly, a 69 year-old female patient was treated for MDS (RAEB) which developed one year after neprectomy for hypernephroma. Disease progression after 6 months of low-dose Ara-C therapy could be postponed for further six months by using the combination therapy with fludarabine until the death due to MDS. The third patient (male, 58 years) suffered from a follicular lymphoma which relapsed 3 times and was finally treated with high-dose therapy and autologous stem cell transplantation plus interferon maintenance therapy. 5 years after high-dose treatment a MDS (RAEB) was diagnosed and the patient received two weeks of low-dose Ara-C therapy . Progression after 2 months could be sucessfully encountered by application of the FA-regimen but the patient died 5 weeks later of MDS. Conclusion: The FA-regimen seems to be an effective palliative treatment in Ara-C refractory MDS-patients which can prolong survival of these prognostically unfavorable patients.

336

Prognostic implications of serum levels of soluble TNF family members Fas (CD95/Apo-1) ligand and 4-1BB (CD137) ligand in patients with myelodysplastic syndromes

H. Salih, V. Nüssler, G.C. Starling, C. Denzlinger, R. Pelka-Fleischer, P.A. Kiener, H. Schmetzer (Tübingen, München, D; Princeton, NJ, USA)

Release of soluble Fas ligand (sFasL) has been hypothesized to participate in the pathogenesis of various diseases. Recently we demonstrated that there is also a soluble form of 4-1BB ligand (s4-1BBL) that is released from hematological cells. In this study we analyzed the serum levels of sFasL and s4-1BBL in patients with myelodysplastic syndrome (MDS) by ELISA and estimated cut-off values with maximum probability for significant differentiation between cases with higher/lower levels of these molecules. We found that high serum levels of sFasL and low levels of s4-1BBL correlated statistically significant with time to progression of disease and vice versa as estimated by the log rank test, both at time of initial diagnosis (n=34) and during the course of disease (n=99). Our results indicate that serum levels of sFasL and s4-1BBL can be used as prognostic factors, which can be useful parameters for treatment decisions in patients with MDS. In addition, these observations may open novel perspectives for therapeutic intervention.

337

Morphology of the 5q-minus syndrome

A.A.N. Giagounidis, U. Germing, M. Aivado, M. Schroeder, C. Aul (Duisburg, Düsseldorf, D)

We analyzed the morphological characteristics of 43 patients with myelodysplasia and 5qminus syndrome. 39 patients had a deletion of the long arm of chromosome 5 as the sole anomaly, whereas 4 patients had additional cytogenetic aberrations. The male to female ratio was 0.59 to 1. Three patients had secondary MDS. The age ranged from 39 to 82 years (median 66). Morphological classification at initial diagnosis according to FAB was RA in 28 cases, RARS in 7, RAEB in 7 and RAEB/t in 1 case. 7 patients eventually transformed to AML. Myeloblast count at diagnosis ranged from 0 to 28% (median 2%). 85% had a blast count of less than 5%. Median erythroblast count in the bone marrow was 26% (range 0 to 63%). 10% of patients had more than 45% of erythroblasts out of all nucleated bone marrow cells. In 10% the erythroblasts stained positive for periodic-acid Schiff. 75% of the patients display a high number of mononuclear megakaryocytes. However, about 15% did not show the typical medium-sized spheronuclear megakaryocytes. Surprisingly, 2 patients had hypersegmented megakaryocytes in the bone marrow, too. Mikromegakaryocytes occurred in 50% of patients. An increase in lymphoid cells was noted in 40% of patients. Dysplastic features within the granulocytic cell lineage occurred in about 25% of cases, while Auer rods were never seen. Median survival of the entire group was 80 months (range 2 to 282), for RA patients 82 months, RARS 80, RAEB 12 and RAEB/t 2 months.

338

First results of a multicentric phase III-trial of amifostine long-term application vs. supportive care in patients with low-/intermediate-I-risk MDS (IPSS<1.5)

D. Haase, W. Gassmann, K. Vehmeyer, R. Steffens, J. Schanz, M. Stitz, F. Griesinger, C. Binder, A. Yilmaz, W.-D.. Ludwig, C. Wuchter, Fliche, S. Hahnfeld, L. Trümper, B. Wörmann (Göttingen, Siegen, Berlin, Jena, D)

It is the aim of our study, to examine the clinical value and the influence on disease progression of amifostine in a randomised comparison with supportive care in patients with low-risk MDS (IPSS<1.5). The treatment-schedule is biphasic with an intensive induction of 500 mg amifostine i.v. 4x/week for 4 weeks, followed by a maintenance of 500 mg once a week for further 48 weeks. Sequential bone marrow biopsies with centralized morphology, cytogenetics, colony assays and FACS-analyses are performed before start of therapy/observation phase and after 4, 25 and 51 weeks. Primary determinants of therapeutic effects are hemoglobin, reticulocytes, neutrophils, platelets and transfusion frequencies. Furthermore, we will search for biologic parameters predicting response to amifostine. In the now opened study, as of yet 28 pts. (16 RA, 9 RARS, 3 RAEB) have been enrolled. Preliminary observations are as follows: In the supportive care-group (14 pts.) 2 pts. died due to infectious complications; two out of 3 pts. displayed progressive cytopenias within 6 months of observation time, one pt. showed a stable course over 12 months. In the amifostine-group (14 pts.) 3

out of 3 pts. displayed a response to the substance within 6 months of observation: One of them had a minor (10 to 50%) increase of leucocytes, another pt. showed a minor three-lineage response, while a further pt. showed a major leucocyte response (+93%). Sequentially colony assays in pts. treated with amifostine revealed a significant increase (+1,150%) of CFU-GM in the bone marrow of one pt. associated with a major leucocyte response, an improvement of hematopoesis with erythroid differentiation (significant increase (+1,380%) of BFU-E in the peripheral blood) associated with a minor three-lineage response in a second pt. and a minor (+30%) increase of CFU-GM in the bone marrow after amifostine induction phase in a third pt.. Sequential cytogenetic data are available for 7 pts. in the supportive care goup and 8 pts. in the amifostine therapy arm. A stable karyotype was observed in 5/8 pts. receiving amifostine, 3/8 pts. displayed a cytogenetic response. Three out of 7 pts. treated by supportive care displayed a cytogenetic progression, 4 pts. had a stable cytogenetic course.

339

Mutation of myeloid transcription factor C/EBP alpha in a patient with MDS (RAEB)

W. Verbeek, A. Kaeferstein, M. Faulhaber, M. Stadler, W.K. Hofmann, H.P. Koeffler, A. Ganser (Hannover, D; Los Angeles, USA)

Mutations in myeloid transcription factors contribute to the molecular pathogenesis of acute myelogenous leukemia (AML). Recently, dominant-negative mutations in the transcription factor C/EBP alpha have been described in AML. We performed a mutational analysis of the C/EBP alpha gene in the myelodysplastic syndromes by PCR SSCP and direct sequencing. No mutations were found in patients with refractory anemia (n=28) or CMML(n=4). In one out of ten patients with RAEB/RAEB-T we detected a four base pair insertion disrupting codon 69. The novel +1 frame shift mutation is predicted to result in the translation of a truncated protein of 106 amino acids which include 68 aminoterminal amino acids of C/EBP alpha and 38 new additional amino acids containing 12 arginine residues. In all cases reported so far (n=5)a +2 frame shift resulted in a different position of the stop codon. The second protein predicted to be translated is the C/EBP alpha isoform p30, which was shown to inhibit C/EBP alpha's DNA binding and transactivation. The heterogeneity of the truncated proteins between cases with frame shift mutation supports the notion that it may not be important for the suppression of C/EBPalpha function. We conclude that dominant-negative C/EBP alpha mutations are a late event in MDS and that their emergence may be associated with blast accumulation.

340

Acute abdomen due to G-CSF treatment in a patient with myelodysplastic syndrome

A. Wölfler, W. Olipitz, C. Tinchon, H. Auner, G. Höfler, W. Linkesch, H. Sill (Graz, A)

In patients with myelodysplastic syndrome (MDS) human recombinant granulocyte-colony-stimulating factor (G-CSF) is routinely used to attenuate MDS-associated neutropenia as well as to accelerate granulocyte recovery after anti-neoplastic therapy. Here we report a patient with refractory anemia with excess of blasts (RAEB) with complex chromosome aberrations involving the short arm of chromosome 5 who developed an acute abdomen after treatment with human met-G-CSF (filgastrim).
In a 59-year-old Caucasian male with RAEB anti-neoplastic therapy consisting of topotecan and high dose cytarabine was applied. This regimen was well tolerated and on day 14 treatment with filgastrim (5µg per kg body weight) was initiated. On day 19 the patient complained about increasing pain in the lower back radiating to the abdomen. One day later the patient appeared seriously ill showing signs of an acute abdomen: abdominal rigidity, rebound tenderness and recurrent vomiting. These signs coincided with an excessive recovery of peripheral blood leucocytes with a left shift (peak value 30.600/µl on day 22) and a marked increase in LDH activity with a predominance of the type 3 isoenzyme (peak value: 7359 U/l on day 20; normal range 120-240). Laboratory indicators of disseminated intravascular coagulation temporarily occurred. Apart of a slight elevation of serum ASAT, γ-GT, AP activity, and creatinine other laboratory parameters including pancreatic enzymes as well as computed tomography of the abdomen and thorax and ultrasound examination of the abdominal vessels were normal. A bone marrow biopsy showed marked myeloid hyperplasia with predominance of dysplastic promyelocytes. G-CSF treatment was immediately discontinued and a second cycle of anti-neoplastic therapy initiated. Within few days the symptoms disappeared and elevated serum parameters as well as peripheral blood leukocyte counts returned to normal.
In this case G-CSF induced excessive dysplastic myeloid hyperplasia resulting in hyperleukocytosis, acute abdomen with peritonitic signs, bone pain and incipient DIC. We suppose that the patient´s disease with its underlying cytogenetic aberrations may have rendered him hyperreactive to G-CSF. However, the molecular mechanisms leading to G-CSF hyperreactivity remain to be determined.

341

Defective DNA-mismatch repair: a potential mediator of leukemogenic susceptibility in therapy-related myelodysplasia and leukemia

W. Olipitz, F. Quehenberger, G. Hopfinger, R.C.T. Aguiar, E. Gunsilius, M. Girschikofsky, C. Bodner, K. Hiden, W. Linkesch, G. Hoefler, H. Sill (Graz, Wien, Boston, Innsbruck, Linz, A)

Therapy-related myelodysplastic syndrome (t-MDS) and leukemia (t-leuk) are severe long-term complications following chemo- and radiotherapy as well as stem cell transplantation for a primary malignancy. However, as only a limited number of cancer patients is affected, it seems obvious to postulate a leukemogenic susceptibility of these individuals. To elucidate the potential role of defective DNA-MMR as a mediator of leukemogenic susceptibility we have analyzed 37 patients with t-MDS/t-leuk (7 t-MDS, 26 t-AML, 3 t-CML, 1 t-acute biphenotypic leukemia). The patients were treated with chemotherapy (n=13), chemo-and radiotherapy (n=17) or radiotherapy alone (n=7) for various primary malignancies, among which non-Hodgkin´s lymphoma (n=13), Hodgkin´s disease (n=7) and breast cancer (n=4) were the most common. The degree of microsatellite instability (MSI), the hallmark of defective DNA-MMR, was assessed using mononucleotide markers BAT25, BAT26, BAT34C4 and dinucleotide markers D2S123, D5S346, D17S250, D13S175, D18S58. Five of 37 (14%) patients displayed high MSI whereas three other patients had low MSI (8%). To address the question of at which stage defective DNA-MMR had developed we analyzed primary tumors of MSI positive patients. As a control group nine primary tumors from microsatellite stable patients were also screened for MSI. Three of four patients analyzed with high MSI in their t-MDS/t-leuk were also instable in the corresponding primary tumor, whereas none of the primary tumors of the low MSI and control group displayed MSI (P = .0182 ; Fisher's exact test). To investigate the biological role of MSI we further looked for mutations in exonic repeats of genes known to be targeted by defective DNA-MMR (BAX, TGFbeta RII, IGFII R, Caspase-5, APC, PTEN, E2F4, MBD4, hMSH6, hMSH3). Primary tumor and t-MDS/t-leuk samples presenting high MSI as well as cell lines serving as positive controls were analyzed by direct sequencing. However, no mutation was found in any of the patients samples whereas mutations in cell lines could be detected unambiguously. In summary the significant association between high MSI in the primary tumor and the corresponding t-MDS/t-leuk indicates that these individuals have a propensity to develop MSI. Defective DNA-MMR is therefore most likely of pathogenetic relevance in the development of these malignancies and in conferring leukemogenic susceptibilty to these individuals. Lack of mutations of repetitive exonic target sequences suggests that there may be other yet unidentified target genes related to the pathogenesis of these disorders.

342

Characterization of the T-cell receptor repertoire in patients with myelodysplastic syndromes before and after immunosuppressive therapy with antithymocyte globulin (ATG)

M. Faulhaber, W. Verbeek, M. Stadler, J. Carls, A. Kaeferstein, A. Ganser, E.M. Mischak-Weissinger (Hannover, D)

Recent studies suggest the involvement of a T cell-mediated autoimmune process in the pathogenesis of the cytopenia in myelodysplastic syndromes. Immunosuppressive therapy with antithymocyte globulin (ATG) improves cytopenia in a subset of MDS patients. To further investigate a T-cell involvement in the pathogenesis of MDS we studied the complementarity-determining region 3 (CDR3) size distribution of the T-cell receptor (TCR) V-beta-chain (Vß) subfamilies in the bone marrow of MDS patients (n=8; median age=62,5yrs.) before and 3-6 month after ATG therapy. To address this issue we used the PCR based technique 'TCR spectratyping' and compared the results with age-matched controls (n=4; median age=68 yrs.). A skewed Vß-profile was defined as a two-fold or more increase of any peak from its predicted position in the Poisson distribution of TCR lengths. Three MDS patients (1 RA, 1 CMML, 1 RAEB) responded to ATG (1 good response [GR], 2 minor responses [MR]), 5 (4 RA, 1 RAEB) showed no response [NR] to ATG. We found skewed Vß-profiles prior to ATG therapy in 4 out of 8 MDS patients (1 GR, 1 MR , 2 NR). Four (1 MR, 3 NR) out of 8 patients showed no skewing prior to ATG. Skewing of at least 2 Vß-subgroups was seen in all age-matched controls. Three MDS patients responded with a normalization of at least 1 TCR-Vß-subtype 3-6 month after ATG therapy (2 responders, 1 non-responder). The emergence of more than two skewed Vß profiles after ATG therapy in the absence of Vß skewing prior to therapy was observed in 2 patients. One of these patients achieved MR shortly after ATG, while the other patient did not respond. In both patients MDS transformed to secondary AML in less than 3 months after the second sample. Two non-responders developed a skewing of 1 Vß region respectively after ATG therapy. One of the non responders showed no change of 4 Vß regions which were skewed prior to therapy. We conclude that TCR Vß skewing is frequent in an elderly, apparently healthy population (control group). Normalization of at least 1 initialy skewed Vß profile after ATG therapy occured in 2 of 3 MDS patients responding to ATG, but it was also found in one patient (1/5) who did not respond. We are currently investigating the possibility that the skewed Vß profiles are equivalent to an oligoclonal expansion of T-cells. The analysis of a potential correlation between a response to ATG and normalization of the TCR repertoire requires a larger study population.

343

Gene expression profiles of mononuclear bone marrow cells in patients with myelodysplastic syndromes

M. Aivado, U. Steidl, O. Modlich, S. Kliszewski, A. Rong, C. Strupp, J. Koch, A. Giagounidis, B. Hildebrand, R. Kronenwett, N. Gattermann, R. Haas (Düsseldorf, Duisburg, D)

Introduction: cDNA-based array technology allows for broad-scale analysis of gene expression. Transcriptional profiles of bone marrow cells have not yet been reported for patients with myelodysplastic syndrome (MDS). We studied the expression of 1185 different genes in mononuclear bone marrow cells (MNBMC) from 4 patients (pts) with MDS. Patients and Methods: Pt 1 had refractory anemia with excess of blasts (RAEB II; with a complex karyotype (KT) 49,XY,+8,+18,+21,t(2;6)), and pts 2, 3 and 4 had refractory anemia (RA; KT pt 2: 47,XY,+8; KT pt 3: 46,XX,20q-, KT pt 4: normal male). The pt with RAEB II had cytotoxic chemotherapy six months before gene expression profiling. Between 0.45 and 3 x 10^7 MNBMC were obtained from each patient. Reverse transcription incorporating 32P-labeled dATP was performed using approximately 4µg of total RNA from MNBMC. cDNA probes were hybridized to a nylon-membrane based array containing 1185 different cDNAs (Human 1.2 Array, Clontech, Heidelberg) and the arrays were read out by a phosphorimager following hybridization for 16 to 20 hours. Results: We clustered 173 genes into the categories „apoptosis", „adhesion „, „transcription", „growth factors", and „differentiation". The mean level of gene expression within these clusters were similar for the pts with RA and the one with RAEB II. After this general view, we looked at the single gene level. Here, the expression of CD95 and its ligand FasL were virtually the same. Also, some other molecules, that are suspected to play a role in MDS, did not yield significant differences, for instances tumor necrosis factor (TNF) alpha (1.6-fold higher in RAEB II) and bcl-2 (1.9-fold higher in RA). Similarly, some moderate differences could be observed for the expression of for instance transforming growth factor beta (2.9-fold higher in RA) and TNF-related apoptosis inducing ligand (TRAIL) (2.8-fold higher in RA). Still, at the single gene level, some clear differences were observed. For 112/1185 genes, expression was found to be 5- to 176-fold higher in RA than RAEB II. The greatest differences were found for angiotensin converting enzyme (176-fold) and inhibin beta A subunit (46-fold). On the other hand, expression of 6 genes was 5- to 17-fold higher in the pt with RAEB II. Significant differences were found for ikappaB kinase complex-associated protein (17-fold) and cyclin T (6-fold). Conclusion: Although MNBMC represent a quite heterogenous cell population, we were still able to recognize some differences in the gene expression profiles between RA and RAEB II. Therefore, further analyses of more patients and healthy controls are warranted. Aside from cluster evaluations, analysis at the single gene level seems to be recommended. We will continue investigation of MDS patients using highly purified CD34-positive stem cell populations and we will compare their gene expression profiles with those of healthy controls.

344

Gemcitabine chemotherapy of acute leukemia evolving from chronic myelomonocytic leukemia (CMML)

A. Kündgen, L. Dietze, C. Strupp, N. Gattermann, U. Germing (Düsseldorf, D)

In March, 1997, a 60-year old female patient was admitted to our hospital with anaemia (11.5g/dl), thrombocytopenia (79000/µl), a normal WBC (9000/µl), weakness and rheumatalgia. In retro-spect, she had a moderate anaemia and a mild of thrombocyto-penia since 1989 at least. The differential blood count revealed a monocytosis of 19%. Bone marrow cytology yielded to the diag-nosis of chronic myelomonocytic leukemia (8% blasts, 16% monocytes). The karyotype was normal. Since this indicated a good prognosis and because there were no signs of disease pro-gression, we did not recommend chemotherapy at this time. In March, 1999, she was admitted again with severe arthritic pain, fever, night sweats, splenomegaly and generalized lympha-denopathy. High-dose therapy with cortisone achieved a signifi-cant clinical improvement. Lymphadenectomy revealed infiltration with plasmocytoid monocytes. In May, 2000, the patient had progressive disease with lympha-denopathy. Since there was no satisfactory response to corticos-teroids, we started treatment with hydroxyurea. This was well-tolerated and improved clinical symptoms. Four months later dyspnea, lymphadenopathy and severe thrombocytopenia (8000/µl) occurred. Treatment with Thalidomide (200mg) had no effect. Transformation into acute myeloid leukemia (FAB M4) was demonstrated on bone marrow cytology two months later. We did not see a chance of cure for this patient but tried to ameliorate her symptoms as well as the massive leukocytosis (95100/µl). Therefore, palliative chemotherapy with gemcitabine was administered. The reduction of lymphadenopathy and WBC was im-pressive (from 95100/µl to 1000/µl after the first dose). Between December 2000 and February 2001 gemcitabine was applied on seven occasions (1800mg). In March and April, 2001, we gave only 1000mg because of a prolonged thrombocytopenia. There was no drug resistance to Gemcitabine during those 5 months. With a dose of 1000mg gemcitabine WBC could be brought down again from 79100 to 4500/µl. In May, 2001, we could not continue this treatment because of fatal infectious complications. To our knowledge, this is the first report of successful cytoreduction with gemcitabine in acute leukemia derived from CMML.

345

Comparative cytogenetics of bone marrow stroma cells in myelodysplastic syndromes (MDS)

D. Haase, K. Vehmeyer, R. Steffens, F. Alves, L. Trümper (Göttingen, D)

MDS are clonal diseases of hematopoietic stem cells. Fifty to 80% of cases display cytogenetic abnormalities which can be demonstrated at the level of CD34+/CD38- immature progenitor cells. The role of the bone marrow stroma in the pathogenesis of MDS is still unknown. Preliminary data suppose perturbed adhesive interactions between stem cells and the bone marrow stroma in at least some cases. As yet, it is not known, whether stroma cells themselves belong to the abnormal MDS clone. Therefore, it was the aim of our study to determine whether stroma cells bear the same chromosome abnormalities as the myeloid population. In four patients with MDS and clonal cytogenetic abnormalities (case 1: del(5)(q13q33), case 2: complex abnormalities with partial loss of 5q, case 3: del(11)(q21), case 4: complex rearrangement of chromosomes 9 and 11 with partial loss of 11q) we were able to establish long-term pure stroma cell cultures. In these cases additional FISH-analyses with informative markers were performed in myeloid cells as well as in stroma cells. In none of the four patients the abnormality demonstrable in the myeloid cells (by classical cytogenetics and FISH) could be observed (by FISH) in the bone marrow stroma cells. Thus, our data strongly suggest that in MDS the bone marrow stroma is not part of the malignant cell clone. It remains to be determined whether stroma cells are clonally altered on a submicroscopic level, and whether the defects of adhesive interactions reside in the MDS-stem cells only or if MDS-stroma cells display functional impairments.

346

Azathioprine induces high-risk secondary myelodysplastic syndromes

U. Germing, C. Strupp, G. Kobbe, S. Knipp, B. Hildebrandt, J. Richter, N. Gattermann, R. Hass (Düsseldorf, D)

Azathioprine (AZA) is an immunsuppressive agent, prescribed in autoimmune diseases and after organ transplantation. We de-scribe the development of treatment-related Myelodysplastic Syndromes (MDS) in ten cases following therapy with AZA for different non-malignant disorders. Three of these patients were treated with AZA because of transplantation of heart or kidney. The other patients were treated with AZA because of a systemic rheumatic disorder (SLE, M. Wegener, psoriatic arthritis, polymyositis, rheumatoid arthritis and Sharp's syndrome). The median cumulative dosage of AZA was 141 g. Two of the pa-tients additionally received Cyclophosphamide because of their non-malignant disorder. Sex distribution was equal. Secondary MDS was diagnosed at a median age of 60 years. Morphological subtypes of MDS at diagnosis were as follows: refractory anaemia (RA) in 3 cases, refractory anaemia with ring sideroblasts (RARS) in 1 case, refractory anaemia with excess of blasts (RAEB) in 3 cases, refractory anaemia with excess of blasts in transformation (RAEB-T) in 1 case, chronic myelomonocytic leukemia (CMML) in 2 cases. At diagnosis, median value of LDH was 250 U/l, median bone marrow blast count was 8 %. Complex chromosomal abnormali-ties were detected in 4 of 6 karyotyped patients, all involving monosomy 7. Median survival time from diagnosis of MDS was 9 months. Five patients developed a secondary AML after a median duration of three month, two patients died of infectious diseases and only one is still alive. This patient received inten-sive chemotherapy after diagnosis of the MDS. Despite a high percentage of patients with a limited medullary blast count, we could observe a remarkably high number of development of secondary AML and an extremely short median survival time. Conclusion: Azathioprine seems to be a drug with a high risk of leading to secondary MDS, characterised by complex karyotype anomalies, especially monosomy 7 and a very poor prognosis. The number of treatment related MDS after long-term use of Azathioprine in patients with benign disorders may rise in the future. Azathioprine thus should be prescribed carefully.

Plenarsitzung:
Best Abstracts

347

VH Mutation status, CD38 expression, chromosomal aberrations and survival in CLL

A. Kröber, T. Seiler, A. Benner, P. Lichter, H. Döhner, S. Stilgenbauer (Ulm, Heidelberg, D)

In chronic lymphocytic leukemia (CLL) the prognostic impact of biological risk factors has recently been shown. We analyzed the immunoglobulin variable heavy chain gene (VH) mutation status (n=300) by PCR and direct sequencing using family specific VH- and JH-consensus primers, the most frequent genomic aberrations with a comprehensive FISH probe set (+3q, 6q-, +8q, 11q-, +12q, 13q-, t(14q), 17p-) (n=300) and the CD38 expression level (n=157) by triple color FACS (CD5, CD19, CD38) in an unicentric cohort of CLL patients (median follow-up time 69 months (m)). In all cases at least one clonal VH rearrangement was amplified. To determine the prognostic influence of different VH mutation rates and the CD38 expression levels we carried out maximally selected logrank statistics. The estimated VH homology rate yielding the highest significance in separating two groups with different survival probabilities was not 98% but 96% (95% confidence interval 95% to 98%, p<.0001). The strongest CD38 expression level cutoff value to separate two groups with different survival probabilities was 7% (95% confidence interval 2% to 71%, p=.02). Applying the 96% germline homology cutoff level VH was mutated in 35% and unmutated in 65% of cases. Comparison of the two groups showed the following characteristics (VH mutated vs. unmutated): 1.) Similar distribution of stage, age and sex; 2.) Different VH gene usage: VH1 (7% vs. 35%), VH3 (61% vs. 47%), VH4 (29% vs. 14%); 3.) CD38+ >7% (30% vs. 63%); 4.) Similar overall incidence of chromosomal aberrations (80% vs. 83%); 5.) Different incidence of high-risk aberrations: 17p- (2% vs. 9%), 11q- (2% vs. 25%). 6.) Different incidence of favorable aberrations: 13q- (70% vs. 48%), 13q-single (56% vs. 25%). The median survival times for the groups with (n=82) or without (n=75) CD38+ > 7% was 79m vs. not reached (p=.0009). In the multivariate analysis the VH mutation status and 17p- were independent adverse prognostic factors. The median survival times for the 17p- (n=20), VH unmutated (n=176) and VH mutated (n=104) subgroups were 30m, 83m and not reached (p<.0001). In conclusion, our data confirm the biological and clinical distinction between CLL with mutated or unmutated VH. High-risk genomic aberrations occurred frequently in VH unmutated CLL. Multivariate analysis identified the VH mutation status and 17p- but not CD38 expression as independent prognostic factors in our series of 300 CLL patients.

348

Global gene expression analysis by oligonucleotide microarray can identify altered cellular pathways in mantle cell lymphoma

W.K. Hofmann, S. de Vos, L.C. Jones, W. Wachsman, G.S. Pinkus, D. Hoelzer, J.W. Said, H.P. Koeffler (Frankfurt/Main, D; Los Angeles, San Diego, Boston, USA)

To expand the understanding of altered cellular pathways in mantle cell lymphoma (MCL), we analyzed gene expression in lymph node preparations from 16 patients with MCL as compared to the pattern in 12 nonmalignant hyperplastic lymph nodes (HL). Oligonucleotide microarray analysis was performed comparing 5 MCL to 4 HL utilizing high-density microarrays (HuGeneFL, Affymetrix, Santa Clara, CA). The expression data were analyzed using the Genespring software (Silicon Genetics, San Carlos, CA). For confirmation, the expression of selected genes was analyzed by real-time PCR in all of the samples. We required for significant changes that the genes classified as „up-regulated", had raw data values of at least 500 in the MCL samples, and to be called „down-regulated" had raw data values of at least 500 in the control. Within those parameters, genes were selected if they were either up- or down-regulated at least 3-fold in 4 of 5 cases of MCL studied by microarray. We selected 50 up- and 42 down-regulated genes in MCL. The gene encoding for the homolog of mouse double minute 2 (MDM2) which controls the inhibition of the transcription factor E2F5 and binds to the retinoblastome protein (RB1) was up-regulated in MCL. The FAS associated via death domain (FADD) gene acts downstream of the FAS cascade as a key gene to induce apoptosis; it was more than 10-fold down-regulated in MCL. Furthermore, the death associated protein 6 (DAXX) gene, the caspase 2 (CASP2) gene and the RIPK1 domain containing adapter with death domain (RAIDD) gene, which are key genes in other pro-apoptotic pathways were also decreased in the MCL samples. We suggest that in addition to the known overexpression of cyclin D1 which drives entry into the cell cycle, disturbances of pathways associated with apoptosis contribute to the development of MCL.

349

Combined immunotherapy (R-FCM) is superior to a fludarabine-containing chemotherapy (FCM) alone in recurrent indolent lymphoma - results of a prospective randomized comparison of the German Low Grad Lymphoma Study Group

W. Hiddemann, R. Forstpointner, F. Fiedler, M. Gramatzki, B. Dörken, H.-J. Illiger, M. Kneba, M. Pfreundschuh, R. Pasold, R. Parwaresch, M.H. Dreyling, M. Unterhalt (München, Chemnitz, Erlangen, Berlin, Oldenburg, Kiel, Homburg, Potsdam, D)

The clinical course of indolent lymphoma is characterized by frequent recurrences of the disease and a steadily declining survival curve. However, even in relapse, the disease is sensitive to conventional chemotherapy in a significant subset of patients. In a national multicenter trial, patients with relapsing indolent lymphomas or mantle cell lymphomas were randomly assigned to receive either a anthrachinon-containing chemotherapy (FCM: fludarabine 25 mg/m^2 d 1-3, cyclophosphamide 200 mg/m^2 d 1-3, mitoxantrone 8 mg/m^2 d 1) alone or in combination with an anti-CD20 antibody (375 mg/m^2). Subsequently, patients responding to cytoreductive induction therapy were randomized for an anti-CD20 antibody maintenance therapy (4 x 375mg/m^2 weekly, twice a year). Since November 1998, 147 patients were randomized; currently, 70 patients are evaluable for response. In the chemotherapy only arm (FCM), only 18 patients achieved a remission (53% OR: 15% CR, 38% PR) whereas in the combined immuno-chemotherapy cohort (R-FCM), 32 remissions were reported (89% OR: 36% CR, 53% PR). These differences were highly significant for the total group of patients in favor of the R-FCM arm (sequential test based on overall remission rate (OR): p=0.000715). Similar improvements of remission rates were detected in follicular (95% vs. 68%) as well as mantle cell lymphoma (77% vs. 27). Both treatment options were associated with hematological toxicities of grades II to IV but were well tolerated with only 1 and 3 treatment related deaths in the R-FCM and FCM arm, respectively (5.6%). In summary, to the best of our knowledge, this is the first prospectively randomized study which confirms the synergistic benefits of a combined immuno-chemotherapy in comparison to one of the most effective regimes of chemotherapy alone in relapsed indolent lymphoma.

Fasturtec® Wirkstoff: Rasburicase. **Zusammensetzung:** 1 Durchstechflasche enthält 1,5 mg Rasburicase; sonstige Bestandteile: Alanin, Mannit, Dinatriumphosphat, Natriumdihydrogenphosphat; Lösungsmittel: Poloxamer 188, Wasser für Injektionszwecke. **Anwendungsgebiete:** Prophylaxe und Behandlung einer akuten Hyperurikämie, zur Verhinderung eines akuten Nierenversagens bei Patienten mit hämatologischen Malignomen mit einer hohen Tumorlast und dem Risiko einer raschen Tumorlyse oder -verringerung nach Beginn der Chemotherapie. **Gegenanzeigen:** Überempfindlichkeit gegen Uricasen oder einem der sonstigen Bestandteile. G6PD-Mangel und andere metabolische Störungen im Zellstoffwechsel, die eine hämolytische Anämie auslösen können. **Nebenwirkungen:** Bei gleichzeitiger Verabreichung von Rasburicase mit zytoreduktiver Chemotherapie, ist eine signifikante Anzahl unerwünschter Ereignisse durch die Grunderkrankung und deren Behandlung zu erwarten. Häufiges Auftreten von Fieber, Erbrechen, Übelkeit, gelegentlich Kopfschmerzen und allergische Reaktionen. **Warnhinweis:** Überwachung der Patienten auf allergische Nebenwirkungen, besondere Vorsicht bei Patienten mit atopischen Allergien in der Anamnese. **Wechselwirkungen mit anderen Medikamenten:** bisher nicht bekannt. **Handelsformen:** N1/ 3 Durchstechflaschen mit Trockensubstanz und 3 Ampullen Lösungsmittel. Hinweise der Fachinformation beachten. Verschreibungspflichtig. Stand: März 2001. **Sanofi-Synthelabo GmbH**, Potsdamer Str. 8, 10785 Berlin.

sanofi~synthelabo

FASTURTEC®
Mehr Sicherheit in der Tumortherapie.

350

Role for the adapter protein Grb10 in Akt activation

S. Urschel, C. Peschel, J. Duyster (München, D)

The serine/threonine kinase Akt plays an important role in growth-factor induced survival and has been shown to suppress the apoptotic death of a number of cell types induced by a variety of stimuli, including growth factor withdrawal, loss of cell adhesion and DNA damage. We have recently identified Akt as a binding partner of Grb10. Grb10 belongs to a family of adapter proteins that are thought to play a role in receptor tyrosine kinase-mediated signal transduction. The function of Grb10 remains uncertain and it is not clear whether Grb10 acts as an activator or an inhibitor of signal transduction. We previously reported that Grb10 associates with BCR-ABL in a phosphotyrosine dependent manner and that Grb10 SH2 domain interacts specifically with c-kit receptor. Deficient activation of Akt by a constitutively activated c-kit mutant lacking the binding site for PI3-kinase could be compensated by overexpression of Grb10. Importantly, high amounts of Grb10 expression increase Akt activation. The distribution of Akt is mainly cytosolic in unstimulated cells. Upon PI3-kinase activation Akt translocates to the plasma membrane, where it becomes activated. Activated Akt then appears to detach from the plasma membrane and translocates through the cytosol to the nucleus. We propose a model in which the recruitment of Akt in stimulated cells is mediated by the adaptor protein Grb10. According, fusion of Grb10 sequence to Akt creating an Akt/Grb10 hybrid showed constitutive Akt activation in vivo. To further elucidate the biological significance of Grb10 for the membrane targeting of Akt, we constructed a farnesylated Grb10 construct, which translocates constitutively to the plasma membrane. To study the regulation of Akt translocation by Grb10 in more detail, we performed a yeast two-hybrid screen using the full length Grb10 as a bait to further analyse the specific role of Grb10 and found binding to a 14-3-3 protein. 14-3-3 proteins are highly conserved, ubiquitously expressed and bind to a variety of proteins involved in signal transduction and cell cycle regulation. Binding of 14-3-3 to a consensus motif of phosphorylated proteins RXXSXP was described. This motif was identified in Grb10. Mutation of this motif in Grb10 does not diminish the binding to 14-3-3. However, phosphatase treatment of Grb10 inhibits the association with 14-3-3 in vitro, demonstrating the necessity of Grb10 phosphorylation for this interaction. Interestingly, the coexpression of Akt inhibits the interaction of 14-3-3 and Grb10. This data supports a model in which Grb10 exists in two different states of phosphorylation. In the unphosphorylated state Grb10 binds to Akt and translocates to the plasma membrane upon stimulation. Grb10 phosphorylation occurs, the Grb10/Akt complex dissociates and Grb10 binds to 14-3-3. This is the first study describing a model for Akt translocation from the cytosol to the plasma membrane involving the adapter protein Grb10 and 14-3-3.

351

Cytogenetic profile of acute myeloid leukemia (AML) and myelodysplastic syndrome (MDS): More similiarities than differences

C. Schoch, S. Schnittger, W. Kern, S. Bursch, W. Hiddemann, T. Haferlach (München, D)

MDS and AML seem to be related diseases. So far distinction between these two groups is based on the percentage of blasts in the bone marrow. According to the new WHO-classification an AML is diagnosed if the blast count exceeds 20%. We asked the question, whether MDS and AML can be characterized better according to cytogenetics. Therefore we analysed 444 MDS, 110 AML evolving from MDS, 91 therapy-related AML and 1513 de novo AML that had been sent to our laboratory for cytomorphological and cytogenetic diagnostics. In MDS and de novo AML the incidence of chromosome abnormalities was the same (55%), but aberrant karyotypes were more frequent in AML evolving from MDS (75%) or in t-AML (80%). Overall balanced rearrangements were more common in AML than in MDS (21% vs. 10%). However, the pattern of the most common unbalanced karyotype abnormalities like del(5q), del(7q), del(20q), -Y,-7, +8 was similiar in MDS and AML with small differences in the incidence. The incidence of complex aberrant karyotypes was 15% in the total group and highest in t-AML, MDS RAEB-T and MDS RAEB (33% vs. 23.5% vs. 19%). No MDS specific karyotype abnormality was observed, which did not also occur in AML. In contrast, t(8;21), t(15;17), inv(16)/t(16;16) and translocations involving the MLL gene were only found in de novo AML and t-AML. In addition we analysed 40 MDS patients and 125 AML patients showing a complex aberrant karyotype in more detail with 24 color FISH and FISH with loci specific probes for 5q31, 7q31 and the p53 gene in addition to chromosome banding analysis. A similiar pattern of abnormalities was observed in MDS and AML: Structural abnormalities leading to the loss of material of 5q, 7q , 17p and 12p were the most frequent karyotype changes observed in both subgroups. The only difference between AML and MDS in this analysis was the significantly higher percentage of cases with a p53 deletion in AML compared to MDS (53% vs. 35%; p=0.049). In conclusion, from the cytogenetic point of view AML and MDS are overlapping entities. The underlying genetic defects will be more important for the definition of biologic subgroups than blast count in the bone marrow.

352

Oxaliplatin (OX) in patients (pts) with relapsed or cisplatin-refractory germ cell cancer (GCT): Final results of a German Testicular Cancer Study Group investigation

C. Kollmannsberger, O. Rick, H.-G. Derigs, N. Schleucher, P. Schöffski, J. Beyer, R. Schoch, H.G. Sayer, A. Gerl, M. Kuczyk, C. Spott, L. Kanz, C. Bokemeyer (Tübingen, Berlin, Mainz, Essen, Hannover, Marburg, Kiel, Jena, München, D)

Treatment of pts with cisplatin-refractory GCT remains a therapeutic challenge. Previously, paclitaxel, gemcitabine and oral etoposide have shown some clinical activity in refractory pts. The present study investigates the activity of OX, a new platinum-analogue with incomplete cross-resistance to cisplatin in-vitro, in pts with metastatic, intensively pretreated and cisplatin-refractory GCT. Patients and methods: Cisplatin-refractory disease was defined as progression or relapse within 4 weeks after cisplatin-based chemotherapy or relapse after salvage high-dose chemotherapy. OX was administered at 2 different dose levels with either 60 mg/m^2 weekly for 3 consecutive weeks (group 1) or 130 mg/m^2 biweekly (group 2). Response was evaluated every 4 weeks. This study with its two treatment groups was considered statistically as two separate, subsequent phase II studies. Results: Thirty-two pts have been enrolled. The initial 16 pts received 60 mg/m^2 weekly and the subsequent 16 pts 130 mg/m^2 biweekly. Median age was 35 [21-61] years. Primary tumor localisation was gonadal/retro-peritoneal/mediastinal in 27/5/0 pts. Pts had been pretreated with a median of 7 (group 1) and 9 (group 2) cisplatin-containing cycles and 12 (75%; group 1) and 13 (81%; group 2) pts had previously failed high-dose chemotherapy with PBSCT. 12/16 pts and 15/16 pts of groups 1 and 2, respectively, were considered cisplatin-refractory or absolutely refractory. Median number of OX applications was 3 [3-9] (60 mg/m^2) and 4 [2-16] (130 mg/m^2). All pts are assessable for response and toxicity. Toxicity was generally acceptable with myelosuppression, nausea/vomiting and cold-induced neuropathy being the most frequent side-effects. NCI-CTC III°/IV° thrombocytopenia/ granulocytopenia occurred in 2 pts each (group 1) and in 3 and 1 pts (group 2). Neurotoxicity III° was only observed in 1 pt (group 2). Two pts developed III° nausea/vomiting. There were no cases of neutropenic fever or treatment related deaths. In group 1, only 1/16 pts achieved a PR+ of 6 weeks duration. In contrast, 3 (19%) and 2 (13%) pts of group 2 achieved PRs or disease stabilisations, respectively. All responding pts were classified as cisplatin-refractory. Conclusions: OX at a dose of 130 mg/m^2 biweekly demonstrates antitumor activity in pts with cisplatin-refractory GCT. The response rate appears to be related to the applied peak dose of OX. Further evaluation in combination regimens is warranted.

353

Bolus 5-fluorouracil (5-FU)/ folinic acid (FA) (Mayo) vs. weekly high-dose 24h 5-FU infusion/ FA + Oxaliplatin (OXA) in advanced colorectal cancer (ACRC). Results of a phase III study

A. Grothey, B. Deschler, H. Kröning, K. Ridwelski, P. Reichardt, A. Kretzschmar, M. Clemens, W. Hirschmann, M. Lorenz, M. Asperger, T. Büchele, H.J. Schmoll (Halle, Magdeburg, Berlin, Trier, Kassel, Frankfurt, D)

5-FU/FA plus Oxaliplatin (OXA) has shown to be superior in response rate (RR) and progression free survival (PFS) in patients with ACRC compared with 5-FU/FA infusion. However, this combination has never been compared to standard bolus 5-FU/FA. Based on the encouraging activity in 5-FU refractory pts and low toxicity (Buechele et al., ASCO 1998), a prospective randomized study was performed comparing a weekly protocol of 5-FU/FA+OXA with standard Mayo regimen. DOSAGE AND SCHEDULE: 5-FU bolus 425 mg/m^2/FA 20 mg/m^2, d1-5, qd29(A) and OXA 2h inf 50 mg/m^2 + 5-FU24hinf 2000 mg/m^2/FA 500 mg/m^2, d1,8,15,22, qd36 (B). 252 pts (A 129, B 123) were randomized by 60 centers in 435 days. Both treatment arms were well balanced for sex, age, site of primary, PS, and number of metastatic sites; however, only 15.4% of pts in Arm B had received prior adjuvant chemotherapy compared with 24.8% in Arm A (p=0.084). RESULTS: Hematologic toxicity was less frequent and severe in Arm B with WHO°3/4 toxicity (Arm A vs B): neutropenia 23.4% vs 6.7% (p=0.0003), leukocytes 12.1% vs 2.5% (p=0.006). Non-hematologic toxicity was comparable in both arms: mucositis 8.9% vs 2.5% (p=0.051), diarrhea 12.1% vs 21.2% (p=0.082), nausea 2.4% vs 6.8%, cardiac 1.6% vs 0.8%. WHO°3/4 sensory neuropathy occurred in 12.7% of pts (B) with a median onset after a cumulative OXA dose of 1000 mg/m^2. RR was significantly increased in Arm B with 47.8% vs 21.8% (p<0.0001) with 8.7% of pts in Arm B achieving CR. After a median follow-up of 20.3 months PFS is significantly different with 5.3(A) vs 8.1(B) months (p<0.0001). Survival data are still preliminary since more than 50% of pts in Arm B are still alive; however, current median overall survival (OS) for both treatment arms are 16.7 (A) vs 20.4 (B) months. No early death due to progressive disease occurred in Arm B compared with 6 cases in Arm A. After 12 (24) months 67.2 (33.4)% vs 78.2 (49.0)% of pts are alive in Arm A and B, resp. More than 70% of pts in both arms received at least second-line therapy, a fact which has conceivably contributed to the long OS achieved in both treatment arms. CONCLUSION: Weekly OXA/5-FU24h inf /FA has superior efficacy vs. the Mayo Clinic protocol with a favorable toxicity profile except cumulative neuropathy. This weekly combination appears to be as effective and conceivably less toxic than bi- and tri-weekly OXA-based protocols. Final results on OS will be presented at the meeting.

Wissenschaftliches Symposium: Indolente Lymphome

354

Allogeneic transplantation in indolent lymphoma

O.J. Stoetzer, M. Schleuning, W. Hiddemann, H.-J. Kolb (München, D)

Allogeneic bone marrow transplantation from HLA-identical siblings meanwhile represents a standard treatment procedere for young patients with relapsed follicular lymphoma in advanced stages. For conditioning therapy mainly total body irradiation is used. Three-year probabilities of overall survival rankes from 49% up to 70%, disease free survival is almost identical. Main causes of dead therefore are transplantation associated. Only few smaller studies were available for patients with mantle cell lymphoma. Depending on disease course and stage overal suvival at three years is 23% up to 55%. Relapses are more frequent than in patients with follicular lymphoma. In our own centre 16 patients with advanced indolent lymphoma underwent allogeneic transplantation from HLA-identical siblings or unrelated donors. Median follow up is 850 days. Overall survival and disease free survival is 67%. 56% of the patients suffered from acute Graft-versus-host disease grade 2. Recently reduced intensitiy protocols for allogeneic stem cell transplantation have been developed (reduced intensitiy stem cell transplantation = RIST). Such protocols are aimed to reduce transplant related toxicity and to make allo-transplantation available for elderly patients or patients with reduced organ functions. Most RIST-protocols are fludarabine based. Preliminary result of several different RIST-protocols in patients with indolent lymphoma demonstrated good response rates (73%-100% CR) and excellent engraftment data (70%-100%). However the median follow up by now is short in most studies. Khouri et al reported a clear superiority of their RIST protocol as compared to standard full-TBI conditioning regimen according to OS (73% vs. 45%) and DFS (73% vs. 43%) at 2 years. In Munich we transplanted 9 patients with advanced indolent lymphoma using a RIST protocol containing Cyclophosphamid, Fludarabine and TBI 2-4 Gy. Most patients have had autologous transplantation before. HLA-identical sibling and unrelated donors were included. Median follow up is now 268 days. OS is 78 % and DFS 67%. 7 of 9 patients suffered from acute Graft-versus-Host disease ≤ grade 2.
Allogeneic bone marrow or stem cell transplantation is a successfull therapeutic option in well selected patients with advanced indolent lymphoma. If reduced intensity conditioning stem cell transplantation is superior to standard bone marrow transplantation has to be evaluated in clinical trials.

355

Role of genomics for immunological target identification in lymphoma

J.L. Schultze (Boston, USA)

The potential power of T cell-mediated cancer immunotherapy has been demonstrated in multiple animal models and clinical trials. However, the lack of adequate tumor associated antigens (TAA) as targets for immunotherapy has hampered its progress and success for most patients. Most currently known TAA such as the MAGE gene family are restricted to a subset of tumors and often are expressed heterogeneously within an individual tumor. These TAA are not essential for tumor growth and therefore tumor escape by immunoselection can occur. Moreover, the identification of TAA by classical approaches is labor-intensive, costly and limited to genes that evoke immune responses in cancer patients. To overcome the current limitations we have developed a strategy to identify more ideal TAA. We hypothesized that the identification of genes that are universally expressed in all cells of all tumor histologies and that are involved in oncogenesis might overcome the limitations of currently known TAA. This platform technology consists of the following steps: Publicly available genomic, proteomic and literature databases are data mined for genes that are expressed in tumors but not in normal tissues and that are involved in oncogenesis. Genes with these characteristics are further analyzed by prediction algorithms for peptides that potentially bind with high affinity to HLA molecules. This is a prerequisite for immunogenicity. Peptide candidates are synthesized and tested in an HLA-peptide binding assay in micro-titer plate format. Peptides with high binding affinity are tested for the induction of peptide-specific T cells from healthy individuals and cancer patients. These T cells are tested for lysis of tumor cell lines and primary tumor tissue of different histologies including carcinomas, sarcomas, melanomas and hematological malignancies as well as normal cells. Using this technology the catalytic subunit of telomerase and the cytochrome P450 1B1 were identified as the first universal tumor antigens. This methodology makes use of the increasing genomic information and links it to immunological information leading to the development of universally applicable immunotherapy for the vast majority of cancer patients including patients with lymphomas.

356

Radioimmunotherapy (RAIT) with 131-Iodine labeled Rituximab in patients with B cell NHL in a myeloablative and non-myeloablative regimen

B. Schmidt, K. Schwarz, K. Scheidhauer, M. Schwaiger, Ch. Peschel, C. von Schilling (München, D)

In ongoing phase I/II dose escalation trials, we treated 41 patients (pts) affected by CD20+ NHL (follicular, mantle cell, marginal zone, lymphoplasmacytoid or transformed subtype) with RAIT using 131-Iodine labeled rituximab: 22 with non-myeloablative and 19 with myeloablative doses, supported by autologous PBSCT. Pts were eligible for RAIT if no standard curative treatment was available and/or if multiple salvage treatments had yielded < PR. Doses were escalated to define MTD: for myelosuppression in the non-myeloablative and non-haematologic organ toxicity in the myeloablative group. In the non-myeloablative group calculated doses were escalated from 45 to 75 cGy total body dose. Pt characteristics: stage II, 2; stage III, 11; stage IV, 9 pts; median number of prior therapies: 4 (1-12). Median time from diagnosis to RAIT 38 (8-192) months (mo). Prior rituximab therapy in 4 pts, prior aPBSCT in 5 pts. 5 pts were treated during rapid progression. Acute toxicity was mild, with 20 % incidence of flu-like symptoms grade II. Nadirs occurred at a mean of 36 days post-treatment. Haematological toxicity: grade 0, 6 pts; grade I, 1 pt; grade II, 5 pts; grade III, 1 pt; grade IV, 5 pts. 4 pt are not (yet) evaluable. The responses at 3 mo post treatment were: CR 5 (duration 6-18+ mo), PR 3 (3+-18+ mo), SD 1 (3 mo), PD 8 pts, not yet evaluable 5 pts. In the myeloablative group activities of 250-390 mCi were applied. Pt characteristics: stage II, 2; stage III, 7; stage IV, 10 pts; number of prior therapies: 4 (2-11). Median time from diagnosis to RAIT was 52 (12-235) mo. 5 pt had bone marrow involvement, 4 pt had previously rituximab and 8 pt had high-dose chemotherapy for one of their previous relapses. Acute toxicity was mild, with 80 % nausea max. grade II. Reinfusion of PBSC was performed 22 (15-35) days after RAIT. There was 1 fatal polymicrobial pneumonia (d+154), 1 fatal pulmonary PCP and aspergillosis (d+104) and 1 fatal paraspinal abscess (d+110). Beside, only 6 febrile episodes (1 port infection, 1 perianal abscess, 2 non-fatal pulmonary aspergillosis, 2 FUO) and 2 mucositis in the neutropenic phase were seen. 2 pt developed pneumonitis. All severe toxicities occurred in pt who had undergone prior aPBSCT. All pts reached a stable ANC > 0.5 G/L and all but one were transfusion-independent by d +15 after PBSCT. In summary, treatment was well tolerated. The responses at 3 months were: CR 4 (duration 3+-18+ mo), PR 3 (3-12+ mo), PD 7 pts, not yet evaluable 5 pts. Despite extensive pre-treatment, RAIT with the 131-Iodine labeled chimeric humanized antibody Rituximab induced an overall response rate of 50% (25% CR) in pt with relapsed or refractory B cell NHL. Response could be predicted by histology (follicular or marginal zone vs. other subtypes), rate of progression at time of RAIT, and extend of pre-treatment. The results are encouraging and comparable to those reported for radiolabeled murine anti-CD20 antibodies. Dose escalation continues.

357

Myeloablative radiochemotherapy with autologous stem cell transplantation in indolent lymphoma - results of a prospective randomized comparison of the German Low Grade Lymphoma Study Group

W. Hiddemann, M. Pfreundschuh, V. Diehl, L. Trümper, R. Hehlmann, P. Koch, H.-J. Illiger, H. Wandt, R. Kuse, B. Dörken, R. Parwaresch, M.H. Dreyling, M. Unterhalt (München, Homburg, Köln, Göttingen, Mannheim, Münster, Oldenburg, Nürnberg, Hamburg, Berlin, Kiel, D)

In a large national multicenter trial, patients < 60 years with advanced low-grade lymphomas were randomly assigned to receive either intensive consolidation (myeloablative radiochemotherapy followed by autologous peripheral stem cell transplantation) or interferon alpha (IFN) maintenance after initial cytoreductive chemotherapy (CHOP or MCP: Mitoxantrone/ Chlorambucil/Prednisone). In patients of the high dose arm, peripheral stem cells were mobilized with DexaBEAM. Subsequently, patients successfully harvested underwent fractionated total body irradiation with 12 Gray plus cyclophosphamide (60 mg/kg) on days -2 and -3 followed by transfusion of unpurged autologous stem cells. Patients of the IFN maintenance arm received 2 additional courses of the cytoreductive chemotherapy followed by $5x10^6$ units interferon alpha 3x weekly until progression. Initiated in April 1996, the study recruited 231 patients evaluable for interim analysis until October 2000 with a maximum follow-up of 4 years. A recurrence of lymphoma was observed in 22 of 103 (21%) patients of the high dose therapy arm, whereas the lymphoma progressed in 45 of 128 patients (35%) receiving IFN maintenance, corresponding to an estimated risk reduction of relapse of 53% in favor of the high dose arm (p=0.00253). Both treatment options were well tolerated with only two treatment related deaths in the high dose therapy arm (<2%). Besides the initial hematological toxicities, fatigue and muscle pain were the most significant treatment related symptoms with similar frequencies in both treatment arms. At present time, no significant differences in overall survival have been observed due to the rather indolent nature of the disease and the optional second line dose intensification. Since the benefits of high dose therapy may be diminished by late toxic side effects (e.g. secondary malignancies), the current study has been extended to define the impact of high dose therapy with autologous stem cell transplantation on overall survival.

358

Meta-analysis of granulopoiesis-stimulating factors in the treatment of malignant lymphoma

J. Bohlius, M. Reiser, A. Engert (Köln, D)

Background: Granulopoiesis-stimulating factors (G-CSF and GM-CSF) are used in clinical practice to prevent febrile neutropenia, to maintain dose intensities and to improve tumor response and overall survival in the treatment of patients with malignant lymphoma. The results from single studies are inconclusive and evidence-based guidelines are predominantly based on studies in solid tumors. The critical question whether G-CSF and GM-CSF improve dose-intensity, tumor response and overall survival in the treatment of malignant lymphoma has not been answered yet. Objective: We undertook a systematic review to determine the effectiveness of G-CSF and GM-CSF in the primary prevention of febrile neutropenia and infections, the maintenance of dose intensity, improvement in quality of life, tumor response and overall survival in the treatment of malignant lymphoma. Methods: The Cochrane, Medline, Embase, CancerLit and smaller databases were searched. Additionally, we searched Internet-databases of ongoing trials, conference proceedings of the American Society of Clinical Oncology and the American Society of Hematology and contacted experts in the field. Randomised controlled trials comparing parenteral application of G-CSF or GM-CSF versus placebo/no treatment in adult and elderly patients with malignant lymphoma undergoing chemotherapy were included in this review. Both study arms had to receive the same chemotherapy and the same supportive care. We included full-text and abstract publications as well as unpublished data. Eligibility and quality assessment, data extraction and analysis were done in duplicate. All authors have been contacted to obtain missing data. Results: We included 10 eligible studies with 1383 randomised patients. The relative risk to suffer neutropenia, febrile neutropenia and infections was statistically significant reduced in the experimental group. The relative risk to suffer bone pain and skin reactions, such as injection site reactions and skin rash, was increased in the group treated with G-CSF or GM-CSF ($p < 0.05$). There was no evidence for G-CSF or GM-CSF to decrease the number of patients who require iv antibiotics, or to improve tumor response, freedom from treatment failure (FFTF) and overall survival (OS). None of the studies evaluated quality of life. Results on infection related mortality and received dose intensity depend on data that is currently being collected and analysed. Conclusion: Based on the data available we conclude that the benefit of G-CSF and GM-CSF in this clinical setting is restricted to endpoints related directly to neutrophil recovery. There is no evidence for G-CSF and GM-CSF to improve tumor control, e.g. tumor response, FFTF and OS. Another 3 studies including further 962 patients will be completed within the next years. This additional information may change the current evidence of the clinical impact of granulopoiesis-stimulating factors.

359

Cyclin D3 (CCND3) is the target gene of t(6;14)(p21.1;q32.3) recurrent in B-cell lymphomas

R. Siebert, T. Sonoki, L. Harder, D.E. Horsman, I. Taniguchi, T.G. Willis, L. Karran, S. Gesk, D. Steinemann, E. Zucca, F. Sole, B. Schlegelberger, A.J. Mungall, R.D. Gascoyne, M.J.S. Dyer (Kiel, D; Kumamoto, J; Vancouver, CDN; Leicester; Bellinzona, CH; Barcelona, E; Hinxton, UK)

Chromosomal translocation t(6;14)(p21.1;q32.3) has been reported as a rare but recurrent event not only in myeloma and plasma cell leukemia, but also in diffuse large B-NHL (DLBCL) and splenic lymphoma with villous lymphocytes (SLVL); however, the nature of the target gene(s) has not been determined. We identified t(6;14)(p21.1;q32.3) in three cases of transformed extranodal marginal zone B-NHL, in one case of SLVL and one case of a low-grade B-cell lymphoproliferative disorder. In a sixth case, a CD5+ve DLBCL, the translocation was identified by molecular cloning, in the absence of detectable cytogenetic change. Two chromosomal translocation breakpoints were cloned using long-distance inverse PCR (LDI-PCR) methods. Comparison with the genomic sequence for chromosome 6p21.1 showed breakpoints approximately 59 and 73kb 5' of the Cyclin D3 (CCND3) gene. Newly developed interphase fluorescent in-situ hybridization (FISH) assays, using BAC clones spanning and flanking the CCND3 locus as well as probes for the IGH locus, confirmed the CCND3/IGH fusion and demonstrated localization of 6p21.1 breakpoints within the same region in all five cases with suitable material available. Furthermore, overexpression of CCND3 was demonstrated at RNA and/or protein levels by northern and western blotting and by immunohistochemistry. These data implicate CCND3 as a dominant oncogene in the pathogenesis and transformation in several histological subtypes of mature B-cell malignancies with t(6;14)(p21.1;q32.3) and suggest that CCND3 over-expression seen in about 10% of DLBCL may have a genetic basis.

Wissenschaftliches Symposium:
Anämien

360

Diagnosis and management of the anaemia of chronic disease

G. Weiss (Innsbruck, A)

Anaemia of chronic disease (ACD) is a very prominent clinical syndrome at least in being the most frequent anaemia in hospitalised patients. ACD develops in patients suffering from chronic inflammatory disorders which involve activation of cellular immunity, such as patients with chronic infections, auto-immune diseases or neoplasm. Although, ACD can be easily diagnosed the underlying mechanisms are not fully understood. A number of different pathways have been shown to contribute to ACD, such as diversion of iron traffic, a diminished erythropoiesis, a blunted response to erythropoietin, erythrophagocytosis or a decreased erythrocyte half life. Nevertheless, ACD is a reflection of an activated immune system and possibly results from a innovative defence strategy of the body in order to withdraw the essential growth factor iron from invading pathogens and to increase the efficacy of cell mediated immunity. Diagnosis of ACD can be assessed by examination of chances in serum iron parameters with low to normal serum iron, transferrin saturation and transferrin concentrations on the one hand and normal to increased ferritin, zinc protoporphyrin IX and cytokine levels on the other side. A ratio of TfR/log ferritin is useful to identify ACD patients with a true iron deficiency f.e. due to bleeding. Therapy of ACD includes the cure of the underlying the disease. Apart from this transfusions for rapid correction of haemoglobin levels, and human recombinant erythropoietin for prolonged therapy (sometimes in combination with iron supplementation) are used. However, response rates to recombinant erythropoietin are sometimes low. Thus, we need predictive variables which help to identify those patients which may respond to erythropoietin therapy. Iron alone should be strictly avoided due to its growth promoting effect towards micro-organism and tumour cells and because of it capacity to inhibited T-cell mediated immune effector pathways. Finally and most importantly, we definitely need prospective clinical trials in order to gain knowledge about the effects of anaemia correction and/or the use of Erythropoietin towards the course of the underlying disease.

362

The role of zinc protoporphyrin in screening iron metabolism

J. Hastka (Mannheim, D)

Zinc protoporphyrin (ZPP) is produced instead of heme when the iron support to erythropoiesis becomes insufficient and zinc instead of iron is incorporated into protoporphyrin IX. According to its fluorescence properties, ZPP can be measured fluorometrically. The development of a front-face hematofluorometer, which allows a rapid determination of the erythrocyte ZPP/heme ratio, made the routine use of this parameter possible. In normal persons without iron deficiency (ID), ZPP is < 40 mmol/mol heme. In simple iron depletion (grade I iron deficiency), ZPP values are still within the normal range, because the iron support to erythropoiesis is still sufficient. ZPP allways detects an iron-deficient erythropoiesis, the first clinically relevant stage of ID. Normal or only slightly increased ZPP values exclude clinically relevant ID and show the physician that, whatever the pathological findings for the given patient, they are not caused by ID. ZPP concentrations are increased not only in real ID, but in all conditions that induce impaired supply to the erythroid marrow. Increased ZPP levels are measured in ACD-anemias, lead poisoning and in myelodysplastic syndrome as well. By recording the derangements quantitatively, ZPP allows monitoring of therapy of these diseases. Thus, ZPP is not a simple parameter of ID but rather a screening parameter of iron deficient erythropoiesis, surveying all steps of the iron metabolism from iron uptake to its utilization. It serves as a kind of „end point control" of the whole iron metabolism, which detects real iron deficiency as well as derangements of iron metabolism in chronic inflammatory and neoplastic diseases, in sideroblastic disorders or lead poisoning.

363

Erythropoietin decreases transfusion requirements in patients with relapsed lymphoma treated with the Cologne high-dose sequential chemotherapy regimen prior to autologous stem cell support

A. Josting, J.P. Glossmann, G. Wassmer, H. Flechtner, Y. Ko, C. Rudolph, H.J. Illiger, B. Dörken, S. Wiedenmann, V. Diehl, A. Engert (Köln, Bonn, Cottbus, Oldenburg, Berlin, D)

Introduction: Patients treated with highly myelosuppressive regimens usually develop severe anemia requiring multiple red blood cell (RBC) transfusions whereas erythropoietin (EPO) has been shown to reduce the number of transfusions in chemotherapy-induced anemia. This study evaluated the ability of EPO to reduce the need of RBC transfusions in patients with relapsed lymphoma treated with the Cologne high-dose sequential chemotherapy (HDSCT) regimen. Patients and methods: Sixty patients (pts) with early or late relapsed Hodgkin's lymphoma (HD) or first relapse of aggressive Non-Hodgkin's lymphoma (NHL) were treated in a randomized trial in 23 centers in Germany. Inclusion criteria were histology (relapsed HD or aggressive NHL), age (18 to 60 years [y]) and a Karnofsky score > 70%. Patients were stratified in both treatment arms according to gender, age (<40 vs. >40 y) and Hemoglobin (Hb) level before therapy (< 10 vs. > 10 g/dl). Both study arms received two cycles DHAP; pts with PR or CR received cyclophosphamide 4g/m^2, followed by PBSC harvest; methotrexate 8g/m^2 plus vincristine 1.4mg/m^2; and etoposide 2g/m^2. The final myeloblative course was BEAM followed by autologous stem cell support. One arm additionally received 10,000 IE EPO (NeoRecormon 3x/week) from the start of therapy until the end of VP-16. Primary endpoint of the study was number of RBC units needed during HDSCT. In addition, Hb levels and quality of life were measured. An interim analysis was planned after 42 pts. Mean values were compared using t-test. Results: The median age of the 29 male and 31 female pts was 37y (19y - 60y). Thirty-nine pts (65%) had HD, 21 pts (35%) had NHL. Eighteen pts had stage I/II diesease vs. 42 pts with stage III/IV. The interim analysis (42 pts) showed no differences in pts age, gender and Hb level before therapy. The mean number of RBC units given in the EPO arm was 4.5 compared to 8.9 in the control arm (p=0.00049). Five pts (25%) receiving EPO needed no transfusion, 1 patient (5%) in the control arm. Twelve pts (55%) in the control arm received = 8 RBC units compared to 4 pts (20%) in the EPO arm. EPO especially reduced the need of tranfusions in the highly myelotoxic cycles (CTX, VP-16). The mean Hb levels during therapy were 10.4 g/dl in the EPO arm and 9.7 g/dl in the control (p=0.022). Quality of life (QLQC30) and fatigue (MFI) assessment showed no significant difference. Conclusion: This analysis indicates that EPO is remarkably decreasing RBC transfusion requirements in patients with relapsed lymphomas receiving HDCT while the mean Hb level seems to be raised. Quality of life and fatigue of pts appear not to be affected.

364

Treatment of severe PNH with allogeneic hemopoietic stem cell transplantation from unrelated donors using 2 Gy total body irradiation, fludarabine and a combination of cyclosporine and mycophenolate mofetil

U. Hegenbart, T. Lange, S. Leiblein, W. Pönisch, M. Deininger, J. Edelmann, F. Kamprath, E. Holler, S. Forman, D. Niederwieser, R. Storb (Leipzig, Regensburg, D; City of Hope, Seattle, USA)

PNH is a clonal hematological disorder with a prognosis ranging from a few months to several years dependent on the occurrence PNH-associated complications. Allogeneic HSCT is the only curative approach; this therapy, however, is associated with a substantial transplant related mortality (TRM), especially with unrelated donors. Here we report on allogeneic HSCT after minimal conditioning in 4 patients with high-risk PNH. Conditioning consisted of 2 Gy TBI before and a combination of MMF and CSP for 96 and 180 days, respectively, after HSCT. FLU 30 mg/m^2/day was given on days -4, -3, and -2. Patients were 25-49 years old. All had a history of severe complications due to PNH including hemolytic crises, hepatic vein thromboses including Budd-Chiari syndrome, thromboses of mesenteric and cerebral veins, and life-threatening infections. Diagnoses were made by postive HAM-test and/or FACS analysis of blood cells showing deficiency of CD55 and/or CD59. Other hematological diseases were excluded using standard methods. Because of the the lack of family donors, HLA-A, B and DRB1 matched unrelated HSCTs were performed between october 1999 and november 2000. Granulocytes decreased posttransplant below 0.5x10^9/L a median of 4.5 days. No platelet transfusions were required. One patient had in the neutropenic period treated with antibiotics. No transplant-related complications except FUO (n=1) and septicemia (n=1) were observed. GvHD grade II occurred in three patients and was successfully treated with corticosteroids. VNTR or FACS chimerism studies of blood on day +56 revealed 94-100% donor cells in whole blood with 75-90% CD3+ cells. FACS analyses on days +56 and +180 showed no evidence of the PNH clone. After a median observation of 254 days (range 188-590) all patients are alive without evidence of PNH. One patient has limited chronic GvHD. We conclude that unrelated HSCT after minimal conditioning is an effective treatment with little toxicity for patients with high-risk PNH.

365

Fluorescently labeled variant of aerolysin (FLAER): A new method for detection of GPI-deficient cells in paroxysmal nocturnal hemoglobinuria (PNH) and aplastic anemia (AA)

H. Schrezenmeier, S. Körper, A. Punnose, K. Krause, M. Rojewski, E. Thiel (Berlin, D)

Flow cytometric measurement of expression of glycosyl-phosphatidyl-inositol anchored proteins (GPI-AP) is the method of choice for the diagnosis of PNH and detection of GPI-deficient cells in bone marrow failure syndromes. However, there are some disadvantages and pitfalls of this test: It does not measure the deficiency of the GPI-anchor itself, but the presence or absence of proteins linked to the GPI-anchor. Since expression patterns of GPI-AP differ between hemopoietic lineages and since expression can change within one lineage during differentiation, a safe flow cytometric diagnosis of a GPI-deficient population requires measurement of several different (at least two) GPI-AP on at least two different lineages. Careful establishment of cut-off values for a GPI-deficient population on each lineage is necessary. It has been recently proposed (Brodsky et al., 2000) that some disadvantages of the antibody-based method, might be overcome by the bacterial toxin aerolysin which binds specifically and with high affinity directly to GPI-anchors. In this study we used an Alexa Fluor 488-labelled variant of aerolysin (FLAER)(Protox, Biotech; Canada). We compared results of a standard protocol (CD58, CD59 on erythrocytes and reticulocytes; CD16, CD66b on granulocytes) with the FLAER-staining of granulocytes in a series of 35 consecutive patients (pts.) which were analyzed to detect a GPI-deficient clone in the context of AA or because of clinical suspicion of PNH. According to the result of the antibody-test we grouped the pts. in group A (normal expression of GPI-AP; n=19), group B (deficiency of GPI-AP on granulocytes only, n=7) and group C (significant deficiency of GPI-AP on erythrocytes, reticulocytes and granulocytes; n= 9). There was a good

Fordern Sie ein Gratis-Probeheft an!

Im Dienste der klinischen Krebsforschung und -behandlung

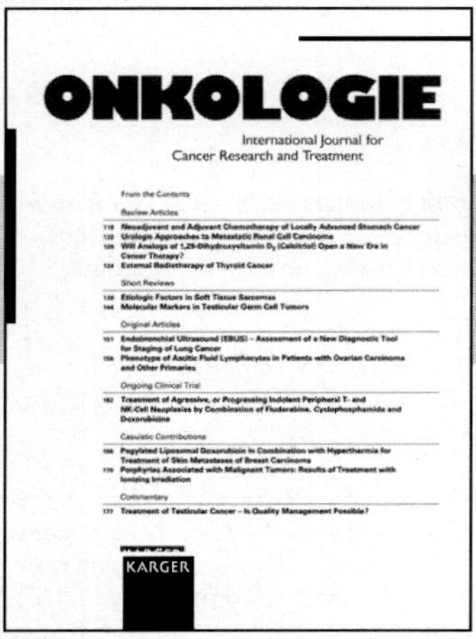

ONKOLOGIE

International Journal for Cancer Research and Treatment

Offizielles Organ der Deutschen Gesellschaft für Hämatologie und Onkologie,
der Österreichischen Gesellschaft für Hämatologie und Onkologie,
der Österreichischen Krebsgesellschaft – Krebsliga unter Fortführung der
«Österreichischen Zeitung für Onkologie»

Schriftleitung
***W. Queißer**, Mannheim*
***W. Scheithauer**, Wien*

Fachschriftleitung
***C. Bokemeyer**, Tübingen*
***Th. Cerny**, St. Gallen*
***R. Hartenstein**, München*
***P. Kreienberg**, Ulm*
***K. Mross**, Freiburg*
***H. Rübben**, Essen*
***P.M. Schlag**, Berlin*
***Th. Wendt**, Jena*

Einsenden von Arbeiten
Wissenschaftliche Beiträge müssen in englischer Sprache abgefasst sein.

Manuskripte sind zu senden an:

S. Karger GmbH
Redaktion Onkologie
Postfach
D–79095 Freiburg (Deutschland)

Lesen Sie's Online:
www.karger.com/journals/onk

Ihr Direktzugang zu Volltext-artikeln, Online-Reprints, Inhalts-verzeichnissen – mit freiem Zugriff auf Abstracts, Probenummern und auf E-mail ToC-Alert Service.

KF 02059S

In Original- und Übersichtsarbeiten behandelt «Onkologie» aktuelle Fragen der Krebsforschung und -behandlung. Als interdisziplinäre Zeitschrift veröffentlicht sie Arbeiten aus der internistischen Hämato-Onkologie, der chirurgischen, neurochirurgischen, gynäkologischen, pädiatrischen und urologischen Onkologie, der Mund-Kiefer-Gesichtschirurgie, Dermatoonkologie, der diagnostischen Radiologie in der Onkologie, Psychoonkologie, Pathologie, klinisch-experimentellen Onkologie, Biometrie und onkologischen Rehabilitation. Weitere Schwerpunkte sind klinische Informationen und die onkologische Fortbildung. In kompakter Form hält «Onkologie» den Forscher und Kliniker auf dem laufenden.

Eine Auswahl repräsentativer Beiträge

Intraoperative Radiotherapy in Patients with High-Grade Glioma in Comparison to Conventionally Treated Patients – a Matched-Pair Analysis: ***Wagner, W.** (Osnabrück); **Rübe, C.** (Münster); **Alfrink, M.** (Osnabrück); **Palkovic, S.; Willich, N.** (Münster)*

Clinical Significance of the Echogenicity in Prostatic Ultrasound Findings in the Detection of Prostatic Carcinoma: ***Manseck, A.; Guhr, K.; Hakenberg, O.; Rossa, K.; Wirth, M.P.** (Dresden)*

Reconstructive Surgery after Tumor Resection in the Oral Cavity and Oropharynx: ***Keiner, S.; Bootz, F.; Flisek, J.; Fuchs, M.** (Leipzig)*

Adjuvant Radio-Chemotherapy with 5-Fluorouracil and Leucovorin in Stage II and III Rectal Cancer: 12 Months vs. 6 Months of Therapy. A Study of the Association for Medical Oncology of the German Cancer Society: ***Queißer, W.; Hartung, G.** (Mannheim); **Kopp-Schneider, A.** (Heidelberg); **Diezler, P.; Hagmüller, E.** (Mannheim); **Baur, A.; Weniger, J.** (Erfurt); **Wojatschek, C.** (Bad Friedrichshall); **Janssen, N.** (Siegen); **Hossfeld, D.** (Hamburg); **Lindemann, H.** (Hagen); **Schnabel, T.** (Ludwigshafen); **Edler, L.** (Heidelberg); **for the Study Group Colon and Rectum Carcinoma** (Mannheim)*

Selection of CD34-Positive Blood Cells for Allogeneic Transplantation: Approaches to Optimize CD34-Cell Recovery, Purity, Viability, and T-Cell Depletion: ***Fritsch, G.; Scharner, D.; Fröschl, G.; Buchinger, P.; Peters, C.; Matthes, S.; Witt, V.; Gadner, H.** (Wien)*

Attitudes and Beliefs towards Disease and Treatment in Patients with Advanced Cancer Using Anthroposophical Medicine: ***von Rohr, E.** (Bern); **Pampallona, S.** (Evolene); **van Wegberg, B.** (Bern); **Cerny, T.; Hürny, C.** (St. Gallen); **Bernhard, J.** (Bern); **Helwig, S.** (Arlesheim); **Heusser, P.** (Bern)*

Onkologie
2002: Band 25 mit 6 Heften
Sprachen: Englisch, Deutsch
ISSN 0378–584X
Aufgeführt in bibliographischen Diensten und in
Current Contents®, Index Medicus, Reference Update, Biological Abstracts, EMBASE/Excerpta Medica.

Preise für Studenten, Ärzte in Weiterbildung und Mitglieder onkologischer Gesellschaften

Abonnement		+ Versandkosten	
CHF	96.50	+ CHF 28.80	Europa
CHF	96.50	+ CHF 42.70	Übersee
EUR	69.–	+ EUR 12.–	
USD	74.00	+ USD 28.80	

+ Online-Abonnement pro Band
CHF 39.– / EUR 28.– / USD 30.00

Abonnementspreise

Abonnement		+ Versandkosten	
CHF	193.–	+ CHF 36.–	Europa
CHF	193.–	+ CHF 53.40	Übersee
EUR	138.–	+ EUR 12.–	
USD	148.00	+ USD 36.00	

+ Online-Abonnement pro Band
CHF 58.– / EUR 42.– / USD 45.00

Preisänderungen vorbehalten
EUR-Preis nur für Deutschland,
USD-Preis nur für USA

Ich abonniere

die Zeitschrift **Onkologie** ab Jahrgang ____________

☐ zum regulären Preis von EUR 138.–

☐ zum Sonderpreis von EUR 69.– für Mitglieder der
☐ AIO ☐ CAO ☐ DEGRO ☐ DGHO ☐ ÖGHO
(ausschliesslich an Privatadresse)

Diese Preise gelten für 2002 und verstehen sich inkl. MWSt., zzgl. EUR 12.– Versand.
Preisänderungen vorbehalten.

☐ Bitte senden Sie mir ein kostenloses Probeheft von **Onkologie**

Ich kann diese Bestellung innerhalb einer Woche durch eine einfache Mitteilung an den Verlag widerrufen.
Das Abonnement gilt zunächst für ein Jahr und verlängert sich um jeweils ein weiteres Jahr, wenn ich die Zeitschrift nicht bis 1. Dezember des laufenden Jahres abbestelle.

S. Karger GmbH
Lörracher Strasse 16A
D–79115 Freiburg (Deutschland)
Tel. +49 761 45 20 70
Fax +49 761 45 20 714
E-Mail: information@karger.de
www.karger.com/journals/onk

KARGER

Ich bezahle ☐ mit Scheck ☐ auf Rechnung

Bitte belasten Sie diese Bestellung meiner Kreditkarte
☐ American Express ☐ Eurocard ☐ Visa

Karten-Nr.: Gültig bis:

Ich ermächtige die S. Karger GmbH bis auf Widerruf, jährlich meine Abonnement-Gebühren im Lastschriftverfahren abzubuchen (nur innerhalb Deutschlands möglich)

bei Bank oder Post:

in: BLZ:

Giro-Konto-Nr.:

Meine Postanschrift lautet (Blockschrift):

Tel.:

Datum/Unterschrift:

correlation between percentage of CD66b-neg. cells and FLAER-neg. cells in all groups: group A: 0.5% CD66b-negative cells, 0.8% FLAER-negative cells (medians), group B: 8.9% and 7.4% negative cells, r=0.98; group C: 56.1% and 64.5% negative cells, r=0.99. Correlation was less good with CD16-staining since in many pts. CD16-negative cells were present which were obviously not GPI-deficient since they bound CD66b and FLAER. Fluorescence intensity of FLAER positive granulocytes was high which made it easy to discriminate positive and negative cells. Since FLAER can be used at low concentration, costs for one test with FLAER is only about 20% of costs for the antibody method (with 2 different antibodies on one lineage). In primary PNH and AA-PNH syndrome, granulocytes are the most affected population. Staining of granulocytes with FLAER therefore can be used as a simple, rapid and cheap screening method for detection of GPI-deficient cells. FLAER might in particular be suitable to monitor the proportion of GPI-deficient cells during treatment or follow-up. Further studies in a larger series of pts. are required to decide whether FLAER can completely replace protocols using antibodies to GPI-AP.

Wissenschaftliches Symposium: Therapeutische Tumorimpfung

366

Anti-tumor treatment with dendritic cells reveals the delicate balance between autoimmunity and tumor immunity

B. Ludewig (Zürich, CH)

Vaccination with dendritic cells (DC) presenting tumor antigens induces primary immune responses or amplifies existing cytotoxic antitumor T cell responses. We recently demonstrated that antitumor treatment with DC may cause severe autoimmune disease when the tumor antigens are not tumor-specific but are also expressed in peripheral non-lymphoid organs Growing tumors with such shared tumor antigens that were, at least initially, strictly located outside of secondary lymphoid organs were successfully controlled by specific dendritic cell vaccination. However, antitumor treatment was accompanied by fatal autoimmune disease, i.e. autoimmune diabetes in transgenic mice expressing the tumor antigen also in pancreatic b-islet cells or by severe arteritis, myocarditis and eventually dilated cardiomyopathy when arterial smooth muscle cells and cardiomyocytes expressed the transgenic tumor antigen. However, the activation of self-reactive CTL by DC after uptake of exogenous antigen was limited firstly, by the short half-life of MHC class I-associated peptides on DC in vitro and in vivo, and, secondly by the rather inefficient MHC class I presentation of cell-associated self-antigens by DC. Taken together, these results reveal the delicate balance between tumor immunity and autoimmunity and therefore point out important limitations for the use of not strictly tumor-specific antigens in antitumor vaccination with DC.

367

Antigen-specific immunotherapy: A long way from antigen identification to clinical application

T. Wölfel (Mainz, D)

Early cancer immunotherapy trials were performed without knowing target antigens on human malignant cells. During the past two decades anti-tumor T cell responses were generated in vitro and some of their target antigens were identified. The expression pattern of these antigens differs considerably. They are over-expressed in malignant versus normal cells and contribute to malignancy, are expressed at similar or even higher levels in normal parental cells (differentiation antigens), can only be found in tumor and testis cells (MAGE-type) or are highly specific for the individual tumor (mutated/genetically altered antigens). They were identified primarily in melanoma, but also in a variety of carcinomas, leukemias and lymphomas. Meanwhile a continuously growing list of peptides is available that are processed in tumor cells from tumor-associated proteins and are presented on the cell surface by various HLA molecules. Their identification fostered a series of phase I trials on therapeutic vaccination against distinct antigens and also on adoptive immunotherapy with T cells of known specificity. While their outcome was quite encouraging, it became soon clear that a more detailed understanding of the interactions between cellular immunity and antigens presented by malignant cells is inevitable for further progress. As most relevant issues in this context appear the heterogeneity and stability of antigen expression in tumor tissues, the rules and exceptions for post-translational modification, processing and transport of peptides, the hierarchy between antigen categories, the mechanisms of tolerance, ignorance and anergy induction and maintenance, the quantitative and qualitative assessment of T cell responses, and the rules for the induction and maintenance of efficient T cell responses both in vitro and in vivo. Their consideration in future trial designs will pace further progress in the field of cancer immunotherapy. In addition, it will be important to pave the way for clinical development strategies, which make biotechnology investments in this area sufficiently attractive and to overcome complex patent issues slowing down development programs.

368

Induction of renal cell carcinoma specific CTL using dendritic cells transfected with tumor-RNA

M. Müller, F. Grünebach, W. Brugger, P. Brossart (Tübingen, D)

Dendritic cells (DC) are professional antigen-presenting cells and play a central role in the induction of antigenspecific cytotoxic T-cells (CTL). DC are currently used in vaccination protocols to treat malignant diseases. However, immunization with defined tumor antigens is limited to a small number of known tumor antigens and certain HLA types. We therefore tested a whole-tumor-approach to induce antigen specific CTL and analized the efficiency of tumor-RNA-transfection into DC using different cationic liposomes and a transferrin receptor based delivery system. To evaluate the sensitivity of these different approaches, pure in vitro RNA-transcripts from the Enhanced Green Fluorescence Protein (EGFP) and b-galactosidase coding vectors were utilized. HeLa cells used a controls and DC transfected with these in vitro transcripts showed a dose dependent expression of the reporter genes. For the experiments using total tumor-RNA a renal cell carcinoma cell line N43 was transfected with a vector coding for EGFP. The transfection of HeLa cells and immature DCs with total cellular tumor-mRNA showed an expression kinetic for the EGFP with a maximum at 30 h post transfection. One restriction of this approach in clinical trials could be the limited number and size of available tumor samples. We therefore developed a protocol for the amplification of total tumor-mRNA. Full length cDNA was synthesized from total isolated cellular mRNA which carries the T7 phage promoter. The resulting cDNA was unspecifically amplified in a PCR. The PCR product was then used as template for RNA in vitro transcription using T7 RNA polymerase in the presence of a 5'-cap analogue. To test the efficacy of the RNA-approach we transfected DC with same amounts of nativ total mRNA and in vitro amplified mRNA derived from a renal cell carcinoma cell line A498. The obtained CTLs specifically lysed the A498 tumor cells in a standard Cr-release assay. However, the nativ mRNA appeared to be more effective than the amplified mRNA product in stimulating CTL. In summary, tumor-RNA transfected DC might be an effective strategy for immunotherapies of malignant diseases like RCC.

369

Dendritic cells as vaccine in BCR-ABL-positive CML: a phase I-study

J. Westermann, J. Kopp, I. Körner, K. Döhner, H. Döhner, B. Dörken, A. Pezzutto (Berlin, Ulm, D)

In chronic myeloid leukemia (CML), in vitro and in vivo data show that distinct bcr/abl fusion peptides are able to elicit T cell responses. Furthermore, other yet undefined antigens resulting from chromosomal abnormalities might also play a role as immunotherapeutic targets. Since DC and leukemic cells of CML patients share a common progeny, there is constitutive expression of putative tumor antigens such as bcr/abl in DC which are derived from leukemic progenitor cells. In vitro these DC are able to elicit an anti-leukemic T cell response. In our institution, a phase I study is being performed in bcr/abl-positive CML patients in chronic phase who did not achieve a major cytogenetic response after a-IFN treatment. DC are generated from monocytes in the presence of GM-CSF, IL-4 and TNF-a under GMP conditions. Vaccination consists of four s.c. injections with $1x10^6$ to $5x10^7$ cells. Keyhole limpet hemocyanin (KLH) is used as adjuvans. Results from the first five patients using monocyte-derived DC show that around 60-70 % of the cultivated cells were DC according to morphology and immunophenotype. In 20 clinical applications the yield of harvested cells/seeded PBMC was in the range of 1-9.5%. 40-70 % of the DC were bcr/abl-positive by fluorescence in situ hybridization (FISH). Vaccination was well tolerated without severe toxicity. DTH could be documented usually after the third DC injection, skin biopsy in one patient showed an infiltrate of CD4+ and CD8+ lymphocytes. FISH analysis of PBMC showed a significant reduction of bcr/abl-positive cells in 2 patients 8 and 12 weeks after vaccination. After DC vaccination, PBMC of all patients had an increasing proliferative response upon stimulation with autologous bcr/abl-positive DC. In one HLA-A2-positive patient having a 78% reduction in bcr/abl-positive PBMC after vaccination, a specific T cell response against a bcr/abl-fusion peptide could be demonstrated by an Elispot assay. In CML, vaccination with DC might possibly play a role in post remission therapy, i.e. after HD-chemotherapy with autologous stem cell support.

370

Production of recombinant individual idiotype vaccines: Interim experience from a phase I trial of active immunotherapy for NHL

H. Veelken, E. Unterbusch, F. Simon, K. Mauerer, L. Berchem, J. Scheele, A. Hafkemeyer, F. Osterroth (Stanford, USA; Freiburg, D)

The idiotype (Id) immunoglobulin of B-cell non-Hodgkin´s lymphomas (NHL) constitutes an individual tumor-specific antigen and a promising target for specific lymphoma immunotherapy. In previous work, we have developed a strategy for bacterial expression of recombinant Id Fab fragments as the basis for active anti-Id immunization of NHL patients (ref.1). Stimulation of peripheral blood T cells from NHL patients in vitro with recombinant Id Fab derived from their tumor leads to efficient induction of Id-specific cytotoxic activity (ref. 2). To assess the feasibility and toxicity of Id Fab-based individual tumor vaccines, a phase I clinical trial was initiated. Up to 5/01, 16 tumor biopsies were obtained from NHL patients to produce recombinant idiotype vaccines. Tumor samples were 9 lymph node (LN) extirpations, 3 fine needle biopsies (2 LN, 1 kidney), 1 bone marrow aspirate, and 3 peripheral blood samples. Histopathological diagnoses were follicular lymphoma in 11 cases, mantle cell lymphoma in 3, and CLL and lymphoplasmacytoid immunocytoma in one patient, respectively. In 9 biopsies that had been analyzed by routine diagnostic flow cytometry, tumor cells constituted between 30 and 95% (median, 80%) of the total cells in the biopsies. Heavy and light chain immunoglobulin (Ig) transcripts (mu = 11, gamma = 4, alpha = 1; kappa = 12, lambda = 4) were amplified by anchored (A-)PCR (ref. 1), cloned, and sequenced. All 308 clones contained the respective amplified Ig gene; no aberrant sequences were observed. Within a single biopsy, the fraction of identical (i.e., clonal) sequences ranged from 0.25 to 1.0 (median, 0.6). Dicistronic expression constructs for periplasmatic expression of recombinant Id Fab fragments (ref. 1) have currently been constructed for 12 cases. Successful Fab protein expression in E. coli was achieved in 10 cases (8 kappa, 2 lambda). As of 31/5/01, large-scale production of Fab was performed for 2 patients who have undergone immunizations according to the clinical protocol. The protein yield was 4 and 5 mg recombinant protein with purities of 95 and 99%, respectively. Endotoxin contamination was in both cases < 50 U/mg. The complete vaccine is formulated with escalating doses of Id Fab mixed with the lipid-based adjuvant MF59 and injected i.d., with GM-CSF injected s.c. at the same site. Both patients tolerated the vaccinations without major side effects. The current experience indicates that the Id cloning strategy by A-PCR and the bacterial expression system permit successful production of individual Id vaccines in the vast majority of cases, even from very small or weakly tumor-infiltrated biopsies. Therefore, this approach should facilitate large-scale clinical testing of active and patient-specific anti-lymphoma immunotherapy. A multi-center Phase II/III trial based on this strategy is currently being designed.

371

Specific immunotherapy of multiple myeloma patients using idiotype pulsed serum free generated dendritic cells

V. Reichardt, C. Milazzo, S. Wirths, G. Stuhler, H. Einsele, W. Brugger, L. Kanz, P. Brossart (Tübingen, D)

Specific immunotherapy of multiple myeloma (MM) patients has become a feasible adjuvant therapeutic option which appears promising for patients who have achieved minimal residual disease after high dose therapy (HDT). Autologous MM antigen pulsed dendritic cells (DC) of different sources have been used by investigators to induce MM specific immune responses in vivo. Idiotypes are MM derived antigens which can serve as tumor specific targets.We first generated functional monocyte derived DC comparing various serum free commercially available tissue culture media. All media allowed the generation of cells with DC morphology, surface expression of DC related molecules (HLA-DR+, CD83+, CD1a±, CD54+, CD80+ and CD86+) and typical DC in vitro function. X-VIVO 15 medium compared favorably to other media tested in functional tests of soluble antigen uptake and induction of antigen specific T cell responses. Serum free generated DC could be used to induce peptide specific CTL in vitro. We have initiated a clinical phase I/II study to evaluate DC based idiotype (Id) vaccination of MM patients after HDT and autologous peripheral blood stem cell transplantation (PBSCT). Id proteins were purified from sera collected prior to therapy. As of 5/01 twelve patients have been vaccinated on our trial. Using serum free culture conditions, we obtained a median number of 4,5 x 10E6 autologous Id pulsed DC (n=24) per 100mL peripheral blood taken 3 months after HDT. A homogenous population of DC pulsed in vitro with individual Id was used twice as cellular vaccine followed by 5 subcutaneous booster injections of conjugated Id/KLH (keyhole limpet hemocyanin) with GM-CSF as adjuvant. All patients tolerated the Id/DC vaccines well with no side effects. The s.c. booster injections caused local and transient reactions in most patients. One patient was in CR, 11 patients were in PR after HDT prior to vaccination. Two of the twelve patients who have completed their series of Id vaccination have developed a measurable Id specific T cell proliferative response, one patient developed an Id specific CTL response. All patients mounted KLH specific T cell and antibody responses proving immunocompetence to a neoantigen early after HDT. Three patients have declining Id spikes during or after Id vaccination. Id vaccination of MM patients is feasible and well tolerated using serum free generated DC. Id specific immune responses can be induced in patients after HDT. In collaboration with the German Multiple Myeloma Study Group specific immunotherapy by Id pulsed DC vaccination is now being offered to MM patients who have achieved a clinical CR after tandem autologous PBSCT.

Wissenschaftliches Symposium: Hämostaseologie

374

Therapy with platelet concentrates and alternatives

E. Seifried, R. Henschler (Frankfurt, D)

Transfusion of blood platelets to overcome the period of thrombocytopenia after blood and bone marrow transplantation (BMT) regimens has been of paramount importance for the feasibility and safety of this therapy. Pooled random donor platelet concentrates (RDPC) from whole blood and apheresis-derived thrombocyte single donor platelet concentrates (ADPC) have been the two competing favourite praparations used for these patients. Recently, however, inline leukocyte-depleted (ILD)-RDPC have been introduced to the market and will replace the conventional RDPC by October 2001 in Germany. A multicenter randomized trial and numerous other studies have so far shown that ILD-RDPC, which contain $<1x10^6$ leukocytes per unit, can reduce the incidence of side effects ascribed to leukocytes such as HLA immunization and refractoriness to platelet transfusion from RDPC down to levels obtained with ADPC. These results encourage a much broader use of ILD-RDPC for platelet substitution therapy even for long term substitution in hematological patients. This development is also driven by the more economical preparation of RDPC compared to ADPC; however the recently developed possibility to prepare two PC from one apheresis in our hands provides an attractive alternative to ILD-RDPC. In some cases of thrombocytopenia corticosteroids can be helpful. In patients with von Willebrand disease or platelet function disturbances DDAVP can be an alternative approach. A novel approach, which is completing development is the photoinactivation of viruses and bacteria potentially present in platelet concentrates by substances like thionin and psoralens. This could be of relevance for hematological patients e.g. in the case of Parvovirus B19 or CMV. A very different, still experimental strategy has been the transfusion of platelet progenitor cells (PPC); however it is too early to draw conclusions regarding its potential for the further improvement of platelet insufficiency after stem cell transplantation.

375

Faktor V-Leiden-mutation and orthotopic liver transplantation

A. Loew, M. Stockmann, M. Bahra, G. Hintz, H Riess (Berlin, D)

Background: FV-Leiden-Mutation (FV:Q506) resulting in APC-resistance is a common risk factor predisposing to venous thromboembolism. Patients undergoing orthotopic liver transplantation (OLT) can either acquire or loose the defect, depending on the genotypes of donor and recipient.
The role of FV:Q506 in recipients as well as the significance of OLT-related APC-resistance is still open to discussion. Methods: We tested blood-samples from 720 patients having undergone OLT for FV:Q506 by PCR-technology, and for the functional defect with medium-specific (APCR, Chro-

Risikoadaptierte Thrombose-Prophylaxe

...einfach in 2 Wirkstärken

Fragmin® P
Dalteparin Natrium 2.500 I.E.

Bei niedrigem und mittlerem Risiko

Fragmin® P Forte
Dalteparin Natrium 5.000 I.E.

Bei hohem Risiko

Fragmin® P/-Forte. 2.500/5.000 I.E.* Wirkstoff: Dalteparin-Natrium. Verschreibungspflichtig. **Zusammensetzung:** 1 Fertigspritze Fragmin® P mit 0,2 ml Injektionslösung enthält Dalteparin-Natrium 2.500 I.E.* anti-Faktor Xa. 1 Fertigspritze Fragmin® P Forte mit 0,2 ml Injektionslösung enthält Dalteparin-Natrium 5.000 I.E.* anti-Faktor Xa. Sonstige Bestandteile: Wasser für Injektionszwecke. Zusätzlich bei Fragmin® P: Natriumchlorid. **Anwendungsgebiete:** Fragmin® P: Zur peri- und postoperativen Primärprophylaxe tiefer Venenthrombosen bei niedrigem oder mittlerem thromboembolischen Risiko, am OP-Tag auch bei hohem Risiko. Fragmin® P Forte: Zur postoperativen Primärprophylaxe tiefer Venenthrombosen bei hohem thromboembolischen Risiko (z.B. orthopädische Chirurgie). **Gegenanzeigen:** Überempfindlichkeit gegen Dalteparin-Natrium, andere niedermolekulare Heparine oder Heparin. Aktueller oder aus der Vorgeschichte bekannter allergisch bedingter Abfall der Blutplättchen (Thrombozytopenie Typ II) auf Heparin oder Dalteparin-Natrium. Verletzungen oder Operationen am Zentralnervensystem, Augenoperationen. Aktuelle aktive Blutungsprozesse. Erkrankungen, die mit einer erhöhten Blutungsbereitschaft einhergehen. Erkrankungen bei denen der Verdacht einer Verletzung des Gefäßsystems besteht. Fragmin® P/-Forte sollte nicht angewendet werden bei: Verdacht auf Malignom mit Blutungsneigung, Nieren- und Harnleitersteinen, chronischem Alkoholismus. Nur unter Vorsicht anwenden bei Leber- und Nierenschäden, Magen-Darmgeschwüren in der Krankengeschichte, bei gleichzeitiger Behandlung mit den Serum-Kalium-Spiegel erhöhenden Arzneimitteln, oralen Antikoagulanzien und/oder Acetylsalicylsäure. Unter der Geburt ist die rückenmarksnahe Narkose bei Schwangeren, die mit gerinnungshemmenden Medikamenten behandelt werden, absolut kontraindiziert. Klinische Erfahrungen mit Kindern liegen nicht vor. **Nebenwirkungen:** Dosisabhängig und häufiger bei Patienten mit zusätzlichen Risikofaktoren können vermehrt Blutungen auftreten. Einzelfälle von schwerwiegenden Blutungen sind berichtet worden. Häufig wird ein Anstieg der Leberenzyme (GOT, GPT, Gamma-GT), sowie der LDL und Lipase beobachtet, der zumeist nach Absetzen der Behandlung reversibel und klinisch nicht bedeutsam ist. Ein Anstieg der Serum-Kalium-Konzentration ist nicht auszuschließen. Gelegentlich tritt zu Beginn der Behandlung mit Heparin eine leichte vorübergehende Verminderung der Zahl der Blutplättchen (Thrombozytopenie Typ I) mit Werten zwischen 100.000/µl und 150.000/µl auf. Komplikationen kommen im allgemeinen nicht vor. Die Behandlung kann daher fortgeführt werden. Selten wird ein allergisch bedingter schneller Abfall der Zahl der Blutplättchen (Thrombozytopenie Typ II) mit Werten deutlich unter 100.000/µl oder weniger als 50% des Ausgangswertes beobachtet. Bei Patienten ohne vorbestehende Überempfindlichkeit gegen Heparin beginnt der Abfall der Blutplättchen in der Regel 6–14 Tage nach Behandlungsbeginn. Bei Patienten mit Überempfindlichkeit gegen Heparin tritt dieser Abfall unter Umständen innerhalb von Stunden auf. Diese schwere Form der Verminderung der Blutplättchenzahl kann verbunden sein mit arteriellen und venösen Thrombosen/Thromboembolien, Verbrauchskoagulopathie, Hautnekrosen an der Injektionsstelle, Haut- und Schleimhautblutungen, flohstichartigen Blutungen und Teerstuhl. Dabei kann die gerinnungshemmende Wirkung des Heparins vermindert sein (Heparin-Toleranz). Bei Patienten, bei denen die genannten allergischen Reaktionen auftreten, muß Fragmin® P/-Forte sofort abgesetzt werden. Bei ihnen darf auch in Zukunft kein heparinhaltiges Arzneimittel mehr angewendet werden. Wegen der genannten gelegentlich auftretenden ungünstigen Wirkung des Heparins auf die Thrombozyten muß deren Zahl, insbesondere zu Beginn der Behandlung mit diesem Arzneimittel, engmaschig kontrolliert werden. Eine Überempfindlichkeit gegenüber Dalteparin sowie anaphylaktische Reaktionen sind selten. Allergische Erscheinungen beinhalten Symptome wie Übelkeit, Kopfschmerzen, Temperaturanstieg, Gliederschmerzen, Nesselsucht, Erbrechen, Pruritus, Dyspnoe, Bronchospasmen und Blutdruckabfall. Nebenwirkungen wie Haarausfall in seltenen Fällen, Osteoporose nach längerer Anwendung, Priapismus und Vasospasmen in Einzelfällen, Hypotonie und Bradykardie, Hypoaldosteronismus mit Hyperkaliämie und metabolischer Azidose, besonders bei Patienten mit eingeschränkter Nierenfunktion und Diabetes mellitus in sehr seltenen Fällen, sind nicht auszuschließen. Lokale Gewebsreaktionen an der Injektionsstelle werden gelegentlich beobachtet. Absterben von Hautgewebe (Hautnekrosen) und Blutungen an der Injektionsstelle sind selten. *) 1 I.E. = 1 Einheit des 1. internationalen Standards für niedermolekulares Heparin. Nicht zu verwechseln mit Heparin I.E.! **Stand:** 04.2000. **Darreichungsform und Packungsgrößen: Fragmin® P:** 5 Fertigspritzen à 0,2 ml Injektionslösung (N1) AVP: 62,70 DM. 10 Fertigspritzen à 0,2 ml Injektionslösung (N1) AVP: 115,05 DM. 20 Fertigspritzen à 0,2 ml Injektionslösung (N2) AVP: 222,77 DM. 35 Fertigspritzen à 0,2 ml Injektionslösung (N3) AVP: 383,00 DM. 100 Fertigspritzen à 0,2 ml Injektionslösung (N3) AVP: 925,– DM. 25 Fertigspritzen à 0,2 ml Injektionslösung (Klinikpackung). **Fragmin® P Forte:** 5 Fertigspritzen à 0,2 ml Injektionslösung (N1) AVP: 109,30 DM. 10 Fertigspritzen à 0,2 ml Injektionslösung (N1) AVP: 209,60 DM. 20 Fertigspritzen à 0,2 ml Injektionslösung (N2) AVP: 419,19 DM. 35 Fertigspritzen à 0,2 ml Injektionslösung (N3) AVP: 698,25 DM. 100 Fertigspritzen à 0,2 ml Injektionslösung (N3) AVP: 1.550,– DM. 25 Fertigspritzen à 0,2 ml Injektionslösung (Klinikpackung). **Stand:** Juni 2001.

Pharmacia GmbH, 91051 Erlangen

mogenix™), and high FV-specific tests (Akzelerininactivation-test Blood 1996; 88[11]: 4398-9) APCR-V, Chromogenix™). In addition to clinical records 552 of 720 patients returned a questionnaire focused on thromboembolic events (TE) before and after OLT. Results: The prevalence of FV-Leiden in the OLT-recipient population was 6,9%. Carriers of the FV:Q506-Mutation suffered more TE before transplantation compared to non-carriers (28% vs.7%,p=0.0005), and especially suffered significant more hepatic vein thrombosis (Budd Chiari Syndrome, BCS) before OLT (10% vs.1,8%, p=0.0005) than patients without the defect. After OLT 49 carriers of FV:Q506 did not show APCR in the functional tests. One patient heterocygous for FV:Q506 stayed APC-resistant after OLT and received obviously the liver from a FV:Q506 -carrier. Overall we detected acquired APC-resistance in 36 cases (5,0%). In one patient the results of functional assays after OLT are compatible with a FV:Q506-homocygous-donor organ. We found a higher incidence of TE after transplantation in patients with acquired APC-resistance in a mean follow up of 5 years (16.6%vs 4,3%;p=0.001) including one life threatening TE in the early postoperative phase. Conclusion: Before OLT APC-resistance due to FV-Leiden-Mutation is a risk Factor for TE especially for hepatic vessel thrombosis like BCS. OLT related APC-resistance has to be considered as a risk factor for venous thrombosis without clearly elevating the acute postoperative complication rate in the early phase after OLT.

376

Dominant inherited giant platelet syndromes are caused by mutations in nonmuscle myosin heavy chain A

L.E. Carlsson, N. Mouraviev, K.E. Heath, J.A. Martignetti, A. Greinacher (Greifswald, D; New York, USA)

Background: Giant platelet syndromes such as May-Hegglin anomaly (MHA), Fechtner syndrome and Sebastian platelet syndrome (SPS) form a group of autosomal dominant inherited platelet disorders sharing the triad of thrombocytopenia, giant platelets and granulocyte inclusion bodies. Fechtner syndrome is also associated with nephritis, cataract, and neurosensorial deafness. Recently, by linkage analysis the defect in giant platelet syndromes has been localized to chromosome 22q11-13.
We report on twelve patients (five different families) with giant platelet syndromes where the genetic defect was localized to the gene encoding nonmuscular myosin heavy chain A (MYH9) and sequence-specific primer (SSP-PCR) methods for rapid diagnosis. Material and Methods: Family A: the original SPS family, four affected members; family B: SPS, three affected members; Family C: SPS or MHA, three affected members; case D: unclassified macrothrombocytopenia; case E: Fechtner-syndrome. Platelet and granulocyte morphology and ultra structure were studied by standard methods. For each missense mutation identified in MYH9 by sequencing a SSP-PCR method was established. Results: Giant platelets and granulocyte inclusion bodies shared typical morphological features of SPS and Fechtner syndrome. Three different mutations in MYH9 were detected. One of the mutations (D1424N) was found in three of the families. Conclusions: The genetic defect in inherited macrothrombocytopenia has been located on MYH9 in twelve individuals of five unrelated families. Importantly, all index cases were initially wrongly diagnosed and treated as chronic autoimmune-thrombocytopenia, three of five patients underwent splenectomy. Using our SSP-PCR methods enables rapid identification of these missense mutations and will further improve differential diagnosis of chronic thrombocytopenia.

Wissenschaftliches Symposium: Zytogenetische und molekulare Diagnostik

377

Cytogenetic findings in non-Hodgkin's lymphomas

I. Wlodarska (Leuven, B)

NHL comprise a complex group of neoplasms with heterogenous clinical, histopathological and genotypic characteristics. Cytogenetic and molecular investigations resulting in the identification of several recurrent chromosomal abnormalities, as well as affected genes controlling cell proliferation, differentiation and apoptosis allowed a better insight into the biology of lymphoma. The most frequent NHL-related chromosomal translocations lead to a deregulation of (onco)genes by their juxtaposition with regulatory sequences of immunoglobulin genes. The best known examples include t(3;14)(q27;q32), t(8;14)(q24;q32), t(11;14)(q13;q32) and t(14;18)(q32;q21) involving the BCL6/LAZ3, CMYC, BCL1/CCND1 and BCL2 genes, respectively. Other, less frequently affected genes include BCL3 (19q13), PAX-5 (9p13), BCL8 (15q11-13), BCL9 (1q21), BCL10 (1p22), and MUM1/IRF4 (6q25), FGFR3 (4p16) and CMAF (16q23), which are recurrently involved in myeloma. In contrast to leukemias, only two NHL-related translocations generate chimeric genes with an oncogenic potential. The first of them, t(2;5)(p23;q35) involving NPM-ALK is associated with a subset of ALCL. The second one, t(11;18)(q21;q21) results in the API2-MLT fusion playing a significant role in MALT lymphomagenesis. In addition to chromosomal translocations other karyotypic aberrations like deletions, trisomies and amplification of specific regions have been non-randomly observed in NHL. The overview of these chromosomal abnormalities, possibilities of their FISH detection and association with histopathological subsets of NHL will be discussed.

379

Multicolor interphase-FISH for the detection of recurrent chromosomal translocations in lymphoid neoplams

J.I. Martín-Subero, S. Gesk, L. Harder, I. Chudoba, B. Luhmann, W. Grote, F.J. Novo, M.J. Calasanz, R. Siebert, Institut für Humangenetik, MetaSystems, Department of Genetics (Kiel, Altlusheim, D; Pamplona, E)

Conventional cytogenetic analysis in lymphoproliferative disorders is frequently hindered by the low mitotic index of tumor cells in vitro and by cytogenetic complexity. Molecular genetic approaches to detect lymphoma-associated translocations, e.g. t(11;14) in mantle cell lymphoma or t(8;14) in Burkitt's lymphoma, are limited due to spreading of the breakpoints over large genomic regions. In contrast, fluorescence in situ hybridization (FISH) using suitable locus-specific probes does not rely on metaphases and is not influenced by scattering of the genomic breakpoints. Here, we present novel FISH assays for the diagnosis of breakpoints in T-cell receptor (TCR) loci and in different oncogene loci involved in B-cell malignancies. After having tested the performance of all these new probes, they were comprehensively combined with previously published and commercially available probes, all labeled with distinct fluorescent dyes. Thus, multicolor locus-specific assays for B-cell non-Hodgkin´s lymphoma (B-NHL) including multiple myeloma (MM) and T-cell non-Hodgkin´s lymphoma (T-NHL) were established. The multicolor B-NHL assay included probes for BCL1 (11q13), BCL2 (18q21), BCL6 (3q27), MYC (8q24) and IGH (14q32) allowing the detection of the diagnostically most important translocations t(11;14)(q13;q32), t(14;18)(q32;q21), t(3;14)(q27;q32) and t(8;14)(q24;q32) in a single experiment. In MM, the multicolor probe pool was composed of probes for the FGFR3 (4p16), IRF4 (6p25), BCL1 (11q13), MAF (16q23) and IGH (14q32) genes for the simultaneous detection of the MM-associated translocations t(4;14)(p16;q32), t(6;14)(p25;q32), t(11;14)(q13;q32) and t(14;16)(q32;q23). The T-NHL multicolor assay included probes flanking the TCRA/D and TCRB loci in 14q11 and 7q35, respectively. The reliability of all multicolor assays was tested in negative and positive controls. Even in the limited number of cases included in this pilot study unexpected findings were obtained with the multicolor assays, e.g. a cytogenetically cryptic translocation affecting the TCRB locus in addition to a TCRA/D breakpoint in a T-NHL or two simultaneous translocations with breakpoint in the IGH locus affecting BCL1 and FGFR3 genes in a case of MM.

380

Early detection of imminent relapse using sequential quantitative chimerism analyses in sorted CD34+ cells from the peripheral blood

K. Lutterbeck, M. Bornhäuser, U. Oelschlägel, M. Kiehl, C. Brendel, U. Platzbecker, A. Neubauer, A.A. Fauser, G. Ehninger, C. Thiede (Dresden, Idar-Oberstein, Marburg, D)

Detection of minimal residual disease (MRD) after allogeneic stem cell transplantation (SCT) is important to schedule early intervention ,e.g. DLI. PCR based procedures, detecting tumor specific genetic aberrations, e.g. chromosomal translocations, have been successfully used for this purpose. However many patients with AML or MDS lack appropriate molecular markers. We asked, whether analysis of donor chimerism in circulating hematological progenitor cells, i. e. CD 34+ cells, might be a feasible method to monitor minimal residual disease in patients after allogeneic SCT. Material and Methods: We prospectively analyzed 87 patients (44 CML, 25 AML, 10 ALL, 6 MDS, 1 CMMOL, 1 NHL), who underwent BM (n = 31) or PBSC (n = 56; T-cell depleted: n = 11) transplantation from related (n = 23) or unrelated (n = 64) donors in two different centers. CD 34+ cells were prepared from peripheral blood samples and sorted (FACS Vantage) after immunomagnetic preenrichment. These data were compared with other cellular subsets (T-, NK-, CD 15+). Quantification of donor chimerism was done using multiplex STR-PCR with fluorescence detection. Subset chimerism data were compared to overall chimerism and results of nested RT-PCR for the BCR-ABL rearrangement in Ph-pos. cases. Results: In the 87 patients a median of 7 analyses (range 3-22) covering a median period of 257 days (range 28-1067) were performed. Secondary graft failure observed in four patients was associated with low donor NK-cell chimerism in those three patients transplanted from an unrelated donor. A total of 26/87 patients (29.8%) died due to graft failure, relapse, multiorgan failure or infections. A hematological relapse was observed in 17/83 engrafting patients (20%) after a median time of 111 days (range: 52-595). In twelve of these patients, donor chimerism in CD34+ cells decreased 12-128 days (median 40 days) before the clinical diagnosis. CD34+ chimerism analysis failed to demonstrate the relapse one week before the diagnosis in a case with CML in blast crisis, although the blasts were shown to express CD34. In the other four patients, the interval between the diagnosis of relapse and the last subset analysis exceeded 100 days. In one patient with CML, molecular relapse was observed after 319 days, which was associated with a decrease of donor CD34+ cells. Reduction of immunosuppression induced clinical GvHD which was associated with an increase of donor derived CD34+ cells and clearance of BCR-ABL positive cells. Conclusion: Sequential chimerism analysis in CD34+ cells represents a feasible and sensitive method for the detection of minimal residual disease. It allows early detection of relapse and facilitates monitoring of therapeutic intervention. This method appears especially useful in patients with high risk disease lacking other suitable markers for the detection of MRD.

381

Chromosomal aberrations in skin fibroblasts following total body irradiation before allogeneic stem cell transplantation

P. Zschieschang, G. Thiel, G. Massenkeil, J. Pross, E. Schümann, R. Arnold (Berlin, D)

Introduction: Radiotherapy and chemotherapy can produce detectable cytogenetic aberrations and are known to be mutagenic in in-vitro and animal test systems. Development of secondary malignancies after radiotherapy and chemotherapy might therefore be caused by DNA alterations through point mutations or chromosomal aberrations. Patients undergoing stem cell transplantation (SCT) often receive total body irradiation as part of the conditioning regimen prior to stem cell infusion. The cytogenetic effects of this therapy were studied in patients before and after SCT. Patients and methods: In a retrospective study cytogenetic investigation of skin fibroblasts was performed in 15 adult patients 23 - 216 months after SCT. Patients (8 female and 7 male) ranged in age from 18 - 52 years at the time of transplantation. Eight patients had CML, three had AML, two had ALL and two had MDS. This study group was compared to 15 control patients, who were analysed before SCT. Skin samples were 4 mm punch biopsies. Samples were cut into small pieces, placed at the bottom of three flasks and cultured according to standard techniques. Cytogenetic analysis was performed with trypsin-Giemsa banding technique. Results: No abnormal mitoses were found in the control patients. In the post-SCT group cytogenetic analysis revealed complex aberrant karyotypes in all 15 patients affecting 43 - 100% of the analysed mitoses. Interestingly, we found exclusively stable chromosomal aberrations (translocations, inversions, deletions), which were clonal in part. Conclusion: Our data indicate that total body irradiation and high dose chemotherapy induce stable chromosomal aberrations in skin fibroblasts in vivo. Arising questions as the influence of clinical factors on the extent of chromosomal abnormalities as well as the possibility to describe risk groups for secondary malignancies will be discussed.

382

Association between mutations in the dihydropyrimidine dehydrogenase gene and severe toxicity of treatment with 5-fluorouracil: A prospective multicenter study

C. Kollmannsberger, C. Bokemeyer, C. Marx, J. Fischer, F. Honecker, M. Schwab, M. Eichelbaum (Tübingen, Stuttgart, D)

5-Fluorouracil (5-FU) is frequently used for the treatment of several solid tumors, especially colorectal cancer. Although usually well tolerated, about 3-5% of pts experience severe or even lethal toxicity. The cause for this toxicity may be deleterious mutations in the gene for dihydropyrimidine dehydrogenase (DPD), the initial and rate-limiting enzyme in pyrimidine catabolism. At least 17 mutations in the DPYD gene on chromosome 1 have been identified either in patients with inborn DPD deficiency resulting in thymine-uraciluria or in pts experiencing severe 5-FU toxicity. A mutation leading to the deletion of exon 14 (DPYD*2) was previously described in several unrelated individuals. It remains unclear, whether these mutations might predict the development of 5-FU related toxicity.
The association between DPYD mutations and 5-FU toxicity is presently assessed in a large prospective multicenter study including pts receiving single agent 5-FU therapy (± folinic acid) as bolus or infusional therapy for colorectal cancer or other tumors. Of 351 patients presently recruited by 36 centres; 210 have been further analysed so far. Of these 133 had no toxicity reported and 77 pts (37%) had developed toxicity (34 pts grade 2, 34 pts grade 3 and 9 grade 4) following 5-FU. To establish high-throughput mutation screening, we applied denaturing high pressure liquid chromatography (DHPLC), a highly automated technique suitable for the detection of known and unknown mutations. DHPLC mutation analysis of DPYD was optimized for PCR amplicons covering all 23 exons and their exon-intron boundaries generated from genomic DNA. The DPYD*2 mutation was detected in 7 of 210 pts, corresponding to an overall allele frequency of 3.3 %. All patients carried the DPYD*2 allele in a heterozygous condition. Three of 9 pts (33%) with grade 4 toxicity and 2 of 34 pts (5.8%) with grade 2 toxicity had exon 14 skip mutations. Overall 2 of 133 pts (1.5%) without toxicity in contrast to 5 of 77 pts (6.5%) with grade 2 - 4 toxicity showed skip mutations at exon 14. Interestingly, 4 pts with exon 14 skip mutation and toxicity had received 5-FU bolus application and only one part as infusional therapy and both unaffected patients with mutations had also received 5-FU as continuous infusion. These preliminary results suggest that heterozygosity for DPYD*2 accounts only for part of the cases of 5-FU induced toxicity. Complete mutation analysis in all patients is ongoing to investigate whether additional mutations are found. Recruitment of up to 800 pts is planned to asess the overall frequency of mutations affecting the DPD pathway in this setting. Supported by grant 01 GG 9846 from the German Federal Ministry of Education and Science (BMBF) and by the Robert Bosch Foundation, Stuttgart, Germany.

Wissenschaftliches Symposium: Nichtkleinzelliges Bronchialkarzinom

383

The potential significance of immunological and molecular aspects for the therapeutic strategy in lung cancer

J. Fischer, A. Neuner, T. Muley, H. Lahm (Heidelberg, D)

Recent insight into the mechanisms of pathogenesis of malignant disease have revealed a number of biological factors that may not only be involved in the development and course of the disease but may also serve as prognostic markers for survival and, perhaps even more important, as predictive markers for the efficiency of a specific therapeutic strategy. Immunosuppressive molecules are not only produced by immunocompetent cells, but also by a whole variety of tumor cells. Therefore, escape from immune surveillance mediated by tumor-derived immunosuppressive factors may be an important step in the development and clinical course of malignant disease. Moreover, an impairment of immune defense and a suppression of cytokine secretion capacity in tumor patients may have clinical relevance and influence survival. In SCLC, secretion of TGFb1 and impairment of IL-2 secretion capacity have been demonstrated. Suppression of IL-2-secretion significantly correlated with reduced survival. In NSCLC, increased production of immunosuppressive IL-10 by the tumor and increased plasma IL-10 concentrations are correlated with reduced survival. In addition, IL-2 secretion was suppressed in NSCLC-patients. Moreover, suppression of IL-2 secretion is predictive for survival of NSCLC patients. With a cut-off value for IL-2 of 1100 pg/ml, survival was different between the group with high IL-2 (n=38, 29 failures) and with low IL-2 (n=52, 47 failures) with a p-value of 0.014 in the whole group of 90 NSCLC-patients. With an observation time from 0.5 to 96.4 mo, 5-year-survival was 24.5% and 6.2% in the group with high and low IL-2, respectively. In surgically treated patients (n=33), survival was different with a p-value of 0.011. In 14 patients with surgical resection of the tumor and high IL-2 at diagnosis and 19 patients with surgical resection, but low IL-2 at diagnosis, median survival was 86.2 mo and 11.3 mo, respectively. Molecular changes in the tumor as well as in peripheral blood may also be good candidates as predictive markers and may even serve as a basis for individualised therapy. Point-mutations, hypermethylations and overexpression of different oncogenes, tumor-suppressor-genes, or genes coding for growth factors or their respective receptors can now be detected by modern screening procedures. For example, b-tubulin mutations predict the response of NSCLC to taxotere. Molecular abnormalities such as hypermethylation or overexpression of DNA-repair-genes may result in resistance to different cytotoxic drugs. Thus, the combined assessment of these markers might serve as a basis for future individualised treatment strategies based on individual biological tumor characteristics.

384

Radio-chemotherapy in stage III non-small-cell lung cancer

M. Stuschke (Essen, D)

Combined radiation and chemotherapy can improve treatment results in prognostically or technically inoperable patients with stage III NSCLC over conventional radiotherapy alone. Three randomised studies showed that sequential cisplatin-based combination chemotherapy followed by radiotherapy improves long term survival over the same radiation schedule alone. Three further studies found that sensitising doses of cisplatin or carboplatin given simultaneously to radiotherapy can improve survival. In a second step of optimisation, a newer study from the West Japan Lung Cancer Group and an immature trial from the RTOG showed that full dose combination chemotherapy is more effective given concurrently instead of sequentially to radiotherapy. Treatment intensity is thereby increased. For the future, there is room for intensified radiotherapy in combination with chemotherapy, either by using accelerated hyperfractionated radiation schedules or dose escalated conformal radiotherapy. In addition, the value of newer drugs given concurrently to radiotherapy is currently tested. In conclusion, radiation chemotherapy is suitable and standard for selected patients with inoperable stage III NSCLC. Further treatment optimisations are underway and bear great promises.

385

Coagulation and fibrinolysis in patients with non small cell lung cancer

U. Epler, T. Blankenburg, S. Schädlich, W. Schütte (Dresden, Halle/Saale, D)

Objective: Patients suffering from malignant disease present with a higher risk of thrombosis due to a subclinical activation of the coagulation-fibrinolysis system. Especially lung cancer is frequently responsible for thromboembolic events. The incidence of thromboembolism in cancer patients caused by malignancy is generally specified up to 11 percent. The aim of our study was to evaluate the thrombin activation and fibrinolysis before and after surgery and to analyse the influence of tumour resection on parameters of hemostasis. Method: We measured the coagulation and fibrinolytic state in 72 patients (7 female, 65 male) suffering from NSCLC before surgery by use of thrombin-antithrombin III complex (TAT) and prothrombin fragment 1+2 (F1+2) as indicators of thrombin activation. Additionally d-dimeres (DD) and plasminogen activator inhibitor (PAI) belonging to fibrinolytic system were analysed. Finally we performed coagulation tests on 30 patients after R0-resection of a NSCLC. Results: TAT, F1+2 and DD were increased before surgical treatment, therefore a concealed activation of coagulation may be postulated. However, a thromboembolic event was not to be seen at all. After R0-surgery all measured parameters of hemostasis were significantly decreased (TAT $p<0.001$; F1+2 $p<0.005$; DD $p<0.005$; PAI $p<0.05$) which could be respected as a normalisation of the hemostasis disturbed by malignancy. A pre-therapeutic increase of DD and decrease of PAI in patients with progressive disease indicate an imbalance of hemostasis caused by predomination of fibrinolysis. Activation of coagulation can be concluded from an increase of TAT and F1+2 which is independent from stage and histological type of NSCLC. Conclusion: Non small cell lung cancer is an important risk factor for thrombosis. Prophylactic application of anticoagulants (for example low molecular weight heparin) in addition to the common perioperative prophylaxis should be generally considered. To evaluate the prognostic value of these parameters concerning the progress of NSCLC or other malignancies further studies are required.

386

Identification of metastasis associated genes that predict survival in early stage non-small cell lung cancer

C. Müller-Tidow, W. Wang, S. Böing, R. Metzger, P. M. Schneider, M. Thomas, W. E. Berdel, H. Serve (Münster, Köln, D)

Non-small cell lung cancer is a leading cause of death. Only complete resection of the tumor offers a chance for cure of the disease. However, even after complete resection of early stage NSCLC, distant metastasis develop in a substantial fraction of patients. The identification of patients at high risk for the development of metastasis could help to develop novel therapeutic strategies. To identify differences in stage I adenocarcinoma that either did or did not subsequently metastasize we used a subtractive hybridization strategy of pooled RNA from different primary tumors. RNA from five stage I adenocarcinomas from male patients that did not show evidence for metastasis after a 5-year follow-up period was subtracted from pooled RNA from four similar patients that did develop metastasis. The resulting subtracted cDNA library was cloned into standard cloning vectors and individual clones were sequenced. In total, 225 sequences were obtained from the subtracted cDNA library with 125 unique sequences. Twenty-six genes were identified more than once. About 20% of the identified sequences were previously described to be associated with cancer and or metastasis. Almost the same percentage of genes was either unknown or did not possess any known function. Real-time quantitative RT-PCR analyses provided evidence for 11 of 12 genes that average expression levels of the identified genes were higher in the four tumor samples that subsequently led to metastasis than in the five others. Further analyses of mRNA expression levels in a cohort of 70 NSCLC patients (stage I to IIIA) demonstrated that the association of the identified genes with metastasis was stage and histology specific. Cox regression analyses identified two genes that were independent prognostic parameters for patients' survival in stage I and II disease. One of them, thymosine b4 has recently also been identified in a metastasis screen for melanoma. The other gene was a so far unidentified transcript from chromosome 11q13 that contained an open reading frame of 51 amino acids. We named this gene MALA1 (Metastasis Associated in Lung Adenocarcinoma). The predictive value of strong MALA1 expression was very high: Within stage I and II patients only 9% (2/22) with low expression of MALA1 died within a 5-year follow up period whereas more than 40% (12/28) of patients with high expression succumbed to their disease ($p=0.01$). Taken together, we have identified genes that predict NSCLC primary tumors' metastatic potential. Also, we have cloned MALA1, a novel gene on 11q13, that was closely associated with patients´ prognosis in early stage NSCLC. These findings could help to identify early-stage NSCLC patients at high risk for the development of distant metastasis.

387

Correlation of FDG-PET response with morphometric tumor response after induction chemotherapy with Docetaxel (D) and Carboplatin (C) and Erythropoietin (EPO) support in locally advanced (stage III) non-small cell lung cancer (NSCLC)

F. Griesinger, J. Leugering, F. Lahmann, C. Criée, H. Schmidberger, B. Herse, J. Schirren, W. Keppler, B. Hemmerlein, H. Leupold, R.-P. Baum (Göttingen, Erfurt, Wiesbaden, Immenhausen, Bad Berka, D)

Background: D and C combination chemotherapy (DC) has shown high response rates in advanced NSCLC. Histologic tumor response after chemotherapy or combined modality induction is strongly associated with systemic tumor control and potentially cure. Metabolic tumor response assessed by FDG-PET after induction chemotherapy with etoposide, ifosfamide and cisplatin (VIP) has been shown to be predictive of outcome in NSCLC. Finally, Epo may prevent the decrease in hemoglobin levels that was seen in our previous study of DC (median drop 2.7 g/dl; Thorac Cardiovasc Surg 1998, 298) and thus may enhance treatment efficacy. The aim of the present study was to correlate FDG-PET scan with morphometric findings after DC induction chemotherapy plus Epo. Methods: 38 patients (pts) with NSCLC stage IIIA (8 pts) or IIIB (30 pts) were enrolled and received treatment with D 100 mg/m^2 d1 and C AUC 7.5 d2 q21 days for 4 cycles. Epo was given at 10,000 IU s.c. three times a week. All pts received adjuvant radiotherapy. Results: Of 38 enrolled patients, 32 were evaluable for response by CT imaging. 22/32 pts (69%) achieved PR. Of the 26 pts are evaluable for repeat FDG-PET. 21/26 pts had a decrease of standardized uptake values (SUV) by 50%, 12/26 had SUV <2.5 (CR). Seven of these 12 pts underwent tumor resection, and specimens were subjected to morphometric analysis. In all 7 PET-CR cases, no vital tumor cells were detected in the specimens. In contrast to our previous study, hemoglobin levels increased by a median of 0.3 g/dl. Conclusion: Morphometric tumor response after induction chemotherapy correlates strongly with metabolic remission by FDG-PET. FDG-PET appears to be a useful non-invasive diagnostic tool to predict pathologic response and potentially long-term outcome in stage III NSCLC.

388

Weekly docetaxel as second line chemotherapy in advanced non small cell lung cancer (NSCLC)

M. Serke, S. Fassheber, N. Schönfeld, R. Loddenkemper (Berlin, D)

The activity and toxicity of weekly docetaxel (D) after platin-based first-line therapy for advanced NSCLC were investigated in a prospective phase II study. A final analysis of 36 patients is presented.
Patients and method: 36 patients were enrolled between 1/99 and 4/00. Most pts. (n = 26) showed progressive disease under first-line therapy. One third (n= 11) of pts. had a „sensitive relaps". Pts were treated with 6 cycles of weekly D (35 mg/m^2) each interrupted by a 14 days break. In total 3 courses were administered. A total of 211 infusions of docetaxel were administered (median 6 weekly infusions). Results: Toxicity: Severe (grade III / grade IV) hematotoxicity was not seen. Other than one grade IV diarrhea, grade III non-hematologic toxicities included nausea (4), asthenia (3), pneumothorax (1), fluid retention (1), arrythmia (1), and nail toxicity (2). Mild cutaneous and nail toxicity occurred in 27 pts, neutropenia grace II in 2 cases, and mild asthenia (grade I and II) in 47 courses, anemia grade II in 3 courses. Response: 34 patients were evaluable for response. Partial response (PR) was observed in 4/34 (12%), stable disease in 7/34 (21%), minor response in 5/34 (15%) and progressive disease (PD) in 18/34 (53%). From 36 pts 33 have died so far. 3 pts are still alive (censored 8,3%). Median survival was 160 days (115-205, 95% CI., Kaplan-Meier-analysis, 3 cases censored) and 1-year-survival rate was 24%. Conclusion: Weekly docetaxel (35 mg/m^2) as second line therapy is a well tolerated and safe regimen without occurrence of grade III/ IV hematotocicity. Response rate (PR=12%), median survival and 1-year-survival rate indicate weekly docetaxel as a very active regimen and confirm the results generated with 3-weekly docetaxel with an improved toxicity profile.

Kompetenznetz „Akute und chronische Leukämien": DRGs in der Hämatologie

390

Cost-evaluation in a hemato-oncology center – Results from University Hospital Frankfurt

A. Böhme, E. Harrison, C. Teubner (Frankfurt, D)

Reducing costs for health services, especially for hospitalised patients, has become a major priority in Germany. However, in the field of Hematooncology new medications such as the monoclonal antibodies, growth factors as well as the expansion of indications for allogeneic stem cell transplantation, such as the kind of diseases as well as patient age, combined with intensive complication management, pose a substantial financial burden on clinical departmental budgets. In addition „high-tech medicine" does not always improve the overall cure rate. It prolongs survival and is cost-intensive. Furthermore, higher standards of supportive care, including consequent antibiotic treatment (e.g. liposomal amphotericine B), major diagnostic tools, such as repeated CT-scans, PETs, PCRs and serological screenings, and particularly substitution therapies utilising blood products and coagulation factors constitutes an ever-increasing part of the total case costs. Given the impending introduction of a DRG (diagnosis related groups)-based payment system, a realistic estimation and experience-based calculation of costs per diagnosis - including medical complications - has become a strategic necessity. In additional to medical costs of service, it is essential that these calculations consider all care-delivery costs, which includes rent and renovation of buildings (because of special public funds for buildings, rent and renovation costs should only be considered in internal, not in external pricing), fees and economic costs for technical services as well as employer-specific costs for clinical staff and hospital administration. In particular, special attention must be paid to the hidden costs of medical personnel, as these costs are widely underestimated due to unpaid working overtime, as overtime is traditionally considered an expression of „normal job demands". A methodology for hematological/oncological case-cost calculations will be presented using experienced-based treatment data from Medical Clinic III / University Ffm. In addition to examples of cost calculations for routine treatments of AML and ALL, examples of „high-expense" cases will also be presented.

Wissenschaftliches Symposium: DNA-Methylierung und genetische Instabilität

393

Multidrug resistance is generated in cancer cells and aneuploid cell lines by reversible chromosome assortments – Irrespective of deletions of multidrug resistance genes

P. Duesberg, R. Stindl, R. Hehlmann (Berkeley, USA; Mannheim, D)

Abstract Cancer cells and aneuploid cell lines can acquire resistance against multiple chemotherapeutic drugs that are over 3000-fold those of normal levels, and can display spontaneous resistances up to 20-fold of normal levels. Two mutually exclusive mechanisms are thought to generate this phenotype, either (i) chromosome reassortments that are catalyzed by cancer- and cell line-specific aneuploidy or (ii) classical mutation of preexisting metabolic and transport genes. To distinguish between these two mechanisms we have asked here whether aneuploid mouse cell lines, from which either two or three multidrug resistance genes (Mdr1a, Mdr1b, and Mrp1) have been deleted, can become drug resistant. It was found that all three mouse lines without these central drug resistance genes could nevertheless mutate to multidrug resistance, although they were initially more drug sensitive than undeleted control cells. But no drug-resistant diploid mouse cells were obtained. We conclude that aneuploid cells can generate resistance to cytotoxic drugs de novo via specific chromosome assortments, independent of preexisting genes. Further we have asked whether in the absence of cytotoxic drugs aneuploid, drug resistant Chinese hamster cells could revert spontaneously to drug sensitivity at the high rates that are typical of chromosome reassortments catalyzed by aneuploidy, or at the very low or zero-rates (ie, deletion) of conventional gene mutation. We found that four drug resistant hamster cell lines reverted to drug sensitivity at rates of about 2 to 3% per generation, while two closely related lines remained resistant under our conditions. Thus, the minor resistances, less than 20-fold, that are commonly observed in untreated cancers and unselected lines, reflect accidental chromosome assortments, and the high resistances, obtained after chemotherapy or multiple drug selections in vitro, reflect complex, specific assortments.

394

DNA methylation in Ph positive leukemias

D. Ben-Yehuda, P. Shteper, Z. Siegfried, F. Asimakopoulos, S. Krichevsky, G. Palumbo, E. Rachmilewitz (Jerusalem, IL)

We previously reported that the ABL1 promoter (Pa) undergoes de novo methylation in the course of CML. The methylation status of the ABL1 promoter was monitored by PCR amplification of the Pa region, following digestion with several methylation-sensitive restriction enzymes. Some 74% of the DNA samples from blood and marrow drawn in the chronic phase were unmethylated, similar to control samples from non-CML patients. The remaining 26% were partially methylated in the ABL1 Pa region. Methylated samples were mostly derived from patients clinically indistinguishable from others but known to have a disease of longer duration (26 months versus 7.5 months, p=0.01). Samples of 30 IFNa-treated patients were sequentially analyzed in the course of treatment. Fifteen patients with no evidence of Pa methylation prior to treatment remained methylation-free. The remainder, who displayed Pa methylation before treatment, reverted to the methylation-free status. The outcome is attributed to IFNa therapy as the Pa methylation status was not reversed in any of the patients treated with hydroxyurea. More recently we have adapted the techniques of methylation-specific PCR and bisulfite sequencing to study the regulatory regions of ABL1 and other genes with a role in DNA repair or genotoxic stress response. In cell lines established from CML blast crisis which only carry a single ABL1 allele nested within the BCR-ABL fusion gene, only methylated ABL1 promoters were detected. By contrast, in clinical samples from patients at advanced stages of the disease both methylated and unmethylated alleles were detectable. To distinguish between allele-specific methylation and a mixed cell population pattern, we studied the methylation status of colonies derived from single hematopoietic progenitors. Our results show that both methylated and unmethylated alleles co-exist in the same colony. Furthermore, ABL1 methylation was detectable in the vast majority of colonies from blast crisis, and only in a small proportion of colonies derived from patients in the chronic phase. Topologically, in both cell lines and clinical samples from acute phase CML uniform hypermethylation of the promoter region was noted. Moreover, we showed that ABL1 methylation does not reflect a generalized process and may be unique among DNA repair/

genotoxic stress response genes. Our data suggest that specific methylation of the Ph-associated ABL1 allele accompanies clonal evolution in CML. We asked ourselves whether methylation of the ABL1 promoter is an epigenetic modification also associated with Ph-positive ALL and if it has any clinical significance. The methylation status of the ABL1 promoter in 18 Ph-positive ALL samples was studied. We found that gene-specific ABL1 promoter methylation is associated mainly with the P210 form of BCR-ABL and not the P190 form. While 6 out of the 7 P210-positive ALL samples had ABL1 promoter methylation, all 11 P190-positive ALL samples were methylation-free. In addition, we estimated the extent and relative abundance of ABL1 promoter methylation in several Ph-positive ALL samples and compared it to the methylation pattern in chronic and acute phase CML samples. We put forth a model that correlates the different types of leukemias with the different levels of ABL1 promoter methylation.

395

Centrosome aberrations are associated with tetraploidization of blastoid variants in mantle cell lymphoma

A. Krämer, S. Schweizer, K. Neben, P. Lichter, U. Krause, C. Giesecke, H.K. Müller-Hermelink, A.D. Ho, G. Ott (Heidelberg, Würzburg, D)

Mantle cell lymphoma (MCL) is characterized by the chromosomal translocation t(11;14)(q13;q32), resulting in the rearrangement and overexpression of the cyclin D1 gene. According to cytomorphology, two variants were recognized: a classical and a blastoid subtype of MCL, with blastoid variants being characterized by elevated mitotic counts and proliferation indices as well as a striking tendency to harbor chromosome numbers in the tetraploid range. Since this feature clearly separates these neoplasms from other types of B-cell non-Hodgkin's lymphoma, it has been argued that the inappropriate expression of cyclin D1 in MCL is the reason for this particular karyotypic alteration. Recently, centrosome aberrations have been described as a possible cause of aneuploidy in many human tumors (Marx J, Science 292: 426-429, 2001). Most interestingly, centrosomes, which assure accurate chromosome segregation during mitosis, undergo duplication precisely once before cell division as a function of cyclin-dependent kinase 2 activity at the G1/S boundary. To investigate whether centrosome aberrations do occur in MCL and are associated with tetraploidization of blastoid variants we examined seven diploid and two tetraploid MCL samples for centrosome abnormalities using antibodies to centrosomal proteins. Evaluation was performed by analyzing at least 400 cells per sample using fluorescence microscopy. Centrosomes were considered abnormal if they were present in more than two copies per cell or if they had diameters > two times the diameter of centrosomes in nontumor control cells. First, we found that all MCL analyzed displayed numerical as well as structural centrosome aberrations as compared to normal tonsillar tissue. Centrosome abnormalities were detectable in $37.18 \pm 16.83\%$ of the cells in MCL but only in $5.25 \pm 0.7\%$ of the cells in normal tonsillar tissue ($p<0.001$). Whereas the diploid MCL contained $29.67 \pm 6.54\%$ cells with abnormal centrosomes, the tetraploid samples harbored centrosome aberrations in $63.48 \pm 15.24\%$ of the cells ($p<0.001$). These results indicate that centrosome defects are a common feature of MCL and suggest that they may contribute to tetraploidization of blastoid variants.

396

Automated screening for genomic imbalances in malignant tumors using Matrix-CGH

S. Wessendorf, C. Schwänen, G. Wrobel, B. Fritz, M. Küppers, S. Joos, H. Döhner, M. Bentz, P. Lichter (Ulm, Heidelberg, D)

Comparative genomic hybridization (CGH) and Fluorescence in situ hybridization (FISH) have significantly contributed to our understanding of recurrent genomic alterations in malignant diseases. The knowledge about these alterations and the understanding of their prognostic implications led to the development of new risk adapted tumor treatment strategies. Based on CGH we here report on a new technology allowing the application of CGH to defined DNA targets immobilized on glass slides (Matrix-CGH). Major advantages of this approach are higher resolution, marker-oriented analysis (e.g. with chromosomal contigs for the evaluation of minimal deleted or amplified regions), copy number analysis of hundreds of defined regions in a single experiment and fully automated evaluation procedures. Therefore we designed a matrix-CGH array containing 188 genomic PAC and BAC DNA fragments, which were selected from known oncogenes, tumor suppressor genes and frequently altered chromosomal regions of different tumor entities. Three cell lines (Colo 320 HSR, HL 60 and NGP) as well as selected high grade Non-Hodgkin lymphoma samples were analyzed with this array and these results were compared to chromosomal CGH. An average of 87% congruent results were found between the two techniques. Compared to findings of Southern blot analysis and Fluorescence in situ hybridization we demonstrated that in a given amplicon (e.g. MYC in Colo 320 HSR or MYCN in NGP), the actual number of DNA-copies within the amplified region is assessed more accurately than with chromosomal CGH. Additionally, in HL 60, analysis of eight contiguous clones covering the MYC region on 8q24 revealed discontinuites which remained undetected by conventional CGH, thus demonstrating the higher spatial resolution of the method. Finally, a distinct amplification on 9p24 and multiple low-copy-number changes (2q25, 4q25,22q12 and Xp22) were detected with this array in a primary mediastinal B-cell lymphoma. Our results underline the potential of matrix-CGH as a powerful technique for the identification of chromosomal aberrations in malignant tumors. Within clinical trials, automated tumor screenings for many tumor entities can be envisioned in order to develop new risk adapted tumor therapies. Currently the array has been extended to> 500 DNA probes and is now beeing used for the characterization of genomic aberrations in a larger series of low and high grade non-Hodgkin's lymphomas.

397

Loss of caspase-8 expression by hypermethylation confers resistance to death-receptor - or drug-induced apoptosis

S. Fulda, M. Küfer, B. Dockhorn-Dworniczak, K.M. Debatin (Ulm, Münster, D)

In addition to genetic alterations, epigenetic modifications such as DNA hypermethylation have recently been implied in loss of gene function and treatment resistance of tumor cells. Here, we report loss of caspase-8 expression by hypermethylation in a variety of resistant neuroectodermal tumor cells including Ewing tumor, neuroblastoma, malignant brain tumors, rhabdomyosarcoma and melanoma. In resistant tumor cells caspase-8 protein and mRNA expression was strongly reduced without caspase-8 gene loss. Methylation-specific PCR revealed hypermethylation of caspase-8 regulatory sequences. Treatment with the demethylation agent 5-Aza-2'-deoxycytidine reversed hypermethylation of caspase-8 resulting in restoration of caspase-8 expression and function thereby sensitizing cells for death receptor-, and importantly, also for drug-induced apoptosis. Similar to demethylation, introduction of caspase-8 by gene transfer sensitized for apoptosis induction. Hypermethylation of caspase-8 was linked to reduced caspase-8 expression in different neuroectodermal tumor cell lines in vitro and, most importantly, also in primary tumor samples. Thus, re-expression of caspase-8 by demethylation or caspase-8 gene transfer might be an effective strategy to restore sensitivity for chemotherapy- or death receptor-induced apoptosis in various resistant tumors in vivo.

398

Methylation of class II trans-activator promoter IV in HNSCC: A novel mechanism of tumour immune escape?

M. Meissner, T. Whiteside, C. Huber, P. van den Elsen, B. Seliger (Mainz, D; Pittsburgh, USA; Leiden, NL)

The expression of the major histocompatibility complex (MHC) class II molecules on the cell surface is required for the presentation of peptides to CD4+ T-lymphocytes. Lack of MHC class II-restricted antigen presentation function might play an important role in the evasion of head and neck squamous cell carcinoma (HNSCC) from immune recognition. The expression of the class II transactivator (CIITA), the major regulator of the MHC class II antigen processing machinery (APM), is controlled by four different promoters. Promoter III directs the constitutive expression, promoter IV the IFN-gamma inducible expression. In this study, the constitutive and IFN-gamma regulated expression of MHC class II APM components of HNSCC cell lines were examined by RT-PCR, Southern blot, gel shift assays, flow cytometry and promoter analysis. HNSCC cell lines revealed an extremely heterogeneous expression pattern of APM components, especially for HLA-DR alpha, -DM alpha, -DM beta and the invariant chain. The IFN-gamma mediated induction of these components appears to be quantitatively controlled by the CIITA expression level. Four HNSCC cell lines lack constitutive MHC class II surface expression. One of the MHC class II deficient cell lines further lacks constitutive and IFN-gamma inducible MHC class II APM component expression. This was caused by hypermethylation of the IFN-gamma responsive CIITA promoter IV, thereby preventing promoter occupancy and IFN-gamma inducible transcription, whereas the factors essential for CIITA expression were present and could activate reporter gene expression. Together, this is the first report of the silencing of CIITA expression in a tumour cell line by methylation of the CIITA promoter IV. Further studies are in progress revealing the frequency and clinical significance of such alterations in different tumor entities.

Wissenschaftliches Symposium: Transfusionsmedizin - Sicherheit von Blutprodukten

399

Risk of prion transmission by blood and blood products

J. Löwer (Langen, D)

It is well established that prion diseases (also known as transmissible spongiform encephalopathies, TSE) can be experimentally transmitted with CNS tissue of animals or humans which are in the clinical stage of the disease or in the third part of the incubation period. In humans, impressive examples are the transmissions of Creutzfeldt-Jakob disease (CJD) with human growth hormone prepared from cadaveric pituitary glands or with dura mater preparations. The question whether or not prion diseases can also be transmitted by blood is not yet completely answered. While there is no indication by epidemiological data that classical forms of CJD have been transmitted by blood or blood products experimental data obtained from laboratory animals hint to the possibility that the agent causing TSE might be present in peripheral blood at low concentrations (approximately 10 IU/ml). A variant form of CJD (vCJD) has been nearly exclusively observed in UK. It is most probably caused by the agent of bovine spongiform encephalopathy (BSE). Since the experience with this disease is not as extensive as with classical forms of CJD and since only in vCJD the pathological form of the prion protein can be detected in peripheral lymphatic tissues (like in laboratory animals) a precautionary approach should be taken assuming that infectivity might be present in the blood of vCJD patients in the clinical stage as well as during the incubation period. A number measures (e.g. exclusion of plasma collected in the UK from fractionation, exclusion of donors who stayed a minimum period of time in UK between 1980 and 1996) has been taken to reduce the theoretical risk. As realised recently BSE is more widely spread in continental Europe than believed before. Therefore, the possi-bility cannot be excluded that also outside the UK individuals may already have been infected with the BSE agent and may develop vCJD. In October 2000, a number of EU regulations (e.g. removal of specified risk mate-rial at slaughtering) came into force which will minimise or even eliminate the risk of BSE transmission from cattle to man. Therefore, it is expected that further primary transmissions will not longer occur. However, it is necessary to consider the possibility that secondary transmissions from man to man can take place. There is no doubt that diagnostic tests which are suitable for screening blood and plasma donors would allow the interrup-tion of such infection chains. As long as they are not available other interventions (e.g. exclusion of transfusion recipients from donation) have to be considered.

400

Bacterial contamination of blood products

P. Morel, M.F. Lecorte des Floris, L.Bardiaux, F. Pouthier, P. Hervé (Dijon, F)

Background: Blood component (BC) contamination may lead to severe Transfusion Reaction due to Bacterial Contamination (TRBC), originating from bacteria growth, the strain, the BC itself or of unknown origin. Origin: BC contamination occurs generally at the donation step. Venous puncture is often responsible for bacteria derived from the skin flora (2-4%), but donor infection and bacteremia are involved in the most severe cases. Mechanism : Bacteria growth represents the main TRBC factor, but a part of growth factors still remains unknown. TRBC severity is correlated to the rate of CFU/mL. Although the rate of donated blood or BC contamination is known, it is still difficult to assess the actual incidence of TRBC, as it is difficult to identify and relate this phenomenon to transfusion. Sepsis during transfusion is seldom related to blood product contamination, it is necessary to confirm 1) the sepsis, 2) BC contamination and 3) demonstrate the presence of the same bacteria in patient blood cultures and BC cultures. Bacterial analysis appear a great source of error and a possible impact of BC transfusion on a septic episode may be suspected. Thus transfusion may be involved in immediate infectious symptoms, related to soluble molecules like HLA class I or Fas-ligand. Findings: Through the haemovigilance system implemented in France in 1994, 185 cases of transfusion reaction due to the bacterial contamination of blood components have been reported over 5 years (1994-1999), thus confirming the frequency and severity of bacterial risks in transfusion medicine. In the French experience the incidence of bacterial contamination is 12.6/million homologous BC, 12.8/million Platelets Concentrates (PC) from whole blood, and 69.1/million apheresis PC versus 10.7/million Red cells Concentrates (RCC). The observed death rate is 0.7/million RCC, 1.1/million PC derived from whole blood and 12/million apheresis PC. In RCC, cocci (staphylococcus and streptococcus) are involved in 58%, Gram-negative bacilli in 32% and others in 10% of the cases. In PC, Gram-negative bacilli represent 36% of cases, cocci 42% and others 22%. Severity: from 1994-1999, 63.8% (118/185) of the cases were considered minor while 26.5% (49/185) represented vital threat. Eighteen (9.7%) had a fatal outcome. Four major observations could be drawn from BC bacterial contamination-related incidents: 1) the findings of the French haemovigilance network confirm the incidence of such reactions; 2) minor reactions are very frequent, prescribers' attention must therefore be drawn on the fact that fever or shivers are sometimes the sole symptoms, as suggested by M. Blajchman; 3) Gram-positive Cocci are very often responsible for RBC contamination: this is consistent with the rate of whole blood contamination by skin flora, as described by C. Bruneau et al; 4) we must improve investigations to study the role of transfusion in septic episodes and document the origin of BC contamination. Conclusion: Activation of the haemovigilance network and awareness-raising among transfusion actors have made it possible to prevent possible BC-related reactions (for several BCs derived from one donation) and improve case reporting. Continuous analysis, in real time, of the reported cases has improved prevention strategies at every stage of the transfusion chain, from collection to transfusion.

401

Yield Of HCV, HBV and HIV-1 Nat ccreening at the Red Cross Blood Transfusion Service Centers in Germany

E. Seifried, W. Roth (Frankfurt am Main, D)

Background: At present the reduction in the residual risk of hepatitis C (HCV), HIV-1, and hepatitis B (HBV) infection due to the introduction of NAT-Screening for viruses is mainly based on theoretical estimates. To define the actual residual risk of transfusion transmitted viral infections a prospective multi-center study on NAT-testing was launched by the research council of the German Red Cross.
Method: The survey is based on the analysis of a questionnaire requesting data on pre-NAT, NAT and post-NAT testing procedures, logistics and results of serological and NAT blood donor screening for HCV, HIV, and HBV. It includes allogeneic blood donations of first time and repeat donors collected at 12 Red Cross Blood Transfusion Service Centers (RCBSC) in Germany from January 1997 through December 2000. NAT testing for HCV, HIV-1 and HBV was performed in 12.7, 10.7 and 10.7 million donations respectively. During the investigation period 27 NAT-positive donations (HIV 1, HCV 11, HBV 15) were identified which were negative in corresponding serological tests (table). HBV follow up of 10 additional NAT-reactive, and seronegative donations is not yet completed. HBV NAT positive HBsAg negative donors were either pre-seroconverters or chronic anti-HBc positive low-level carriers. In contrast to the RCBSCs, many blood banks of the German Association of Blood Transfusion Services (GABTS) perform only HCV NAT since April 1999. Out of 2.3 million donations, 17 PCR-only positives could be detected, resulting in a rate of positives of 1/137.884 and a prevalence of 7.3 per million (IABs-Meeting Juni 2001). Conclusion: PCR contributes to the reduction of viral transmission by transfusion of blood components. However, the yield of contaminated units detected in the window period by NAT-screening of the RCBSCs is smaller than estimated by earlier calculations. The benefit-cost ratio of NAT at the RCBSCs of Germany is very low. It is imperative that the difference in the yield between the donation population of the German Red Cross and the GABTS must be be studied further.

402

Pathogen inactivation of blood components

H. Klüter (Mannheim, D)

Beside significant improvement in the evaluation of blood donors and serological and nucleic acid testing of blood products prior to transfusion, blood components continue to carry a substantial risk of transmitting pathogens. A new photochemical treatment (PCT) using psoralen S-59 in combination with a low dose of UVA light is now demonstrating an encouragingly effective inactivation of high levels of enveloped and non-enveloped viruses,

gram-positve and –negative bacteria, protozoa and leukocytes without affecting non-nucleiated blood components. To this extend platelet concentrates (PCs) and plasma are mixed with S-59 which reversibly intercalates into DNA and RNA. Upon UVA illumination S-59 reacts with nucleic acid, forming covalent monoadducts and crosslinks. After illumination, residual S-59 and free photoproducts are removed by a compound adsorption device. The results of several studies demonstrate, that PCT of platelet concentrates has a negligible effect on platelet function and viability compared with standard products. Clinical studies carried out in healthy volunteer and thrombocytopenic patients after chemotherapy showed no significant difference in post transfusion platelet recovery, platelet life span, template bleeding time and hemostasis when compared with a control group. Further studies are currently undertaken in the US and in Europe to further investigate the therapeutic efficacy and safety of psoralen treated blood components.
Furthermore, studies evaluating the effect of psoralen S-59 and UVA in fresh frozen plasma have been started, while a similar method for pathogen inactivation of red blood cell units (RBC) using psoralen S-303, is under development. This new photochemical inactivation technology, which can be easily incorporated into blood center manufacturing processes, may set a new standard in transfusion medicine.

403

Preemptive granulocyte transfusions (GTX) during neutropenia in patients with high risk of life threatening infections after stem cell transplantation

S. Hermann, M. Duchscherer, S.A. Klein, H. Bialleck, G. Bug, D. Hoelzer, H. Martin (Frankfurt, D)

Background: Preexisting fungal and other infections have been associated with a high risk of mortality in patients treated with stem cell transplantation. Aims: To reduce morbidity and mortality in pts. with preexisiting infections, we initated a prospective phase II study using preemptive granulocyte tranfusions (GTX) during posttransplant neutropenia. Patients and methods: Sixteen allografted pts. (14 HLA-id. sib + 2 MUD) and 1 autografted pt. (median age 57, range 34-63, years) received preemptive GTX. Sixteen pts. had high risk or advanced leukemia (10 AML, 4 ALL, 1 biphenotypic ALL, 1 CML-BC), and 1 pt. NHL. Eleven pts. had proven, 1 suspected previous fungal lung infections, 1 mucormucosis of sinuses, 1 assumed osteomyelitis, 2 older pts. with advanced leukemia received GTX with no evidence of preexisting infections but with anticipated high risk of transplant-related toxicity. Eight pts received dose-reduced conditioning regimens, 2 pts. standard-dose conditioning and 6 pts. additional 188-Rhenium-labelled anti-CD66 MoAbs. Amphotericin B and G-CSF was given during neutropenia. GTX were collected from relatives or friends after stimulation with G-CSF 5/µg/kg. Results: Patients received a median of 6 (range 2-12) GTX containing median 4.6 (range 0.2-22) x 10E10 neutrophils from median 2.5 (range 1-3) donors. GTX also contained medium 3 (range 0.9-6.4) x 10E11 platetets, corresponding to an average platelet transfusion. No severe adverse effects during GTX were documented. One patient (5%) died due to infection (candida sepsis) on d+9 during neutropenia. In all other pts. preexisting infections were controlled by GTX. Serial CT-scans showed decreasing sizes of lesions in all but one of 12 patients with radiololgical evidence of fungus. Nonrelapse mortality on d+100 was 3/17 (17%): in addition to the septic death on d+9 two pts. died on d+52 due to idiopathic IP and on d+85 due to GvHD. Late non-relapse mortality was seen in 2 pts due to bacterial sepsis on d+440 and d+531(non-compliance due to alcoholism). Eight pts. relapsed after a median of 96 (range 32-765) days and 4 pts. with preexisting pulmonary fungus are in CCR after a follow-up of median 781 (range 109-857) days. Conclusions: Preemptive GTX in combination with antifungal therapy prevents progression of fungal lesions during posttransplant neutropenia, resulting in an acceptable d+100 mortality. In addition, granulocyte transfusions save platelet transfusions. Long-term outcome in this high risk group of older patients with advanced leukemia appears to be related to other factors than preexisting fungal infections.

Wissenschaftliches Symposium: Gentherapie

404

Selection of gene-modified hematopoietic cells

C. Baum, Z. Li, J. Duellmann, B. Schiedlmeier, H. Klump, E. Will, J. Meyer, B. Fehse, A. Zander, C. Stocking, W. Ostertag, A. Ganser (Hannover, Hamburg, D)

Gene transfer into hematopoietic stem and progenitor cells offers promising perspectives in the treatment of malignant, infectious, or metabolic disorders. Improved techniques for retroviral transduction now allow efficient gene transfer in >10% of repopulating cells. However, the persistence and functional potential of individual transduced clones after engraftment and the maintenance of transgene expression within a clonal progeny is insufficiently investigated. Moreover, the inability to control the expansion of gene-modified cells in vivo, after transplantation, constitutes a significant limitation. In some applications, targeted elimination of transduced cells will be desirable in case of inappropriate side effects. Given this background, our work is concentrating on the development of retroviral vectors that mediate adequate levels of transgene expression in primitive hematopoietic cells and their differentiating progeny. We have shown that appropriate combinations of cis-regulatory elements can be defined for retroviral vectors to achieve a step-wise improvement of transcription and post-transcriptional transgene processing. A second goal of our research is to determine the conditions for and consequences of genetic engineering for controlled selection of transduced hematopoietic cells. We have developed suitable vectors to investigate the utility of MDR1 to confer hematopoietic resistance to a spectrum of cytotoxic agents during cancer chemotherapy, and to express HOXB4, the latter providing a spontaneous selective advantage. We have also introduced tCD34 for the phenotypic marking and selective enrichment of CD34 negative cells such as gene-modified T-cells. Long-term observations in mice suggest that tCD34, MDR1 or HOXB4 can be functionally expressed in hematopoietic cells, without severe alterations of multilineage hematopoiesis. Using tCD34, we could demonstrate that immunoaffinity-selection allows engraftment with gene-modified hematopoietic stem cells that show persistent multilineage gene expression in vivo. However, myelodysplasia with progression to AML M4 was observed in mice in association with high ectopic expression of another gene proposed for cell-surface marking, truncated LNGFR. Taken together, our data suggest that (1) mechanistic insights into vector-host interactions in the hematopoietic system can be obtained and (2) it may be possible to define a therapeutic window for some genetic marking and selection strategies which, however, requires long-term observations in appropriate in vivo models.

406

Retroviral mediated transgene expression after in vitro differentiation of mobilized human peripheral blood progenitor cells

H.G. Eckert, C. Lindemann, A. Schilz, A.A. Fauser, C. Baum, K. Kühlcke (Idar-Oberstein, Hannover, D)

Retroviral gene transfer is at present most frequently used to transduce hematopoietic cells for ex vivo gene therapy. In our previous work we have established a protocol for the genetic modification of human peripheral blood stem cells that is efficient (Schilz et al., Mol. Ther., 2000) and in compliance with current GMP requirements (certified by German authorities in February 2000). In in vitro studies we now evaluated EGFP expression mediated by a set of different retroviral vectors during lineage committed differentiation. Separated human CD34+ peripheral blood stem cells were transduced, sorted for transgene expression and subsequently differentiated to dendritic cells, granulocytes, monocytes and erythrocytes. We found a substantial decrease of EGFP positive cells during dendritic cell differentiation. Significant amounts of CD83+/EGFP+ cells could only be detected after transduction with vectors previously optimized for high expression in early hematopoietic stem/progenitor cells. In contrast, loss of transgene expression in other differentiated cell populations was only low. EGFP expression was detected in up to 97% of CD15+ granulocytes, in 94% of CD14+ monocytes as well as in 71% of the erythroblasts. In summary we show evidence for differently regulated transgene expression from identical vector backbones in different myeloid lineages. Results from these in vitro differentiation studies might help to determine the suitability of vectors for expression in relevant target cells.

407

Gene transfer of MGMT-P140K into hematopoietic cells: Hematoprotection from alkylating agents and selection of transduced cells

U.R. Sorg, M. Jansen, S. Ragg, D.A. Williams, M. Flasshove, S. Seeber, T. Moritz (Essen, D; Indianapolis, USA)

Gene transfer of drug resistance (CTX-R) genes has been advocated as a means to reduce chemotherapy induced myelosuppression. One CTX-R gene is methylguanine DNA-methyltransferase (MGMT), encoding a DNA repair protein specific for O6-alkyl lesions of guanine. O6-DNA alkylating agents used in anti-cancer chemotherapy (CTX), such as chloroethylnitrosotype drugs (ACNU, BCNU) or tetrazene derivatives (temozolomide (TMZ), dacarbazine) often have dose limiting hematologic side effects. While hematopoietic cells express wildtype (wt) MGMT only at low levels, some tumors show high endogenous levels of wtMGMT. Therefore, transfer of mutants resistant to O6-benzylguanine (BG), a specific inhibitor of wt-MGMT, such as MGMTP140K, look especially promising for use in hematoprotective strategies that combine MGMT gene transfer and application of BG and O6-alkylating agents. We evaluated this approach in a murine gene transfer model in which lethally irradiated mice were transplanted with bone marrow cells transduced with a retroviral vector expressing MGMTP140K in combination with green fluorescent protein, or with mock transduced cells. After hematological reconstitution mice were treated with 5 cycles of combined BG/ACNU, BG/BCNU or BG/TMZ application at weekly intervals (BG: 30mg/kg; ACNU 6mg/kg; TMZ: 70mg/kg; BCNU: 10mg/kg). While 26/26 control mice died of progressive myelosuppression during CTX, nearly all animals transplanted with MGMTP140K transduced cells and treated with BG/ACNU (12/15), BG/BCNU (5/5) or BG/TMZ (10/10) survived. After an initial decline, hematologic parameters in these animals improved already during CTX and pretreatment levels were reestablished during or shortly after CTX. In all animals enrichment of transgenic granulocytes (14-50fold) and lymphocytes (11-40fold) was detected in the peripheral blood. Also, the bone marrow showed substantial enrichment of transduced granulocytes (87.2%) and lymphocytes (67.7%), compared to 1.6% and 3.6% in untreated animals. In the spleen 63.3% transduced lymphocytes were detected in CTX-treated animals compared to 0.8% in untreated animals. Transduction of long term reconstituting cells in this model was demonstrated by serial transplantation i.e. transfer of chemotherapy resistance to recipients of secondary and tertiary transplants. Thus, these data demonstrate (i) that stem cells retrovirally transduced with MGMTP140K offer efficient protection from the hematotoxicity induced by different O6-alkylating agents and (ii) substantial enrichment of transduced cells occurs after combined BG/CTX treatment. This approach may be helpful for designing improved CTX protocols for tumors responsive to O6-alkylating and allow in vivo selection of transduced hematopoietic stem cells.

408

Epstein Barr Virus (EBV) and Adeno-Associated Virus (AAV) based vectors as novel tools for CLL-targeted gene therapy

C. Wendtner, C. Kurzeder, D. Kofler, H. Theiss, W. Hammerschmidt, H. Buening, M. Hallek (München, D)

Gene therapy for B-cell chronic lymphocytic leukemia (B-CLL) remains a challenge due to the low transduction efficiency and lack of cellular targeting of currently available vector systems. Because EBV has a natural tropism for B cells, we explored the characteristics of recombinant EBV vectors for transducing B-CLL cells. A packaging cell line (TR-2/293) which contains the entire EBV genome but lacks the terminal repeats (TR) allows encapsidation of nontransforming recombinant EBV vectors in the absence of EBV helpervirus. Recombinant EBV vectors encoding EGFP (EBV/EGFP) were generated after induction of the lytic cycle with an infectious titer up to $2x10^6$ per ml. Freshly isolated cells from fourteen patients diagnosed with B-CLL were successfully transduced with EBV/EGFP at low multiplicity of infection (MOI 0.2) resulting in transgene expression between 13.8 and 85.5% of cells 48hrs later. Transduction could be specifically abrogated by the monoclonal antibody (mAb) 72A1 which is directed against the viral envelope glycoprotein gp350/220. Due to its B-cell tropism and its high packaging capacity up to 150 kb, this EBV-based vector system holds promise for the development of cellular vaccines for B-CLL patients. Furthermore, recombinant adeno-associated virus (rAAV) was explored for CLL-targeted gene transfer. Using an optimized adenovirus-free packaging system, rAAV vectors coding for EGFP (AAV/EGFP) and CD40 ligand (AAV/CD40L) were packaged and highly purified resulting in genomic titers up to 3x1011 per ml. Cells obtained from 15 patients with B-CLL were infected with AAV/EGFP or AAV/CD40L at an MOI of 100 resulting in transgene expression in up to 48% of cells as detected by flow cytometry. Transduction with AAV/CD40L induced upregulation of the costimulatory molecule CD80 on noninfected bystander leukemia B cells, while this effect could be specifically blocked by an anti-CD40L mAb. Vaccination strategies using B-CLL cells infected ex vivo by rAAV vectors seem now possible in the near future.

409

Identification of target genes shared by histone deacetylase recruiting fusion proteins in AML using an oligonucleotide microarray approach

B. Steffen, T. Cauvet, C. Müller-Tidow, M. Ruthardt, B. Sargin, J. Worch, S. Klümpen, S. de Vos, H. P. Koeffler, W. E. Berdel, H. Serve (Münster, Frankfurt, D; Los Angeles, USA)

Acute myeloid leukemia (AML) is associated with fusion proteins that result from chromosomal translocations. These fusion proteins are thought to mediate one of the major features of acute myeloid leukemia - a block in myeloid differentiation. The translocation products AML1-ETO, PML-RAR-alpha and PLZF-RAR-alpha encode DNA-binding proteins that recruit co-repressors and histone deacetylases (HDACs) to their target genes. In order to gain insight into the mechanism of the differentiation block in AML, we sought to identify shared target genes of these fusion proteins. DNA microarray technologies allowed us to look at the global gene expression changes that result from a single abnormality. We used oligonucleotide arrays (HuFL 6800) to analyze the expression of more than 6500 genes in U937 cells inducibly transfected with AML1-ETO, PML-RAR-alpha or PLZF-RAR-alpha. Expression patterns obtained from U937 cells were compared to those from cells expressing the three fusion proteins with or without exposure to Trichostatin A (TSA), an inhibitor of histone deacetylases. 30 genes were significantly upregulated by all three fusion proteins. The expression of 60 genes was repressed by all three fusion proteins. Repression of some of these genes was relieved after HDAC inhibition by TSA. We used quantitative real time RT-PCR technology to confirm the gene expression data for 60 of the identified genes. Within the group of downregulated genes some are known to play a role in the process of myeloid differentiation and some upregulated genes are known to function as oncogenes. We will further analyze the functional effects of regulated genes that are involved in cell differentiation and proliferation, signal transduction, apoptosis and other pathways. Our results provide further evidence that all three fusion proteins share common target genes that lead to a block in differentiation and contribute to transformation in acute myeloid leukemia. Supported by German José-Carreras Leukemia Foundation.

Wissenschaftliches Symposium:
Spätkomplikationen nach Tumortherapie

410

The biological basis of normal tissue reactions to radiotherapy

W. Doerr (Dresden, D)

Normal tissue responses to radiotherapy are classified as acute effects or chronic sequelae, This classification is based on the time course, with acute reactions occurring within 90 days from onset of treatment. However, the majority of acute side effects on one hand, and of late responses on the other hand, follow common pathogenetic principles. Acute effects are typically seen in turnover tissues, like epidermis, oral mucosa or bone marrow, where normally proliferation precisely balances cell loss. Radiation impairs cell production, while differentiation and cell loss remain largely unchanged. This results in progressive hypoplasia and eventually complete depletion of the functional cell compartment. Restoration is based on surviving cells within the irradiated volume, or on cells from outside the field. During fractionated radiotherapy a compensatory tissue response is induced. These „repopulation" processes are based on a complex reorganisation of the proliferative tissue organisation. This includes loss of the division asymmetry, with production of two proliferating (stem) cells instead of one proliferating and one differentiating daughter, accelerated proliferation, but also residual abortive divisions of damaged cells. Based on these repopulation processes, the predominant factor affecting the severity of acute effects is overall treatment time. Accelerated radiotherapy, aiming at minimisation of tumour repopulation, is associated with aggravation of acute side effects. The pathogenesis of late radiation effects is more complex. Target cell populations not only include parenchymal cells, like type II pneumocytes in the lung, or kidney tubular cells, but also fibroblasts and vascular endothelial cells. Sterilisation of parenchymal cells results in impairment of organ function. Radiation damage to fibroblasts induces differentiation into postmitotic fibrocytes, which deposit substantially more collagen than fibroblasts. The increase in extracellular matrix in consequence further affects organ function. Moreover, vascular endothelial changes lead to impairment of blood supply and hence contribute to organ dysfunction. The relative importance of these three processes varies between individual organs and tissues. Late radiation effects are progressive and irreversible. Dose fractionation, i. e. size of dose per fraction, is the most prominent parameter of radiation tolerance for late responding tissues, because of their marked potential for recovery of radiation damage. This is the biological basis for hyperfractionated radiotherapy. Interaction between acute and late effects of radiotherapy result in consequential late effects. In these cases, the acute response results in breakdown of a barrier against mechanical and/or chemical stress. This inflicts additional trauma to target structures of the (generic) late effects. In some organs, e. g. urinary bladder, the consequential component is the predominant mechanism of the late response.

411

Long-Term sequelae after the treatment for Hodgkin's lymphoma

A. Engert, V. Diehl, K. Breuer (Köln, D)

Hodgkin's lymphoma has become a curable disease after the introduction of improved radiation techniques and combination chemotherapy regimen such as MOPP, COPP or ABVD. Depending on the stage at diagnosis and other risk factors, 60 – 90 % of patients achieve long-term remission. However, numerous sequelae such as hypothyreosis, immune dysfunction, pneumonitis, infertility, fatigue and secondary neoplasias can develop as long-term consequence of the initially successful treatment. The risk of secondary neoplasias has been of particular concern, since most secondary neoplasias are incurable with standard treatments. After 15 year of follow up, a total of 17,6 % of patients developed secondary malignancies in a survey from 1985. Of these, 13,2 % were solid tumors, 3,3 % acute leukemias and 1,6 % Non-Hodgkin's lymphoma. Risk factors for the development of secondary neoplasias include age, smoking, combined modality, the use of alkylating agents and topoisomerase II inhibitors. One area of particular recent concern is the gonadal toxicity which is related to the type of chemotherapy given and the number of cycles applied. The infertility rate varies between 33 % (ABVD) and 100 % (MVPP) with recovery rates from 0 – 100 %. Most recent data from the German Hodgkin Study Group indicate that recoveries are possible as late as five years after the initial chemotherapy. Interestingly, only 29 % of 158 male patients with diagnosis of Hodgkin's lymphoma show a normal semen analysis in their pre-treatment fertility status.

412

Therapy-related myelodysplastic syndrome (t-MDS) and leukemia (t-AML) following chemotherapy treatment of patients (pts) with potentially curable malignancies

C. Kollmannsberger, C. Bokemeyer (Tübingen, D)

T-MDS and t-AML are considered to be the most serious late complications of cytostatic treatment in pts originally cured of their primary disease. Whereas secondary solid tumors are predominantly related to radiation therapy, chemotherapy is mostly associated with t-MDS and t-AML. Regardless of the primary illness, t-AMLs exhibit certain characteristics which differentiate them from spontaneously occurring leukemias. Two main forms of t-AMLs can be distinguished, one following therapy with alkylating agents, e.g. oxazaphosphorins, melphalan but also cisplatin, and one following treatment with topoisomerase-II inhibitors, e.g. epipodophyllotoxins and anthracyclins. T-AMLs usually occur within the first 5-10 years after therapy, while incidence rates plateau thereafter. The risk for developing t-AML appears to be related to the cytostatic agent used, the cumulative dose applied and to the application schedule. Larger investigations on t-AMLs have been performed in pts treated for Hodgkin´s disease, breast cancer, germ cell cancer as well as non-Hodgkin´s lymphoma (NHL). Pts treated for Hodgkin´s disease have an increased risk of 20-40 fold depending on the cytostatic agents applied. Adjuvant cyclophosphamide/methotrexate/5-FU (CMF) for high-risk breast cancer has a low risk of t-AML (5/10000 pts treated), while an increased incidence has been observed after anthracycline-based chemotherapy particularly when mitoxantrone is used. In addition, a high incidence of 3% has been reported after sequential, dose-intensified anthracyclin-based chemotherapy. A total dose of etoposide ^{2} 2g/m^2 is associated with a cumulative incidence of t-AML of 2% at 5 years in pts with germ cell cancer, whereas the incidence is only 0.5% with cumulative doses < 2 g/m^2. High cumulative incidences up to 24% at 43 months for developing t-MDS/t-AML have been reported following high-dose chemotherapy (HD-CT) with autologous stem cell support for NHL. Several investigations indicate that the intensive conventional chemotherapy prior to HD-CT with autologous stem cell support may be the most important factor in the etiology of t-AML following HD-CT for NHL. The molecular mechanisms in the development of t-AML are still not completely understood. The few available data suggest different genetic pathways and the involvement of p53 mutations, microsatellite instability, altered glutathione-S-transferase levels and polymorphism of cytochrome p4503A in the leukemogenesis of t-AML. Pts suffering from t-AML exhibit a very unfavorable prognosis and long-term survival is unlikely. Allogenic bone marrow transplantation appears to be the only potentially curative treatment option and should be considered early in the course of the disease. In the future, a more precise insight into the molecular mechanisms of leukemogenesis may identify pts at increased risk already at the time of primary tumor treatment. This may allow the design of individual therapies or preventive measures.

413

Occurrence of leukemia / MDS in donor cells after allogeneic stem cell transplantation - an evaluation of the European Group for Blood and Marrow Transplantation

B. Hertenstein, L. Hambach, D. Niederwieser, T. Ruutu, T. Ruutu, T. Ruutu (Hannover, Leipzig, D; Helsinki, FIN)

The occurrence of leukemia or myelodysplasia in donor cells (DCL) after allogeneic transplantation is only rarely described, but may be more common than reported. To evaluate the characteristics, frequency, and treatment options of DCL a questionnaire evaluation within the EBMT was performed. 91 centers responded and a total of 12 cases of DCL occurring between 1982 and 2000 were identified. 5 of these cases were previously published and 7 „new“ cases were reported. 7/11 cases were AML (org. disease: AML 3, ALL 1, SAA 2, CML 1), 2 ALL (AML 1, ALL1), 2 MDS (AML) and 1 CML (CML). The donor cell origin of the ''new'' leukemias was demonstrated by cytogenetic analysis in 6 cases. In 3 of these cases male cells with additional cytogenetic abnormalities were found in female recipients, in two patients the Y chromosome was directly demonstrated by FISH in blast cells in a female recipient. In 7 patients the donor cell origin was demonstrated by VNTR or STR analysis. Median age of the patients at diagnosis of DCL was 28 ys. (13-54). The interval between SCT and diagnosis of DCL ranged between 4 month and 14 years (median 13 m). The allografts were BM in 11 and PBPC in one case. In 10 cases the donors were HLA-id. siblings, in 1 case a HLA-id. cousin and in one patient DCL occurred after transplantation from a matched unrelated donor. The occurrence of leukemia in the donors was not seen in any case. 5 patients received MTX as part of their GvHD prophylaxis and one patient received 6 doses of intrathecal MTX after transplant. No additional treatment with cytotoxic drugs was reported in the other patients. All but one patient received treatment for their DCL. 5/9 patients entered a remission after chemotherapy, 2 died of complications and two of progressive disease. One patient remained in remission after chemotherapy alone, one patient relapsed. Five patients underwent a second transplantation. Two of these patients are alive 5 and 11 months after a second transplant from an alternative sibling and unrelated donor, respectively. From this data it can be concluded that DCL is an extremely rare event after allogeneic SCT. However, treatment of this complication with conventional chemotherapy and retransplantation might be successful.

414

Long term toxicity and health-related quality of life in patients treated with hematopoetic stem cell transplantation for solid tumors and hematologic malignancies

A. Oetzel, C. Oetzel, S. Martin, H. Einsele, W. E. Aulitzky (Stuttgart, Tübingen, D)

Aims: To evaluate long term toxicity, health-related quality of life (HRQOL) and Fatigue in patients in continuous complete remission after high-dose chemotherapy followed by autologous stem-cell transplantation (ASCT) for solid tumors and hematologic malignancies. Patients and Methods: A retrospective analysis was performed on patients treated with high-dose chemotherapy followed by autologous stem-cell transplantation for cancer and hematologic malignancies and who had no relapse of their disease at least two years after transplantation. Toxicities and late complications were evaluated by review of patients' charts 0.5, 1, 1.5 and 2 years

after transplantation. In addition, patients were asked to complete the questionnaire of the European Organization for Research and Treatment of Cancer QLQ-C30 and the Multidimensional Fatigue Inventory (MFI). Results: 112 Patients, 58 with solid tumors and 54 with hematologic malignancies were included in the study. 84 (75%) of the 112 Patients, 47 (81%) with solid cancer and 36 (66%) with hematologic malignancy completed the questionnaire. The median time between transplantation and response to the questionnaires was 2.9 years (range 1.6 to 7.2). At two years after transplantation few toxicities according to the World Health Toxicity criteria were noted: 72/112 (75%) had no signs or symptoms. Major complaints in the remaining patients were depression/fatigue 9/112 (8%) myalgia/arthralgia 10/112 (8.9 %), dry mouth/eyes 3/112 (2.6%) and alopecia 2/112 (1.7%). 1/112 (1%) patient reported sexual dysfunction. In contrast, the QLQ-C30 and MFI questionnaire showed a pronounced impairment of the emotional functioning (score 65, score range 1-100) than of the physical functioning (score 81, score range 1-100) . These values were considerably lower than those reported for normal individuals of the same age groups and comparable to those reported for other cancer patients. The most pronounced alterations were seen in items related to sexual functioning with scores in the range of 50. 87% patients reported disease- and treatment-related changes in their sexual functioning.
Conclusion: High dose chemotherapy followed by stem-cell transplantation for solid tumors and hematologic malignancies is well tolerated causing few severe long term complications. Patients in complete remission report ongoing impairment of emotional and sexual functioning in QOL questionnaires. In addition, our data indicate that sexual dysfunctions are frequently missed in routine history taking in out-patients clinics.

Wissenschaftliches Symposium:
Morbus Hodgkin

415

New aspects in biology and pathology of Hodgkin's lymphomas

S. Poppema, E. Maggio, A. van den Berg (Groningen, NL)

Hodgkin lymphomas (HL) are characterized by the presence of a small proportion – usually less than 1% - of neoplastic cells, the so called Reed-Sternberg cells and their mononuclear variants (R-S cells) and a majority of reactive cells, mostly lymphocytes admixed with histiocytes, plasma cells and eosinophils. As such, HL constitutes an extreme example of a tumor with an immune response that is not effective and may even aid the tumor cells. The R-S cells are generally considered to be clones of 'crippled' - germinal center derived - neoplastic B cells that have many features of antigen presenting cells. There are two HL entities, nodular lymphocyte predominance HL (NLPHL) and classical HL, that includes mixed cellularity (MC), nodular sclerosis (NS), lymphocyte rich and lymphocyte depleted subtypes. In NLP HL small polyclonal B-lymphocytes predominate and the R-S cells are surrounded by CD4+, CD57+ T cells. The NS subtype is characterized by predominance of CD4 T-cells and by the presence of collagen rich sclerotic bands, whereas in the MC subtype also CD8 T-cells are present and sclerotic bands are lacking. In both subtypes an admixture of eosinophils, histiocytes and plasma cells is typically present. Recent studies provide evidence that R-S cells produce factors that may explain the characteristic inflammatory infiltrate in the affected tissues of Hodgkin lymphoma. R-S cells of classical HL produce TARC as well as MDC, two chemokines that attract activated Th2 lymphocytes that express the CCR4 chemokine receptor. This finding may explain the preferential influx of TH2 type T-cells. In addition, R-S cells frequently produce TGF-beta and/or IL-10, immunosuppressive cytokines, that suppress Th1 type responses. The production of the various cytokines and chemokines may result from the upregulation of NF-kB in the R-S cells. The immunophenotype of the lymphocytes surrounding the R-S cells is indeed consistent with anergic and/or TH2 type T-cells. Therefore, these cells do not support a cytotoxic anti tumor response. Since the R-S cells are neoplastic B-cells, the cytokines produced by these T-cells may in fact help their growth and/or survival. Moreover, the TH2 type cytokines produced by the R-S cells may induce the lymphocytes to produce similar cytokines, leading to an autocrine cytokine loop. The production and induction of various other cytokines may also explain the influx of eosinophils (IL-5, eotaxin) and plasma cells (IL-6). Differences in chemokine and cytokine production may be responsible for the differences between the histological subtypes. In conclusion, by secreting potent cytokines and chemokines R-S cells may not only promote their own growth and evade immune surveillance, but also cause the characteristic histology and symptoms of HL

417

Treatment of Hodgkin's lymphoma: The EORTC studies

E. Noordijk, P. Carde, M. Henry- Amar, J. Meerwaldt, J. Thomas, H. Eghbali, J. Raemaekers, C. Fermé, B. Aleman, J. Kluin-Nelemans (Leiden, NL)

Since 1964, the EORTC Lymphoma Group has been involved in developing treatment strategies in Hodgkin's lymphoma (HL). In the first four consecutive controlled trials for stages I and II, called H1, H2, H5 and H6, the main objective was to improve overall survival (OS). The respective roles of adjuvant vinblastine, splenectomy, spleen irradiation, (sub) total nodal irradiation, MOPP, ABVD, and staging laparotomy were prospectively assessed. Prognostic factors were identified, which were used to stratify patients into different prognostic subgroups called favourable and unfavourable. Treatment intensity was adapted to those prognostic groups. From 1988, trial strategies changed. The main objective became the reduction of (late) treatment toxicity, replacing 'standard' radiotherapy by smaller fields and lower doses, and reducing intensity and duration of chemotherapy. In the following three consecutive trials, called H7, H8 and H9, the subjects studied were: minimal or no radiotherapy, EBVP, MOPP/ABV and BEACOPP. Despite reduction in treatment aggressiveness in the second period, event-free survival (EFS) and overall survival (OS) rates have been higher in every consecutive EORTC trial for stage I and II:
8-yr EFS: H1 48%, H2 66%, H5 68%, H6 77%, H7 78%. H8 (5-yr) 86%
8-yr OS: H1 71%, H2 83%, H5 84%, H6 88%, H7 89%, H8 (5-yr) 94%
Since 1981, stages III and IV HL have also been studied in two consecutive trials called HIIIB/IV and H34. In the first trial (1981-1986), 8 courses of MOPP proved inferior to 2 MOPP / 2 ABVD / 2 MOPP / 2 ABVD. In the second trial (1989-2000) the usefulness of involved field radiotherapy (IF RT) after 6-8 courses of MOPP/ABV hybrid was tested. The 6-yr EFS was 76% and the 6-yr OS was 82%, both for patients in CR after chemotherapy randomised for no further treatment or IF RT and for (irradiated) PR patients. We conclude that, despite reduction in treatment aggressiveness, EFS and OS rates have improved for all stages of HL. However, after almost four decades of clinical research there is still no standard therapy for patients with either favourable, unfavourable or advanced disease. Clinical trials with very long follow-up are needed to establish the best treatments with optimal survival rates and minimal late toxic effects in each of the different prognostic subgroups.

418

New approaches to the treatment of relapsed Hodgkin's disease

A. Engert, A. Josting, V. Diehl (Köln, D)

In order to evaluate new treatment strategies for patients with relapsed Hodgkin's disease, we conducted a large phase II study, in which more than 100 patients with relapsed and refractory Hodgkin's lymphoma were enrolled in a multicentre setting. Treatment consisted of two cycles of DHAP, followed by a sequential highdose (hd) chemotherapy in those patients responding (hd cyclophosphomide, hd methotrexate, hd etoposide). This was followed by one cycle of BEAM supported by peripheral stem cells. The latest interim analysis of this study included 69 patients of which 16 were primary refractory and 53 were relapsed. The overall response was 80 % in relapsed patients and 50 % in those with primary refractory disease. Toxicity was acceptable with mucositis being the major non-hematological toxicity. Taken together, these data demonstrate that the protocol is feasible and very effective in a large multicentre setting. Since the results of the HDR1 study of the German Hodgkin Study Group demonstrated superior tumor free survival in relapsed chemosensitive patients treated with two cycles of chemotherapy followed by BEAM, these data formed the rationale for the current prospectively randomised relapsed study (HDR2). This study is conducted together with the EORTC and some EBMT centres. Patients with early and late relapsed Hodgkin's disease are randomised between the described protocol and a reduced version consisting of two cycles of DHAP followed by BEAM. The study has started recruiting patients earlier this year and has enrolled 24 patients by June 2001.

419

Incidence, treatment and prognosis of secondary leukemia and secondary myelodysplastic syndrome in patients treated for Hodgkin's disease in the German Hodgkin Study Group (GHSG)

A. Josting, S. Ansén, M. May, J. Franklin, M. Sieber, J.P. Glossmann, J. Wolf, A. Engert, V. Diehl (Köln, D)

Purpose: Depending on stage and risk factor profile, more than 80% of patients with Hodgkin`s disease (HD) can be cured with front-line treatment. However, long-term survivors of HD are at risk for late complications. Among these, secondary malignancies are the most serious since they are often fatal. The most widely reported secondary malignancies are solid tumors and acute leukemia. The aim of this retrospective analysis is to assess the incidence of secondary leukemias and myelodysplastic syndromes in Hodgkin disease patients who were treated in the last 20 years according to the protocols of the German Hodgkin Study Group (GHSG). Patients and Methods: From 1981-1998, the GHSG conducted three generations of clinical trials for the treatment of primary early, intermediate and advanced stages HD involving a total of 5411 patients (HD1-HD9). Data on incidence, treatment and outcome of secondary AML/MDS were updated in June 2000. Results: Altogether 45 patients were documented with secondary AML/MDS within 5411 patients, treated in the above mentioned trials. The median age at leukemia diagnosis was 47 years (range 22-79 years). Primary therapy was as follows: radiotherapy alone n=3; ABVD n=1; COPP/ABVD-like n=30; BEACOPP baseline n=2; BEACOPP escalated n=9, 12 pts. developed AML/MDS after salvage therapy for relapsed HD including HDCT/ASCT in 4 pts.HDCT/ASCT. 35 of the secondary malignancies were AML/ALL, 10 were MDS. The majority of secondary AML/MDS occurred in the first 5 years after completion of the initial therapy for Hodgkin's disease. After a median observation time of 55 months (range 0-160 months) incidence of secondary AML/MDS was 1 % (95% CI. 0.6-1,2%). Treatment for secondary AML/MDS was as follows: high-dose Ara-C containing regimens (TAD/HAM, HAM, Ida-Ara-C, Ida-Flag and ICE+HAM) n=11, TAD-chemotherapy n=5, several other regimens n=3, palliative oral chemotherapy n=3, allogeneic SCT n=9. After 24 months of observation FFTF was OS was 2% and 8%, respectively. There was no difference in the overall survival rate between patients who developed AML versus patients who developed MDS. Conclusion: The incidence of secondary AML/MDS in the GHSG trials HD1-HD9 is 1% at 55 months. The prognosis of these patients is dismal with only 1 patient surviving more than 5 years.

Wissenschaftliches Symposium: Zellzyklusregulation und Apoptose

420

CD95(Apo-1/Fas)-mediated apoptosis: Signaling and diseases

P. Krammer (Heidelberg, D)

CD95, a member of the tumor necrosis factor (TNF) receptor superfamily induces apoptosis upon receptor oligomerization. The receptor and its ligand are important for apoptosis of peripheral T cells, for downregulation of an immune response and most likely, at least in part, also for peripheral T cell tolerance. In Aids, apoptosis mediated by this system might contribute to the depletion of T helper lymphocytes. Likewise, in diseases in which liver cells are destroyed the CD95 system might play a major role. In a search to identify the intracellular signalling pathway of CD95 several molecules coupling to oligomerized CD95 were immunoprecipitated from apoptosis-sensitive human leukemic T cell and lymphoblastoid B cell lines. The following binding molecules were only associated with aggregated and not with monomeric CD95: phosphorylated FADD (MORT1) and caspase 8. Thus, caspase 8 was identified as the most CD95 receptor proximal protease which starts the cascade of protease reactions important for CD95-mediated apoptosis. Association of FADD and caspase 8 with CD95 was not observed with C-terminally truncated non-signalling CD95. FADD and FLICE did also not associate with a CD95 cytoplasmic tail carrying the lprcg amino acid replacement. FADD and caspase 8 form a death-inducing signalling complex (DISC) with the CD95 receptor and are, thus, the first CD95 associating proteins of a signalling cascade mediating apoptosis.The function of the DISC is discussed in detail, particularly with respect to its role in sensitivity and resistance to apoptosis.

The CD95 death system plays a role in destruction of liver tissue. In hepatitis cytotoxic T lymphocytes might use the CD95 system to kill infected hepatocytes. In M. Wilson copper overload leads to upregulation of the CD95 ligand that may finally contribute to acute liver failure. In HCC from patients treated with chemotherapeutic drugs the CD95 receptor and ligand are upregulated and may contribute to apoptosis of the tumor or, dependent on the drug sensitivity of the tumor, to the status of the tumor as an immunoprivileged site.

421

Genetic control of apoptosis and long-term cell cycle arrest in B-cell lymphoma in vivo

C. Schmitt (Berlin, D)

Treatment failure and ultimately death from cancer, particularly from hematological malignancies, are - at least in part - determined by genetic programs encoding cellular mechanisms of growth control. Since most anticancer agents can induce cell cycle arrest or apoptosis, mutations of genes involved in cell cycle regulation and programmed cell death might contribute to chemoresistance. Moreover, because apoptosis and senescence are cellular fail-safe mechanisms to counter oncogenic transformation, mutations in these programs occur frequently in tumor development, hereby potentially co-selecting for treatment resistance even prior to any drug encounter. Mutations of the tumor suppressor gene p53 or deletions of the complex INK4a/ARF locus are among the most frequently detected alterations in B-cell neoplasia. p53 can induce apoptosis, cell cycle checkpoints, cellular senescence, differentiation, and can suppress angiogenesis. The INK4a/ARF gene product p16 is an indirect inhibitor of the retinoblastoma protein-mediated G1/S phase progression and has been implicated in cellular senescence. ARF, encoded in alternate reading frame by the INK4a/ARF locus as well, acts as a positive upstream regulator of p53. Consequently, loss of p53 function can result in improved cellular survival, increased proliferation, immortalization, genomic instability, neo-angiogenesis and metastasis. Mutations of genes acting in the p53 pathway or controlling similar biological functions might be able to phenocopy to some extent the contribution of p53 mutations to tumorigenesis and treatment outcome. We sought to address the impact of mutations in the p53 and INK4a/ARF locus on tumor biology and chemosensitivity in a murine lymphoma model in vivo. The Eμ-myc transgenic mouse develops B-cell lymphoma with genetic and histological similarity to the human Burkitt's lymphoma. Crossbreeding to 'knock-out' mice or retroviral gene transfer into lymphoma cells or hematopoietic stem cells provide the opportunity to study the role of defined genetic lesions in tumorigenesis and therapy. Using a genetic approach to disrupt programmed cell death by strictly anti-apoptotic lesions such as Bcl-2 or a dominant-negative Caspase 9, we were able to dissect p53's effector functions in tumor suppression. While disruption of apoptosis provides the selective pressure to mutate p53 during lymphoma development, aneuploidy and cell cycle checkpoint dysregulation seem to be nonessential byproducts of p53 inactivation. In contrast, not only compromised apoptosis, but loss of p53's ability to control cell cycle arrest is critical for chemoresistance. In the presence of wild-type p53 genes, the two INK4a/ARF gene products p16 and ARF cooperate to control drug-induced apoptosis and long-term arrest. Genetically dissected in the Eμ-myc lymphoma model in vivo, loss of ARF primarily inactivates apoptosis whereas additional loss of p16 dramatically reduces treatment outcome presumably due to impairment of a drug-mediated arrest program. We expect that a genetic dissection of the functions of p53 and INK4a/ARF gene products with particular emphasis on apoptotis and cell cycle arrest in vivo will help to develop novel treatment strategies targeting essential features of tumor viability, progression and treatment resistance.

422

The Rb-pathway of cell cycle regulation and the mitochondrial apoptosome: Relevance for tumor biology, resistance to therapy and clinical prognosis

P. Daniel (Berlin, D)

The development of cancer is a multifactorial process and the result of alterations that affect different cellular machineries resulting ultimately in deregulated proliferation. Therapeutic strategies aim to eliminate all cancer cells. Virtually all antitumor therapies (except surgery) act via the induction of apoptosis and utilise to this end the endogenous death machinery. It is therefore not surprising that most therapeutic strategies fail to ultimately cure the malignant disease due to the selection of resistant clones. Deregulation of cell cycle control by the Rb-pathway is a frequent event in cancer. Transgenic mice with cyclin D1 overexpression and knock-out mice with inactivation of p53, p16INK4a, or p14ARF or other key genes in this signalling network show an increased rate of spontaneous tumor development. Multiple signaling events closely intercalate cell cycle regulation and cell death control. In cytotoxic cancer therapy, both cell cycle programmes and cell death signalling cascades are targeted which ultimately converge at the level of the mitochondrial apoptosis signaling cascade. A prominent example is the p14ARF/mdm-2/p53/Bax/APAF-1 signaling pathway which is a key regulator of cell death induced by cytotoxic therapies. The mitochondrial activation during apoptosis is inhibited by Bcl-2, Bcl-xL and other apoptosis-preventing members of the Bcl-2 gene family. In contrast, apoptosis-promoting members of this gene family, such as of the Bax-subfamily (Bax, Bak and Bok) have been demonstrated to be direct activators of the mitochondria. Both upregulation of Bcl-2 or loss of Bax is associated with development of therapy resistance in many types of cancer. Members of the Bcl-2 family, the mitochondrial protein cytochrome c and the cytosolic APAF-1 recruit the initiator caspase procaspase-9 and thereby form the mitochondrial apoptosome complex to activate the downstream effector caspase cascade and execution of cell death. In human cancer, a variety of studies have identified single components of the Rb pathway as prognostic factors. In esophageal squamous cell carcinoma, we previously observed that intact p16INK4a gene expression is associated with good outcome (J Clin Oncol 2001,19:2272). The detailled analysis of genes such as pRb, p16INK4a, p27KIP1, Cyclin D1, cyclin E, and others allows to refine the discrimination between patients at risk for early death and patients with a good prognosis after therapy. The prognostic power can be further increased by the inclusion of cell death regulators like p53 and Bax in such a multigene analysis (J Clin Oncol 1999,17:1364). Thus, the analysis of these pathways can be useful for individualisation of risk-adapted therapeutic strategies in the future. The analysis of these molecular events is the basis not only for the understanding of tumor pathogenesis but also of resistance mechanisms which impair the response to cytotoxic therapies. Furthermore, this also provides the basis for the development of novel therapeutic targets in cancer treatment. To this end, strategies for a genetic approach in the therapy of resistant carcinoma cells will be presented.

423

A transcription-independent pathway for p53-induced apoptosis

M. Schuler, D.R. Green (Mainz, D; San Diego, USA)

The understanding of the molecular pathways of p53-mediated apoptosis is still incomplete. In several systems the involvement of transcription-dependent and –independent mechanisms has been described. Here we show in cancer cell lines and in primary fibroblasts from wild type, p53-deficient and bax-deficient mice that p53 induces apoptosis by death receptor-independent caspase activation. This activity requires pro-apoptotic Bcl-2-family members, such as Bax, which upon activation by p53 alters its conformation and translocates to the mitochondria to trigger cytochrome c release. p53-induced Bax translocation and cytochrome c release can occur in the absence of protein synthesis, in cytoplasts devoid of nuclei, and in a cell-free reconstituted system encompassing isolated murine liver mitochondria. Using primary fibroblasts from p53 mutant „knock in" mice, or fibroblasts retrovirally transduced to express p53 mutants, we find that this transcription-independent pro-apoptotic activity of p53 in vivo is invariantly regulated by a transcription-dependent and Mdm-2-dependent mechanism. In conclusion, the interplay between transcription-dependent and –independent activities of p53 contributes to the engagement of the apoptotic machinery in response to stress stimuli.

424

SMAC sensitizes resistant tumor cells for death receptor or drug-triggered apoptosis

S. Fulda, K.M. Debatin (Ulm, D)

Despite aggressive therapies resistance of many tumors to current treatment protocols still remains a major concern in cancer therapy and may be caused by defective apoptosis programs. Here, we report that Smac can bypass the Bcl-2 inhibition and sensitizes resistant tumor cells for death receptor or drug-induced apoptosis. Transfection-enforced overexpression of Bcl-2 inhibited release of cytochrome c, AIF or Smac from mitochondria into the cytosol thereby blocking apoptosis following treatment with anti-CD95, TRAIL, TNFa or cytotoxic drugs. Interestingly, Bcl-2 overexpression did not prevent cleavage of caspase-8 or Bid, but inhibited cleavage of caspase-9, -7, -3 or PARP indicating that apoptosis was blocked at the level of effector caspases, e.g. by inhibitor of apoptosis proteins such as XIAP. Overexpression of a cytosolic form of Smac without mitochondrial targeting sequence bypassed the Bcl-2 inhibition and sensitized for death receptor and drug-induced apoptosis by neutralizing the inhibitory function of XIAP. In addition, a N-terminal peptide of Smac similarly sensitized for apoptosis. Sensitization for apoptosis was found in various different tumor cell lines and primary tumor cells derived from neuroblastoma or malignant brain tumors. Thus, Smac gene transfer or N-terminal Smac peptides may be novel strategies to bypass resistance and to sensitize resistant tumor cells for death receptor and, most importantly, also for drug-induced apoptosis.

425

Cloning of FOCA1 (Friend of cyclin A1), a novel cyclin A1 interacting protein

S. K. Metzelder, T. Cauvet, W. Wang, A. Westermann, G. Idos, S. Diederichs, M. Möller, H. Serve, W. E. Berdel, C. Müller-Tidow (Münster, D)

Human cyclin A1 is highly expressed in testis and acute myeloid leukemia. It is thought to play an important role in cell cycle progression by interacting with Rb, E2F-1 and B-myb. cyclin A1 overexpression was shown to induce leukemia in transgenic mice. To get more insight into the functions of Cyclin A1, we screened $2x10^6$ independent cDNA clones for interacting partners using a yeast-triple-hybrid approach with conditional CDK2 expression and Cyclin A1 as bait. Several partner candidates were identified and a novel protein - Friend of Cyclin A1 (FOCA1) - was confirmed to interact with cyclin A1 in vitro and in vivo. The full length cDNA was cloned by 5'RACE. The mRNA consists of 1225 base pairs and contains a single ORF that encodes a 242 amino acid protein. The protein is not homologous to any other known protein. The foca1 gene is located on human chromosome 17, spans a 10kb region and contains 6 introns. FOCA1 is most highly expressed in the testis while lower expression was found in ovary and spleen. Nuclear localization of FOCA1 was found by fluorescence microscopy after transient transfection of a GFP-FOCA1 fusion construct. A CpG island was found directly upstream of the transcriptional start site. By cloning the upstream fragments into luciferase-reporter vectors we demonstrated that this region possesses promoter activity. Further experiments showed that the FOCA1-promoter depends on the Sp1 transcription factor. Currently, we are performing experiments to further elucidate the functional role of the FOCA1 and cyclin A1 interaction.

Wissenschaftliches Symposium: Biologie und Behandlung von metastasierenden Krebskrankheiten

426

Pathology of CUP-syndrome

C. Wittekind (Leipzig, D)

The CUP-syndrome is defined as metastasis of a primary tumour which cannot not be localised by clinical routine examinations. This syndrome is considered a special disease since specific diagnostic and therapeutic strategies are necessary for adequate treatment of the patients. 2 - 4% of patients with malignant tumours present with a CUP-syndrome and even 5 - 10% of patients in oncological centres. It is most frequent in the 6th decade. At the time of diagnosis the CUP-syndrome corresponds to a disseminated disease (80% of patients in stage IV, 20% with a solitary metastasis [50% solitary lymph node metastasis, 50% solitary organ metastasis]. In the pathohistological diagnosis it is of utmost importance to delineate tumours which are potentially curable from those which are not. The former group comprises germ cell tumours and malignant lymphomas. Especially in the diagnosis of tumours in the liver, the precise diagnosis of intrahepatic cholangiocarcinoma is important, since it could be treated with curative intention. In the second group there are some tumours which may have a fairly good prognosis, e.g., prostate or breast carcinoma, and other which will lead to a rapidly downhill course of the patient. The diagnosis of a primary tumour has been facilitated by using immunohistochemical methods which in certain tumours will enable pathologists to exactly indicate the origin of the primry tumour (e.g., thyreogobulin in differentiated carcinoma of the thyroid, prostate specific antigen in adenocarcinoma of the prostate). The use of other antibodies, e.g. cytokeratins, will at least show a certain direction for clinical search of the primary tumour. Some criteria must be considered by the pathologist when dealing with CUP-syndrome patients: localisation of metastasis, histology (malignant melanoma versus sarcoma versus carcinoma with different subtypes, e.g., adenocarcinoma, squamous cell carcinoma), immunohistology, age, sex, history, preceding diseases). Furthermore, one has to bear in mind that lung and pancreatic carcinoma are the most tumours causing CUP-syndrome. Otherwise frequently occurring tumours are rarely the tumours of origin in CUP-syndrome, i.e., carcinomas of breast, colorectum and stomach. It is postulated that in about 10% of CUP-syndrome the primary tumour is detected during lifetime, while in 90% it is found after autopsy. In only a minority of cases, by using the above mentioned sophisticated methods it will not be possible to detect a primary tumour at autopsy (less than 1%).

427

Cancer of unknown primary site – diagnostic and therapeutic aspects

D. Hossfeld (Hamburg, D)

Cancer of unknown primary site is defined as a biopsy proven malignancy, where the anatomic origin remains unidentified after history and physical examination, laboratory studies, computertomography of thorax and abdomen, and mammography in women. Considerable controversy surrounds the optimal evaluation of such patients. Less experienced physicians have a tendency to initiate an exhaustive evaluation which rarely results in sound therapeutic consequences. The most important aspects of the evaluation are the patient's previous history, a careful clinical examination, and the biopsy of a metastatic site. Subsequent laboratory studies and imaging examinations should be directed by results of the histological examination of the biopsy. The management of a patient with cancer of unknown primary site should be determined by the general condition of the patient, histology, localisation of metastases, whether metastatic disease is isolated or diffuse. To give two examples: A patient with cervical adenopathy and histology of squemous carcinoma should be handled as if he had primary head-neck cancer; the management of a female patient with an isolated axillary lymph node should be based on the treatment of stage II breast cancer. The management of patients with metastases to the lungs, bones, liver, CNS will be discussed.

428

Long-term survival in patients with cancer of unknown primary (CUP) and bulky disease

K. Jentsch-Ullrich, M. Koenigsmann, T. Kalinfke, A. Roessner, A. Franke (Magdeburg, D)

Introduction: Cancer of unknown primary (CUP) is defined as the presence of metastatic cancer documented in the absence of an identifiable primary tumor site. This entity represents approximately 2% to 4% of all new cancer diagnoses. The duration of the overall median survival is 3 to 11 months. Only 5% to 15% of patients with CUP survived more than 5 years. Patients: We describe two cases with complete remission of CUP to radiochemotherapy administration and survival of more than 7 years. A 46-year-old man and a 65-year-old woman with CUP were admitted to our hospital. They had lymphadenopathies (man with bilaterally enlarged cervical, supraclaviculary and paraaortal bulky disease, woman with right axillary bulky disease and enlarged right supraclaviculary lymph nodes). History, physical examination, laboratory studies, chest radiography, computed tomography, mammography, endoscopy, tumor markers and pathological analysis failed to identify the primary tumor site. Histology of the lymphatic metastasis showed a „poorly differentiated, papillary carcinoma“ and an „anaplastic carcinoma“. Both patients were treated with cisplatin-containing combined radio-/chemotherapy (FACP: 5-fluorouracil/adriamycin/cyclophosphamid/cisplatin). After therapy patients demonstrated a clinically ongoing complete remission over 7 years. Conclusion: CUP-patients have a poor median survival duration. One important task for the physician is to recognize those subgroups of patients who may have a better prognosis and thus may be treated in a different way. Therapy-responsive subgroups are adenocarcinoma in women with peritoneal carcinomatosis and axillary node metastasis, in men with blastic bone metastasis or high serum PSA or positive PSA tumor staining, poorly differentiated carcinoma/adenocarcinoma (neuroendocrine tumors, tumor location in mediastinum, retroperitoneum, lymph nodes) and squamous cell carcinoma with cervical and inguinal lymphadenopathy. In spite of bulky disease in our two cases of poorly differentiated carcinoma an ongoing complete remission over 7 years could be achieved with a platin-derived protocol.

429

Kinetics and phenotype of disseminated breast cancer cells

W. H. Krüger, F. Tögel, N. Kröger, H. Renges, F. Gieseking, F. Jänicke, A. R. Zander (Halle/Saale, Hamburg, D)

Tumour cell (TC) contamination of stem cell collections (PSC) from breast cancer (BC) patients is frequent but with varying findings. Although so-called co-mobilization of TC has been hypothesized, the origin of TC contamination in PSC is still unknown. A total of 47 G-CSF mobilised PSC grafts from patients with nodal-positive (n=30), chemosensitive metastatic (n=11), and five women with inflammatory BC was evaluated for TC by immunocytochemistry (IC). Additionally, 40 bone marrow (BM) aspirations and 23 peripheral blood (PB) samples collected prior to apheresis and after one to two cycles of conventional chemotherapy were examined. TC contamination of PSC correlated best with pre-harvest PB state. This was valid when the nominal (positive/negative) presence of TC in PB was compared to the nominal presence of TC in PSC and when it was correlated to the tumour cell load of PSC samples (TCL = tumour cells per 10^6 nucleated cells investigated). The correlation between PB and PSC was better than that between BM and PSC despite the larger sample size of BM aspirations. The presence or absence of TC in PSC could not be safely predicted by the presence or absence of TC in BM or PB alone. Diagnostic specificity seems to improve from a combination of results. No correlation was found in quantitative analysis of TC contamination between BM and PB. 194 data sets consisting of BM and PB samples obtained prior and after HDT and of aliquots of PSC products were searched with IC and rtPCR technique for disseminated TC. Presence of TC in BM is frequent prior and after HDT, but HDT reduces the amount of TC in BM significantly. In contrast, there was no effect on the number of circulating TC. Reinfusion of contaminated PSC products was surprisingly associated with a low number of malignant cells in BM after high-dose therapy and vice versa. Disseminated TC from BM, PB, and PSC have been investigated by double-stain technique for uPA-R expressing cytokeratin-positive cells. uPA-R+/CK+ cells could be found in all qualities of samples, however, significantly less in PSC compared to samples of other provenance (p=0.02). Disseminated TC had no prognostic impact so far. In conclusion the results suggest that PB and BM represent different compartments for TC and that contaminating TC in PSC may be derived from the blood and/or marrow compartment. The TC contamination of PSC can not be safely predicted by a preceding PB or BM analysis. HDT leads to a significant but not complete eradication of disseminated TC. Populations of disseminated TC are phenotypically heterogeneous. Epithelial cells of malignant phenotype occur in PB, BM and PSC of BC patients. Reduced uPA-R expression on TC from leukapheresis samples might suggest a less aggressive nature of these cells compared cells found in BM. Furthermore, the data suggest that the phenotype of TC contamination in PSC products differs significantly from those of disseminated cancer cells in bone marrow or blood.

430

Gemcitabine and treosulfan (GeT): a chemotherapy combination with activity in a variety of frequently chemoresistant cancers

U. Keilholz, A. Schmittel, M.E. Scheulen, J. Siehl, G. Hütter, J. Baumgart, E. Thiel (Berlin, Essen, Hamburg, D)

Recently, it has been discribed that gemcitabine and treosulfan have synergistic activity in vitro against ocular melanoma cell lines. Based on this in-vitro synergy, we initiated a clinical phase I/II study and administered chemotherapy containing gemcitabine and treosulfan to patients with metastatic cancers either as first line treatment in frequently chemoresistant histologies, or as salvage regimen in patients refractory to standard chemotherapy. The treatment regimen consisted of gemcitabine, $1g/m^2$ and treosulfan, 2.5-3.5 g/m^2 on days 1 and 8. Cycles were repeated every 4 weeks for a maximum of 6 cycles. To date, 35 patients have been enrolled. After the phase I part, the recommended dose for treosulfan was established as 3.5 g/m^2 on days 1 and 8 in conjunction with $1g/m^2$ of gemcitabin. Subjective toxicity was mild and objective toxicity was mainly hematological. Only 4 patients experienced grade IV hematotoxicity. There was no objective response in 23 patients with melanoma, although 6 out of 14 patients with ocular melanoma experienced rewarding disease stabilisation for up to 2 years. Out of 5 patients with renal cell carcinoma there were 2 PR and 3 SD, in four patients with ovarian cancer refractory to at least one standard chemotherapy regimen there were 3 PR and 1 CR. None of the 3 patients with colorectal cancer responded. In conclusion, this well tolerated regimen of gemcitabine and treosulfan should be further evaluated in patients with ocular melanoma, renal cell cancer, and ovarian cancer.

FEMARA® – JETZT ZUGELASSEN FÜR
DIE FIRST-LINE THERAPIE
des hormonabhängigen fortgeschrittenen Mammakarzinoms bei postmenopausalen Frauen
• 50% mehr Remissionen*
• Signifikante Verlängerung der Wirkdauer*
• Signifikant bessere Wirksamkeit*
FEMARA® Letrozol
* Im Vergleich zu Tamoxifen. Quelle: Henning Mouridsen, et al., Letrozole is superior to tamoxifen as first-line hormonal treatment of post-menopausal women with locally advanced or metastatic breast cancer (bc) (on behalf of the Letrozole P025 Breast Cancer Study Group), Annals of Oncology, Vol.11, Suppl. 4, Oct. 2000, p.155
FEMARA® Wirkstoff: Letrozol. Zusammensetzung: Eine Filmtablette enthält: Arzneilich wirksamer Bestandteil: 2,5 mg Letrozol; sonstige Bestandteile: hochdisperses Siliciumdioxid, mikrokristalline Cellulose, Lactose-Monohydrat, Magnesiumstearat, Maisstärke, Poly (O-carboxymethyl)stärke, Natriumsalz, Methyl-hydroxypropylcellulose, Macrogol 8000, Talkum, Farbstoffe E 171 und E172. Anwendungsgebiete: First Line Therapie des hormonabhängigen fortgeschrittenen Mammakarzinoms bei postmenopausalen Frauen. Außerdem ist Femara zur Behandlung des Mammakarzinoms im fortgeschrittenen Stadium nach Rezidiv oder Progression der Erkrankung bei Frauen angezeigt, die sich physiologisch oder nach einem künstlichen Eingriff in der Postmenopause befinden und die zuvor mit Antiöstrogenen behandelt wurden. Bei Patientinnen mit östrogenrezeptornegativem Status ist die Wirksamkeit nicht belegt. Gegenanzeigen: Überempfindlichkeit gegenüber dem Wirkstoff oder einem der Hilfsstoffe. Prämenopausaler Hormonstatus, Schwangerschaft, Stillzeit. Nebenwirkungen: Endokrines System: häufig Hitzewallungen. Nervensystem: gelegentlich Kopfschmerzen, Schwindel. Vaskuläres System: gelegentlich Thrombophlebitis. Respiratorisches System: selten Atemnot. Gastrointestinaltrakt: häufig Übelkeit, gelegentlich Erbrechen, Dyspepsie, Appetitlosigkeit, Obstipation. Haut und Anhangsgebilde: häufig Ausdünnen der Haare, gelegentlich Alopezie, Hautausschlag. Bewegungsapparat: gelegentlich Myalgien. Urogenitalsystem: gelegentlich Vaginalblutung, Fluor vaginalis. Veränderungen der Laborwerte: selten Anstieg der Aminotransferasen, in Einzelfällen Veränderungen der Lymphozytenzahl. Verschiedenes: gelegentlich Gewichtszunahme, Gewichtsverlust, generalisierte oder periphere Ödeme, Müdigkeit. Wechselwirkungen: Zum gegenwärtigen Zeitpunkt liegen keine Hinweise auf Wechselwirkungen bei der Anwendung von Femara in Kombination mit anderen Krebsmitteln und anderen häufig verschriebenen Arzneimitteln vor. Die gleichzeitige Anwendung von Femara mit Arzneimitteln, deren Verfügbarkeit hauptsächlich von den Isoenzymen CYP2A6 und 2C19 abhängt und die eine enge therapeutsiche Breite haben, sollte mit Vorsicht erfolgen. Dosierung sowie Art und Dauer der Anwendung: Einmal täglich 2,5mg Letrozol. Die Behandlung mit Femara sollte solange durchgeführt werden, bis eine Progression der Tumorerkrankung festgestellt wird. Bei älteren Patientinnen ist keine Anpassung der Dosierung erforderlich. Leber- und/oder Niereninsuffizienz: Bei Patientinnen mit Niereninsuffizienz und einer Kreatinin-Clearance von > 30 ml/min ist keine Anpassung der Dosierung erforderlich. Eine Dosierungsempfehlung in Fällen von Niereninsuffizienz mit einer Kreatinin-Clearance von < 30 ml/min oder bei Patientinnen mit schwerer Leberinsuffizienz kann aufgrund zu geringer Daten nicht gegeben werden. Weitere Hinweise: siehe Fachinformation. Verschreibungspflichtig. Handelsformen und Preise: 30 Filmtabletten (N1) = DM 452,70; 100 Filmtabletten (N2) = DM 1.432,24. Stand: Februar 2001. Novartis Pharma GmbH, 90327 Nürnberg. www.pharma.de.novartis.com
NOVARTIS
ONCOLOGY

431

Radio-chemotherapy against carcinoma in combination with hyperthermia

P. Wust, B. Hildebrandt, B. Rau, W. Tilly, G. Sreenivasa, H. Riess, R. Felix, P. Schlag (Berlin, D)

Regional hyperthermia is performed in pelvic malignancies since more than 10 years employing annular phased array systems (e.g. SIGMA 60-applicator of the system BSD 2000). Recently, a randomised study of the Rotterdam group has shown a statistically significant benefit for patients with locally advanced cervical cancer, if regional hyperthermia was added to a conventional radiotherapy regimen (The Lancet 355, 2000, van der Zee et al.). Other randomised studies have been initiated (soft tissue sarcomas, rectal carcinoma) and various phase II-studies were performed. The results are encouraging enough to further evaluate and technically develop this modality. Firstly, a randomised study with a preoperative regimen (radiochemotherapy ± regional hyperthermia) for locally advanced uT3/4 has been started 1995 at Charité Medical Centre, Berlin. We performed an interim analysis on 113 patients on 6/2000. We found a trend of higher effectiveness (down-staging and WHO-remission) in the hyperthermia arm with equal toxicity and perioperative morbidity in both arms. Survival rate is excellent in both arms (3 years overall survival 85 %) and superior to previous study results. Median follow-up time of about 2 years is too short to evaluate group related survival curves. However, we found a clear dose response relationship in favour for patients with better heating conditions. Secondly, we performed a phase II-study in locally advanced prostate cancer (cT3/4 pN0 M0). We treated 15 patients with primary cancer and 7 with recurrencies since 1997 with the median follow-up time of 3 years. We found a PSA-control after 3 years of 66 % and an overall 3 years survival of > 90 %. These results are equivalent or even better than data achieved with radiotherapy and neoadjuvant/adjuvant androgene ablation achieved in a similar patient group. Further observation is required, however, additional hyperthermia appears to enhance a lot the effectiveness of radiotherapy alone. Thirdly, for preirradiated rectal cancer recurrencies we performed a phase II-study combining oxaliplatinum (50 mg/m^2), 5-FU (1,000 mg/m^2), and folinic acid (500 mg/m^2) with regients. In 7/8 patients the CEA-level decreased more than 50 %. Toxicity was acceptable. Longer follow-up is required to determine time to progression and survival. Based on this encouraging results a phase II/III-study has been established to compare chemotherapy alone (oxaliplatinum, capecitabine) vs. hyperthermic chemotherapy with respect to palliation and time to progression. All in all, the oncological potential of regional hyperthermia appears quite high. Therefore, further clinical evaluation in well designed phase III-studies is strongly needed. In parallel, significant technological progress has been achieved concerning control of power deposition patterns (multi antenna applicators) and online monitoring of temperature/perfusion distribution (based on MRI imaging). In the view of this progress, we hope to develop regional hyperthermia as a useful and effective treatment modality in oncology for pelvic as well as for abdominal locally advanced (or recurrent) tumour diseases.

432

Response to neoadjuvant thermochemotherapy predicts long-term survival for adult patients with retroperitoneal and visceral high-risk soft tissue sarcomas

C. Wendtner, S. Abdel-Rahman, J. Baumert, W. Hiddemann, R. Issels (München, D)

Purpose: Patients with retroperitoneal or visceral (RP/V) soft-tissue sarcomas (STS) have a poor prognosis with a median survival of 20 months in retrospective analysis. Since local recurrence without distant metastasis often results in death for patients with RP/V tumors, efficacy of a combined modality treatment including regional hyperthermia (RHT) was evaluated in this high-risk group in two consecutive phase II studies. Methods: From 1991 to 1997, 58 patients with high-risk STS (grade II/III; size >5cm; extracompartmental) at RP/V sites were prospectively treated with 4 cycles of etoposide, ifosfamide and adriamycin (EIA) combined with RHT followed by surgery, another 4 cycles of chemotherapy and external beam radiation. Twenty-one patients presented with primary, 18 with local recurrent and 19 with inadequately resected STS. Results: Objective response assessable in 40 patients was 13% (5PR) after neoadjuvant thermochemotherapy. Including eight patients with minor radiographic responses resulted in an overall response rate of 33%. Pathologic response (complete necrosis=pCR, >75% necrosis=FHR) evaluable in 26 patients after surgical resection was 42%. After completion of protocol treatment, 25 patients (43%) had no macroscopic tumor left. Patients responding to neoadjuvant thermochemotherapy associated significantly with freedom from macroscopic disease (P = .008). Median overall survival (OS) was 31 months for the entire study cohort. At a median observation time of 74 months, 5-year probabilities (estimated according to Kaplan-Meier) of only local failure-free, any local failure-free, distant metastasis-free, event-free and overall survival were 53%, 25%, 51%, 20% and 32%, respectively. At 5-year follow-up, probability of OS was in favor of patients responding to neoadjuvant thermochemotherapy (60% vs. 10%; P = .0014). Conclusion: Response to neoadjuvant thermochemotherapy which includes the evaluation of pathologic and minor radiographic responses is predictive for a better long-term survival in patients with high-risk STS at unfavorable RP/V sites. Chemotherapy combined with RHT seems to be an effective new treatment modality for this high-risk group with impact on local tumor control and survival.

433

Selective inhibition of tyrosine kinases - a new therapeutic principle in oncology

A. Hochhaus, S. Kreil, T. Lahaye, U. Berger, G. Metzgeroth, R. Hehlmann (Mannheim, D)

Tyrosine kinases are enzymes that regulate mitosis, differentiation, migration, neovascularization, and apoptosis. Their spectrum and association with specific malignancies offer multiple targets for therapeutic intervention. Variability of the ATP binding pocket allows the design of small molecules acting as selective inhibitors. Due to the knowledge of the molecular etiology, chronic myelogenous leukemia (CML) represents an ideal target for a therapy using a selective inhibitor of the BCR-ABL tyrosine kinase. The 2-phenylpyrimidine derivative STI571 was rationally designed to inhibit ABL and BCR-ABL tyrosine kinase activity through competitive ATP binding pocket interactions. Phase II data demonstrate stable hematologic responses in 91% interferon refractory chronic phase (CP), 69% accelerated phase (AP) and 29% blast crisis (BC) patients. Major cytogenetic remissions have been observed in 55% CP, 24% AP and 16% myeloid BC patients. However, long term observation is needed to confirm that response data result in prolongation of survival. Acute side effects are common but mild in most patients; data on long term adverse effects are still missing. Various molecular and cytogenetic mechanisms of primary and secondary resistance have been demonstrated in vitro and in vivo. Monotherapy with STI571 may select for clones with acquired resistance, such as mutations of the ATP binding site. In vitro data suggest synergistic or additive efficacy of combinations of chemotherapeutic drugs with STI571. Combination trials in CP and BC are ongoing. The efficacy of STI571 is being investigated in other malignancies, including leukemias characterized by expression of alternate molecular forms of BCR-ABL and those expressing protein tyrosine kinases with ATP binding pockets structurally similar to ABL, e.g. c-kit and PDGF-R. Gastrointestinal stromal tumor (GIST) cells overexpress the stem cell factor receptor CD117, the product of the proto-oncogene c-kit. Mutations of c-kit that cause constitutive activation of tyrosine kinase activity play a central role in the pathogenesis of the disease. Inhibition of c-kit in vivo results in an immediate metabolic change of the tumor cells, detectable by positron emission tomography. Since c-kit overexpression is inhibited in small cell lung cancer cell lines a study with STI571 as second line therapy of c-kit positive small cell lung cancer is ongoing. Mastocytosis is based on c-kit overexpression. Since the sporadic form of mastocytosis is linked to the Asp816Val mutation resulting in primary resistance patients should be tested for their individual response to STI571. Clinical studies are ongoing in malignancies accociated with an enhanced activity of the PDGF-R, such as high grade glioma, prostate cancer, pancreatic cancer and leukemias with rearrangements of PDGF-R. A particular target might be idiopathic myelofibrosis characterized by increased secretion of PDGF ligand. The development of selective tyrosine kinase inhibitors is considered a promising approach for the design of new drugs. Clinical responses to STI571 in various malignancies may stimulate greater interest in the clinical use of tyrosine kinase inhibitors.

435

cDNA expression profiling of peripheral blood lymphocytes in cancer patients treated with whole body hyperthermia

D. Schoeler, B. Hildebrandt, F. Ringel, T. Kerner, O. Ahlers, P. Wust, S. Srock, H. Riess, F. Schriever (Berlin, Wartenberg, D)

Background: The molecular basis of the anti-neoplastic properties of whole body hyperthermia (WBH) is still poorly understood. Aims: In the current study, we generated complex arrays of cDNAs that were differentially expressed in peripheral blood lymphocytes (PBL) of cancer patients during WBH as part of a systemic Cancer Multistep Therapy (sCMT). This treatment modality consists of WBH, induced hyperglycemia and hyperoxemia, and chemotherapy (Wust et al., Int. J. Hyperthermia 2000; 16(4):325-39). Methods: Subtracted cDNA libraries were constructed from PBL of 4 patients with metastatic solid tumors (2x colorectal, 1x head & neck, 1x ovarian). Blood samples were obtained before start of WBH and after a heating period of more than 180 min, including a 60 min plateau phase of 42°C. cDNA clones were screened by Dot-Blot and semi-quantitative RT-PCR (sq-RT-PCR) for upregulation. Results: Among 192 clones, 39 cDNAs were found to be significantly upregulated during WBH. Sequencing results revealed the genes encoding: heat shock protein (hsp) 60, hsp 90 and hsp 105, the cell cycle associated calcyclin binding protein, hemoglobin beta-chain and Mip-1b. Upregulation of mRNA in these genes was confirmed by sq-RT-PCR. Further analyses with sq-RT-PCR demonstrated that hsp 70 was also upregulated by WBH. Among the clones of cDNAs that were significantly downregulated we identified genes encoding for actin related protein 2/3 and eukaryotic translation elongation factor 1. Conclusions: The molecular effects of WBH in cancer patients involves the upregulation of heat shock proteins and proteins involved in cell cycle regulation. We are currently investigating which lymphocyte subpopulations are responsible for the cDNA pattern induced by WBH.
(Supported by DFG-SFB 273, Wilhelm-Sander-Stiftung and Deutsche Krebshilfe).

Wissenschaftliches Symposium: Chronische myeloproliferative Erkrankungen

436

PRV-1, a molecular marker for the diagnosis of polycythemia vera

H. Pahl (Freiburg, D)

Polycythemia vera (PV) is a clonal stem cell disorder characterized by hyperproliferation of the erythroid, myeloid and megakaryocytic lineages. Unlike the related disorder CML, for which a pathognomonic alteration has been described, to date, the molecular pathogenesis of PV remains unknown. In order to investigate the molecular defects underlying PV, we used subtractive hybridization to isolate a novel gene, subsequently named PRV-1, which is highly expressed in granulocytes from PV patients (n = 65), but not detectable in normal control granulocytes (n = 45). Moreover, PRV-1 is not expressed in granulocytes from patients with secondary erythrocytosis (n=8), or in mononuclear cells from patients with CML (n=8) or AML (n = 5). Interestingly, of 29 patients with essential thrombocythemia (ET), 16 express PRV-1while 13 do not. The 16 PRV-1 positive ET-patients form endogenous erythroid colonies (EECs) in peripheral blood cultures, while the 13 PRV-1-negative ET-patients do not. Whether PRV-1 positive ET-patients show distinct clinical features, distinguishing them from PRV-1 negative ET-patients or whether they proceed to develop PV during the course of their disease is currently being investigated. In order to determine the cause of PRV-1 overexpression in PV patients, we have cloned 3.4 kb of the PRV-1 promoter. Deletion analysis shows that 98 bp of the proximal promoter region are sufficient to direct high-level, myeloid specific transcription of a reporter gene. Within this 98 bp segment, three protein binding sites are required for activity: a SP-1-site as well as two CCAAT-like sequences. While we have demonstrated SP-1 binding to its cognate site in the PRV-1-promoter, the identity of the CCAAT-factor binding protein remains to be determined. It is clear that both sites are bound by the same protein, which, however, is immunologically distinct from members of the CEBP family of transcription factors.

437

Pharmacotherapy of Ph-negative chronic myeloproliferative disorders: General strategies and recent advances

G. Finazzi (Bergamo, I)

Philadelphia-negative chronic myeloproliferative disorders include Polycythemia Vera (PV), Essential Thrombocythemia (ET) and Idiopathic Myelofibrosis (IF). The cornerstone of treatment in PV patients is phlebotomy, with the goal of keeping the hematocrit value below 0.45. No additional therapy may be needed in stable patients who are at low risk for thrombosis (age below 60 yrs, no history of thrombosis). In patients at high risk of thrombosis or who develop progressive thrombocytosis or splenomegaly, the choice of myelosuppressive agent depends on the patient's age. Older patients can be managed with radiophosphorus, low-dose busulfan or pipobroman. Hydroxyurea (HU) has a role in the management of patients at all ages despite concerns about its potential leukemogenicity. Interferon or anagrelide should be considered in younger patients, but controlled clinical studies are urgently needed to assess their efficacy and safety on hard clinical end-points, such as survival, major thrombosis and hematological progression. In all patients with cerebrovascular, coronary or peripheral arterial ischemia, the association of low-dose aspirin is recommended. The treatment of ET patients should be based on the expected risk of major thrombotic complications. Young asymptomatic subjects with platelet counts in the range of 600-1,000 x 10^9/l are at lower risk and can be followed untreated. The risk/benefit of low-dose aspirin in the primary prevention of thrombosis in these patients remains to be tested in appropriate clinical trials. For „high-risk" patients, HU (plus aspirin in the case of ischemia and/or thrombosis) is the treatment of choice because its efficacy in preventing thrombotic complications has been demonstrated. The possible long-term leukemogenicity of this drug, as well as that of other effective cytoreductive agents such as busulphan and pipobroman, remains a major concern. Anagrelide and interferon could overcome this worry but their efficacy has been hitherto demonstrated only in lowering the platelet count, and clinical studies aimed at documenting additional clinical benefit in ET are needed. At present, ablation of the abnormal hemopoietic clone with high-dose chemotherapy and allogeneic bone marrow transplantation is the only chance for achieving a cure in IM patients. In the few cases reported in the literature, engraftment was prompt in most, transplant-related mortality low, and resolution of marrow fibrosis with disease remission a frequent outcome. No medical therapy has proven to be active on overall patient survival and therapeutic strategies are mainly aimed at maintaining quality of life by correcting anemia, thrombocytosis and symptomatic hepato-splenomegaly. New potential targets for therapy in IM patients include antifibrotic and antiangiogenic drugs. Pilot studies on pirfenidone and thalidomide are currently ongoing.

438

The pathogenesis of hereditary chronic myeloproliferative diseases

R. Skoda (Heidelberg, D)

Studying hereditary myeloproliferative disorders (MPD) has the advantage that genetic linkage analysis can be applied to test the involvement of a candidate gene or to define a chromosomal region where the mutated gene must be located. Such analyses led to the identification of mutations in the erythropoietin receptor (EPO-R) in a subset of familial polycythemia and mutations in the thrombopoietin (TPO) in several families with hereditary thrombocythemia. However, in many MPD families neither EPO-R nor TPO is altered and it follows that mutations in other genes must be responsible for the MPD phenotypes. Interestingly, in some of these families more than one MPD phenotype (e.g. polycythemia and/or thrombocythemia and/or fibrosis) can be observed in the same pedigree and in some affected female individuals clonal hematopoiesis can be found. Thus, genetic analysis in these families may allow us to define genes relevant for acquired forms of MPD. We are studying three such MPD pedigrees, a Spanish, Italian and US family. The current status and an overview of the field will be presented.

439

Angiogenesis is increased in myeloproliferative disorders

A. Willer, J. Hastka, P. LaRosée, S. Saussele, C. Kuhn, A. Reiter, A. Hochhaus, R. Hehlmann (Mannheim, D)

Angiogenesis is crucial for both growth and metastasis of solid tumors, but very likely is also involved in the pathogenesis of a variety of hematological diseases. Thus, an increased bone marrow angiogenesis has been found in AML, ALL, multiple myeloma, CML (chronic phase), polycythemia vera (PV) and osteomyelofibrosis (OMF). Basic fibroblast growth factor (bFGF) and vascular endothelial growth factor (VEGF) are major factors that can induce angiogenesis. Therefore we wanted to investigate, i) if patients with advanced CML (accelerated phase, and blast crisis) or essential thrombocythemia (ET) have an increased bone marrow angiogenesis, ii) if treatment with interferon-alpha(IFN-alpha) that can inhibit angiogenesis in hemangiomas results in a decrease of bone marrow angiogenesis in patients with myeloproliferative disorders, and iii) if bFGF and VEGF play a role in ET angiogenesis. In total, we studied 124 patients with myeloproliferative disorders that included 12 patients with CML-accelerated phase, 7 patients with CML-blast crisis and 23 patients with ET, as well as 30 patients with CML-chronic phase, 23 patients with PV, 6 patients with OMF, and 30 normal individuals as controls. The bone marrow angiogenesis was determined by counting the microvessels in bone marrow biopsies thin sections after microvessel staining. The mRNA expression of bFGF and VEGF was quantified by real-time PCR. The median microvessel density in the bone marrow of normal individuals was 4.0 per mm^2. In the bone marrow of all myeloproliferative disorders, a significantly increased bone marrow density was detected: in CML-chronic phase, the median microvessel density was 8.7 per mm^2 ($p<0.0001$), in CML-accelerated phase 8.5 per mm^2 ($p<0.0001$), in CML-blast crisis 9.3 per mm^2 ($p<0.0001$), in PV 8.0 per mm^2 ($p<0.0001$), in ET 7.5 per mm^2 ($p<0.0001$) and in OMF 10.3 per mm^2 ($p=0.0015$). 3/4 patients with CML-chronic phase, who received 42 MU of IFN-alpha per week, had a significantly lower microvessel density ($p=0.024$) when compared to patients with CML-chronic phase at diagnose. When compared to healthy individuals, we observed a difference only of bFGF mRNA expression in patients with ET (6/6) ($p=0.0004$). We conclude that i) all myeloproliferative disorders are associated with an increased microvessel density, ii) a medium dose of Ifn-alpha seems to inhibit angiogenesis in patients with CML-chronic phase and iii) the increased level of bFGF mRNA in patients with ET could contribute to increased angiogenesis. The increased microvessel density in myeloproliferative disorders may be involved in the pathogenesis of these diseases.

440

The impact of hemostatic parameters in relation to the clinical presentation of essential thrombocythemia

M. Griesshammer, M. Grünewald, M. Hafner, C. Güthner, H. Beneke, A.E. Brommer (Ulm, D)

Essential thrombocythemia (ET) is a chronic myeloproliferative disorder characterized by an increased incidence of thromboembolic, microcirculatory and, less frequently, bleeding complications. In order to evaluate the value of determining hemostatic parameters in relation to the clinical presentation we prospectively investigated 37 newly diagnosed, untreated patients (pts) with ET over a period of 7 years. The following parameters were determined: aPTT, TT, PT, fibrinogen, factor VIII-complex, factor XII, protein C and S, AT III, t-PA, PAI I, FgDP, FbDP, TDP, TAT, F1+2, bleeding time (BT), platelet aggregation studies, platelet activation markers CD 62p, CD 63 and thrombospondin (TSP). The results were compared with 41 healthy controls. An acquired von Willebrand's disease was detected in 25% of pts and was associated with bleeding complications ($p=0.024$). Bleeding time and platelet aggregation studies could not predict bleedings. In comparison with healthy controls, ET pts had both an activation of the plasmatic system (TAT, F1+2) and an increased platelet activation (CD 62p, TSP). An elevation of prothrombin-fragments F1+2 was significantly correlated with the presence of ET-related complications ($p=0.048$). Pts with ET also had activated fbrinolysis with increased t-PA and FbDP levels. Thromboembolic and microcirculatory disturbances were positively correlated to increased FbDP values ($p=0.018$). In order to determine a bleeding diathesis, pts with ET should be screened for an acquired von Willebrand's disease. The correlation of bleeding time and platelet aggregation studies with bleeding symptoms is poor. The activation parameter F1+2 or an increased FbDP are positively correlated with thrombotic events. These hemostatic parameters should be helpful for risk stratification in prospective clinical trials.

Postersitzung:
Akute myeloische Leukämien - Pathophysiologie

441

The CGRP receptor Ramp-1 is downregulated during myelopoiesis

M. Harzenetter, U. Keller, S. Beer, M. Wimmer, C. Riedel, C. Peschel, B. Holzmann (München, D)

Growth of early hematopoetic cells is impaired in patients with spinal cord injuries or in mice following experimental bone marrow denervation. Possibly, this reflects cellular defects caused by the malfunctioning neuronal regulation of immune and bone marrow function. Since neuropeptides are known to be crucial for myelopoesis, we focused on the neuropeptide CGRP, which is released from nerve fibres in lymphatic organs, around blood vessels, and in bone marrow. The present study examines the differentiation-dependent expression and the regulation of the CGRP-receptor during myelopoesis. Using semiquantitative PCR we can show that RAMP-1, component of the CGRP receptor, is expressed in human CD34+ stem cells but not in peripheral blood granulocytes. In vitro myelopoesis experiments confirmed these results and directly demonstrated that RAMP-1 is down-regulated during the differentiation of CD34+ stem cells into granulocytes. Additionally, we analysed the expression of the other receptor components CRLR and RCP. By functional cAMP assays we showed that the RAMP-1 expressing CD34+ cells were CGRP responsive, whereas granulocytes exhibited no specific cAMP response. These results confirm the concept that CGRP plays a role during myelopoesis. Based on the known effect of CGRP on hematopoetic cells, it is conceivable that CGRP regulates the adhesion and/or migration of CD34+ cells in the bone marrow microenvironment.

442

Mandatory expression of either c-kit or FLT3 in adult AML

M. Zühlsdorf, O. Albrecht, K. Kratz-Albers, H. Serve, Th. Büchner, W.E. Berdel (Münster, D)

Stem cell-associated antigens like CD34, AC133, CD117 (c-kit) and CD135 (flt3) are expressed on AML blasts with variable frequencies. Published data range from 50% for CD34 or AC133 to 66% for CD135 and 70% for CD117. We have reassessed these frequencies using three colour immunofluorescence for the validaton of a diagnostic panel and for patient evaluation before tyrosine kinase-directed therapy. The antibodies used were anti CD33-PerCP and CD34-PerCP, CD45-FITC and -PerCP, and CD117-PE (Becton-Dickinson), AC133-PE (Miltenyi Biotec) and CD135-PE (Pharmingen). Bone marrow aspirates were pretreated with NH4Cl-lysis. Blasts were gated by SSC / CD45 density and the positivity of the other antigens was measured against isotype controls. Positive samples expressed the antigen on größergleich 20% of blasts (30 000 acquired). The table shows the numbers and frequencies of positive samples in %. CD34 and AC133 were coexpressed in 48 samples. The frequencies for CD34 and AC133 and the proportion of positive blasts within each sample were significantly lower than those for the receptor tyrosine kinases. 71 samples coexpressed CD117 and CD135, independently of FAB type. The clone used here for anti-CD135 detects 96% of AMLs as positive, identical to a published frequency for flt3-mRNA. Either c-kit or flt3 was expressed in all samples but one. In conclusion, a panel with CD33 and either CD117 or CD135 and CD45 would detect >95% of AML blast populations.

443

Expression of cytokine receptors on AML blasts

M. Graf, S. Danhauser-Riedl, C. Schoch, S. Schnittger, T. Haferlach, H. Schmetzer (München, D)

Haematopoetic cytokines regulate haematopoetic cell functions via specific cell surface receptors. There is evidence to suggest, that those receptors could have a role in leukemia. We have studied the expression of cytokine receptors on mononuclear bone marrow (BM-) cells of 93 patients with acute myeloid leukemia (AML) at first diagnosis by FACS – analysis using directly Fluorescein conjugated antibodies: CD 114 (hG-CSF-R), CD 116 (hGM-CSF-R), CD 117 (hSCF-R), CD 123 (hIL3-R), CD 130 (gp130 subunit), CD 135 (hFL-R). Our findings indicate, that most of patients with

Dienstag, 2. 10. 2001

FAB-type M4/M5 express CD116, whereas most of the patients with FAB-type M2 express CD114. CD117 and CD 123 were found in nearly all patients with FAB-type M0-M5. Most of patients with FAB-type M1, M4 and M5, but only a part of patients with FAB-type M2 and M3 expressed CD 135. Separating our patient cohorts in cytogenetic risk groups we could not detect differences in the cytokine receptor profiles of the 'bad risk group' compared with the 'good' and 'intermediate' risk groups. We can conclude, that cytokine receptor expression in AML patients is very variable. This reveals the great diversitiy of immunophenotypes in AML and might contribute to identify individual blast phenotypes in order to detect residual disease in remission. On the other hand the detection of specific cytokine receptors on leucemic cells might be important for individual cytokine therapies in AML-patients. The prognostic value for the clinical follow up of the patients remains to be evaluated.

444

AML1/ETO inactivates the transcription factor PU.1 in acute myeloid leukemia

R. Vangala, M. Neumann, J. Rangatia, S. Singh, A. Elsässer, V. Reddy, N. Singhal, C. Schoch, P. Pelicci, D. Tenen, W. Hiddemann, G. Behre (München, D; Milan, I; Boston, USA)

A reciprocal translocation, t(8;21) (q22;q22), observed in 40% of patients with M2 subtype of Acute Myeloid Leukemia (AML), forms a fusion protein AML1/ETO. Here we report a negative functional impact of AML1/ETO on the transcriptional activity of PU.1, an important transcription factor for normal myeloid development, as knockout mice exhibit complete block in normal myelopoises. c-Jun is shown to co-activate PU.1 by binding to ß3ß4 domain but AML1/ETO interacts at same region and displaces c-Jun. However this interaction does not recruit any TSA sensitive co-repressors nor change the DNA binding capacity of PU.1. The cell surface markers regulated by PU.1 or its own expression levels in AML M2 patients with t(8;21), were observed to be lower than in M2 without t(8;21), suggesting that PU.1 regulated genes are down-regulated by this mechanism. Expression of PU.1 in mouse bone marrow progenitor cells causes decrease in proliferation and differentiation to macrophage/granulocytic lineage which is blocked by AML1/ETO. Over-expression of PU.1 in Kasumi-1 cells having t(8;21) differentiates them to the monocytic lineage. Thus PU.1 function is altered in presence of AML1/ETO whereas over-expression of PU.1 would restore normal differentiation.

445

Identification of a protein domain of ETV6 which interacts with the MYST domains of TIP60, MOZ (ZNF220) and MORF

S.K. Bohlander, J. Putnik, C.D. Zhang, S. Bartels, M. Kickstein (München, Göttingen, D)

Rearrangements of human chromosomal band 12p13 are found in a variety of human leukemias (e.g.: acute myeloid leukemia (AML), acute lymphoblastic leukemia (ALL), myelodysplastic syndrome). 12p13 is involved in balanced translocations with numerous partner bands. The ETV6 (formerly called TEL) gene was identified as the gene that is affected by most of these balanced translocations. ETV6 was originally cloned as the partner gene of PDGFRB in a (5;12) translocation in a patient with chronic myelomonocytic leukemia. Several other fusion partners of ETV6 have been identified including CBFA2, ABL, ARG, MN1, EVI1-MDS1, JAK2, BTL and STL. An ETV6/CBFA2 fusion mRNA is observed in up to 30% of childhood B-cell ALLs. The non-rearranged allele of ETV6 is frequently deleted in childhood ALL cells. These observations suggest that alterations of ETV6 play a pivotal role in the development of different types of leukemia. In order to analyze the normal function of ETV6 we performed a yeast two hybrid screen to identify proteins that would interact with ETV6. One of the interacting proteins was the beta isoform of TIP60, the HIV Tat interacting protein. TIP60 is a histone acetyltransferase (HAT) containing a 200 amino acid long protein domain called the MYST domain which is highly conserved in evolution. The interaction between ETV6 and TIP60 could be confirmed in GST-pulldown experiments. Mapping of the TIP60 interaction domain of ETV6 showed that the regions between amino acids 249 and 287 and the region between amino acids 282 and 345 were each sufficient for interaction with TIP60beta. However, when we tried to define the ETV6 interaction domain of TIP60 we were unable to construct a deletion mutant of TIP60 that was able to interact with full length ETV6. ETV6 interacting TIP60 deletion mutants could only be identified when the previously defined TIP60 interaction domain of ETV6 (aa282 to 345) was used. It could be shown that aa282 to 345 of ETV6 specifically interact with the C2HC zinc finger (aa 253-287) of the TIP60 MYST domain. Since MYST domains are also found in the human proteins MOZ and MORF. Both MOZ and the related MORF have been identified as fusion partners with CBP in cases with AML and t(8;16) and t(19;16) translocations, respectively. We tested whether aa282 to 345 of ETV6 would also interact with the MYST domain of these proteins. This turned out to be the case. These findings suggest that alterations in the function of MYST domain HATs play an important role in leukemogenesis.

446

Acute myeloid leukemia: Disease of which stem cell?

R. Henschler, I. Junghahn, G. Bug, M. Becker, I. Fichtner (Frankfurt, Berlin, D)

The discovery of stem cell plasticity has led to new questions regarding the ontogenic derivation of various organs and organ systems. To more closely characterize the putative stem cells giving rise to acute myeloid leukemia (AML), human AML cells from patients with various FAB forms (M2, M4, M5) were injected intravenously into NOD/SCID immunodeficient mice. After 2 to 12 weeks, human cells were detected in blood samples taken from the mice using a human chromosome 17-specific PCR. Positive mice were sacrificed, and their bone marrow, liver, spleen, lung and brains were prepared either as tissue sections, or cell suspensions. Immunohistochemical analysis using a panel of human-specific antibodies directed against tissue-specific markers for human hematopoietic cells (CD45, CD33, CD38, CD15, CD68), endothelial cells (KDR, CD31/PECAM-1), or mesenchymal cells (5B5, AS02) was performed with material from the AML-mice and non-transplanted controls. Positive samples contained not only human hematopoietic cells (up to 88% of bone marrow cells), but also human endothelial and mesenchymal cells at frequencies of 0.2 to 2%. Human-specific RT-PCR confirmed the data from the immunohistochemical analyses. Furthermore, human endothelial cells and fibroblasts could be grown after isolation from the AML-mice in cultures supported by NOD/SCID stromal cells for several weeks, as assessed by staining of the cultures by immunohistochemistry or by RT-PCR for human tissue-specific markers. Importantly, transplantation of clonal human AML cell lines KG-1 and EOL-1 led to similar results, indicating that both mesenchymal- and hematopoietic-type cells in the mice arise from a common stem cell. These data show that AML, upon transplantation into mice, displays an as yet underestimated plasticity. Moreover, AML thus may descend not from a hematopoietic, but from an ontogenetically more primitive stem cell. These results have implications for the concept of the malignant transformation leading to leukemia, and for the development of treatment strategies against human leukemia.

447

Resistance to drug-induced cytotoxicity in acute myeloid leukemia is significantly correlated between inhibitors of topoisomerases I and II and nucleoside analogues but is independent from the effect of FAS-ligand

I. Zimmermann, M. Fiegl, P. Schneiderat, C. Schoch, W. Kern, W. Hiddemann, J. Braess (München, D)

Resistance to cytotoxic therapy in AML may be the consequence of A) pharmacokinetics or intracellular metabolism, B) cytokinetic features esp. cell cycle status or C) imbalances of pro- vs. antiapoptotic factors of the apoptotic machinery. Whereas the effects of A) have a diverse impact on different classes of drugs B) and C) are likely to influence most of the presently used induction agents collectively. The present study in 53 patients with AML at various stages of their disease (primary diagnosis n=38, relapse following chemotherapy n=7, relapse following allogeneic BMT n=8) intended to establish whether ex-vivo sensitivity to agents of the major groups used in AML induction chemotherapy [AraC (ARC), daunorubicine (DNR), idarubicine (IDA), etoposide (VP16), mitoxantrone (MIT), topotecan (TOP)] was independent of other classes [indicating A-type resistance] or correlated [indicating B- or C-type resistance]. As a potential effector of cell-mediated cytotoxicity and as a postulated mediator of drug-induced cytotoxicity the cytotoxic effect of FAS-ligand was also assessed. Drug sensitivity was measured by a tetrazolium salt based (WST-) assay at 6 different pharmacokinetically relevant concentrations for each drug. A 4-parameter logistic equation $(y=[(A-D)/(1+(x/C)B]+D)$ was fit to the empirical data, from which the lethal concentration 50% (LC50%) was derived. The maximum slope of the curve was best approximated by the parameter B of the fitting equation and was used to indicate the heterogeneity of the response (steep slope indicates homogeneous response = high B values). LC50% values (µg/ml) after 96h exposure were as follows [median, (variation coeffi-

Dienstag, 2. 10. 2001

cient)]: ARC 1.04 (2.02) ; VP16 5.26 (1.16); DNR 0.09 (4.41); IDA 0.03 (4.40); MIT 0.05 (3.34); TOP 0.13 (2.47). When compared with the sign-rank test the LC50% values were highly correlated (p<0.01) excepting TOP/DNR. Since these drugs are different in their pharmacokinetics and mechanisms of action these findings call into question the sole relevance of specific A-type mechanisms of resistance. Differences in cell cycle status as a B-type reason for heterogeneity in response was suggested by significantly increasing B-values during increasing exposure times (4-96h) for TOP and DNR. This was not observed in the other agents which stresses the relevance of a fully functional apoptotic machinery as a sine-qua-non for a response to most cytotoxic agents in AML. In contrast, FAS-L was unable to induce significant cytotoxicity in AML casting doubts on its relevance for mediating cytotoxic effects of some antileukemic agents and stressing the role of the alternative perforin/ granzyme B system for mediating cell-based antileukemic activity.

448

Detection of cyclooxygenase-2 in acute myeloid leukemic blasts - Impact on multidrug resistance?

U. Puhlmann, D. Reinhardt, A. Klein-Franke, C. Ziemann (Münster, Göttingen, D)

Background: Multidrug resistance (MDR) was identified as a prognostic factor in acute myeloid leukemia (AML). However, endogenous mechanisms of MDR regulation in tumor cells are still unclear. Overexpression of the cyclooxygenase-2 (COX-2) enzyme, which catalyses the rate-limiting step in prostaglandin synthesis, has been demonstrated in different tumors or during inflammation. Methods/Results: Studies with primary rat hepatocyte cultures, characterized by a time-dependent mdr1b overexpression, pointed to a direct role of COX-2 in mdr1b gene (homologue of the human MDR1 gene) regulation. Functionally active mdr1b overexpression was induced by arachidonic acid and the prostaglandins PGE2 and PGF2a and decreased by COX-2 inhibition (NS-398, meloxicam), leading to enhanced intracellular accumulation of the mdr1 substrate rhodamine 123. Therefore we raised the question whether the COX-2 enzyme might also participate in MDR1 regulation in AML- blasts. In children, who suffered AML (n=9), including relapse (n=3), MDR1 and COX&#64979;2 mRNA expression, MDR1-type P-glycoprotein (Pgp) expression and functional rhodamine 123 efflux were analysed. Morphological subtyps were AML FAB M0 (n=2), M1 (n=1), M4/M5 (n=6). In Northern blot analyses with gene-specific oligonucleotide probes (neuroblastoma cell line GIMEN served as positive control and internal standard) all patients tested exhibited significant MDR1 (about 3-fold variation) and COX-2 (about 2-fold variation) mRNA levels. Pgp expression and rhodamine efflux were assessed by multiparametric flowcytometry, validated by testing cell lines CADO ES1, ET.1, ET.2.1, VH 64, and HL 60. Pgp-expression was found positive in eight of nine patients. However, effective rhodamine123 efflux ranged between 9 and 94% and not always correlated with level of Pgp expression, suggesting that other MDR-mechanism might also be involved. Summary: In summary, the expression COX-2 and MDR1 mRNA associated with a functional drug efflux could be demonstrated in AML blast cells. Further studies should analyse whether the COX-2 enzyme indeed participates in regulation of multidrug resistance. Supported by Deutsche José-Carreras-Stiftung (SP 01/03) and by a grant of the University of Göttingen, Faculty of Medicine.

449

Establishment of a new c-kit+ acute myeloid leukemia (AML) cell line with immuno-suppressive properties

U. Erben, A. Bittroff-Leben, G. Bochert, I. Fichtner, C. Schoch, E. Thiel, M. Notter (Berlin, München, D)

Primary culture of blast cells derived from a patient with AML-M5a relapsing following unrelated stem cell donor transplantation gave rise to a permanent c-kit+ cell line, subsequently referred to as CS-1 with overexpression of the stem cell factor (SCF) receptor c-kit. The karyotype of primary leukemic cells as well as of the CS-1 cell line, was 45,XX, t(2;11) (q31;p13),-7,t(10;12)(q24;q24) as identified by cytogenetics and 24-color fluorescence-in situ-hybridisation. CS-1 cells express myeloid surface markers (CD13, CD33). By flow cytometry they show an aberrant phenotype including co-expression of lymphatic antigens (CD4: 48.8%; CD7: 54.4%), and the asynchronous expression of progenitor-associated antigens (CD34: 99.5%; CD38: 36.0%; HLA-DR: 97.4%; c-kit: 99.9%). In contrast to non-malignant progenitor cells which are AC133hi+ only 10.5% of CS-1 cells expressed this early progenitor antigen. CS-1 cells were propagated for more than 1.300 days in continuous in vitro culture without any exogenous cytokines. The generation time was 57.6+10.8 hours. The cells do not secrete SCF as determined by an enzyme-linked-immunosorbent assay arguing against an autocrine SCF/ c-kit-loop. Addition of soluble SCF, however, caused dose-dependent growth stimulation with saturating concentrations (100 ng/ ml) resulting in a 2.2-fold higher plateau of 3H-thymidine incorporation compared to medium control. To investigate the immunogenicity of CS-1 cells, mixed-lymphocyte-tumor-cultures were set up using allogeneic responder T cells and soluble interleukin-2 (25 U/ml). Stimulation with EBV-B cells, but not with CS-1 efficiently activated T cells as reflected by a 5.8±0.6 vs. 1.2+0.1-fold expansion of CD25+ T cells on day 6. By weekly restimulation, T cell cultures could be maintained beyond 32 days when EBV-B cells were present, but only until day 21 using CS-1 stimulators. In 51Chromium-release assays, maximum specific lysis from stimulator cells by EBV-B-stimulated T cells was 82.4±0.3% on day 32 comparing to 24.2±0.7% by CS-1 stimulated cultures on day 21 (effector: target ratio 30: 1). These findings indicate that CS-1 cells induce profound T cell anergy. The immunosuppressive effects appear at least partially being related to down-regulation of co-stimulatory molecules (CD58: 35.6 vs. 99.7%; CD80: 8.0 vs. 99.9%; CD86: 14.8 vs. 99.8%) and of HLA-DR (2.28-fold) and HLA class I (4.86-fold). Cultured CS-1 cells injected subcutaneously into NOD/SCID mice produced tumors that disseminated into the bone marrow and spleen. Tumor cells were passed twice in NOD/SCID mice with consistent distribution and phenotype. In summary, a novel c-kit+ factor-independent AML cell line is described that may be useful to identify as of yet unknown tumor-related immunosuppressive mechanisms and to study immunotherapeutic strategies in vitro and in vivo. In addition, it is a model to evaluate the efficacy of novel small compounds, such as STI571, targeting the c-kit pathway in AML.

450

Relevance of caspase activation in drug-induced versus FAS-ligand induced cell death in acute myeloid leukemia

P. Schneiderat, I. Zimmermann, M. Fiegl, C. Schoch, W. Hiddemann, J. Braess (München, D)

Acute myeloid leukemia (AML) has changed from an invariably fatal disease to a potentially curable condition. The differences in survival are associated with distinct cytogenetic abnormalities of the leukemic clone. However, differences in prognosis in the overall group of AML are not due to potential variations in the natural course of the disease but are a consequence of differential susceptibilities to the induction of cytotoxicity. At present two main antileukemic treatment modalities are used in the clinic: 1) drug-induced cytotoxicity by chemotherapeutic agents and 2) cell-mediated antileukemic effects by cytoxic T-cells. Cytotoxic T-cells can induce cell-death by FAS-ligand/FAS interaction and via the perforin/ granzyme B system. In contrast the various drugs used in leukemia therapy are directed against diverse intracellular targets but the main effect and common denominator is the induction of single- and double strand breaks in the DNA. However, even though the nature of these death inducing signals is widely diverse they all involve the induction of programmed cell death.

The main clinical problem in AML therapy is refractoriness of the leukemic clone to the cytotoxic effects of chemotherapy or of cytotoxic cells (following allogeneic bone marrow transplantation). The aim of the present study in 31 patients with first diagnosis of AML was to investigate the nature of cell-death of leukemic blasts induced by VP16 and cytosine arabinoside (AraC) as the core agents in the treatment of AML and the FAS-L system as one of the effector mechanisms of cytotoxic cells. Early events such as the apoptotic volume decrease (AVD) were measured by the micro-culture kinetic (MicK) assay, downstream effects were assessed by measurement of caspase 3 activation and final loss of viability (LOV) was measured by a tetrazolium based assay (WST). Exposure to VP16 resulted in a prominent AVD and strong caspase 3 activation. Early cytoreduction at 24h was clearly caspase dependent and could be partially prevented by caspase inhibitors. However following AraC a similar extent of cell death could be induced with substantially less AVD and caspase activation. Moreover viability at 96h was clearly not associated with caspase activation and could not be prevented by caspase inhibitors for both drugs indicating that only the rapid initial cytoreduction following VP16 is clearly a caspase dependent process. Interestingly FAS-L was unable to initiate caspase activation and subsequent cells death in AML blasts even though the majority of AML blasts are CD95 positive suggesting that other mechanisms such as the perforin/ granzyme B system are relevant for cell-mediated antileukemic effects

451

Conditional expression of the AML1/ETO fusion protein in human monocytic U-937 cells: a model to study the role of the leukemia specific chimaeric transcription factor

M. Fliegauf, J. Burger, R. Claus, M. Stock, F. Otto, M. Lübbert (Freiburg, D)

The AML1 gene is fused to the ETO gene in a frequent, nonrandom chromosomal translocation t(8;21) in AML with maturation. Thus far only a limited number of AML1/ETO target genes have been identified and human hematopoietic cell line models with high level ectopic expression of the fusion protein have not been intensively studied. We established a model of conditional AML1/ETO expression in human myeloid leukemic U-937 cells using the ecdysone inducible expression system. Individual clones were generated and checked for high level, time- and dose-dependent conditional expression of AML1/ETO mRNA and protein after induction with hormone (Muristerone A or Ponasterone A). We analyzed the possible in vivo modulation of known or putative target genes of AML1 and/or AML1/ETO by Northern blotting. We found that the expression level of lysozyme mRNA was clearly reduced following 24 hours of hormone treatment. The lysozyme gene, which previously has not been reported as an AML1 target gene, contains 5 perfect matches of the AML1 consensus DNA binding site in a 2.2 kb promoter upstream region, providing a useful model to study the effect of AML1/ETO on transcriptional repression. The influence of AML1/ETO upon cell growth was determined by flow cytometry. Cells were stained with DiOC6/propidium iodide and viability was analyzed. Cell cycle analysis was performed by DNA content measurements of propidium iodide stained cells. After induction of AML1/ETO expression for 24 hours, the percentage of cells in G1-phase was increased, consistent with the previously reported shortening of the G1-phase by wild-type AML1, indicating a dominant negative inhibitory effect of AML1/ETO over wild-type AML1 function. Following AML1/ETO expression for 3 days, viability was reduced up to 20% and the population of hypodiploid cells (sub-G1) was markedly increased, both indicating induction of apoptosis. Prolonged expression of the chimaeric protein for 10 days led to nearly complete inhibition of cell growth. In order to identify differentially expressed genes, cDNA representational difference analysis (cDNA-RDA) was performed following 48 hours of AML1/ETO induction. Several bands were isolated and subcloned and individual cDNA fragments were identified by sequencing and BLAST search. The differential expression is now being analyzed by Northern Blot hybridization and the putative AML1/ETO target genes are scanned for AML1 binding sites and other DNA binding motifs in their regulatory sequences. In conclusion, conditional expression of AML1/ETO has effects on cell proliferation, cell cycle and the rate of apoptosis of U-937 cells. It provides a model to identify differentially expressed genes and for studying its interaction with target genes such as lysozyme.

452

Proteasome activation by poly-ADP-ribose-polymerase(PARP) represents a possible target within chemotherapeutic defense mechanisms developed by human myelomoncytic leukemia cells

Ö. Ciftci, R. Hass, O. Ullrich (Berlin, D)

Certain anthracyclines, including adriamycin, develop oxygen radicals to derail nuclear protein functions i.e. by oxidation of histones and damage of DANN. Thus, a variety of leukemic cell lines and leukemic patients, which resemble certain states of differentiation and proliferation, were used in the present study to investigate the adriamycin-induced activation of the nuclear proteasome and a possible involvement of the PARP-associated proteasome activation in developing adriamycin-resistance. Treatment of different human leukemia cell variants with the anthracycline adriamycin was associated with a rapid activation of the proteasome. Thus, proliferating U937, TUR and retrodifferentiated U937 cells exhibited a 4.3-fold, a 5.8-fold, and a 4.3-fold proteasome activation within 15min after adriamycin treatment, respectively. In contrast, little if any proteasome activation was detectable in a growth-arrested differentiated U937 population following adriamycin treatment.

Further analysis of this mechanism revealed a significant reduction of adriamycin-induced proteasome activity after inhibition of poly(ADP-ribose) polymerase (PARP), a nuclear enzyme that catalyzes the synthesis and transfer of poly(ADP-ribose) from NAD+ molecules to various nuclear acceptor proteins, by 3-aminobenzamide (3-ABA) in the proliferating leukemic cell types., These findings suggested that PARP is involved in the regulation of drug-induced proteasome activation. Indeed, anti-PARP immunoprecipitation experiments of adriamycin-treated cells revealed increasing levels of co-precipitated, enzymatically active proteasome particularly in the proliferating cell variants in contrast to the differentiated U937 cells with a maximum after 15min and sensitivity to PARP inhibition by 3-ABA. The specific role of the PARP was investigated in U937 and TUR cell clones stably transfected with a constitutively active antisense PARP (as-PARP) vector. Thus, asPARP TUR cells developed a 25-fold increased sensitivity to adriamycin treatment. Furthermore, we investigated leukemic blasts isolated from AML patients and obtained a similarly enhanced proteasome activity after adriamycin treatment, which was dependent on the PARP and thus, could be co-precipitated with anti-PARP antibodies. Transient transfection of leukemic blasts with the asPARP vector significantly reduced the adriamycin-induced proteasome activation. These data suggest that the PARP-associated nuclear proteasome activation represents a potential target within chemotherapeutic defense mechanisms developed by leukemia cells.

453

In-vitro characterisation of secondary resistance in AML using the CCC-assay

T. Schinköthe, M. Gierling, A. Puppe, V. Diehl, P. Staib (Köln, D)

Only about 30% of patients with acute myeloid leukemia (AML) have a relapse free survival of more than five years. Generally patients with a relapsed AML had a poorer prognosis then patients with a primary diagnosis. The initial therapy surviving leukemic cells are cause of the relapse. It is obviously that these cells are one of the central problems in the treatment of AML. To characterise these cells, we established the in-vitro CCC-assay. The main target of this assay is to simulate a chemotherapy in-vitro and to characterise the therapy surviving cells. Therefore the leukaemic cells from patients with AML were incubated for 48 hours with chemotherapeutics. For controls probes of each patient were incubated without chemotherapeutics and not incubated at all. The surviving cells were isolated and the proteins of these cells were prepared afterwards. The identification and quantification of proteins of interest were effected by Western-Blotting. In addition to these cell isolation we carried out a chemosensitivity-assay (e.g. DiSC-assay) to define the grade of sensitivity or resistance of the surviving cells. Results: The mononuclear cells from bone marrow samples of 51 patients were available. In 29 of these 51 samples (57%) the cell count was sufficient to perform the in-vitro incubation with Ara-C (0.56µg/ml), Daunorubicin (0.125 µg/ml) and Gemcitabine (0.2 µg/ml) in separate attempts. The expression levels of Bcl-2 and Bcl-x after in-vitro incubation could be measured in all 29 cases. An incubation control without any drug was carried out to determine the degree of expression changes. The relationship between the protein expression of the cells surviving the cytotoxic drugs and the cells of the incubation control was computed by linear regression. In case of Bcl-2a we observed a very strong correlation with ara-c (R^2=0.91) and gemcitabine(R^2=0.95). The expression of Bcl-2a in the ara-c surviving cells was increased by 32%, and in the gemcitabine surviving cells it was nearly unchanged. In contrast, the expression of Bcl-2b was increased by 130% after gemcitabine incubation and nearly unchanged after incubation with ara-c. Conclusions: For comparison of the initial protein level with the results achieved by the in-vitro chemosensitivity assay the data are too preliminary and do not allow any conclusions. Data from the recent literature suggest a strong correlation between the expression of e.g. Bcl-2 and the appearance of in-vitro drug resistance. Thus, the CCC-assay may be an useful tool for understanding the induction of secondary resistance in acute leukemia. Further investigation using the CCC-assay may help to develop strategies to overcome secondary resistance.

454

Dual signal targeting against aberrant antigen combinations of acute myeloid leukemia (AML) blasts induces selective T cell stimulation and breaks AML-induced T cell anergy

M. Notter, U. Erben, A. Bittroff-Leben, G. Bochert, E. Thiel (Berlin, D)

Myeloid leukemias prove T cell-sensitive diseases. Raising an autologous transplant rejection type T cell response is therefore an intriguing goal to improve therapy in residual disease states following remission induction and consolidation. Its realization asks for identification of effective T cell epitopes on AML blasts and breaking tumor-induced T cell anergy. We tested the hypothesis that aberrant antigen expression on AML blasts is sufficient to selectively activate T-cells. The asynchronous expression of the stem cell factor (SCF) reced on SCF-immunoglobulin (IgG1) (b) and interleukin (IL)-2-IgG1 as templates, we constructed a SCF-IL-2 fusion cytokine with an apparent molecular weight of 50 - 66 kDa, and demonstrated its specific binding to the c-kit+ AML cell line CS-1 and to CD25+ T- cells. Fresh c-kit+ CD64+ AML-M1 blasts, but not donor CD34+ non-malignant

Dienstag, 2. 10. 2001

progenitor cells, precoated with SCF-IL-2, caused HLA-matched donor T cells carrying submitogenic concentrations of the CD3 MAB OKT3 on the membrane to proliferate in 3-H-thymidine incorporation assays. Simultaneous binding of SCF-IL-2 and OKT3 to AML-M4 blasts and autologous T cells preactivated by PHA, respectively, also resulted in selective T cell stimulation. In mixed-lymphocyte-tumor-cultures (MLTC) with soluble IL-2 (25 U/ml), EBV-transformed B cells caused dramatic expansion of CD25+ cytotoxic T cells (CTL). The c-kit(hi) CD64(lo) AML cell line CS-1 induced profound T cell anergy in this system as reflected by virtually no expansion of CD25+ T cells and induction of only minimal CTL activity against CS-1 targets as assessed by 51-Chromium-release assays. When compared to medium (M)- and murine IgG (mIgG)-pretreated control MLTCs containing soluble IL-2 (25 U/ml), binding of SCF-IL-2 to CS-1 cells not only re-induced proliferation of OKT3-coated T cells on day 14 but, most importantly, restored their capacity to differentiate into CTL cultures with lytic activity against CS-1 blasts. In conclusion, dual signal targeting using CD3 MAB and novel fusion cytokines against aberrant antigens on AML cells represents a novel immunotherapeutic strategy to selectively stimulate autologous T cells and to overcome leukemia-induced T cell anergy.

455

Differentiation of promyelocytic leukemia: Alterations in Fas (CD 95/Apo-1) and Fas ligand expression

H. Salih, G. Starling, S. Brandl, R. Pelka-Fleischer, T. Haferlach, W. Hiddemann, P. Kiener, V. Nuessler (Tübingen, München, D)

Prolonged survival of leukemic blasts contributes to the pathological mechanism of acute promyelocytic leukemia (APL). Whilst treatment of APL with retinoic acid (RA) is a model of differentiation therapy, little is known about possible effects of this treatment on the Fas/FasL system. Investigation of APL cells from patients undergoing differentiation therapy with RA and of HL-60 and U-937 cells cultured with RA resulted in a reduction of surface expression of both Fas and its ligand. Accordingly, the sensitivity of the cells to anti-Fas induced apoptosis decreased proportionally and the reduced expression of FasL resulted in a decreased ability of the leukemic cells to induce apoptosis in T cells. Our findings demonstrate that there are significant changes in Fas and FasL expression during RA treatment of APL, which likely have consequences for the interaction between host immune and leukemia cells and thus may be involved in the beneficial effects of differentiation therapy.

Postersitzung:
Allogene Stammzelltransplantation - Klinik

456

Progressive multifocal leucoencephalopathy (PML) caused by Jc virus: A rare complication of the late period after allogeneic stem cell transplantation

M. Prumbaum, U. Lörcher, S. Bräuninger, A. Burchardt, D. Judith, S. Weier, F. Trostdorf, H. Baurmann, R. Schwerdtfeger (Wiesbaden, Frankfurt/Main, D)

Introduction: PML is a demyelinating disease caused by papovavirus (JC) in immunocompromised hosts. Despite a high prevalence of papovavirus infections in recipients of allogeneic stem cell transplants and a severe and prolonged immunosuppression in this group, PML has very rarely been reported. Case report: A 41-year-old male with MDS (CMML) underwent allogeneic bone marrow transplantation from a one-HLA-antigen mismatched unrelated female donor in October 1999. Conditioning regimen consisted of busulfan 16 mg/kg, cyclophosphamide 200 mg/kg and ATG (Merieux) 10 mg/kg. He received CSA and short-course methotrexate as GvHD prophylaxis. GvHD of the upper gastrointestinal tract (UGT) occured and responded to methylprednisolone 1 mg/kg. He experienced three episodes of CMV reactivation due to a high risk constellation without any evidence of CMV disease. In December 1999 a coronary angiography revealed vasospastic angina without coronary stenosis. The further course was uneventful. Regular analysis of T cell subsets showed severe cellular immunodeficiency, though gradually improving by time. Due to a chronic GvHD of the UGT the patient remained on CSA and low-dose steroids. On day + 406 he was hospitalized because of headache, diplopia, hemianopia and skotoma. MRI revealed multiple white matter lesions, mostly in the parieto-occipital region, with minimal space – occupying effect and only a slight contrast enhancement. CSF analysis showed no pleocytosis, PCR testing for HHV 6, HSV, VZV, CMV, EBV and Toxoplasma genome was negative, also PCR for JCV-DNA was negative. A broad spektrum antimicrobial therapy (including itraconazole, meropenem, foscarnet) was initiated and CSA therapy was stopped. For the next two weeks the clinical symptoms remained unchanged, but MRI findings showed a further deterioration. Then the patient reported improvement of headache and diplopia, MRI was stable. A stereotactic brain biopsy was performed, histological findings showed the typical pathological features of PML, JC virus was demonstrated by in-situ-hybridization. Unfortunately, the patient developed ventricular fibrillation after the neurosurgical procedure, resuscitation was unsuccessful. Conclusions: 1. Although PML is a very rare complication, it should be kept in mind also for the late period after allogeneic bone marrow transplantation. 2. Stereotactic brain biopsy – though potentially dangerous and even deleterious as in our case - might be inevitable to differentiate PML from other diagnosis. 3. No proven antiviral treatment is available. If cessation of CSA therapy might be helpful in improving immunocompetence and thus controlling infection, remains an open question.

457

Conditioning with fractionated total body irradiation and high-dose cyclophosphamide (fTBI/Cy) does not result in any cardiotoxicity as shown by serum cardiac troponin T (cTnT) levels, ventricular repolarisation indices and echocardiography

H. W. Auner, C. Tinchon, M. Eibl, R. I. Brezinschek, S. Sormann, R. Schmon-Kampel, C. Maizen, G. Halwachs-Baumann, W. Linkesch, H. Sill (Graz, A)

Myocardiocyte necrosis and heart failure have been reported after various chemotherapy regimens including high doses of cyclophosphamide (Cy). Recently, we and others have shown that cardiac troponins are highly sensitive biomarkers of early and subclinical chemotherapy-induced myocardial damage. To evaluate the cardiotoxic effects of conditioning with fTBI/Cy, we performed serial measurements of serum cTnT levels. In addition, we determined systolic and diastolic echocardiographic parameters and analysed the QT-dispersion (QTD), which may serve as an indicator of pathologic ventricular repolarisation. A total of 26 patients with various hematologic malignancies (AML, n=10; NHL, n=14; CML, n=2) received conditioning with fTBI/Cy according to standard protocols (12Gy divided in 6 fractions on days 1 to 3 and 120mg per kg body weight given in two doses on days 4 and 5) . None of the patients had a known cardiac disease, and 22 had received prior treatment with anthracyclines at moderate cumulative doses. Serum concentrations of cTnT were determined by immunoassay with a detection limit of 0.01ng/ml. Levels were considered elevated if they were greater than or equal to 0.03ng/ml, which represents the functional sensitivity of the assay. Starting from treatment day 1, a median of 13 cTnT levels (range 5-18) per patient were determined until treatment day 18 (median, range 10-29). One patient had an increase of cTnT on day 7 (0.37ng/ml), which coincided with a hypertensive crisis and returned to normal within 3 days. No further cTnT elevations were noted. In 17 patients, echocardiography was performed before treatment initiation and at a median follow up of 6 months (range 2-15 months). Left ventricular ejection fraction (LVEF), fractional shortening (FS), E/A-ratio and deceleration time of the E wave (DTE) were determined. Median values at baseline did not differ significantly from results at follow up (LVEF 63% vs. 66%, p=0.42; FS 34% vs. 36%, p=0.107; E/A-ratio 0.99 vs. 1.06, p=0.244; DTE 217ms vs. 194ms, p=0.38). Ventricular repolarisation indices were measured from standard 12-lead ECGs in 17 patients immediately before treatment with Cy and within 7 days of treatment completion. No significant differences were noted for median QTD (20 ms vs. 20 ms, p=0.524) and median corrected QTD (22.5 ms vs. 23 ms, p=0.579). None of the 26 patients developed clinically overt heart failure during the observation period. These results indicate that both early and late cardiotoxic effects of conditioning with fTBI/Cy are marginal in patients without preexisting cardiac dysfunction. Myocardial damage observed after chemotherapy regimens including high-dose Cy may rather be due to additive cardiotoxicity of different antineoplastic substances than to Cy alone.

458

Improved disease-free survival after PBSCT as compared to BMT from HLA-identical unrelated donors in patients with 1st chronic phase CML

A.H. Elmaagacli, S. Basoglu, R. Peceny, R. Trenschel, H. Ottinger, A. Lollert, V. Runde, D.W. Beelen, U.W. Schaefer (Essen, D)

PBSCT from HLA-A,-B,-DR compatible unrelated donors (n=34) were compared to BMT (n=52).The recovery of both neutrophils and platelets was faster with PBSCs than with BM (p<0.05). PBSCT had an improved immune reconstitution compared to BMT with elevated peripheral blood counts of naive and memory helper T cells at 3 months and 12 months post transplant (p<0.03). The cumulative incidence of acute (grade II-IV) and chronic GVHD were similar. But patients after BMT had a slightly higher incidence of cumulative severe acute GVHD (grade III+IV)with 27% compared to 13% (p<0.09).

A molecular relapse, as defined by two consecutive positive BCR-ABL - PCR assays with a 4-week interval was detectable in 11 of 44 evaluable patients (25%) after BMT and 3 of 34 (9%) after PBSCT (p=0.06). Cytogenetic relapse developed 5 of 52 patients after BMT (10%), but in none of the 34 patients after PBSCT (n.s.). Median follow-up time was 15 months after PBSCT and 29 months after BMT. 16 of 52 patients died after BMT compared to 2 of 34 patients after PBSCT. The predominant cause of death in the BMT group was mainly associated with fatal infections and severe acute GVHD. The estimated probablity of TRM and DFS at 1000 days after transplant were 31% and 66% in the BMT group and 7% and 93% in the PBSCT group (p<0.03). Overall survival 1000 days after transplant was estimated in the BMT group with 68% versus 93% after PBSCT. In the multivariate analysis only acute GVHD had significant influence on TRM (p<0.01).

459

Successful allotransplantation from unrelated and related donors in patients >60 years of age with active myeloid malignancies

H. Bertz, K. Potthoff, D. Behringer, J. Finke (Freiburg, D)

The widely accepted upper age limit for related or unrelated donor transplantation is 60 and 55 years, respectively. But AML and sAML/MDS occur mainly in elderly patients. Here we present our results of 19 patients with a median age of 63 (range 60-70) years, transplanted from an unrelated (n=12) donor or a matched sibling (n=7). Patients diagnosis were AML, AML/MDS, MDSRA/RAEB-T and OMF. Except for 1 pt each in CR1 and CP1 respectively, all patients had active disease with primary refractory AML (n=6)or untreated AML/MDS (n=11). Median blasts in the bone marrow at transplantation were 50% (range 2-90). For conditioning our reduced intensity conditioning regimen („FBM“) with fludarabine 5x30mg/m^2, BCNU 2x150mg/m^2 and melphalane 110mg/m^2 was used. GvHD prophylaxis consisted of CSA and mainly MMF and in unrelated transplants patients received ATG-S (Fresenius) 40-60 mg/kg additionally. All pts. engraftend with leukocytes >1x10e9/l day +11 and platelets >20x10e9/l day +16 in median. Rates of acute GvHD °II-°IV were 54% (80% in related and 41% in MUD transplants). TRM was 16% in the majority due to fungal infection in this heavily pretreated and advanced disease population. 17 pts. (88%) achieved CR, two pts. are not evaluable. OS and EFS is 78% and 60%, respectively with a median follow up of 305 days (range 107-514). Three patients relapsed, one achieving again cCR after donor lymphocyte infusions. Additionally, lymphocyte reconstitution data will be presented. Conclusion: Age as such or high bone marrow blast infiltration should not be used as exclusion criteria for allogeneic transplantation. Unrelated donor transplantations are save and effective. TRM is determined by disease state and pretreatment. FBM as reduced intensity conditioning regimen has high activity in active myeloid malignancies.

460

Intensification of the conditioning regimen with Re - 188 labeled Anti - CD66 monoclonal antibody and the use of T Cell depleted stem cell grafts may improve the outcome of allogeneic stem cell transplantation in patients with high - risk AML /MDS

D. Bunjes, I. Buchmann, C. Duncker, U. Seitz, M. Wiesneth J. Kotzerke, M. Stefanic, D. Dohr, G. Glatting, S. Von Harsdorf, A. Buck, W. Grimminger, R. Schlenk, G. Munzert, T. Karakas, H. Döhner, L. Bergmann, S. Reske (Ulm, D)

Patients (pts) with high - risk AML in 1.CR and pts with AML > 1.CR have a poor outcome after allogeneic stem cell transplantation (SCT) due to both a high relapse rate and a high transplant related mortality (TRM). Over the past 3 years we have evaluated a strategy of intensifying standard - dose conditioning regimens with a Re - 188 labeled monoclonal anti - CD66 antibody and giving patients T cell depleted allografts in 38 pts with high - risk AML / MDS. The patient cohort consisted of 16 males and 22 females with a median age of 45y (range 17 - 63y). 34 pts were suffering from AML and 4 from MDS. The stage of disease of the AML pts at transplant was as follows: 1.CR (n=8, high - risk), and > 1.CR (n=26, 2.CR.n=7. For the purposes of the following analysis pts were further divided into those with a low tumor burden i.e. < 15% marrow blast (n=25) at transplant and those with a high tumor burden ie. >15% blasts (n=13). All pts had a favourable dosimetry and the radiolabled antibody provided a mean marrow dose of 15,3 Gy. All pts were subsequently treated with standard - dose conditioning consisting either of TBI 12Gy (n=26) or Bu 12,8mg /kg (n=12)plus CTX 120mg / kg with additional Thiotepa 10mg / kg in 6 cases.PBPC were the stem cell source in 32 pts, bone marrow in 4 pts. The donor was a HLA -identical sibling in 16 cases, a matched unrelated donor in 13 cases, a haploidentical family donor in 6 cases, a mismatched family donor in 2 cases and an identical twin in 1 case.All 38 pts achieved rapid and stable engraftment. The day +30 mortality was 3%, the day +100 mortality 6%. After a median follow - up of 23 mo 50% of pts are surviving free of disease. The actuarial risk of relapse was 25 %, the risk of TRM was also 25%. There was a striking and statistically highly significant difference in disease - free survival (DFS) between pts with a low tumor burden and those with a high tumor burden at transplant. The actuarial probability of DFS is 71 % for those with a low tumor burden compared with 14% of those with a high tumor burden. The respective risks for relapse and TRM are 18% vs 38% and 14% vs 38%. The tumor burden was the only factor predictive of outcome, there no differences btween the groups with respect to radiation dose to the marrow, conditioning, type of graft or incidence of GvHD.

461

Chimerism of B lymphoid precursor cells after allogeneic transplantation: Comparison bone marrow to G-CSF mobilized stem cells

J. Winkler, A. Biller, W. Rösler, K. Schäfer-Eckart, H. Wandt, C.N. Kluender, R. Schwerdtfeger, J. Hahn, E. Holler, M. Gramatzki (Erlangen, Nürnberg, Wiesbaden, Regensburg, D)

The engraftment and prolonged repopulation of B lymphoid progenitor cells is dependent on pluripotent hematopoietic stem cells capable of multilineage reconstitution and engraftment of all hematopoietic lineages. Little is known about B cell repopulation after transplantation of G-CSF-mobilized stem cells. We compared the B lymphopoiesis in the bone marrow after allogeneic transplantation from recipients who recieved bone marrow (BM) (n=20) or peripheral blood stem cells (PBSC) (n=28). All patients had hematologic malignancies and were conditioned with high-dose chemotherapy and/or fractionated total body irradiation. PBSCs were obtained from donors after mobilization with standard dose filgastrim from day 1 until day 4 (23 HLA identical siblings, 5 unrelated donors). BM was harvested from 12 HLA identical siblings and 8 unrelated donors. After myeloid engraftment, B lymphoid progenitor cells were analysed in the bone marrow. CD19+CD10+ pro-B/pre-B cells were detected in all recipients of BM 1 month to 12 years after transplantation. The percentage of CD19+CD10+ cells ranged from 9 to 98% (mean 36.4%) of all B lymphocytes. In recipients of PBSC analysed 1 month to 4 years after transplantation, we found CD19+CD10+ preB cells in 25 of 28 patients. Those patients without detectable precursor B cells relapsed (n=1) or had a graft failure (n=2). The percentage of pro-B/pre-B cells in PBSC recipients ranged from 1 to 98% (mean 43.2%) of B cells. A hyperactive B lymphopoiesis with high percentages of precursor B cells was observed in both BM recipients (mean 85.3%) and PBSC recipients (mean 65.7%) during the first 6 months after transplantation. To analyse whether the precursor B cells were donor derived, CD19+CD10+ pro-B/pre-B cells from 20 PBSC recipients and from 14 BM recipients were sorted by flow cytometry. DNA isolated from puri-

fied populations was analysed for polymorphic short tandem repeat markers by PCR. In all BM and PBSC transplanted patients pro-B/pre-B cells were derived exclusively from the donor.
Thus, B-lymphopoiesis is repopulated from donor derived stem cells to a similar extent after allogeneic transplantation of BM or PB source. These data provide evidence that G-CSF mobilized PBSCs contain hematopoietic stem cells capable of long term reconstitution of the B cell compartment.

462

Sensitive detection of HCMV-specific CTL-responses by flow cytometry in healthy individuals and patients following allogeneic stem cell transplantation

H. Hebart, S. Stevanovic, S. Daginik, U. Grigoleit, G. Rauser, C. Sinzger, G. Jahn, J. Löffler, L. Kanz, H. Einsele (Tübingen, D)

Reconstitution of HCMV-specific cytotoxic T cells (CTLs), predominantly directed against the tegument protein pp65, has been found to provide protective immunity for the development of HCMV disease after allogeneic stem cell transplantation (SCT). By computer-based epitope prediction, pp65-derived epitopes presented by the HLA alleles A*0101, A*0201, A*1101, and B*0702 were identified. The frequency of CTLs in healthy HCMV-seropositive individuals as assessed by flow cytometry after peptide stimulation varied between ~0.1-3.3% of all CD8+ T-cells, whereas no peptide-specific CTL's were detectable in HCMV-seronegative individuals. HCMV-specific immune reconstitution was assessed in 17 patients after allogeneic SCT. Eight patients received a transplant from a matched unrelated, 2 from a mismatch family, and 7 from an HLA-identical family donor. Six out of the 17 patients were treated with a non-myeloablative conditioning regimen. In 7 out of 10 recipients of an allogeneic SCT from an HCMV-seropositive donor, HCMV peptide-specific CTLs and/or HCMV-antigen specific CD4+ T-lymphocytes were documented until day 100 post-transplantation. All patients showing peptide-specific T-cell responses also demonstrated lymphoproliferative responses to HCMV, PHA and IL-2 and showed rapid clearance of HCMV infection upon antiviral therapy. In contrast, only 1 of 5 HCMV-seropositive patients receiving a transplant from a seronegative donor developed HCMV-specific T cell responses until day 100. Absence of CMV-specific T cell responses was observed in 3 patients with documented CMV disease, with 1/3 dying developing fatal CMV-IP. In conclusion, prospective monitoring of HCMV-specific cellular immune reconstitution can be performed by flow cytometry after specific stimulation with HCMV peptides and proteins. The introduction of these highly sensitive and rapid techniques as a diagnostic marker will have a major implication on the clinical management of HCMV infection after allogeneic SCT.

463

Unrelated donor transplantation: Better survival rates after peripheral blood stem cell transplantation (PBSCT) compared to bone marrow transplantation (BMT)

K. Potthoff, H. Bertz, J. Finke (Freiburg, D)

Allogeneic stem cell transplantation has the highest curative potential for malignant diseases of the lympho-haematopoietic system. In recent years cytokine-mobilized peripheral blood stem cells (PBSC) have been increasingly used instead of bone marrow (BM) as stem cell source even for allogeneic unrelated donor transplantation. Here we report on a large single-center retrospective analysis of consecutive patients (pts.) who underwent HLA-identical allogeneic unrelated PBSCT (n=46) or BMT (n=98) following myeloablative conditioning comprising TBI or busulfan (n=121) or fludarabine-based low intensity conditioning (n=23) between November 1995 and April 2001 at our institution. For all patients included it was the first stem cell transplantation. Diagnoses in the PBSCT-group were acute leukemias/MDS (n=39), CML (n=3) or others (n=4), in the BMT-group acute leukemias/MDS (n=52), CML (n=41) or others (n=5). The remission at SCT was defined as early disease or standard risk, i.e. CR1/CP1, in 10/46 pts. (21.7.5%) in the PBSCT- and in 35/98 pts. (35.7%) in the BMT-group. All other pts. had advanced disease, the majority primary refractory or relapsing. The median age was 48.2 years (PBSC, 19.0-70.5) vs. 36.7 years (BM, 17.3-62.6). GvHD prophylaxis was performed with cyclosporine A (CSA) and MTX (PBSC: 28/46; BM: 88/98) or mycophenolate mofetile (PBSC: 17/46; BM: 7/98) or prednisone (PBSC: 1/46) or CSA alone (BM: 3/98). Additionally all pts. received ATG-S (Fresenius, Bad Homburg) prior to transplantation. Results: Engraftment was rapid in both groups (PBSC/BM (median): WBC>1000/µl at day +13.5/+16 (9-23 days/1-29 days), platelets >20000/µl at day+17/+22.5 (8-63 days/9-150 days), respectively). Severe acute GvHD >°II developed in 7/46 (15.2%) PBSCT-pts. and in 5/98 (5.1%) BMT-pts., extensive chronic GvHD in 7/35 PBSCT-pts. (19.4%) and 14/76 (18.4%) BMT-pts.. With a median follow up of +353 days in the PBSC-group (49-1534 days) and +1163 days in the BMT-group (69-2015 days), 36/46 PBSCT-pts. and 53/98 BMT-pts. are alive. Comparing the risk groups, for advanced disease patients the estimated overall survival (OS) is 72.4 % in the PBSCT-group vs. 34.2 % in the BMT-group. For early disease or standard risk patients the estimated OS is 83.3 % (PBSCT) vs. 78.1 % (BMT). Thus, for patients with advanced disease the overall survival (OS) after PBSCT is significantly superior to BMT (p=0.0274). Conclusion: For advanced leukemias the use of PBSC instead of BM as stem cell source seems to be superior concerning overall survival in allogeneic unrelated transplantation. Thus, PBSCT should be considered to be first choice in patients with advanced disease.

464

Identification of hematopoiesis-specific minor-histocompatibility antigens by SEREX technology in patients with leukemia

D. Després, T. Wehler, Ö. Türeci, U. Sahin, M. Pfreundschuh, C. Huber, G. Derigs (Mainz, Homburg, D)

Transplantation of bone marrow and peripheral blood stem cells in patients with hemtopoietic malignancies is often complicated by graft versus host disease (GVHD). In case of HLA-identity between donor and recipient GvHD is caused by alloreactive T-cells against minor histocompatibility antigens (mHag). mHags are immunogenic peptides derived from polymorphic cellular proteins and are presented to T-cells as peptide fragments associated with class I or II-HLA molecules. Tissue expression of MH varies, in that some of them are expressed ubiquitously whereas others are restricted to hematopoietic and leukemic cells. Hematopoiesis-specific mHags could also contribute to graft versus leukemia effect. Known mHags were identified by different assays using alloreactive T-cells. Since these are difficult to establish, we sought for an easier way to identify new mHags.
In the SEREX approach (serological identification of antigens by recombinant expression cloning) a cDNA library is constructed from tumor cells packaged into lambda-phage vectors and expressed recombinantly in Escherischia coli. Recombinant proteins are transferred onto nitrocellulose membranes, and identified as antigens by their reactivity with high-titer IgG antibodies present in the patients serum. They are detected by an enzyme conjugated secondary antibody specific for human IgG. Positive clones are subcloned to monoclonality and subjected to further molecular analysis.
We adapted this method and constructed a cDNA library consisting of mononuclear cells from a patient with acute myeloid leukemia in first remission after two cycles of conventional chemotherapy. Serum collected from the patient four months after peripheral blood stem cell transplantation from his HLA-identical sibling was used for immunoscreening. At that time the patient had mild GVHD and was treated with steroids. Screening more than 1 Mio recombinants, ten clones were identified, one of them belonging to the family of defensins. Circulating autoantibodies to this clone were detected up to 17 months after transplantation. Ongoing studies are focussed on the typing of polymorphism of this clone between donor and recipient and the assessment of their potential to constitute a new mHag
We conclude that SEREX could serve as a versatile strategy to identify new mHags.

465

Evidence by T cell epitope mapping and allele typing that the proteinase 3 variant region may be a candidate as minor histocompatibility antigen

B. Khattab, B. Eiz-Vesper, A. Ganser, B. Hertenstein, R. Blasczyk (Hannover, D)

Minor histocompatibility antigens (mHags) are thought to be targets for graft-versus-host disease (GvHD) as well as for a graft-versus-leukemia (GvL) effect after allogeneic, HLA-identical stem cell transplantation (SCT). Recently, the valine/isoleucine (V/I) polymorphism at position aa119 in proteinase 3 (PR3) was identified. PR3 is mainly localized in the primary, azurophilic granules of differentiating myeloblasts. PR3 expression was also observed on the plasma membrane of neutrophils. In myeloid leukemias PR3 is often overexpressed. To assess the PR3 polymorphism as a possible mHag, we screened the distribution of the aaV119I polymorphism by sequence-specific primed PCR in healthy blood donors (n = 221). The allele frequencies for PR3-119V and PR3-119I were 0.56 and 0.44, respectively. To determine, whether peptides derived from the PR3-V119I variation may act as mHags, we investigated the ability of synthetic peptides to induce peptide-specific T lymphocytes (CTLs) in vitro. According to the peptide binding motif of HLA-A*0201 decamers were chosen. HLA-A*0201 positive peripheral blood mononuclear cells (PBMCs) of PR3-V119V, PR3-I119I and PR3-V119I positive donors were stimulated with the individual peptides (spPR3-119V and spPR3-119I). Intracellular detection of IL-2 (icIL-2) by flow cytometry was used as an indicator of peptide-specific CTL activation. PBMCs of PR3-I119I homozygous individuals showed a significant CTL induction, when stimulated with the discordant spPR3-119V peptide. Likewise, an increase of icIL2 expression of CTLs could be detected, when stimulating homozygous PR3-V119V PBMCs with spPR3-119I. These findings indicate that PR3 decamers are capable of inducing peptide-specific CTLs in vitro. The distribution and frequencies of the diallelic polymorphism were analyzed in 104 HLA-identical SCT donor / recipient pairs. 11 % of the recipients from 36 HLA-A*0201 sibling SCT pairs had a PR3-119I disparity in GvH direction, and 8% showed a PR3-119V disparity in GvH direction. Since the respective discordant peptides were immunogenic for either PR3-V119V or PR3-I119I homozygous HLA-A*0201 positive, healthy blood donors, an overall disparity in GvH direction of 19% was calculated for HLA-A*0201 sibling SCT pairs. Finally, 36% of the recipients from 14 HLA-A*0201 MUD SCT pairs showed a PR3-199I disparity in GvH direction. These data provide functional as well as structural evidence, that PR3 has the potential to act as a minor histocompatiblity antigen.

466

Transplantation of highly purified peripheral blood CD34+ cells from sibling donors in chronic myeloid leukemia in first chronic phase

R. Peceny, A.H. Elmaagacli, G. Kummer, H. Ottinger, R. Trenschel, V. Runde, U.W. Schaefer, D. Beelen (Essen, D)

Since September 1998, we are studying prospectively the feasibility of transplantation using G-CSF-mobilized highly purified peripheral blood CD34+ cells from HLA-identical sibling donors.
Twenty-six adult patients with chronic myeloid leukemia in first chronic phase were transplanted in this program up to date. CD34+ cells were enriched by large-scale magnetic-activated cell separation, which resulted in grafts containing a median of 0.6 (range 0.1 to 8.5) x 10^4 CD3+ T lymphocytes per kg of recipient body weight. No prophylactic immunosuppression was given with one exception. This patient had to receive additionally an unmodified peripheral blood stem cell graft because the donor did not mobilize a sufficient number of CD34+ cells to counteract the inherent losses of the enrichment procedure. All patients engrafted rapidly, neutrophils increasing above 500 /µL after 9 days median. Isolated acute graft-versus-host disease (GvHD) of the skin stage I to II developed in 3 patients, but resolved completely after phototherapy alone (n=2) or in combination with low-dose steroids (n=1). The second patient developed a secondary graft failure and was treated successfully with an unmanipulated blood stem cell graft from the same donor. One patient died of metastatic prostate cancer 27 months after transplant. Eleven patients received donor T-cell infusions because of molecular or cytogenetic relapse or progressively disappearing donor chimerism. Six of 11 patients are in sustained molecular remission and stable complete chimerism again, all others without further progression being treated with increasing doses of donor T-cells. No hematologic relapse has been seen up to date. Twenty-five of 26 patients are surviving between 1 and 29 months (median 10 months) after transplantation. In conclusion, transplantation of highly purified peripheral blood CD34+ cells from sibling donors appears to be safe, is effectively preventing acute GvHD and is capable to induce complete hematologic and molecular remissions. This transplant approach with very low mortality and markedly reduced transplant related morbidity has to be further evaluated also in the light of new developments in conventional treatment with allogeneic transplantation still being the only curative therapy for chronic myeloid leukemia.

467

Treatment of Langerhans cell histiocytosis refractory to chemotherapy (study LCH-II) by allogeneic stem cell transplantation

A.M.J. Peters, H.U. Kontny, B. Strahm, J. Thomas, O. Haas, C.M. Niemeyer (Freiburg, D; Wien, A)

Langerhans cell histiocytosis is characterized by clonal proliferation of bone marrow derived Langerhans cells. Typical infiltrates can be found in various organs. The clinical course is variable, but multisystem involvement in young children is associated with a poor prognosis despite chemotherapy or immunosuppressive therapy. We report here on a two year old girl with generalized Langerhans cell histiocytosis involving bone marrow, skin, liver, spleen, gastrointestinal tract, and central nervous system (diabetes insipidus). After chemotherapy according to study LCH-II including prednisone, vinblastine and etoposid skin and gastrointestinal findings improved while the other organ systems showed progressive disease activity. Pancytopenia due to bone marrow involvement was followed by recurrent infections and required daily transfusions. Due to progressive disease, 4 months after diagnosis a non-myeloablative peripheral blood stem cell transplantation from the HLA-identical brother was performed according to the Seattle protocol. Fludarabine and total body irradiation (2 Gy) were used for conditioning. 13 x 1,000,000 CD34/kg body weight positive cells were transplanted after T cell depletion. Cyclosporine A and mycophenolate mofetil were given for immunosuppression and GvHD prophylaxis. After stem cell transplantation a mixed chimerism developed, monocytes and granulocytes being mostly of donor origin while lymphocytes being mostly of recipient origin. With increasing numbers of donor NK cells, disease activity resolved, neutrophil numbers increased and transfusions could be stopped on day +73. Treatment for diabetes insipidus could be discontinued. On day +132, after tapering immunosuppression, hemophagocytosis and the need for desmopressin treatment reermerged, so that immunosuppression was intensified. We conclude that a „graft versus histiocytosis“ effect in combination with immunosuppression were effective in treating multi-organ histiocytosis in this two-year-old girl.

468

Influence of the conditioning therapy on bone marrow progenitor cells after stem cell transplantation

C. Kunert, H.G. Sayer, K. Schilling, C. Kasper, L-O. Mügge, H.J. Fricke, K. Höffken (Jena, D)

Various risk-adapted conditioning regimens are currently used prior to allogenic or autologous stem cell transplantation (SCT). To investigate effects of different conditioning regimens (classic conditioning with total body irradiation/ cyclophosphamide followed by allogenic SCT : 6 pt, modified risk-adapted conditioning with fludarabine/ busulfan/ATG followed by allogenic SCT : 13 pt and myeloablative chemotherapies followed by autologous SCT : 7 pt) on hematopoiesis we studied the growth of different progenitor cells before and after SCT. Clonogenic tests for CFU-GM (colony forming unit-granulocyte/macrophage), BFU-E (burst forming unit-erythroid) and CFU-E (colony forming unit-erythroid) were performed using a methylcellulose assay (Methocult). Also, we analyzed the grade of differentiation of individual colonies. With respect to normal controls (CFU-GM: 100.5 ±40.8 BFU-E: 40.3±10.7 CFU-E: 159,0±80.1 prc 1x 10e5 MNC) growth of progenitor cells from pt before SCT was reduced in all three conditioning groups. One month after SCT we found in the group of modified conditioning higher numbers of progenitor cells (CFU-GM: 72,4±45,9 BFU-E: 10.4±9,5 CFU-E: 90.6±73,4) than in the other groups: TBI containing conditioning (CFU-GM: 51,3±34,8 BFU-E: 6,1±6,7 CFU-E: 36,3±6,1) and autologous regimens (CFU-GM: 50.8±24,2 BFU-E: 12,9±5,8 CFU-E: 73,8±48,9). As a result of these findings, it may be postulated that growth of erythroid progenitors is a very sensitive parameter of influences of conditioning therapies. By contrast, we found no relationships between inoculum of CD34+ cells and numbers of progenitor cells. When investigating cellular composition of individual colonies, we have seen a good differentiation. Longer follow ups regarding long term observation of hematopoiesis after SCT needs to be done to find out long-lasting differences of the conditioning regimens.

469

Allogeneic peripheral blood stem cell transplantation induces prolonged oligoclonality of the T cell receptor repertoire

L.O. Mügge, H.G. Sayer, C. Kunert, J. Clement, K. Höffken (Jena, D)

Peripheral blood stem cell transplantation (PBSCT) has been shown to reduce transplantation related mortality and improve disease free survival when compared to conventional bone marrow transplantation (BMT). Strikingly, acute graft-versus-host-disease (aGVHD) seems not to be increased in PBSCT patients even though the amount of T cells in the peripheral stem cell transplant is about 10-fold higher than in the conventional bone marrow transplant. We have investigated the development of the T cell receptor repertoire in patients with chronic myelogenous leukemia (CML) after allogeneic peripheral blood stem cell transplantation. While granulopoiesis recovers relatively fast after PBSCT, the relative and total count of CD3+ T cells remains severely impaired for a prolonged period of time. A more detailed analysis of the T cell receptor repertoire using the spectratyping method reveales profound imbalances with expansions of very few T cell clones in the early period after PBSCT. The recovery of a diverse T cell receptor repertoire with Gaussian distribution within the Vbeta-subfamilies needs about one year to develop. These results are comparable to those available from transplantations using conventional bone marrow. It predicts a comparable incidence of viral infections after allogeneic transplantation procedures employing PBSCT versus BMT.

470

Degree of telomere shortening after haploidentical megadose stem cell transplantation is not correlated with the number of CD34+ cells infused

T. Brümmendorf, H. Wolke, A. Wahl, H. Einsele, P. Lang, P. Bader, P. Schlegel, C. Bokemeyer, L. Kanz, R. Handgretinger, D. Niethammer (Tübingen, D)

Telomeres are specialized structures at the end of eukaryotic chromosomes that consist of hundreds to thousands of tandem repeats of the sequence TTAGGG in vertebrates. Telomere length measurements in short lived peripheral blood myeloid cells (i.e. granulocytes and monocytes) can be used as a surrogate marker for hematopietic stem cell turnover in normal hematopoiesis and after allogeneic stem cell transplantation (SCT). In previous studies, telomere shortening was shown to be restricted to the first year post SCT. In the current study, we used quantitative fluorescence in situ hybridisation and flow cytometry (Flow-FISH) to analyze lymphocytes and myeloid cells from recipients and their respective donors after haploidentical megadose SCT in order to investigate the role of the infused stem cell dose on the degree of telomere reduction post transplantation. In total, we have analyzed mononuclear cells (MNC) and peripheral blood leukocytes (PBL) from 2 adult and 15 pediatric recipient-donor pairs at an average of 434 (range: 28-1487) days post transplant. The number of transplanted CD34+ cells per kg body weight ranged from 0.4 to 4,0 x 10E7 (mean 2 x 10E7). Two patients had to be excluded from analysis due to incomplete chimerism (<95% donor cells). Results were expressed as difference in molecular equivalents of soluble fluorochrome (kMESF) units between donor and recipient (deltaTel) both for myeloid cells and total lymphocytes. We found an average (mean + S.E.) telomere shortening of 1,0 + 0.6 kMESF in myeloid cells (deltaTELmyeloid). Interestingly, telomere length in myeloid cells of 3 recipients after SCT was found to be elongated compared to their respective donors. No significant correlation between deltaTelmyeloid and the number of CD34+ cells infused was observed. These data suggest that the number of HSC actively contributing to hematopoietic recovery after haploidentical Megadose-SCT is not only restricted to the number of transplanted CD34+ cells but also to recipient parameters, e.g. number of stem cell niches available, immune phenomena and/or other factors. DeltaTellymph in lymphocytes was variable (-0.34 + 0.4 kMESF), probably reflecting the contribution of different lymphocyte subpopulations to the peripheral blood lymphocyte pool at different points in time post SCT. Future studies will aim at a separate analysis of telomere length in phenotypically defined leukocyte subpopulations post SCT by combining Flow-FISH with cell surface staining.

471

Allogeneic transplantation of patients with primary or secondary cardiac amyloidosis

J. Casper, C. Köhne, A. Lück, B. Steiner, D. Wolf, M. Freund (Rostock, D)

Primary or secondary cardiac amyloidosis is a life threatening condition with a median survival of approx. 13 month (primary systemic amyloidosis) after diagnosis. Since amyloidosis is frequently accompanying multiple myeloma, treatment approaches using autologous or allogeneic transplantation have been attempted. Using conventional transplantation approaches treatment results in patients with cardiac involvement of their amyloidosis were quite poor so far. We have used a toxicity reduced conditioning regimen consisting of treosulfan and fludarabine prior to allogeneic transplantation of these high risk patients. 3 patients with a median age of 50 years (49-52), 2 suffering from multiple myeloma with secondary amyloidosis (pulmonary and cardiac involvement) and 1 pat. with a primary cardiac amyloidosis were transplanted after an interval of 1, 2 and 4 years after first symptoms. 2 pts. had been pretreated with conventional chemotherapy (CX) and the third pat. with autologous transplantation as well. Treosulfan 10g/m^2 i.v. was given from d-6 to d-4, fludarabine 30mg/m^2 i.v. d-6 to d-2. 3 matched related donors donated peripheral blood stem cells. The 2 pat. with previous conventional CX did not drop below 0.5 Gpt/L WBC and recovered >1 GpT/l between d +7 and +10. Platelet recovery (>50 Gpt/l) occurred between d+8 and +15. Extramedullary toxicity was mild not exceeding CTC °1 for creatinine, ASAT/ALAT, bilirubine or mucositis. All three pts. reached a complete donor chimerism by d +28. After a median observation time 183 d (range 62-294) aGvHD °1 occurred in one pat. only. cGvHD was developed by 2 pts. (1 limited, 1 extensive) but could be controlled by immunosuppression. One pat. with an accompanying severe cardiac amyloidosis died on d+61 due to a sudden cardiac death/septic shock. The other two pts. are quite well. The walking distance of one pat. with repeated angina pectoris, confined to only a few hundred meters before transplantation has returned to normal and the pat. is now free of angina pectoris. An amyloid involvement of the mammae has decreased considerably and the pat. wants to return back to work. No signs of further progress of her multiple myeloma were observed. In the other pat. with primary amyloidosis the septum thickness did not increase any longer but seems to be decreasing (15 -> 14 mm). At this point we feel, that a toxicity reduced allogeneic transplantation might be successfully performed in this high risk population and that these patients might benefit from allogeneic transplantation after a toxicity reduced conditioning regimen with treosulfan and fludarabine.

472

Allogeneic stem cell transplantation (SCT) in de novo AML and AML secondary to MDS

H. Baurmann, S. Bräuninger, M. Prumbaum, S. Tonndorf, D. Judith, A. Burchardt, A. Yaman, C. Klünder, R. Schwerdtfeger (Wiesbaden, D)

Introduction: In conventional treatment of AML previous myelodysplasia is an important adverse factor for overall prognosis. It is unknown whether this negative impact is maintained after allogeneic SCT. We compared the results of allogeneic SCT for primary (pAML) and secondary AML following MDS (sAML) in 98 consecutive patients (pts) treated in our institution. - Pts and Methods: 70 pts with pAML and 28 pts with sAML were grafted with a related (n = 43) or unrelated (n = 55) donor. Median age was 38 for the pAML and 52 for the sAML group. In pAML 36% of pts were in CR1, 30% in CR2 and 34% beyond CR2 after a median of 4 chemotherapy cycles. In sAML all pts had had chemotherapy prior to SCT for a median of 2 cycles. 29% of them achieved CR. 6/70 patients in the de novo and 12/28 patients in the secondary group were conditioned with reduced intensity. In MUD SCT ATG (Thymoglobuline® Imtix-SangStat 10 mg/kg BW) was added before grafting. PBSC were used for a comparable portion of pts with pAML and sAML as graft source (n = 42). - Results: At a median follow up of 365 days probability of overall survival (POS) for pAML is 28% without difference between MRD (28%) and MUD (29%) transplants. Infections (n = 10) and toxicity (n = 4) were the main causes of death. At a median follow up of 334 days POS for sAML is 56% with a clear survival advantage for MUD (69%) over MRD (38%) pts. Relapse (n = 3) was the leading cause of death after MRD SCT, wheras no relapse has occured after MUD SCT so far. - Conclusions: In this cohort results of allogeneic SCT in sAML - despite the higher pt age -are superior to those of pAML. Possible explanations are less intensive prior therapy and a higher proportion of reduced conditioning. For MUD SCT in sAML a stronger GvL effect and/or a disease specific modulation by ATG could explain these encouraging results.

473

Allogeneic stem cell transplantations in patients with idiopathic ulcerative colitis or Crohn's disease and leukemia

A. Berkenfeld, G. Kummer, D.W. Beelen, V. Runde, U.W. Schaefer, A. Elmaagacli

Three patients with inflammatory bowel disease and chronic myeloid leukemia (CML) in 1st chronic phase who underwent allogeneic stem cell transplantation from a healthy donor were studied clinically for the outcome of their inflammatory bowel disease after transplant. Between July 1994 and June 2000, two patients with Crohn's disease and one patient with idiopathic ulcerative colitis and CML were transplanted with bone marrow (n=1) or blood stem cell transplantation (n=2) from a HLA-identical sibling donor (n=2) or unrelated HLA-matched donor (n=1). All three patients had no active disease of their inflammatory bowel disease before transplantation. At time of this analysis, all patients are alive and were follow up for 1, 3 and 7 years after transplantation. Six months after transplant one patient suffered from relapse of CML, but did not develop any signs of her Crohn's disease again. She achieved a remission of CML with DLI and an immuntherapy with interferon again and remained free of symptoms of her Crohn's disease for 6 years up to date. Another patient showed signs of his idiopathic ulcerative colitis 6 months after transplant. Multiple biopsy specimen of the gut did not prove a relapse of idiopathic ulcerative colitis. The symptoms of ulcerative colitis diminished after 6 months and this patient remained free of any signs of inflammatory bowel disease up to date with a follow up time of two years (3 years after transplant). The third patient remained free of Crohn`s disease with a follow up time of 1 year now. All patients are also in complete remission of CML. We concluded that these observations suggest that host immune dysregulation plays a role in the perpetuation of Crohn's disease that may be corrected by allogeneic hematopoietic cell transplantation.

474

Fludarabine and cytosine arabinoside as salvage therapy in patients with relapsed acute and chronic leukemia after BMT or allogenic stem cell transplantation

A. Lollert, V. Runde, D. Beelen, U. Schaefer, A. Elmaagacli

Background : Chemotherapy with fludarabine and cytosine arabinoside is an effective treatment option for relapsed acute lymphoblastic leukemia (ALL). The objective of this study was to evaluate its use in relapsed acute and chronic leukemic disease after allogenic stem cell transplantation.
Patients and Methods : 14 patients (10 male, 4 female) with relapsed ALL (n=5), AML (n=4) and CML (n=5) after BMT (n=3) or PBSCT (n=11) were included in this study. The median age of patients was 35 years (range 19-57 years). Patients with CML were transplantated in 1st chronic phase (n=2), acceleration (n=2) or blast crisis (n=1) and relapsed in blast crises (n=4) or stable chronic phase (n=1). Patients with acute leukemia underwent transplantation in 1st CR (n=1), 2nd CR (n=1), 1st relapse (n=1) or 2nd relapse (n=5). Relapse after transplant occurred at a median time of 386 days (range 59-1773 days) Chemotherapy consisted of fludarabine 30 mg/m^2 and Ara-C 100 mg/m^2 for 5 days. Post chemotherapy 10 of 14 patients received also unfractioned DLI. Results : Thirteen (93 %) patients achieved a complete remission. But 9 of 13 (69%) patients relapsed again. Duration of complete remission ranged from 80 to 781 days with a median at 314 days. Up to date, seven of 14 patients (50%) are still alive. The median survival time after occurrence of relapse post transplant was 362 days. Four of the 7 patients who are alive are without signs of relapsed disease, the median survival time in this group is 473 days. The main side effect was hematotoxicity and worsening of graft-versus-host-disease of skin and liver. Conclusions : Chemotherapy with fludarabine and cytosine arabinoside is an effective antileukemic regimen in patients with advanced relapsed leukemia after stem cell transplantation, which can induce long lasting complete remissions.

Postersitzung:
Antikörpertherapie

475

A novel recombinant bispecific single-chain antibody, Wue-1xCD3, induces T-cell mediated cytotoxicity directed against autologous primary human myeloma and immunocytoma cells

D. Hönemann, F. Riechert, M. Chatterjee, K. Bommert, S. Lentzsch, I. Körner, P. Kufer, G. Riethmüller, A. Greiner, H.-K. Müller-Hermelink, A. Wolf, P. Bäuerle, B. Dörken, R. Bargou (Berlin, München, Würzburg, Martinsried, D)

Introduction: The development of antibody-based strategies for the treatment of malignant plasma cell disorders has been hampered by the fact that suitable plasmacell-specific surface antigens have been missing so far. Although normal and malignant plasmacells express a number of well characterized surface markers they all have turned out to be not plasmacell specific. However, very recently a novel monoclonal antibody, designated Wue-1, which specifically binds to normal and malignant human plasma cells has been generated. Therefore, Wue-1 represents an interesting and promising candidate for the development of novel immunotherapeutic strategies for the treatment of multiple myeloma and immunocytoma. An interesting variant of antibody design is the generation of bispecific antibodies. In particular, recombinant bispecific single-chain antibodies are interesting candidates that show exceptional biological properties. We cloned and expressed a recombinant bispecific single-chain antibody, Wue-1xCD3 (bscWue-1xCD3), consisting of two different single-chain antibody fragments joined by a glycine-serine linker. One specificity is directed against the CD3 antigen of human T-cells, and the other antigen-binding site engages the recently described plasma cell specific Wue-1 Antigen. Results: We analyzed the biological activity of the recombinant bscWue-1xCD3 antibody using freshly isolated primary myeloma or immunocytoma cells derived either from the peripheral blood or bone marrow aspirates of patients. Specific binding to the Wue-1 antigen and CD3 was demonstrated by flow-cytometry. In (10/12) cases analyzed so far the bscWue-1xCD3 antibody induced efficient plasma cell-directed cytotoxicity by autologous PBMCs derived from the peripheral blood or random allogeneic donors. The bscWue-1xCD3 mediated cytotoxicity is antigen specific, demonstrated by the following findings:
(1) a control antibody of different target specificity but same molecular structure, bsc17-1AxCD3, fails to lyse myeloma cells.
(2) the antibody fails to lyse B cells from a patient with B CLL which belong to the B lineage, but do not express the Wue-1 antigen.
(3) cytotoxicity can be blocked by the parental Wue-1 monoclonal antibody.
Conclusions: Due to its plasma cell specific expression pattern the Wue-1 antigen may be a potential target for antibody-mediated immunotherapy of malignant plasma cell disorders. A recombinant bispecific antibody Wue-1xCD3 shows specific cytotoxicity against primary myeloma and immunocytoma cells and therefore might be an interesting candidate for immunotherapy.

476

Chimeric IgA antibodies for lymphoma therapy

M. Dechant, G. Vidarsson, B. Stockmeyer, R. Repp, M. Glennie, M. Glennie, J.G.J. van de Winkel, M. Gramatzki, T. Valerius (Erlangen, D; Utrecht, NL; Southampton, UK)

Background: Antibody therapy has become a new treatment option for lymphoma patients. Antibodies against HLA class II variants - such as 1D10, or Lym-1 - are actively investigated in clinical trials. However, antibodies against „classical" HLA class II were associated with systemic complement activation and severe toxicity in primate studies. Therefore, we aimed to generate novel antibody constructs with reduced complement activating capacity, and - at the same time - improved effector cell recruitment. Promising candidates to fulfill these criteria were IgA antibodies, because IgA was documented not to activate the classical complement pathway, and FcalphaRI (CD89)- directed bispecific antibodies effectively recruited polymorphonuclear cells - the most populous effector cells in human blood. Therefore, we generated and functionally tested mouse/human chimeric lymphoma antibodies of different IgA and IgG isotypes. Material and Methods: cDNA coding for the variable regions of murine monoclonal antibodies F3.3, Lym-1, and Lym-2 was amplified from the original hybridomas by PCR, and cloned into vectors coding for the constant regions of human kappa light chain, and human alpha1, alpha2, or gamma1-4 heavy chains, respectively. After expression in BHK cells, panels of antibodies were obtained with identical antigen- specificities, and different human constant regions. Integrity of antibodies was proved by ELISA and SDS- PAGE, binding to target antigens was shown by immunofluorescence. Functional activity of antibodies was tested in 51Cr- release assays against ARH-77 mature B- cells, and freshly isolated tumor cells from patients with B-CLL. As effector source, whole blood was compared with plasma (containing human complement), mononuclear (MNC) or polymorphonuclear (PMN) effector cells to identify the lytic mechanism of antibodies. Results: Functional studies with the panel of chimeric F3.3 and Lym-1 variants revealed that killing of malignant human B cells by the IgG1 construct was mainly triggered by human NK cells and complement. In contrast, both the IgA1 and IgA2 variants mediated effective lysis by PMN, but did not activate complement, or recruit NK cells. Importantly, IgA antibodies were similar effective as respective IgG1 antibodies in killing freshly isolated human CLL cells. In conclusion, IgA antibodies may be an attractive alternative to commonly used human IgG1 antibodies.

477

Efficient primary lymphoma B-cell depletion by autologous T cells in the presence of a recombinant bispecific CD19xCD3 single-chain antibody

A. Löffler, M. Grün, C. Wuchter, F. Schriever, P. Kufer, T. Dreier, F. Hanakam, P. A. Bäuerle, L. Karawajew, B. Dörken, R. C. Bargou (Berlin, München, Martinsried, D)

The limited options in the treatment of therapy-refractory B-NHL have stimulated the search for immunotherapeutic strategies. Antibodies and antibody conjugates against B lymphocyte differentiation antigens play an increasing role among these attempts. In a previous study, we have shown that a novel recombinant bispecific single chain antibody (bsc), CD19xCD3, induces rapid and high lymphoma directed cytotoxicity mediated by unstimulated T-lymphocytes. So far functional analysis of the bscCD19xCD3 has been done exclusively with B lymphoma cell lines and data about the killing mechanism of this novel molecule are lacking. Here we analyzed the activity of the antibody using T-cells and autologous primary lymphoma B-cells from patients with B-cell non-Hodgkin's lymphoma (B-NHL). We could show that bscCD19xCD3 induced T-cell mediated depletion of primary malignant cells in 22 out of 26 cases analyzed (24 B-CLL, 1 immunocytoma, 1 mantle cell lymphoma). This effect could be observed at very low E:T ratios (e.g. approx. 1:30) and in the majority of cases without any additional T-cell pre- or costimulation. To our knowledge a comparable in vitro activity of biscpecific CD19xCD3 antibodies has never been described before and therefore discriminates this molecule from other bispecifics generated until now. The addition of the molecule to peripheral blood mononuclear cells of lymphoma patients resulted in an upregulation of activation-specific cell surface antigens on T cells. Target B-NHL cells disclosed elevated levels of CD95 but did not become susceptible to apoptosis induced by the agonistic anti-Fas-antibody CH11. Additionally these cells upregulated cell adhesion molecule CD54 but not the activation or differentiation related antigens CD25 and CD38 respectively. The bscCD19xCD3 induced lymphoma cell depletion is mediated at least in part by the apoptotic caspase pathway since addition of different caspase inhibitors could partially block cell death of the malignant target cells. In conclusion, the properties of the bispecific single-chain antibody CD19xCD3 make this molecule a promising agent for the treatment of B cell malignancies.

478

Single dose of gemtuzumab ozogamicin induced complete remission in second relapse of a patient with AML

M.P. Radsak, H.R. Salih, M. Sökler, L. Kanz, C. Denzlinger (Tübingen, D)

Patients with recurrent or refractory acute myeloid leukemia (AML) have a poor prognosis unless they can be treated with an allogeneic stem cell transplant. However, this option is restricted to patients below the age of 65 in good clinical condition and to donor availability. Alternative treatment options are needed for the majority of elderly patients unfit for this intensive approach. Gemtuzumab ozogamicin (CMA-676) is an immunotoxin consisting of an anti CD33 humanized murine monoclonal antibody conjugated with the potent tumor antibiotic calicheamicin. The CD33 antigen is broadly expressed on most AML blasts as well as on normal hemotopoetic progenitor cells, but not on CD34 positive stem cells. Several clinical trials are currently evaluating CMA-676 for first or second line treatment of AML. We report a case of a 74 year old female patient presenting with second relapse of AML who was successfully treated with a single dose of CMA-676. Initial diagnosis of AML M5 with a t(1;22) translocation was October 1998. Induction treatment was conducted with cytarabine, daunorubicine and PSC833 as multi-drug resistance modulator. Complete remission was obtained and consolidated with two cycles of cytarabine combined with daunorubicine and PSC833 or mitoxantrone and etoposide. 15 months later the first relapse was diagnosed and the patient was treated with mitoxantrone, etoposide and cytarabine, again resulting in complete remission. Consolidation consisted of one course of idarubicine, etoposide and cytarabine. However, six months later the patient had a second relapse. Oral low dose melphalan, given for four weeks, was ineffective. Finally, the patient was treated with a single standard dose of CMA-676 (9 mg/m^2) resulting in complete remission of AML without major extramedullary side effects or toxicity. The clinical course was only complicated by one episode of fever during neutropenia two weeks after treatment. CT scans showed a pulmonary infiltrate and anti-infective therapy was initiated. Hematologic reconstitution was prolonged: white blood counts > 3 000/µl were achieved 44 days, and platelets > 20 000/µl 54 days after treatment with CMA-676. In an attempt to decrease the risk of a third relapse the patient is being treated on an outpatient basis with interleukin 2, 4 million units subcutaneously on five consecutive days every month. The impact of this maintenance strategy is currently under investigation. After more than six months of follow-up after treatment with CMA-676, the patient is still in complete remission and doing very well. This case demonstrates that CMA-676 can be a highly active and fairly safe treatment option in elderly patients with recurrent AML. A single dose of CMA-676 may be sufficient to achieve complete remission even in second relapse after two regimens of conventional combination chemotherapy.

479

Successful treatment of systemic lupus erythematosus with rituximab

R. Weide, J. Heymanns, H. Köppler (Koblenz, D)

Background: SLE is a chronic, inflammatory autoimmune disease which may involve multiple organ systems. Treatment consists of immunosuppression, cytotoxic treatment, plasmapheresis and immunoglobuline therapy. Treatment of patients refractory to standard treatment approaches is difficult and results are poor. Methods: We describe a 39-year old patient with SLE suffering from grand mal epilepsia due to cerebral vasculopathy with positive lupus anticoagulant, who was refractory to standard treatment modalities. The patient was treated with the anti-CD20 monoclonal antibody rituximab (375 mg/m^2 x 4, repeated in weekly intervals).One rituximab-application was delivered in 10/00 and a second application in 03/01. Results: The patient showed a major improvement of her clinical condition, 6 months after the beginning of the rituximab therapy she is free of any symptoms. Inflammation parameters, ANA and lupus anticoagulant declined significantly after the treatment.
Conclusion: Rituximab may be a new effective therapy in SLE.

480

Treatment of a patient with cold agglutinin disease (CAD) with rituximab

A. Jakob, D. Messerschmidt, B. Weber, E. Stöckl, F.W. Hirsch (Offenburg, D)

Introduction: CAD is characterized by the production of IgM antibodies directed against antigens of red blood cells (RBCs). Most cases are associated with lymphoproliferative or infectious diseases. Cold agglutinin-mediated hemolysis occurs at low temperatures and is rarely severe. However, anemia caused by CAD is difficult to treat. Case report: We report on a 60-y-old woman presented to our hospital on march this year. In 1990 laborytory studies had shown a positive direct coombs test and finally an idiopathic cold agglutinin disease was diagnosed. There were only mild signs of hemolysis with normal hemoglobin level requiring no specific therapy. During the year 2000 Raynaud's phenomenon worsened and the Hb level decreased. When she was adimitted to our hospital the hemoglobin level was 10.0 g/dl, the hematocrit was 31,1 %, the reticulycyte count was 5,7 % , haptoglobin level was < 20mg/dl and the LDH level was elevated to 583 U/l. Physical examination showed no adenopathy or enlargement of liver or spleen. The direct coombs test was positive attributable to C3d (4+). Serum protein electropheresis revealed a monoclonal protein defined by Immunofixation as IgM kappa. A bone marrow aspirate and biospsy was performed and showed erythroid hyperplasia and lympho-plasmocytic infiltrates (10%). Immunophenotyping was done and demonstrated an abnormal B-cell phenotype. A computed tomographic (CT) scan of the abdomen showed no adenopathy and normal liver and spleen size. First we recommended a treatment with cyclophosphamide but the patient refused to take a chemotherapeutic regimen. Therefore we decided to treat her with the chimeric anti-CD 20 monoclonal antibody rituximab (375 mg/m^2 iv weekly for 4 weeks). Five weeks after the last infusion the hemoglobin level had increased to 12,1 g/dl, the reticulocyte count decreased to 2,5 % and the LDH level decreased to 335 U/l. The therapy was well tolerated, no side effects occured. Conclusion: A patient with cold agglutinin disease associated with a low grade B-cell lymphoproliferative disease was successfully treated with rituximab. No side effects occured and no additional immunosuppressive or chemotherapeutic agents had to be administered.

481

Monoclonal antibody therapy with CAMPATH-1H: Response and toxicity in B-cell non-Hodgkin's lymphoma patients

M. Uppenkamp, A. Engert, V. Diehl, D. Bunjes, D. Huhn, G. Brittinger (Ludwigshafen, Köln, Ulm, Berlin, Essen, D)

18 patients (pts.) with B-cell NHL were treated in 4 centers in Germany with the monoclonal antibody CAMPATH-1H (CP-1H). 16 pts. suffered from a low grade and 2 from a high grade NHL. All pts. had received chemotherapy before and had either relapsed or were refractory to conventional therapy. 2 pts. received CP-1H in a dose range finding trial once weekly and 16 pts. as a fixed dose of 30 mg three times weekly. 12 pts. completed their therapy and were restaged. 8 out of 18 pts. (44%) achieved a partial remission (PR). 2 (11%) had stable and 5 had progressive disease. One pt. with PR progressed on continuation of CP-1H therapy. 4 pts. could not be evaluated for response, because of death (2 pts.) possibly related to

CP-1H therapy and serious adverse events (2 pts.). All pts. with response to CP-1H had a low grade NHL. Non-hematological toxicity was severe in 2 pts. who suffered from bronchospasm grade III/IV WHO. Common acute adverse events grade I-II WHO included fever, chills, rigor, urticaria, nausea, and vomiting. The main side effects were infections. 11 pts. suffered from bacterial or viral infections; some had recurrent infections (5). A total of 12 different infections were reported. The most frequent infections were caused by Herpes virus (7). 3 pts. developed pneumonia and 1 pt. suffered from bronchitis. Hematological toxicity included thrombocytopenia in 4 and lymphocytopenia in 7 pts. Although the antibody is humanized, the non-hematological toxicity is substantial and probalbly due to a cytokine release syndrome. Furthermore CP-1H induced leukocytopenia (absolute lymphocytopenia) with severe infections especially infludarabine pretreated pts. Prophylactic treatment of the side effects is strongly recommended for pts. treated either with CP-1H alone or in combination with chemotherapy.

482

Rituximab in the treatment of idiopathic thrombocytopenic purpura (ITP)

A.A.N. Giagounidis, J. Anhuf, P. Schneider, D. Söhngen, M. Heinsch, C. Aul (Duisburg, Düsseldorf, Köln, D)

We evaluated the effects of Rituximab treatment in 10 patients with chronic ITP refractory to front-line therapy. Pretreatment patient (pt) characteristics: We treated 4 men and 6 women, age 24 to 80 years: 8 patients had earlier splenectomy, but 2 had remaining accessory spleens. Two splenectomized women (pts A+B) had relapsed and were unresponsive to steroids, chemotherapy, and immunoglobulins. One woman relapsed after steroids, splenectomy and chemotherapy, unresponsive to repeated steroid treatment and immunoglobulins. She had additional autoimmune hemolytic anemia (pt. C). One women had chronic ITP with no improvement to steroids, splenectomy and chemotherapy, but responded to immunoglobulins (pt. D). One splenectomized male patient refused further chemotherapy at relapse, although vincristine led to sustained increase in thrombocytes. He was unsuccessfully subjected to steroids but responded to immunoglobulins (pt. E). One woman relapsed after splenectomy and improved to steroid treatment. She refused further corticosteroids after an increase of 40 kg body weight, any chemotherapy or the removal of an accessory spleen, which was identified at abdominal ultrasound before Rituximab treatment. (pt. F). Patient G relapsed after splenectomy and didn't respond to steroids and immunoglobulins, but had an accessory spleen. Patients H and I were irresponive to steroids and immunoglobulins, but not splenectomized. Patient K (male) was unresponsive to immunoglobulins and had additional active endocarditis. Steroids, chemotherapy or splenectomy were judged contraindicated. Response criteria: Complete remission (CR)was defined as complete normalisation of the thrombocyte count over a period of at least 4 weeks. Partial remission (PR) was defined as sustained (> 4 weeks) increase in thrombocyte count to >15.000/µl without further therapy. Treatment protocol: Patients received Rituximab 375 mg/m^2 day 1,8,15,22, unless thrombocyte normalisation occurred earlier. Results: Overall response (CR+PR)was 70%. Pts B, H and I failed to improve. Patient D had partial remission (thrombocytes between 19.000 and 41.000/µl). Patient A, C, E, F, G and K had sustained normalization of thrombocyte counts. Patient K was successfully operated on his aortic valve. Patient F developed thrombocytosis of > 1.000.000/µl but autoimmune hemolysis failed to improve, however, this was successfully treated with steroids afterwards. Median follow-up for responders is 3 months. There were no adverse events noted as a result of Rituximab administration. We conclude that Rituximab is an effective salvage therapy in patients with chronic refractory ITP. To further evaluate this treatment option, we present a protocol for a multi-center clinical trial.

483

The Rituximab-induced inhibition of proliferation and induction of apoptosis depend on CD20 expression

L. V. von Engelhardt, U. Lohmann, S. Serke, J. P. Deans, I. Schwaner (Berlin, D; Vancouver, CDN)

Therapy of CD20-expressing low-grade and follicular lymphomas with the chimeric anti-CD20 monoclonal antibody rituximab has been well established in the last few years. Therapeutic trials in patients with B-cell non-Hodgkin's lymphoma (NHL) have shown significant efficacy with a primary response rate of about 50%, and a secondary response rate of about 44% in prior responders. Recently a patient was identified who developed a transformed NHL that lost CD20 expression after two courses of therapy with rituximab. Because many patients relapsed after anti-CD20 antibody therapy and have not been biopsied to identify clones with down-regulated CD20 antigen, the frequency of this phenomenon is unknown. Furthermore it is unknown whether rituximab-effects depend on the membraneous density of the CD20 antigen. In our study, we investigated whether the density of CD20 expression may influence the effects of rituximab on lymphoma cells. Therefore we used three differently CD20-expressing mutants of the T lymphoblastoid cell line Molt-04, designated M20, ND50 and ND252. M20 was transfected with wild type CD20. Different deletions of the cytoplasmatic domain induced an impaired CD20-expression in ND252 and ND50 cell clones. The differences in CD20 expression were confirmed using flow cytometry. In these clones inhibition of proliferation induced by different rituximab-doses (100 ng/ml, 300 ng/ml, 1 mg/ml, 10 mg/ml, 100 mg/ml) in the presence of human complement was investigated. Proliferation was measured by the modified tetrazolium salt (MTT) assay. Apoptosis was measured using the flow cytometric Annexin V assay. Results: In non-transfected Molt-04 cells rituximab was without effect. The CD20-expressing clones, ND50, ND252 and M20, showed substantial differences in the dose-dependency of the observed antiproliferative effect. Similarly, induction of apoptosis was different. The antiproliferative effect and induction of apoptosis increased according to the ascending membraneous expression of the CD20-antigen. We conclude, that cytotoxic effects of rituximab are dependent on CD20 expression. Thus, differences in the CD20 expression in different lymphoma subtypes or downregulation of CD20 during treatment may influence treatment outcome.

484

Successful use of an anti-CD 20 antibody (rituximab) for the treatment of refractory immune-mediated thrombocytopenia in a patient with chronic graft-versus-host disease

A. Helwig, J. Freiberg-Richter, M. Bornhäuser, G. Ehninger, A. Haack, K. Hölig (Dresden, D)

We report the case of a 33 year old male with chronic GVHD of the skin and severe immune-mediated thrombocytopenia refractory to any treatment modality. The patient had undergone BMT for CML in first chronic phase three years ago. He developed a chronic GVHD of the skin and remained steroid dependent. At the onset of the immune-mediated thrombocytopenia he received prednisolone 20mg/day, mycophenolate mofetil 2g/day ans cyclosporine A in therapeutic levels, all as treatment for chronic GVHD. Bound autoiantibodies against platelet antigens could be detected with the indirect MAIPA test. Despite treatment with steroids, gamma globulin and splenectomy, only short periods of platelet increases could be achieved. Finally treatment with rituximab, an anti-CD 20 chimeric monoclonal antibody, was started at a dose of 375mg/sqm weekly for 4 weeks. The infusions were performed on an outpatient basis and were tolerated well. The therapy resulted in a depletion of B-cells in the peripheral blood and decreased levels of platelet antibodies. The increased plateled count (up to 300 x 10e9/L)persisted over 6 months despite tapering of immunosuppressive therapy for chronic GVHD. Single cases of successful tratment with rituximab for immune-mediated thrombocytopenia associated with GVHD have been reported in the last years. The autoimmune-mediated Thrombocytopenia in chronic GVHD seems to be due to B-cell dysfunction, resulting in the production of autoantibodies to antigens. The pathological B-cell clones can be efficiently eliminated with the anti-CD 20 antibody. The use of rituximab may be warranted in autoimmune syndromes associated with chronic GVHD.

485

Complement cascade activation before and during rituximab-infusion shows no prognostic relevance in lymphoma patients undergoing rituximab treatment

L. V. von Engelhardt, I. Schwaner (Berlin, D)

Therapy of CD20-expressing low-grade and follicular lymphomas with the chimeric anti-CD20 monoclonal antibody rituximab induce objective responses in 50 % of patients with low-grade and follicular B-cell lymphoma. The antilymphoma effects of rituximab are probably due to CDCC, ADCC (complement and antibody-dependent cellular cytotoxicity) and apoptosis. The mechanism of resistance to rituximab-treatment is yet unknown. Neoplastic cells, like other nucleated cells, are usually resistant to complement-mediated killing. This is mainly owing to complement regulatory proteins and cell repair mechanisms. Recently we have shown, that infusion-related adverse events of rituximab are induced by an activation of the complement cascade (Schwaner et al., submitted for publication). In the present study, we investigated whether an activation of the complement cascade is important for treatment outcome. Therefore, in 34 Non Hodgkin's lymphoma patients (15 FCL, 8 MCL, 4 MM, 6 B-CLL, 1 IC) response (NCI criteria) was evaluated one year after start of rituximab treatment. The status of complement activation was investigated before and after Rituximab-treatment and correlated to response. To investigate the complement cascade activation before and during rituximab infusion, we measured the degradation product C3a-desArg as a sensitive marker of complement cascade activation. Measurements were performed before infusion start and 1, 2, 4, 6, 9 and 24 hours after infusion start. At any time point the median C3a-desArg concentration of 16 responder (9 FCL, 4 MCL, 2 B-CLL, 1 IC) was significantly different compared to 18 non-responder (6 FCL, 4 MCL, 4 MM, 4 B-CLL), $P > 0.05$, respectively. Thus, the status of complement cascade activation - before, during and after rituximab treatment - is not different between respondern and non-respondern. In regard to these data, in vivo complement cascade activation did not influence treatment outcome.

486

Quantification of CD20-positive cells by FACS analysis in peripheral blood for prediction of side effects to treatment with rituximab

C. Tinchon, W. Linkesch, D. Strunk (Graz, A)

Patients with very high lymphocyte counts in peripheral blood (PB) are prone to have severe side effects to treatment with rituximab, including fatal events. Cytokine release from destroyed lymphocytes has been proposed as the main mechanism for these reactions (Winkler et al. Blood 1999,94,2217-24). However there are also cases of severe adverse events in patients with lower cell counts. We questioned wether FACS analysis of CD20-immunoreactivity on circulating tumor B cells allows for a better prediction of side effects. In a series of 19 patients (7 follicular lymphoma, 6 B-CLL, 1 B-PLL, 3 diffuse large cell lymphoma, 1 immunocytoma, 1 primary mediastinal large cell lymphoma) the number of CD20-positive cells in PB and the intensity of antibody reactivity were assessed by FACS analysis. Rituximab was given at 375 mg/m^2 as a 1mg/ml infusion beginning with 50 ml/hour, increasing the dose every 30 minutes, except in one leukemic patient in whom 100mg in 4 hours was given on day 1 and the rest of the dose on day 2. Patients were monitored for adverse events according to National Cancer Institute criteria. All events observed were of grade 1 or 2 except one case of arterial hypertension which had to be treated, therefore classified as grade 3. Nine of eleven patients with more than 2000*10^6/l CD20-positive cells showed side effects. In contrast only one of eight patients with a CD20 cell count less than 2000*10^6/l had side effects (p=0.005). Absolute white blood cell counts also significantly correlated with adverse events (> 50000*10E6/l: p=0.008, >20000*10^6/l: p=0.047). Our data confirm that the interaction of rituximab with CD20-positive cells in peripheral blood is related to manifestation of side effects, but we also suggest that adverse events may be predicted more accurately by measuring CD20-positive cells in PB.

487

Advanced non-Hodgkin's lymphomas (NHL) with central nervous system (CNS) involvement after treatment with rituximab - Report on four cases

A. Jakob, I. Dresel, B. Weber, E. Stöckl, M. Heiland, M. Hofstetter, F.W. Hirsch (Offenburg, D)

Introduction: CNS involvement is uncommon at presentation but ultimately occurs in about 10 percent of patients with lymphoma. The prognosis is usually poor in those patients.
Case 1: A 64-y-old woman with secondary high grade NHL (stage IV with bone marrow involvement) was treated with 4 cycles of CHOEP in ´94. A good partial remission was achieved. In the further course of her disease she received interferon alpha, steroids and cyclophosphamide. 04-05/98 we treated her with rituximab (4 x 375 mg/m^2). The lymph nodes regressed but shortly thereafter (06/98) multiple intracerebral lesions were detected. CNS involvement of the NHL was confirmed by biopsy. The patient received a total brain irradiation, however, died one month later.
Case 2: A 70-y-old man presented with a generalized lymphadenopathy and an enlargement of the left testis. A high grade follicular NHL was diagnosed and he was treated with combination chemotherapy. Three months later there was a cervical relapse and an involved field irradiation was performed. One month later the NHL progressed with a generalized lymphadenopathy and a chemotherapy with cytarabine and idarubicin was started. There was no response, so we treated the patient with bendamustin and rituximab. A partial response was achieved, but one month later the patient progressed with CNS involvement and died 3 months thereafter.
Case 3: 09/98 a 65-y-old man presented with mantle cell lymphoma (stage IIIb). The patient was treated with six cycles combination chemotherapy (FND) and a complete remission was achieved. 05/00 the NHL relapsed and we started a therapy with bendamustin and rituximab. A complete remission could be induced, but seven months later CNS involvement was diagnosed and treated with repeated intrathecal MTX injections. Up to now the patient is well without evidence of disease.
Case 4: A 33-y-old man received high dose chemotherapy with peripheral stem cell transplantation because of an advanced mantle cell NHL in ´94. He relapsed 1^2 years later and was treated with fludarabine. Two years after first relapse the lymphoma further progressed with intraspinal involvement (D7-D9). A radiation therapy was performed; afterwards rituximab was administered. Only a few weeks later the patient presented with paralysis of the left n. abducens. MRT showed a menigeal infiltration of the basis cerebri. The patient died shortly thereafter.
Conclusion: There are reports on successful treatment of primary CNS lymphomas with rituximab. However, these four examples suggest that rituximab cannot prevent CNS involvement in advanced NHL.

488

Fc receptor- directed single chain bispecific antibodies for lymphoma therapy

Y. Wachter, J. Brünke, K. Barbin, K. Schreiter, M. Peipp, M. Gramatzki, T. Valerius, G. Fey, R. Repp (Erlangen, D)

Fc receptors are important for the clinical efficacy of therapeutic antibodies. Bispecific antibodies (bsAb) are immunoglobulin- conjugates with two different binding specificities. BsAb, produced by chemical coupling of one antibody against a tumor cell surface antigen with another against an Fc receptor, mediate effective interactions between effector and target cells. Here, genetically coupled bispecific single chain Fv (bsscFv) were produced, directed against one of the effector cell antigens FcgammaRI (CD64), FcgammaRIII (CD16) or FcalphaRI (CD89), and against HLA class II - a known effective target for effector cell- mediated lysis of malignant human B- lypmhoid cells. The two component scFv were fused via a flexible 20aa linker. ScFv fragments were generated by producing phage libraries from corresponding hybridomas, and screening the libraries with antigen- positive cells. Recombinant scFv against HLA class II, CD16, CD64 and CD89 were thus obtained from the hybridomas F3.3, 3G8, 22 and A77, respectively. Expression of functional bsscFv in E. coli was limited by inefficient transport of recombinant proteins to the periplasmic space. This problem was overcome by expression in eukaryotic cells. Functional bsscFv were expressed and secreted both by insect and mammalian cells, and were purified via Nickel chelate chromatography. Purified bsAb reacted with HLA class II- positive target cells and with Fc receptor- expressing effector cells, and triggered potent ADCC reactions in Cr-51 release assays. For example, in conjunction with the [FcgRIII x HLA class II] bsscFv, human mononuclear cells mediated effective killing of LLM cells - pro B-leukemia cells with the t(4;11) translocation, which are resistant to apoptosis induced by most chemotherapeutic agents. Thus, these recombinant scFv may allow the design of highly specific and effective antibody derivatives for the treatment of B-lymphoid malignancies.

489

Development of new and less immunogenic fusion protein for the treatment of malignant lymphoma

B. Matthey, M. Huhn, S. Sasse, P. Borchmann, S.M. Rybak, S. Barth, A. Engert (Köln, D; Frederick, USA)

The clinical use of current immunotherapeutic strategies with plant and bacterial toxins is limited due to 1) their side effects and 2) their immunogenicity. In order to overcome these problems, we developed a new human fusion-protein consisting of the CD30 ligand and angiogenin. Angiogenin is cytotoxic after incorporation into the cell. The completely human fusion gene was inserted into a pET-based expression plasmid. Transformed E.coli BL21(DE3) were grown under osmotic salt stress conditions in the presence of compatible solutes. After IPTG induction, the 37 kDa His10-tagged Ang-CD30L was directed into the periplasmic space and functionally purified by a combination of metal ion affinity followed by enterokinase cleavage of the His10-Tag and molecular size-chromatography. The characteristics of the recombinant protein were assessed by ELISA and flow cytometry showing specific activity against CD30+ Hodgkin-derived L540cy cells. Ang-CD30L showed RNase activity in vitro and exhibited specific cytotoxicity against L540cy cells (IC50 = 8 ng/ml) as determined by cell proliferation assays. Thus, a new class of targeted therapeutic agents containing a recombinant completely human fusion proteins with angiogenin as a toxic moiety can be designed that potentially will be less immunogenic and with lower side effects than immunotoxins available currently.

490

Anti-CD20 antibody (IDEC-C2B8, Rituximab) enhances drug-induced apoptosis in vitro. Effects of complement, cytokines and caspases

K.U. Chow, W.D. Sommerlad, S. Boehrer, B. Schneider, D. Hoelzer, P.S. Miltrou, E. Weidmann (Frankfurt, D)

In the present study we investigated effects on CD20 expression per cell, apoptosis and cytotoxicity of lymphoma cells under influence of rituximab in combination with complement, standard and new chemotherapeutics (doxorubicin, mitoxantrone, cladribine, bendamustine). The studies were performed in CD20 positive lymphoma cell lines DOHH-2, WSU-NHL, Raji and ex vivo cells of patients with CLL(n=17) and leukemic low-grade lymphomas(n=9). In a further series of experiments influence of rituximab, cytotoxic drugs and complement as single agents and in combination on main events of apoptosis (Annexin, JC-1, western blot analysis of pro-caspase-3 and 7 PARP, +inhibitors) were determined. Third aim of the study was to determine the effect of various cytokines (IL-2,IL-4,IL-6,IL-13,GM-CSF,TNFalpha) on expression of CD20 molecules per cell. We could demonstrate that: i) rituximab at a dose of 10µg/ml sufficiently blocked CD20 molecules per cell in all tested cell populations. ii) complement (rabbit serum) at a dose of 25µl/ml showed the maximum of intrinsic activity (cytotoxicity in a range of 15-25%) in DOHH-2 and WSU-NHL cells. To reach this intrinsic activity in Raji cells higher doses (100µl/ml complement) were needed. Subtoxic doses (no lysis of the cells by intrinsic activity) were obtained at doses <25µl/ml complement in the cell lines. iii) Maximum of cytotoxicity (60-70%) was achieved in cell lines after 4-10 hours using 10µg/ml rituximab in combintion with 25µl complement. Single application of rituximab or complement did not induce higher cytotoxicity than 25% of the cells. iv) simultaneous combination of rituximab with cytotoxic drugs was able to decrease significantly the IC30- and IC50 dosages (on apoptosis) of all tested chemotherapeutic drugs on all studied cell populations. The additional application of complement did not further decrease the necessary dose to reach the IC30- or IC50 effect. v) Experiments with addition of caspase-inhibitors DEVD-CHO or ZVAD-fmk despite determining apoptosis and expression of PARP, caspase 3 and 7, showed that in these settings caspase-7 may function as main executioner caspase in rituximab sensitized apoptosis. vi) preincubation of CLL and lymphoma cells with all studied cytokines did not increase the level of CD20 expression per cell after 24 or 48 hours. A reduction of the rate of apoptotic cells was observed after incubation with IL-2 and IL-4. From these data we conclude, that rituximab significantly sensitizes lymphoma cells for drug-induced apoptosis by cytotoxic agents and that this effect is possibly mediated caspase-7 and independent from complement. Pretreatment with cytokines may not result in an improved efficacy of rituximab, because CD20 expression was unaffected and apoptosis was inhibited by some cytokines.

491

Rituximab induced clinical remission and maintenance therapy in variant hairy cell leukemia resistant to chemotherapy

M. Montemurro, F. Schüler, C. Hirt, T. Kiefer, G. Dölken (Greifswald, D)

Hairy cell leukemia usually responds very well to treatment with purine analogues, e.g. 2-chlorodeoxyadenosine (2-CdA/Cladribin), yielding high rates of long term remission and overall survival. Here we present a case of a variant hairy cell leukemia (CD25 neg.) with high peripheral blood leukocyte counts (>400,000/µl, >90% hairy cells). Only a partial remission of very short duration was achieved after treatment with 2 cycles 2-CdA. Alpha-interferon, 2 cycles Dexa-BEAM, vindesine, VACOP-B and VIP were tried without success to reduce the high peripheral blood counts and the extensive bone marrow infiltration of hairy cells associated with severe thrombocytopenia (<20,000/µl). All treatment periods were complicated by severe infections (pneumonia, septicemia). Because of splenomegaly and severe thrombocytopenia splenectomy was carried out leading to a very short remission regarding the low platelet count. Therefore an experimental therapy with the monoclonal CD20 antibody Rituximab (375mg/m^2, weekly, 4x) was started. During the first infusion the patient developed due to the extremely high number of circulating CD20 pos. hairy cells the expected cytokine release syndrome with fever and severe bronchoconstriction (WHO IV). A complete morphological remission (cytology, FACS-analysis) was achieved after four infusions of the antibody lasting for about a year, but a CDR3 clone-specific quantitative real-time PCR always detected circulating lymphoma cells near its detection sensitivity limit (1 lymphoma cell/10^6 PBMC). The cytologically detectable relapse was successfully treated with one infusion of rituximab leading to a complete morphological remission (<0.1% hairy cells), but again a very low number of circulating clonal cells (1 hairy cell/1x10^6 PBMC) was detected by PCR. The next relapses occurred 7 and 4 months later respectively and were successfully treated with a single infusion of rituximab again. A further shortening of the relapse free interval can be expected. To our knowledge, this is the first report of a prolonged clinical remission of variant hairy cell leukemia by treatment with rituximab. Rituximab seems to be a very useful option for treating hairy cell leukemia resistant to 2-CdA.

492

Targeting properties of an anti-CD16/anti-CD30 bispecific antibody in an in-vivo system

C. Renner, I. Stehle, M. Pfreundschuh, F. Lee, C. Hall, B. Catimel, E. Nice, A. Mountain, A. Rigopolous, A. Scott (Homburg, D; Victoria, AUS)

Bispecific antibodies are currently entering clinical trials in increasing numbers in the areas of breast cancer, prostate cancer, Non-Hodgkin's and Hodgkin's lymphoma. We have previously performed two clinical trials in patients with Hodgkin's disease with an anti-CD30/anti-CD16 bispecific antibody and demonstrated a 30% response rate in a cohort of patients otherwise resistant to standard therapeutic modalities. However, no surrogate marker could be defined in these trials indicative of optimal antibody dosing/scheduling or predictive for favorable response. In order to accurately evaluate the potential biodistribution properties of bispecific antibody in patients, we have performed a detailed analysis of the binding properties and animal model in-vivo characteristics of these constructs. For this purpose, the parental antibodies (anti-CD30 and anti-CD16) and the bispecific antibody (anti-CD30/anti-CD16) were radiolabeled with either 125I or 111In. Antibody integrity and binding properties after labeling were confirmed by Scatchard plot and Lindmo analysis. 111In labeled antibodies revealed superior targeting properties in a standard SCID mouse tumor model. Both, the bivalent parental anti-CD30 monoclonal antibody and the monovalent anti-CD30/anti-CD16 bispecific antibody showed excellent uptake in CD30+ tumors which did not differ significantly between the two (max. uptake 16.5% ± 4.2% vs. 18.4% ± 3.8% injected dose/gram tissue). The equivalent targeting properties of the bispecific antibody compared with the parental anti-CD30 antibody encourages the further clinical development of this bispecific antibody and might help to explain the clinical responses seen with this antibody so far in patients suffering from Hodgkin's disease.

Postersitzung: Chronische lymphatische Leukämie

493

Meningeal leukemia as a rare event complicating chronic lymphcytic leukemia of B-cell type

B. Weiß, M. Hoffmann, C. Anders, K. Lowitzsch, M. Bohrer, P. Hellstern, M. Uppenkamp (Ludwigshafen, D)

We report about a 60 years old male patient with chronic lymphocytic leukemia of B-cell type who was submitted to our hospital with physical detoriation, intermittend disorientation, signs of depression, and weight loss. He reported of a brief period of double vision, which was diagnosed as possible signs of transient ischemic attacks. Neurological examination revealed no specific disorder. An MRI scan of the brain detected periventricular defects of the white matter in frontal areas. Laboratory work up showed asymptomatic mild lymphocytosis which was in accordance with chronic lymphocytic leukemia diagnosed in 1995. The patient presented with painless minimal peripheral lymphadenopathy in both inguinal regions and mild splenomegaly. From clinical features and stage no cytoreductive treatment was indicated at that point. A lumbar puncture revealed meningeal leukemia with 70% of reactive T-lymphocytes and 30% clonal B-lymphocytes on flow cytometry. Repeated intrathecal therapy (Mtx, Ara-C, Dexa)not only led to a reduction of tumor cells from the liquor but als improved his clinical condition.

494

Heterogeneity of B-CLL: Relation of serum parameters, distribution of lymphocyte subsets and new immunological features to clinical course

A. Dorn-Beineke, V. Zunftmeister, J. Hastka, T. Nebe (Mannheim, D)

Previous studies have shown antigenic heterogeneity of B-CLL, however its correlation to the clinical course is still a matter of controversity. Therefore, we reinvestigated this phenomenon in 76 B-CLL patients by determining the serum parameters lactate dehydrogenase (LDH), total protein, immunoglobulins and creatinine. Furthermore we analyzed the distribution of the lymphocyte subsets. Additionally a panel of monoclonal antibodies against B-cell antigens was used and the number of antibody binding sites of CD5, CD10, CD11c, CD22, CD23, CD25, CD38, CD103 and FMC-7 were evaluated. The clinical course was judged according to the RAI criteria, additional adverse clinical signs and laboratory parameters. Univariate analysis identified the increase of LDH and the decrease of total protein and immunoglobulins to be significant in most of the clinical signs of progression. Our results concerning tumor cell mass, the distribution of natural killer (NK)-cells and the T-cell subsets underline their significance in the immunological control of B-CLL. We substantiated the immunophenotypic heterogeneity of B-CLL and obtained significant results for decreased CD5- and CD103-expression and increased CD25- and CD38-density concerning several clinical signs of disease progression. Multivariate logistic regression analysis ruled out several serum and immunological parameters to be independent for eleven response variables. Two models, generated by a general linear procedure, showed that a combination of several serum and immunological markers can improve the significance levels. Our results underline a) that beside the serum parameters LDH, total protein and immunoglobulins especially NK-cells and activated T-cells are excellent markers of disease progression, b) the usefulness of dividing B-CLL patients into a CD38low and CD38high expressing group for a better assessment of individual risk and c) the advantage in monitoring B-CLL by the combined use of conventional serum markers as well as by the immunological parameters described in this work to the sole application of the RAI or Binet classification system in order to estimate the clinical course of the disease as shown in multivariate analysis. Furthermore, this is the first report of CD103 as an progression marker in B-CLL.

495

Fludarabine monotherapy versus combination of fludarabine plus epirubicin in the treatment of chronic lymphocytic leukemia – preliminary results of a randomized phase III study

M. Rummel, S. Stilgenbauer, H. Gamm, T. Karakas, C. Wojatschek, A. Rost, M. Grundheber, E. Lengfelder, P. Mitrou, H. Döhner, D. Hoelzer, L. Bergmann (Frankfurt, Ulm, Mainz, Bad Friedrichshall, Darmstadt, Trier, Mannheim, D)

Background: In our previous phase-II study using the combination of fludarabine plus epirubicin in previously untreated patients with CLL we reached an overall response rate of 92% with a high rate of complete remissions of 40%. Based on these promising results we initiated a multicenter randomized phase-III study to compare efficacy and toxicity of the combination regimen fludarabine plus epirubicin versus monotherapy with fludarabine as first-line therapy or therapy of first relapse. In addition, cytogenetic aberrations will be investigated in all patients to evaluate, if special chromosomal abnormalities are associated with prognosis and response to fludarabine. Methods: Treatment schedule: Fludarabine 25 mg/m^2 day 1-5, epirubicin 25 mg/m^2 day 4+5 for a maximum of 6 cycles every four weeks. Results: 144 patients entered the study so far, 104 are evaluable for response. The overall response rate (ORR) was 82% with a CR-rate of 22%. 63 patients were previously untreated in which CR was achieved in 19 cases (30%) and PR in 37 cases (59%). In 41 patients entering the study in their first relapse the overall response rate was 72% with a CR-rate of 10%. Patients in stage Binet B achieved a better CR rate with 30% in contrast to patients with Binet C with an CR rate of 11%. The overall response rates (ORR) for patients in stage Binet B was 84% compared to the patients in stage Binet C reaching an ORR of 80%. Conclusions: From our preliminary results so far we can corroborate the high activity of fludarabine in the tratment of CLL. It is of particular interest if a higher response rate and a longer duration of achieved remissions will be observed for the combination therapy in this ongoing study.

496

CD38 expression is an independent prognostic marker in chronic lymphocytic leukemia

J. Dürig, M. Naschar, U. Schmücker, K. Renzing-Köhler, A. Hüttmann, U. Dührsen (Essen, D)

Employing a multicolour flow cytometry assay 133 B-chronic lymphocytic leukemia (B-CLL) cases were analysed for surface expression of CD38. Based on a cut-off value of 20%, CLL patients were categorised into a CD38 positive (>20%, n=55) and a CD38 negative subgroup (<20%, n=78) and separately analysed for clinical and laboratory parameters. Patients in the CD38 positive cohort were characterised by an unfavourable clinical course with a more advanced disease stage at study enrollment, poor responsiveness to chemotherapy, short time to initiation of first treatment and shorter survival. In contrast, the CD38 negative group required minimal or no treatment, remained treatment-free for a longer time period and had prolonged survival ($P<0.05$). There was an unequal sex distribution between the two groups with a male-female ratio of 3.6 and 1.3 in the CD38 positive and CD38 negative cohorts, respectively ($P<0.05$). CD38 expression in individual patients was a robust marker in that it was stable over time and not significantly influenced by chemotherapy. In conclusion, our data confirm recent studies suggesting a role of CD38 as a predictor of clinical outcome in patients with B-CLL.

497

Induction of apoptosis in B-CLL cells by bendamustine in vitro

S. Srock (Berlin, D)

Background: Bendamustine is a well known chemotherapy that is used since several years for the treatment of low grade non-Hodgkin's lymphoma. Bendamustine probably combines the features of alkylating agents and purin analoga. However, few data are available on the in vitro effects of this drug on malignant B cells. Aims: The aim of this study was to determine the in vitro activity of bendamustine alone and in combination with dexamethason or fludarabine in previously untreated and pretreated patients with B-CLL (chlorambucil responsive or refractory). Methods: B-CLL cells were separated from heparinized blood and incubated at 37° C for 24-48 hours alone, with bendamustine, fludarabine, dexamethason or in combinations of the three agents. Bendamustine concentrations of 0.0005-2 mM were tested. Dexamethason with concentrations of 100 ng/ml and fludarabine with 500 ng/ml were used. To determine the induction of programmed cell death (apoptosis) of the different drugs, staining with Annexin-FITC and propidium iodide was determined by flow cytometry. In vitro assays were correlated with clinical data. Results: B-CLL cells of 28 patients were analysed for induction of cell death by bendamustine. Apoptosis occurred in a dose and time dependent manner. The IC50 for bendamustine was 0.125 mM and 0.035 mM after 24 and 48 hours respectively. The combinations of bendamustine, fludarabine and dexamethason induced higher cell death rates than the single agents, but no synergistic effects were found. Responsiveness to bendamustine did not correlate with clinical parameters, e.g. stage, time to progression, leucocytes, hemoglobin, thrombocytes, LDH, â2-microglobulin, age, sex, or number of prior chemotherapies or clinical resistance to chlorambucil.

Conclusions: Bendamustine has significant in vitro activity against B-CLL cells. The amount of apoptosis induced by bendamustine in vitro is not influenced by clinical parameters of the disease. Future studies are needed to determine whether bendamustine can serve also in vivo as a valid option for treatment of progressive or non-responsive B-CLL.

498

Ex-Vivo expanded NK T cells induce apoptosis in primary CLL-cells

M. Kornacker, G. Moldenhauer, E. Weilguni, M. Herbst, M. Hensel, A.D. Ho (Heidelberg, D)

Natural killer T cells (NK T cells) are defined by the coexpression of T- and NK-markers. Two types of NK T cells exist: One type is selected by CD1d, expresses a skewed T-cell receptor repertoire and is found primarily in thymus and liver. A second type is independent of CD1d, has an unbiased T-cell receptor repertoire and is found in spleen and bone marrow. A cell population similar to CD1d independent NK T cells can be expanded from peripheral blood mononuclear cells by the timed addition of g-IFN, IL-2 and OKT-3. In vitro culture leads to a 900 fold expansion of CD3+56+ NK T cells in three weeks. MHC independent cytolysis of various syngeneic tumor cell lines could be demonstrated in a mouse model (Baker et al., Blood 97(10), 2001). Here we show that human chronic lymphocytic leukemia (CLL) cells are targeted by ex-vivo expanded NK T cells. Freshly isolated CLL-cells can not be labelled sufficiently with 51Cr and are therefore not suitable as targets in a standard 51Cr-release cytotoxicty assay. We measured apoptosis in CLL targets by flow cytometry using annexinV and propidiumiodide. NK T effector cells were labelled with the red membrane dye PKH26 and were excluded from analysis. After 4 hour incubation with NK T cells and CLL cells at an effector to target ratio of 10 : 1, specific apoptosis (annexin and/or PI positivity) in CLL cells was 29%. Cell death in target cells could be further increased by using a bispecific antibody crosslinking CD3 on NK T cells and CD19 on CLL cells. When 100 ng/ml of CD3xCD19 bispecific antibody was added to the assay, specific apoptosis in CLL cells was 67%. The antibody alone had no effect on target cell survival. These results could be reproduced in 2 other CLL patients. In conclusion, NK T cells induce apoptosis in primary CLL cells without antigenic priming. Retargeting of NK T cells with a bispecific antibody is able to further increase target cell death. The fact that these ex vivo activated and expanded effector cells did not cause GVHD in a cross major transplantation model (Baker et al.) holds promise for their use in cellular therapy after allogeneic stem cell transplantation in CLL and lymphoma.

499

Quantitative molecular monitoring of residual tumor cells in chronic lymphatic leukemia

H. Schulz, T. Pfitzner, M. Reiser, S. Barth, P. Borchmann, T. Schinköthe, F. Oberhäuser, M. Tur, V. Diehl, A. Engert (Köln, Sulzbach, Aachen, D)

Rationale: Most likely all patients with Chronic Lymphocytic Leukemia (CLL) in CR treated with standard chemotherapy relapse, due to minimal residual disease. Therefore, it has become a major focus of interest to quantify the exact number of residual tumor cells remaining after treatment. This approach becomes even more interesting in B-CLL, since molecular remissions have been observed after more aggressive treatment approaches such as high-dose chemotherapy (HDCT) and allogeneic bone marrow transplantation. Thus, new improved techniques to monitor residual tumor cells on a molecular basis in CLL are warranted. We designed a real-time quantitative allele-specific-oligonucleotide-PCR (ASO-PCR) for patients with B-CLL. The PCR-assay was standardised with identical cycling parameters, as well as primer-, probe- and MgCl2-concentration for each patient. Results: We monitored ten patients with DNA samples obtained at 52 time points (median 5,2 per patient). The median follow up per patient was 11,4 months. 9/10 patients had PCR-detectable residual tumor cells in the peripheral blood after therapy. One patient became PCR negative with a combination of Fludarabine and Rituximab after the end of treatment. The MRD-levels in patients with detectable disease ranged from 0.002 % to 10.1 % after therapy. Conclusion: Real-time quantitative ASO-PCR can be utilised for quantitative molecular monitoring of MRD in B-CLL patients in CR, new effective treatment approaches such as combined Chemo/Immunotherapy can render CLL patients PCR negative and different MRD-levels in PCR positive patients were observed warranting further investigation into possible correlation with clinical outcome.

500

Clinical markers of disease progression in chronic lymphocytic leukemia do not correlate with the proliferation rate of CLL cells in the peripheral blood

T. Bochtler, W. Beisker, M. Nüsse, G. Meinhardt (München, Neuherberg, D)

Chronic lymphocytic leukemia (CLL) is a clonal B-cell disorder with a variable pattern of disease progression. CLL is characterized by an accumulation of largely quiescent cells arrested in the G0/G1 cell cycle phase. In patients with rapid disease progression, clinical markers suggest a higher proliferation rate of the malignant cells.We examined in 24 patients, whether the heterogeneity of disease progression is reflected by a high variability of proliferative activity of CLL cells in peripheral blood. We compared predictors of disease progression like leukocyte count, thymidine kinase and beta 2-microglobulin levels with the rate of cells in the S- and G2/M-phase of the cell cycle. Leukocyte count ranged from 11,000 to 320,000/ul with a mean of 91,400/ul (SD of ± 63,500/ul). DNA replication was determined by FACS analysis utilizing an intravital stain with 3µg/ml of the dye Hoechst 33342. The rate of cells in the S- and G2/M-phase ranged from 0.08% to 0.45% with a mean of 0.26% and a SD of ± 0.14%. Statistical analysis demonstrated no significant correlation between the number of proliferating CLL cells and the leukocyte count, thymidine kinase or beta 2-microglobulin levels, respectively. Interestingly, the proliferating fraction of mononuclear cells from peripheral blood samples consisted entirely of CLL cells. Taken together, quantification of cells in S- and G2M-cell cycle phase by FACS with Hoechst 33342 stain seems to have no prognostic value.

501

Fludarabine versus fludarabin plus cyclophosphamide in first line therapy of advanced chronic lymphocytic leukemia (CLL): First report of a phase III study (CLL4 protocol) of the German CLL Study Group (GCLLSG)

B. Schmitt, M. Bergmann, R. Busch, C. Wendtner, R. Pasold, O. Sezer, O. Anders, C. Steinbrecher, U. Söling, B. Emmerich, M. Hallek and the GCLLSG (München, D)

Introduction: The purine analogue fludarabine (F) and the combination fludarabine plus cyclophosphamide (FC) are very effective treatments of pretreated and untreated CLL with response rates of 60% to 90%. In a phase III study F was shown to induce higher remission rates and longer progression free survival than chlorambucil in first line therapy of advanced CLL (Rai et al., 2000). At present it is not clear, whether the FC combination improves the outcome of untreated CLL patients (pts). The CLL4 study, a phase III study, therefore was conducted, to evaluate the effect of F versus (vs) FC in first line therapy of patients with advanced CLL. Results From 05/1999 to 2/2001 105 pts were included; 53 pts were randomized to FC and 52 to F. So far 165 cycles of chemotherapy have been documented. The main side effects in both arms were myelosuppression, which occcured slightly more frequently in the FC arm than in the F arm. Frequency and severity of infections were equally distributed in both arms. Three therapy related deaths were documented: two pts died in the F arm and one pt in the FC arm. One pt died because of severe autoimmune hemolytic anemia one pt because of gastrointestinal bleeding and autoimmune thrombocytopenia, which was refractory to immunosuppressive therapy. Another pt died because of sepsis during neutropenia. Two deaths due to progressive disease were reported. At the moment, it is still too early to evaluate the effects of FC vs F with regard to the treatment response. The overall response rate, which was ducumented in 32 patients, is 81% (26 of 32 pts), with 3 complete remissions and 23 partial remissions. Conclusion: F and FC are very effective regimens in first line therapy of CLL, but both regimens show severe side effects, with myelosuppression and infections. Two deaths due to autoimmne hemolytic anemia and autoimmune thrombocytopenia were reported.

502

The kinase inhibitor rottlerin, previously characterized as inhibitor of caspase-dependent apoptosis in various cell systems, is a potent inducer of cell death in chronic lymphocytic leukemia

G. Meinhardt, T. Bochtler (München, D)

Chronic lymphocytic leukemia (CLL) is an incurable malignant clonal B-cell disorder. Although therapeutic alternatives such as monoclonal antibodies or high-dose chemotherapy with autologous stem cell transplantation have emerged recently, all patients will ultimately relapse. We identified the kinase inhibitor rottlerin as a potent cytotoxic compound in CLL cells. CLL cells from peripheral blood samples of seven patients with Binet stages A to C were exposed to different concentrations of rottlerin for 1, 2 and 4 days. Cell death was assayed for hypodiploid DNA content (subG1 cells) by FACS analysis of fixed, propidium iodide stained cells. Beforehand, we established that hypodiploid DNA content correlated with Annexin V positivity in unfixed cells. Cell death in untreated control cells was 3.2% after 4 days. Treatment with 3ug/ml of fludarabine for 1, 2, and 4 days increased the amount of subG1 CLL cells to 3%, 5%, and 14% (±6.7%), respectively. Incubation of CLL samples with 3, 10, and 40 uM rottlerin resulted in an elevated amount of cells in the subG1 phase that extended to 6 (±3)%, 10 (±3)%, and 33 (±25)% after 2 days and to 10 (±3)%, 13 (±6)%, and 29 (±16)% after 4 days. For the individual patients, a clear dose-response relationsship could be observed, although the degree of sensitivity between individual patients varied. Sensitivity to fludarabine was not correlated with a comparable susceptibility to rottlerin treatment. Furthermore, a combination of both rottlerin and fludarabine did not show synergistic effects. In contrast, the curves plotting subG1 cells for rottlerin and fludarabine+rottlerin treated samples were superimposable. This is the first report, that the kinase inhibitor rottlerin, with its attributed selectivity for protein kinase C isoforms, induces cell death. Rottlerin has previously been demonstrated to inhibit apoptosis in various cell systems. Thus, modulators of certain kinases could have therapeutic potential in the treatment of chronic lymphocytic leukemia.

503

Association of CD38 antigen expression with other prognostic parameters in early stages of chronic lymphocytic leukemia

D. Heintel, I. Schwarzinger, C. Chizzali-Bonfadin, R. Thalhammer, J. Schwarzmeier, M. Fritzer-Szekeres, A. Weltermann, I. Simonitsch, K. Lechner, U. Jäger (Wien, A)

The expression of the surface molecule CD38 on B cell chronic lymphocytic leukemia (B-CLL) cells has recently been described as a prognostic marker for patient survival. We have analyzed CD19/CD38 expression in 81 patients with predominantly early stages of B-CLL (69 Binet A, 7 Binet B, 5 Binet C). 62 patients (77%) had less than 30% CD38+/CD19+ cells, while 19 (23%) had > 30%. There was a significant association between Binet stages (A vs. B+C, p<0.0001), Rai stages (0-II vs. III+IV, p<0.001) and CD38 expression, confirming the published cut-off level of 30%. A particularly strong association between CD38 expression was found with serum CD23 (sCD23) levels of > 2000 U/ml (p<0.0001). and ß2-microglobulin serum levels of >3 mg/L (p<0.0001) indicating that CD38 is a marker of tumor mass as well as disease progression. A borderline association was found with lymphocyte doubling time (LDT) <12 months (p=0.05) due to low patient numbers, while there was no association with age, sex or immunoglobulin deficiency. Discordant results were obtained in a number of patients: 10 of 67 patients (15%) with Binet A had a CD38 >30% while 3 of 7 patients with Binet B had a CD38 <30%. In these two subgroups CD38 and other prognostic factors gave discrepant results. Due to the early stage and short median observation time (12 months, range 1-24 months) calculations concerning patient survival were not performed. However, our data show a strong association between CD38 and other known prognostic factors. The results also suggest that this factor is not always reliable in Binet A patients.

504

Fludarabine induces secondary resistance in CLL

T. Schinköthe, R. Henrichs, T. Siebert, D. Voliotis, M. Reiser, B. Lathan, V. Diehl, P. Staib (Köln, D)

Fludarabine, prednisone and vincristine are frequently used in the treatment of CLL. Patients often receive these standard drugs repeatedly. We investigated whether pretreatment with these drugs leads to secondary resistance analyzing by in-vitro chemosensitivity testing. Material and Methods: For the in-vitro chemosensitivity testing we used the differential staining cytotoxicity (DiSC) assay. Before performing the DiSC assay, we incubated fresh leukemic cells with and without various drugs for four days (37°C, 5%CO2). Drugs were tested in duplicate samples and 5 different concentrations. The dose range was chosen around the clinical steady state plasma level. We computed the area under the curve for analyzing the dose-response-relationship. The fresh leukemic cells were isolated from peripheral blood samples from 37 patients with the established diagnosis of CLL. 10 patients had received no therapy before the time of carrying out the in-vitro test. 23 patients were pretreated with prednisone, 8 patient with fludarabine and 6 patients with vincristin (most patients ha received poly-chemotherapy). Results and Discussion: We compared the in-vitro sensitivity of cells from patients pretreated vs. not pretreated with the analyzed drug. Pretreatment with prednisone or vincristine did not show any significant changes in the in-vitro sensitivity (prednisone p=0.98; vincristine p=0.69). However, cells from patients pretreated with fludarabine had a 3.9-fold decreased sensitivity (p=0.008). We conclude that patients develop a drug induced resistance after treatment with fludarabine, thus fludarabine does not seem to be suitable for repeated therapy in patients with CLL.

505

Severe hematological and non-hematological toxicity after treatment with Bendamustine (B) in pretreated patients with relapsed or progressive chronic lymphocytic leukemia (CLL)-preliminary results of a phase I study of the German CLL study group GCLLSG

M. Bergmann, M. Herold, B. Emmerich, M. Wilhelm, L. Böning, M. Hallek (München, Erfurt, Würzburg, D)

Introduction: Bendamustine (B) is an antineoplastic agent combining a purine-like benzimidazol and a bifunctionally alkylating nitrogen mustard group. Though B is widely used as third-line therapy in relapsed CLL patients(pts), no recommendations are established with regard to optimal dosing and treatment intervals. Therefore, a phase I study was initiated to determine the MTD and the DLT of B in order to prepare future phase II studies. Patients and methods: From Oct 2000 to May 2001 10 pts with relapsed or progressive CLL were included. The median age of pts was 68 years. Pts were pretreated with at least two treatments, usually chlorambucil and fludarabine. Results of previous phase II studies suggested to start with a dose of 100 mg/m^2 on days 1 and 2 every 3-4 weeks. In the absence of any DLT in the first 3 pts after the first treatment course, the dose was escalated by 10 mg/m^2/d. If a DLT occurred in 1/3 pts, 3 additional pts were included at the same dose level. If a DLT occurred in 2 pts at this dose level, the dose was decreased for the next 3 to 6 patients by 10 mg/m^2/d. Toxicity was evaluated by the common toxicity criteria (CTC). One pt was not evaluable for toxicity.
Results: The first 2 pts treated with 100 mg/m^2/d on d1 and d2 showed a CTC grade 4 leukopenia after the 1st course. One pt developed an erysipela with fever in neutropenia requiring antibiotic treatment. In both cases treatment was discontinued. The next pt was treated with 90 mg/m^2 on d1 and d2. He showed a leukopenia CTC grade 4, treatment was continued at a 50% dose reduction. Because of the severe side effects occuring at this dose, it was decided to decrease the dose to 80 mg/m^2/d. The subsequent 5 pts were treated with 80 mg/m^2/d. One pt showed progressive disease after the 1st course and died one month later. Another patient developed acute renal failure CTC grade 4 after administration of the 2nd course, treatment was discontinued. 2 pts developed a CTC grade 4 pneumonia requiring hospitalization and i.v. antibiotics after the 1st course of therapy. One of these pts died because of a HSV-pneumonia (diagnosed by BAL) and renal failure after the 1st course. Due to these 2 grade 4 toxicities at the 80 mg/m^2/d dose level, the dose was again decreased by 10 mg/m^2/d. The study currently continues at this dose. Conclusion: In contrast to reports recommending a higher dose of B for treatment of low grade NHL, this study found a very high toxicity in pretreated pts with CLL, even at a dose of 80 mg/m^2/d on d1 and d2 every 3-4 weeks. Grade 3 and 4 toxicity occured in most patients. Myelotoxicity and severe, life-threatening infections were the most frequent side effects. The study is ongoing using a B dose of 70 mg/m^2/d on d1 and d2 every 3 weeks. Preliminary analysis of treatment efficacy shows that B seems to be very effective in pretreated pts with CLL. However, it must be cautioned that B might be much more toxic in pretreated CLL patients than previously thought.

Postersitzung:
Dendritische Zellen

506

Activated CD8+ T-lymphocytes induce differentiation of monocytes to dendritic cells and restore the stimulatory capacity of IL-10 treated antigen presenting cells

S. Wirths, J. Reichert, F. Grünebach, L. Kanz, W. Brugger, P. Brossart (Tübingen, D)

Dendritic cells (DC) and T-cells can interact with each other and CD4+ positive T-Helper lymphocytes were shown to play an important role in the regulation of DC function via the expresion of the TNF/TNF receptor family molecules on the cell surface. In our study we analyzed the influence of purified CD8+ T-cells on DC function and differentiation. Incubation of purified blood monocytes with activated but not resting CD8+ T-cells resulted in the differentiation of monocytes into DC with a CD1a+, CD83+ and CD14- phenotype thus demonstrating that CD8+ T-cells can productivelly interact with antigen presenting cells (APC) and influence their phenotype and function. In a previous report we demonstrated that the addition of IL-10 to the cultures containing GM-CSF and IL-4 completely inhibited the differentiation and T-cell stimulatory capacity of APC generated from peripheral blood monocytes. These cells have a macrophage-like CD14+, CD1a- phenotype. Addition of activated CD8+ lymphocytes induced down regulation of CD14 expression within 24 hours on IL-10 treated monocytes. Furthermore, it resulted in the expression of DC-related CD1a on the cell surface and upregulation of mRNA transcripts for RelB, IL-12 p40 and the MCP-2, MIP-1alpha and ELC chemokines within 48-72 hours. These DC were potent stimulatory cells in MLR and induced primary CTL responses against a Her-2/neu derived peptide antigen. Blocking antibodies and soluble proteins specific for CD137, TRANCE and CD40L had no effect on the interaction of activated CD8+ T-cells with APC in our experiments. Transwell experiments revealed that beside the receptor/ligand interactions soluble factors independent of TNF-a may contribute to these observations. Blocking of IFN-a and its receptors resulted in only marginal inhibition of CD1a expression, whereas mAbs against IFN-g completely abolished the CD1a induction on IL-10 treated APC. These results demonstrate that CD8+ cells can efficiently interact with and provide stimulatory signals to APC to induce their differentiation into DC and to increase their immunostimulatory capacity. According to our results this interaction does not have to be cognate.

507

Induction of clinical and immunological responses in cancer patients after vaccinations with peptide pulsed dendritic cells

P. Brossart, J. Wierecki, C. Gouttefangeas, S. Wirths, M. Radsack, L. Kanz, W. Brugger (Tübingen, D)

Pulsing of dendritic cells (DC) with tumor antigens and their applications in cancer patients was shown to be effective in immunotherapies of malignant diseases. We analyzed the clinical and immunological responses in patients with malignant diseases using autologous mature DC derived from peripheral blood monocytes. DC were pulsed with HLA-A2 binding peptides deduced from Her-2/neu and/or MUC1. For the recruiting and activation of CD4+ T-helper lymphocytes DC were furher incubated with the PAN-DR binding peptide PADRE. DC vaccinations were performed sc every two weeks four times and repeated afterwards monthly until tumor progression. After the four DC injections patients additionally received 3 injections/week of low dose IL-2 (1Mio IE/m^2) sc. The vaccine-induced enhancement of T cell precursor was monitored using intracellular IFN-g staining and ELISPOT assay. Furthermore, we tested the ability of postvaccination PBMC to produce cytokines in response to challenge with vaccine-related epitopes by quantitative real-time PCR. 16 patients with advanced malignancies were included so far in this study. In one patient with metastatic renal cell carcinoma (RCC) a partial remission of the metastatic sites was induced after the first 4 vaccinations with MUC1 peptides pulsed DC, that was ongoing after the next cycles. Vaccine-induced peptide specific T-cell responses in vivo were detected in the PBMNC of this patient and in peptide specific DTH reactions. Another RCC patient developed a regression of a pulmonary metastasis after four vaccinations while being stable with abdominal leasions. In a patient with two metastatic malignancies (breast and ovarian cancer) a mixed response with complete remission of lymph node metastases but progression of abdominal manifestations was observed. This study demonstrates that peptide pulsed DC can be effective in cancer patients and might induce significant clinical and immunological effects, especially in patients with RCC. Further data on clinical and immunological responses from this ongoing trial wiil be presented.

508

Signal regulatory molecule SIRPalpha interacts with CD47 and is involved in T-cell activation by dendritic cells

P. Brossart, M. Seifert, C. Cant, M. Cella, M. Colonna, W. Brugger, L. Kanz, A. Ulrich, H.-J. Bühring (Tübingen, D)

Signal-regulatory proteins (SIRPs) represent a new family of inhibitory/activating receptor pairs which consist of three highly homologous Ig-like domains in their extracellular regions, but differ in their cytoplasmic regions suggesting different biological functions. Recently, we demonstrated that SIRPalpha binds to CD47 that is expressed on T-lymphocytes. To analyze the expression, function and ligand binding capacity of SIRPalpha and SIRPbeta molecules on DC, soluble fusion proteins consisting of the extracellular domains of SIRPalpha1, SIRPalpha2, and SIRPbeta1, as well as SIRPalpha- and SIRPbeta-specific monoclonal antibodies (mAbs) were generated. The specificity of generated mAbs was verified using transfectants and western blot analysis. In contrast to SIRPalpha1 and SIRPalpha2, no adhesion of SIRPbeta1 to cellular expressed and soluble CD47 could be detected. Using deletion constructs, the epitope responsible for SIRPalpha1 binding to CD47 could be confined to the N-terminal Ig loop. FACS analysis with SIRPalpha- and SIRPbeta-specific mAbs revealed a strong SIRPalpha expression on primary myeloid and plasmocytoid DCs isolated from peripheral blood and on in vitro generated monocyte derived DCs. In contrast, SIRPbeta expression was higher on plasmocytoid DC. Analysis of the T-cell stimulatory capacity of in vitro generated DC in the presence of soluble SIRPalpha1 fusion proteins as well as SIRPalpha- and CD47-specific mAbs revealed a significant reduction of T cell proliferation in mixed lymphocyte reaction and inhibition of induction of primary T-cell responses under these conditions whereas blocking of SIRPbeta had no effect on DC function. Our data suggest that SIRPalpha and CD47 play an important role during T cell activation and induction of antigen specific CTL responses by DCs and demonstrate a novel regulatory parhway in DC-T-cell interaction.

509

Enhancement of cytokine-induced killer (CIK) cell cytotoxic activity against multiple myeloma cells by coculturing with idiotype-pulsed dendritic cCells

A. Märten, S. Renoth, M. von Lilienfeld-Toal, P. Buttgereit F. Schakowski, T. Sauerbruch, I. Schmidt-Wolf (Bonn, D)

Background and objective: There are numerous reports of in vitro and in vivo usage of dendritic cells (DC) pulsed with idiotype, the tumor-specific antigen of MM, for immunotherapy of multiple myeloma (MM). Data suggest that not only T cells, but also the innate immune system reacts against MM. Here, we examined the cytotoxic activity of cytokine-induced killer (CIK) cells against myeloma cells. This heterogenous effector population consists of T-, NK- and NKT cells. Design and Methods: CIK cells generated from buffy coats or blood from patients with MM were cocultured with autologous idiotype-pulsed DC. The cytotoxic activity was investigated in LDH release assays against cell lines or autologous CD138 positive cells from bone marrow. Results: CIK cells were able to lyse MM cells at low effector to target ratios. This effect could be significantly enhanced by coculturing with specifically pulsed DC (54.4% lysis at an effector to target ratio of 6:1). High cytotoxic activity was also shown in a completely autologous setting against enriched CD138+ cells from a patient with MM. Interestingly, there was no cytotoxic activity against the CD138- fraction of the bonemarrow. Using interferon gamma secreting MACS separation assay, the cytotoxic activity of CIK cells could be shifted to still lower effector to target ratios (99.9% at an effector to target ratio of 5:1). Interpretation: Using a heterogenous population of effector cells, we are able to activate the innate and the adoptive immune-system against myeloma cells. Conclusion: CIK cells showed high lytic activity against MM cells, which could be enhanced by coculturing with antigen-specific pulsed DC.

510

Transfection of immunostimulatory CpG-sequences in immunologic effector cells and dendritic cells to augment anti-tumor activity

M. Gorschlüter, F. Schakowski, P. Buttgereit, B. Schöttker, C. Ziske, M. v. Lilienfeld-Toal, A. Märten, S. Renoth, I. Schmidt-Wolf (Bonn, Köln, D)

Bacterial DNA sequences and oligodesoxynucleotides (ODN) containing unmethylated CpG dinucleotides in a certain base context are able to stimulate a variety of immune cells such as antigen-presenting cells, B cells, T cells and NK cells very efficiently. CpG-ODN are promising agents in immunology-based antitumor strategies. Recent studies demonstrated successful use as adjuvants in tumor antigen vaccination or monoclonal antibody therapy in animal models. CpG-ODN probably elicit their effects by cellular uptake and subsequent activation of transcription factors in the cytoplasm of immune cells. Surprisingly, the effect of transfection of CpG-ODN on their immunological capabilities has not yet been investigated. We established dendritic cells (DC) and a CD3+/CD56+ rich cytotoxic effector cell population from peripheral blood of healthy donors. We optimized conditions of CpG-ODN transfection of DC and cytotoxic T cells which are generally known to be relatively resistant to nonviral gene transfer. We applied electroporation and cationic complexes (e.g. branched polyethylenimine, lipofectamine) and various CpG-ODN. Transfection efficiency as assessed by flow cytometry after transfer of fluorescein-labeled CpG-ODN could be increased up to 99% of cytotoxic T cells. The average rate of necrosis of cytotoxic T cells under this condition was 23%. Furthermore, we present analyses of expression of cell surface molecules. First results of chromrelease assays of transfected cells and colon carcinoma cells as targets showed no significant increase of cytotoxicity. Further cytotoxicity assays are in process. In conclusion, our results show efficient CpG-ODN transfection in cytotoxic T cells and dendritic cells and demonstrate perspectives of this approach for cancer immunotherapy. Supported by the H.W. & J. Hector-Stiftung, Weinheim.

511

Idiotype-KLH pulsed dendritic cell vaccine before and after autologous hematopoietic cell transplantation for patients with multiple myeloma

S. Auffermann-Gretzinger, K. Stockerl-Goldstein, C.J Benike, B. Taidi, A. Liso, R. Rajapaksa, A.v. Beckhoven, E.G. Engleman, K.G. Blume, R. Levy (Palo Alto, USA)

Idiotype-based vaccination of multiple myeloma patients has been shown to be feasable and well tolerated, but Id-specifc immune responses are rarely achieved. We therefore investigated a new vaccination strategy administering vaccine prior to and after autologus hematopoietic cell transplantation (AHCT). The regimen consisted of an Id-KLH pulsed dendritic cell (DC) vaccination before stem cell collection. The patients then received one intraveneous infusion of DC pulsed with Id-KLH at 6 month following AHCT and 5 subcutaneous boosts of Id-KLH and GM-CSF. 21 patients have been included in this trial. 9 patients went off study after the first DC vaccine. Two more patients dropped out of the study due to progressive disease before having received half of the vaccines. Out of the remaining 10 patients 5 have completed the series and 5 are still undergoing the vaccinations. 9 out of these 10 patients generated an idiotype-specific proliferative immune response and all 10 patients attained a KLH-specific response. We also measured the cytokines TNF-a, IFN-g, IL-2, IL-4, IL-5, IL-10 secreted by PBMC after stimulation with Idiotype and KLH and detected mainly TH1 type immune responses. In 2 patients tested we detected idiotype specific cytotoxic T cells. The high rate of induced idiotype-specific immune responses seems to favor this new schedule of the vaccination. We conclude that the approach of idiotype vaccination prior to and after AHCT can induce Id specific cellular responses with a high efficiency. An update of the data will be presented at the meeting.

512

Anti-tumor immune therapy of pediatric patients suffering from solid neoplasias with tumor protein loaded dendritic cells

T. Felzmann, W. Holter, V. Witt, H. Gadner (Wien, A; Erlangen, D)

In spite of combination chemotherapy and dose escalation solid tumors of childhood still have a poor prognosis. Many centers, therefore, search intensively for new treatment concepts. Anti-tumor immune therapy (ATIT) is one of the most promising new strategies. Particularly the increasing understanding of the biology of antigen presenting dendritic cells (DC) as the central regulators of an immune response gave first encouraging results in therapeutic applications. We are currently conducting a clinical pilot trial for the ATIT treatment of patients suffering from solid pediatric malignancies with tumor antigen loaded DCs. So far eight patients are included into this trial (3 osteosarcomas, 3 hepatocellular carcinomas, 1 renal cell carcinoma, 1 Ewing's sarcoma) and five have been treated. Three patients completed the whole treatment whereas two others dropped out of the protocol after the first cycle. No major toxic side effects of the treatment were observed. In the patients who completed the treatment we found a positive response to the tracer antigen keyhole limpet hemocyanine that is included into the tumor vaccine. We conclude that ATIT with DCs in a feasible approach for tumor treatment with only minor side effects. More patients will be included into this pilot trial that will be followed by a phase I/II study to determine efficacy.

513

Human cytomegalovirus efficiently infects dendritic cells and suppresses DC function and cell-mediated immunity

U. Grigoleit, H. Hebart, G. Jahn, C. Sinzger, S. Riegler, F. Gruenebach, L. Kanz, H. Einsele (Tübingen, D)

Human cytomegalovirus (HCMV) infection is one of the most important and life threatening infections in patients undergoing allogeneic stem cell transplantation (SCT). In addition, HCMV infection seems to induce severe immunosuppression and is often followed by other life-threatening infectious complications, e.g. invasive aspergillosis. To better understand the effects of HCMV on dendritic cells (DC), the most professional antigen presenting cells, DC of healthy HCMV sero-negative donors were infected with endothelotropic HCMV strains in vitro. Twenty-four hours after virus infection, the efficiency of infection was determined by immunocytochemical staining for immediate early antigen (IEA). In addition, the cytokines IL-1, IL-6, IL-10, IL-15, IL-18 and TNF-alpha were analysed in the DC culture supernatant. Furthermore the expression of CD1a and CD83, MHC class I and II as well as the co-stimulating molecules CD 80, CD 86 and CD 40 were determined on infected and uninfected DC by flow cytometry. For their immunostimulatory function, the DC were tested in a mixed lymphocyte reaction (MLR). As a negative control mock infected DC of each donor were cultured and analysed under the same conditions. After 24 h of infection, 60%-80% of the DC infected with endothelotropic HCMV strains expressed HCMV specific protein, e.g. IEA. Whereas the expression of CD1a and CD83 was not significantly different between non-infected and infected DC, MHC class I, CD80 and CD40 were significantly reduced in surface expression 24 h after HCMV infection, while a significant reduction of MHC class II was detected 72 h post infection. The proliferation rate with infected DC as stimulator cells in the MLR was 2-3 fold lower compared to the one with non infected DC as stimulator cells. In conclusion, endothelotropic HCMV strains efficiently infect dendritic cells, reduce the expression of MHC class I and costimulatory molecules and thus suppresses cell-mediated immunity. This might be one explanation of HCMV mediated immunosuppression.

514

Quantitative and qualitative in vitro assessment of phagocytosis by dendritic cells

A. Eil, P. Reinhardt, B. Maccari, S. Hübsch, M. Wiesneth, M. Schmitt (Ulm, D)

BACKGROUND: Phagocytosis and intracellular digestion by oxygen-dependent mechanisms are essential for host defense against bacteria, viruses and tumor cells by dendritic cells (DC). Several immunotherapy trials using DC preparations for patients with tumors or leukemias have been initiated and might be useful for the treatment of infectious diseases. The assays described here are helpful methods for the in vitro quality control of DC preparations. MATERIAL AND METHODS: Monocytes were collected by leukapheresis, separated by adherence and cultured in standard medium supplemented with IL-4 and GM-CSF for 5 days with a half medium change on day 3. On day 6 and 7, the medium was supplemented additionally by TNFa. The phenotype of the cells was determined by immunofluorescence analysis (FACS) on day +1 and day +7 of culture using a panel of monoclonal antibodies and the double-staining technique. Phagocytosis and oxidative burst were measured on day 0 (naive monocytes), day +1 (immature DC) and day +8 (mature DC) after in vitro culture.
RESULTS: After selection by adherence and after one day of culture the cells showed a phenotype compatible with a predominance of monocytes/macrophages: CD 11c (99%), CD 14 (86%), CD 83 (13%), CD 86 (98%), HLA-ABC (98%) and HLA-DR (98%). On day +8 of culture, the phenotype of the cells was characteristic for DC: CD 11c (75%), CD 14 (0%), CD 83 (56%), CD 86 (95%), HLA-ABC (67%) and HLA-DR (98%). In comparison to the naive monocytes of day 0, the immature DC of day +1 after culture showed a 30% median reduction of the phagocytic capacity and a 50% median decrease of the oxidative burst. On day +8 of culture, the mature DC had no ingestive and digestive function anymore. CONCLUSION: Immature, but not mature DC are characterized by ingestive and processing functions, suggesting that early coincubation with proteins or apoptotic tumor cells would be preferable for pulsing of DC with tumor antigens.

515

Laser-assisted optoporation allows high transfection efficiency of a green fluorescent protein (GFP) vector into dendritic cells (DC)

M. Schmitt, A. Hendinger, A. Eil, P. Reinhardt, S. Hübsch, M. Wiesneth, R. Sailer, H. Schneckenburger (Ulm, Aalen, D)

Recently, DC gained high interest in the field of cellular therapy as cancer vaccines. They are the most professional antigen presenting cells (APC) with the potential to induce strong cellular and humoral immune responses to kill tumor cells. For specific immune therapy it would be highly desirable to transfer DNA or proteins of tumor antigens into DC. Like other hematopoetic cells DC are difficult to transfect with conventional methods. Optoporation is a novel technique to open up biological membranes of cells or cell organelles. To our knowledge, this is the first report on the use of this novel method for the transfection of DC with DNA. Monocytes were prepared from leukapheresis product of healthy volunteers, and cultured in standard medium containing IL-4 and GM-CSF. Flow cytometry revealed the phenotype of DC at day 8. We developed a device where adherent cells or cells fixed by optical tweezers (Nd:YAG laser of 1064 nm) were submitted to optoporation using an argon ion laser (AIL) of 488 nm and a monomode glass fiber. Successful optoporation was deduced from an influx of trypan blue into the cell. When using the focused beam of the AIL and phenol-red at a concentration of 40 µM as an absorbing dye, small dark spots (1-2 µm) and an interference pattern were observed by interference contrast videomicroscopy immediately after the optoporation of mature DC on day 8 with 1 MW/cm^2 over 1 to 10 s, indicating changes of membrane geometry or fluidity. These changes were reversible and disappeared within about five minutes. According to a colony forming assay, the viability of these cells after optoporation was maintained to about 80 % when using laser energy doses of 1 MJ/cm^2 or less applied for one second. Following laser irradiation a rapid release of the cell marker calcein at a concentration of 5 µM into the surrounding medium was observed by fluorescence microscopy, thus proving the applicability of the method. In a further experiment, the cellular uptake of a green fluorescent protein (GFP) coding plasmid dissolved in the incubation medium, was studied: brightly fluorescent GFP with an emission maximum around 510 nm was observed after 24 hours. Cell transfection rates with laser-assisted optoporation were 40 %, and thus higher than with the well established plasmid transfection methods of lipofection (10 %) or electroporation (10 %). Further increase of the transfection rate appears possible after optimization of experimental parameters. Optoporation might be an effective transfection method for hematopoetic cells difficult to transfect with conventional methods, and can be performed on a large cell number scale with an appropriate cell flow apparatus. Higher transfection rates and safety in comparison with adeno- or retroviral vector systems could finally be the main advantages of laser optoporation as compared with existing methods of cell transfection.

516

Blast-derived dendritic cells in AML show a severe defect in the expression of functionally important cytokines and cell surface receptors

K. Kratz-Albers, J. Schwäble, B. Massmann, J. Kienast, W.E. Berdel, H. Serve (Münster, D)

Recent evidence suggests that dendritic cells (DC) derived in vitro from AML blasts (BDDC) might be used to specifically enhance immune recognition of AML. Although BDDCs could be readily obtained by many groups, reports about an immune response from cytotoxic T cells conditioned on these DCs have been sparse, and our own previous results indicated a severe functional defect of BDDC. Dendritic cell function is dependent on several cytokines and membrane proteins, one of which is the recently identified ICAM-3 receptor DC-sign. In order to evaluate the immune function of BDDC, we measured the mRNA expression of a large panel of cytokines and of DC-sign in these cells and compared it with monocyte-derived DC (MDDC) from healthy donors. We generated BDDC from five patients with high bone marrow infiltration by AML (>80%) and MDDC from five healthy donors. Cells were cultured with Granulocyte-Macrophage-Colony Stimulating Factor (GM-CSF), Interleukin (IL) -4 and Tumor Necrosis Factor-alpha for 9 days and then analyzed by flow cytometry (CD14, CD83, CD1a, CD80, CD86, HLD-DR, HLA-ABC). Cells were sorted according to their CD80 and CD86 expression by fluorescence-activated cell sorting (FACS) to a purity of > 98%. Real-Time-PCR was used to quantify the mRNA-content of cytokines and other functionally important molecules: Interferon (IFN) gamma, GM-CSF, Transforming Growth Factor (TGF)-beta, IL -6, IL-7, IL-10, IL-8, IL-12p35, IL-12p40, IL-15 and DC-Sign. Expression values were normalized to GAPDH expression. There was a significantly lower or no expression of IFN-gamma mRNA in BDDC from all five patients if compared with MDDC. Also, GM-CSF expression was significantly reduced in most BDDC preparations. Most significantly, no, or very low levels of DC-Sign could be detected in BDDC, while MDDC expressed high levels of this functionally important ICAM-3 receptor. In summary, the transcriptional profile of functionaly important genes in FACS-sorted DC populations reveals a severe defect that possibly explains the failure of these cells to elicit a sufficient immune response in vitro.

517

Dendritic cell generation in AML and MDS - comparison of different FCS-free culture conditions

H. Schmetzer, R. Pelka-Fleischer, T. Kröll, W. Hiddemann, V. Nüssler (München, D)

Insufficient presentation of tumor antigens may result in an insufficient elimination of leukemic cells by the immune system. Dendritic cells (DC) are professional antigen presenting cells and induce a T-cell mediated immune response.DC can be successfully generated from AML-cells presenting leukemic genotypes and antigens as well as DC markers .The aim of our study was to find optimal FCS-free culture conditions for the generation of DC from AML and MDS-PB-samples.We cultured mononuclear PB-cells from 2MDS and 8AML patients and 4 healthy donors in parallel in two different FCS-free media (X-vivo 15, Cellgro DC), supplemented with GM-CSF,IL-4 ,TNF-alpha ,plusminus FL, plusminus 10% autologous plasma. The DC- and leukemic antigen expressions were evaluated at day (D) 0,7 or 14 using monoclonal DC markers (e.g. CD1a, CD83, HLA-Dr), costimulatory antibodies (e.g. CD80, CD86) together with myeloid and blast markers (e.g. CD33, CD117, 7.1, CD34) (Table 1).

Our data show, that DC can be grown in FCS-free conditions from normal or leukemic PB-samples with amounts depending from diagnosis and culture conditions:

1.) comparison Cellgro/X-vivo: amounts of DC and antigen profiles were similar ,however with higher proportions of dying cells under Cellgro cultivation.

2.) culture plusminus FL: whereas in AML and healthy PB samples amounts of DC were not significantly different in the parallel culture , in MDS higher DC proportions could be found if FL was added to the media.

3.) culture ± autologous plasma: in most of the AML and healthy samples studied an inhibitory effect on DC generation was detectable, if plasma was added to the culture.

4.) proof of leukemic origin of DC: in AML patients with a high percentage of blasts positive for CD117, CD34 or 7.1 (marker for 11q23 aberrations) a coexpression of those antigens with DC antigens could be demonstrated pointing to the fact, that DC generated derived from clonal, leukemic cells.

We can conclude that optimal generation of DC in AML and healthy donors is possible with X-vivo without FL and autologous plasma; in cases with MDS the addition of FL and autologous plasma might improve the harvest of dendritic cells.

518

Dendritic cells fused with acute myelogenous leukemia (AML) blast cells induce activation of antitumor cytotoxic lymphocytes

N. Usluoglu, K. Ihlow, S. Hoppmann, O. Christ, M. Hoeck, H.B. Pralle, G.A. Banat (Giessen, D)

Several observations indicate that T-lymphocytes can recognize AML blasts and thereby mediate antileukemic reactivity. For example, it is documented that T cells are involved in the pathogenesis of antileukemic graft versus host disease. Moreover, AML blasts often have chromosomal abnormalities which could encode for leukemia specific antigenic peptide sequences probably presented in the context of self MHC molecules. In most patients AML blasts express both MHC class I and II molecules for antigenic presentation. The expression of costimulatory molecules on AML blasts is very heterogenous and could be downregulated during one cell cycle, suggesting the ability to initiate an antileukemic T cell response which differs not only between individual patients, but may also result in T cell anergy. The coculture of AML blasts and T lymphocytes as well in an autologous as in an allogeneic manner showed that a T cell stimulation could be induced in only a very low percentage of the cases. To overcome this failure we fused in this study AML blasts with autologous DC by a combination of polyethylene glycol and electroporation. The fusion cells were positive for MHC class I and II, B7-1, B7-2 and adhesion molecules. Most of the fused cells exhibited typical DC morphology with veiled processes and dendrites. In a mixed lymphocyte hybrid reaction we could demonstrate that the fusion cells retained the functional potency of DCs and were able to induce proliferation of autologous T cells compared to the most naive AML blasts. Moreover, the autologous T lymphocytes could be primed by the fusion cells to induce cytotoxicity against autologous AML blasts up to 35% in a effector/ target ratio of 10:1 where as the autologous MLR was without any effect indicating the specificity of the T cells. Antibody blocking assays could show that the lysis is mediated by positive T lymphocytes in a MHC class I dependent reaction.

520

A stratified, randomized monocentric comparison of ceftriaxone/gentamicin versus cefepime/gentamicin in neutropenic sepsis

O. Cornely, U. Bethe, B. Salzberger, H. Seifert, M. Schrappe, G. Fätkenheuer (Köln, D)

Background: Neutropenic sepsis is a frequent and life threatening adverse event of chemotherapy. Once-daily ceftriaxone/aminoglycoside regimens are a simple and cheap option. But the safety of ß-lactams with no antipseudomonal activity is still debated. Methods: A prospective randomized monocentric trial was performed on the efficacy and safety of ceftriaxone 2g qd + gentamicin 5mg/kg qd in comparison to cefepime 2g tid + gentamicin 5mg/kg qd in the treatment of neutropenic fever. In case of fever (orally >38.5°C once or >38.0°C twice within 24 hours) and granulocytopenia (below 500/µl when expected to fall below 500 within 72 hours) patients with hematological malignancies or solid tumors were assigned to ceftriaxone or cefepime, each with gentamicin. The primary endpoint was defined as defervescence on day 4 to 6 followed by at least 7 afebrile days. Secondary endpoints were overall response, defined as defervescence on day 28 and toxicity. Results: 211 episodes were included. Fever of unknown origin accounted for 124 episodes (58.8%), microbiologically defined infection for 39 (18.5%), clinically defined infection for 25 (11.8%), both clinically and microbiologically defined infection for 19 episodes (9%). On an intent-to-treat basis 207 episodes were evaluable for the primary endpoint. Ceftriaxone plus gentamicin and cefepime plus gentamicin were successful in 49.5% and 51%, respectively. Overall response was achieved on study day 28 in 92.5% and 91%, respectively. Diarrhea was more frequent with ceftriaxone/gentamicin (6.5% vs 17%), while nausea/vomiting was less (12.1% vs 5%). Conclusion: Once-daily ceftriaxone plus gentamicin was not inferior to cefepime tid plus gentamicin qd in the empirical treatment of neutropenic fever.

521

Abdominal infections in acute leukemia: Interim analysis of a prospective study combining sonography and microbiology

M. Gorschlüter, S. Baumgartner, R. Clarenbach, C. Hahn, C. Ziske, U. Mey, R. Schiemann, T. Sauerbruch, G. Marklein, I.G.H. Schmidt-Wolf, A. Glasmacher (Bonn, D)

In retrospective studies we and others found abdominal infections in neutropenic patients to be associated with high mortality rates. However, very limited prospective data are available concerning epidemiology and adequate diagnosis of abdominal infections in long-term neutropenic patients outside the setting of allogeneic transplantation. Therefore, we initiated an ongoing monocentric prospective study (started January 2001) in patients with acute leukemia in which chemotherapy with an expected duration of neutropenia of at least 10 days was administered. All patients underwent abdominal sonography (liver, gallbladder, gut) before start of chemotherapy, on day 2 to 4 after chemotherapy, and again in case of fever, diarrhea or abdominal pain. This regimen was routinely combined with a standardized clinical examination and a broadened spectrum of microbiological examinations (including non-resident intestinal bacteria, fungi, intestinal viruses, CMV, Clostridium difficile and Cryptosporidium spp.. As of May 2001, 56 neutropenic episodes were included and 45 episodes were evaluable. In two patients neutropenic enterocolitis developed which represents an incidence rate of 4.4%. Both patients died. Marked bowel wall thickening could be detected in both, but in no other patient. In three patients (6.7%) Clostridium difficile enterocolitis developed. All could be managed with oral metronidazole without complications. Abdominal infection caused by intestinal viruses, CMV, and Cryptosporidium spp. were not observed. These results confirm and extend retrospective analyses of our patients between 1993 to 2000. We conclude, that in neutropenic patients with acute leukemia and cytotoxic chemotherapy (1) abdominal infections are life-treatening but relatively infrequent, (2) bowel wall thickening is not a feature of chemotherapy-induced mucositis, but should be always considered as sign of infectious enterocolitis, (3) Clostridium difficile enterocolitis can be efficiently managed with oral metronidazole, and that (4) virusus, classic enteric pathogens (Salmonella, Shigella, Yersinia, Campylobacter spp.) and Cryptosporidium spp. are probably irrelevant pathogens in patients with acute leukemia undergoing intensive induction or consolidation chemotherapy. Supported by the Deutsche José Carreras Leukämie-Stiftung e.V.

522

Late infectious complications after high dose therapy (HDT) and autologous stem cell transplantation (ASCT)

B. Metzner, R. Grüneisl, W. Gebauer, D. Reschke, E. Ost, R. Husmann, M. Klee, D. Reichert, B. Rosien, F. del Valle, I. Zirpel, K.P. Kohse, F. Schunter, H.J. Illiger (Oldenburg, D)

Clinical data of 192 consecutive patients treated with HDT and ASCT in a single institution (from 9/1993 to 12/1999) were analyzed to identify factors predictive for late infectious complications. 25 patients were not evaluable (4 therapy-related early deaths, 17 relapse-related deaths within the first 6 months, 4 lost to follow-up). The diagnoses of 167 evaluable patients were: 61 lymphomas, 17 myelomas, 4 AML, 44 breast cancer, 23 germ cell cancer, 4 ovarian cancer, 11 Ewing's and soft tissue sarcomas, 3 others. The following HDT-regimens were used: TBI and/or busulfan containing chemotherapy (n=16), other aggressive regimens like BEAM (n=139) or less aggressive regimens like VIP or Melphalan (n=95). 57 patients received 2 or more (up to 4) HDT-courses. Altogether 250 courses of HDT were performed. For 25 patients enriched CD34+ cells (Ceprate system) were used. For Pneumocystis carinii prophylaxis inhalation with pentacarinate or oral cotrimoxazole was used for 3 to 4 months following ASCT. The patients were examined and interviewed on infections every 4 to 12 weeks after ASCT. Results: In the first 6 months following ASCT after discharge we saw on average 1 infectious episode per patient (median, range 0-7), usually afebrile not requiring antibiotic therapy (median 0 episode, range 0-3, mostly banal upper airway infections) or febrile and oral antibiotic therapy requiring infections respectively (median 0, range 0-6). 14 patients were treated for infection and recovered in hospital, 6 with active tumor (2 solid tumors, one with active tumor). 3 infection-related deaths occurred in ICU (1 patient with myeloma due to infection with adenovirus, aspergillus and probably CMV, 2 patients with malignant lymphoma following relapse due to Pneumocystis carinii pneumonia one of whom had partly omitted oral prophylaxis). In the second half of the first year after ASCT median 1 (0-6) infection occurred with similar characteristics, 5 patients had to be hospitalized. No further infection-related death occurred, apart from one patient with NHL as a result of Pneumocystis carinii pneumonia 16 months after ASCT. During the first 12 months after HDT 31 patients developed zoster (21 % of 150 evaluable patients with a follow-up > 12 months). They were usually treated orally without any severe course. One infection with parvovirus B 19 occurred in a patient with Hodgkin's disease resulting in transfusion-dependent anemia 9 months after ASCT. He recovered completely as a result of high-dose immunoglobulin treatment. Severe infections occurred only in patients with hematologic malignancies. No other predictive factors (such as age, amount of reinfused CD-34+ cells, CD-34+ cell selection, type of HDT, lymphocyte subsets or immunoglobulin levels) could be identified. Conclusion: Late infectious complications after ASCT are usually mild. Patients with hematologic malignancies have a slightly increased risk of more serious infections and should be observed carefully.

523

Preemptive adoptive immunotherapy for cytomegalovirus (CMV) reactivation after allogeneic transplantation

E.M. Weissinger, A. Ganser, B. Hertenstein (Hannover, D)

Reactivation of viral infections is a major cause of morbidity and mortality after allogeneic hemopoietic stem cell transplantation (HSCT). More then 70% of the patients reactivate CMV after allo-HSCT. The reactivation is treatable in most cases with ganciclovir, but recurrent infections remain a problem. The patients do not develop immunity against CMV. The HSCT leads to the development of tolerant chimeras, which allow donor lymphocyte transfusions(DLT).Unmanipulated DLT has been applied with success to control leukemic relapses after BMT, the control of EBV-lymphoproliferative disease and even for CMV-reactivation. Although the results are promising, one major complication of this treatment is the induction of severe GVHD. We propose the use of CMV-specific T cell lines (CD8 and CD4 T cells), generated ex vivo in order to maximize chances of CMV- prevention and minimize the risk of GVHD development. So far we have generated CMV-specific T cell lines from 15 CMV-sero+ indivduals using the cytokine secretion assay from Miltenyi and we have analyzed these lines for function. We have also initiated screening for CMV-specfic T cells in vivo in patients , who reactivated CMV after allo-HSCT and are currently treated with Ganciclovir or Foscarnet. So far, none of these 5 patients developed CMV-reactive T cells under treatment. In a first clinical trial we use a broad spectrum of CMV-specific T-cell lines generated after stimulation with CMV-lysate and enrichment with MACS (immunomagnetic cell separation).

524

Antifungal prophylaxis with itraconazole reduces proven invasive fungal infections in neutropenic patients: A meta-analysis of 2181 patients

A. Glasmacher, M. Gorschlüter, B. Djulbegovic, C. Hahn, E. Molitor, G. Marklein, T. Sauerbruch, I. Schmidt-Wolf (Bonn, D; Tampa, USA)

BACKGROUND: Although several randomized studies have been performed, the efficacy of itraconazole prophylaxis is not established. Therefore, a systematic review of the available data is warranted. METHODS: Published, randomized, controlled studies with itraconazole for antifungal prophylaxis in neutropenic patients were identified through Medline, the Cochrane database and reference lists. The statistical analysis used the relative risk (RR, 95%-confidence intervals) and a fixed effects model. Identical results were obtained when Peto's odds ratio was used. The primary endpoint was the incidence of proven invasive fungal infections (IFI, non-disseminated sinonasal infections excluded). The criteria for proven IFI in the individual studies were adapted to conform with the EORTC/MSG consensus where possible. RESULTS: Seven studies with 2181 patients/episodes were included (four studies evaluated itraconazole capsules and three oral solution). The relative risk for proven IFI was reduced to RR=0.55 (0.35-0.87) in the itraconazole arm (P=0.01). The relative risk for mortality from IFI was significantly reduced by 83% in studies using itraconazole solution (RR=0.17; 0.05-0.66; P=0.01), but not in studies using itraconazole capsules (RR=1.38; 0.65-2.93; P=0.4). The overall mortality showed a non-significant trend towards a reduction in patients with itraconazole prophylaxis (RR=0.82; 0.60-1.12; P=0.2). Two studies reported a higher incidence of hypokalemia in the itraconazole arm (RR=2.05; 1.47-2.87; P<0.001) and the rate of drug discontinuation was higher in a study that tested cyclodextrin oral solution against fluconazole suspension (RR=3.99; 2.22-7.18; P<0.001). This rate was comparable to control in studies with capsule or cyclodextrin placebo. No other adverse events different from control were reported. CONCLUSION: Itraconazole prophylaxis effectively prevented 45% of proven invasive fungal infections in neutropenic patients. The oral itraconazole solution reduced the mortality from invasive fungal infections by 83% in these patients.

525

Computer-assisted selection of anti-emetic strategies

M. Hartmann, D. Schulz, R. Kath, K. Höffken (Jena, Essen, D)

Recommendations on the appropriate use of corticosteroids (cort) and HT-3 antagonists (HT3A), alone or in combination with benzodiazepines (benz) and dopamine receptor antagonists (dopA) vary between 3 and 9 different anti-emetic strategies. We retrospectively compared the use of anti-emetics in our clinical practice with the consensus recommendations of Peruggia (Eur J Cancer 35: 351-79, 1999). We analyzed all patients of a single departement unit receiving chemotherapy on an in-patient basis during February 2000. 135 days of chemotherapy in 38 patients (20 - 78 years, male/female ratio 1,7/1) suffering from various malignant diseases, were included. Five patients (13.1%) received anti-emetic therapy for other reasons than preventing nausea and vomiting (e.g. cortocosteroids as comedication with paclitaxel).We grouped the chemotherapy in „very low risk" (n=5), „low risk" (n=44), „moderate risk" (n=16), „high risk" (n=40) and „severe risk" (n=30) protocols, according to the Peruggia guidelines. We calculated the total theoretical consensus-guided costs for anti-emetic treatment to be 2079.04 Deutsch Marks (DM) and the real costs to be 2024.93 DM (hospital supplier prices). The differences for each group of anti-emetics are indicated below:

	HT3A	cort	benz	dopA
Theoretical costs	1757	181,26	85,2	55,58
Real costs in DM	1968	55,44	0	1,49
% Costs difference	+12	-70	-100	-98

In conclusion, we found a considerable drug-group related difference between consensus guided anti-emetic strategies and clinical practice. However, the total expenditures did not vary significantly. We assume, that the differences observed are due to non-chemotherapy but patient related factors like history of nausea and emesis, tumour stage, tumour localisation, sex, age, co-morbidity and alcohol consumption. A computer-assisted anti-emetic decision model (Emeto-Help) will be presented which includes both the drug related emetogenic potential and the individual patient's risk.

526

Procalcitonin-reduced sensitivity and specificity in heavily leucopenic and immunosuppressed patients

M. Svaldi, J. Hirber, A. Lanthaler, O. Mayr, M. Mitterer (Bozen, Meran, I)

PCT has proven to be a very sensitive marker of sepsis for non-leucopenic patients. Little is known about its relevance in immunosuppressed and leucopenic adults. 475 PCT determinations were carried out in 73 hematological patients: in 221 occasions the WBC value was below 1000/µl and in 239 occasions it exeeded 1000/µl leucocytes. Patients were classified as: non systemic infected controls (n =280), patients with bacteremia (n =32), sepsis (n =30), severe sepsis (n =3), septic shock (n =3) and SIRS (n =62). When the WBC count was more than 1000/µl s, gram-negative bacteria induced higher PCT levels (median 9,4 ng/ml) than gram-positives (median 1.4 ng/ml). In cases with less than 1000 WBC/ul PCT levels were comparable for gram-negative and gram-positive bacteria (1.1 ng/ml vs 0.85 ng/ml). Regardless of the leucocyte count, the median PCT level in bacteremia always remained below 0.5 ng/ml. In heavily leucopenic situations, PCT levels never exceeded 2 ng/ml even in sepsis, severe sepsis/septic shock, while in the presence of more than 1000/µl WBCs the median PCT values were 4.1 ng/ml and 45 ng/ml, respectively. The positive predictive value for sepsis (cut-off 2 ng/ml) was 93% in the case of more than 1000/µl WBCs, but only 66% in leucopenic conditions. The negative predictive value (cut-off 0.5 ng/ml) was 90% when more than 1000/µl WBCs were counted and 63% in leucopenic conditions. Procalcitonin is an excellent sepsis marker with a high positive and negative predictive value in patients with more than 1000 WBC/µl, but does not work satisfactorily below this leucocyte count.

527

Predictive value and clinical relevance of surveillance Aspergillus antigen tests in the serological diagnosis of invasive aspergillosis in high-risk neutropenic patients with acute myeloid leukemia

M. Starck, S. Zewen, C. Waterhouse, C. Nerl (München, D)

Background: Invasive pulmonary aspergillosis is one of the most life-threatening infectious complications in high-risk neutropenic patients. Early diagnosis is an important basis for proper therapy. Many investigators have suggested that detection of Aspergillus antigen is useful for early diagnosis. We investigated the predictive value of a serological Aspergillus antigen test in neutropenic patients with acute myeloid leukemia. Patients and methods: Twice weekly serological surveillance monitoring of Aspergillus antigen using an enzyme immunoassay for early diagnosis of invasive pulmonary aspergillosis was evaluated retrospectively in 62 neutropenic episodes in 25 patients receiving myelosuppressive induction or consolidation chemotherapy for acute myeloid leukemia. The results were compared with clinical, radiological and microbiological examinations. Results: In 12/62 neutropenic episodes invasive pulmonary aspergillosis was suspected by clinical and radiological signs. In all these 12 cases clinical and radiological improvement was seen after antifungal treatment, and therefore invasive pulmonary aspergillosis was assumed. In 1/62 neutropenic episodes invasive pulmonary aspergillosis was proven by microbiology and histopathology. A questionable positive test for Aspergillus antigen was only documented in 2 of the 12 suspected cases. A positive Aspergillus antigen test was never observed in clinically and radiologically unsuspected cases. In both cases of positive antigen test clinical and / or radiological signs conclusive for invasive pulmonary aspergillosis were additionally found, and hence a positive antigen test never was the sole criterion for introduction of systemic antifungal treatment. In none of the cases a positive test was first indication of invasive aspergillosis. Sensitivity, specifity, positive predictive value, and negative predictive value were 15%, 100%, 100%, and 82%, respectively. Conclusion: In contrast to other reports, in our retrospective analysis serological surveillance monitoring of Aspergillus antigen did not yield sufficient positive results for invasive pulmonary aspergillosis in high-risk neutropenic patients. The test was insufficiently sensitive to diagnose invasive pulmonary aspergillosis before other symptoms are apparent and therefore was unsuitable for earlier initiation of systemic antifungal treatment.

528

Totally implantable venous access systems for patients undergoing chemotherapy for hematological malignancies: A prospective controlled clinical observation trial

C. Harter, G. Egerer, G. Nöldge, H.J. Salwender, H. Goldschmidt, H.J. Hansmann, G. Kauffmann, A.D. Ho (Heidelberg, D)

PURPOSE: To determine prospectively the outcome of radiological placement of central venous access devices in patients with haematological malignancies we prospectively evaluated the performance and rate of short- and long-term complications of the Z-Port-F-8 system (Zambon Biomedica, Antibes, France), a totally implanted vascular access system. MATERIALS AND METHODS: In 86 patients (pts.) with haematological malignancies chest port systems were placed radiologically. The median follow up was 9.5 months (range 6 – 15 months). Success and complication rates (immediate and long-term) were recorded. RESULTS: The median age of the pts. was 53 years (range, 16 - 78). No extravasation, skin necrosis or leakage occurred after 16.536 days of use. Perioperative complications were recorded in 6 pts. (6.9%): pneumothorax in 2 pts. (2.3%), bleeding in 2 pts. (2.3%), and hematoma in 2 pts. (2.3%). In 7 pts. (8.13%) late complications occurred: catheter disconnection in 1 pts. (1.16%), thrombosis in 3 pts. (3.4%) and subcutaneous port-infections in 3 patients (3.4%). Late complications were higher in patients who had received chemotherapy prior to 24 hours following implantation. Patency of the port system was maintained using a regular flushing schedule once every 28 days. CONCLUSION: The results of this study suggest that the Z-Port-F-8 system can provide a safe and reliable method for venous access in patients requiring intermittent or prolonged intravenous chemotherapy. To reduce the rate of late complications a 24 hour interval after implantation of the Z-Port-F-8 system should be achieved.

529

Increased risk of complicated CMV infection after the use of mycophenolate mofetil in allogeneic stem cell transplantation

L. Hambach, M. Stadler, E. Dammann, A. Ganser, B. Hertenstein (Hannover, D)

Mycophenolate mofetil (MMF) is increasingly used for prophylaxis and therapy of GvHD in allogeneic stem cell transplantation. In some recent reports on use of MMF in solid organ transplantation a high incidence of CMV diseases in patients has been described. We therefore evaluated the frequency and course of active CMV infection in patients who received MMF compared to those who did not receive MMF in our institution. We retrospectively analyzed 49 adult patients who consecutively underwent unmanipulated allogeneic bone marrow (n=15) or peripheral stem cell transplantation (n=34) from HLA-compatible family donors (n=30) or unrelated donors (n=19) from 02/97 to 09/00. Only patients who were evaluable for the first 100 days were included in this analysis. MMF was given prophylactically in 11 patients and therapeutically in 7 patients. The number of patients with GvHD > II was not significantly different in patients receiving MMF (MMF+ group) and patients receiving no MMF (MMF- group) (9/18 vs. 12/31). CMV antigenemia was monitored by CMV pp65 antigen twice weekly.

CMV viremias (n=15) were almost only observed in CMV IgG+ recipients (14/23, 61%), whereas in CMV IgG - patients only one case was observed (1/26). 9/23 CMV IgG+ patients received MMF for GvHD prophylaxis and/or therapy, whereas 14/23 received no MMF. CMV IgG + / MMF+ patients developed a high number of CMV antigenemias (8/9, 89%) compared to the CMV IgG+ / MMF- patients (6/14, 43%), $p<0.05$ (Fisher's exact test). Moreover CMV viremia was persistent (still CMV pp65 Ag positive after 14 days of gancyclovir) or recurrent in 5/7 patients who received MMF continuously compared to only 1/6 MMF- patients with CMV antigenemia. No patient in either group developed CMV disease or died on CMV related complications. Multivariate analysis including the use of MMF, sibling donor vs. unrelated donor or the presence vs. absence of acute GvHD revealed MMF to be an independent risk factor for complicated (persistent or recurrent) CMV-infection ($p<0.05$).

There appears to be a trend towards a higher incidence and number of complicated CMV viremias in patients receiving MMF for GvHD prophylaxis or therapy. This has however to be confirmed in a larger patient number.

530

Development and evaluation of the nuclisens basic kit NASBA for the detection of RNA from yeast frequently resistant to antifungal drugs

H. Hebart, J. Loeffler, C. Dorn, P. Cox, S. Magga, H. Einsele (Tübingen, D)

Invasive Candida infections are often life-threatening diseases with increasing incidence in immunosuppressed patients. Early diagnosis is mandatory for appropriate, successful therapy. The aim of this study was to develop and evaluate a NASBA protocol (Nucleic Acid Sequence Based Amplification) for the detection of RNA from yeasts which frequently show resistance to azoles (Candida krusei, Candida glabrata, Candida inconspicua, Candida dubliniensis, Candida norvegensis) or amphotericin B (Candida lusitaniae) and to compare this test with a PCR assay based on slot blot hybridization with species-specific probes. Serially diluted cells of C. krusei, C. glabrata C. inconspicua, C. dubliniensis, C. norvegensis and C. lusitaniae were used for development and evaluation of the new molecular-based method. Primers and specific probes for the 6 species were designed within the fungal 18S rRNA gene region. Blood samples from healthy donors (n=90) spiked with fungal cells or -RNA and specimens from bone marrow transplant recipients (n = 24 patients) were compared subsequently by PCR and NASBA. NASBA showed a detection limit of 1 CFU/ml blood (PCR: 10 CFU/ml). The 6 probes did not show any cross-reaction to mould-, viral- and human RNA. In the presence of a specific RNA protection buffer, fungal RNA was detectable up to 24 h post-spiking demonstrating the robustness of the assay. In clinical specimens, both assays showed identical positive and negative results demonstrating the feasibility of the assay for clinical routine.

The NASBA Basic Kit technology is a valuable tool for sensitive, specific, fast and reliable detection of Candida-RNA with potential for routine diagnosis.

531

Prospective comparison of PCR-based versus late mRNA-based HCMV-directed preemptive antiviral therapy in patients after allogeneic stem cell transplantation: interim analysis

H. Hebart, P. Ljungman, T. Klingebiel, D. Wolf, J. Löffler, I. Lewensohn-Fuchs, L. Barkholt, P. Schlegel, J. Lunenberg, L. Kanz, G. Jahn, H. Einsele (Tübingen, D; Huddinge, S; Boxtel, NL)

Prophylactic and preemptive antiviral therapy based on home-made PCR and pp65 antigenemia assays have been shown to significantly reduce HCMV-associated mortality following allogeneic stem cell transplantation (SCT). Both approaches are associated with a high rate of over-treatment. In contrast to the detection of HCMV-DNA, the detection of HCMV-specific messenger RNA (mRNA) directly reflects active viral gene expression. Recently, an internally controlled and standardized assay for the detection of HCMV pp67 mRNA, a true late gene product of HCMV, has been commercialized (NucliSens CMV pp67, Organon Teknika, Netherlands). A prospective randomized multi-center trial was initiated comparing PCR-based with NucliSens CMV pp67-based preemptive HCMV-directed antiviral therapy in patients after allogeneic SCT. Study objectives are to assess 1) safety of the mRNA screening approach, and 2) to analyze whether this approach can help to avoid over-treatment.

The study is intended to randomize a total of 100 patients, an interim analysis was performed after completion of 40 patients, 20 in each arm. Viral screening was performed once weekly until day 100. Antiviral therapy was initiated either at the second consecutive positive PCR result or the first detection of late mRNA. If clearance of HCMV DNA or late mRNA was demonstrated after 14 days of antiviral therapy, therapy was stopped. If HCMV infection persisted, antiviral therapy was continued in a reduced dose. The two arms were well balanced with regard to age, gender, HCMV-serostatus, donor type, underlying disease and disease stage, conditioning regimen and stem cell source. Twenty of 40 patients received antiviral therapy prior to day 100 after transplant. Therapy based on the detection of late mRNA was found to be similar to PCR-based therapy with none of the patients in either arm developing HCMV disease. This is the first study that prospectively compares two screening tests for the initiation of preemptive therapy and monitoring of the efficacy of antiviral therapy after allogeneic stem cell transplantation. The interim analysis clearly indicates that NucliSens CMV pp67-based therapy is similar to PCR-based preemptive antiviral therapy, and no patient developed CMV disease.

532

Cidofovir for CMV infection and disease in stem cell transplant recipients

S. Günzelmann, G. Lentini, N. Basara, A.A. Fauser, M.G. Kiehl (Idar-Oberstein, D)

Viral infections after allogeneic stem cell transplantation are still a major complication. Especially cytomegalovirus (CMV) infection is associated with a substantial morbidity and mortality for patients receiving mismatched or unrelated transplants. Database analysis reported by Ljungman et al (Blood 2001;97:388-392) demonstrate some beneficial effects of cidofovir in the treatment of CMV infection. We herein report first data of our prospective study on cidofovir for first line preemptive therapy in allogeneic stem cell transplant recipients. So far, we treated 21 patients (5 mg/kg bodyweight give once a week for the first 2 weeks followed by one application every other week). Therapy was continued until CMV PCR and/or pp65 antigenemia was negative for at least 3 consecutive weeks. Patients were routinely monitored for CMV by PCR and pp65 2 times a week. Failure of preemptive therapy was defined as continued presence of PCR positive signal and/or pp65 antigenemia. In only 1 patient CMV PCR conversion to a negative signal remaining negative during maintenance and at least for 3 consecutive weeks after study drug discontinuation was observed. 15 patients became CMV PCR negative after two weeks of therapy but a positive signal in the CMV PCR was again observed 2 – 3 weeks later during maintenance therapy. 5 patients did not respond to cidofovir. In the not responding patients as well as in the patients converting from a negative signal to a positive signal during maintenance antiviral therapy was changed to ganciclovir. In 14 of these 20 patients a conversions to a negative CMV PCR signal was observed during ganciclovir therapy. Two patients remained CMV PCR positive despite change of antiviral therapy to foscarnet. Out of these 21 patients 11 are still alive, whereas 10 died (4 leukemia relapse, 1 generalized toxoplasmosis, 5 refractory graft versus host disease and pneumonia). Cidofovir was well tolerated and only one patient experienced a renal toxicity. From these data we conclude that cidofovir is not a useful alternative for ganciclovir or foscarnet as first line CMV preemptive therapy. On the other hand it is important to note that we treated 2 additional patients not responding to ganciclovir as single agent because of a U97L-mutation known as ganciclovir resistance with the combination of ganciclovir and cidofovir. Both patients became CMV PCR negative after 3 weeks of combination therapy. Therefore, we conclude that cidofovir can be given with a low risk of toxicity even in combination with ganciclovir.

533

Central venous catheter (CVC)-associated bloodstream infections in patients treated for acute leukemia (AL)

M. Karthaus, T. Doellmann, S. Weber, G. Heil, A. Ganser (Bielefeld, D)

Background: Pts with AL generally receive noncuffed multilumen CVC's for administration of antineoplastic, antimicrobial or parenteral nutritional therapy. Fever during aplasia has been reported in up to more than 90% of pts with AL. Although the origin of fever is often assumed to relate to a CVC-infection, data regarding the incidence of CVC-bloodstream infections (BSI) are rare.
Methods: The incidence of CVC-related infections was analyzed in 100 pts (45 m/55 f) treated for AL within 227 consecutive chemotherapy cycles from 01/96 to 12/99. All pts received a multilumen uncoated CVC (Arrow). Insertion sites were inspected daily and covered with a dry dressing and local administration of PVP. Proven CVC-related BSI was defined as the isolation of the same species from peripheral blood culture and CVC-tip (>15 CFU). Results: Within 227 chemotherapy cycles a total of 324 CVC (4859 CVC days; mean 15; SD 9.1; CVC range 1-4/cycle) were analyzed. All patients had a severe neutropenia (<0.1/µl), with a mean duration of aplasia of 15.85 days (SD 8.2)/cycle. Fever was noticed in a total of 160 out of 227 cycles (70.5%). Inflammation at the CVC insertion site was documented in 153 of 324 inserted CVC (47.2%), with fever in 59 out of these CVC (38.6%). Colonized CVC-tips were documented in a total of 45 CVC (13.9%). Fever was observed in 23 out of these 45 CVC, however, proven CVC-related BSI were the origin of fever in 9 cycles (4%) only. All these CVC-related BSI were attributed to coag.-neg. staphylococci, in one episode with Corynebacter sp. in addition. The incidence of proven CVC-related BSI was 1.85/1000 CVC-days. Conclusion: Although febrile neutropenia was observed in 70.5% of all treatment cycles for AL, CVC-related infections were a rare origin of fever (4%). The incidence of proven CVC-related BSI was low (1.85/1000 d). Local inflammation at insertion site was common, however, proved to be an unspecific sign for CVC-related BSI.

534

Catheter-related blood stream infections in hematological and oncological patients - a surveillance study

M. Leithäuser, C. Kahl, K. Neuber, S. Wilhelm, D. Kunst, J. Rutjes, A. Podbielski, M. Freund (Rostock, D)

A major cause of infectious morbidity in hospitalised patients are catheter-related blood stream infections. Hematological and oncological patients often need central venous catheters (CVC) for a long period of time. Only little data on CVC-related blood stream infections (CVC-BSI) is available in this group of patients. We used the criteria for definite central venous catheter infections published by the infectiology working group of the DGHO in a retrospective analysis of 156 catheters from 91 patients. Our current preventive policy includes maximal sterile barrier precautions during insertion and daily changes of infusion sets in „sterile" technique. Non-tunneled multi lumen catheters were used in all patients.The insertion site were subclavian in 130 and jugular veins in 26 cases, respectively. The mean duration of catheter stay was 23 days leading to 3650 device days. In 17 of 156 CVC a definite CVC-related BSI was detected. Overall, the rate of CVC-BSI was 11.5% and 4.7 per 1000 device days. Further analysis of subgroups revealed 14.8% and 5.1/1000 of 61 CVCs in leukemic patients and 8.3% and 4.1/1000 of 36 CVCs in solid tumor patients, respectively. Analysis of 56 CVCs used for autlogous stem cell transplantation showed similar results with 10.7% and 4.8 CVC-BSI/1000 device days. Analysing the 36 CVCs with a duration of stay of more than 28 days (mean 48 days; range 30-93 days) the CVC-BSI rate was 8.3% and 1.75/1000, respectively. The causative organism found on the catheter tip and corresponding blood cultures were coagluase-negative staphylococci in 15 of the 17 cases of CVC-BSI. No case of CVC-BSI-related mortality was seen in this group of patients. Of the remaining 139 CVC 26% were found to be colonised on removal.
The rate of CVC-BSI found in our unit is within the expected range, although other groups have reported lower rates. This might be due to the amount of teaching committments in CVC insertion and daily care. At present, the available data does not support the need for changes in our preventive policy, for example use of coated catheters or longer intervals of infusion set usage. Prolonged use of non-tunneled CVC above the recommended time of 28 days does not lead to an increased rate of CVC-BSI, further data might reveal risk factors leading to early CVC-infection.

Postersitzung:
Mammakarzinom / Ovarialkarzinom

535

Treatment of advanced breast cancer (ABC) with docetaxel and gemcitabine ± human granulocyte colony-stimulating factor (G-CSF)

G. Kornek, K. Haider, W. Kwasny, B. Schüll, W. Fiebiger, D. Depisch, E. Kovats, F. Lang, W. Scheithauer (Wien, Wr.Neustadt, Baden, Neunkirchen, A)

Purpose: A multicenter phase II trial was performed to investigate the efficacy and
tolerance of docetaxel, gemcitabine ± recombinant human granulocyte colony-stimulating factor (G-CSF) in patients with metastatic breast cancer. Patients and Methods: Between July 1999 and November 2000, fifty-two patients participated in this trial, 51 of whom are evaluable for response. Thirty-eight patients received this combination as first-line and 14 patients as second-line chemotherapy, including 8 patients who had failed anthracyclines. Therapy consisted of gemcitabine 1500mg/m^2 on days 1 + 15 and docetaxel 50 mg/m^2 on days 1 + 15 every 4 weeks. Depending on the absolute neutrophil counts (ANC) on the day of scheduled chemotherapeutic drug administration, a 5-day course of granulocyte colony-stimulating factor (G-CSF) 5 mg/kg/d was given. Results: The overall response rate was 60.5% in patients receiving docetaxel plus gemcitabine as first-line chemotherapy, including 4 complete (10.5%) and 19 partial remissions (50%); 9 patients (24%) had disease stabilization, and only 5 (13%) progressed. Second-line treatment with this regimen resulted in 6/14 (43%) objective responses, five had stable disease and 3 PD. The median time to progression was 8.5 months in the first-line, and 6.6 months in the second-line setting, respectively. After a median follow-up time of 15 months, 37 patients (71%) are still alive with metastatic disease. Myelosuppression was commonly observed; WHO grade 3 or 4 neutropenia both occurred in 15 (29%) patients, and was complicated by septicaemia in 3 cases; grade 3 or 4 thrombocy-

topenia was not observed, and grade 3 anemia in only 1 patient (2%). Severe (grade 3) nonhematologic toxicity except for alopecia, was rarely observed, and included nausea/vomiting in 3 (6%), and stomatitis in 2 (4%), peripheral neuropathy, skin toxicity, and raise of liver enzymes each in one patient. Conclusion: Our data suggest that docetaxel and gemcitabine ± G-CSF is an effective and fairly well tolerated regimen for the treatment of advanced breast cancer. It might be particularly useful in patients previously exposed to adjuvant or palliative anthracyclines and/or alkylating agents.

536

Second-line hormonal treatment in advanced breast cancer breast cancer patients with soft tissue metastases

P. Schmid, M. Wischnewsky, K. Possinger (Berlin, Bremen, D)

Introduction: Letrozole 2.5 mg has previously shown superiority over megestrol acetate (MA) as second-line therapy in postmenopausal women with advanced breast cancer (ABC). The purpose of this paper is to analyze the efficacy of letrozole and MA with respect to response, duration of response (DR) and overall survival (TTD) for pts with only soft tissue metastases using machine learning techniques. Methods: 150 ABC-patients with soft tissue metastases [Mean age: 66.2; disease free interval > 24 months: 61.3%; no previous chemotherapy: 73.3%; objective response to prior therapeutic antiestrogen: 25.3%; performance status WHO grade 0: 66.0%;] were randomly assigned to receive letrozole 2.5 mg (=L2.5; n =44), letrozole 0.5 mg (=L0.5; n = 57), or megestrol acetate 160 mg (=MA; n = 49) once daily in a double-blind, multinational, peer reviewed trial until progression of disease or any other reason requiring discontinuation of study treatment. Results: Response, duration of response and overall survival are shown in table 1. L2.5 is significantly more effective than L0.5 or MA with respect to objective response and clinical benefit. Whereas duration of response is significantly higher for L2.5 compared to MA (p=0.003), there is no difference between L0.5 and L2.5.
Using KDD, the most important prognostic factors for OR are (C4.5-algorithm): Treatment, disease free intervall, age (OR is increasing with age), receptor status and previous chemotherapy (negative impact). In contrast, conventional statistics (Logistic regression) delivers only treatment as prognostic factor with 64.4% correct classification. Conclusion: Letrozole 2.5 mg is highly active in patients with soft tissue metastases. Letrozole 2.5 mg is significantly more effective in terms of response and duration of response than MA. There is a dose effect with respect to response favoring Letrozole 2.5 mg over Letrozole 0.5 mg. However, duration of response is comparable between Letrozole 2.5 mg and 0.5 mg.

537

Rupture of the abdominal aorta due to metastatic infiltration from breast cancer

B. Röhne, N. Dreier, M. Schroeder, A.A.N. Giagounidis, R. Erlemann, H. Müller-Wiefel, C. Aul (Duisburg, D)

Metastatic infiltration of the aorta in cancer is a rare event. It has mainly been described in lung cancer patients, where the primary tumor extended to, and infiltrated, the vessel wall. In metastatic breast cancer this complication has only been described once before. We report the case of a 65 year old patient with breast cancer first diagnosed in 02/1998. Initial staging was pT3 pN2 pM0. Primary treatment consisted of modified radical mastectomy and lymph node excision followed by chemotherapy (4xEC plus 4xCMF) and irradiation. c-erb-2 expression was highly positive (+++). Metastatic disease was first diagnosed in 08/2000, and treated with trastuzumab and paclitaxel. Further disease progression was noted with liver and pulmonal metastases, and additional treatment with vinorelbine, paclitaxel and trastuzumab was given. In 04/2001 the patient presented with severe lower back pain caused by disseminated infiltration of the lumbar vertebrae. Palliative irradiation was performed. In 05/2001 rupture of the abdominal aorta led to a retroperitoneal hematoma extending to the right thigh. Operation revealed metastatic infiltration of the aortic vessel wall with an adenocarcinoma expressing the identical immunohistochemical characteristics as the primary breast cancer. After successful insertion of an aortoiliacal prothesis the patient is now undergoing further palliative chemotherapy with gemcitabine and cisplatinum.

538

Microsatellite instability is an independent and negative predictor of poor prognosis in patients with breast cancer treated by high dose chemotherapy and autologous stem cell transplantation

P. Wild, A. Reichle, R. Andreesen, F. Hofstädter, A. Hartmann (Regensburg, D)

Aim: The treatment of patients with advanced breast cancer using high dose chemotherapy and autologous stem cell transplantation (HDCT-ASCT) is presently subject to controversial discussions. The aim of this study was to define molecular alterations which could predict a likely response of the tumor to high dose chemotherapy. Methods: 39 patients with breast cancer and lymph node and/or systemic metastases were selected from a prospective study of sequential HDCT-ASCT. After laser microdissection (PALM), microsatellite analysis was done using 15 microsatellite markers selected for detection rates of MSI in breast cancer according to literature. Exon 5-9 of the p53 gene were directly sequenced. Expression of p53, HER2-neu, hMSH2 and hMLH1 was evaluated by immunohistochemistry. Results: MSI of at least 3 markers (>=20%) was detected in 13 of 39 patients (33%). All MSI-positive tumors showed normal expression of the mismatch repair proteins hMSH2 and hMLH1. There were 6 p53 mutations in 33 cases with complete sequencing results (18%). Overexpression of HER2-neu and p53 was detected in 7 (27%) and 12 (46%) of 26 evaluated patients. The presence of microsatellite instability was highly associated with shorter overall (p=0.0004) and progress free survival (p=0.04). None of the clinical factors (e.g. ER, PR, initial tumor size, number of involved lymph nodes, localization of metastases) or any of the other investigated genetic alterations correlated significantly with the outcome of the patients in a univariate analysis. Using various multivariate Cox regression models, MSI remained a highly significant and negative risk factor for overall survival. Conclusions: MSI is frequent in advanced breast cancer and is not associated with alterations of hMSH2 and hMLH1. MSI could be an indicator of chemotherapy resistance and poor outcome in breast cancer patients treated with HDCT-ASCT. Large prospective studies are needed to evaluate the prognostic significance and the underlying mechanisms of MSI in breast cancer.

539

Anti-HER2 antibody Herceptin enhances cytolytic activity of alass I-restricted her2-specific T lymphocytes against her2-overexpressing tumor cells

C. Meyer zum Büschenfelde, C. Hermann, B. Schmidt, C. Peschel, H. Bernhard (München, D)

The human epidermal growth factor receptor 2 (HER2/neu/c-erbB2) represents an appealing target for antigen-specific humoral and cellular immunotherapy. HER2 is expressed at high levels in a variety of human cancers. Overexpression of HER2 contributes to the malignant phenotype of the tumor. Therefore, HER2-overexpressing tumors might not be able to easily escape HER2-targeted immunotherapy through immunoselection of antigen-loss variants. Antibody-based immunotherapy regimens have been developed to target the HER2 molecule. In clinical trials, infusion of the humanized anti-HER2 monoclonal antibody trastuzumab (Herceptin) induced tumor regressions in some patients with HER2-overexpressing breast cancer. The underlying mechanism of action is based on receptor internalization upon binding of Herceptin to HER2. Receptor internalization abolishes signal transduction via HER2 that further results in suppression of tumor cell proliferation. Antibody-induced internalization of HER2 leads to the degradation of HER2 and subsequently to the generation of HER2-derived peptides. We asked the question whether cytolytic activity of HER2-specific T lymphocytes is augmented by Herceptin due to the induction of newly synthesized peptide-MHC-complexes on the tumor cell surface. HER2-specific, class I-restricted cytotoxic T lymphocytes (CTL) were generated from different donors by repetitive stimulation with autologous DC either transduced with a HER2-encoding retrovirus or loaded with HER2-derived peptides known to be naturally presented with HLA-A2. The HER2-overexpressing breast cancer cell line SKOV3tA2 was used as target for HER2-specific CTL in a standard chromium release assay. SKOV3tA2 cells expressing the class I molecules HLA-A2 and HLA-A3 were lysed by HLA-A2-restricted and by HLA-A3-restricted CTL clones directed against HER2. Lysis of SKOV3tA2 cells by these CTL clones was greatly enhanced when tumor cells were preincubated with anti-HER2 antibody Herceptin. Lytic activity of anti-HER2 CTL clones was not altered when SKOV3tA2 cells were treated with an irrelevant monoclonal antibody recognizing transferrin receptor. In conclusion, HER2-specific CTL using different restriction elements lyse more efficiently HER2-overexpressing tumor cells following incubation with anti-HER2 antibody Herceptin. The synergistic activity of HER2-directed antibodies and T cells in vitro encourages the further development of a HER2-targeted immunotherapy using a combination of inhibitory antibodies and cytotoxic T cells for patients with HER2-overexpressing tumors.

540

Salvage chemotherapy with docetaxel, oxaliplatin, gemcitabine in platin-pretreated advanced epithelial ovarian cancer - Preliminary results of a phase II study

L. P. Mueller, G. Seliger, O. Kellner, T. Buechele, T. Kegel, B. Schaedlich, K. Jordan, D. Wunderlich, H.-G. Strauss, A. Grothey, H. Koelbl, H.-J. Schmoll (Halle, D)

RATIONALE: The strategy for a salvage treatment of advanced epithelial ovarian cancer (EOC) after carboplatin chemotherapy remains an issue of ongoing investigation. For new drugs like gemcitabine (GEM) a promising single-agent activity has been demonstrated. Combination therapy with taxanes and platin-derivates might result in synergistic activity and could help to prolong survival with decreased effects on the quality of life. PATIENTS/DESIGN: 17 female patients (median age 57.6 yrs; 27-76 yrs) with relapsed or platin-refractory EOC have been enrolled to date. All patients had received a median of 3 (2-6) chemotherapies for advanced disease including at least one (1-3) platin-containing regimen. The majority of patients had received a previous therapy with paclitaxel. Chemotherapy was performed in an outpatient setting as a biweekly (q d15) regimen of docetaxel (55 mg/m^2, 60 min, d1) followed after 1hr by gemcitabine (500 mg/m^2, 30 min, d1) and on the subsequent day oxaliplatin (70 mg/m^2, 120 min, d2) until maximal response or disease progression. Response, progression free survival, overall survival, toxicity and quality of life were evaluated. RESULTS: 14 patients are evaluable. A CR was observed in 2 patients (14.3%) and a PR in 4 patients (28.6%) resulting in a total response rate of 42.9%. A stable disease was seen in 5 patients (36.0%) and 3 patients (21.4%) presented with primary progressive disease. Treatment resulted in a median progression free survival of 3.5 months (0.1-11.8) and an overall survival of 7.8 months (2.8-17.2). Of the 6 patients with remission 5 were evaluable for follow up. All of those are alive to date with an overall survival of 11.4 months (5.4-16.3) and a progression free survival of 11.4 months (5.4-16.3). WHO 3° toxicity was seen in 5 patients with 3 cases of diarrhea/nausea and 1 case of myelosuppression and stomatitis respectively. CONCLUSION: The combination therapy of docetaxel/gemcitabine/oxaliplatin demonstrated an objective response activity in heavily pretreated patients with relapsed and/or platin-refractory EOC. This regimen is feasible in an outpatient setting and results in only moderate impairment of the patients quality of life. Analysis regarding the efficacy in correlation to response to previous carboplatin therapies and especially response rates in platin-refractory EOC as well as detailed toxicity data will be presented at the meeting.

541

Deregulation of the p16/p53/Bax cell cycle and apoptosis pathway in ovarian cancer: Comparison between adenomas, borderline tumors and carcinomas

S. Hermann, S. Hauptmann, I. Sturm, P. Hemmati, D. Stuckenberger, B. Dörken, P. Daniel (Berlin, D)

P16(ink4a/CDKN2) is a cyclin dependent kinase inhibitor (CDKI) known to mediate arrest at the G1 restriction point of the cell cycle. P53 was shown to cooperate with p16 in the induction of cell death and to be a transcriptional activator of the BAX gene. Recently, we showed that BAX and high p16(ink4a/CDKN2) are prognostic indicators in patients with oesophageal squamous cell carcinomas (Sturm et al, J Clin Oncol 2001,19(8); 2272-81). Since high BAX expression is a positive prognostic factor in ovarian cancer we analysed p53, BAX and p16(ink4a/CDKN2) in ovarian carcinomas as compared with borderline tumours and adenomas. 60 ovarian tumours (32 carcinomas, 10 adenomas, 18 borderline tumors) were investigated for protein expression by immunohistochemistry. In addition, we analyzed the tumor tissues for p53 and BAX mutations by single-strand conformation polymorphism-polymerase chain reaction (SSCP-PCR). Highest nuclear p53 expression was observed in the carcinomas while the expression of p53 was lower in the borderline tumours (p=0.004) and lowest in the adenomas (p=0.01). In the same line, the CDKI p16 (ink4a/CDKN2) levels were increased in the carcinomas as compared with the borderline tumours (p<0.0007). No differences were observed between borderline tumours and adenomas (p=0.7). Interestingly, BAX expression was increased in both the borderline tumors and the carcinomas (serous: n=19), non-serous: n=13) when compared with the BAX levels in the adenomas. Lower levels of BAX were detected in the serous carcinomas when compared with the non-serous carcinomas (p=0.035) which is in line with the poor prognosis of the serous tumors. Thus, deregulation of p16, p53 appears to be a frequent event in malignant transformation in ovarian cancer, especially in view of the borderline tumors. While BAX may be expressed at higher levels in some tumors, the differences in BAX expression may be related to the tumor biology and prognosis of histopathological subtypes, i.e. serous versus non-serous carcinomas.

542

Efficiency of the regional intra-arterial and endolymphatic neoadjuvant chemotherapy in the treatment of the destructive forms of a breast cancer

G.V. Bondar, I.E. Sedakov, V.N. Smirnov, D.V. Trukhin (Donetsk, UKR)

From 1990 to 2000 years 85 patients with local advanced breast cancer, accompanied by the tumour disintegration and secondary infection were treated and studied. The purpose of our investigation was to study the efficience of various types of complex therapy of these patients. In cases of the local-advanced breast cancer use of a neoadjuvant polychemotherapy at the first stage of medical treatment is conventional. In 38 cases the neoadjuvant chemotherapy were carried out in regional intra-arterial regimen through the basin of a. thoracica interna. The catheterization of an a. thoracica interna were accomplished through the a. epigactrica superior, outside of a tumoral and bacterial infiltration zone. In 30 cases the neoadjuvant chemotherapy were carried out in endolymphatic regimen, after the catheterization of femoral lymphatic vessels. For 17 patients the medical treatment were begun from mastectomy and adjuvant regional chemotherapy. The average age of the patients were: in the first group – 53.3 years, in the second – 55.6 years, in third – 64.7 years. The secondary intoxication caused by a bacteriemia was stoped in the first group by application of regional intra-arterial antibacterial therapy, in the second group – by the endolymphatic antibacterial therapy. In the third group the antibacterial regional therapy were carried out in the early postoperative period through the basin of a. thoracica interna, which catheterization were fulfilled through it perforans branches in 2 – 3 intercostal interval during the mastectomy. For the patients of the first and second group 2-3 courses of neoadjuvant chemo- and antibacterial therapy were carried out, that allowed us to continue a treatment in complete volume and after the radiotherapy course to execute a radical mastectomy. The five years' survival rate in the group of patients with regional chemotherapy was 43.3 %, in the group with endolymphatic chemotherapy – 51.3 %, in the group of patients, which treatment was begun from a sanitary mastectomy – 27.5 %. Use of an a. thoracica interna basin for neoadjuvant chemotherapy in treatment of the local-advanced and destructive forms of a breast cancer essentialy removes terms of relapses and metastasises occurence. The presence «of lymphatic dump» from wound and tumor surface favours to high efficiency of endolymphatic chemo- and antibacterilal therapy method. Application of neoadjuvant methods for treatment of the local–advanced and destructive forms of a breast cancer increase the efficiency of radiotherapy and promotes to execute a radical mastectomy with low relapses probability, that leads to increase of a five years' survival rate.

543

Ligands of PPARg and VDR induce apoptosis of breast cancer cells in vitro

O. Doudina, W. Zang, M. Wächter, J. Reichardt, K. Possinger, E. Elstner, L. Binderup, B. Afanassiev, H.P. Koeffler (Berlin, D; Ballerup, DK; Saint-Petersburg, RUS; Los Angeles, USA)

A new class of antidiabetic drugs termed thiazolidinediones (TZDs) has been developed for insulin-resistant diabetes mellitus. TZDs are specific ligands for the peroxisome proliferator-activated receptor gamma (PPARg). Data from us and other groups have shown that TZDs can inhibit the proliferation of a variety of cancers. Pioglitazone (PGZ, sold as ACTOS by TAKEDA) is a member of TZDs with no evidence of drug-induced hepatoxicity and with high selectivity for PPARg. The aim of this study is to evaluate the therapeutic effect of the combination of potent 20-epi-vitamin D3 analogs (CB1093, KH1060, MC1288) with PGZ in the breast cancer cell line MCF7. Dose-response studies with PGZ showed that PGZ alone is a weak inhibitor of clonal proliferation of MCF7 cells. In contrast, 20-epi-vitamin D3 analogs are potent inhibitors of clonal proliferation of these breast cancer cells (ED50 10^{-9}-10^{-10} M) with CB1093 being the most effective(ED50 10^{-10} M). Furthermore, the combination of CB1093 with PGZ (10^{-5} M) synergistically inhibits their clonal growth (ED50 10^{-11}-10^{-12} M) and induces apoptosis in 27% of MCF7 cells as measured by TUNEL assay. Interestingly, the combination of ligands for PPARg and VDR markedly decreased the level of bcl-2 protein and increased the level of the tumor suppressor PTEN as measured by western blot. In conclusion, these ligands of nuclear hormone receptors when used together, may have an adjunctive role in the therapy of breast cancer.

544

Dynamic surface tensiometry of serum from patients with uterine carcinoma - a diagnostic tool to monitor radiotherapy

R. Miller, U. Pison, D. Trukhin (Golm, Berlin, D; Donetsk, UKR)

Dynamic surface tension of biological liquids was studied for patients with malignant neoplasm of the brain and spinal cord, the lungs, liver, stomach, breast and reproductive organs before, during and after a respective medical treatment. Dynamic surface tension characteristics depend on tumour localisation and change during treatment. Measurements were performed with the maximum bubble pressure tensiometer MPT2 and the pendant drop tensiometer PAT1 (from LAUDA and SINTECH, Germany). Dynamic surface tension of serum obtained from patients suffering from uterine carcinoma (n = 46) was lower compared to healthy controls. During radiotherapy these values normalized. Major complications of radiotherapy include radiation proctitis and cystitis. Significant differences in the serum dynamic tensiograms and mean values of the tensiometric parameters were found for patients who developed such inflammatory complications (n = 14) as compared to the cases without these problems. Because the clinical symptoms of inflammation are the final consequences of extensive tissue alterations, one can expect to detect systemic changes, i.e. in serum, long before inflammation becomes clinically evident. Dynamic interfacial tensiometry, being quite sensitive to such changes, has the potential to predict the development of complications caused by radiotherapy already at early stages of the treatment. Indeed, the development of radiation proctitis or urethritis could be predicted from dynamic surface tension measurements in 85 % of all patients who developed these complications before it became clinically evident. Comprehensive prophylactic actions aimed to prevent complications of radiotherapy are most useful if performed early. Interfacial tensiometry can predict complications of radiotherapy and therefore is a useful tool to monitor radiotherapy in cancer patients.

545

Prognostic significance of Ep-Cam and Her-2/neu overexpression in breast cancer

J. Spizzo, P. Obrist, C. Ensinger, I. Theurl, M. Dünser, A. Ramoni, E. Gunsilius, G. Eibl, G. Mikuz, G. Gastl (Innsbruck, A)

To assess the frequency and prognostic impact of Ep-CAM overexpression in patients with breast cancer and to determine its relationship with other prognostic markers including Her-2/neu, we enrolled in this retrospective study 205 breast cancer patients with a median follow-up of 10.8 years. Overexpression of Her-2/neu and Ep-CAM in tumor tissue samples was assessed by immunohistochemistry. Ep-CAM and Her-2/neu overexpression was found in 35.6% and 23.9% of the tumor samples, respectively. Ep-CAM as well as Her-2/neu overexpression was predictive for poor disease-free and overall survival. Concurrent Ep-CAM and Her-2/neu overexpression was present in 14.1% of tumor specimens and had an additive negative impact on disease-free and overall survival ($p<0.001$). These patients had a median time to relapse of 40 months, while the median was not reached in the patient population without overexpression of the antigens. By multivariate analysis Ep-CAM overexpression proved to be an indicator of poor prognosis, independent of tumor size, histologic grade, hormone receptor expression and Her-2/neu overexpression. Overexpression of Ep-CAM and Her-2/neu complement each other as predictors for poor prognosis in patients with invasive breast cancer. Determination of both tumor markers should help in assigning breast cancer patients to one of three distinct risk categories and provide a rationale for targeting strategies with monoclonal antibodies.

546

Circulating tumor cells can be detected in the peripheral blood of patients with localized breast cancer

O. Doudina, J. Fritz, M. Fenner, M. Wächter, K. Possinger, E. Elstner, B. Afanassiev (Berlin, D; St. Petersburg, RUS)

Several studies have shown the prognostic value of detecting tumor cells in the bone marrow of breast cancer patients, including those with localized disease. The invasiveness of the procedure limits the more widespread use of this approach, including serial measurements during and after treatment. We therefore started a pilot study with the aim to detect circulating tumor cells (CTC) in the peripheral blood of breast cancer patients. After obtaining written informed consent, 40 ml of peripheral blood was drawn from a total of 24 breast cancer patients. We used immunocytochemistry after immunomagnetic enrichment of cancer cells by human epithelian antigen (HEA) magnetic microbeads (Miltenyi Biotec) for the detection of CTC. The results are normalized for the number of mononuclear cells, the number of CTC per 3 x 10^7 mononuclear cells is reported. The sensitivity of the procedure was at least 1 CTC per 1 x 10^7 mononuclear cells. A total of 43 samples were obtained from 24 breast cancer patients, including 13 patients with newly diagnosed localized breast cancer and 11 patients with metastatic disease. CTC could be detected in all 24 patients. There was no difference in the median number of CTC between patients with localized (median 28, range 6-249 cells) vs. metastatic (median 39, range 1-4290 cells) breast cancer. Only two patients had more than 250 CTC (739 and 4290 cells), both patients had advanced metastatic breast cancer and died 15 and 12 weeks after sample collection, respectively. We were able to morphologically separate intact and apoptotic tumor cells, the number of intact tumor cells was higher than the number of apoptotic tumor cells in most of the patients. CTC were measured before and/or after therapy in 13 patients, including 7 patients with localized cancer receiving preoperative chemotherapy and 6 patients with metastatic disease. In this limited number of patients, no obvious correlation between clinical response and changes in CTC could be observed. This pilot study is one of the first reports that demonstrates the feasibility of detecting CTC in the peripheral blood of breast cancer patients. Measurements in more patients are needed to correlate these data with the clinical, radiological, laboratory and pathological markers of the disease and to evaluate the value of this interesting new tool in monitoring breast cancer patients during and after treatment. The isolation of tumor cells from peripheral blood of breast cancer patients provides an unique opportunity for to study these cells in vitro and gain new insights into the metastatic process.

547

Low incidence of mammaglobin mRNA detection in periferal blood of non-metastic breast cancer patients

O. Zach, H. Kasparu, M. Girschikofsky, O. Krieger, D. Lutz (Linz, A)

To elucidate the incidence of tumor cells in peripheral blood (pB) of breast cancer (BC) patients, samples were tested for mammaglobin (hMAM) expression via a nested RT-PCR assay consisting of four PCR setups per sample. BC patients were classified into three clinical subgruops: at diagnosis (Dx), with no evidence of disease (NED, with and without any adjuvant treatment) and with metastatic disease (MD). In addition, samples from patients with benign lesions of the breast were investigated. Results of the hMAM nested RT-PCR assay were as follows: only one or two positive PCR setups were detected in 6/143 (4%) patients with benign lesions of the breast, in 17/123 (14%) BC patients at Dx, 24/187 (13%) with NED and 39/157 (25%) with MD. On the contrary, no sample from patients with benign lesions of the breast resulted in three or four positive PCR setups, but 1/123 (1%) BC patients at Dx, 4/187 (2%) with NED and 46/157 (29%) with MD. All 5 BC patients at Dx or NED having pB samples with at least three positive setups relapsed shortly afterwards, in contrast to 22/171 (13%) hMAM negative patients and 5/41 (12%) patients with one or two positive setups. On account of these results we suggest that samples with at least three hMAM positive PCR setups are likely to contain tumor cells with metastatic potential, hence only these patients are evaluated as hMAM positive. We conclude that the incidence of hMAM positive samples is low in BC patients at Dx or NED (1-2%) compared to MD (29%). However, all patients with such samples at Dx or NED relapsed shortly afterwards, indicating a possible prognostic relevance of the hMAM nested RT-PCR assay.

548

Multiple myeloma and the use of 18F-FDG positron emission tomography

M. Bommer, H. Schirrmeister, A. Buck, I. Buchmann, D. Bunjes, H. Döhner, S. Reske, M. Bentz (Ulm, D)

The incidence of multiple myeloma among the population of western countries is high. Staging procedures include bone marrow biopsy, planar radiography and measurement of immunoglobulin levels. Since most of the manifestations of multiple myeloma are restricted to the bone or bone marrow, a correct evaluation of the bone is one of the most important goals in myeloma staging. It has been shown, that a technetium polyphosphonate bone scan is of no value in detecting myeloma associated osteolytic bone lesions. Therefore there is a need for other, more specific diagnostic tools. Few myeloma-patients have extramedullary manifestation of the disease such as soft tissue tumors. These tumors are often difficult to detect but are very important for the choice of therapy and the patients prognosis. Evaluation of myeloma therapy is mainly based on the measurement of serum immunoglobulin levels. A small number of patients suffer from non-secretory myeloma and therefore are very difficult to monitor. Positron emission tomography (PET) using the glucose analogue F-18 fluorodeoxyglucose (FDG) has the unique ability to reveal medullary and extramedullary manifestations of myeloma in a single diagnostic procedure. Aim of the present study was to determine wether PET is a technique that can be used for staging and therapy evaluation in myeloma patients. We examined 43 myeloma-patients using the PET - technique. 37 patients had active disease, 6 patients were in complete remission. 10 patients had solitary plasmocytoma and 5 had extramedullary lesions. PET scans were positive in all patients with extramedullary lesions and 9 of 10 patients with isolated bone manifestation. All patients with focal uptake had active disease. Compared to standard staging procedures used in myeloma patients, we were able to detect additional bone lesions in two patients which were missed on planar radiography, changing their diagnosis from single plasmocytoma to multiple myeloma. In two patients 5 previously unknown soft tissue tumors were found. In one patient the only manifestation of myeloma was a large abdominal bulky tumor. In this case after radiation of the tumor the immunoglobulin-levels were normal and the PET scan showed no metabolic activity. Our results indicate, that 18FDG-PET is a useful technique to detect active lesions of myeloma that even seems to be superior to routine diagnostic procedures in some cases. Especially in cases were other diagnostic tools fail (e.g. non-secretory myeloma) it may be the only non-invasive technique for monitoring the activity of multiple myeloma.

549

Deletion of 22q is a prognostic parameter in multiple myeloma independent of chromosome 13q

E. Urbauer, H. Kaufmann, J. Ackermann, N. Worel, T. Nösslinger, A. Weltermann, E. Krömer, U. Jäger, H. Gisslinger, N. Zojer, H. Ludwig, R. Heinz, H. Huber, J. Drach (Wien, A)

Since genes located on chromosome 22q have been implicated in the pathogenesis of various malignancies, we evaluated the chromosome 22q status in 138 patients (pts) with newly diagnosed multiple myeloma (MM) using interphase fluorescence in situ hybridization (FISH) and probes for 22q11.2 (N25-locus and TUPLE 1-locus, respectively) and 22q13 (arylsulfatase A-locus). Durie & Salmon stages of pts were as follows: Stage I, 17 pts; stage II, 28 pts; stage III, 93 pts. FISH revealed a deletion of (del) 22q in 43 pts (31.1%), with simultaneous del of both 22q regions in 17 pts; in the remaining cases, either the 22q11.2 region (24 pts) or the 22q13 region (2 pts) was deleted. Del 22q occurred in a median of 28% of cells (range, 10% to 90%; cut-off level 10%) and were always observed to be monoallelic. Del 22q by FISH was confirmed by classical cytogenetics in the 8 cases with informative metaphases. There were no significant correlations of del 22q with standard MM parameters (Ig-isotype, hemoglobin, creatinine, CRP, beta-2-microglobulin, LDH, plasma cell morphology, Ki-67), whereas a significant correlation of del 22q with del 13q14 (present in 45% of pts) was observed (p= .0005). All symptomatic patients (n = 120), who were treated either by conventional-dose chemotherapy (melphalan-based, 84 pts; VAD, 23 pts) or high dose chemotherapy (22 pts), were studied for response to treatment and survival. Pts with del 22q had an inferior response rate (55% vs. 77%; p = .02) and shorter median overall survival (OS, 29 vs. 53 months; p = .0008) compared to pts with normal chromosome 22q. On multivariate analysis, del 22q was an independent prognostic parameter for survival (p=.027) besides del 13q14 (p=.01), beta-2-microglobulin >4 (b2M) (p=.026) and immature plasma cell morphology (p=.034). Based on this analysis, a prognostic model for MM was established according to the number of risk factors (del 22q and/or del 13q14, b2M>4, immature morphology): Pts who presented with none of those factors had superior OS (median not yet reached) compared to pts with one, two and three risk factors (median OS, 49.2, 22.5 and 9.8 months), respectively (p<.0001). In summary, del 22q is a new recurrent chromosomal abnormality in MM, which provides prognostically relevant information in addition to chromosome 13q14.

550

Detection of 13q14 and p53 deletions in patients with multiple myeloma by fluorescence in situ hybridization (FISH)

C. Hahn, A. Glasmacher, T. Urban, J. Mezger, C. Maintz, C. Losem, T. Sauerbruch, I. Schmidt-Wolf (Bonn, Karlsruhe, Würselen, Neuss, D)

Background: In patients with multiple myeloma (MM) partial or total deletion of chromosome 13 and 11q abnormalities have been reported to be an independent adverse prognostic factor. However, conventional cytogenetics are successful in only 20 to 60% of patients. It was shown that the use of FISH improves cytogenetic results in MM. Using FISH, we studied deletions of 13q14 and p53, a tumor suppressor gene the importance of which in MM has been discussed controversially. Methods: FISH analysis was performed on bone marrow slides obtained from 58 patients (pts.) (30 females, 28 males) with MM at different stages of disease (stage I 10 pts, stage II 5 pts, stage III 36 pts). In 36% of cases the specimen was obtained at the time of initial diagnosis. Locus-specific probes for bands 13q14 (Rb-1) and 17p13.1 (p53; Vysis, Bergisch Gladbach, Germany) were used according to standard protocols and analysis was successful in 51 pts. each. Results: Deletion of 13q14 (13q-) was detected in 16/51 patients (31%). At this time point, survival was not yet different in patients with 13q- vs. patients with a normal 13q result: median survival from time of bone marrow puncture: 266 vs. 276 days, p=0.45; median survival from time of initial diagnosis: 600 days vs. 2837 days, p=0.097. All patients with -13q had stage III disease vs. 55% with normal 13q (p=0.006). Deletion of p53 was found in 7/51 patients (14%). In three of these cases, deletion of p53 was associated with a loss of the whole chromosome 17. Again, survival was not statistically different in patients with -p53 vs. patients with a normal p53 result (median survival: 112 vs. 284 days, p=0.47). All patients with -p53 were in stage III of their disease. Conclusion: (1) Deletions of 13q14 or p53 were only found in patients with stage III myeloma. (2) We found a lower frequency of p53 deletion than reported in one study but confirm the results of two other studies. (3) The evaluation of the prognostic value of 13q14 and p53 deletions will require further follow-up. Supported by Leukämie-Initiative Bonn e.V.

551

Recurrent partial trisomies of chromosome 11q with a terminally located commonly gained segment as delineated by CGH and FISH in multiple myeloma

C. Wendl, P. Liebisch, K. Roth, N. Baßermann, A. Viardot, M. Bentz, S. Stilgenbauer, H. Döhner (Ulm, D)

Introduction: Various recurrent aberrations involving the long arm of chromosome 11 (11q) have been identified in multiple myeloma (MM). A translocation t(11;14)(q13;q32) is detectable in about 15-20% of newly diagnosed patients (pts.) by FISH. Although highly recurrent trisomies of chromosome 11 (+11) have been described in both, cytogenetic as well as molecular cytogenetic studies, partial trisomies of 11q (+11q) were not reported to be frequent in the disease. Deletion of 11q (11q-) seems to be a rare genetic event in MM. In a series of 46 pts. examined by comperative genomic hybridization (CGH) at our institution, gains involving 11q were the second most frequent chromosomal gains detectable. Moreover, a commonly gained segment comprising bands 11q23-qter could be delineated. To screen a large number of myeloma specimen for 11q abnormalities by FISH, we used a PAC probe from our disease-specific DNA probe set recognizing chromosome band 11q22.3-q23.1. A third hybridization signal was present in 46/126 pts. (36.5%), consistent with a partial or complete +11q. Loss of one hybridization signal, consistent with an 11q-, was detectable in only 5/126 cases (3.9%). Thus, +11q was far the most frequent trisomy in our series, whereas 11q- seems to be a very rare genetic event in MM. To further characterize +11q in MM, we used a preliminary set of DNA probes specific for chromosome 11. Results: We analyzed 14 pts. that exhibited three hybridization signals when studied by FISH and the 11q screening probe (11q22.3-q23.1) using four additional BAC/PAC probes mapping to chromosome bands 11p15.4, 11q13, 11q23 (mapping distally from our screening probe as shown by FISH on metaphase spreads), and 11q25. 10/14 pts. exhibited findings consistent with a +11, while 3/14 pts. showed complete +11q. In one case a partial +11q located distally of 11q13 was identified. More interestingly, as our CGH data suggested a commonly gained genomic segment at 11q23-qter, we also studied 28 pts. that have been shown to be disomic at band 11q22.3-q23.1 with the probes mapping to chromosome bands 11q23 and 11q25. While most of the pts. (26/28) were also disomic at these two loci, two pts. exhibited a third hybridization signal each, suggesting a rather small partial trisomy located distally to band 11q22.3-q23.1. Conclusions: Partial or complete +11q is one of the most frequent genetic abnormalities in MM. According to our CGH data that indicated a commonly gained chromosomal segment comprising bands 11q23-qter, FISH with a preliminary panel of 5 different DNA probes confirms the terminal location of the minimally gained +11q segment in MM and suggests a critical region that maps distally of chromosome band 11q22.3-q23.1. Further studies with a larger panel of DNA probes mapping to 11q22.3-q23 are needed to delineate the region of interest on chromosome 11qter more precisely.

552

Atypical radiological appearence of extramedullary plasmocytoma

T. Heß, G. Egerer, A.D. Ho, H. Goldschmidt, G.W. Kauffmann (Heidelberg, D)

PURPOSE: The radiological staging of multiple myeloma includes projection radiography of the spine, the pelvis, the femura and humeri. This is sufficient to define the sceletal status regarding osteolyses for the classification of Durie and Salmon. There are, however, manifestations of this disease that require additional imaging, mainly CT and MR. This is especially true for the extramedullary forms of the plasmocytoma. We want to present rare extramedullary manifestations of this disease. METHOD/MATERIALS: 5 Patients out of a group of 500 outpatients of the university of Heidelberg with multiple myeloma and plasmocytoma were identified to have extramedullary manifestations of the disease. The radiological appearence on chest x-ray, computed tomography and magnetic resonance imaging is described. RESULTS: 2 of the 5 patients had mediastinal and hilar tumors, one of these tumors showed calcifications identified on CT. The other of the two patients had a pleural tumor without involvement of the adjacent rip. One patient presented with a nodular pulmonary amyloidosis. The lesions remained stable in size over the last 2 years. Another patient showed multiple focal liver lesions which were hypointens on T1w MR images and slightly hyperintens on T2w images. One patient had a peritoneal involvment that was identified indirectly by detection of ascites. All cases were histologically proven. CONCLUSIONS: Primary extramedullary plasmocytoma is a very rare manifestation of the disease. It can however mimic a lot of different tumors and thus has to be taken into account if there are clinical hints for multiple myeloma or plasmocytoma.

553

Molecular monitoring of the tumor load predicts progressive disease in patients with multiple myeloma after high dose therapy with autologous peripheral blood stem cell transplantation

F.W. Cremer, E. Lipinski, A.D. Ho, H. Goldschmidt, H. Goldschmidt, M. Moos (Heidelberg, D)

The clinical relevance of the assessment of minimal residual disease (MRD) in patients with multiple myeloma (MM) to predict disease recurrence has not been proven, yet. In the present study, we retrospectively analyzed the tumor load in peripheral blood (PB) and bone marrow (BM) samples of 13 patients with MM both in remission after high dose therapy (HDT) with autologous PBSC transplantation (PBSCT) and at the time of progressive disease (PD). For 6 patients, subsequent samples obtained in remission could be included in the study. Tumor cells were assessed by means of quantitative PCR assay with allele-specific oligonucleotides (ASO-qPCR) based on the method of limiting dilutions. PD was documented with ASO-qPCR in BM samples (median concentration of tumor cells in remission versus at PD: 0.18% vs. 4.6%) representing a significant increase by a median factor of 8.7. In PB, the median tumor load was 0.0005% in remission and 0.012% at PD. With a median factor of 47, the increase was much more pronounced. Comparing the results of the molecular monitoring in PB with those of the determination of the monoclonal protein, routinely applied as parameter for the course of the disease, revealed a superiority of the molecular monitoring because of the significantly higher increase in the tumor load. Analyzing the subsequent remission samples showed an increase of the malignant cells in 4 out of 6 PB samples and in all 4 BM samples available, indicating the potential of ASO-qPCR for an early PD recognition.

554

Cloning of translocations into the immunoglobulin heavy chain locus in patients with multiple myeloma

H. Szelenyi, Pl. Bergsagel (New York, USA)

Translocation into the immunoglobulin heavy chain (IgH) locus are a hallmark in multiple myeloma (MM). Occurring in nearly 100% of myeloma cell lines the most common chromosomal partners are FGFR3/MMSET, cyclin D1, and c-maf and their respective breakpoints have been cloned. In MM patients (pts) translocation into the IgH locus are seen in about 50-70% but chromosomal partners are identified in a limited number of samples nor are translocation breakpoints cloned. Here we aimed to identify and clone IgH translocation in order to find IgH translocation of known or unknown oncogenes in MM pts. We examined bone marrow (BM) samples of MM pts with our comprehensive Southern blot method using restriction enzymes and pairs of probes flanking each class of switch region to identify rearrangements of IgH switch regions. Restriction fragments hybridizing to two switch probes (i.e. involving 2 switch regions) were classified as legitimate switch recombination fragments. Fragments hybridizing to only one switch probe were classified as illegitimate switch recombination fragments (ISRF). BM samples from 30 pts were analyzed. Pts Ig idiotype was: IgG 24, IgA 4, IgM 1, light chain 2 pts. In 13 pts (43%) we found at least one ISRF, 8 pts (27%) having two ISRF. The frequency of ISRF might be underestimated because we did not select for (malignant) plasma cells and the number of BM plasma cells was low (<15%) in 10 pts (33%). Ten ISRF were seen in the switch mu region, 10 in the switch gamma region and 1 in the switch alpha region. In a first step to clone the ISRF in the patients samples we used inverse PCR (iPCR). In MM cell lines where all ISRF were shown to be translocations into IgH we successfully cloned ISRF up to a size of 4 kB by iPCR. Surprisingly, iPCR products of the pts samples cloned turned out to be germline IgH fragments or productive IgH switches, including one inversion. Unlike to the cell lines we were not able to clone translocations from respective ISRF by iPCR in the pts samples, especially in the switch gamma region. Therefore we are currently constructing and screening DNA libraries of the respective pts samples to clone the ISRF seen. Updated data on the translocations including breakpoints will be presented.

555

A combination of morphology and FISH techniques in multiple myeloma

C. Nickenig, N.K. Lang, C. Schoch, W. Hiddemann, T. Haferlach (München, D)

Multiple myeloma (MM) is a B-cell neoplasm characterized by the clonal accumulation of malignant plasma cells in the bone marrow. Although the predominant cell type in MM is the plasma cell, the target cell for transformation is considered to be a more immature B-lineage cell. Using modern cytogenetic techniques the aberration rate in MM is higher than 70%. Unfortunately, up to now only limited informations are available about the significance of the chromosomal aberration and especially about the timepoint of the neoplastic transformation. Fluorescence in situ hybridization (FISH) techniques combined with standard cytomorphological May-Grünwald-Giemsa (MGG) staining or immunophenotyping seemed to be the most appropriate methods to answer these questions.
When infiltration rate with plasma cells in the bone marrow was <50% FISH was performed on cytospins after Magnetic Cell Sorting (MACS) with the antibody anti-CD 138. In patients (pts.) with >50% infiltration rate FISH was applied to original bone marrow smears. The FISH panel consisted of: centromer specific DNA-probes for chromosome 1,7,9,11,15 and 17, DNA-probes for 5pq, Rb-gene locus (13q14), Cyclin D1 (11q13) and p53 (17p13). Using a combination of MGG, FISH and fluorescence immunophenotyping and interphase cytogenetics as a tool for investigation of neoplasms (FICTION) with the antibody CD 34 we furthermore investigated chromosomal aberrations on single cell level. MACS was performed on bone marrows of 57 pts. and led to $1x10^6$ to $3.5x10^8$ CD138+-cells. In all cases purity for plasma cells was >85%. Using FISH we detected aberrations in 48/63 (76%) of cases. Afterwards the combination of MGG and FISH was performed in 8 pts. with chromosomal aberrations detected before by interphase FISH: a total of 2623 bone marrow cells were morphologically identified and investigated for their specific chromosomal aberra-

tions. 61 cells of the erythropoietic lineage, 698 cells of the granulopoiesis, and 168 lymphocytes showed no chromosomal aberrations in any cell. 417/1619 plasma cells were also not aberrant, but the other plasma cells beared the aberration. Fiction was applied on bone marrow smears in 10 pts.: 78 CD34-positive precursor cells did not show any specific aberration detected before in the plasma cells.
Using FISH techniques combined with standard MGG staining or with FICTION in 8 pts. with MM we were able to demonstrate that only plasma cells beared the chromosomal aberration, whereas neither granulopoietic, erythropoietic nor other lymphoid cells were aberrant. In addition FICTION demonstrated that all CD34-positive precursor cells were not involved into the specific chromosomal aberration. Therefore, the malignant transformation has to start on a pre-plasma cell level and is followed by uncontrolled proliferation. Further investigations on molecular levels are necessary to identify specific chromosomal events in MM that initiate the malignant transformation.

556

Effects of proteasome inhibition in multiple myeloma cell lines

I. Zarvrski, M. Schweigert, M. Becker, U. Heider, C. Jakob, K. Niemöller, K. Possinger, O. Sezer (Berlin, D)

Background: The proteasome is a ATP-ubiquitin dependent protease which has a central role in degrading cytosolic proteins. Most of its substrates are short living proteins critical for the regulation of the cell cycle progression and apoptosis. A coordinated expression and degradation of regulatory proteins is necessary for cell survival. The inhibition of degradation leads to accumulation and activation of the cell death program. Having shown the effects of proteasome inhibitors on leukemic cells and hematopoietic progenitor cells (Eur J Haematol 2000; 65: 221-236), we tested a cell-permeable proteasome inhibitor MG-132 on multiple myeloma cell lines IM-9 and RPMI8226 in this study. Methods 0.5x10^6/ml cells were cultured in 12-well flat-bottom plates and subjected to DMSO control, 1µM or 10µM MG-132 for 0, 8, 12, 16, 20, 24 and 48 hours. Apoptosis was studied using Annexin-V. Cell-cycle analysis was performed using 7-AAD and flow cytometry. Results: In vitro experiments on multiple myeloma cell lines (IM-9, RPMI8226), which are mutant for p53, showed a dose and time dependent externalisation of phosphatidylserine and either an arrest in G0/G1 or in G2/M, which both increased nearly linearly with the time of exposure. For RPMI8226 the following data could be obtained: after 8 hours 6 percent of cells incubated with 1µM MG-132 and 12 percent incubated with 10µM MG-132 underwent apoptosis. After 48 hours the specific apoptosis rate rose to 49 percent in the 1µM MG-132 group and to 84 percent in the 10µM MG-132 group. The cell-cycle arrest occurred in G0/G1. The obtained cell counts showed a growth inhibition. IM-9 cells showed already after 24 hours a phosphatidylserine externalisation of 84 percent in the group treated with 1µM MG-132 and 89 percent in the group treated with 10µM MG-132. Conclusion: MG-132 was able to induce apoptosis in multiple myeloma cell lines on a much higher level in comparison to hematopoietic progenitor cells. We conclude that proteasome inhibitors may act as potent anti-myeloma drugs.

557

Anti-gp130 antibodies inhibit growth of human malignant plasma cells

M. Gramatzki, R. Burger, F. Bakker, A. Günther, W. Baum, E. Claret, J. Wijdenes (Erlangen, D; Besancon, F)

Previous studies have shown that the growth of malignant plasma cells is stimulated by cytokines of the gp130-family. In particular, IL-6 plays an important role as growth and survival factor. A xenograft tumor model with our IL-6-dependent human plasma cell line INA-6 was established to study myeloma cell growth as well as therapeutic intervention strategies. In contrast to the in vitro situation, INA-6 cells produced IL-6 during growth in SCID mice. However, explanted tumor cells were not able to survive in vitro without cytokine supplementation. Remarkably, expression of other gp130 cytokine receptors was induced in certain tumor sublines resulting in growth responsiveness to additional gp130-cytokines, including LIF and a novel member of the gp130-family, BSF-3. When tumor sublines were reinjected i.p. into SCID mice, an engraftment rate of >90% was achieved with a mean engraftment time of 37 days. Initial spontaneous in vitro proliferation of explanted tumor cells could be almost completely inhibited with neutralizing anti-gp130 antibodies, while blocking IL-6 or the IL-6 receptor (IL-6R) resulted only in partial reduction of DNA synthesis. Furthermore, in contrast with anti-IL-6R antibody treatment of INA-6 bearing SCID mice, anti-gp130 antibodies completely prevented tumor development. This effect was also achieved by using anti-gp130 F(ab)2 fragments underlining the central role of the gp130 signal transducer for plasma cell growth. The results provide evidence that INA-6 growth in vivo may be dependent on various gp130 cytokines and that gp130 signaling may be effectively inhibited by anti-gp130 antibodies. Despite certain limitations of a xenograft myeloma model, such an approach may be considered a valuable option for treatment of patients with multiple myeloma. Supported by the Wilhelm Sander-Foundation and DFG.

558

Prediction Of MHC I restricted peptides within the variable regions of Ig heavy chains from patients with multiple myeloma using computer assisted analysis

O. Christensen, S. Belle, F. Cremer, A.D. Ho, H. Goldschmidt, M. Moos (Heidelberg, D)

Introduction: The idiotype of the monoclonal paraprotein from patients with multiple myeloma can be regarded as a specific tumor antigen. Several groups reported about the activation of CD4 and CD8 positive idiotype specific T cells in myeloma patients using different kinds of vaccination strategies. Recently Li et al. (Blood 96, 2828-33, 2000) demonstrated that human myeloma cells are lysed by autologous idiotype specific T cells. Such an activation of cytotoxic T cells seems to be a promising tool for immunotherapeutic protocols. Up to now little is known about the nature of MHC I restricted peptides within the variable region of Ig heavy chains. In this study the sequences of the hypervariable Ig regions (Complementarity determining regions (CDR) 1-3) of patients with multiple myeloma were analysed using a HLA prediction software. Methods: Bone marrow samples with high plasma cell infiltration rates were used to identify sequences of the CDR 1-3 regions of 8 patients with multiple myeloma. Sequences were validated with commercially available DNA alignment programs.The prediciton of MHC-I restricted peptides within the CDR and frame work (FR) regions was performed using the HLA Peptide Binding Predictions software [BioInformatics & Molecular Analysis Section, National Institutes of Health (BIMAS), USA] and the peptide motif search software syfpeithi [Departement of Immunology, University of Tübingen, Germany]. For each residue position, the BIMAS-program examines which amino acid is appearing at that position. The running score is then multiplied by the coefficient for that amino acid type, at that position, for the chosen HLA molecule. The resulting running score is multiplied by a final constant to yield an estimate of the half time of disassociation.The prediction of the syfpeithi-program is based on published motifs (pool sequencing, natural ligands) and takes into consideration the amino acids in the anchor and auxiliary anchor positions, as well as other frequent amino acids. This score is calculated according to the following rules: The amino acids of a certain peptide are given a specific value depending on whether they are anchor, auxiliary anchor or preferred residue. Ideal anchors will be given 10 points, unusual anchors 6-8 points, auxiliary anchors 4-6 and preferred residues 1-4 points. Amino acids that are regarded as having a negative effect on the binding ability are given values between -1 and -3. Results: The amino acid sequences of the heavy chain CDR and FR regions of 8 multiple myeloma patients with HLA type -A2 or –A3 were depicted. Scanning these sequences with the HLA prediction software revealed a majority of the nonapeptides predicted to bind to MHC I molecules were found within the framework regions. A „focus" with a predicted peptide/MHC-I dissociation half-life of >10 min (BIMAS-program) was found for 7 of 8 patients in the framework region 3. Only for 2 patients peptides with a peptide/MHC-I dissociation half-life of > 10 min were found for the most variable CDR3 region. Analysis with the syfpeithi-program showed similar results.Conclusion: The peptide prediction analysis of the idiotype of 8 patients with multiple myeloma patients indicate that the most favourable peptides for binding to MHC I molecules are located within the framework regions of the variable Ig heavy chain region. In vitro analyses are in progress to evaluate the functional properties of the predicted peptides.

559

Differentiation of MGUS and multiple myeloma using flow cytometric characteristics of plasma cells

U. Heider, M. Schweigert, I. Zavrski, K. Niemöller, C. Jakob, J. Eucker, K. Possinger, O. Sezer (Berlin, D)

Background: Monoclonal gammopathy of undetermined significance (MGUS) and multiple myeloma (MM) are the most frequent forms of monoclonal gammopathies. For differentiation between both, various classification systems exist, based on a combination of clinical criteria. Especially in

early MM stages, the differential diagnosis according to these systems may be uncertain. The aim of our study was to evaluate whether flow cytometric parameters are able to distinguish between MGUS and MM.
Methods Patients were classified as having MM or MGUS according to the criteria of Durie and Salmon. They were divided into three groups. Group A (n=13) consisted of multiple myeloma patients with bone marrow plasma cell infiltration 10- 30%, a low serum M-protein (IgG $\leq$3.5g/dl or IgA $\leq$2g/dl) and without evidence of osteolytic bone lesions in conventional radiography at the time of flow cytometric analysis. Group B included 53 MM patients with bone marrow plasma cell infiltration >30%, a serum M-protein IgG > 3.5g/dl or IgA > 2g/dl or osteolytic bone lesions. 17 MGUS patients were included into group C. Bone marrow (BM) and peripheral blood (PB) were analyzed by flow cytometry. Plasma cells (PC) were identified by their CD38/CD138 (B-B4)/CD45 expression. Coexpression of CD19 and CD56 was analyzed to distinguish between normal (CD19+/CD56-) and atypical (mostly CD19-/CD56+ or CD19-/CD56-) plasma cells. Furthermore, DNA analysis of bone marrow plasma cells was performed to study the percentage of plasma cells in S-phase and their DNA content.
Results The ratio of immunophenotypically normal to all bone marrow plasma cells (BMPC), the number of peripheral blood PC (PBPC), the percentage of BMPC in S-phase and the DNA content of BMPC were analyzed. All of the MGUS patients in this study had residual normal PC in their bone marrow, but only 39% of group A and 28% of group B. The median proportion of phenotypically normal to all BMPC was 0% in MM group A and B and 60% in MGUS. In this study, all individuals with MGUS and no patient with MM had a percentage of $\geq$20% phenotypically normal BMPC to all BMPC. This ratio was highly significant for the differential diagnosis between MGUS and MM, both for group A and B (p<0.0005). The median of monoclonal PBPC was 1/µl (range 0-30/µl) in group A, 2.2/µl (0-211/µl) in group B and 0/µl (range 0-2/µl) in group C (MGUS). The median percentage of BMPC in S-phase was 1.6% (0.4%-2.5%) in group A, 3% (0.3%-10%) in group B and 1.6% (0.7% - 3.0%) in MGUS. Aneuploidy was found in 11% of group A, 41% of group B and 12% of group C. These parameters were significantly different between MGUS and group B, but not between MGUS and multiple myeloma in group A. Conclusion: The ratio of immunophenotypically normal to all BMPC was the only flow-cytometric parameter for the differentiation of MGUS and MM with low bone marrow infiltration, low serum M-protein and without bone lesions. The number of PBPC, the percentage of BMPC in S-phase and the DNA content of the BMPC are of less importance for differentiation.

560

Comparison of the bone resorption parameters serum carboxy-terminal telopeptide of type-I collagen (ICTP), urinary amino-terminal collagen type-I telopeptide (NTx) and deoxypyridinoline (Dpd) in MGUS and multiple myeloma

C. Jakob, M. Schweigert, K. Niemöller, U. Heider, I. Zavrski, J. Eucker, K. Possinger, O. Sezer (Berlin, D)

Bone resorption caused by increased osteoclast activity is a major feature during the disease course in multiple myeloma. As shown by histomorphometric studies, an increased bone resorption can be present in multiple myeloma patients before radiographic abnormalities can be observed. Excessive osteoclast-mediated bone resorption was found also in some individuals with monoclonal gammopathy of undetermined significance (MGUS). To increase the diagnostic sensitivity in estimating the activity of bone resorption, new laboratory parameters, which reflect bone metabolism accurately, are urgently needed. In this study we analyzed three parameters of osteoclastic bone destruction, i.e. urinary deoxypyridinoline (Dpd) and amino-terminal collagen type-I telopeptide (NTx) and serum carboxy-terminal telopeptide of type-I collagen (ICTP) in 72 individuals with monoclonal gammopathy of undetermined significance (n=12) or multiple myeloma (n=60). ICTP serum levels increased parallel to the stage of the disease and differed significantly (p=0.002) between MGUS, myeloma stage I and myeloma in stages II-III according to Salmon and Durie. Serum levels of ICTP were significantly (p=0.003) elevated in patients with multiple myeloma compared to individuals with MGUS, while in contrast to ICTP no significant difference was found for urinary NTx and Dpd. Furthermore there was a significant difference in ICTP serum levels between MGUS and myeloma patients in stage I. To evaluate the prognostic relevance of bone resorption parameters ICTP, NTx and Dpd levels were compared in myeloma patients in this study for the first time. In the Kaplan-Meyer analysis ICTP was a prognostic factor for overall survival in myeloma patients (log-rank test: p=0.03). In addition urinary NTx in myeloma patients was also found to be a significant (p=0.04) prognostic factor in the survival analysis. For Dpd no prognostic significance was found. These data show that serum ICTP levels increase parallel to the progression of the disease and give the first report on a significant difference in serum ICTP levels between individuals with MGUS and patients with myeloma in stage I. In the comparative evaluation of these bone resorption parameters for survival analysis we identified serum ICTP and urinary NTx as prognostic factors for survival in multiple myeloma, while Dpd had no prognostic relevance.

561

Natural T lymphocyte response against HLA class I epitopes of telomerase reverse transcriptase and mucin-1 in patients with multiple myeloma

A. Schmittel, C. Scheibenbogen, D. Nagorsen, H. Goldschmidt, S. Bauer, E. Thiel, U. Keilholz (Berlin, Heidelberg, D)

The tumor antigens telomerase reverse transcriptase (TERT), Mucin-1 (Muc-1) and MAGE-3 are overexpressed in multiple myeloma. Since these antigens are potential candidates for vaccination strategies in myeloma, we analysed, whether a natural T cell response to HLA-A2 binding peptides derived from these antigens already exists in myeloma patients. We used the IFN-g ELISPOT assay for direct ex vivo detection of ciculating T cells in peripheral blood mononuclear cells (PBMC). In 42 HLA-A2 positive myeloma patients the T cell response to the following HLA-A2-restricted peptide epitopes was analysed: ILAKFLHWL derived from TERT, LLLLTVLTV and STAPPVHNV both derived from Muc-1 and FLWGPRALV derived from MAGE-3. Eight patients had T cells specifically secreting IFN-g in response to one up to three peptide epitopes derived from TERT and Muc-1 (N=6 TERT; N=4 Muc-1). T cell responses could be detected in stage I and III patients before or in the course of chemotherapy. No patient had a natural T cell response to MAGE-3 antigen. The results of this study indicate that a natural T cell response to telomerase reverse transcriptase and mucin-1 exists in myeloma patients. Furthermore this is the first demonstration of a natural T cell response against telomerase in cancer patients.

562

Down-regulation of survivin in myeloma cells by antisense leads to increased apoptosis and sensitivity to chemotherapeutic drugs

I. Tamm, D. Höhnemann, M. Chatterjee, B. Dörken (Berlin, D)

Survivin blocks apoptosis and is detected in most human cancers but not in normal cells. Survivin expression is 40x increased during G2/M-phase of the cell cycle. Survivin associates with the mitotic spindle and binds to p21. To further investigate the function of survivin, we employed two survivin antisense oligonucleotides targeting different regions of the survivin mRNA to down-regulate survivin expression in plasmocytoma cell lines. Both constructs downregulated mRNA up to 60%, one day after start of lipofection. This effect was dose-dependent and accompanied by a reduction of cell viability as determined by DNA fragmentation and AnnexinV-staining. Cell death following treatment involved increased caspase-3 cleavage as measured by immunoblot. A non-sense as well as sense-construct did not alter the viability of the cells and did not activate caspase-3 suggesting that the effect was specific. In addition, inhibition of survivin expression increased apoptosis as well as sensitivity to chemotherapeutic drugs in vitro. These findings indicate that survivin is an interesting target for a specific cancer therapy.

563

High-producer genotyp of interleukin-10 is associated with an increased risk for multiple myeloma (MM)

K. Neben, J. Mytilineos, T.M. Möhler, A. Preiss, A. Krämer, A.D. Ho, H. Goldschmidt (Heidelberg, D)

Interleukin-10 (IL-10) is a pleiotropic cytokine that was found to have strong immunosuppressive effects via inhibition of Th1 type cytokines, including interferon-gamma and interleukin-2. Studies in monozygotic twins have shown that 75% of the variations in IL-10 production is genetically determined. The aim of this study was to examine the effect of polymorphic variations in IL-10 genes on the predisposition to MM. The single base pair polymorphisms at position –1082, -819 and –592 of the IL-10 promoter gene was studied using a sequence-specific polymerase chain reaction assay. Blood samples were analysed from 255 pts. with MM who presented at our institution. According to Salmon and Durie, 32 pts. presented with stage I, 36 pts. with stage II and 187 pts. with stage III disease at diagnosis. As a control, the genotype distribution was also studied in 96 healthy, local blood donors. Comparison of the extended IL-10 haplotype in the myeloma pts. and controls showed a significant excess of high-producer alleles in myeloma pts. (p=0.03). This genotype was present in 23.5% of myeloma pts. but in only 11.5% of the controls (p=0.004). Given the central role of IL-10 within the cytokine network, we conclude that individuals with polymorphism associated with a high production of IL-10 are at a significantly increased risk of developing MM. Since IL-10 is a growth factor for myeloma cells and has a potent stimulating effect on B-cells, this finding emphasizes the testing of immunomodulatory agents in MM.

564

Treosulfan is a potent inducer of cell death in myeloma cell lines

G. Meinhardt, E. Eppinger (München, D)

Despite recent advances utilizing high-dose chemotherapy followed by autologous stem cell transplantation, multiple myeloma remains an incurable disease to date. Median survival after relapse is usually short with myeloma cells being largely resistant to conventional therapeutic strategies. Thus, we attempted to identify cytotoxic compounds that could represent alternatives to current mephalan-based front-line or salvage treatment strategies in MM. The alkylating agent treosulfan, with has been used in the treatment of ovarian cancer for over twenty years, demonstrated strong cytotoxicity against myeloma cells in in vitro assays. NCI-H929 and U-266 myeloma cells were exposed to various concentrations of melphalan or treosulfan. Cell death was independently assayed by FACS using unfixed cells (Annexin V/PI staining) or fixed cells (hypodiploid DNA content). Increasing concentrations and exposure time of treosulfan strongly altered the ratio of viable (Annexin-/PI-) vs. apoptotic/necrotic (Annexin+) cells. Wheras this ratio was approximately 72:25 in untreated NCI-H929 control cells, 30µM and 100µM treosulfan lead to a ratio of 41:50 and 1:83 after a two day exposure, respectively. Moreover, a four day incubation resulted in more than 61% cells being apoptotic or necrotic with treosulfan concentrations of 10µM and 89% with 30µM. Using subG1 phase analysis of fixed, propidium iodide stained cells, the amount of dead cells was 15% and 47% after 2 days and 40% and 72% after 4 days of treatment with 25µM or 100µM treosulfan, respectively. In comparison, a two day incubation with 10µM or 30µM melphalan resulted in 37% or 56% and a four day course in 47% and 67% cell death using subG1 phase analysis of fixed cells. Similar results were obtained with U-266 cells. Melphalan, used as reference compound, induced cell death with a ratio of 77:21 (viable vs. apoptotic/necrotic) and 56:41 after a 2 day incubation with 10 and 30µM melphalan, respectively. In comparison, analysis of subG1 phase cells demonstrated 8% and 41% cell death after 2 days and 15% and 42% cell death after 4 days of 10 and 30µM melphalan, respectively. Treosulfan treatment resulted in 6% and 18% subG1 phase cells after 2 days and 20% and 50% after 4 days of a dose level of 25µM and 100µM. Moreover, treosulfan induced the appearance of 19% and 21% Annexin V positive cells after a 2 day incubation and of 43% and 63% after a 4 day incubation with 10µM and 30µM treosulfan, respectively. In this context, it has to be noted, that very high plasma levels of treosulfan can be achieved at a conventional dose level of e.g. 8g/m^2, although recently the mean tolerated i.v. dose was determined to be 47g/m^2. Studies are underway to compare the effects of treosulfan combinations. Preliminary data also suggest high cytotoxic efficacy in myeloma cells from bone marrow samples of patients. Thus, treosulfan has promising potential as a candidate drug for testing in myeloma patients.

565

Inhibition of the gp130-JAK-STAT-pathway does not affect bone marrow stromal cell mediated survival, but sensitizes multiple myeloma cells towards drug induced apoptosis

M. Chatterjee, D. Hönemann, R. Savino, K. Bommert, R. Burger, M. Gramatzki, B. Dörken, R. Bargou (Berlin, Erlangen, D; Pomezia, I)

The bone marrow stromal cell (BMSC) derived cytokine IL-6, which activates the gp130-JAK-STAT-signal tranduction pathway, is supposed to act both as survival and drug resistance factor for multiple myeloma (MM) cells. Therefore inhibition of this pathway should be a reasonable therapeutic strategy. To investigate the significance of the gp130-JAK-STAT-pathway for BMSC mediated cell survival and drug resistance of MM cells we blocked signal transduction at different levels. The effects of the IL-6 receptor antagonist SANT7, anti gp130 mAb and JAK tyrosine kinase inhibitor tyrphostin AG490 were evaluated on IL-6 dependent cell line INA-6 as well as on primary MM cells. In the absence of viable BMSCs inhibition by SANT7, anti gp130 mAb and AG490 resulted in suppression of cell proliferation and induction of apoptosis. However, in the presence of BMSCs INA-6 and primary MM cells were markedly less sensitive to blockade of the gp130-JAK-STAT- pathway. Furthermore, we examined, if activation of gp130-JAK-STAT is linked to BMSC mediated drug resistance of dexamethasone and all trans retinoic acid. Interstingly, SANT7, anti gp130 mAb and AG490 could overcome BMSC mediated drug resistance. These data suggest that in the presence of BMSCs the IL-6 activated gp130-JAK-STAT-pathway alone is not essential for survival of MM cells, but pharmacological targeting of this pathway with the intention to overcome BMSC mediated drug resistance could of potential therapeutic interest.

566

Chromosome aberrations in multiple myeloma: a systematic comparison of Comparative genomic hybridization (CGH) and Fluorescence in-situ hybridization (FISH)

N. Baßermann, A. Viardot, P. Liebisch, C. Wendl, K. Roth, H. Döhner, M. Bentz (Ulm, D)

In multiple myeloma (MM), there are accumulating data supporting a biological and prognostic importance of chromosomal aberrations. Up to now, only few studies using comparative genomic hybridization (CGH) have been performed in MM. In addition, no systematic comparison of CGH and other molecular-cytogenetic methods has been published. Therefore, we analysed bone marrow samples of 46 patients (34 were male and 12 female) with MM by CGH. Imbalanced chromosomal aberrations were identified in 61% of the cases. The most frequent changes (present in >10% of all cases) were gains on chromosome arms 1q (11 cases), 5p (5 cases) , 9p (5 cases), 9q (6 cases), 11q (7 cases), 15 (5 cases) and 18q (5 cases), as well as losses on 1p (6 cases), 6q (6 cases), 8p (6 cases), Xp (5 cases) and Xq (5 cases). In 27 cases, fluorescence in-situ hybridisation (FISH) results were available for comparison. Using locus-specific probes mapping on 6q21, 6q27, 11q22.3, 13q14 and 17p13, we found a high concordance between CGH and FISH data. For the different chromosomal regions, identical results were obtained in the following percentages of cases: 6q21: 77%; 6q27: 77%; 11q22.3: 77%; 13q14: 76%, and 17p13: 72%. 19 aberrations were found by FISH which were not confirmed by CGH. These discrepances were mostly due to sampling errors with a low proportion of tumor cells in the bone marrow samples. In 4 cases, positive results of CGH were not confirmed by FISH. These data show that using CGH chromosomal aberrations can be detected with high reliability in MM. In contrast to FISH, new recurrent chromosomal aberrations (e.g. losses on chromosome arm 1p and 8p as well as gains on 18q) can be identified even in cases, where metaphase preparations for banding analysis is not possible.

567

Plasma concentrations of soluble endothelial specific tyrosine kinase receptors (Stie2/Sflt-1) correlate with disease activity markers and functional magnetic resonance imaging parameters in multiple myeloma

T.M. Möhler, H. Goldschmidt, K. Neben, B. Kasper, H. Hawighorst, G. van Kaick, A. Benner, P. Reusch, A.D. Ho, D. Marme, J. Drevs (Heidelberg, Freiburg, D)

The endothelial specific tyrosine kinase receptors Tie2 and vascular endothelial growth factor receptor (VEGF) 1 (FLT-1) play an important role in vasculogenesis and angiogenesis. Shedding of these receptors is induced upon activation of endothelial cells. Induction of angiogenesis and activation of bone marrow endothelium was recently discovered in multiple myeloma (MM). We investigated plasma levels of soluble forms of Tie2 (sTie2) and FLT-1 (sFLT-1) in 48 patients with advanced MM using a two-step ELISA system. We correlated our findings to clinical parameters as well as bone marrow microcirculation analysis by contrast-enhanced dynamic magnetic resonance imaging dMRI in a subset (n=29) of patients. STie2 and sFLT-1 plasma concentration were above the upper limit of healthy controls in 3 (6.25%) and 11 (23%) pts, respectively. Using a spearman´s correlation analysis a statistically significant positive correlation was found between sTie2 and sFLT-1 plasma levels (corrsTie2: 0.53, p<0.001; CorrsFLT-1 : 0.29, p=0.04) and serum LDH as adverse prognostic tumor marker in MM. In addition, we found that Amplitude A (parameter of maximum signal increase) of dMRI significantly correlated with sTie2 plasma levels (corr: 0.44; p=0.02). Our study indicates that plasma levels of soluble endothelial specific receptor kinases (sTie2 and sFLT-1) might be useful to further elucidated the pathopysiologic role of angiogenesis and endothelial activation in MM. Moreover studies on the prognostic impact of circulationg sFLT-1/sTie2 levels in MM are warranted and are currently under way in our departments.

568

Role of CD82 in normal and malignant hematopoiesis

A. Burchert, M. Schmidt, S. Wölfl, T. Härle, S. Pankuweit, M. Jänike, M. Ritter, B. Reckzeh, A. Wilke, A. Neubauer (Marburg, D)

CD82 is a transmembrane protein, which belongs to the tetraspanin family (TM4SF). TM4SF members were initially identified as mediators of adhesive properties in non-hematological tissues. Since adhesion is essentially required to regulate hematopoiesis, we studied CD82 expression of CD82 on primitive normal progenitor cells and leukemic blasts and found a differential expression.
Aim of our subsequent studies was to investigate the biological relevance of CD82 expression in normal versus malignant hematopoiesis. In absence of a functional ligand to activate CD82 signaling, we developed a model system with immobilized or suspension-CD82-specific monoclonal antibodies to mimic the effects of a putative CD82 ligand. Ligation of CD82-specific monoclonal antibodies on CD34+ progenitor cells in adhesion assays, long term culture experiments, phosphorylation assays, and colony forming unit assays uncovered significant effects of CD82-activation on the biology of stem cells. It resulted in the induction of adhesion and morphology changes, inhibition of myelopoiesis, and signal transduction events. Furthermore, we found a significantly altered responsiveness of primary leukemic blasts compared to normal progenitors. Using cDNA array, we explored regulated targets of an antibody induced CD82 signaling. Conclusions: Our work established CD82 as a regulator of normal hematopoiesis and suggested genetic targets of CD82-signaling in immature cells. In contrast to normal primitive progenitor cells, leukemic CD34+/CD82+ blasts appear to be deficient in their ability to adequately mediate CD82-signaling. Manipulation of progenitor biology with tetraspanin-specific antibodies may be of therapeutic use in the ex vivo manipulation of progenitor cell growth and survival.

569

Characterization of bone marrow-derived mesenchymal stem cells using monoclonal antibodies recognizing neuronal progenitor cells

W. Vogel, H.-J. Bühring, F. Grünebach, A. Wahl, A. Marxer, L. Kanz, W. Brugger (Tübingen, D)

Mesenchymal stem cells (MSC) are multipotent adherent, fibroblastoid cells that express CD13, CD90, and CD105 and can give rise to myocytes, chondrocytes, adipocytes and osteogenic cells. Recently, the differentiation of MSC into nonmesenchymal cells including neuronal and glial cells has been demonstrated. We investigated bone marrow-derived MSC (n=6) for the presence of neuronal and glial progenitor cells using a variety of unique monoclonal antibodies (mAbs) generated in our lab against the retinoblastoma cell line WERI-RB-1 along with mAbs against different CD antigens. Human bone marrow mononuclear cells were cultured in RPMI 1640 medium supplemented with 10% FBS. After 14 days of culture and removal of non-adherent cells, a morphologically uniform MSC population with fibroblast-like appearance was obtained. Flow cytometry analysis revealed that these cells expressed CD10, CD13, CD61, CD90, CD105 (endoglin), CD140b (PDGF-RB), CD164, and CD172a (SIRP alpha), but were negative for CD45, CD34, and CD133. In contrast to the majority of the studied CD antigens, many of the new antibodies, including W4A5, W6D3, W7C5, W8C3, W8B2, W12D1, and 57D2, reacted with distinct MSC subpopulations, indicating that MSC are phenotypically heterogeneous. The expression profile of MSC depends also on the passage number. A gradual decrease of W7C5 antigen expression to about 10% of the original value was recorded from the first to the fifth passage (14-61days). To analyze neuronal antigen expression, APAAP staining of intracellular and extracellular matrix proteins as well as surface antigens was performed on MSC grown in Lab Tek chamber slides. The differentiation antigens consisted of vimentin, neurofilament, GFAP, laminin beta-2 chain, Neu-N, NSE, N-CAM (CD56), and our own markers defined by the mAbs W4A5, W6D3, W8B2, W8C3, W12D1, and 57D2. Of the tested markers, only Neu-N, laminin beta-2 chain, W4A5, W6D3, W8B2, W8C3, 57D2, and CD56 reacted with MSC subpopulations. Interestingly, W4A5, W6D3, and 57D2 mAbs also reacted with neuronal progenitor cells, and W4A5, and W6D3 additionally reacted with astrocytes obtained from CellSystems, St. Katharinen, Germany. Currently, we are analyzing MSC cultures for GFAP, Nestin, NSE and other neuronal markers by RT-PCR. Nestin mRNA expression was found in such MSC cultures. Our data suggest the presence of neuronal precursor/progenitor populations within the MSC cultures. To address this question, sorted MSC fractions will be analyzed for their differentiation capacity into neuronal and glial cells.

570

Generation and expansion of endothelial cells from CD34+ cord blood, bone marrow and peripheral blood cells

U. Keller, K. Götze, R. Oostendorp, M. Wimmer, A. Dwenger, C. Peschel (München, Freiburg, D)

It has been proposed that endothelial and hematopoietic progenitor cells are derived from common stem cells called hemangioblasts. Recent studies provide strong evidence for the existence of endothelial progenitor cells (EPC) in CD34+ peripheral blood cells of patients with malignancies mobilized with G-CSF. It also has been shown that normal adults have a small number of circulating endothelial progenitor cells in peripheral blood The origin of these circulating EPC remains unclear. Since endothelial cells (EC) provide the physical interface between blood and surrounding tissue and play a central role in pathologic conditions like myocardial infarction and autoimmune disease, they could soon become of therapeutical interest. In the present study, CD34+ cells and CD34-lin- cells from different sources (peripheral blood after chemotherapy and G-CSF treatment, bone marrow, cord blood) were isolated and tested for their potential to form endothelial cell layers in vitro. Isolated CD34+ and CD34-lin- cells were cultured in vitro with different growth factor combinations in the presence of extracellular matrix proteins. Endothelial cells could be successfully established in about one half of all started cultures from CD34+ cord blood and CD34+ mobilized peripheral blood cells. EC differentiation was tested by immunohistochemistry, flow cytometry and RT-PCR. Cultured cells from passage 3-5 show endothelial-like morphology very similar to human umbilical vein endothelial cells (HUVEC), which were used as a control in all experiments performed. Immunohistochemically, EPC/ EC cultured from CD34+ cells show high expression of vWF, KDR-2, CD31 and Tie-2 and moderate expression of CD34. These cells show typical morphologic characteristics of endothelial cells and stain negative for CD45 and CD14. Using RT-PCR, the expression of Flk-1could be shown in EC and control cells (HUVEC). Functional integrity was evaluated by tube formation in Matrigel. To evaluate the potential of these EC/ EPC to seed into ischemic areas of blood vessels, experiments with immunodeficient animals will be performed. In conclusion, the present study shows the generation of human EPC/ EC from different CD34+ sources. Development of ex vivo expansion techniques for angioblast-like endothelial precursor cells will be useful to promote vascular healing, could provide suitable coating for vascular grafts or could be used to deliver toxins to tumor vascular cells.

571

Amplification of human primitive hematopoietic stem cells by overexpression of the homeobox transcription factor HOXB4

M. Feuring-Buske, C. Buske, D. Hogge, C. Eaves, K. Humphries (München, D; Vancouver, CDN)

Multiple members of the clustered HOX family of transcription factors are preferentially expressed in the most primitive hematopoietic stem cell (HSC) populations thus implicating these genes in critical early regulatory processes. Furthermore, constitutive overexpression of the homeobox gene HOXB4 in the murine model leads to marked increases in the regenerative capacity of long term competitive repopulating cells (CRU) without altering lineage determination. These findings suggest HOXB4 as an attractive candidate for the amplification of genetically modified human HSC. To test directly the hypothesis that HOXB4 can amplify early HSC growth, we retrovirally expressed the gene in human highly purified CD34+/GFP+ cord blood cells and analyzed the impact of constitutive HOXB4 expression on differentiation and proliferation in vitro and in vivo. When long-term culture initiating cell (LTC-IC) numbers were determined 2-9 days after transduction by limiting dilution assays HOXB4 induced a 3-30fold increase (n=6) of the progenitor frequency compared to the GFP control (p=0.01). This effect was confirmed using the limiting dilution CRU assay in a NOD/SCID mouse model which revealed a 4fold increase in the CRU frequency of HOXB4 transduced CB cells (p=0.02). HOXB4 also had a profound effect on committed progenitors: analysis of the hematopoietic progenitor content of suspension cultures after 6 weeks showed a 100-fold increase in the frequency of colony forming cells (CFC) (p<0.0005). In addition, HOXB4 transduced progenitors had greatly increased proliferative capacity as indicated by an 8-fold increase in 2° CFC formation (n=6, p=0.04) in serial replating experiments. Interestingly, a greater proportion of HOXB4 derived secondary colonies were of erythroid type (81% versus 16% for controls (p<0.0001)) suggesting that HOXB4 might have some impact on lineage commitment. This was further supported by analysis of 6 wk in vitro liquid cultures in which the yield of erythroid committed progenitors was 4.x104 in HOXB4 transduced cultures, compared to none in the control. Furthermore, HOXB4 overexpression promoted early B cell development in vitro in a feeder dependent culture system with an 18fold increase in the proportion of CD34+/CD19+ pro-B cells. The data provide evidence that expression of HOXB4 is a critical nodal point in human stem cell self-renewal / proliferation decisions. These studies will help to gain more insight into the regulatory mechanisms controlling HSC expansion and differentiation.

572

Analysis of adherent hematopoietic progenitor cells on modified polymer substrates by laser scanning cytometry

T. Köhler, F. Kobe, T. Pompe, J. Oswald, R. Plettig, K. Salchert, M. Bornhäuser, C. Werner (Dresden, D)

Adhesion to stroma cells or artificial surfaces is considered to influence the fate of hematopoietic progenitor cells upon in vitro manipulations. Thus, the preparation of polymer scaffolds with well-balanced bioadhesive characteristics and the investigation of adherent progenitor cell cultures deserve attention. Cytokine driven cultures of CD 34+ selected peripheral blood stem cells were performed in direct contact with modified polymer substrates exhibiting biomimetic characteristics. For that aim thin films of poly(ocatdecen alt maleic anhydride), a versatile reactive copolymer, were covalently bound onto glass carriers and subsequently coated with different components of the extracellular matrix (fibronectin, heparansulfate, collagen I, collagen IV). Results of the ongoing study indicate pronounced adhesion of the progenitor cells onto coatings of collagen I, escpecially in early stages of cell culture. Current experiments are dedicated to the characterization of the (immuno)phenotype of the adherent cells after different culture periods. The established analytical method for the detection of hematopoietic progenitor cells is based on flow cytometry using antibodies labelled with fluorochromes (FACS). However, flow cytometry requires a considerable quantity of suspended cells and cannot provide any microscopic information about phenotype and shape of the analysed populations. Therefore, Laser Scanning Cytometry (LSC) and Laser Scanning Microscopy (LSM) were adapted for the detailed characterization of adherent hematopoietic progenitor cells on artificial substrates. The LSC is a microscope linked instrument that measures fluorescence and forward scatter of cells on a standard microscope slide. In addition to multicolour cytometric analysis equivalent to flow cytometry, this configuration offers the advantage to permit accurate relocation and microscopic visualization of analysed cells. Laser Scanning Microscopy can complement LSC data by high-resolution imaging of individual cells. Both techniques were found to be applicable to adherent cells on thin organic layers on glass carriers (i.e. cover slits). Fluorescence staining had to be performed with special care to prevent ablation of the adherent cells. In order to evaluate the sensitivity of LSC different samples of CD34/CD38 labelled hematopoietic stem cells were analysed by FACS and by LSC in parallel. The results show a good correlation between FACS and LSC data even if the analysed number of cells was substantially lower in the LSC analysis. In consequence, LSC can be recommended as a useful analytical tool for adherent or infrequent cell populations which might be important to facilitate engraftment in primary recipients after ex vivo culture.

573

Migration of normal and malignant hematopoietic cells induced by stroma-derived factor-1 alpha is associated with increased podia formation and downregulation of chemokine receptor gene expression

S. Fruehauf, K. Srbic, R. Seggewiss, J. Topaly, A.D. Ho (Heidelberg, D)

Normal and malignant hematopoietic cells extend multiple podia of yet unknown function. To study the role of different forms of podia in locomotion we investigated podia formation and migration of normal primitive hematopoietic cells using scanning electron microscopy and a time-lapse camera system. Subsequently, we investigated the influence of stroma-derived factor-1-alpha (SDF-1-alpha) on podia formation, transmigration and transcriptional regulation of putatively relevant genes in leukemic cell lines. The impact of BCR-ABL expression on these parameters in a chronic myelogenous leukemia cell line was studied using the ABL kinase inhibitor STI571. Scanning electron microscopy revealed the formation of uropodia at the one pole and lamillopodia at the opposite, i.e. on the fibronectin-coated (fragment CH-296) contact surface of CD34+ cells, FACS-sorted from umbilical cord blood. Locomotion of CD34+ cells was always associated with cells demonstrating short podia (uropodia and lamillopodia with the latter always representing the leading edge) as shown by time-lapse video microscopy. Following exposure to SDF-1-alpha, podia formation was significantly increased in five of six leukemic cell lines studied (BCR-ABL+: BV173, K562, 32Dp210; BCR-ABL-: HL60, 32D, not KG1a). In an indicator cell line (BV173) SDF-1-alpha lead to a significant 8-fold increase in transmigration ($p < 0.05$) and to a downregulation of the gene expression of the chemokine receptors CCR4, CXCR4, CXCR5, indicating a negative feedback regulation. SDF-1-alpha-induced transmigration and podia formation were independent of BCR-ABL expression as both were not affected by the BCR-ABL inhibitor STI571. Our data are compatible with the hypothesis that formation of specific, fibronectin-dependent podia by hematopoietic cells is associated with egression of these cells from the bone marrow.

574

Application of a new PCR technique allows to study individual retrovirally marked human hematopoietic stem cells with marrow repopulating ability

B. Gentner, S. Laufs, K.Z. Nagy, K. Kühlcke, W.J. Zeller, B. Schiedlmeier, A.D. Ho, S. Fruehauf (Heidelberg, Idar-Oberstein, D)

Gene-transfer protocols using e.g. the multidrug-resistance-1 (MDR1) gene to protect blood cells from chemotherapy related toxicity are being introduced into clinical testing. However, little is known about the lifespan and clonal diversity of human hematopoietic stem cells in steady state and post-transplantation hematopoiesis, information that is important to evaluate the success of such protocols. We transplanted NOD/SCID mice with retrovirally marked human CD34+ peripheral blood progenitor cells (PBPC). We applied a ligation-mediated PCR (LM-PCR) method to study bulk cell populations of chimeric mouse bone marrows with regard to the number of different human cell clones present after 5-13 weeks of engraftment. This method allows simultaneous single tube amplification of different integration site fragments from whole bone marrow. Therefore, we digested the DNA, thus creating a cell clone-specific restriction fragment length polymorphism (RFLP), followed by amplification of the genomic DNA- retroviral vector LTR junction by LM-PCR and sequenced the informative DNA regions. The detection threshold for a clone was in the range of 1 retrovirally marked cell per 100-200 unmarked cells. In 5 chimeric NOD/SCID mouse bone marrows we were able to detect between 6-17 integration sites. When RFLP's were generated with different restriction enzymes we were able to pick up identical integration sites confirming the reliability of our protocol. On the other hand, each digest also contained singular integration sites, indicating that more clones may become detectable with the analysis of further RFLP patterns. In a blast search analysis of our integration site DNA sequences in the currently available human genome databases we found identical regions to human clones in 28%-60% (mean: 46%) of cases which validates this LM-PCR method. In four cases known human gene sequences /genes could be identified. The protocol described here allows to track individual human hematopoietic stem cells marked with murine-derived retroviral vectors even in a background of mouse genomic DNA. This opens the door for a multitude of studies on the biology of bone marrow transplantation.

575

Cysteinyl leukotrienes enhance beta 1-integrin-mediated adhesion of CD34+ hematopoietic progenitor cells and may play a role as paracrine mediators in the bone marrow microenvironment

A. Boehmler, F. Bautz, C. Denzlinger, L. Kanz, R. Möhle (Tübingen, D)

Hematopoietic stem cell homing is a prerequisite for the establishment of hematopoiesis after bone marrow transplantation. However, our current understanding of the factors and pathways involved in the adhesion and migration of hematopoietic cells is still limited. We have recently shown that the cysteinyl leukotriene receptor CysLT1 is highly expressed and functionally active in CD34+ hematopoietic progenitor cells (HPC) [Blood 97:3433-3440, 2001]. Since the production of leukotrienes has been observed in the normal bone marrow, and CysLT1 is structurally related to chemokine receptors (G protein coupled 7-transmembrane receptors), CysLT1 might modulate adhesion and migration of HPC similar to the SDF-1 receptor CXCR4. We analyzed adhesion of mobilized human CD34+ HPC to endothelial cells (EC), immobilized adhesion molecules, and extracellular matrix (ECM) proteins in vitro. Preincubation with cysteinyl leukotriene D4

(LTD4) rapidly upregulated adhesion of HPC to IL-1 beta-activated EC in a time- and dose-dependent manner by up to 81.14 ± 17.87 % (mean ± SD, n=3), with maximum adhesion being observed in cells treated with 1 μM LTD4 for 15 minutes. LTD4-triggered adhesion to endothelium could be blocked by an integrin alpha 4 beta 1 (CD49d/CD29: VLA-4) antibody (mAb), but not by CD11a (LFA-1 alpha) or CD18 (beta 2 integrin) mAbs. Adhesion of HPC to the VLA-4 ligand VCAM-1 was more than two-fold increased by LTD4 (2.10 ± 0.17, n=3), including both primitive CD34+/CD38- cells and lineage-committed clonogenic progenitors. The LTD4-induced activation of VLA-4 could be suppressed by the specific CysLT1 receptor antagonist MK-571. In addition, LTD4 also upregulated both alpha 4 beta 1 and alpha 5 beta 1 (CD49e/CD29: VLA-5) dependent HPC adhesion to fibronectin. We conclude that activation of CysLT1 by cysteinyl leukotrienes results in increased activity of VLA-4 and VLA-5 by inside-out signalling. Cysteinyl leukotrienes may therefore play a role as paracrine or autocrine mediators supporting adhesion of HPC to endothelium, stromal cells and ECM, and may thus be involved in bone marrow homing in a way similar to peptide chemokines. These observations link cysteinyl leukotrienes to physiological effects apart from their impact in allergy and inflammation.

576

Human peripheral blood-derived CD34-negative stem cells with mesenchymal and hematopoietic characteristics

K. Thalmeier, C. Conrad, S. Moosmann, S. Sagebiel, I. Wiest, K. Öttrich, R. Huss (München, D)

To study the stem cell potential of cells undergoing the hematopoietic as well as the mesenchymal differentiation pathway, we have established a model system, where we isolated two CD34-negative, peripheral blood-derived, monoclonal cell lines, V54/1 and V54/2. Both cell lines show a fibroblast-like morphology in monolayer cultures and can be stimulated to produce non-adherent, round shaped progeny in-vitro. Analysing the hematopoietic potential in classical CFU-assays showed that the non-adherent cells gave rise to single colonies, while the non-adherent layer were negative for colony formation. Previous studies have shown that the adherent growing cells were predominantly quiescent. Differences in the expression of surface markers between adherent and non-adherent could be shown by immunohistocemical stainig: while the adherent cells did not express the SCF-receptor c-kit (CD117), the non-adherent cells were almost all CD117-positive. FACS analysis revealed positive staining for CD10 in the adherent cells of both cell lines, but no staining for CD45, CD117 and HLA-DR. While V54/1 cells showed less than 0.5% CD34-positive cells among the total population, 1-2% CD34-positive cells were found in the V54/2 cell population. There is also evidence that V54/2 cells contain a small side population (SP cells) as demonstrated by Hoechst 33342 staining. As shown by immunohistochemistry, V54/1 cells have a collagen expression pattern (I and V positive, II and III neagtive) typically found in cells of mesenchymal origin. Both cell lines are also positive for the osteogenic marker osteocalcin, but negative for alkaline phosphatase.
In summary and as previously shown in the murine and canine model, we haved also isolated CD34-negative cell lines from human peripheral blood cells, which prove to be a potential source for adult hematopoietic and mesenchymal stem cells, a possible alternative for somatic stem cells. Whether those peripheral blood-derived and possibly totipotent stem cells might also replace the need for embryonic stem cells will be shown in further experiments.

577

Co-cultivation of bone marrow mononuclear cells and hepatocytes of one donor in a 3-D bioreactor system

S.W. Reif, G. Pless, N. Obermayer, Y.-I. Kelnar, U. Neumann, C. Müller, P. Neuhaus (Berlin, D)

Purpose: Optimization of cultivation conditions for bone marrow mononuclear cells (BM-MNC) alone, or in co-culture with autologous hepatocytes. Material and Methods: Human primary bone marrow cells from one donor aspirated from sternum and from vertebra alone, or in combination with autologous hepatocytes were cultured in a 3-D bioreactor perfusion system and, as a control, under conventional culture conditions. For all assays the same cultivation conditions and medium (CellGro(R)SCGM (Cell Genix, Germany) + 10% fetal calf serum) over a time period of 27 days was used. Every day biochemical data were analysed (glucose, lactate, lactate-dehydrogenase, urea, ammonium). Cells - aspirated after different cultivation periods - were characterized by a) growth-kinetic, b) viability, c) morphology, d) immunophenotype and d) clonogenic potential. Results: In the bioreactor system BM-MNC could be cultured with a stable viability of 45% over the whole cultivation period. Clonogenic ability could be verified even after 26 d cultivation. Granulocytopoesis dominated morphologically, after day 12 macrophages proliferated. The co-culture of BM-MNC with autologous hepatocytes showed a high amount of pluripotent precursor cells (d8: CD34: 1,36%, CFU: 0.6 ‰) during the first 2 weeks of cultivation. Afterwards the maturation of granulocytopoiesis took place, macrophages occured. Cultivation of hepatocytes in the bioreactor perfusion system resulted in an increase of t-lymphocytes, early hematopoietic development took place. Biochemial analysis in all bioreactor-experiments showed an uptake of glucose and a production of lactate. Hepatocytes metabolized ammonium. After day 24 the biochemical activity decreased. Best biochemical values could be obtained in the co-culture of BM-MNC with hepatocytes. Conclusion: The bioreactor perfusion system proved to be suitable for the cultivation of human primary hematopoietic cells as well as liver cells. The co-culture of bone marrow cells with hepatocytes is favorable for early hematopoietic cell cultivation. To investigate more aspects of stem-cell-plasticity, further studies are required.

578

Ex-vivo expansion of CD34+ hematopoietic progenitor cells with SCF, FL, IL-3, TPO and HGF: Phenotypic and functional characterization

J. Grassinger, G. Müller, M. Zaiss, S. Rose-John, L.A. Kunz-Schughart, R. Andreesen, E. Holler, A. Reichle, B. Hennemann (Regensburg, Kiel, D)

Ex-vivo expansion of hematopoietic cells can be used to amplify defined cell populations for specific clinical purposes in the setting of allogeneic and autologous stem cell transplantation. In this study, the effect of hepatocyte growth factor (HGF), stem cell factor (SCF), flt-3 ligand (FL), interleukin (IL)-3, hyper-IL-6 (HIL-6) and thrombopoietin (TPO) on the expansion of CD34+ cells, colony forming cells (CFC), long-term culture initiating cells (LTC-IC) and the immunophenotype was investigated. First, single sorted CD34+38- cells isolated from the blood of healthy volunteers were incubated in serum free medium (SFM) with SCF, FL, IL-3 (SF3) with and without HIL-6, TPO, HGF and combinations thereof. On day 8, the highest number of clones was found after stimulation with SF3+TPO+HGF (42±10%, SF3TH) and was significantly increased as compared to SF3 (26±5%). Then, 5x104 lineage depleted cells from G-CSF mobilized tumor patients (PAT) and healthy donors (VOL) were incubated for 7 days in SMF plus SF3 and SF3TH whereby the expansion of the total cell number after 7 days was 13±3 (1.1±0.3) and 16±4 (1.8±0.3) for VOL (PAT), respectively. Accordingly, the number of CD34+ cells increased 11±3 fold and 14±7 fold for VOL derived cells but decreased for PAT derived cells. Mean CFC numbers increased 10±3 and 17±7 fold for VOL and 7±2 and 12±6 fold for PAT. The number of VOL derived CD34+ cells and CFC were additionally measured after 4 days and increased 2±0.6 (SF3) and 4±1 fold (SF3TH) and 4±2 and 7±3 fold, respectively. The number of VOL derived LTC-IC decreased form 1012±437 per 105 CD34+ cells seeded to 152±65 and 239±111 on day 4 and to 86±33 and 228±113 on day 7. The proportion of CD34+ cells decreased from 85±5% on day 0 to 60±6% (SF3) and 63±10% (SF3TH) on day 7. The percentage of myeloid (CD33+) cells increased from 46±26% to 73±22% and 73±24%, CD15/66b+ granulocytes increased form 5±3% to 20±5% and 22±7% whereas the percentage of monocytes and T-lymphocytes remained unchanged. However, a significant increase of B-lymphoid cells form 7±3% to 21±7% and 23±2% was detectable.
In summary, our data indicate that mobilized hematopoietic progenitor cells from healthy volunteers can be expanded more efficiently than cells from tumor patients. After 7 days of culture, the proportion of early hematopoietic progenitor cells is reduced and the majority of the generated cells consists of differentiated myeloid and B-lymphoid cells.

579

Stromal cells from mouse embryonal AGM subregions support long-term maintenance of hematopoietic stem cells

R. Oostendorp, K. Harvey, N. Kusadasi, M. de Bruijn, C. Saris, A. Medvinsky, A. Medvinsky, A. Medvinsky, A. Medvinsky, A. Medvinsky, A. Medvinsky, E. Dzierzak (München, D; Rotterdam, NL; Thousand Oaks, USA; Edinburgh, UK)

The aorta-gonads-mesonephros (AGM) region autonomously generates the first adult repopulating hematopoietic stem cells (HSC) in the mouse embryo. HSC activity is initially localized to the dorsal aorta, vitelline and umbilical arteries. Thereafter, HSC activity rapidly increases in the urogenital ridges (UG) and the liver. To examine which embryonic microenvironment best supports HSCs growth and maintenance we established stromal cell lines from mid-gestation mouse embryos transgenic for a temperature sensitive immortalizing gene or a marker gene. Several embryonic stromal clones were found to maintain adult hematopoietic progenitors and HSCs to a similar degree as, or better than, previously described murine adult and fetal stromal clones. Interestingly, the stromal clone most highly supportive of HSCs is derived from the UG subregion of the AGM, suggesting that the AGM provides its own potent microenvironment for growth and maintenance of the first HSCs during mid-gestation.

580

Efficient isolation and characterization of AGM subregion-derived stromal cell lines from novel temperature sensitive SV40 T antigen transgenic mice

R. Oostendorp, A. Medvinsky, A. Medvinsky, N. Nakayama, N. Kusadasi, C. Orelio, T. Covey, R. Ploemacher, R. Ploemacher, C. Saris, E. Dzierzak (München, D; Edinburgh, UK; Thousand Oaks, USA; Rotterdam, NL)

The temperature sensitive form of the simian virus 40 (SV40) large tumor antigen (TAg) has been widely used for the isolation of permanent cell lines. Previously, hematopoietic stromal cell lines have been established through in vitro transduction of the SV40 Tag gene by retroviral vectors, thus requiring some ex vivo cell growth before the immortalizing gene is expressed. To isolate stromal cell lines that may be more representative of the in vivo hematopoietic microenvironment, particularly during ontogeny, we generated transgenic mice with the temperature sensitive form of the SV40 TAg gene (tsA58) under the transcriptional control of constituitive and ubiquitously expressing promoters for the beta-actin and PGK genes. We report here the generation of tsA58 transgenic mice and the isolation of hematopoietic stromal cell populations and clones from the embryonic hematopoietic microenvironments of such transgenic animals. We demonstrate that growth and cloning efficiency of stromal cells is increased with presence of the SV40 Tag transgene. Furthermore, our results show that the large panel of stromal populations and clones isolated from the different hematopoietic subregions of the mouse embryo exhibit functional heterogeneity in their ability to promote hematopoietic differentiation and in hematopoietic growth factor gene expression profiles. Interestingly, a few of these stromal clones express a recently described chordin like gene that is preferentially expressed in cells of the mesenchymal lineage

581

Preferential expression of human Fc gamma receptor II on human hematopoietic progenitor cells committed to the erythroid lineage

M. Zaiss, A. Rehm, C. Hirtreiter, L. Kunz-Schughart, A. Boyd, R. Andreesen (Regensburg, D; Brisbane, AUS)

Fc gamma receptors (FcgR), a subclass of the large family of Fc receptors, are expressed on almost all hematopoietic cells. Binding of IgG-antibodies to these receptors triggers or suppresses immune responses. Recently, the expression of the murine FcgR II/III was shown to be specific for granulo-monocytic progenitor cells. We investigated the expression of the FcgR IIA (CD32) on human hematopoietic progenitor cells isolated from peripheral blood. 20 to 30 % of the CD34+ progenitor cells from non-mobilized peripheral blood but only 2 to 5 % of G-CSF mobilized peripheral blood progenitor cells were positive for CD32. In colony assays in semisolid agar CD34+CD32+ cells gave rise to 192±10 erythroid, 37±16 myeloid and 10 mixed colonies per 1000 cells seeded, indicating that human CD34+CD32+ are mainly erythroid progenitor cells. To determine if the CD34+CD32+ subpopulation contains primitive progenitor cells we performed liquid long-term culture assays. In these assays CD34+CD32- cells had a much higher proliferative potential compared to CD34+CD32+ cells. Similar results were observed for the number of secondary CFC. 1000 CD34+CD32- cells gave rise to almost 7000 CFC, while the same number of CD34+CD32+ cells formed only 100 CFC over the six week culture period. Using RT-PCR we investigated the expression of transcription factors in CD34+CD32+ cells. In samples from small cell numbers (10 cells) we were able to detect mainly GATA-2 expression. PU.1 and SCL was also expressed in some of these samples, but always together with GATA-2. In single CD34+CD32+ cells GATA-1 and GATA-2 was expressed in the majority of the cells, suggesting a commitment along the erythroid or megakaryocytic differentiation pathway. No expression of C/EBP alpha was detectable. In summary our results show that in humans the expression of CD32 on CD34+ progenitor cells from peripheral blood is not restricted to the granulo-monocytic cell line. Instead the FcgR IIA identifies a population of committed CD34+ progenitor cells predominantly of the erythroid lineage.

582

In vivo phenotype of human mesenchymal stem cells (MSC)

S. Kaiser, B. Hackanson, Y. Guo, M. Follo, K. Geiger, E. DeLima-Hahn, M. Engelhardt, U. Kapp (Freiburg, D)

Human Mesenchymal Stem Cells (MSC) can be isolated from bone marrow and umbilical cord blood and were shown to differentiate in vitro into osteocytes, chondrocytes, adipocytes, tendon and myocytes. To obtain MSC, mononuclear cells are isolated from bone marrow by ficoll separation and cultured in culture dishes at a density of 50.000-100.000 cells/cm2. The culture medium is Alpha-MEM containing 2% fetal calf serum, 1% penicillin/streptomycin, PDGF, EGF, dexamethason, insulin, ascorbic acid, transferrin and LDL. MSC attach to the culture dishes after several hours. The early attachment can be used to separate these cells from all the other mononuclear cells. Non-attaching cells are completely removed after two to three medium changes. Although much is known about the phenotype of MSC in vitro (e.g. cell surface markers) the character of the MSC progenitors in vivo still remains to be seen. MSC stain positive for the Thy-1 (CD90) antigen that is also expressed on hematopoietic progenitors and T-cells, but in vitro do not express CD45, a common marker for hematopoietic cells. We stained mononuclear cells from human bone marrow with fluorescence-conjugated CD45 and CD90 antibodies and sorted them using a MoFlow high-speed cell sorter. Afterwards the cells were cultured under MSC-selecting conditions. MSC growth was restricted to the CD90+ fraction. In a double staining CD45/90 adherent cells could not be cultured from the double negative fraction but grew in most experiments from the double positive fraction, indicating a possible change of the CD45 expression from an early stage positive cell to a CD45 negative phenotype of MSC in vitro. These results imply that the character of early MSC in vivo might be different from their progeny in vitro due to the artificial culture conditions. This changed phenotype is possibly accompanied by a diminished differentiation potential, which is under investigation in our ongoing studies. Furthermore cell separations will be performed with antibodies for other early stem cell markers and by staining with the fluorescent DNA-binding dye Hoechst 33342, a method for obtaining side population (sp) cells, to define new isolation and culture conditions for earlier cells with more differentiation capacity than in vitro cultured MSC.

583

Influence of TGF-beta on hematopoietic stem and progenitor cells in vitro

A. Wiesmann, S. Schneider, W. Brugger, L. Kanz (Tübingen, D)

Hematopoietic stem and progenitor cell (HSPC) transplantation is an increasing therapeutical option for many diseases. Cell numbers of transplanted HSPC is still one limiting factor for autologous and allogeneic transplantation. Therefore, in vitro expansion of HSPC without loss of stem cell function would overcome this problem. One possible reason for reduced homing or proliferation capacity of HSPC after in vitro culture is the increase of cells in S and G2 phase as primitive stem cells are known to be quiescent. We have shown recently, that transforming growth factor (TGF)-beta1 is able to induce cell cycle arrest in short-term culture of mouse HSPC but nevertheless did not increase repopulation potential in vivo. This observation might be due to TGF-beta induced apoptosis. To answer this question, human CD34+ HSPC were isolated from apheresis products of healthy stem cell donors by magnetically labelling (MidiMACS). Purity of positive cell fraction was always about 95%. Isolated HSPC were then cultured under serum-free conditions with the addition of SCF, IL-6 and FLT3L for up to 10 days. TGF-beta1, TGF-beta3 or both cytokines were added to the culture in a second step 24-72h before end of culture. Cultured cells were then monitored for cell number, cell-cycle status, percent apoptotic/dead cells, CD34 expression and cloning efficiency (colony-forming-unit cells). TGF-beta1 was most potent in inducing cell cycle arrest, increasing G0/1 cells from about 65% to 85% on day 3 of culture and from about 80% to 95% on day 10 of culture. On the other hand, TGF-beta1 and TGF-beta3 also increased the number of apoptotic/dead cells in any culture condition. In 5 independent experiments with 10 days in vitro culture, apoptotic/dead cells increased from 12,8 + 4,5% to 22,7 + 7,2% if TGF-beta1 was added on day 8. Furthermore, TGF-beta reduced the proliferation rate as measured by total cell numbers but did not show any influence on CD34 expression. Cloning efficiency after 10 days in culture was decreased from 146 + 18,6% to 86 + 20.0% if TGF-beta1 was added on day 8. Cloning efficiency of uncultured cells was 66%. In summary, TGF-beta1 and TGF-beta3 induce cell cycle arrest of in vitro cultured human HSPC, but increase the fraction of apoptotic/dead cells, reduce proliferative capacity in vitro and reduce cloning efficiency. Therefore it seems to be unlikely to increase stem cell potential of HSPC after in vitro culture in the presence of TGF-beta.

584

Migration of hematopoietic progenitor cells into a novel 3-D homing model is controlled by rho family small GTPases and cytokines

G. Bug, T. Roßmanith, R. Henschler, B. Schröder, L. Kunz-Schughart, M. Kampfmann, D. Hoelzer, O.G. Ottmann (Frankfurt, Regensburg, D)

Homing of hematopoietic progenitor cells (HPC) is a multistep process that is still poorly defined. To mimic the in vivo cell-cell and cell-matrix interactions involving a complex array of cytokines, chemokines and adhesive molecules, we established a multicellular spheroid coculture model by initiating the murine stromal cell line M2-10B4 to form aggregates of about 250 µm diameter in liquid overlay prior to addition of umbilical cord blood (UCB) derived CD34+ cells or the HPC line TF-1. Entry of hematopoietic cells into spheroids was quantified by flow cytometric analysis of trypsinised spheroids after 6-72 hours. TF-1 cells migrated into the spheroids in a time dependent manner, resulting in a maximum proportion of 35.6±5.7% TF-1 cells per spheroid after 48 hours (n=7). Migration of UCB derived CD34+ cells was enhanced from 24.6±4.6 to 43.9±11.9% (n=3) by preincubation in the presence of SCF, FL, TPO and IL-3 for at least 24 hours, according to what we have found previously for the SDF-1-induced transendothelial migration of freshly isolated and cultured CD34+ cells. Surprisingly, entry of CD34+ cells into the three-dimensional network of bone marrow stromal cells proved to be independent of both CXCR4 and CXCR3, as blocking antibodies had no effect. Family members of the Rho-like GTPases including Rho, Rac and Cdc42 have been identified as key regulators of both cell-cell adhesion and cell motility. We therefore investigated the impact of signalling through small GTPases on the migration of HPC into stromal cell spheroids. Pertussis toxin, an inhibitor of Gi/o GTPases, reduced the proportion of TF-1 cells entering the spheroids from 12 to 4% after 12 hours (n=3). A comparable reduction (60-70%) resulted from a panel of bacterial toxins selectively targeting different Rho proteins (kindly provided by Klaus Aktories, Freiburg, Germany), i.e. the C2/3 Fusion Toxin for specific inactivation of Rho A and Lethal Toxin for Rac and Cdc42 (n=2). Inhibition of the PI3 kinase pathway by wortmannin or LY294002 did not impair migration of TF-1 cells, although Rac- and Cdc42-induced cell motility and invasion have been shown to be mediated by PI3 kinase. Blocking of all Rho GTPases by Toxin B completely abrogated migration of TF-1 cells without increase of apoptosis or impaiment of proliferation. Flow cytometry analysis showed that treatment with Toxin B did not alter VLA-4 and VLA-5 integrin expression on TF-1 cell surface. Our data show that the in vitro migratory capacity of UCB derived CD34+ cells may be improved by culture in presence of appropriate cytokines. Stromal cell spheroids provide a novel model to study signal transduction cascades involved in the regulation of HPC migration besides the well-known SDF-1/CXCR4 signalling, e.g. the Rho family small GTPases.

585

Telomere length dynamics in feline blood cells point to a rapid turnover of hematopoietic stem cells in the first year of life

T. Brümmendorf, J. Mak, G. Baerlocher, K. Sabo, J. Abkowitz, P. Lansdorp (Tübingen, D; Vancouver, CDN; Seattle, USA)

Telomeres in all vertebrates are composed of TTAGGG repeats and associated proteins. In somatic cells, aproximately 50-200 base pairs of telomere repeats are lost with each cell division. Previous cross-sectional studies of the telomere length in normal human peripheral blood granulocytes have pointed to substantially increased turnover of hematopoietic stem cells in the first year of life (J. Exp. Med. 190:157-167). In view of the large variation in telomere length at any given age, longitudinal studies are required to confirm these results. Because such studies are difficult to perform in humans, we analyzed the telomere length in serial blood samples from cats. Lymphocytes and granulocytes from peripheral blood of two newborn kittens, a four year old cat, a ten year old feline recipient of a double autologous stem cell transplant and a ten year old control animal were analyzed at 2-week intervals over a one year period. Long telomeres were found in granulocytes and lymphocytes from the two kittens (mean ± S.D. = 59.2 ± 2.8 TFU and 63.2 ± 3.4 TFU resp.) compared to the four year old cat (53.3 and 62.3 TFU, resp.) and from the two adults (46.2 ± 0.3 TFU and 45.3 ± 1.6, resp.). The rate of telomere shortening in both granulocytes and lymphocytes was most rapid in the kittens (slope: 15.0 ± 0.6 and 12,8 ± 2.5 TFU/year, resp.). Compared to humans, telomere shortening with age in cats is strikingly more rapid. To investigate whether this reflects differences in cellular turnover or differences in the number of repeats lost per cell division, we expanded T-lymphocytes from peripheral blood with IL-2 and PHA. An average rate of telomere attrition of 450 + 42 bp/pd was found for lymphocytes from both the kittens and the adult animals, approx. 5-fold higher than the rate observed with human cells. Based on the assumption that the telomere length in granulocytes reflects the telomere length in the stem cell compartment, we estimate that the number of population doublings (pd) in the feline stem cell compartment in the first year of life was resp. 15.4/0.55 = 28 pd and 14.6/0.49 = 29.8 pd for the two kittens. These results are in good agreement with our previous estimate of the turnover of stem cells in human newborns (15 pd in the first 6 months). Species specific differences in the rate of telomere shortening per pd need to be considered when telomere length dymanics are linked to estimates of cellular turnover.

586

Stimulation with the monoclonal antibody anti-HPCA-1 induces receptor-mediated endocytosis of the CD34 molecule on hematopoietic cells

J. Krauter, M. Hartl, L. Hambach, A. Kohlenberg, E. Gunsilius, A. Ganser, G. Heil (Hannover, D; Innsbruck, A)

CD34 is a highly glycosylated 115kd transmembrane protein belonging to the sialomucin family. It is constantly found on early progenitors of both normal and malignant hematopoiesis. In addition to the full-length CD34 isoform, a truncated CD34 lacking most of the intracellular part of the molecule is generated by alternative splicing. Evidence was found that CD34 most probably acts as a receptor molecule, whereas its precise function remains to be elucidated. To further analyse the mechanisms of CD34 mediated signal transduction we wished to track the intracellular pathway of CD34 after ligand-binding by immuno-electron microscopy. Therefore, cells of normal bone marrow (NBMPC), acute lymphoblastic and acute myeloid leukemias and the KG1a cell line were incubated with the anti-CD34 antibody HPCA-1 and subsequently with a secondary antibody coupled to 30nm gold colloids at 4°C. Thereafter, the cells were incubated in medium for various time intervals at 37°C and then analysed by transmission immuno-electron microscopy. In parallel, the expression of the RNA of the two different CD34 isoforms (full-length and truncated) in these cells was quantified by isotype-specific real time PCR. Before stimulation, CD34 was evenly distributed over the surface of the cells. After binding of the anti-HPCA-1 antibody to CD34 and labelling with 30nm immunogold, a rapid capping of CD34 and a subsequent internalisation from the cell surface via clathrin coated pits and coated vesicles was observed. The percentage of internalised CD34/immunogold complexes ranged from 8-80% in the NBMPC and the leukemic blasts whereas KG1a cells showed an internalisation rate of only 0.4%. Moreover, in the KG1a cells, the CD34/immunogold complexes were excluded from the coated pits. These differences in CD34 internalisation did not correlate with the mRNA expression for the full-length or truncated CD34 as assessed by isotype specific real time PCR. In conclusion, evidence was found that CD34 is a surface receptor molecule which is modulated by receptor-mediated endocytosis. CD34 on KG1a cells appears to have a functional and/or structural defect preventing modulation of this epitope.

Postersitzung:
Präklinische und klinische Erfahrungen mit STI571

587

Gene expression profiles of leukemic cells isolated from patients with chronic myeloid leukemia (CML) and Philadelphia-positive acute lymphoblastic leukemia (ALL) receiving the Abl-kinase inhibitor STI571

N. von Bubnoff, F. Schneller, Ch. Peschel, J. Duyster (München, D)

We examined changes in gene expression of leukemic cells isolated from patients with chronic myeloid leukemia (CML, accelerated phase(AP) or blast crisis (BC)) or Philadelphia-positive acute lymphoblastic leukemia (Ph+ ALL) receiving the abl-kinase inhibitor STI571 using cDNA microarray analysis. Expression of BCR-ABL in freshly isolated leukemia cells was confirmed by western blot. So far, 16 Patients have been included. cDNA array analysis has been done for five Patients comparing expression profiles before onset of therapy and during therapy with STI571, and has been completed including analysis at the time of refractoriness towards STI571 for 2 Patients (Ph+ ALL one patient, CML/myeloid blast crisis one patient). Five patients (myeloid BC two patients, Ph+ ALL two patients, and lymphatic BC one patient) have been analysed so far at the time of

STI571-refractory disease. cDNA analysis revealed profound changes in gene expression of leukemic cells comparing expression profiles before onset of therapy, during therapy and at the time of STI571-refractory disease. We were able to discriminate different clusters of gene expression profiles over time. There were genes characterized by a decrease of expression after onset of therapy and an increase at the time of STI-refractoriness. Among those were genes which are known to be transactivated by BCR-ABL, e.g. c-myc. In contrast, there were genes which were induced upon inhibition of abl-kinase by STI571 and whose expression level decreased at the time of relaps. Moreover, there were multiple genes which were induced only at the time of refractoriness towards STI571, among them oncogenes (e.g. c-jun), genes for transcription factors, detoxification proteins (e.g. glutathion-S-transferase) and genes for DNA repair proteins. Several of those genes have been implicated in mechanisms of resistence towards cytostatic agents. Thus, in view of the work done so far, philadelphia-positive leukemia cells seem to induce different groups of genes when resistancy towards STI571 develops. Gene expression analysis, as presented in this work, will give insights into both the molecular pathogenesis of Philadelphia-positive leukemias and the mechanisms of refractoriness towards the abl-kinase inhibitor STI571.

588

Response to treatment with the selective tyrosine kinase inhibitor STI571 (Glivec) in a patient with a metastatic gastrointestinal stromal tumor (GIST)

G. Metzgeroth, S. Kreil, T. Lahaye, J. Hastka, U. Haberkorn, W. Back, H. Gschaidmeier, R. Hehlmann, A. Hochhaus (Mannheim, Heidelberg, Nürnberg, D)

Gastrointestinal stromal tumors (GIST) are mesenchymal neoplasms that arise from precursors of the connective tissue of the gastrointestinal tract. GIST cells express the growth factor receptor with tyrosine kinase activity c-kit. This receptor can be detected by immunohistochemical staining for CD117, which is considered the most specific diagnostic criterion for the diagnosis of GIST. c-kit mutations result in constitutive tyrosine kinase activity and uncontrolled cell proliferation. Prognosis of patients with non resectable tumors and/or metastatic disease is limited with chemotherapy.
Therefore we used the selective inhibitor of c-kit tyrosine kinase STI571 (Imatinib, Glivec) in a GIST patient pretreated with chemotherapy. A 78 year old male patient presented in 1/98 in good performance status with advanced unresectable GIST of the stomach with abdominal masses. Abundant c-kit expression was documented by immunostaining of the primary tumor for CD117. Despite palliative adriamycin monotherapy between 10-11/00 tumor size progressed.
Therapy was changed to STI571 400 mg p.o. daily in 3/01. Toxicity, blood count and blood chemistry were assessed weekly for the first two months, then monthly. Positron emission tomography with 18F-fluorodeoxyglucose (FDG-PET) was performed 2 hours prior and 24 hours after the first STI571 dose. The FDG uptake was determined using a region of interest (ROI) analysis and normalization of the values to the injected dose and the body weight. At baseline four different regions of tumor manifestations with increased FDG accumulation were seen. 24 hours after initiation of therapy FDG uptake was significantly reduced in all lesions. Two months after initiation of STI571 therapy CT scan and abdominal ultrasound showed a clear regression of tumor size. No new lesion appeared. The abdominal discomfort disappeared completely. No severe side effects of STI571 were observed. Mild nausea (WHO grade 1) disappeared when the drug was taken after a meal. No hematologic, hepatic or renal toxicity was noted. We conclude that (i) the selective tyrosine kinase inhibitor STI571 is an effective and well-tolerated therapy for unresectable or metastatic GISTs. (ii) Treatment is well tolerated with minor side effects only after three months. (iii) 18F-FDG-PET is suggested as an adequate method to evaluate the very early biological response of STI571 on GISTs.

589

Combination of different signal transduction inhibitors results in strong synergistic activity on chronic myelogenous leukemia cells

J. Topaly, S. Laufs, I. Herr, W. Zeller, A.D. Ho, S. Fruehauf (Heidelberg, D)

The tyrosine kinase activity of the BCR-ABL oncoprotein results in reduced apoptosis and thus prolonged survival of chronic myelogenous leukemia (CML) cells. The ABL-specific tyrosine kinase inhibitor STI571 was shown to selectively suppress the proliferation of the BCR-ABL-positive cells. However, resistance may occur. We were interested whether the farnesyltransferase inhibitor (FTI) L-744,832 is still active in STI571-resistant CML cells and whether the combination of both compounds is synergistic in primary CML cells or CML cell lines. A tetrazolium based MTT assay was used to quantify growth inhibition of CML cell lines after 48 hrs of exposure to FTI alone and in simultaneous combination with STI571. The drug interactions were analyzed using the median-effect method of Chou and Talalay. The combination index (CI) was calculated according to the classic isobologram equation; CI values of 1, <1 or >1 indicate additivity, synergism or antagonistic effects, respectively.
L-744,832 produced a strong synergistic effect (CI=0.53±0.21 at a 50% growth inhibition level (IC50); CI=0.47±0.18 at IC75) in combination with STI571 in the BCR-ABL+ EM-3 cell line. Both drugs were shown to be potent inducers of apoptosis as shown by annexin V staining and TdT-mediated dUTP nick end labeling (TUNEL). The CML cell lines K562 and BV173 were less susceptible to L-744,832-induced growth inhibition than EM-3. Nevertheless, L-744,832 potentiated the cytotoxic effect of STI571 in these cell lines. No cross-resistance to L-744,832 was found in BV173-R and K562-R cell lines resistant to 5 μM and 0.7 μM of STI571 respectively. The influence of both drugs on CFU-GM colonies formed by primary CML progenitor cells obtained from patients was investigated and a mutual enhancement of inhibitory effects was observed. Combination of STI571 with FTIs and possibly other signal transduction inhibitors represents a promising direction for developing new clinical trials addressing treatment of CML in advanced phases of the disease.

590

Determination of alpha-1 acid glycoprotein (orosomucoid) in patients with Ph+ chronic myeloid leukemia under treatment with STI571

P. le Coutre, K.-A. Kreuzer, J. Lupberger, I.-K. Na, M. Hohlhoff, M. Schwarz, C. Appelt, C. Müller, U. Platzbecker, G. Ehninger, C.A. Schmidt (Berlin, Dresden, D)

The tyrosinkinase activity of the oncogenic fusion protein bcr/abl is required for the transformation in almost all patients with chronic myeloid leukemia and some patients with acute lymphatic leukemia. The tyrosinkinase inhibitor STI571 effectively inhibits the bcr/abl and abl kinase activity and therefore inhibits growth and viability of bcr/abl positive cells, representing a novel therapeutic approach. In previous investigations in a nu/nu mouse model the binding of STI571 to elevated levels of the plasmaprotein alpha-1 acid glycoprotein (AGP) was identified as an in-vivo mechanism of resistance that inactivates STI571 by reducing the free and active plasma fraction of the tyrosinkinase inhibitor. In this report we provide data on the role of AGP in 39 CML patients under therapy with STI571. Patients either recieved 400 or 600 mg STI571. Apart from clinical parameters we determined AGP and CRP plasma levels, quantitative expression of both bcr/abl and AGP mRNA in peripheral blood cells and correlated these parameters to clinical response to STI571. To further investigate a common mechanism of resistance we also determined MDR1 mRNA expression in a subset of our patients. In conclusion we demonstrated that during the initial weeks of therapy: 1) Plasma AGP levels in CML patients correlate with white blood cell count (wbc) and progressive disease. 2) Elevated AGP levels at the initiation of therapy did not obstruct STI571 efficacy. 3) Treatment with STI571 normalizes elevated AGP and CRP plasma levels although mRNA levels of AGP in peripheral blood cells remained stable. 4) Initially normal levels of plasma AGP remained in the normal range under treatment with STI571, indicating that STI571 does not trigger AGP expression in humans. 5) MDR1 gene expression during an effective treatment with STI571 is not altered.

591

Effects of STI 571 on neutrophil alkaline phosphatase and whole blood levels of histamine and tryptase in patients with CML

H. Agis, A. Hauswirth, W. Sperr, V. Sagaster, A. Tueni, R. Thalhammer, I. Schwarzinger, H. Pirc-Danoewinata, O. Haas, K. Lechner, P. Valent, K. Geissler (Wien, A)

STI 571 a specific inhibitor of the BCR-ABL protein is going to have a major impact on treatment strategies for CML. Several biochemical parameters have been found to be altered in CML. A low neutrophil alkaline phosphatase (NAP) staining score is usually found in CML and has been used as a diagnostic mean in the premolecular era. Recently our group has found that whole blood levels of histamine and tryptase are elevated in untreated CML and may serve as disease related markers in these patients. Therefore, we performed serial determinations of NAP, whole blood histamine and whole blood tryptase in CML patients treated with STI in order to investigate the effects of STI in CML patients not only on hematological and molecular/cytogenetic markers but also on biochemical parameters. Between

October 2000 and May 2001 we have treated 24 chronic phase and 8 accelerated phase CML patients with STI 571 within the expanded access program (study 113 and study 114) of the Novartis company. Patients with chronic phase CML received 400 mg/day and accelerated phase patients 600 mg/day STI 571, respectively. Up to now 18 patients are evaluable with regard to hematological and cytogenetic effects at 3 months. The median WBC count of these patients was 15400/µL (range 2160-89470) prior to STI treatment and 4340/µL (1530-7530) at 3 months. In 10 patients a major cytogenetic response was observed. The suppressive effect of STI 571 on hematopoiesis was associated with a dramatic and significant reduction in both whole blood histamine and tryptase levels (median histamine level at STI start 1086ng/mL [83-7108], normal value <200ng/mL, and after 2-4 months 60ng/mL [12-345]; whole blood tryptase: 126ng/mL [3.4-864] vs 3.6ng/mL [0.9-40.9], normal value <20ng/mL). In contrast, NAP levels increased during STI treatment (7 [1-120] vs 49 [3-267]). There was no difference in the suppression of histamine and tryptase levels in patients achieving a major cytogenetic response and those who did not, however, NAP levels only increased in patients with major cytogenetic response. Our data show that hematological response in CML patients receiving STI 571 is associated with major changes in biochemical paramaters. A longer follow up of these patients is required to investigate the significance of these biochemical parameters in the prediction of disease progression.

592

In vivo effects of STI 571 on the compartment of hematopoietic progenitor cells in patients with CML

V. Sagaster, H. Agis, E. Kabrna, M. Kollars, S. Stengg, R. Thalhammer, I. Schwarzinger, H. Pirc-Danoewinata, O. Haas, K. Lechner, K. Geissler (Wien, A)

STI 571 a specific inhibitor of the BCR-ABL protein is going to have a major impact on treatment strategies for CML. CML is characterized by a massive expansion of the hematopoietic progenitor cell compartment, particularly in peripheral blood. STI 571 has been demonstrated to suppresses the growth of CML progenitor cells in vitro. Clinical trials have shown that STI 571 can normalize the number of white blood cells in most patients with chronic or accelerated phase CML but the effect of this drug on progenitor cells has not been reported. Between October 2000 and May 2001 we have treated 24 chronic phase and 8 accelerated phase CML patients with STI 571 within the expanded access program (study 113 and study 114) of the Novartis company. Patients with chronic phase CML received 400 mg/day and accelerated phase patients 600 mg/day STI 571, respectively. Serial determinations of circulating hematopoietic progenitor cells have been performed during treatment in order to study the in vivo effect of STI 571 on the progenitor cell compartment. Up to now 18 patients are evaluable with regard to hematological and cytogenetic effects at 3 months. The median WBC count of these patients was 15400/µL (range 2160-89470) prior to STI treatment and 4340µL (1530-7530) at 3 months. In 10 patients a major cytogenetic response was observed. The suppressive effect of STI 571 on hematopoisis was not restricted to mature blood cells but also involved hematopoietic progenitor cells. Beyond 1 month of STI treatment marked and significant reductions in progenitor cell numbers were observed for boths, myeloid (CFU-GM) and erythroid (BFU-E) colony forming units (median progenitor cell numbers [ranges] /mL at STI start: CFU-GM 2083 [25-43523], BFU-E 695 [0-31914], and after 2-4 months CFU-GM 28 [7-203], BFU-E 70 [0-330]). There was no difference in the suppression of the progenitor cell pool in patients achieving a major cytogenetic response and those who did not. Our data show that hematological response in CML patients receiving STI 571 is associated with a marked reduction in the hematopoietic progenitor cell pool. Molecular studies are ongoing to assess the origin of residual progenitor cells in these patients.

593

Safety and efficacy of STI-571 in patients with bcr/abl-positive chronic myelogenous leukemia (CML) after autologous peripheral blood stem cell transplantation

C. Reifenrath, G. Hess, T. Fischer, S. Kreil, P. Meinhardt, B. Schuch, J. Beck, H. Gschaidmeier, R. Hehlmann, A. Hochhaus, C. Huber (Mainz, Mannheim, Nürnberg, D)

Background: Chronic myeloid leukemia (CML) is caused by expression of the BCR-ABL fusiongene product. Phase-II studies with STI-571 (Glivec), a selective inhibitor of the BCR-ABL tyrosine kinase, showed substantial activity in patients with Philadelphia-chromosome (Ph)-positive CML. We examined the safety and efficacy of this agent in patients status post autologous peripheral blood stem cell transplantation (PBSC-T). Patients and Methods: 21 patients, who had received mobilization chemotherapy were enrolled in the multicenter CSTI571 trials 102, 109, 113, 114, and 115. Autologous PBSC-T had been performed in 19 patients. At study entry, 5 patients presented in blast crisis (BC), 5 in accelerated phase (AP), and 11 in chronic phase (CP). Patients were evaluated for hematologic response. duration of response, survival and toxicity. Tumor load was quantitated by cytogenetic analysis, FISH and quantitative RT-PCR. Results: In general, STI-571 was well tolerated in this heavily pretreated group of patients with a non-hematologic and hematologic toxicity profile similar to that observed in a previous phase-I trial at comparable doses. 6/10 patients with CML in transformation (AP, BC) were evaluable for hematologic response. Of the 6 patients, 2 had transient reductions in WBC and blasts and 4 patients achieved a sustained hematologic response lasting at least 4 weeks (2 patients: return to chronic phase; 1 patient: no evidence of leukemia; 1 patient: complete hematologic response). Cytogenetic analysis in these patients showed numerical and/or structural responses. In 11/11 CP-patients, STI-571 induced a complete hematologic response. In this group, major cytogenetic responses were observed in 5/9 evaluable patients including 4 complete cytogenetic responses. Conclusions: Taken together, this report indicates that STI-571 is a safe and effective drug in patients post autologous PBSC-T. No apparent additional side effects were noted in this patient cohort. In CP-patients, STI-571 was able to induce a high rate of complete hematologic and cytogenetic responses, combined with good quality of life. This is even more remarkable, as intensive treatment approaches plus IFN-alpha failed to be efficient in achieving long term stabilization of the underlying disease in this patient cohort. Further clinical studies are warranted to explore the feasibility and efficacy of STI-571 for in-vivo purging of PBSC followed by autologous PBSC-T plus STI-571.

594

STI571 induced apoptosis correlates with inhibition of STAT5 activation in BCR-ABL positive cells

T. Kindler, T. Stevens, F. Breitenbücher, B. Carius, R. Kühn, H. Gschaidmeier, C. Huber, T. Fischer (Mainz, Nürnberg, D)

STI571, a newly developed inhibitor of BCR-ABL tyrosine kinase exhibits its effects on BCR-ABL expressing cell lines and primary CML cells due to strong downregulation of constitutive STAT5-tyrosine phosphorylation. This results in inhibition of proliferation and colony formation. In recent reports, it has been demonstrated that STAT5 inhibition correlates with reduced expression of Bcl-XL, a potent inhibitor of apoptosis. Here, we explored the molecular mechanisms of growth inhibition by STI571 treatment in K562 cells, BCR-ABL transfected 32D cells and in primary CML cells. Our results show that STI571 specifically inhibits proliferation of BCR-ABL positive cells. As apoptosis seems to be a potential mechanism, we investigated BCR-ABL positve cells treated with STI571 using TUNEL- and annexin assays as well as cell cycle analysis. STI571 treatment resulted in induction of apoptosis in K562 cells, BCR-ABL positive 32D cells and in primary CML blasts. As revealed by Western blot analysis, induction of apoptosis involves specific and early cleavage of caspase-3 targets as PARP and RFCp140. Further, STI571 induced apoptosis is partial inhibited by selective inhibition of caspase 3. In addition, we analysed the effects of the chemotherapeutics Ara-C and hydroxyurea, which are both effective in CML therapy, on inhibition of STAT5 activation. K562 and primary CML blast were treated with Ara-C for 12 to 48 hours. Western blot analysis demonstrated a specific decrease in STAT5-tyrosine phosphorylation in a time dependent manner. Currently we explore the molecular mechanisms involved in Ara-C mediated inhibition of phosphorylation. In conclusion, our results show that STI751 treatment specifically induces apoptosis in BCR-ABL expressing cells. Further, our data suggest that Ara-C, a widely used drug in CML therapy, exhibits its effects in part by downregulating STAT5-tyrosine phosphorylation. This inhibition of STAT5 activation may result in the reported synergistic effects of STI571 and Ara-C.

595

A tyrosine to phenylalanine-mutant of the major-autophosphorylation site of BCR-ABL shows a higher sensitivity against STI571 in vivo

C. Miething, C. Mugler, R. Grundler, J. Hoepfl, Ch. Peschel, J. Duyster (München, D)

The causative genetic abnormality of chronic myelocytic leukemia (CML) is the constitutive activation of the Abl-tyrosine kinase through a translocation fusing parts of bcr to sequences from the Abl-gene. The recent development of a specific Abl-kinase inhibitor has led to a new promising therapeutical approach in the treatment of CML. The crystal structure of STI571 in complex with Abl has been resolved recently, showing an important role for the tyrosine 393 of the Abl gene. When phosphorylated, this tyrosine-residue strongly reduces the affinity of STI571 binding to the Abl activation loop in vitro. In order to examine the effects of this phosphorylation induced conformational change on STI571 induced Abl kinase inhibition in vivo, we have transfected the factor-dependent murine pre-B-cell line Ba/F3 with wild-type BCR-ABL and a tyrosine to phenylalanine mutant of the tyrosine 393 (AblY393F), mimicking a constantly unphosphorylated amino-acid residue. Both cell lines grew factor-independent with similar growth curves. After adding different concentrations of STI571, the AblY393F transformed Ba/F3 cells showed higher inhibition already at concentrations of 0.1µM STI571 than wt BCR-ABL transformed Ba/F3 cells (40% vs. 89% cells compared to control after 72h). Thus, phosphorylation of Abl at tyrosine 393 indeed modulates the response of Abl transformed cells to STI571. Interestingly, the addition of 1µM sodium-orthovanadate, an unspecific phosphatase-inhibitor, was able to restore cell growth of the BCR-ABL transformed Ba/F3 cell line inhibited by STI571. This may also implicate that increased phosphorylation of BCR-ABL is responsible for reduced STI571 activity. Taken together, the results suggest that the amount and activity of kinases and phosphatases influencing the phosphorylation of BCR-ABL at tyrosine 393 may play an important role in the response of a BCR-ABL transformed hematopoetic cell to STI571 and may represent a further mechanism of resistance in patients with Abl-expressing leukemias.

596

Prognostic impact of immunophenotype and additional cytogenetic aberrations in patients with myeloid blast crisis treated with Glivec (STI 571)

T. Lahaye, S. Kreil, C. Schoch, S. Schnittger, P. Paschka, M.C. Müller, P. Worst, U. Berger, J. Hastka, H. Löffler, A. Fabarius, T. Haferlach, H. Gschaidmeier, R. Hehlmann, A. Hochhaus (Mannheim, München, Nürnberg, D)

Response to chemotherapy is limited in blast crisis (BC), the terminal phase of chronic myeloid leukemia (CML). The causative event in the initiation of CML is the Philadelphia (Ph) translocation t(9;22) resulting in the chimeric BCR-ABL oncogene. Evolution of the disease is frequently characterized by additional cytogenetic and molecular aberrations. STI571 (Glivec) is an inhibitor of the BCR-ABL tyrosine kinase with efficacy in BC. We investigated the role of the individual immunophenotype and of additional cytogenetic aberrations in 59 patients (27 m., 32 f.) treated with STI571 for myeloid BC of CML in multicenter phase II (0102) and expanded access (0115) protocols. Immunophenotype was determined by FACS analysis and/or immunostaining (APAAP method) of peripheral blood (PB) and/or bone marrow (BM) blasts in 48 cases. 30 patients (51%) showed only myeloid markers, six (10%) expressed additional lymphatic, seven (12%) megakaryoblastic, and eight (14%) monoblastic markers. Hematologic response was observed in 9/30 patients with myeloid markers only, 2/6 patients with coexpression of lymphoid markers, 1/7 megakaryoblastic crises, and 3/8 monoblastic crises. Median survival was 270 days in all patients, 346 days in patients with myeloid markers only, 270 days in patients with lymphatic coexpression, 120 days in megakaryoblastic and not yet reached in monoblastic crises. There is a significant difference in survival between patients with myeloid markers only and patients with additional megakaryoblastic markers (p=0.026). Cytogenetic analysis of BM and/or PB metaphases was performed in 53 cases. 22 Ph+ patients lacking additional aberrations had a median survival of 270 days, in 31 cases additional aberrations have been observed (median survival 176 days, n.s.), being aneuploidy, n=19, 132 days; trisomy 8, n=8, 276 days; second Ph chromosome, n=4, 99 days; aberrations of chromosome 17, n=11, 98 days. A significantly impaired survival has been observed in patients with a second Ph chromosome (p=0.022). We conclude that immunophenotype of myeloid blasts and specific additional cytogenetic aberrations may be associated with response to STI571. These parameters should be considered in multivariate analyses of prognostic factors and in the preparation of future combination studies with STI571 and chemotherapy.

597

Molecular monitoring of response to the therapy with the tyrosine kinase inhibitor STI571 (Glivec) in CML patients

P. Paschka, S. Kreil, T. Lahaye, C. Schoch, A. Weißer, P. La Rosée, K. Merx, M.C. Müller, C. Kuhn, U. Berger, T. Haferlach, H. Gschaidmeier, R. Hehlmann, A. Hochhaus (Mannheim, München, Nürnberg, D)

Cells transformed by BCR-ABL are specifically inhibited by the tyrosine kinase inhibitor STI571 (Glivec). We investigated 75 peripheral blood (PB) and bone marrow (BM) sample pairs from 37 patients with CML who achieved a partial (1-34% Ph+ metaphases, n=8), or complete (n=29) cytogenetic remission (CCR) after 3-9 months of STI571 therapy in three multicenter phase II trials and in an expanded access program. At starting of STI571 therapy 30 patients were in chronic phase, 3 in accelerated phase and in myeloid blast crisis respectively, and 1 in lymphoid blast crisis. Response to therapy was determined by cytogenetic analysis of BM metaphases, hypermetaphase and interphase fluorescence-in-situ-hybridization (HM-FISH, IP-FISH) of BM cells, and by real time quantitative RT-PCR (Q-PCR) for BCR-ABL transcripts. For Q-PCR 20 ml PB and 4 ml BM have been processed. Total ABL and glucose-6-phosphate-dehydrogenase (G6PD) have been quantified as internal controls. In complete cytogenetic responders the median proportion of BCR-ABL+ metaphases by HM-FISH (median 126 metaphases evaluated, range 14-504) was 0% (range 0-17.1%). IP-FISH (median 300 nuclei evaluated, range 73-414) showed a median proportion of 4.7% BCR-ABL+ BM cells (range 0-10.1%, background level 5%). All patients had evidence of residual disease by RT-PCR; the median ratio BCR-ABL/ABL was 1.7% (range 0.012-67%) in PB and 1.8% (range 0.012-51%) in BM; the median ratio of BCR-ABL/G6PD was 0.052% (range 0.0006-2.5%) in PB and 0.072% (range 0.00034-3.4%) in BM. The median level of BCR-ABL transcripts was 1.1 times higher in BM compared to PB, which was not significantly different. Ratios BCR-ABL/ABL and BCR-ABL/G6PD (r=0.86, p<0.0001) and Q-PCR results from PB and BM (r=0.70, p<0.0001) showed a good correlation. A significant correlation was seen between HM-FISH results, IP-FISH (r=0.62, p<0.0001) and ratios BCR-ABL/ABL in PB (r=0.63, p<0.0001) or in BM (r=0.56, p<0.0001). Correlation was worse between IP-FISH in BM and ratios of BCR-ABL/ABL in PB (r=0.48, p<0.0001) and BM (r=0.43, p=0.0003). After a median treatment period of 0.7 years in complete cytogenetic responders the best molecular response to STI571, expressed as the ratio BCR-ABL/ABL, ranged between 0.012-8.2% (median 0.33%); at present this level of minimal residual disease is higher compared with complete cytogenetic responders after IFN alpha therapy (median time of therapy 4.7 years) or allogeneic stem cell transplantation. In patients with complete cytogenetic response to STI571, HM-FISH, and Q-PCR from PB or BM may reveal BCR-ABL+ cells. A limitation of IP-FISH is the variant proportion of false positive cells. Longer follow up is required to determine, if STI571 is able to induce molecular remissions in patients with CML.

598

Therapy of resistant chronic myelogenous leukemia (CML) with STI571 (Glivec): Experience from the expanded access program

A. Hochhaus, S. Kreil, T. Lahaye, U. Berger, M.C. Müller, A. Weißer, K. Merx, P. Paschka, P. Worst, J. Hastka, C. Schoch, A. Louis, B. Heinze, H. Gschaidmeier, R. Hehlmann (Mannheim, München, Ulm Nürnberg, D)

Chronic myelogenous leukemia (CML) is caused by expression of the BCR-ABL tyrosine kinase oncogene, the product of the t(9;22) Philadelphia translocation. STI571 (Imatinib mesylate, Glivec) is an orally administered selective inhibitor of the BCR-ABL tyrosine kinase. In a phase I study, STI571 induced hematologic responses at daily doses of 300-1000 mg in 54/54 patients with chronic phase and 21/38 patients with myeloid blast crisis. Between July 2000 and May 2001, a total of 190 patients (112 m., 78 f.; median age 60.3, range 16.3-81.6 years) with Ph and/or BCR-ABL positive CML who had failed prior therapy with interferon alpha (IFN)-based regimens or had progressed to advanced disease were recruited into three expanded access protocols: 0113, IFN resistant chronic phase (CP) n=82; 0114, accelerated phase (AP) n=67; 0115 myeloid blast crisis (BC), n=40 and lymphoid BC, n=1. BC was pretreated with chemotherapy in 13 cases. Therapy was started with 400 mg STI571/day in a single oral dose in chronic phase patients and 600 mg STI571 in advanced disease. Patients were treated in close cooperation with the referring center. In case of hematologic resistance, dose escalation was permitted to 600 mg or 2x400 mg, respectively. Patients will be evaluated for induction of hematologic response, major cytogenetic response, molecular response, time to progression, survival, and safety. Median duration of therapy was 208 days (range 8-334) in CP, 158 days in AP (4-335), and 74 days in BC (2-278 days). All CP patients are still in CP and alive, eight (11.9%) AP patients progressed to BC and died, 13 (31.7%) BC patients died from progressive disease. Non-hematologic toxicity was manageable and consisted of nausea, fluid retention, muscle cramps, diarrhea, rush and dermatitis. Grade 3 or 4 non-hematologic toxicity was infrequent. In advanced disease, thrombocytopenia was common. Patients received platelet transfusions in case of bleeding signs and/or platelet counts <10/nl. Neutropenia was frequent but rarely accompanied by fever and/or infection. Only two patients (1.1%) discontinued treatment due to drug-related adverse events, and no treatment-related deaths occurred. STI571 induced hematologic and cytogenetic responses in CML patients with CP who had failed prior IFN-based therapy and in patients with advanced disease, with manageable toxicity. Our experience suggests that STI571 does significantly improve treatment options of patients with CML in advanced disease and of chronic phase patients who failed prior therapy with IFN. The results are a basis for trials combining STI571 with regular and pegylated IFN, chemotherapy, and/or stem cell transplantation.

599

STI-571 (Glivec) treatment in a case of AML-M0 with secondary acquisition of the Philadelphia chromosome and BCR-ABL fusion product

T. Flohr, T. Kindler, G. Heß, H. Gamm, H. Gschaidmeier, C. Huber, T. Fischer (Mainz, Nürnberg, D)

Late appearance of the Philadelphia chromosome is a rare event in acute myelogenous leukemia. We here present the clinical course of a patient with M0 subtype of AML, and a 46, XY del(20q) karyotype at initial diagnosis. 29 months after initial diagnosis, the patient developed a t(9,22) including appearance of a Philadelphia chromosome and BCR-ABL fusion transcript. The patient had no acceptable blood stem cell donor. Therefore, he received STI-571 treatment, which led to reduction of leukemic blasts in peripheral blood and bone marrow. Primary diagnosis of the AML–M0 was made in 07/1998. Cytogenetic analysis revealed a 46, XY del (20q) karyotype. Subsequent induction treatment (AraC/Ida, Hovon Sakk trial: AML 29) resulted in first cytologic and cytogenetic remission. After a second induction cycle (HDAraC/Amsacrine) the patient received autologous stem cell transplantation after conditioning with busulfan and cyclophosphamide in 12/98. Due to incomplete recovery of peripheral blood counts post transplant, repetitive bone marrow biopsies were performed without sign of cytologic or cytogenetic relapse. In 06/99 cytogenetic analysis revealed reoccurrence of the leukemic cell clone without change of karyotype. Due to a oligoblastic nature of the relapsed leukemia and with no donor for allogeneic transplantation available, the patient received supportive therapy. First morphological evidence of blasts in the marrow was detected in 02/00 and cytogenetics at this timepoint still showed the initial 46, XY del(20q) karyotype. In the beginning of 2001 transfusion frequencies were rising and increasing number of leukemic blasts were present in peripheral blood. Subsequent bone marrow analysis revealed additional acquisition of the Philadelphia chromosome and a changed immunophenotype of the leukemic clon. Clonogenic evolution was confirmed when backup samples of the autografts obtained 3 months after initial diagnosis were tested negative for the BCR-ABL fusion product by RT-PCR technique. The patient was recruited into the multicenter CSTI5710115 trial and received 600mg STI-571 daily. Upon 8 weeks, this treatment resulted in clearance of peripheral blasts and marked decrease of leukemic cells in the bone marrow. Results of cytogenetics, FISH and PCR are pending. Additional in-vitro analysis showed constitutive STAT-5 tyrosine phosphorylation in the leukemic cells which was downregulated upon in- vitro incubation with STI-571. In summary, we here present the rare event of late appearance of the Philadelphia chromosome in a case of AML. Furthermore, the clinical course and in -vitro analysis revealed susceptibility of the evolved cell clone to STI-571.

600

Efficacy of Glivec in patients with chronic myeloid leukemia (CML) relapsing in advanced stage disease after allogenic stem cell transplantation (alloSCT)

H. Pfeifer, B. Wassmann, U. Scheuring, C. Thiede, M. Bornhäuser, J. Atta, P. Brück, H. Gschaidmeier, D. Hoelzer, O.G. Ottmann (Frankfurt/Main, Dresden, Nürnberg, D)

Background: The prognosis of patients suffering from CML recurring after alloSCT is poor. Therefore, the use of Glivec (formerly STI571), a selective ABL tyrosine kinase inhibitor shown in multicenter phase II trials to have clinical efficacy in BCR-ABL/Ph+ acute leukemias, is a promising agent in CML patients relapsing after prior alloSCT. Methods: We studied 13 patients who had previously received an alloSCT and relapsed in blast crisis (n=9) including 1 patient with a chloroma, in accelerated phase (n=3) or in chronic phase (n=1). The patients were treated in consecutive multicenter phase II clinical trials of Glivec. The median time interval between allo SCT and relapse was 6.5 months (2-50 months). One patient had also been treated with Glivec prior to SCT. Marrow cytology, cytogenetic and FISH analysis and determination of donor chimerism were used to assess efficacy. Results: A complete hematologic response was obtained in 7 of 13 patients (4/9 in BC, 2/3 in AP and 1 in CP) within a median time of 1 month (0.5 –3 months) of Glivec. 6 patients were refractory to treatment (5 in BC, 1 AP), although one of these patients had clearance of peripheral blood blasts. 2 of 7 CR patients (1 myeloid and 1 bilineage BC) relapsed after 6 weeks (initially an isolated meningeal relapse) and 2.5 months. After a median of 11 months (2-20 months) CR ist still ongoing in 5 patients, all of whom also showed a complete or major cytogenetic response, which persists after a median of 10.5 months. In responding patients, donor chimerism increased from a median of 50.5% at study entry to above 98% during the first 12 weeks of therapy. In general, adverse effects of Glivec were mild to moderate, the most frequent being gastrointestinal discomfort and peripheral and facial edema. 2 patients refractory to Glivec who were thrombocytopenic developed a subdural hygroma /hematoma, which required surgical decompression in one patient. Conclusions: Our results indicate that Glivec may induce a complete hematologic and even cytogenetic response in a significant proportion of patients with Ph+CML who relapse with advanced disease after allogenic stem cell transplantation, with reestablishment of donor chimerism which may be prolonged. Our findings suggest that Glivec is a promising agent in the post transplant setting in CML, although its efficacy and safety still need to be established in prospective clinical trials, some of which are ongoing.

601

Isolated central nervous system relapse in a patient with chronic myeloid leukemia with lymphoid blast crisis under STI571 due to low penetration of STI571 into the cerebrospinal fluid

A. Petzer, E. Gunsilius, M. Hayes, G. Stockhammer, H. Duba, F. Schneller, K. Grünewald, W. Poewe, G. Gastl (Innsbruck, A)

We report on a 53-year old man with Ph+ chronic myeloid leukemia who was treated with STI571, a selective inhibitor of the enzymatic activity of BCR-ABL, after he had developed lymphoid blast crisis. The patient responded excellently to a daily dose of 600 mg STI571 obtaining a complete cytogenetic remission after 3 months of therapy (25/25 BM-metaphases Ph-, <5% of BM-interphases BCR-ABL+). Although remission in the bone marrow was sustained the patient developed an isolated central nervous system (CNS) relapse, presenting with neoplastic meningitis after 5 months on STI571 therapy. Subsequent analyses of STI concentrations in the cerebrospinal fluid (CSF) revealed ~2log lower CSF levels of STI571 compared to corresponding plasma levels. Measured CSF levels were below the concentration known to be required for 50% inhibition of the BCR-ABL tyrosine kinase. Predominantly normal albumin ratios in the CSF (below 8) indicate a low STI571 penetration at a normal blood-brain barrier. At the time of CNS relapse the percentage of myeloblasts in the bone marrow was 0.5 and all mature progenitor cells (CFC) and primitive progenitor cells (LTC-IC) in the BM were BCR-ABL-. CNS relapse was successfully treated with intrathecally administered metothrexate. These are the first data demonstrating a low penetration of orally administered STI571 into the CSF in humans. STI571 treatment of lymphoid blast crisis in CML should therefore be combined with additional antileukemic CNS prophylaxis using either inthrathecal chemotherapy or CNS-irradiation.

602

Evaluation of STI571 for treatment of Hodgkin's disease

D. Re, C. Wickenhauser, V. Diehl, J. Wolf (Köln, D)

STI571 is a selective tyrosine kinase inhibitor that interacts specifically with BCR-ABL, c-abl, PDGFR and c-kit. In chronic myeloid leukemia, proliferation of malignant cells expressing the oncogenic BCR-ABL encoded tyrosine kinase can effectively be reduced by administration of STI571. The aim of the presented study was to evaluate the therapeutic potential of STI571 in Hodgkin´s disease (HD). Based on data showing expression of c-kit on Hodgkin and Reed Sternberg cells in approximately 50% of cases of HD (Pinto et al., Blood 1994), we analysed expression of c-kit and its ligand SCF in six HD derived cell lines and 20 cases of classical HD. A strong c-kit expression and soluble SCF were detected in the cell line L1236 by FACS analysis and ELISA, respectively, but not in the five other HD derived cell lines. C-kit expression was absent in 20 primary cases of classical

HD as determined by immunohistochemistry. A potential functional role of the c-kit/SCF interaction in vitro was tested by performing MTT proliferation assays with L1236 cells and controls (M07e for positive control and a B-lymphoblastoid cell line and a c-kit negative HD derived cell line for negative control). Cells were incubated with high amounts of SCF, with neutralizing anti-SCF antibodies and with STI571, respectively. Proliferation of the SCF dependent c-kit positive cell line M07e but not of the L1236 cell line was dependent on these factors. As this observation could be explained by mutations leading to a consitutive activation of c-kit, we sequenced several c-kit hot-spots in L1236 cells. Sequence analysis of exons 11 and exon 17 did ecxlude genomic c-kit mutations in L1236 cells. We conclude that activation of c-kit is not involved in the control of proliferation in L1236 cells. In summary, our results suggest that patients with HD may not benefit from administration of STI571.

603

Complete cytogenetic remission to STI571 in a patient with myeloid blast crisis of CML characterized by simultaneous occurence of philadelphia translocation and inv(16)

T. Lahaye, S. Kreil, C. Schoch, S. Schnittger, U. Berger, J. Hastka, T. Haferlach, M. Bornhäuser, H. Gschaidmeier, R. Hehlmann, A. Hochhaus (Mannheim, München, Dresden, Nürnberg, D)

Chronic myeloid leukemia (CML) in myeloid blast crisis was diagnosed in a 63 year old male patient in January 2000. Bone marrow karyotype showed the typical Philadelphia translocation in 25/25 metaphases with a subclone of 7/25 metaphases with additional inversion 16. Molecular analysis revealed typical b3a2 BCR-ABL and rare type E MYH11/CBFbeta transcripts. By morphology and immunophenotyping, blasts were of M1 type; the M4Eo phenotype was not observed. Therapy with STI571 (Glivec, 600 mg/d) was commenced according to the 0102 multicenter phase II protocol leading to a complete hematologic and cytogenetic response after three months. Remission was accompanied by a significant and parallel decrease of BCR-ABL and CBFbeta-MYH11 transcript levels determined by quantitative RT-PCR. After six months of treatment, hematologic relapse occured with reappearance of both the Ph+ clone and the subclone with additional inv(16). STI571 therapy was stopped and high dose chemotherapy (thioguanine, cytarabine, daunorubicin; TAD protocol) was administered, followed by low dose cytarabine (40mg s.c./day for seven days) after hematologic recovery. Transient hematologic but no cytogenetic response was observed. STI571 therapy was restarted, but, after five weeks blasts reappeared. Non-myeloablative HLA identical related stem cell transplantation was considered. The patient died from septicemia in frank myeloid blast crisis. We conclude that (i) in myeloid blast crisis therapy with STI571 may result in hematologic and cytogenetic remission even in cases with clonal evolution. (ii) Resistance of blasts to STI571 may be maintained even in hematologic remission after chemotherapy. (iii) Due to high risk of relapse allogeneic stem cell transplantation should be considered early after remission. (iv) Parallelity of BCR-ABL and CBFbeta/MYH11 transcript levels demonstrate the usefullness of real-time PCR on the follow up of patients during therapy.

Postersitzung:
Transplantationsmodelle und -organisation

604

Do we have to protect stem cell harvests from X-ray-hand-luggage-control systems at international airports?

H. G. Speth, E. Gunsilius, E. Hoflehner, D. Nachbaur, G. Gastl, A. L. Petzer (Innsbruck, A)

Matched unrelated donor transplantation has become an increasing transplantation procedure for patients with hematologic malignancies and marrow or peripheral blood stem cell (PBSC) harvests are commonly transported from the harvesting center to the transplantation center by aircrafts. So far the stem cell (SC) crafts have to be protected from X-ray-hand-luggage-control-systems according to IBMTR/EBMT-guidelines. To address the question whether SC harvests have indeed to be prevented from this security procedure, we performed a series of experiments, where we exposed aliquots form mobilized PBSC harvests 0, 1, 2, 3, 5 and 10 times, respectively, to a widely used X-ray-hand-luggage-control-system (HEIMANN HI-SCAN 6040 A, Wiesbaden, Germany, radiation dosage 1,5±0.6 µSV/run) at Innsbruck International Airport. Cell viability as well as growth of mature progenitor cells (Colony Forming Cells, (CFC), including BFU-E, CFU-GM, CFU-Meg) and primitive progenitor cells (longterm culture-initiating cells, (LTC-IC)) were analyzed thereafter and compared to control cells not exposed to irradiation. Cell viability was 96%±1 in control cells (n=11) and did not decrease after 1, 2, 3, 5 or 10 passages (P), respectively, through the X-ray scan (98%±1, 97%±1, 96%±1, 96%±2, 97%±1). The growth of mature progenitor cells (CFC, n=11) was 98%±6 (1P), 103%±7 (2P), 95%±9 (3P), 106%±13 (5P) and 109%±10 (10P) compared to controls (set 100% at 0 P). Primitive LTC-IC (n=11) did not decline either, if passed through the X-ray scan: 102%±12 (1P), 93%±167 (2P), 121%±31 (3P), 153%±30 (5P) and 116%±25 (10P) compared to controls. In addition no significant change in the expansion-potential of MNC and CFC was found after 5 passsages through the X-ray scan compared to non X-ray exposed cells (MNC 100%±12, CFC 94%±19; n=3). In conclusion, these data indicate, that passages up to 10 times through X-ray-hand-luggage-control-systems do not seem to harm SC grafts. It has also to be considered, that the cells are exposed to a radiation dosage of ~60 µSv through cosmic radiation during a 10 hour flight, comparable to ~40 hand luggage controls at the airport. Further investigations with this aspect should be performed to eliminate potential conflicts with security standards at international airports.

605

The development of allogeneic stem cell transplantation in Mecklenburg/Vorpommern (MV) from 1991 to 2000

D. Wolff, B. Krammer-Steiner, V. Lakner, J. Casper, I. Richter, G. Doelken, M. Freund (Rostock, Greifswald, D)

Allogeneic stem cell transplantation (alloSCT) are increasingly used in treatment of hematological disorders. New indications, the use of unrelated donors as well as the extension of the age barrier caused a significant increase of alloSCT during the last decade. In order to document the effect of the latter factors as well as the effect of the foundation of local transplant centers, data from all alloSCT performed on patients living in MV during 1991 and 2000 were evaluated regarding date and location of the alloSCT, diagnosis and age of the patient as well as donor type.
Results: AlloSCT in patients from MV increased 6 fold from n=5 in 1991 to n=30 in 2000. Due to the lack of local transplant centers until 1997 alloSCT of patients from MV were exclusively performed at other locations, whereas in 2000 only 2 of 30 alloSCT were performed in transplant centers outside MV. The implementation of local centers adjusted the frequency of alloSCT of patients in MV to the frequency of Germany(1991 Germany 4,17/10E7 inhabitants, MV 2,67/10E7 inhabitants, 2000 Germany 17,4/10E7 inhabitants, MV 16,7/10E7 inhabitants). The increase of the number of alloSCT was mainly due to an increased use of unrelated stem cell donors, which increased from n=0 1991 to 12 out of 30 in 2000. An other important factor was the increase of the age barrier with a mean age of 32 years in 1991 and 43 years in 2000 (upper range 60 years). In addition indications for alloSCT changed from exclusively acute leukemia and chronic myeloid leukemia in 1991 to NHL in 8 out of 30 alloSCT in 2000. Conclusion: The data reflect the dramatic increase of alloSCT during the last decade caused by the factors listed above. In addition the development of the frequency of alloSCT of patients in MV stresses the need for local transplant centers since only the implementation of local transplant centers equalized the frequency of alloSCT between Germany and MV. This can not be explained by the lack of access to transplant centers but by the missing infrastructure for donor search, long distances to transplant centers as well as the absence of a close interaction between local hematologists and transplant centers necessary for optimal care for patients with hematologic malignancies.

606

Marginal benefit of large-volume peripheral blood stem cell collection: comparison of processing four and five times the patients` blood volume and analysis of CD34+38- and CD34+49d+ subsets during apheresis

P. Bojko, S. Seeber (Essen, D)

Marginal benefit of large-volume peripheral blood stem cell collection: comparison of processing four and five times the patients` blood volume and analysis of CD34+38- and CD34+49d+ subsets during apheresis. Purpose: We investigated whether increasing the patients` processed blood volume (BV) during peripheral blood stem cell collection (PBSCC) from four to five times leads to a greater yield of CD34+ cells. We also studied the kinetics of CD34+38- and CD34+49d+ subsets and compared the amount of transfused cells with engraftment. Methods: All patients (n = 20) received chemotherapy followed by G-CSF for PBSC mobilization. Samples from the patients` peripheral blood and the PBSC harvests were taken after processing 1-, 4- and 5 times the patients` calculated BV. Results: The mean total yields of CD34+, CD34+38- and CD34+49d+ cells were 15.69-, 1.13- and 4.17 x E6/kg body weight, respectively. The mean increase for these subsets between 4 and 5 BV was 10%, 8% and 21%, respectively. Based on the mean number of 2.25 (range 2-3) planned courses of high-dose chemotherapy (HDC) per patient, the mean yield of CD34+ cells per kg body weight and intended course of HDC after 4- and 5 BV was 6.31- and 6.97 x E6. Twenty HDC were evaluable for engraftment. There was some correlation between the number of transfused CD34+ and CD34+38- cells and WBC engraftment (r = -0.66 and 0.69) and CD34+ cells and platelet engraftment (r = -0.56). No toxicity occurred during PBSCC. The mean platelet count decreased by 42% and 50%, respectively, after processing 4 and 5 BV. Conclusion: Processing 4 BV is sufficient to collect >5 x E6 CD34+ cells/kg body weight and intended course of HDC in most patients. Extension to 5 BV increases the total yield of CD34+ cells only marginally.

607

Use of recombinant human DNase for processing of a thawed umbilical cord blood transplant

C. Beck, F. Bernhard, P. Bugert, H. Klüter, H. Eichler (Mannheim, D; Montpellier, F)

Purpose: To prevent loss of hematopoietic progenitors during thawing procedure of frozen umbilical cord blood (UCB) transplants, we introduced a modified washing protocol using a recombinant human deoxyribonuclease I (rhDNase) containing solution. We report an UCB transplantation to a child with relapsed ALL where a rhDNase containing solution was successfully used to process the thawed unit prior to transplantation. Case report: A 6-year old male received an unrelated 2/6 HLA antigen mismatched UCB transplant from the Mannheim Cord Blood Bank for the treatment of a high risk Burkitt type ALL. The UCB was provided as a frozen buffy coat sample with a final volume of 36 mL. Thawed cells were washed with a solution containing 4 percent rhDNase. No clotting of the resuspended unit could be observed, and the patient was transplanted with 2.9 x 10 E7 nucleated cells/kg body weight without any acute or chronic side effects due to rhDNase. On day +35, PCR analysis of bone marrow aspirate showed complete chimerism, and the child engrafted with an ANC >0.5 x 10 E9/L on day +47. Platelet transfusion independence was achieved on day +120.
Conclusion: The supplementation of rhDNase to a thawed cell suspension of UCB is an effective and safe procedure to reduce cell loss and to enhance the transplanted cell dose by preventing cell clotting phenomena due to released DNA.

609

Isolation and characterization of hematopoietic stem cells via Hoechst sorting from murine bone marrow and human cell sources

Y. Guo, M. Follo, S. Kaiser, K. Geiger, U. Kapp, M. Lübbert, M. Engelhardt (Freiburg, D)

Instead of using monoclonal antibodies, Goodell et al (J Exp Med 183:1797, 1996; Nat Med 3: 1337, 1997) selected cells based on the rapid efflux of the fluorescent DNA-binding dye Hoechst 33342 and identified a special cell population termed side-population (SP) cells. Murine BM SP cells had long-term repopulating activity and proved negative for the CD34 antigen. Such SP cells were detected in rats, pigs, monkeys and human cell sources. Within SP cells from human umbilical cord blood, CD7+ CD34- lymphoid progenitor cells were recently described. SP cells from other human hematopoietic cell sources, such as BM from normal donors and patients with hematological malignancies, or mobilized PB from donors and patients, however, have not been characterized extensively. We therefore sought to characterize SP cells from these human cell sources, analyze their CD34 expression, CFU-, LTC-IC-, and repopulation potential in NOD/SCID mice. We detected murine SP with a median frequency of 0.03%. When verapamil was present during staining, the frequency of SP cells significantly decreased, confirming that the SP profile is due to the efflux of Hoechst through the multidrug resistance (mdr)-like protein. There was an increase of CFUs yielded from murine SP cells as compared to unfractionated cells with a median enrichment of 52- fold. The frequencies of human SP cells were highest in AP (donor AP: median 0.04%; patient AP: median 0.03%), followed by CB (median 0.02%) and lowest in BM. Most human SP cells were CD34- and the majority of CD34- cells proved CD34-CD38-. In contrast to murine SP cells, the CFU potential of human SP cells was not increased compared to unfractionated cells. Most likely the immaturity of SP cells led to their low CFU potential, since a significant increase in CFU numbers was observed after short-term expansion in suspension culture and when replated in methylcellulose. Of interest was that a considerable increase in LTC-IC of a median of 49-fold was perceived with SP as compared to unsorted cells. The highest LTC-IC increase was detected from donor AP (median 184-fold), followed by CB (53-fold), and lowest from patient AP or donor BM, in line with the finding that the number of LTC-ICs in unfractionated AP cells is higher than that in BM and CB. Our data demonstrate that primitive hematopoietic cells can be isolated via Hoechst staining. Ongoing experiments, subdividing the human Lin- SP population into CD34+ and CD34- cells, will allow us to compare their LTC-IC potential and in vivo repopulating capacity in NOD/SCID mice, thereby aiding in the discussion of whether HSC are CD34+ or CD34-.

610

Concessions for the clearance of 'healthy' donors of G-CSF mobilized hematopoietic progenitor cells

P. Reinhardt, S. Nowak-Harnau, A. Eil, S. Goldmann, B. Kubanek, M. Wiesneth (Ulm, D)

Purpose: Reflexions about clearance criteria of related and unrelated stem cell donors based on pre-harvest check up's over the last year. Introduction: As the indication for allogenic stem cell transplantation grows, so does the need for healthy donors of hematopoetic progenitor cells (HPC). Sometimes significant medical concessions must be made to make a stem cell collection and transplantation possible, while disregarding the usual guidelines for blood donation. Methods: This is a report on our experience in examining 60 „healthy", allogenic stem cell donors for pre-harvest clearance. All donors were submitted to a thorough clinical examination, ECG, chest radiograph, ultrasound of the abdomen, blood smear, blood chemistry and microbiological analysis. Results: Sixty healthy allogenic stem cell donors (29 male, 31 female) were checked for pre-harvest clearance. The median age was 37, and ranged from 18 to 70 years. 45 out of 60 were family related donors the other 15 were unrelated, registered stem cell donor volunteers. Donor clearance could not be given to 4 donors (3 related, 1 unrelated) for clinical reasons, all diagnosed for the first time at check up. One related donor had a so far asymptomatic monoclonal gammopathy, another was diagnosed with a round pulmonary infiltrate, another with coronary heart disease. The unrelated donor was discovered to have a so far asymptomatic severe mitralis valve insufficiency. Of the remaining 56 donors 31 had relative contraindications such as a newly diagnosed thalassaemia minor, severe hypertension, cystic nephropathy or varicosis to either peripheral blood HPC-donation or bone marrow donation. Among the most severe was a chronic hepatitis B infection in a related HPC-donor and a case of untreated hypertension with a blood pressure of up to 240/130 mmHg in an unrelated HPC-donor. Acute anti hypertensive therapy allowed for an otherwise normal and incident free HPC-collection. Extensive clinical data and reasoning as well as virologic parameters of HBV under G-CSF-stimulation will be presented. Conclusions: Extensive examination of healthy stem cell donors is necessary to recognize potential hazards for both the donor and the recipient. Pre-harvest donor clearance of family related stem cell donors does often demand concessions that may increase the involved risk for either the donor or the transplant recipient. Both donor and transplant center physicians must be involved in the decision for such concessions. Unrelated donors will rarely be given clearance if an increased risk is to be expected. Joint efforts for a standardized risk evaluation should be made, to facilitate future decisions and reduce unnecessary risks for non-healthy donors.

611

Ex vivo expansion of normal progenitor cells from acute myeloid leukemia (AML) cell-contaminated CD34+ peripheral blood progenitor cells after mafosfamide purging

S. Scheding, R. Stilz, F. Grünebach, P. Bader, W. Vogel, L. Kanz, W. Brugger (Berlin, Tübingen, D)

The rationale for purging of autologous AML grafts is to eradicate contaminating leukemic cells that might contribute to relapse. However, in-vitro purging generally delays post-transplant hematopoietic recovery, thus increasing treatment-related complication rates. Theoretically, this prolonged aplasia might be shortened by the additional transplantation of ex-vivo generated progenitor cells. Therefore, we investigated whether non-leukemic progenitors could be ex-vivo expanded from AML cell-contaminated CD34+ peripheral blood progenitor cell (PBPC) preparations. Non-leukemic CD34+-selected PBPC and AML cells (Kasumi-1, KG-1, primary

AML blasts) were cultured in cytokine-supplemented liquid culture for up to 19 days. Cells were used either unmanipulated or following in-vitro purging with mafosfamide (30, 50, 75 µg/ml). Ex-vivo generated cells were assessed by flow cytometry, progenitor cell assays, and polymerase chain reaction. Without prior purging, ex-vivo culture markedly amplified AML cells as well as non-leukemic CD34+ PBPC (day 12: Kasumi-1, 18.5 ± 0.6-fold; KG-1, 52.2 ± 2.6-fold; CD34+, 74.1 ± 5.6-fold). Co-culture with leukemic cells did not affect CD34+ cell growth and vice-versa. Following in-vitro purging, CD34+ PBPC were expanded even at the highest mafosfamide dose (day 19: 25 ± 15-fold), whereas leukemic cells were markedly depleted (approx. 1.5 log). Furthermore, normal CFU could be effectively recovered (day 19: 10 ± 3.1% of pre-purging input CFU), whereas CFU-L were depleted to undetectable levels in six of seven experiments. Finally, leukemic cells were undetectable following ex-vivo co-culture of purged cells (CD34+ PBPC plus 10% Kasumi-1 cells or primary blasts), but were clearly detectable without purging.
Taken together, these data demonstrated that ex-vivo expansion of normal progenitors from mafosfamide-purged AML cell-contaminated grafts might be feasible.

612

Engraftment of retrovirally transduced human cells from umbilical cord blood (UCB) in nonobese diabetic mice with severe combined immunodeficiency (NOD/SCID)

P. Ebeling, P. Bach, T. Trarbach, H. Quang, A. Schneider, S. Niesert, T. Moritz, S. Seeber, M. Flasshove (Essen, D)

In order to investigate optimal conditions for retroviral transduction of human hematopoietic stem cells, we used the NOD/SCID transplantation model known to be a valid assay system for the stem cell compartment. Mononuclear cells from UCB were enriched for CD34+ cells by immunomagnetic separation and exposed to viral particles containing an MSCV-based retroviral vector with the enhanced green fluorescence protein (EGFP). Retroviral supernatant was produced by the cell line PG13. Three different protocols were used for transplantation of transduced cells into sublethally irradiated NOD/SCID mice: (i) CD34+ enriched cells were prestimulated for 48h with Flt3-ligand, thrombopoetin, G-CSF, and stem cell factor and exposed twice to retroviral supernatant for 6h within the following 48h in the presence of fibronectin and cytokines; finally, 1.5 – 4 x 10^6 cells per mouse were injected i.v. (n=5); (ii) Using identical cytokines, CD34 enriched cells were again prestimulated for 48h, but exposed to retroviral supernatant three times for 6h within the following 96h; 1 – 2.5 x 10^7 cells were injected per mouse (n=3); (iii) Prestimulation was shortened to 24h, retroviral supernatant was added three times for 6h within the following 72h; 1 x 107 cells were injected per mouse (n=2). Four weeks after transplantation, bone marrow cells were analyzed by flow cytometry for human CD45+ cells – representing engraftment of human hematopoietic cells – together with CD34 and EGPF expression. Transduction protocol (i) failed to show significant engraftment of human cells and/or expression of EGFP probably due to insufficient numbers of transplanted cells. Prestimulation and transduction according to protocol (ii) resulted in a total of 36.8% EGFP+ cells, 9.7% stained positive for CD34 and 2.2% were CD34+/EGFP+ before transplantation. Engraftment was achieved with up to 2.6% (2.0-2.6%) huCD45+ cells , and up to 10.4% (0-10.4%) of these human cells expressed EGFP indicating successful retroviral transduction of NOD/SCID repopulating cells. Using transduction protocol (iii) we observed a total of 9.7% EGFP+ cells and obtained 72.3% of CD34+ cells with 5.9% CD34+/EGFP+ cells before transplantation. Four weeks after transplantation engraftment rates of 21.8%/36% huCD45+ cells were better than for protocol (ii). 3.4%/2.3% of these cells stained positive for EGFP. According to our data this model requires transplantation of at least 1x107 cells. A shorter transduction protocol seems to result in better engraftment of human retrovirally transduced cells. This observation needs to be confirmed by further investigations.

613

Efficient engraftment of mobilized CD34+ peripheral blood cells in immundeficient (NOD/SCID) mice

T. Trarbach, H. Quang, A. Schneider, S. Grehl, M. Flasshove, S. Seeber, T. Moritz (Essen, D)

Transplantation of human cells into nonobese diabetic mice with severe combined immunodeficiency disease (NOD/SCID) has proven to be a valid assay system to investigate stem cell biology and gene transfer efficiency into human progenitor and stem cells. In order to examine the engraftment of mobilized peripheral blood cells in this model, we injected immunomagneatically separated CD34+ cells (purity: 80-95%) i.v. into sublethally (3.75 Gray) irradiated NOD/SCID mice. Engraftment of human cells was determined after 4-6 weeks in the murine bone marrow using FACS analysis to differentiate human CD45+ versus murine CD45+ expressing cells. To establish the model 5x10^6 CD34+ cells were injected and animals were treated with human GM-SCF (10µg) and SCF (6µg) i.p. three times a week. All (4/4) mice engrafted with 52±17% (mean±SD) human cells. To evaluate minimal requirements for this model subsequent experiments were performed with reduced cell numbers and without growth factor application. While 9/11 mice transplanted with 0.7-1.5x10^6 CD34– cells engrafted showing 7.1±5% human cells (range:2-17%), transplantation of 3x10^6 CD34+ cells resulted in an engraftment of 37±40% (range:17-83%) in 3/3 animals. Engraftment observed after simultaneous injection of murine stroma cell lines expressing soluble or membrane-bound human SCF was as follows: While 6/10 animals engrafted when 1.0x10^6 stroma cells where cotransplanted i.v. with 0.7-1.5x10^6 CD34+ cells (19±14%; range:2-39%), after transplantation of 3x10^6 CD34+ cells 1/1 animal engrafted with 56% human cells. Similar data were observed after i.p. injection of 1x10^6 stroma cells. Here 6/8 mice transplanted with 0.7-1.5x10^6 CD34+ cells engrafted (19±21%; range:4-59%), and 8/8 animals engrafted after transplantation of 3x10^6 CD34+ cells (34±30%; range:4-78%). In addition, in control experiments in which non-human SCF expressing murine stroma cells where cotransplanted i.v. or i.p. 5/6 mice engrafted after transplantation of 0.7-1.5x10^6 CD34+ cells (8±8%; range:2-20%) and 2/2 animals after transplantation of 3x10^6 CD34+ cells (7%; range:4-10%). Subpopulations of engrafted cells using myeloid (CD33), B-cell (CD19), T-cell (CD3) and monocytic (CD14) markers showed multilineage differentiation of human cells in all engrafted mice. Even more important a high percentage (10-20%) of primitive CD34+ cells was present in all mice 4-6 weeks after transplantation. In summary these experiments prove effective engraftment of mobilized peripheral blood stem cells with or without growth factor application. This model should be useful for characterization of peripheral blood stem cells and evaluation of gene transfer efficiency in these cells.

614

Survey of blood stem cell transplantation activities in Germany (1998 – 2000): A report from the national registry „DRST“

H. Ottinger, C. Müller, N. Schmitz, B. Kubanek (Essen, Kiel, Ulm, D)

The German Registry for Blood Stem Cell Transplantations („Deutsches Register für Stammzell-Transplantationen , DRST)“ collects clinical data of all hematopoietic blood stem cell transplantations (HSCT) performed in Germany since January 1998. All 106 German transplant units have meanwhile signed a „registry contract“ clarifying the rights and duties of the DRST and the centers, and the national public health insurances agreed to reimburse the expenses of treatment only to those transplant units co-operating efficiently with the DRST. The present paper was facilitated by the DRST and gives an overview of the changing scene of transplant activities in Germany during the last three years. The main trends for allogeneic first tansplants are: (1) Transplant fre-quencies are still increasing (year 1998: n = 1.118; year 2000: n = 1.438), (2) The percentage of unrelated transplants is growing (1998: 37 %, 2000: 47 %). (3) Peripheral blood (PB) is now preferred over bone marrow (BM) as graft source (1998: 57 % BM, , 43 % PB; 2000: BM 27 %, PB: 73 %), (4) About 90 % of the patients are transplanted for hematologic malignancies (leukemias, lymphomas), whereas solid tumors are a rare indication (5) Chronic myelogenous leukemia continuous to be a major indication for allotransplantation despite the availability of STI (1998: n = 243, 2000: n = 237), (6) In 2000, a total of 427 patients received a conditioning regimen with reduced intensity („mini“ or „micro“ transplants). (7) With 17.3 allotransplants/1 Million inhabitants Germany holds a leading position with regard of transplant frequency among the European countries. The main trends for autologous first tansplants are: (1) Transplant frequencies are significantly declining (1998: n = 2.444, 2000: n = 2.105). (2) This decline can be attributed to patients with breast cancer, (1998: n = 490, 2000: n = 169), whereas, in contrary to the trend, transplant frequencies for multiple myeloma are increasing (1998: n = 422, 2000: n = 580). (3) With 27.1 autotransplants/ 1 Million inhabitants transplant frequencies are lower in Germany than in many other European countries (eg. Spain: 38.8 transplants/ 1 Million inhabitants). We conclude that the DRST might be of help in recognising trends in HSCT medicine, which produces costs of more than 250.000.000 Euro per year.

615

A new cell separator for the collection of peripheral blood progenitor cells: comparison of two leukapheresis programs

N. Schwella, S. Scheding, O. Rick, N. Ahrens, S. Serke, A. Salama (Berlin, D)

Purpose: We evaluated two leukapheresis programs (LP) on a new cell separator (COM.TEC, Fresenius HemoCare, Dreieich, Germany) for the collection of peripheral blood progenitor cells (PBPC). Methods: PBPC were either collected with the LP MNC (LP-MNC) or LP PBSCLYM (LP-PB-SCLYM) following chemotherapy and G-CSF in 42 patients with malignant diseases or after G-CSF alone in 11 healthy individuals. Results: We performed 33 collections on LP-MNC in 27 donors and 35 harvests on LP-PB-SCLYM in 26 donors with a median of a single 11-L-leukapheresis per donor (range: 1-3). Preleukapheresis blood cell counts, collected MNC/kg and CD34+ cells/kg were comparable in both groups. Postleukapheresis counts revealed a higher platelet (PLT) loss with 35% (19-66) for LP-MNC (p=0.022) than 29% (4-52) for LP-PBSCLYM (p>0.05). The median collection efficiency for CD34+ cells was significantly better in LP-MNC collections than in LP-PBSCLYM harvests (p<0.000): 72% (35-99) vs 58% (20-91). However, leukapheresis products of LP-PBSCLYM had a significantly higher mononuclear cell (MNC) purity (p<0.000) and a higher CD34+ cell content (p=0.031) than concentrates harvested by LP-MNC: 92% (69-99) vs 72% (35-98) and 1.2% (0.1-5.9) vs 0.6% (0.2-6.1), respectively. Conclusions: The higher CD34 collection efficiency of LP-MNC is hampered by a lower MNC and CD34+ cell content of PBPC concentrates collected. Leukapheresis-related PLT losses of donors are greater during procedures on LP-MNC.

616

Characterization of acute myeloid leukemia (AML) engraftment in NOD/SCID mice

I. Junghahn, G. Bug, M. Becker, I. Fichtner, R. Henschler (Berlin, Frankfurt, D)

The development of new lines of therapy for AML primarily is short of methods to study survival and growth of AML either in vitro or in animal models: Dexter cultures are not suitable to detect small amounts of AML stem cells and to expand AML cells at all, and transplantation into immunodeficient mice leads either to quick death or no permanent engraftment. We therefore aimed to develop a method to isolate human AML cells engrafted at relatively low levels in Non-Obese Diabetic / Severe Combined Immuno-Deficiency (NOD/SCID) mice. After transplantation of human KG-1 AML cells, a dose- and time-dependent engraftment of AML was seen in the bone marrow and spleens of the animals by flow-cytometric or immunohistochemical analysis using human-specific markers. When bone marrow of transplanted mice was seeded on NOD/SCID mouse stromal cells in cultures supplemented by human cytokines, we detected AML in mice receiving an up to 10 fold lower inoculation dose than via conventional analysis. When primary human AML cells were transplanted, 7/10 AMLs engrafted as detected by conventional methodology. Comparison of their phenotypes by analysis of CD45, CD33, CD38, CD15, CD68 revealed that the original phenotype could be recovered from the marrow of transplanted mice in most cases. In AMLs, cultures of bone marrow from transplanted mice on NOD/SCID stroma revealed outgrowth of human leukemia both in the adherent layer and in the supernatant of the cultures. In one case, retransplantation of passaged AML was tried, and led to reproducible engraftment of the AML in secondary and tertiary recipients, demonstrating the presence of AML stem cells in the cultures. Also, direct culture of human AML on NOD/SCID stroma without previous transplantation revealed the supportive role of this environment, however AML in vitro growth was stimulated by previous passage of the AML through mice. Our data indicate that the NOD/SCID hematopoietic microenvironment can support human AML growth and survival both in vitro and in vivo, which can be exploited for the improved detection and characterization of human AML.

617

Update of results of the Dresden allogeneic cord blood bank (DNS-BB), supported by the DKMS

K. Petzold, K. Hölig, R. Ordemann, S. Mauersberger, B. Schaffer (Dresden, Tübingen, D)

The DNS-BB cord blood bank in Dresden was founded 1997 and started collecting, processing and cryoconservation of umbilical cord blood (UCB) in August 1997. With informed consent of the mothers the collection is performed in collaboration with hospitals of Dresden and hospitals in Berlin, Bautzen, Erlabrunn and Wolfsburg. UCB units are collected and processed in a closed top and bottom system (Macopharma, Langen, Germany). Volume reduction is performed when the collected netto volume is >50 ml. The UCB units are cryoconserved when TNC-count is > 5 x 10^8 after volume reduction. All UCB units are evaluated for TNC and MNC count, for CD34+ cells by FACS analysis and bacterial contamination after the collection and during the processing, ABO/Rh(D) and CFU-GM after volume reduction and the erythroblast count (blood smear) from the native cord blood. The HLA-typing class I and II is performed by PCR-SSOP. The maternal blood is tested for infectious disease markers (HbsAg, anti-HBc-Ak, anti-HCV, HCV-PCR, anti-CMV-IgG/-IgM, anti-HTLV I/II, anti-HIV I/II, TPHA, GPT) and also for HLA-type class I. Results of infectious disease markers are only accepted, when the HLA-type class I of the mother is haploidentical with the cord blood. From August 1997 to March 2001 more then 2979 samples were collected, 1087 units could be frozen. Results: mean netto volume: 78.31 ml , after volume reduction: TNC: 7.28 x 10^8, MNC: 3.90 x10^8, CD34+:2.08 x 10^6, CFU-GM: 56.0 x 10^4. Reasons for discarding UCB-units were: low volume (38%), low TNC after volume reduction (20%), bacterial contamination (2.9%), positive infectious disease markers (1.4 %), others (5%). Efforts are continued to improve the quality of the collected UCB to provide sufficient transplants for unrelated recipients.

618

Relevance of CD34+ mobilization in PB and apheresis products of autologous patients and allogeneic donors

M. Engelhardt, D. Wider, E. Samek, H. Bertz, M. Lübbert, J. Finke (Freiburg, D)

We and others have previously shown that CD34+ cells in peripheral blood (PB) correlate with apheresis (AP) products in tumor patients after chemotherapy (CT) plus G-CSF (BMT 19:529, 1997) and in allogeneic donors after G-CSF (G-) priming alone (JCO 17: 2160,1999). Since the analysis of PB CD34+ cells has recently been determined as mandatory for apheresis guidance, we sought to study within a large group of patients and donors, whether: 1. the mobilization potential of various CT regimens and G-doses differ, 2. differences after CT- plus G- vs. G-mobilization exist, 3. PB and AP closely correlate and can be used for efficient AP scheduling, 4. prior CT- and radiation therapy effect PB and AP results and 5. potential differences in PB and AP products are observed with subsequent aphereses. We evaluated 248 autologous patients and 129 allogeneic donors in a retrospective analysis between 6/1996-1/2001. The underlying diagnoses leading to autologous transplantation (Tx) were lymphoma (n=169), solid tumors (n=67), leukemia (n=9) and skleroderma (n=3); and for allogeneic Tx mostly leukemia. The median age and body weight were comparable in patients and donors with 46 vs. 43 years and 72 vs. 75 kg, respectively. Significantly higher WBC and platelet numbers were observed in donors as compared to patients after CT and G-treatment, nevertheless considerably higher PB-CD34+, AP-CD34+ and CFU numbers were obtained in patients, demonstrating once again the better mobilization efficacy after CT plus G-, rather than G-priming alone. Within this large patient and donor cohort, we observed a strong correlation between PB and AP CD34+ cells, thereby allowing us to accurately predict AP results. With subsequent aphereses, we detected little variation between PB-, AP-CD34+ cells and AP CFU-GM numbers, suggesting a predetermined mobilization capacity in patients and donors. With various tumor diagnoses in autologous patients, considerable differences in PB and AP products were apparent, with highest CD34+ numbers in solid tumor, less in lymphoma and least in leukemia patients. This suggests that predominantly the amount and intensity of pretreatment regimens are responsible for the decreased mobilization capacity, declined cell expansion, slow hematopoietic reconstitution and short telomeres as previously reported (Blood [Suppl] 90: 211a, 1997). In summary, our data demonstrate that 1. a prediction of autologous and allogeneic AP results is possible due to a strong correlation of PB and AP products which is of importance to facilitate PBSC mobilization in patients with high pretreatment, older age and short telomeres, 2. correct timing and schedule are essential for efficient mobilization and apheresis, 3. PBSC mobilization seems 'determined', however, can be increased with CT agents, higher G-doses and large volume apheresis, and 4. PBSC mobilization should be scheduled early in the disease course.

619

Prediction of CD34+ mobilization and individually adjusted performance of PBSC harvests in autologous patients and allogeneic donors

M. Engelhardt, D. Wider, E. Samek, R. Herzog, H. Bertz, M. Lübbert, J. Finke (Freiburg, D)

After having evaluated a large cohort of autologous patients (n=248) and allogeneic donors (n=129) in a single center retrospective analysis between 6/1996-1/2001, we observed a strong positive correlation between peripheral blood (PB) and apheresis (AP) CD34+ cells (for autologous patients according to: f(x)=0.003038*x+0.030256, for allogeneic donors: f(x)= 0.00263*x+0.029957). Since this correlation allows to accurately predict AP results, we sought to estimate these upfront and thereby to individually adjust PBSC harvests in autologous patients and allogeneic donors. In this ongoing prospective analysis, AP results were calculated after determination of the patient's/donor's PB CD34+ counts according to the above correlation analysis of PB and AP products. According to these predicted AP results, the leukapheresis performance was individually adjusted in each patient or donor and later compared to the actually obtained PBSC product. So far, 13 autologous patients and 13 allogeneic donors have been analyzed. Their age and body weight (bw) were comparable with 46 vs. 43 years and 77 vs. 74 kg, respectively. Mean PB WBC, hemoglobin and platelet counts before AP were 15/ml, 9g/dl and 77/ml in patients and 47/ml, 13g/dl and 162/ml in donors, respectively. Mean PB CD34+ counts (in % and /μL) of 0.57% and 67/μL in patients vs. 0.1% and 47/μL in donors were observed. According to PB CD34+ cells and predicted harvest results, aphereses were performed with a mean volume of 15 L. Per single apheresis, mean CD34+ cells of $3.13x10^6$ in patients and $3.1x10^6$/kg bw in donors were obtained. These PBSC results were very similar to our upfront predicted values which were a median of 1.37-fold lower for autologous and 0.94-fold different for allogeneic PBSC harvests. These results demonstrate that 1. a prediction of autologous and allogeneic AP results is possible due to a strong correlation of PB and AP products. This is important to facilitate PBSC mobilization, especially in donors and patients who poorly mobilize, e.g. due to older age and/or high pretreatment, 2. predicted CD34+ results well match actually obtained PBSC results, 3. high PB CD34+ cells allow to perform shorter AP procedures, lower PB CD34+ counts need more extensive PBSC harvests, and very low PB CD34+ counts (<10/μL) suggest to perform no aphereses at all, and 4. the prediction of AP results with our model may prove a valuable tool to best and most efficiently perform PBSC harvests in autologous patients and allogeneic donors.

620

In-vivo / ex-vivo T-cell depleted peripheral blood progenitor cell transplants: A single centre experience of 160 consecutive cases

S. von Harsdorf, C. Duncker, M. Wiesneth, M. Stefanic, W. Grimminger, G. Munzert, R. Schlenk, T. Karakas, D. Dohr, I. Buchmann, J. Kotzerke, S. Reske, L. Bergmann, H. Döhner, D. Bunjes (Ulm, D)

Over the past 10 years the Ulm BMT group has pursued a strategy of in - vivo / ex - vivo T cell depletion and the preemptive use of donor lymphocyte infusions in patients (pts) with low risk leukemias and a HLA -identical sibling donor. In 1996 we switched to peripheral blood progenitor cells (PBPC) as stem cell source and have extended the strategy to all categories of pts except those with refractory acute leukemia.Here we report on the outcome of 160 consecutive cases transplanted between 4 / 96 and 12 /00. The cohort consists of 87 men and 73 women with a mediaan age of 42y (range 17 - 63y). 109 of these pts were considered to be at high risk of treatment failure because of advanced disease and / or the use of an alternative donor, 51 pts were considered to be low - risk pts. The conditioning consisted of TBI 12 Gy or Bu 12,8 mg / kg +CTX 120mg / kg ± Thiotepa 10 mg / kg, 52 pts at high risk of leukaemic relapse were additionally treated with a Re - labelled anti - CD66 monoclonal antibody. The following donors were used : HLA - identical family donors (n=80), matched unrelated donors (n=45),mismatched family donors (n=21),haploidentical family donor (n=14). In - vivo T cell depletion was performed with either rabbit ATG or Campath 1H. Three methods were used for T cell depletion of the grafts : CD34+selection with the CEPRATE device + CD2/CD3 depletion (n=47), CD34+selection with the ClinMACS device (n=61) and Campath 1H in the bag (n=52). Rapid and stable engraftment was achieved in 96% of pts. The actuarial risk of acute GvHD grade II was 23 %, no pt suffered grade III or IV acute GvHD. The actuarial risk of chronic GvHD was 47%. After a median follow - up of of 24 mo (range 6 - 59 mo) 54% of pts are alive, 65% in the low risk group and 49% in the high risk group. Transplant- related mortality was the cause of death in 35% of pts, relapse was the cause of death in 11%.The most frequent single causes of death were sepsis,aspergillus infection and relapse. Approx. 15% of pts developed autoimmune diseases such autoimmunethyroiditis,AIHA or ITP. Overall there were no significant differences between the different T cell depletion modalities with respect to engraftment or GvHD control, but given the higher risk of GvHD in the CliniMACS group, this device probably provides the best GvHD control. In summary, T cell depleted PBPCT is associated with a low risk of graft failure, a low risk of severe acute GvHD, a significant risk of chronic GvHD and a moderate risk of autoimmune disease. No increased risk of relapse was observed.

621

Selection of peripheral blood CD34+ cells using the immunomagnetic (Clinimacs) selection device for autologous and allogeneic transplantation

M. Engelhardt, D. Wider, E. Samek, H. Bertz, D. Behringer, M. Lübbert, J. Finke (Freiburg, D)

In order to efficiently deplete either tumor cells in autologous, or T cells in allogeneic grafts, the selection of CD34+ peripheral blood progenitor cells (PBPC) is widely pursued. To determine the efficacy of an immunomagnetic (Miltenyi, Clinimacs®) device to select CD34+ PBPC for autologous and allogeneic transplantation (Tx), a total of 107 patients scheduled to receive CD34+ selected products were analyzed. One-hundred autologous patients with lymphoma (n=64), solid tumors (n=28) and others (n=8) were mobilized with chemotherapy plus rhG-CSF, while 7 healthy donors received G-CSF alone. Median age was 46 years in patients and 42 years in donors, median body weight (bw) in both groups 66 kg. A total of 59 male vs. 41 female patients and 6 male vs. 1 female donor were included. The Clinimacs device allowed to obtain a median CD34+ purity of 97% in patients and 98% in donors, and a CD34+ recovery of 78% and 86%, respectively. Median CD34+ cells before selection were 6.1 and $4.2x10^6$/kg and after selection 5.7 and $2.6x10^6$/kg bw in patients and donors, respectively. To obtain target CD34+ cells of >4x10e6/kg bw, a median of one immunomagnetic procedure was performed. The T- and B-cell depletion was 4-log (median 11000-fold: 10068x in auto and 26746x in allo Tx) and 3-log (1200-fold: 1090 in auto and 1526 in allo Tx), respectively. These results demonstrate that this selection device yields highly CD34+ pure cell products. The transplantation of these cells, depleted of 'by-stander' cell subsets such as T- and B- cells, did permit rapid hematopoietic engraftment so far in 56 patients (n=51 autologous, n=5 allogeneic patients) reaching WBC>1000/μL on day+11, platelets>20,000/μL on day+12. Four out of 56 patients (7%) died during the transplant course (n=3 autologous: 6%, 1 allogeneic: 20%), the autologous due to pneumonia and sepsis, the allogeneic due to secondary graft failure. In order to further improve the transplant course and avoid severe infections post-transplant, longer antibiotic prophylaxis and lymphocyte subpopulation monitoring are currently under clinical investigation.

Postersitzung:
Tumoren des oberen Gastrointestinaltraktes

622

Effective combination chemotherapy with paclitaxel and cisplatin with or without human granulocyte colony-stimulating factor and/or erythropoietin in patients with advanced gastric cancer

C. Gedlicka, W. Scheithauer, M. Raderer, B. Schüll, W. Fiebiger, H. Ulrich-Pur, B. Schneeweiss, D. Depisch, F. Lang, G. Kornek (Wien, Kirchdorf, Wr.Neustadt, Neunkirchen, A)

A phase II trial was performed to determine the antitumor efficacy and tolerance of combined paclitaxel and cisplatin with or without hematopoetic growth factor support in patients with advanced gastric cancer. Forty-five patients with histologically confirmed advanced gastric cancer were entered in this trial. Treatment consisted of 2-weekly courses of paclitaxel 160

mg/m^2 and cisplatin 60 mg/m^2 both given on days 1 and 15. Depending on absolute neutrophil counts on the days of scheduled chemotherapeutic drug administration (1000-2000/µL), a 5-days course of human granulocyte colony-stimulating factor (G-CSF) 5mg/kg/d was given subcutaneously (s.c.); in addition, if hemoglobin was <12.0mg/dl, erythropoietin 10.000 IU was administered s.c. three times per week. The confirmed overall response rate (intent-to-treat) was 44%, including 5 complete (11%) and 15 partial remissions (33%). Twelve patients had stable disease (27%), eleven (24%) progressed while on chemotherapy, and 2 patients were not evaluable. The median time to response was 3 months, the median time to progression 7.0 months, and the median survival time was 11 months with 12 patients currently alive. Hematologic toxicity was common, though WHO grade 4 neutropenia occurred in only 6 patients (13%). Apart from total alopecia in 16 patients (36%), severe nonhematologic adverse reactions included grade 3 peripheral neuropathy in 7 (16%) and anaphylaxis in 3 patients. In addition, there was one patient each who experienced grade 3 diarrhea, stomatitis, and infection, respectively.
Our data suggest that the combination of paclitaxel and cisplatin with or without G-CSF and/or erythropoietin has promising therapeutic activity in patients with advanced gastric cancer.

623

Outpatient treatment in advanced pancreatic cancer

M. Duru, H. Köppler, J. Heymanns, J. Schimke, G. Jacobs, M. Grundheber, B. Rendenbach, R. Weide (Koblenz, Saarbrücken, Trier, D)

Objective: Evaluation of feasibility, effectivity and toxicity of palliative chemotherapy/supportive care in patients with advanced pancreatic cancer, who were treated completely on an outpatient basis.
Methods: Retrospective analyses of 127 consecutive, unselected patients with advanced pancreatic cancer treated between 6/95-6/2000 in 4 community based oncology group practices.
Results: Median age was 62 years (range 34-88) and performance status ranged from 0-3. 77 were male (60%) and 50 were female (40%). 43 patients (34%) had locally advanced disease and 84 patients (66%) had distant metastases. 94 patients (75%) received cytotoxic treatment during the course of their disease, 33 patients (25%) received best supportive care (BSC) only. First line treatment consisted of gemcitabine (1000 mg/m^2, day 1, 8, 15 q day 29) in 81 patients (86%) and 5-FU in 8 patients (9%), radiochemotherapy in 4 patients (4%) and radiation only in 1 patient (1%). Second line therapy was performed in 30/94 patients (21 5-FU, 6 gemcitabine, 1 etoposid / cisplatin, 1 radiochemotherapy, 1 radiotherapy). 51 patients received gemcitabine treatment only. A total of 1501 gemcitabine-applications were performed during the study period. The observed toxicity was mild. A reversible grade 3 and 4 haematotoxicity was seen after 24 applications (1,3%). A grade 3 or 4 asthenia syndrome was seen in 12 applications (0.7%). Therapy associated hospitalisation was seen in 4 patients (3%). 111 patients (88%) died during the observation time (61 at home, 50 in hospital). The median survival of all patients is 35 weeks (range 4-238). The median survival of patients receiving cytotoxic treatment is 41 weeks (7-238). The median survival of BSC-patients is 21 weeks (4-88). The 6, 12, 24, 36 months survival is 65%, 32%, 14% and 7% respectively. Conclusion: Palliative treatment in advanced and metastatic pancreatic cancer can be performed completely on an outpatient basis. The observed therapy associated toxicity was mild and therapy associated hospitalisation was low. Median survival in this unselected group of patients is high compared to historical controls.

624

Microencapsulated cell-mediated treatment of inoperable pancreatic carcinoma

G. Faulmann, A. Hoffmeyer, J. Kröger, A. Holle, J. Hain, R. Saller, W. Günzburg, W. Günzburg, B. Salmons, B. Salmons, S. Liebe, K. Hauenstein (Mannheim, Rostock, München, D; Wien, A)

Background: The treatment of pancreatic cancer patients in an advanced tumor stage with conventional chemotherapy shows weak positive effects. High toxic side effects of substances like ifosfamide limited the employment for therapy. Hypothesis: The local placement of encapsulated CYP2B1 expressing cells into the tumor leads to a high concentration of the active metabolites of ifosfamide, phosphomustard and acrolein, and a low systemic level of these toxic drugs. Rational: Transfection of CYP2B1 in cells with cellulose sulfate microencapsulation. Experimental Work: In vitro and in vivo experiments showed a stable enzyme activity of encapsulated CYP2B1 expressing L293 cells. After injection intraarteriell into the tumor in xenotrasplanted pancreatic carcinoma in nude mouse a significant antitumor effect was demonstrable (Gen Therapy 1998, 5:1070). Experimental angiographic placement of capsules into the pancreas of pigs showed the feasibility of this method (Ann NY Acad Sci 1999, 88:376; J Mol Med 1999, 77:393; AGO Protocoll No 23). Patients, Material and Methods: CYP2B1 transfected L293 cells were encapsulated with cellulose sulfate on a diameter of 0.8 mm under GCP conditions and packed sterile. Capsules were intraarteriell injected during angiographie in a tumor supporting vessel in patients with inoperable adenocarcinoma of pancreas. After a monitoring of 48 hours, to exclude allergic reactions, the patients were treated with ifosfamide at 1000 mg/m^2 BS on day 1-3 and day 21-23. The patients were followed-up for 5 months. Results: The study protocol started in 7/98 and 17 patients were enrolled. In 14 patients the capsules were instilled as planned and the treatment was tolerated without any complications. The administration of the capsules was technically impossible in 1 patient and two patients had acute infections prior to the procedure. Complications, like allergic reactions or pancreatitis was not encountered. Treatment with ifosfamide was uneventful. Stable disease was observed in 12 of 14 patients and survival time was prolonged to more than 38 weeks in comparison to 24 weeks in the control group. Conclusion: The intratumoral activation of ifosfamide by microencapsulated cells in treatment of pancreatic cancer patients provides a significant better antitumor effect than conventional strategies and reduces the strong toxicity of high systemic concentrations.

625

Value of new tumormarkers P53-antibody, M2PK and tissue factor for detecting pancreatic carcinoma

F. Hummel, H. Hinner, S. Liebe, M. Löhr (Mannheim, Rostock, D)

The value of tumor markers is often overestimated. Nevertheless they are often examined and especially in differenciating chronic pancreatitis (cP) versus pancreatic carcinoma (pa-ca) every clue for optimizing diagnosis is wellcome. Intention of this study was to evaluate p53-Antibody (p53-Ab), M2-Pyruvatkinase (M2PK) and Tissue Factor (TF) in comparism with established markers for pa-ca under special consideration of cP. 300 male indivuals were included. 67 with histologic proven pancreatic adenocarcinoma, 114 with chronic pancreatitis, 79 with other gastrointestinal tumors and 40 healthy individuals (control group). M2PK and TF were messured by a 2-Step-Sandwich-ELISA, p-53-Ab by a common Sandwich-ELISA. Ca19-9, CA72-4, CA242, CEA and AP were messured in routine laboratory assays. Sensitivity of p53-Ab was 21%, specifity was 80% vs. CP and 100 % vs control group. M2PK-levels were lowered in pa-ca but also in cP. For a demanded specifity of 95% vs. control group the sensitivity was 95%, but M2PK could not discriminate vs. cP. Sensitivity of TF 46% for pa-ca for a demanded specifity of 95% vs. control group, but sensitivity decreases to 31 % vs. cP. CA19-9 was the âbestÔ old marker with a sensitivity of 77% (48% for other carcinomas, 25% for cP) with a cut-off for 95% specifity. Results for CA 242 were not much worse. P53-Ab, M2PK and TF are no useful markers for pancreatic carcinoma because they are not able to discriminate pa-ca vs. cP, which is the most relevant clinical question, although normal M2PK-Levels exclude pa-ca. CA19-9 and CA242 show the best profiles for sensitivity and specifity in differentiating pa-ca vs. cP.

626

Results of two phase II studies: Docetaxel and cisplatin combination chemotherapy for advanced gastric cancer

K. Ridwelski, J. Fahlke, U. Keilholz, E. Kettner, H. Lauschke, H. Kröning, R. Kuhn, K. Eichelmann, H. Lippert (Dessau, Magdeburg, Berlin, Bonn, D)

We conducted two phase II studies aiming at investigating the activity and tolerability of combination chemotherapy with docetaxel (Doc) and cisplatin (DDP) in the treatment of patients (pts) with advanced or recurrent gastric cancer. After completion of a single center study using Doc 75 mg/m^2 d1 and DDP 75 mg/m^2 d1 repeated every 21 days for a maximum of 9 cycles (ECCO 10, # 565, 1999), a multicenter study was initiated in 12/98 using the same treatment regimen. Patient characteristics have been comparable for both studies. Thus far, a total of 101 pts have been enrolled in the second trial. To date 71 pts (median age, 60 years; 11 with locally advanced disease, 36 with metastatic disease, 24 with recurrent disease after gastric resection) have been evaluable for toxicity and 72 for response so far.

Response	pts	CR	PR	SD	PD	ORR
Multicenter	52	5	13	19	15	34,6%
Single center	39	4	12	11	12	41%

Toxicity	pts	Median cycles	Myelotox.	GI toxicity	Neurotox.	Alopecia
WHO Grade 3						
Multicenter	72	4	57.2%	2.4%	5.5%	58%
Single center	39	7	17.0%	16.3%	2.3%	6.9%
WHO Grade 4						
Multicenter	57	4	30.9%	0%	0%	0%
Single center	39	7	2.3%	0%	0%	0%

30 patients have died in the multicenter study, and estimated median survival is 9 months compared with 10.4 months in the single center study. In conclusion, the multicenter study confirmed that Doc/DDP is a well tolerated and active regimen for the treatment of advanced gastric cancer. A phase III trial comparing the combination of Doc/DDP with the most frequently used 5FU/DDP regimen.

627

A randomized study of granulocyte colony-stimulating factor (G-CSF) in esophageal carcinoma (EC)

G. Mansmann, G. Gross, V. Hübbel, L. Hägele, G. Wassmer, U. Zorn, B. Ulrich, A. H. Hölscher, H. Schäfer, A. Engert (Köln, Dortmund, Düsseldorf, D)

Patients undergoing esophagectomy for EC are at high risk for severe postoperative infection. We performed a randomized, placebo-controlled, double-blind multicenter study evaluating the efficacy of perioperative G-CSF to prevent postoperative infection in nonneutropenic patients with resectable EC. Primary endpoint of the study was occurrence of infection (pneumonia, sepsis, mediastinitis, peritonitis, wound infection) within the first 10 days after esophagectomy. Mortality, duration of antibiotic therapy, and length of in-hospital and ICU stay have been selected as secondary endpoints. Eligible patients (EC UICC stage 0-III, ECOG 0-2, age 18-75 years) were stratified by study center and preoperative risk score and randomly assigned to receive either recombinant human G-CSF (filgrastim), depending on body weight 300 or 480 µg/d s.c., or placebo for a maximum of 10 days beginning 2 days before surgery. Patients were not eligible if suffering from immunosuppression (myelosuppressive therapy, hematological malignancy, leukopenia <2000/µl HIV infection) or apparent infection at study inclusion. Administration of the study medication was stopped if the primary endpoint was met, or interrupted if leukocytes exceeded 75 x 10^9/l until a leukocyte count of 37.5 x 10^9/l was reached. A total of 150 patients from 5 centers have been randomized. Results of the final data analysis, which is currently underway, will be presented in regard to the primary and secondary endpoints with subgroup analysis by histology, realization of neoadjuvant therapy, and stratification criteria. Supported by a grant of the Deutsche Krebshilfe. Filgrastim was provided free of charge by Amgen GmbH.

628

Mutated Ras-transfected, EBV-transformed lymphoblastoid cell lines as a model tumor vaccine: a pilot trial in pancreatic carcinoma patients

B. Kubuschok, R. Schmits, F. Hartmann, R. Breit, C. Cochlovius, L. Trümper, C. Renner, M. Pfreundschuh (Homburg/Saar, D)

In order to develop a tumor vaccine not limited by MHC restriction, we chose the transfer of a gene coding for a tumor antigen into antigen presenting cells as a promising strategy. The model tumor antigen selected was the point mutated (Codon 12) Ki-Ras p21 oncogene (muRas), which is frequently found in pancreatic adenocarcinoma (about 85%). It has been previously shown to be specifically recognized by CD8+ and/or CD4+ T-lymphocytes from mutated Ras-bearing cancer patients. As antigen presenting cells we used as an alternative to the limited available dendritic cells autologous lymphoblastoid cell lines (LCL), which were established in vitro by spontaneous outgrowth from peripheral blood of pancreatic carcinoma patients and were transfected with an episomal EBV-based expression vector encoding muRas (muRas-LCL). MuRas-LCL were able to elicit a strong muRas-specific cytotoxic T-lymphocyte (CTL) response in vitro. In vivo, vaccines (4x10^6 muRas-LCL) were given subcutaneously once a week for a total of 8 vaccinations in patients with locally advanced pancreatic carcinoma. 6 of 7 patients showed a delayed-type-hypersensitivity reaction and an increase in mutated Ras-specific precursor CTLs measured by IFN-g ELISPOT in peripheral blood. There was no significant toxicity associated with the immunizations or with the generation of a muRas-specific immune response. In 4 of 7 patients stable disease was achieved, although this was a temporally limited clinical effect. In conclusion, human muRas-LCL can serve as an easily obtainable and unlimited source for an autologous tumor vaccine and are able to induce specific immune responses in patients with muRas-positive pancreatic carcinoma.

629

A biweekly schedule of docetaxel, gemcitabine and oxaliplatin for treatment of pancreatico-biliary tumors

O. Kellner, L.P. Müller, T. Kegel, K. Jordan, D. Wunderlich, R. Zamoryn, B. Deschler, T. Büchele, A. Grothey, H.J. Schmoll (Halle, D)

Purpose: Metastatic pancreatico-biliary tumors are associated with short survival and poor quality of life. The current standard therapy with gemcitabine or 5-fluorouracil results in only minor improvement of this situation. Based on first data of combinationtherapy with gemcitabine and oxaliplatin in pancreatic cancer and gemcitabine and taxane in biliary tumors, the efficacy of a therapy with docetaxel (DOC), gemcitabine (GEM) and oxaliplatin (OXA) in patients with metastatic pancreatico-biliary tumors was evaluated in a phase II study. Methods: 12 patients (5 female, 7 male), age 48 - 72 yrs (median 61 yrs) with metastatic pancreatic (8) or biliary (4) carcinoma were enrolled. 3 patients were pretreated for advanced disease with one chemotherapy protocol. Chemotherapy was given in a biweekly (q d 15) schedule of DOC (55 mg/m^2, 60min iv d1) followed after 1 h by GEM (500 mg/m^2 30min iv d1) and, on the subsequent day OXA (70 mg/m^2 120min iv, d2). Results: Currently 9 of 12 patients (6 pts. with pancreatic, 3 with biliary carcinoma) are evaluable for response and toxicity. One partial remission (11%) with a progression-free survival of 12 months in a patient with pancreatic carcinoma was observed. This patient is still alive 18 months after initiation of therapy. In 7 patients (78%; 5 pancreatic, 2 biliary carcinoma) treatment resulted in stabilization and/or clinical improvement of previously progressive disease. To date, these patients show a progression-free survival of 5 months (2 - 10+ months) and a overall survival of 14 months (11 - 18+ months). Progressive disease was only seen in one patient suffering from biliary carcinoma. WHO °3 toxicity was observed in 3 patients with one case of dermatitis, diarrhea and leukopenia, respectively. No WHO °4 toxicity occurred. Conclusion: Despite the low response rate the combination therapy of DOC/GEM/OXA demonstrates a conceivable effect on overall survival in patients with metastatic pancreatico-biliary tumors. It can be performed in an outpatient setting and confers only moderate toxicity without compromising quality of life. Based on these encouraging results a subsequent phase III study is planned.

630

Radiochemotherapy with 5 FU combined with gemcitabine chemotherapy in locally advanced pancreatic cancer

M. Würfel, C. Haag, M. Baumann, T. Herrmann, G. Ehninger (Dresden, D)

We treated 11 patients (8 male, 3 female, median age 50 years) with locally advanced, inoperabel pancreatic cancer with a combination of radiochemotherapy and gemcitabine chemotherapy . Treatment plan: 1 cycle gemcitabine (1000 mg/m^2 i.v., day 1,8,15). 2 to 3 weeks after we continued with the radiochemotherapy: 45 Gy total dose with single doses of 1,8 Gy. During the first and fourth week of radiation therapy 5-FU was given in a dose of 450 mg/m^2 i.v. day 1-3. After finishing the radiochemotherapy gemcitabine monotherapy was planned for at least 3 additional months in the same dose as above. Results: all patients received the first chemotherapy and the radiochemotherapy. All patients received the following gemcitabine chemotherapy, but 4 patients less than 50 % of the planned dose, 5 patients between 50 and 100 % and only 1 patient the full dose. In one patients the data of the dose of gemcitabine and toxicity are not available. The reasons for stopping chemotherapy was toxicity, stable disease or wish of patient. 7 patients continued the gemcitabine chemotherapy more than 3 months. Toxicity: hematological: 4 of 11 patients developed grade 1 and 2 leukopenia, 6 patients grade 3 leukopenia. 6 patients developed grade 1 thrombocytopenia, 3 patients grade 2 and 1 patient grade 3 thrombocytopenia. Non-hematological complications during and after therapy: 1 patient developed an inflammation of M. psoas, which could be treated non surgical. 1 patient suffered from bilioma, 1 patient was treated with stent implantation because of occlusion of common bile duct, 1 patient developed thrombophlebitis and 1 patient pleura effusion and pericard effusion. 1 patient is suffering of duodenitis 10 months after radiochemotherapy.

Survival: After a follow up of 14 months (8-24 months) 6 of 11 patients are alive with stable disease and 5 of 11 patients died after 8-23 months due to progressive disease. Median survival of all patients is 17 months. Conclusions: this combination of gemcitabine before and after radiochemotherapy is a feasible therapy of inoperable, locally advanced pancreatic cancer with a better survival rate than chemotherapy alone or radiochemotherapy alone. The hematological toxicity of gemcitabine chemotherapy after radiochemotherapy was high and lead to dose reduction in most of the patients.

631

Analysis of the p53/Bax/p21 apoptosis pathway identifies patients with good prognosis in esophageal squamous cell carcinoma

I. Sturm, D. Stuckenberger, S. Hermann, H. Petrowsky, S. Hauptmann, P. Hemmati, V. Budach, M. Lorenz, B. Dörken, P. T. Daniel (Berlin, Frankfurt/Main, D)

BAX, a member of the BCL-2 family, is a central downstream apoptosis effector. We have previously shown that loss of BAX epression is a negative prognostic factor in metastatic colorectal cancer (J Clin Oncol 1999, 14:1364). In the present study we addressed the prognostic relevance of BAX and its upstream regulator p53 in squamous cell carcinoma of the esophagus. Retrospective analysis of 53 patients was done for protein expression of BAX, p53, and p21Cip/Waf by immunohistochemistry. All patients underwent R0-resection of squamous cell carcinoma of the esophagus with curative intent to treat at a single institution between 1985 and 1995. In addition, tumor DNA was screened for BAX frameshift mutations by fragment length analysis and for p53 mutations by SSCP-PCR. Overall median survival was 14.0 months. Patients with high BAX protein expression had a median survival of 19.5 months versus 8.0 months with low BAX expression ($p<0.005$). Low BAX expression was an independent negative prognostic marker in multivariate regression analysis with the parameters nodal status, tumor size and grading, non-surgical therapy, and p53 mutation (Cox Proportional Hazard model, relative risk=3.1, $p<0.005$). The worst survival was noticed for patients with p53-mutated/low BAX; the best for p53-wild type/high-BAX (J Clin Oncol 2001, 19: 2272-2281). Neither p53 mutation nor p53 or p21 protein expression alone showed prognostic significance in this tumor entity. We conclude that low BAX expression is an independent negative prognostic marker in patients with squamous cell carcinoma of the esophagus, whereas p53 mutational status or p53/p21 protein overexpression per se have no prognostic relevance. In contrast, the inclusion of the cell cycle regulators p16ink4a and cyclin D1 (which are known prognostic factors) and p27/Cyclin E into the analysis dramatically improved the identification of patients with good prognosis and long-term survival. Thus, such a multimarker model of cell cycle and apoptosis pathway components could be useful for the evaluation of risk-adapted therapeutic strategies in the future, and we therefore suggest prospective studies to confirm these results.

632

Endoscopic ultrasonography (EUS) in assessment of response to chemotherapy in patients with advanced, non-resectable esophageal cancer

D. Arnold, N. Hoepffner, A. Adler, U. Settmacher, C. Stroszczynski, H. Oettle, H. Riess (Berlin, D)

Endoscopic ultrasonography (EUS) of the esophageus has widely been used to assess the extent of circumferential involvement of the esophageal carcinoma with the surrounding structures and lymph node status in preoperative staging for resectable patients. Although it is well-known that computed tomography (CT) scans are less accurate in evaluation of localized tumor depth or/and the degree of adjacent tissue- or lymph node involvement, there is hardly any other data on EUS in follow-up of patients (pts.) with conservative treatment modalities available so far. Methods: Pts. with advanced, inoperable esophageal cancer received chemotherapy in a phase I/II-study of our institution. The chemotherapy regimen consisted of a combination of Gemcitabine, Cisplatin, Folinic Acid and infusional 5-FU. According to protocol, EUS and/or spiral CT-scans were performed in short time intervals of 6-12 weeks to evaluate tumor response for further treatment modalities (resection vs. palliative chemoradiation). Results: 13 pts. (11 male, median age 62 [46-71] yrs., 9 squamous cell- / 4 adenocarcinoma) with non resectable but localized, non obstructive esophageal cancer with lymph nodal involvement (stage III/IV UICC; 10 uT3N1M0, 3 uT4N1M0), underwent a median of 4 [3-6] EUS-assessments within a median period of 9 [7-18] months. 8 pts. showed a reduction of tumor size and/or lymph node involvement of at least 25%. Lymph nodal down staging (from N1 to uN0) was observed in 3 of those 8 pts. while another pt. presented an upward stage migration from UICC-Stage III (uT3N1) to IIa (uT2N0-1) leading to a total of 4 pts. being suitable to surgery. In contrast, CT scans failed to demonstrate measurable response to chemotherapy in 5 of those 8 pts. 3 of 5 pts. without tumor size regression presented progressive disease in further endoscopic/endoultrasound follow-up and were consecutively sent to radiotherapy. CT-scans of those pts. were assessed as „stable disease“. Conclusion: According to our experience, follow-up with EUS seems to be promising in assessment of pts. with localized esophageal cancer receiving chemotherapy. However, the numbers are too small to make any definitive conclusions so far. Prospective comparison with CT-scan is performed in the ongoing phase II setting in terms of tumor response and survival.

633

Phase I study of gemcitabine (GEM) in combination with cisplatin (CDDP), 5-fluorouracil [24 h] (FU) and folinic acid (FA) in patients with advanced esophageal cancer

M. Kern, D. Arnold, J. Stieler, N. Hoepffner, U. Settmacher, H. Riess, H. Oettle (Berlin, D)

Prognosis for advanced, non resectable esophageal cancer remains poor with a median survival of 9-12 months and 5-year-survival rate rarely exceeding 10%. Combination chemotherapy of CDDP and 5-FU is considered to be standard treatment for either squamous cell or adenocarcinoma. Recently GEM has shown activity against various solid tumors as single agent or in combination with CDDP or 5-FU. This study was designed to determine the maximal tolerated dose (MTD) of GEM in standard dosage (1000 mg/m^2 30min) in combination with CDDP and FU (750mg/m^2 24h) / FA (200mg/m^2; 60min). Combination therapy was given on days 1 and 8 of a 3-weekly cycle (q day 22). CDDP dose was escalated according to toxicity at dose levels I/II/III; 25/ 30 /35 mg/m^2 respectively. Pts. characteristics: 16 consecutive chemo-naive patients at stage III or IV (2f/14m; 9 squamous cell- / 7 adenocarcinoma; median age 60 [42-71] yrs., median Karnofsky PS 80% [70-100]) entered the study. 6, 7 and 3 pts. have been enrolled in dose level I/II and III respectively. Toxicity: More than 90% of drug administrations could be given on an outpatient basis. Alopecia has been almost universal. The DLT of this schedule was haematological with febrile neutropenia (IV°) in 2/3 and thrombocytopenia
IV° in 1/3 pts. at dose level III. Dose level III is defined as MTD. Conclusion: This combination appears to be generally well tolerated with 30 mg/m^2 CDDP for pts. with esophageal cancer on an outpatient basis. The study is ongoing in phase II at dose level II.

634

Her2/neu overexpression in pancreatic adenocarcinoma

C. Hempel, H. Oettle, M. Koch, D. Lüftner, A. Winkler, D. Koeneke, J. Langrehr, C. Radtke, H. Schramm, G. Schaller, H. Riess (Berlin, Gera, D)

Her2/neu is a transmembrane receptor tyrosine kinase of the epidermal growth factor receptor (EGFR) family without a known ligand. Its intrinsic tyrosine kinase activity results from heterodimerization with another ligand bound EGFR family member. Overexpression of Her2/neu is associated with poor overall survival and chemoresistance in adenocarcinoma of the breast. Immunohistochemical overexpression of the Her2/neu in adenocarcinoma of the breast shows an incidence of approximately 30% with good correlation to gene amplification proved by fluorescence in situ hybridization (FISH).

So far, the reported rates of immunohistochemical Her2/neu overexpression in adenocarcinomas of the pancreas are inconsistent and range from 1% to more than 60% in small numbers of patients. We determined the rate of Her2/neu overexpression by immunohistochemistry (IHC) in histologically confirmed ductal adenocarcinomas of the pancreas from 117 patients who underwent Whipple's procedure. IHC was performed with DAKO polyclonal cerbB2 antibodies on 4μm sections of paraffin embedded tumor tissue samples. According to the evaluation of two independent pathologists strong overexpression (3+) was seen in one specimen and moderate overexpression (2+) in 6 specimen ; the entire overexpression rate was 7 out of 117 (6 %). For further analysis so far 20 out of 117 specimen were tested for Her2/neu gene amplification by FISH-technique. Interestingly, only one specimen showed a low amplification of 5-10 Her2/neu gene copies but did not demonstrate any overexpression of Her2/neu in the immunohistochemical staining. On the other hand, none of the specimen demonstrating overexpression Her2/neu in the immunohistochemical method showed a gene amplification. Conclusions: In accordance to Koeppen et al., 2001, the overexpression rate of Her2/neu in adenocarcinoma of the pancreas in a group of 117 patients is much lower than reported so far. In a small group (20 patients) out of the 117 patients a correlation between overexpression of the her2/neu receptor (IHC) and gene amplification of the her2/neu gene (FISH) could not be demonstrated.

635

A phase II study of weekly paclitaxel, weekly 24-hour continuous infusion (CI) 5-fluorouracil, folinic acid (FA) and cisplatin in patients (pts) with advanced gastric cancer

F. Honecker, C. Kollmannsberger, D. Quietzsch, C. Haag, J.T. Hartmann, W. Baronius, V. Hempel, L. Kanz, C. Bokemeyer (Tübingen, Chemnitz, Dresden, D)

Background: The present study evaluates the toxicity and efficacy of a combination chemotherapy regimen with weekly paclitaxel, threeweekly cisplatin and weekly 24h c.i. of 5-FU/FA in pts with unresectable, locally advanced or metastatic gastric adenocarcinoma. This combination has been previously evaluated using paclitaxel 175 mg/m^2 administered at weeks 1 and 4. Paclitaxel given on a weekly basis has been suggested to be associated with a lower toxicity but equal antitumor activity. Patients and methods: Twenty-one chemotherapy-naive pts (10 m. and 11 f.) with a median age of 48 years [range 31-72] were enrolled. 5-FU 2 g/m^2 was given weekly over 24h c.i. preceded by FA 500 mg/m^2 as 2h infusion. Paclitaxel 80 mg/m^2 was administered weekly as a 1 hour infusion prior to folinic acid, and cisplatin 50 mg/m^2 as 1h infusion on days 8 and 29. One cycle consisted of six weeks of therapy (days 1,8,15,22,29,36) followed by 2 weeks rest. Results: A median of 3 cycles [range 1-3] were administered to 20 pts assessable for response and survival. One pt (5%) received only 3 weeks of therapy (nausea/vomiting, worsening of performance status) and was therefore not assessable for response. Two pts (10%) obtained a CR and 7 pts (35%) a PR [ORR 45%; 95%-CI: 23-67%]. Seven pts (35%) had stable disease and 4 pts (20%) progressed during therapy. All 21 pts were assessable for toxicity. Neutropenia WHO III°/IV° occurred in 2 pts (10%) with only 1 pt having grade IV. Additional non-hematological WHO III°/IV° toxicities included nausea/vomiting in 2 (10%), alopecia in 9 (43%), and diarrhea in 2 pt (10%). Paclitaxel was discontinued in 2 pts due to allergic reactions. Dose reductions and/or treatment delays were necessary in 12 (23%) and 9 (17%) cycles, respectively, mainly due to nausea/vomiting and neutropenia (grade II/III). There were no therapy-related deaths. All pts were treated on an outpatient basis. Conclusions: The combination of weekly paclitaxel, cisplatin and c.i. 5-FU/FA represents an active regimen for the treatment of pts with advanced gastric cancer. While the overall low toxicity allows its use in the palliative setting, it may also be a promising schedule for neoadjuvant or adjuvant treatment. Patient accrual is ongoing.

Postersitzung:
Urologische Tumoren

636

Cisplatinum/5-FU as palliative option for patients with metastatic hormone-refractory carcinoma of the prostate

M. Schroeder, A.A.N. Giagounidis, M. Heinsch, C. Aul (Duisburg, D)

Patients with progressive prostate cancer developing visceral metastases under treatment with hormones usually have poor prognosis. 15 pts (median age 63 years, range 52-79 years) were treated with Cisplatin 100 mg/m^2 d1 and continuous infusional 5-FU 1000 mg/m^2 d1-5 every 4 weeks. In patients presenting with a performance score > 1 the total Cisplatin dosage was splitted into 5 equal doses of 20 mg/m^2/d. In 2 cases of renal insufficancy patients were treated with Oxaliplat 140 mg/m^2 d1. All pts had been pretreated with hormones and 5/10 pts with cytotoxic drugs and showed signs of tumor progression with development of visceral metastases. Site of metastases was liver in 6 cases, lung in 4 cases, bone marrow in 3 patients, and meningeal involvement in 2 patients. The number of chemotherapy courses applied ranged from 2-13, median 4. Best response was 8 PR, 4 NC and 3 PD. The median response duration was 6 months for PR's and 4+ months for NC. Applying conventional antiemetic support the toxicity was mild with nausea grade I 40 % and no grade II hematotoxicity .
From these data we conclude that the combined treatment with cisplatin and continuous infusional 5-FU is active and well tolerated in patients with metastatic hormone-refractory prostate cancer. Responding patients benefited with an improvement of quality of live and prolongation of survival time.

637

Atypical metastatic lesion of renal cell carcinoma in the palatal tonsil

C. Röllig, J. Babatz, M. Haack, G. Geißler, G. Ehninger, M. Bornhäuser (Dresden, D)

Background: Renal cell carcinoma is characterized by early hematogenic metastasis. Metastatic lesions are most frequently located in the lung, bones, liver or brain. Case report: A 45 year old man suffered from prolonged blurred vision and had a seizure in October 2000. A CT scan showed a solitary tumor in the right occipital part of the brain. Histologic studies from the removed tumour revealed metastastic tissue of a renal cell carcinoma . At the same time the primary lesion in the left kidney and two metastatic pulmonary lesions were diagnosed. In December 2000 the primary tumour could be resected completely by total nephrectomy, the local lymph nodes were free of carcinoma. Three months later we admitted the patient to the hospital for allogeneic transplantation of peripheral blood stem cells (PBSC) from his HLA-identical sister. Controll CT scans showed progression of the pulmonary metastases but no local retroperitoneal or cerebral relaps. On admission, we saw a moderately enlarged right palatal tonsil. Microbial smears from the tonsil showed colonisation with hemolytic strecptococci and antibiotic treatment was initiated. Still the size of the tonsil increased, crossing the midline and causing swallowing troubles, snoring and sleep apnea. Therefore the tonsil was removed by a total tonsilectomy. Tissue sections showed typical clear cells and identified the tumour as metastasis of the known renal cell carcinoma. Conclusion: Inspite of radical removal of the primary tumour, a new metastasis of a renal cell carcinoma was found at an atypical site in the palatal tonsil.

638

Generation of renal cell carcinoma-specific CD8+ T cell clones derived from blood lymphocytes of allogenic HLA class I-matched healthy donors

A. Dörrschuck, A. Schmidt, S. Pötsch, E. Schnürer, T. Gareis, V. Lennerz, A. Lifke, M. Karas, T. Wölfel, W. Herr (Mainz, Frankfurt/Main, Freiburg, D)

Clinical observations suggest that Renal Cell Carcinoma (RCC) is susceptible to immunological control. However, autologous T cell responses directed against RCC have been rarely detected in blood lymphocytes of RCC patients in vitro. One possible explanation for this phenomenon might be that RCC-reactive T lymphocytes are deleted or anergized in the T cell repertoire of these patients. As an alternative approach we tried to generate RCC-specific CD8+ T cells from blood lymphocytes of allogenic HLA class I-matched healthy donors. In a mixed lymphocyte tumor-cell culture (MLTC), CD8+ T cells purified from a buffy coat of healthy donor 1000860 were weekly stimulated against the RCC line ClearCa-3 previously established from patient MZ1257. While serological analysis had shown an identical HLA class I type for donor lymphocytes and RCC cells, genomic HLA sequencing revealed as shared suballeles HLA-A*0201, -A*0301, -B*4402 and -Cw*0501, but detected a suballele mismatch at the HLA-B7 locus (donor: HLA-B*0704; ClearCa-3: HLA-B*0702). On day 25 of MLTC, specificity of responder lymphocytes was analysed in an interferon-gamma (IFN-g) ELISPOT assay. While MLTC responders preferentially recognized ClearCa-3 cells, we also observed significant spot production against an EBV-transformed B-lymphoblastoid cell line generated from patient MZ1257 (MZ1257-EBV). No reactivity was found against the natural killer target K562. After limiting dilution cloning of MLTC responders, we obtained 26 CD8+ T cell clones that recognized in IFN-g ELISPOT analyses both, the ClearCa-3 and the MZ1257-EBV cell line and, consequently, should be directed against the HLA-B7 mismatch or minor histocompatibility antigens. Further 18 CD8+ T cell clones were specifically reactive against ClearCa-3 cells and did not recognize MZ1257-EBV. Spot production of all T cell clones was completely inhibited by the addition of a monoclonal antibody (mAb) directed against total HLA class I molecules. Detailed blocking studies using mAbs specific for single HLA class I alleles identified HLA-A2, -A3, -Bw6 (-B7) or a different HLA-B or HLA-C allele as restriction elements of 11 ClearCa-3-specific CD8+ T cell clones. We presently focus our efforts on the identification of the naturally processed RCC antigens recognized by 3 different HLA-A*0201-restricted ClearCa-3-specific T cell clones using biochemical purification of HLA-A*0201-associated peptides from a ClearCa-3 cell lysate, and sequence analysis of candidate peptide epitopes by mass spectrometry. Our study should ultimately lead to the identification of target antigens recognized by allogenic CD8+ T lymphocytes in HLA-matched settings, such as nonmyeloablative allogenic stem-cell transplantation currently suggested as a treatment modality for refractory metastatic RCC disease.

639

Expression of functional protease-activated receptor 1 (PAR-1) in cultured human prostate cancer cell lines and modulation of cell proliferation by thrombin

J. Liu, M. Steiner, P. Schuff-Werner (Rostock, D)

The functional protease-activated receptor 1 (PAR-1) is physiologically expressed by a variety of different cells and mediates most of the cellular effects exerted by thrombin. Moreover, a growing body of evidence suggests its involvement in mediating cellular effects of thrombin to malignant cells of different origin. Circumstantial evidence predicts the expression of thrombin receptors by prostate cancer cells as calcium transients are elicited by thrombin. In addition, a bimodal effect of thrombin on cell growth has recently been demonstrated. The aim of the present study was to identify protease-activated receptor(s) expressed by cultured human prostate cancer cell lines DU145, LnCAP, and SV40-immortalized human prostate epithelial cells PNT1A. In addition, the effects of thrombin on cell proliferation were assessed. Using RT-PCR techniques, we demonstrated the expression of the prototypic protease-activated receptor PAR-1 in all cell lines investigated whereas none of the cell lines expressed the other identified thrombin receptors PAR-3 and PAR-4. Treatment of growth-arrested cells with various concentrations of thrombin demonstrated dose- and time-dependent effects. At low concentration (<0.5 U/ml), thrombin induces proliferation of all prostate-derived cell lines studied. Thrombin at higher concentration (1.0 U/ml) initially stimulates PNT1A and LnCAP cells to proliferate (time of culture 24 and 48h) but inhibits their growth when assessed after 72h of incubation. In contrast, 1.0 U/ml thrombin causes earlier inhibition of DU145 proliferation starting at 48h of incubation. To confirm that the effects of thrombin on proliferation of prostate-derived cell lines are dependent on its proteolytic activity and cleavage of PAR-1, the experiments were repeated using PPACK-pretreated thrombin thus abolishing protease activity. PPACK pretreatment totally neutralized cell growth-modulating effects of thrombin on prostate-derived cell lines. Our results suggest that PAR-1 mediates proliferation-modulating effects of thrombin on prostate cancer cells thus contributing to the molecular pathogenesis of prostate cancer. (Supported by DFG grant number GK 19-3/01)

640

Immune-reconstitution after conventional- and high-dose chemotherapy ± amifostine in patients with germ cell tumor

O. Rick, J. Beyer, S. Serke, N. Schwella, W. Siegert (Berlin, Marburg, D)

Purpose: To assess the influence of amifostine on immune-reconstitution after conventional-dose (CDCT) and high-dose chemotherapy (HDCT) followed by autologous peripheral blood progenitor cell (PBPC) rescue in patients with germ cell tumor (GCT). Patients and Methods: Fourty patients were treated with one day cycle of 175 mg/m^2 paclitaxel, 5 g/m^2 ifosfamide (TI) plus 10 µg/kg/day granulocyte-colony stimulating factor (G-CSF) to mobilize PBPC, followed by 3 cycles of CDCT with 175 mg/m^2 paclitaxel, 6 g/m^2 ifosfamide, 100 mg/m^2 cisplatin (TIP) plus 5 µg/kg/day G-CSF and one course of HDCT with 1,5 g/m^2 carboplatin, 2.4g/m^2 etoposide, 450 mg/m^2 thiotepa (CET) plus PBPC rescue and 5 µg/kg/day G-CSF. All patients were randomized to receive an absolute dose of 500 mg amifostine (group A, n= 20) on each day of chemotherapy or no amifostine (group B, n= 20). Immune status was measured prior each cycle of CDCT, prior HDCT, after hematologic engraftment from CET as well as 6 weeks and 3 month after PBPC rescue. Immune reconstitution was determined by the assessment of CD3, CD4, CD8, CD16, CD19, CD45R0, CD45RA and CD56 positive lymphocyte subpopulations and the CD4/CD8 ratio. Results: Between the two study arms no statistically significant differences could be observed concerning reconstitution of lymphocyte subpopulations. Throughout treatment with TIP lymphocyte counts and their subpopulations remained low without severe clinical complications. Delayed reconstitution of the CD4+ cell compartment after CET could be observed in both study arms but did not result in any severe or atypical infections. Conclusion: In our study, treatment with amifostine did not significantly influence the immune-reconstitution. Furthermore, low numbers of total lymphocytes during TIP and the delayed reconstitution of CD4+ cells and other lymphocyte subpopulations after HDCT followed by PBPC rescue had no clinical relevance for patients with GCT.

641

Curative external beam radiation therapy alone in patients with localized prostate cancer. A single center experience

A. Wellenhofer, R. Rohloff, F. Abedinpour, C. Nerl (München, D)

Purpose: The majority of patients with prostate cancer is treated with radical prostatectomy. External beam radiation therapy is one of the alternative therapies especially for elder and morbid patients. The results from eight years (1990-98) were analyzed for overall survival and biochemical relapse free survival (bRFS) of all patients and of subgroups (T-Stage, grading, initial Prostate specific antigen (iPSA) and age. Patients and methods: A total of 100 consecutive localized prostate cancer cases treated with external beam radiotherapy alone between 1/90 and 12/98 were reviewed. Most of the cases had available pretreatment parameters including iPSA, grading (22 x G1, 67 x G2, 9 x G3), T-Stage (15 x T1, 48 x T2, 29 x T3, 5 x T4). The median age was 70.8 years. The median iPSA-level was 10.6 ng/ml (range 1,1-126). The median radiotherapy dose was 67 Gy (4-field or rotation-technique). The median-follow up for overall survival was 43 months and for bRFS 36 month, respectively. The biochemical failure was recorded by a PSA level > 1 ng/ml using the ASTRO-consensus (3 times increasing of PSA after a nadir) and was calculated according to the method of Kaplan and Meier. Results: The 5-year overall-survival was 91%. The 5-year biochemical relapse free survival (bRFS) was 61%, and 76% for the group with favourable prognostic factors (G<2, T<2, PSA 1-<10), respectively. 5-year bRFS results for subgroups were: T1 78%, T2 55%, T3 65%, T4 40% (p = 0.014); G1 67%, G2 62%, G3 44% (p = 0.024); iPSA 1-10 ng/ml 67%, 10–20 ng/ml 78%, >20 ng/ml 34% (p = 0.37). There was no significant difference between patients < or > 70 years. 16% of the patients had side effects requiring treatment (5% RTGO III°, 1% RTGO IV°). 43% of the patients showed an erectile dysfunction after radiotherapy. Conclusion: In this retrospective analysis we found good results for overall survival. With a relative low median radiation dose the bRFS after 5 years was comparable to other studies and methods, including radical prostatectomy. Side effects were acceptable. The percutaneous external radiation therapy can be noticed as a good alternative to surgery or other methods. In the future there may be better results by high dose three-dimensional conformal radiation therapy.

642

Differential activation of caspases in sensitive and resistant testicular tumor cells after cisplatin treatment

T. Müller, H. Simon, W. Voigt, A. Grothey, H.-J. Schmoll (Halle, D)

Most anticancer agents exert their cytotoxicity by inducing apoptosis after formation of distinct drug-target complexes in tumor cells. Susceptibility to apoptosis plays a pivotal role in the sensitivity of tumor cells to chemotherapeutic treatment. Testicular tumor cells are usually prone to undergo apoptosis and this correlates closely with their exceptional sensitivity to chemotherapy. However, the underlying molecular mechanism remains obscure. We have previously shown that primary cisplatin resistance in the human testicular tumor cell line 1411HP was associated with a higher intrinsic threshold for induction of apoptosis (Müller et al., 2000; Onkologie 23, S7). A ten-fold amount of DNA-platin adducts, which can only be achieved in vitro, was nessesary to initiate the apoptotic program, which then proceeded with the same characteristics as in the sensitive testicular tumor cell line H12.1. Our in vivo studies in nude mice using cisplatin treatment in clinically relevant dosage demonstrated that regression of H12.1 derived tumors was associated with strong occurrence of apoptosis. In 1411HP-tumors, which responded only with growth retardation, apoptosis was observed to a similar extent as in control tumors. In the current in vitro study, we analysed the cleavage of caspases 2, 3, 7, 8, 9 and 10 in sensitive H12.1- and resistant 1411HP-cells after cisplatin treatment of equitoxic doses. Western blot analysis revealed cleavage of all tested caspases in sensitive and resistant cells during apoptosis. Caspases 2, 8, 9 and 10 may theoretical act as initiator-caspases as they contain large prodomains. Using substrate cleavage assays, the activity of these caspases were measured in treated but still adherent growing cells, which might represent the cell fraction in an early apoptotic stage. Interestingly, in sensitive H12.1 all tested caspases were active whereas in resistant 1411HP only activity for caspase 2 was observed. This observation correlated with results of a cytotoxicity assay (SRB) which showed that cisplatin treatment was affected by inhibitors for caspase 2, 8 and 9 in H12.1 but solely by caspase 2 inhibitor in 1411HP. These results demonstrate that differential caspase activation exists during initiation of apoptosis in our sensitive and resistant testicular tumor cells. Lack of early activation of possible initiator-caspases 8, 9 and 10 cannot block apoptosis but is associated with a higher threshold to commit 1411HP-cells to the apoptotic cascade. Possibly, sensitive H12.1-cells are able to initiate apoptotic cascade by more than one way which could be associated with their low apoptotic threshold. However, further investigations are nessesary to elucidate the exact sequence of caspase activation in our sensitive and resistant testicular tumor cell lines.

643

Mammaglobin expression in gynecological malignancies and prostate cancer detected by nested reverse transcription PCR

K. Grünewald, M. Haun, E. Müller-Holzner, M. Fiegl, C. Marth, G. Gastl, H. Feichtinger (Innsbruck, Wien, A)

Various molecular markers have been used for the detection of circulating tumor cells in blood by reverse transcriptase-polymerase chain reaction (RT-PCR). RT-PCR for human mammaglobin (hMAM) was recently shown to provide a sensitive method for assessing micrometastatic breast cancer. In contrast to other molecular markers such as epidermal growth factor receptor and cytokeratin 19, hMAM mRNA expression showed a correlation with clinical parameters such as nodal status, metastasis, and CA 15-3 serum levels, but was not detectable in peripheral blood cells from healthy volunteers. Although hMAM expression was described to be confined to the mammary gland we evaluated the specificity of this assay by analyzing hMAM expression with nested RT-PCR in 152 normal and malignant epithelial tissues and in 33 mesenchymal tissues. hMAM expression was not restricted to the mammary gland and breast carcinoma, but also detectable in most specimens of benign and malignant epithelial tissue from the ovary (69/72) and uterus (39/39). Notably, hMAM expression was also found in tissues from the normal uterine cervix (10/11) and benign prostatic hyperplasia (5/11) as well as in squamous cell carcinoma of the cervix uteri (9/10) and in prostate cancer (5/9). A significantly lower expression rate was found in normal mesenchymal tissues (4/33). Confirming our previous data, hMAM expression was absent in all of the 124 control samples of peripheral blood and bone marrow from healthy volunteers and patients with hematological malignancies. To verify these data, the size-specific PCR products were sequenced. All of them showed complete alignment with the published hMAM sequence. In addition hMAM amplification products were blotted and hybridized under stringent conditions with a specific probe for hMAM. All Southern blots showed the expected specific hybridization signal. Non-specific amplification of target cDNAs other than hMAM (members of the uteroglobin family, mammaglobin B) therefore can be excluded. Thus, hMAM expression in peripheral blood or bone marrow samples from patients with epithelial cancer is strongly suggestive of circulating tumor cells or micrometastatic disease. In conclusion, hMAM expression assayed by a sensitive nested RT-PCR provides a useful molecular marker for detecting epithelial cancer cells originating from the mammary gland, the ovary and the uterus. This might become most relevant for the diagnosis of micrometastatic disease. Prospective studies have been launched at our Institution to evaluate the predictive value of hMAM expression in peripheral blood and bone marrow from patients with breast carcinoma and other gynecological malignancies.

644

The role of primary apoptosis in the classification of disseminating renal cell carcinoma

J. Atzpodien, E. Richter, K. Oevermann, N. Büntig, S. Sel, S. Stöckel, T. Wandert, J. Grosse, I. Dallmann (Hannover, Bonn, D)

The purpose of the present study was the identification of novel biologic markers of prognosis in primary histopathologic specimens of patients with disseminating renal cell carcinoma. Therefore, apoptotic indices (TUNEL method), proliferation rates (Ki-67 antigen), p21 (WAF1/cip1) expression, and CD95 (APO-1/Fas) expression were determined in paraffin embedded nephrectomy specimens from 73 patients with histologically confirmed progressive metastatic disease. Kaplan-Meier survival analyses, log-rank statistics, and two-proportional Cox regression analyses were employed to identify new risk factors in addition to conventional classification criteria and to demonstrate statistical independence. Multivariate analysis indicated that primary tumor apoptosis is a novel independent predictor in patients with disseminating renal cell carcinoma. This predictor alone or in conjunction with the time interval from diagnosis to metastatic disease showed high predictive impact on overall survival from first diagnosis. Our findings lead to a new classification of disseminating renal cell carcinoma upon first diagnosis.

645

Therapy and outcome of patients with testicular germ cell tumors (GCT): A single center experience

M. Hentrich, C. Göbel, H. Pohlmann, N. Brack, L. Lutz, T. Bergsdorf, H. Schorer, R. Hartenstein (München, D)

The records of all patients (pts) with testicular GCT diagnosed and/or treated at our institution between 1976 and December 1997 were reviewed with regard to tumor histology, staging, treatment, and outcome. In case of incomplete follow-up a questionnaire was send if adresses were available. After completing primary therapy 13 out of 242 pts with GCT (89 seminoma, 153 nonseminoma) were lost to follow-up. The median follow-up for the remaining 229 pts was 67 months (range 4-232). 3 pts evidenced a simultaneous, 4 pts (1,8%) a metachronous bilateral tumor occuring after an interval of 26 to 185 months. 5 pts. suffered primary extragonadal GCT. 66 of 89 (74%) pts with seminoma (S) presented with stage I disease. The relapse rate after adjuvant radiotherapy was 9% and all pts could be successfully salvaged. After radiotherapy all stage IIa S remained in CR but 2 pts with stage IIb S relapsed. One of them died of neutropenic sepsis after salvage chemotherapy (CT) representing the only case of S associated death in this series. Elevation of ß-HCG occured in 25% of S and was associated with an increased risk of relapse. All advanced stage S (IIc–IV, n=16) were „good prognosis" according to IGCCCG. The 5-year survival rate of all pts. with seminomatous GCT was 99%. Primary retroperitoneal lymphadenectomy (rLA), performed in 83% of stage I nonseminoma (NS), resulted in 8% relapses. By contrast 4 of 7 stage I NS managed by surveillance relapsed within an interval of 2 years. Persistent AFP elevation (stage I M) was a negative prognostic factor (3 of 4 NS developed progessive disease resistant to CT). The 10-year survival rate of 41 pts with stage IIa/b NS was 100%. 36/41 of them received primary rLA with (n=27) or without (n=9) adjuvant CT. The majority of advanced stage NS (n=48) received inductive platinum based CT. The 5-year survival rate for pts with „good", „intermediate" and „poor prognosis" according to IGCCCG was 92%, 80% and 48%, respectively. Secondary surgery was performed in 27 of 31 pts with PRm-. Results were mature teratoma in 11, necrosis/fibrosis in 11 and vital tumor in 5 cases, respectively. 24 NS (all stages) relapsed, 21 of them received salvage CT and 3 salvage surgery. The 5-year survival rate of all NS in 1st relapse was 74%. Only 2 of 7 pts with prior inductive CT remained in durable CR after salvage CT ± surgery. 12/21 NS and 2/3 S developed a 2nd relapse frequently localized in the CNS. 3 of them died of progressive disease, 3 of neutropenic sepsis and 3 of other causes (5-year survival rate 36%). 11 of 188 pts (6%) suffered late relapses, 7 of them after prior CT. At a median follow-up of 27+ months (range 6-89) 5 of 11 pts are currently alive without disease. All in all 2 secondary malignancies (1 AML, 1 pancreatic carcinoma) were observed. In conclusion our results compare favourable with data reported in the literature.

646

Induction of a bladder cancer-specific T-cell response by mRNA-transfected dendritic cells

W. Schmitt, M. Stassar, M. Little, B. Cochlovius (Heidelberg, Ladenburg, D)

To design a tumor-specific immunotherapeutic strategy for treating tumors for which no specific antigens are described (such as bladder urothelial carcinoma), we attempted to activate tumor-specific T-cells by dendritic cells transfected with tumor-derived mRNA. Dendritic cells were generated from a patient's peripheral blood and loaded with mRNA derived from the urothelial carcinoma tissue of the same patient. Autologous T-cells were incubated twice on these dendritic cells and tested for their ability to lyse tumor cells. RESULTS: Dendritic cells transfected with tumor-derived mRNA were able to activate T-cells that recognized autologous tumor cells. Cytotoxicity was around 26% for an effector:target ratio of 50:1. Tumor-infiltrating lymphocytes did not kill the autologous tumor cells in vitro, but after a single stimulation with the transfected dendritic cells, they induced tumor cell lysis of 35.7% at an effector:target ratio of 50:1. These results indicate that dendritic cells transfected with tumor mRNA containing messages for one or more tumor antigens could serve for the ex vivo activation of effector T-cells or directly as vaccines for a wide range of human neoplasias.

647

Prostate cancer presenting as leukocytoclastic vasculitis

L. Floro, A. Kiss, K. Horvat-Karajz, L. Sreter, G. Muzes, Z. Tulassay (Budapest, H)

A 85-year-old man was admitted to our department. He complained of fatigue, weakness and low back (bone) pain. Symmetrical papules were found on his neck, abdomen and extremities. The laboratory examinations showed mild normocytic anemia and high erythrocyte sedimentation rate. Biopsy of his skin lesions revealed leukocytoclastic vasculitis There was no sign of extracutaneous manifestations. We couldn't find any infection and he hasn't taken any known vasculitis causing drug. However, the bone scan showed multiplex foci with increased activity in the lumbal region and the PSA level was elevated. The prostate biopsy confirmed prostate cancer so the diagnosis was prostate cancer with multiplex bone metastases, and paraneoplastic cutaneous leukocytoclastic vasculitis. Vasculitits is characterised by inflammatory changes and necrosis of the vessels. There are primary and secondary types of vasculitis. The secondary type has been linked to antigens, for example bacterial antigens, drugs, tumors etc. According to Sanchez-Guerrero et al., cancer is found in about 5% of patients with vasculitis. The majority (about 90%) of patients with malignant neoplasm-associated vasculitis who have been described had hematologic neoplasms. However, a lesser fraction of the patients have solid tumors, most frequently lung cancer. Leukocytoclastic vasculitis is an immuncomplex vasculitis of the small vessels, usually veins. This type of vasculitis is the first report of non-hemorrhagic type leukocytoclastic vasculitis associated with prostate cancer. Altough this type of vasculitis is a very rare paraneoplastic sign of solid tumors, we should never miss such an underlying disease.

648

Systemic mast cell disease in relapse of a primary mediastinal germ cell tumor

T. Südhoff, C. Petz, C. Teschendorf, P. Valent, W. Schmiegel, U. Graeven (Bochum, D; Wien, A)

In 1/00 a 37-year-old male patients with primary mediastinal nonseminomatous germ cell tumor underwent sequential PBSCT following 3 courses of HD-VIP (cisplatin, iphosphamide, etoposide). After normalization of b-HCG and AFP a residual mediastinal mass was resected showing no vital tumor tissue. On follow-up investigation in 11/00 a relapse of the mediastinal tumor and multiple hypodense hepatic and splenic lesions were diagnosed. Although b-HCG and AFP were negative, mediastinal relapse of germ cell tumor was histologically confirmed by EUS guided fine-needle biopsy. Because of severe thrombocytopenia a bone marrow biopsy was performed which showed a mast cell infiltration with focally varying degree (10-80%). Typical clinical symptoms of systemic mastocytosis were not present. Radiation of the mediastinum was initiated due to hypoventilation of the left lung. Since there is no standard therapy for aggressive mast cell disease treatment with steroids followed by interferon alfa-2b (1 MU/d) was started but did not translate into reduced bone marrow infiltration or improved platelet counts. The clinical course was increasingly complicated by relevant bleeding events. Therefore, splenectomy was suggested as ultima ratio. Within 24 h after splenectomy the patient died due to severe GI bleeding. Autopsy revealed multiple infiltrations of the small intestine by the germ cell tumor as source of the bleeding. The coincidence of hematologic disorders and primary mediastinal nonseminomatous germ cell tumors has been described, including two cases linked to mast cell disease. However, it is unclear why some patients with mediastinal germ cell tumor develop hematologic disorders. Cytogenetic instability has been suggested to represent a possible factor and high incidence of i(12p) have been found in these patients. Cytogenetic investigation disclosed a regular karyotype in our patient. Mast cell proliferation tiggered by mutation of the c-kit receptor which has been found in patients with systemic mast cell disease was thought to offer an alternative explanation. Thus, bone marrow samples were analyzed for point mutation of c-kit (Val-816 for Asp), but was found to be negative in our patient. Although, the molecular basis of mast cell disorder complicating germ cell tumor could not be disclosed, the case presented is typical in terms of the short interval between diagnosis of germ cell tumor and hematologic disorder and underlines that hematological disorders associated with germ cell tumor generally show a aggressive clinical course.

649

Metastatic seminoma mimicking primary pancreatic cancer

M. Wehrschuetz, H. Stoeger, F. Ploner, P. Krippl, G. Hofmann, H. Samonigg (Graz, A)

A case of seminoma clinical stage III C arising from the right testis pretending a pancreatic malignancy is reported. A 57 year old male patient presented with obstructive jaundice. He was complaining about recurrent abdominal pain for more than four months and significant weight loss. Computertomographical findings showed a tumor in the head of the pancreas, thrombosis of the vena mesenterica superior due to tumor masses and multipe pathologically enlarged lymphnodes. In the lab findings there were signs of cholestasis and an elevation of the tumor marker Ca 19-9. Endoscopically a stent in the ductus choledochus was implanted and during a laparascopic surgery a biopsy of the head of the pancreas and of a lymphnode was done. Histopathological examination unexpectedly was reported as a seminoma. Laboratory findings showed an elevation of the human placental alkaline phosphatase and LDH in normal range. The patient underwent semicastration of the right suspected testis; subsequent histopathological examination confirmed a seminoma. He received four cycles of standard chemotherapy with platinum, etoposide and bleomycine (PEB) and achieved complete response that is still ongoing. To our best knowledge there is no case reported in literature of germ cell carcinoma spreading to the pancreas. This case provides an example of the possibility of uncommon seminoma metastases and underlines the importance of exact pretherapeutical radiological and histopathological examination.

650

Successful treatment of intracranial germinoma with intracerebral metastases, meningeosis carcinomatosa and bone lesions with primary methotrexate chemotherapy

T. Fietz, W.U. Knauf, G. Stoltenburg, C. Baldus, I.W. Blau, E. Thiel (Berlin, D)

Primary CNS germinoma is a rare diseases that tends to occur primarily at two sites: the pineal gland and the hypothalamic region. Spinal seeding is uncommon, and distant metastases outside the central nervous system hardly ever occur. Due to excellent radiosensitivity localized CNS germinomas are routinely treated with radiotherapy to the primary tumor bed in curative intention. However, mental and pituitary hormonal dysfunctions are major drawbacks of radiotherapy, and especially for patients with spinal seeding or extracranial disease radiotherapy treatment is not sufficient. We report on a 27-year-old female with a history of grand mal seizures since the age of 12; valproic acid has been used successfully as prophylactic regimen over the last 6 years. The patient now presented with vertigo, headache, emesis and progressive ataxia. A spinal tap showed 752/3 cells with a total protein of 1327 mg/L. On MRI scan a prominent tumor in the pineal gland with suprasellar and cerebellar metastases could be localized. A thorough analysis of the tumor cells led to the diagnosis of a germinoma; alpha-fetoprotein and beta-HCG blood and cerebrospinal fluid levels were not elevated. Further evaluation revealed bone metastases to the spinal column (vertebra 2 and 3). The patient was initially treated with intrathecal methotrexate (10mg) for 3 times with a rapid improvement in CSF cell count (33/3 cells) and amelioration of nausea, emesis and headache. We then started a systemic chemotherapy with high-dose methotrexate 4g/m^2 for 4 hours with leucovorin rescue after 24 hrs. The therapy was well tolerated. However, due to reversible hepatic toxicity the methotrexate dose was reduced to 4g total for treatment continuation. After 3 cycles of chemotherapy the patient was in total remission with a Karnofsky index of 100% and two further consolidation cycles were administered. The patient is now 10 months out of therapy with no evidence of disease and normal hormonal functions. This is the first report to show that methotrexate is a potent drug in treating pineal gland germinoma. Long-term side-effects of radiotherapy like reduced mental function and hypopituitarism can probably be avoided. Especially in a more advanced tumor stage with spinal seeding and extracranial disease, single-agent high-dose methotrexate may provide high efficacy with a limited number of adverse-effects.

651

A 55 year old patient with dyspnea and a history of left kidney carcinoid tumor

Th. Heinicke, C. Ziske, P. Albers, N. Morakkabati, R. Kleinschmidt, T. Sauerbruch, I. Schmidt-Wolf (Bonn, D)

History: We report of a 55 year old patient who was admitted to our hospital in December 1999 with dyspnea and suspected pulmonary embolism. A perfusion and ventilation scintiphotography showed perfusion defects in the left lung. In 1991, a carcinoid tumor (pT3aN0M0G2) of the left kidney was resected, but no nephrectomy was performed at that time because of agenesis of the right kidney. A single hepatic metastasis was resected in 1996. Recurrent multiple hepatic metastases were diagnosed in 1997 but no treatment was initiated at that time. Metastatic bone disease was found in late 1998 and bisphosphonate therapy was initiated. In December 1999, a routine octreotide scintigraphy revealed a highly intensive epigastric lesion and further diagnostic tests were recommended. Additional diagnoses at the time of admission were atrial fibrillation, arterial hypertension, mild chronic renal failure, mild normocytic anemia and suspected coronary heart disease. Findings: On physical examination the heart rate was between 60 and 80 beats per minute, the blood pressure was normal, no dyspnea. Respiration rate was normal with no evidence of hypoxemia. CT scan of the central pulmonary arteries revealed no pulmonary embolism. However, an obstructing embolus was seen in the vena cava inferior extending from the hilus of the left kidney up to the right cardiac atrium. This finding was confirmed by echocardiography which showed no signs of elevated pulmonary arterial pressure. Clinical course: Efficient anticoagulation was initiated using intravenous unfractionated heparin. However, recurrent episodes of acute dyspnea persisted. After extensive discussion informed consent was obtained and laparotomy and thoracotomy were performed. The left kidney had to be resected and the tumor was removed from the vena cava inferior as well as from the right atrium. At the end of the procedure the patient was free of tumor. However, disseminated intravascular coagulopathy developed which could not be managed and the patient died the following day. Histologic examination revealed a trabecular as well as rosette like growing malignant tumor which strongly stained positive with a monoclonal antibody directed against chromogranine A. Diagnosis: Recurrent carcinoid of the left kidney. Conclusion: A rare case of recurrent carcinoid of the left kidney was diagnosed nine years after resection of the primary tumor. Most likely, because the initial surgical procedure could not be performed in a radical way, recurrent disease occurred. Continuing growth in the venous system up to the right atrium lead to suspected recurrent pulmonary embolism. Using octreotide scintigraphy combined with CT-scan and MRI the diagnosis of recurrent carcinoid tumor had already been suggested prior to surgery.

Kompetenznetz „Pädiatrische Onkologie und Hämatologie"

656

The Medical Research Network Pediatric Oncology and Hematology

G. Henze, U. Creutzig, R. Herold (Berlin, D)

Pediatric Oncology in Germany: About 95% of children with cancer are treated uniformly according to protocols of the Gesellschaft für Pädiatrische Onkologie und Hämatologie (GPOH). The structure of the GPOH includes the Children's Cancer Registry, the Children's Tumor Registry, the Cooperative Pediatric Stem Cell Transplant Registry, reference laboratories for immunophenotyping, cytogenetics and molecular genetics, the central business office of the GPOH, as well as more than twenty trial offices where children with different malignant diseases are registered. Treatment protocols are designed by pediatric oncologists according to current scientific knowledge and distributed to about 100 participating institutions. About 65% of the children are treated in 25 centers (at least 25 new patients per year). Hence, a structured patient care organized in multicentric trials has been established, and these trials are now instruments of high level quality control as well as clinical research. These institutions have cooperated over the past 30 years in order to develop the mentioned structures, to coordinate research, and thereby to improve the outcome of the affected children. Cure rates for children with cancer are approaching 75%. Aims of the Network: In face of the high cure rates, the complexity and requirements of modern protocols in terms of diagnostic and therapeutic measures, the necessity to collect data on delivered doses and side-effects of therapy, and the fact that about 22,000 survivors of childhood cancer are living in Germany, it deemed necessary to establish an improved, intensified, and structured networking of all these institutions. Therefore, the Medical Research Network Pediatric Oncology and Hematology has been created in late 1999. Specific questions and problems are addressed in ten multicentric projects. Results: A coordination and management group is working as a service unit for all partners involved in research and treatment. The new profession of a research and study assistant has been created and an educational program started in order to support the large treatment centers. A computer based application system for documentation and therapy planning has been developed and is being tested. Solutions for problems of data protection and security have been worked out. Pilot set-ups making use of telemedicine for trial specific purposes are being developed. Three tumor banks with a central tumor data base were established. A national registry of projects in immuno- and gene therapy has been established and treatment modalities were documented. A data base has been compiled for a nationwide analysis on the frequency and predisposing factors for the occurrence of 2nd malignant neoplasms. Standardized follow-up recommendations to measure somatic and neuropsychological long-term effects together with quality of life have been formulated and were integrated as amendments into treatment protocols for children with leukemias and embryonal tumors. Studies on minimal residual disease have been incorporated into clinical trials. Conclusions: The network has initiated numerous cooperations between pediatric and adult institutions. This is to be continued and intensified during the second phase of funding.

657

Telemedicine communication in the GPOH Pediatric Oncology Competence Net: Evaluation of expectations and infrastructure

M. Paulussen, N. Graf, T. Huf, T. Ganslandt, G. Henze, H. Jürgens (Münster, Homburg, Berlin, D)

Background: The Competence Net Pediatric Oncology of the German/Austrian Gesellschaft für Pädiatrische Onkologie und Hämatologie (GPOH) with support from the German Bundesministerium für Bildung und Forschung (BMBF) aims to promote research and to further increase standards of care in pediatric oncology/hematology by enhancing communication standards. Implementation of a tumor bank, standardized patient data acquisition and electronic exchange, an internet based knowledge-server, and telemedicine communication are integral parts of this network project. Methods: The telemedicine project has started by sending out a questionnaire to the 54 largest pediatric hematology/oncology units in Germany. Institutions were asked for their experience, motivation, existing infrastructure, and anticipated benefits and obstacles regarding the implementation of telemedicine in patient care and research. Results: Of the 54 largest German Pediatric Oncology institutions asked, 46 completed the questionnaire. The majority expected practical advantages from telemedicine communication, especially in teleradiological expert consultation but also in telemicroscopy. It was anticipated that telemedicine will increase quality and likely reduce costs in medical care within the next years. Data security and standardization, transfer speed and transmission quality were regarded major issues. There were six institutions (14%) with established telemedicine equipment. Thirty-seven of 46 institutions were prepared to invest in the implementation of telemedicine; the lack of available medical informatics competence and manpower was regarded as the most important obstacle. The need for further detailed information on implementation and for help in technical realization of telemedicine applications was expressed by all participants. Discussion and Outlook: This analysis demonstrates that the use of telemedicine is expected to become standard in pediatric oncology, while the existing infrastructure and status of information regarding this subject at present are insufficient. The most pressing practical need for telemedicine applications is seen in the field of electronic expert consultation. Hence, the Telemedicine-Project of the GPOH Competence Net will focus on this aspect first. A survey of commercially available teleradiology applications was completed recently, and the evaluation of some of the most promising telemedicine applications for use in the GPOH network has been started between several GPOH institutions.

658

Use of an internet based knowledge server for the Medical Research Network Pediatric Oncology and Hematology: http://www.kompetenznetz-paed-onkologie.de/

R. Herold, U. Creutzig, K. Pommerening (Berlin, Mainz, D)

Aims: A long-term working and centrally positioned Internet based Knowledge server was to be newly established for contents related to pediatric oncology and hematology and as a project of the respective BMBF funded, GPOH initiated Medical Research Network. Strategies for content manage-

ment, publication techniques, and long term consolidation were to be developed. Efficacy with respect to various groups of interest and thus to vertical networking was to be evaluated.
Methods: A standard server type computer was set up in mid 2000. Unix server software includes OpenSSL for certificate based data security and integrity, Cocoon for XML-based dynamic page rendering, and data bases for various content types. Pages templates were designed and propagated to participating projects, including navigational elements, upload possibilities, and a frameless scriptless browser-independent HTML syntax. The editorial board (presently, the network's Coordination and Management group) produced relevant content, either from own investigations or from external contributions, and reviewed page statistics. Server usage has been promoted on scientific meetings, during teachings, and by producing a regular paper newsletter which comprises similar contents and may be replaced by the Internet services. Concomitantly, various email distribution lists for special interest groups have been established. Results: Initially, our independent survey disclosed that most of 54 pediatric oncology centers expect the Knowledge Server to reduce searching time, information redundancy and ambiguity. From its start on, the Knowledge server covered topics such as time schedules, organisational documents, educational and general information. Now, contents such as quality standards, guidelines, and relevant literature in the field of pediatric oncology and hematology are available. Security rules were implemented in form of authentication requirements for downloading trial protocols, some data bases, and documents e.g. related to quality control. Daily, about five hundred page hits with about 12 MB data transfer are registered. Announcements and the start page for therapy studies were viewed most often.
Conclusions: Since one year, a working web server was established that already offers a considerable body of information. Daily access statistics indicate a general interest into this Knowledge server. The Knowledge server shall become the central new medium for the emerging vertical networking in pediatric oncology and hematology and should be useful in daily medical practice. Further plans include to convey access to the Cochrane library and to provide substantially more information for patients and relatives.

659

Minimal residual disease in childhood leukemia and lymphoma

J. Harbott (Giessen, D)

During the last years it could be shown by two large studies (Cave et al (1998) N.Engl.J.Med. 339,591; van Dongen et al (1998) Lancet 352,1731) that MRD is an important independent prognostic factor and can be used for therapy stratification in childhood ALL. However, for other types of malignant diseases the clinical relevance of MRD is not yet clear. Therefore, a multicenter project was created under the umbrella of the German competence network „Pediatric Oncology“ to analyze MRD in childhood ALL, ALL-relapse, AML and lymphoma with different techniques. The aims of this project are the comparison of results gained by different methods to find the most suitable technique for each type of malignancy. Furthermore the best time points for therapy control should be evaluated.
Different MRD markers are analyzed in 8 laboratories in Germany: leukemia specific fusion genes in childhood AML (Harbott, Gießen), ALL relapses (Seeger, Berlin) and lymphomas (Reiter, Gießen), leukemia associated immunophenotypes for AML (Griesinger, Göttingen; Reinhardt, Münster), IgH and TCR gene rearrangements for ALL (Schrappe, Hannover; Bartram, Heidelberg for the ALL-BFM therapy trial and Kabisch, Hamburg for the CoALL study group) and ALL relapses (Seeger, Berlin). All results will be compared with clinical data provided by the therapy studies ALL-BFM (Schrappe, Hannover), AML-BFM (Creutzig/Ritter, Münster), ALL-REZ (Henze, Berlin), CoALL (Janka-Schaub, Hamburg), and NHL-BFM (Reiter, Gießen). Although a total of more than 12,000 samples were examined during the past 18 months, no final results can be presented due to the long follow-up of MRD studies. However, some examples may elucidate the preliminary results: Analysis of IgH/TCR and fusion genes of ALL relapse patients showed that the level of MRD at distinct time points might be important for this group as well. Comparing MRD results of therapy protocols ALL-BFM and CoALL based on the IgH/TCR marker the influence of medication became obvious. Whereas only 9% of the BFM patients remained positive after induction (larger than 10-2), 26 percent of the children treated by CoALL had still this high number of leukemic cells. This might be due to asparaginase, given late during consolidation. Fusion transcripts of AML1/ETO persisted in all children after induction, but nearly half of them became negative after consolidation, whereas most of the adults remain positive during treatment. In contrary, the MLL/AF9 transcripts disappeared during the first month of treatment.
All results will be compared to clinical data and finally this project may lead to a more successful therapy and an effective network including medical centers, laboratories, and therapy trials.

660

Functional genomics and immune and gene therapy in pediatric oncology

S. Burdach, G. Hansen, U. Hattenhorst (Halle, D)

Marshalling the immune response to reject human tumors is a daunting challenge. In the past decade, however, the convergence of information from basic studies of cellular and molecular immunology, the molecular identification of tumor antigens, their immunodominant epitopes, and the T cell receptors that recognize them have placed studies of tumor immunology and immunotherpay in the mainstream of immunologcal research. The induction or redirection of T cell-mediated immune responses followed by a protective immunity in malignancies characterized by poor immunogenicity is a most challenging goal for adjuvant immunotherapy. One possible approach in this regard is the tumor-specific delivery of immunomodulators, which achieve effective concentrations in the tumor microenvironment and effectively stimulate cellular immune responses against syngeneic malignancies. Gene delivery systems facilitate the continuous expression of these immunomodulators within or around the tumor and are therefore a very effective method of achieving high intra- or peritumoral concentrations of the immunomodulators while -at the same time- reducing problems like severe side effects due to high systemic concentrations. The efficacy of such approach was confirmed recently by our own group, demonstrating that vaccination with irradiated IL-2-transfected Ewing tumor cells significantly reduced Ewing tumor growth in NOD SCID mice. This antitumor effect was mediated by activated T lymphocytes and natural killer cells. The identification of tumor-specific or tumor-associated proteins is crucial for the development of novel therapeutic strategies in treatment of cancer. Applying high-density DNA-microarrays, we investigated the gene expression patterns in different tumor species profiling them on Affymetrix custom high-density arrays which display approximately 40,000 full length genes and EST's on one single array. Multivariant analysis of all samples and data points allowed uninformed queries of genes upregulated or downregulated. Bone marrow samples from patients with pediatric common acute lymphoblastic leukemia (cALL) were compared to sorted B-cells expressing CD19 and CD10 and to whole bone marrow of healthy donors. Moreover, the gene expression profile of adeno and squamous cell carcinoma samples was compared to normal lung tissues and –so far- twelve tumor-specific genes and their product suitable for therapeutic approaches could be identified. These tumor-specific genes and their proteins are potential targets for therapeutic antibodies delivering drugs selectively to the tumor cells. Moreover, peptides, which are shown to bind to HLA-molecules and which are derived from these target proteins, can be produced synthetically and presented to the patient's immune system by potent antigen-presenting cells like dendritic cells.
A new era of cancer immunotherapy has arrived based on the molecular identification of human cancer antigens.

Wissenschaftliches Symposium: Angiogenese

661

Pathological angiogenesis: Molecular aspects and therapeutic approaches

D. Marmé (Freiburg, D)

Growth and metastasis of solid tumors depend on the formation of new blood vessels which originate from the existing vascular system. These blood vessels grow into the tumor and thus provide the necessary nutrients and growth factors for tumor progression. At the same time, the newly formed blood vessels allow tumor cells to disseminate and form metastases in distant organs. Normally, vascular homeostasis is regulated by a balance of angiogenic and anti-angiogenic mechanisms. Tumor-induced angiogenesis is mainly sustained by the production and secretion of angiogenic factors originating from tumor and stroma cells. The most prominent angiogenic factor is the vascular endothelial growth factor VEGF. Recently, additional angiogenic factors and their respective receptors have been identified and related to tumor angiogenesis. Among these, the angiopoietins and their receptor TIE-2 have been investigated to some detail. Angiopoietin-1, which binds to and activates TIE-2, is obviously responsible for the stabilization of vessels under homeostatic conditions. Angiopoietin-2 binds to the same receptor as angiopoietin-1 but is antagonistic with respect to angiopoietin-1. It destabilizes blood vessels and under appropriate conditions induces complete regression. In the similar situation angiopoietin-2 induces the destabilization of blood vessels, and the angiogenic factor VEGF produced by the tumor induces the massive formation of new vessels. When human melanoma cells A375 are stably transfected to produce the soluble variant of the angiopoietin receptor TIE-2 (sTIE-2) they show a substantial inhibition of tumor growth on nude mice. Similar effects have been seen with the soluble variant of the VEGF-receptor FLT-1 (sFLT-1). In both cases, the vessel density of the tumors is significanty reduced. These experiments show that the inhibition of the angiopoietin/TIE-2-system similar to the inhibition of the VEGF/VEGF-receptor-system has an anti-tumoral effect, most probably due to the inhibition of tumor angiogenesis. Thus, inhibition of both signalling systems seem to be a valid strategy for the development of novel anti-angiogenic therapies. Recently, the inhibition of the VEGF-receptor tyrosine kinase by the compound PTK787/ZK222584 has been shown to substantially inhibit tumor growth and metastases formation, this compound has now entered clinical trials at the Tumor Biology Center in Freiburg. A preliminary evaluation of phase I shows a very promising clinical outcome.

662

Angiogenesis in hematologic malignancies

R. Mesters (Münster, D)

The importance of angiogenesis for the progressive growth and viability of solid tumors is well established. Emerging data suggest an involvement of angiogenesis in the pathophysiology of hematologic malignancies, as well. Recently, we have reported increased angiogenesis in the bone marrow of patients with acute myeloid leukemia (AML) and normali-sation of bone marrow microvessel density when patients achieved a complete remission (CR) after induction chemo-therapy (Padró et al. Blood 2000). Tumor angiogenesis depends on the expression of specific mediators that initiate a cascade of events leading to the formation of new microvessels. Among these, VEGF (vascular endothelial growth factor), FGF (fibroblast growth factor) and angiopoietins play a pivotal role in the induction of neovascularisation in solid tumors. These cytokines stimulate migration and proliferation of endothelial cells and induce angiogenesis in vivo. Recent data suggest an important role for these mediators in hematologic malignancies, as well. Isolated AML blasts overexpress VEGF and VEGF-receptor 2. Thus, the VEGF/VEGF-R2 pathway can promote the growth of leukemic blasts in an autocrine and paracrine manner. Therefore, neovascularization may be a promising target for anti-angiogenic and anti-leukemic treatment strategies. Thalidomide inhibits angiogenesis in animal models. Moreover, it has significant activity in refractory multiple myeloma. Therefore, we evaluated the safety and efficacy of thalidomide in patients with AML not qualifying for intensive cytotoxic chemotherapy. Twenty patients aged 58 – 85 years (median 69 years) were recruited to this phase I/II-study and were treated with a dose of 200 – 400 mg per os per day for a dura-tion of 1 – 40 weeks (median 6 weeks) dependent on the individual tolerability of the drug. In 4 patients we observed a partial response (PR – defined as more than 50 % reduction in leukemic blast infiltration in the bone marrow). This was accompanied by an increase in platelet counts and hemoglobin values. An additional patient showed a significant im-provement of peripheral blood counts without fulfilling the criteria of a PR. In parallel, we observed a significant de-crease in microvessel density in these five patients during treatment with thalidomide. In conclusion, thalidomide seems to have anti-angiogenic as well as anti-leukemic activity in AML. Since the VEGF/VEGF-R2 pathway seems to play an important role in AML, receptor tyrosine kinase inhibitors like SU5416 or SU6668 are currently evaluated in the con-text of phase II-studies in AML. Recently, we were able to induce a stable remission in a patient with second relapse of her AML refractory towards chemotherapy by administration of SU5416 (compassionate use), a tyrosine kinase inhibi-tor of VEGF-R2 and c-kit (Mesters et al. Blood 2001). Current and future studies will clarify the role of anti-angiogenic treatment strategies in AML and other hematologic malignancies.

663

Applications of angiogenesis research

M. O'Reilly (Houston, USA)

Specific angiogenesis inhibitors including angiostatin, endostatin, and antiangiogenic antithrombin. Angiostatin is an internal fragment of plasminogen and endostatin is a C-terminal fragment of collagen XVIII. A change in conformation of antithrombin after cleavage of the molecule's reactive C-terminal loop confers antiangiogenic activity. The discovery of angiostatin and antiangiogenic antithrombin provide further evidence that the clotting and fibrinolytic pathways are directly involved in the regulation of angiogenesis. In vivo, systemic therapy with these agents induces a virtual complete blockade of angiogenesis and potently inhibits tumor growth and no evidence of resistance to therapy has been demonstrated even after prolonged administration. Prolonged therapy with high doses of angiostatin, endostatin, or antiangiogenic antithrombin leads to regression of established tumors and can induce tumor dormancy. The tumor dormancy is defined by a high rate of tumor cell proliferation balanced by apoptosis and a virtual complete blockade of angiogenesis. Further, the antiangiogenic and antitumor activities of human recombinant angiostatin and endostatin are strongly synergistic in vivo. No toxicity has been observed in any mice treated even after prolonged therapy at high dose. When these antiangiogenic agents are combined with cytotoxic or radiation therapy improved anti-tumor efficacy and diminished toxicity is observed. To determine if angiogenesis suppression would induce drug resistance, mice with several types of cancer were treated with cycled endostatin therapy. Mice were treated for several months without any evidence of resistance to therapy or toxicity. After several cycles of therapy, tumor dormancy of lung and prostate carcinomas, fibrosarcomas, melanomas, and leukemia persisted indefinitely off therapy. Prolonged continuous therapy with high dose endostatin also induced self-sustained tumor dormancy. In a model of spontaneous dormancy of metastatic disease, mice implanted with melanoma were apparently cured by surgical resection of a primary tumor. However, histological analysis of their lungs revealed persistent microscopic metastases that remained dormant for the normal life span of the mice. However, the dormant tumors can be reactivated by repeated locoregional injury, transplantation of the tumor to another site on the same mouse, or by systemic treatment with transforming growth factor beta 1 and basic fibroblast growth factor. Further studies of these and other models may help to identify the mechanism of tumor dormancy and to determine why recurrence can be observed in cancer patients after a prolonged disease free survival and may allow for the design of better therapeutic strategies. Based upon these and other studies, the endothelial cell can now be considered as one gatekeeper capable of regulating tumor growth. Endostatin and angiostatin are currently in Phase I clinical trials in cancer patients and antiangiogenic antithrombin is in preclinical development.

664

Inhibition of angiogenesis in vitro using alpha V integrin-directed antisense oligonucleotides

R. Kronenwett, T. Gräf, M. Weber, W. Nedbal, U. Steidl, U. Rohr, S. Stahn, T. Möhler, R. Haas (Düsseldorf, Heidelberg, D)

The inhibition of neoangiogenesis is a promising approach for the treatment of cancer and inflammatory diseases. Since the integrin alpha V beta 3 plays a central role in capillary formation it is a potential target molecule to inhibit angiogenesis. In this study, we examined 3 different phosphorothioate-modified antisense oligodesoxyribonucleotides (AS-ODN) directed against the mRNA of the alpha V chain of the integrin in cell culture models of angiogenesis. The AS-ODN and the respective control ODN (concentration: 0.05 µM) were transfected into human umbilical vein endothelial cells (HUVEC) using cationic lipids. To induce alpha V expression cells were stimulated with PMA 8 hours following transfection. Analysis of gene expression and functional experiments were performed after further 40 hours. As assessed by flow cytometry, transfection of the alpha V-directed AS-ODN inhibited the upregulation of alpha V completely. This was reflected by an 8-fold reduced alpha V mRNA level in antisense transfected cells using a real-time RT-PCR method. The antisense-mediated inhibition of alpha V expression was associated with an 4-fold increase of the apoptosis rate of the endothelial cells and with a reduction of migration through „Matrigel"-coated transwells of 50% in comparison with control cells. Moreover, transfection of AS-ODN inhibited the formation of tube-like structures of HUVEC in „Matrigel" by 65%. For the analysis of the antiangiogenic properties of the alpha V-directed AS-ODN we used a cell culture model of human angiogenesis. In this assay, human endothelial cells (EC) were co-cultured with stroma cells which produced an extracellular matrix. The EC proliferated and migrated through the matrix to form capillary-like tubules which could be stained for the surface marker PECAM-1. Sequential transfections of the EC/stromal cell co-cultures with the AS-ODN every third day over a period of 2 weeks resulted in an inhibition of formation of tubules depending on the AS-ODN between 55% and 90% in comparison to control cells. In conclusion, AS-ODN are potent inhibitors of alpha V expression with induction of apoptosis, reduction of migration of HUVEC and inhibition of angiogenesis in vitro. Therefore, AS-ODN might be a therapeutic alternative to antagonists which directly bind to alpha V beta 3 and might be useful for antiangiogenic therapeutic approaches in the treatment of solid tumors or hematological malignancies.

665

Efficient tumor therapy by lentiviral overexpression

T. Kessler, A. Pfeifer, W.E. Berdel, I. Verma, D.A. Cheresh
(San Diego, USA; Münster, D)

Anti-angiogenesis for cancer therapy is based on the concept that solid tumors cannot grow and probably also hematological malignancies cannot spread without the generation of new blood vessels. Angiogenic endothelial cells recruit matrix metalloproteinase -2 (MMP2) to the cell surface by binding to integrin avb3 thereby facilitating the necessary remodeling of the extracellular matrix. PEX is the inhibitory C-terminal non-catalytic fragment of MMP2 that prevents this association of the integrin to active MMP2. In this study we investigated lentiviral delivery of PEX in different in vitro and in vivo angiogenesis models.Third generation HIV-based lentiviral vectors were generated to deliver FLAG-tagged PEX in different in vitro and in vivo angiogenesis or tumor models. Transgene expression was driven by either the CMV promoter or endothelial-specific promoters of the TIE family. Transduction efficiency in human endothelial cells (HUVECs) was ~90% and these cells secreted large amounts of PEX into their culture media. The resulting PEX expressing HUVECs were significantly impaired in matrigel invasion and tube formation in vitro. In addition, lentiviral delivery of PEX in vivo inhibited the angiogenic response to basic fibroblast growth factor in both the chick chorioallantoic membrane and mouse matrigel models. This was due to lack of MMP2-activation and lack of association to integrin avb3 as shown by gelatin zymography and collagen release assay. Significantly, vectors with endothelial-specific promoters expressed the transgene preferentially in endothelial cells in vivo. Finally, tumor angiogenesis and growth was significantly reduced by a single injection of PEX-lentivirus in the vicinity of subcutaneous melanoma tumors (M21L) in athymic mice. We conclude that PEX is an effective inhibitor of tumor-induced angiogenesis and that lentiviral delivery is not only a feasible approach to provide sustained, high concentrations of anti-angiogenic agents, but that endothelial-specific expression can be achieved by using promoters of the TIE-family. Selective overexpression of anti-angiogenic agents in endothelial cells is a promising therapeutic concept of diseases associated with increased neovascularization, most notably cancer.

Wissenschaftliches Symposium:
Melanome

666

Vakzination beim Melanom – Eine Übersicht

D. Schadendorf (Mannheim, D)

Jährlich werden ca. 8000 Neuerkrankungen des Melanoms in Deutschland registriert, von denen etwa 20% fernmetastasieren werden. Da in diesem fortgeschrittenen Krankheitsstadium keine Heilung mehr möglich ist, werden derzeit tumorimmunologisch-basierte Behandlungskonzepte entwickelt und geprüft. Beim Melanom gelang es so in den letzten 10 Jahren mehrere melanom-assoziierte Antigene mit Hilfe spezifischer T-Zellen zu identifizieren. Die Erkennung durch die spezifischen T-Zellen erfolgt via HLA-Molekül, die intrazellulär prozessierte Peptide präsentieren, und dem T-Zell-Rezeptor. Die Verfügbarkeit derartiger Peptide erlaubt die Entwicklung verschiedener interessanter Vakzinationsprotokolle beim malignen Melanom mit dem Ziel, eine zytotoxische T-Zellantwort gegen den Tumor zu erzeugen. Verschiedene Wege können hierzu beschritten werden: 1. Verwendung dendritischer Zellen (DC), 2. Peptidvakzination in Kombination mit Adjuvantien, 3. Einbringen der Tumorantigene mittels viraler Vehikel, 4. Einbringen der Tumorantigene mittels bakterieller Vehikel oder 5. Immunisierung mit der genetischen Information, die für das Tumorantigen kodiert („genetische Immnunisierung", „nackte DNS"), wobei sich Ansätze 1., 2. und 5. derzeit bereits in der klinischen Erprobung befinden. Dendritische Zellen lassen sich gut aus Blutmonozyten mittels eines Zytokin-Cocktails über 7 Tage in der Kulturschale generieren. Die Spezifität der Zytotoxizität wird durch die Beladung der dendritischen Zellen mit definierten Melanom-assozierten, synthetisch hergestellten Peptiden gewährleistet. Gemäß diesem Konzept sind bis heute über 50 Patienten behandelt worden. Dabei konnten bei geringen Nebenwirkungen bereits einzelne deutliche Tumorregressionen beobachtet werden. Derzeit wird in einer Phase II/III Studie bei 240 Patienten mit metastasierten Melanom (First-Line) geprüft, ob eine Vakzination mit peptid-beladenen DC in ihrer Wirksamkeit vergleichbar ist mit herkömmlichen Chemotherapie mit DTIC. Eine weitere, weniger aufwendigere Therapie ist die direkte Immunisierung mit Peptiden. Allgemein wird die Immunogenität der verwendeten, synthetisch hergestellten Eiweißfragmente als gering erachtet. Aus diesem Grund erfolgt die intradermale Peptidinjektion in der Regel zusammen mit einem Adjuvans, wobei derzeit eine Reihe von Untersuchungen laufen, um das Adjuvans mit der höchsten Effektivität bezüglich der Induktion peptid-spezifischer Immunität zu selektionieren. Zusammenfassend läßt sich feststellen, daß die Tumorimmunologie in der letzten Dekade einer deutlichen Aufschwung genommen hat mit der Identifikation von Molekülen im Tumor, die vom menschlichen Immunsystem erkannt werden. Parallel haben die wachsenden Erkenntnisse über die Arbeitsweise des Immunsystems Möglichkeiten eröffnet, Strategien zu erarbeiten, um auch Tumoren zu bekämpfen.

667

Gene therapy for melanoma: What is the current status?

R. Dummer (Zürich, CH)

Metastatic melanoma remains an aggressive neoplasm. No standard therapy is worldwide accepted. Advances in immune biology and molecular biology of melanoma have resulted in the identification of specific targets for gene therapy approaches. Many patients with metastatic melanomas suffer from extensive skin metastases. These skin metastases are an excellent target for direct injection of various drugs. We have used an adenoviral vector to transduce tumor cells with wild type p53 after the detection of increased nuclear immunoreactivity for p53. Molecular biology studies clearly showed that the transgene was transcribed in the tumor tissue despite the presence of high titer neutralizing antibodies in the serum of the patients. In another protocol we used canary pox viruses to transduce tumor tissue with either the gene for Interleukin-2 (IL-2) or the gene for GM-CSF. We were able to compare the composition of the cellular immune response after injection with these vectors and demonstrated that the transgene determines the composition of the induced infiltrates. There was a predominance of lymphocytes with IL-2 as transgene and a predominance of macrophages and eosinophiles with GM-CSF as transgene. Another approach was the use of fibroblasts, expressing the gene for human IL-2 (VERO-IL-2 cells). These cells were injected in skin metastases. Intensive local inflammatory responses, but no clear tumor regressions were noted. Naked (plasmid) DNA is another elegant approach to introduce immune stimulatory genes into tumor tissue. We currently have activated a protocol using plasmid DNA encoding for both subunits of the human IL-12. In animal models including mice and gray horses, impressive regression of injected metastases were observed.
Until now there is no prove that local immune modulation induces a systemic immune response that is able to target melanoma lesions in visceral organs. Therefore, it is necessary to develop vaccines that mount a potent and persistent immune response. Currently, we have two protocols with this aim. One protocol is using autologous tumor cells transduced by adeno-associated virus with the transgenes B7.2 and GM-CSF and the other protocol is using plasmid DNA encoding for mini-epitopes of 5 different HLA-A2 and HLA-A1 restricted peptides. Today, we can state that gene therapy is a safe and efficient tool to modify the molecular biology and the immune response in melanoma patients. However, there is still a long way to go until this yet experimental approach will have clear indications in patient care.

668

Palliative therapy - Standards and perspectives

U. Keilholz (Berlin, D)

Since 1978, 11 randomized trials have been performed investigating various cytotoxic agents in patients with stage IV melanoma. None of these trials has shown a survival benefit, but from these trials active drugs can be defined. These include DTIC, ist analogue temozolomide, cisplatin, and BCNU. Cyclophosphamide, anthracyclines, bleomycin, procarbazine and tamoxifen have no convincing activity. Another group of randomised trials has investigated the role of the two cytokines IFNa and IL-2. The response rate achieved in larger phase II and in phase III trials with IL-2 as single agent is small (5 - 20%). The combination of IL2 and IFNa has resulted in response rates between 10 and 40%. Only the combination of cytotoxic agents and cytokines, termed chemo-immunotherapy or bio-chemotherapy, has resulted in higher response rates in phase II trials, and this only in case the three drugs IL2, IFNa and cisplatin were included in the regimen. Therefore chemo-immunotherapy is currently the preferred treatment for patients with stage IV melanoma in many institutions. This preference is, however, not based on results from randomised studies. However, already in early phase II trials a proportion of patients with prolonged complete responses to IL2-based treatments were observed. These observations have been the rationale for large phase III biochemotherapy trials in the EORTC and ECOG, the results of wich are awaited. Meanwhile, two other biologic approaches have entered phase III trials: (a) the combination of histamine with IL2, whis was promising in the first trial in patients with liver metastases A large second randomized trial specifically in this patient subgroup is bein initiated to confirm the results. (b) Two other randomized trials currently investigate the impact of peptide vaccines (peptides with adjuvants or peptides loaded onto dendritic cells). Melanoma patients should be entered into these trials wherever possible. Otherwise, palliative treatment needs to be symptom-oriented and based on the characteristics of the few active drugs, knowing that survival benefit is unlikely.

669

Identification of new peptides derived from melanosomal proteins tyrosinase and gp100 that are recognized by melanoma-reactive T cells in association with HLA I molecules

V. Lennerz, M. Fatho, C. Wölfel, W. Herr, A. Lifke, T. Wölfel (Mainz, D)

We set out to facilitate the cDNA expression cloning procedure for T cell-defined antigens by applying instead of long-term cultured CTL clones bulk lymphocyte populations, enriched for tumor-reactive T cells by MLTC (mixed-lymphocyte-tumor-cell culture). As a sensitive readout we applied IFN-gamma ELISPOT assays. - In the MZ7 melanoma model CD8+ T lymphocytes derived from PBLs of different years were weekly stimulated with irradiated autologous tumor cells and were frozen after three to six weeks. MLTC responders were thawed, kept in medium with IL-2 for 2 days and were then tested for recognition of COS-7 cells co-transfected with HLA I cDNAs cloned from MZ7 melanoma cells (coding for HLA-A3.1, -A26.1, -B7.2, -B38.1, -Cw12.3) and cDNAs encoding known melanoma-associated antigens (MAGE-A1, -A2, -A3, -A4, Melan-A/MART-1, tyrosinase, gp100). While reactivity to MAGE antigens and to Melan-A/MART-1 was not detectable in any of the MLTCs, some of MLTCs contained responders to tyrosinase in association with HLA-A26.1 and to gp100 in association with HLA-B7.2. The tyrosinase/A26 and the gp100/B7 peptides were identified via cDNAs fragmentation and subsequent testing of synthetic peptides predicted by database algorithms. CTL clones isolated from MLTCs specifically recognized these peptides and at the same time lysed MZ7 melanoma cells, which formally proved internal processing of both peptides in tumor cells. – These results demonstrate that short-term antigen-driven expansion of MLTCs can generate effector populations that allow to identify tumor antigens. The continued identification of new peptide epitopes presented by different HLA molecules on human tumors cells increases the range of potential targets for T-cell directed immunotherapies and thereby counteracts tumor escape mechanisms.

670

Adoptive transfer of in vitro generated melan-A specific cytotoxic T-lymphocytes in patients with malignant melanoma: A phase I study

A. Mackensen, N. Meidenbauer, S. Vogl, J. Heymann, M. Laumer (Regensburg, D)

Tumor antigens that might serve as potential targets for adoptive T cell therapy have been defined in different tumor entities, especially in malignant melanoma. We have established culture conditions to induce primary T cell responses against HLA-A*0201-restricted melanoma-associated peptides and to allow further expansion of peptide-specific T cells for adoptive transfer. CD8+ purified T cells from melanoma patients were stimulated with modified Melan-A (aa26-36)-pulsed autologous dendritic cells (DC). After 3-4 weekly stimulation cycles, we were able to induce a strong peptide-specific cytotoxic T lymphocyte (CTL) response in vitro. Here we report on a phase I clinical trial of adoptive T cell therapy in advanced melanoma patients using Melan-A-specific CTL. Up to date, five patients received at least three i.v. infusions of 120-900x10^6 T cells at 2 week intervals together with low dose IL-2 (3x10^6 IE s.c./d for 6 days). MHC-peptide tetramer staining demonstrated that the transferred T cell product contained between 10% to 90% Melan-A specific CTL. Clinical side effects were mild and consisted of low-grade fever (WHO grade I-II). In order to investigate the distribution and localization of infused CTL, cells were labeled with 111indium. Results of in vivo tracing and immunological responses will be presented. Altogether, our study demonstrates that we have developed an efficient method for generating sufficient numbers of antigen-specific T cells in vitro that can be used for adoptive T cell transfer in tumor immunotherapy.

671

In vitro anti-melanoma activity of apoptosis-resistant human gamma-delta T cells

K.A. Hollmig, B. Guo, R.D. Lopez (Birmingham, USA)

Whereas the majority of mature human T cells express an alpha beta T cell receptor (TCR) heterodimer, only a small proportion express the gamma delta TCR heterodimer. Evidence suggests that gamma delta T cells possess important innate antitumor activity. However, lacking methods for the isolation of sufficient numbers of viable gamma delta T cells, studies at the preclinical or clinical levels examining their immunotherapeutic potential have until now, not been possible. As gamma delta T cells are extremely sensitive to mitogen-induced apoptosis, methods to simply expand fresh or resting gamma delta T cells do not hold promise. We recently identified a CD2-mediated, IL-12 dependent signal that inhibits apoptosis in mitogen-stimulated human gamma delta T cells, permitting the large-scale expansion and subsequent isolation of viable, functional gamma delta T cells, until now, not possible. Here we investigate the antitumor properties of these apoptosis-resistant human gamma delta T cells, measured against human melanoma cell lines. After 20 days in culture, human gamma delta T cells and control alpha beta T cells were sorted from PBMC cultures derived from separate individuals. Tumor lysis was assessed by standard chromium-release assay where effector gamma delta- and alpha beta T cells were added at effector-target (E:T) cell ratios of 1:1, 5:1, 10:1, 20:1. Alpha beta T cells isolated from the same healthy donor served as a control for MHC- restricted alloreactivity. Percent specific lysis (mean; range) of triplicate assays at an E:T ratio of 10:1 are shown in Table.

	Gamma Delta T Cell induced Lysis (%)		Alpha Beta T Cell induced Lysis (%)	
Cell Line	Mean	Range	Mean	Range
LD-1 (n= 5)	14.7	4.1 - 32.2	4.4	1.3 - 10.9
Mel-1 (n= 5)	35.1	9.7 - 65.9	14.3	2.3 - 33.5
Mel-5 (n= 5)	8.2	6.8 - 8.5	0.2	0 - 0.3
Mel 21 (n= 5)	38.9	19.9 - 70.6	2.8	1.3 - 6.3
Mel-28 (n= 5)	17.8	5.5 - 36.3	1.6	0.1 - 4.6
Mel-624 (n= 5)	23.1	10.8 - 41.9	5.5	0 - 23.9

These results demonstrate significant MHC-unrestricted, innate antitumor activity of ex-vivo expanded, apoptosis-resistant human gamma delta T cells as measured against human melanoma cell lines. These findings would suggest that human gamma delta T cells should be examined for their potential utility in new approches to the immunotherapy of melanoma.

Wissenschaftliches Symposium:
Hämatologische und onkologische Probleme im Alter

672

Geriatric problems of the old cancer patient: Experiences and solutions

G. Kolb (Lingen, D)

The situation of the old cancer patient is frequently accompanied by several problems and misunderstandings as well. Thus, elderly cancer patient are often inadequately diagnosed and treated at all. This is because of a wide spread prejustice that age itself is a risk factor and associated with higher toxicity and low therapeutic outcome. In contrast, several metaanalysis´ and few prospective studies have shown that clinical results will not differ from younger individuals when taking into account the specific age related modified organ functions especially of renal and hepatic parameters. Furthermore, several age related compromising changes in motor function, in functional status and social conditions are framing the context of treatment for malignant diseases in elderly patient and do also play a significant role for the prognosis and treatment outcome. Validated geriatric assess-ment instruments have been developed to identify these specific factors, but the assessment should be followed by systematic interventions and effective symptom control. Supporting patients in their activities of daily living (ADL) and in their motor and cognitive functions is clearly related to success of anticancer treatment in elderly patients. The psychosocial status of elderly cancer patients has a significant influence together with patient´s wish of being treated on the base of the individual prognosis and the quality of life during the treatment process. With respect to the overall importance of „Geriatric oncology topics" more research will hopefully be directed into this field in the future. Interdisciplinary training and common understanding is important for diagnosis, therapeutic decision and achieving therapy goals in elderly cancer patients. However, the cooperation between oncology and geriatrics in Germany is still not sufficiently developed. Thus, a transfer of knowledge between geriatric medicine and clinical hematology/ oncology is urgently needed and was previously proven to be highly effective(1,2). The present situation has formed the background to initiate an interdisciplinary working group on „Geriatric Oncology“ which has been established by the „German Society of Hematology and Oncology“ (DGHO) and the „German Society for Geriatric medicine“ (DGG) in 1999/2000.

673

Particular considerations for the treatment of eldery patients with solid tumours

U. Wedding (Jena, D)

The number of elderly people will increase within the next years in all western countries. Incidence and mortality of malignant tumours increases in an age-dependent manner. The results is an increase of about 50% in the number of cancer patients within the next three decades. Only very few elderly patients with cancer participate in clinical trials. Those who participate are in many cases a selected group of quite healthy elderly patients. They do not represent the majority of elderly patients with cancer. Oncologist are used to make their decisions based on the prognostic factors of the tumour, its histology, stage, grading and other tumour dependent factors. Elderly patients are a very heterogeneous group of patients. Some are in a very good condition, others present with severe comorbidities and are dependent in their functional status. Thus treatment decisions for elderly patients with cancer have to consider the prognostic factors of the patients in addition to the prognostic factors of the tumour. Decisions should not be based just on chronological age, but have to be based on organ function, prevalence and severity of comorbidities, and functional status. A geriatric oncological assessment is a structured way to assess the prognostic factors of elderly patients with cancer. Within geriatric medicine tools for geriatric assessment exists. They have been tested for decision-making in geriatric patients. So far they are not validated for decision-making in patients with solid tumours. To find tools which enable us to answer the following questions is of major importance for our patients in the future. Will this individual elderly person experience the advantage of screening for cervical, colon or breast cancer? Will this individual elderly patients experience the advantage of adjuvant therapy of his colon or breast cancer? Is this patient likely to tolerate palliative therapy for his e.g. cancer of the stomach without increase in toxicity? This decisions should not be based just on the chronological age, but on the biological age of a patients, which is a function of functional status, comorbidities and organ function. Geriatric oncological assessment will enable us to form three groups of elderly patients with cancer. Those within good general conditions, that can be treated according to protocols established for younger patients. Those with some comorbidities and some dependence of their functional status, who need special protocols. And those who are in a very compromised health status for whom supportive care for the control of symptoms is the major treatment goal.

674

Special considerations for the treatment of hematological malignancies in elderly patients

A. Glasmacher, I. Schmidt-Wolf (Bonn, D)

The demographic development in Western Europe will lead to a considerable increase of elderly people in our society. As the incidence of haematological malignancies rises rapidly with age the proportion of elderly patients needing care for these diseases will become much larger in the next ten to twenty years. The haematological community is, however, not yet optimally prepared to meet this challenge. At present, clinical trials, textbooks and teaching focus on patients below 60 years. AIM: To define the heterogeneity of the patient population, to identify the most important medical and sociological problems, to delineate areas that need further study, and to give an overview of current therapeutic concepts. PATIENT POPULATION: Its heterogeneity is not only determined by the nature of the disease (lymphoma, leukemia etc.) but also by a preponderance of unfavourable prognostic parameters (esp. cytogenetics) in older age. Furthermore, comorbidity and age-related changes in pharmacokinetics and –dynamics contribute to the diversity of the patient population. Also life-style, expectations towards further life and the preparedness to cope with severe illnesses vary considerably among elderly patients. With these factors treatment intentions will vary between curative intentions, palliation with the intention to prolong live with good quality or palliation with the intention to control symptoms only. THERAPEUTIC PROBLEMS: (1) To assess prognostic information and comorbidity most accurately as a basis for decision making in elderly patients, (2) to determine individual expectations and intentions towards the treatment and to offer adequate psychosocial support during these processes, (3) to develop therapy regimes with adaptation rules for individual differences in pharmacokinetics, (4) to develop supportive concepts that reduce the impact of comorbidity on treatment complications, (5) to create awareness for the increased need of new social support systems that are able to cope with problems like shorter hospital stays and more outpatient treatment (that will create special difficulties for elderly patients), reduced family sizes and less capacity to support elderly and fragile family members. CURRENT THERAPEUTIC CONCEPTS: These are currently centred on dose-reductions, non-cytostatic drugs (like monoclonal antibodies) and supportive measures (e.g. use of haematopoetic growth factors). Some studies try to evaluate more dose intensive concepts but often selection factors are not addressed and therefore many studies are not representative for the patient population. There is still a great need for age-adapted concepts that take individual factors like comorbidity etc. and the differing treatment intentions into account.

675

Equal prognosis of elderly and non-elderly patients with metastatic colorectal cancer receiving 5-FU treatment. A retrospective analysis

C.-H. Köhne, W. Scheithauer, D. Cunningham, D. Cunningham, D. Cunningham, F. DiCostanzo, B. Glimelius, E. Aranda, P. Rougier, M. Palmer, M. Palmer, J. Wils, H. Hecker, G. Hartung, M. Freund (Dresden, Rostock, Hannover, Mannheim, D; Wien, A; Sutton; Terni, I; Uppsala, S; Cordoba, E; Paris, F; Macclesfield, UK; Roermond, NL)

Introduction: There is uncertainty as to what extent systemic adjuvant or palliative chemotherapy should be offered to elderly patients with colorectal cancer. This fact is related to the unfortunate underrepresentation or even exclusion of fit elderly patients from clinical studies and also to the absence of studies specifically in the unfit elderly, who represent at least 50% of the geriatric population in major hospitals. Methods: We analysed a database of 3825 patients treated in different European trials to compare the prognosis of elderly (> 70 years, n= 711) and non-elderly patients (<= 70 years, n=3114). Results: There was a difference in the percentage of elderly patients treated in the different European countries: the highest rate in Austria (38%) and lowest in the UK (3%), while Germany had 13% and overall 19% of patients were above 70 years of age. The distrubution of age groups was: 70-75 years 370 pts, 75-80 years 116 pts, > 80 years 20 pts (only 1 patient above 85 years). Elderly patients had a higher incidence of rectal primary (p=0.01), more frequent weight loss (p=0.03) and more prior adjuvant chemotherapy (p<0.0001). Significant prognostic factors (Köhne et al. ASCO 2000) such as WBC, thrombocytosis, alkaline phosphastase and LDH all were in favour of elderly patients, however the distribution into good (52% vs. 53%), intermediate (27% vs. 26%) and poor risk (21% vs. 21%) of non-elderly and elderly patients did not differ significantly. Objective responses were more likely to occur in elderly patients (26% vs. 21%, p=0.005). Relapse free survival (median 5.3 vs. 5.8 months , p<0.0001) and overall survival (median 11.1 vs. 12.0 months, p<0.001) were all statistically in favour of elderly patients albeit with insignificant clinical benefit. Survival of elderly vs. non-elderly patients in the three previously defined risk groups did not differ. Older age was also not an independent prognostic factor. Conclusions: Fit elderly patients should not be excluded from clinical trials. Studies in unfit elderly patients are warranted and are currently being discussed by The International Society of Geriatric Oncology (SIOG) and DGHO elderly group. Elderly patients need more attention regarding their functional, social and mental status.

676

Low-dose decitabine in elderly patients with myelodysplastic syndrome: Single-center treatment results

M. Lübbert, P. Wijermans, H. Müller-Berndorff, J. Pitako, M. Daskalakis (Freiburg, D; Den Haag, NL)

Purpose: In high-risk MDS of the elderly, intensive treatment options with curative intent are often limited by performance status and comorbidity of these patients (pts). Therefore treatment appproaches directed against the cytopenia and blast expansion which allow patients to spend most of their remaining time at home are desirable. Low-dose intravenous (i.v.) 5-aza-2´-deoxycytidine (Decitabine, DAC) has shown a ~50% response rate in phase II studies in elderly pts with high-risk MDS (Wijermans et al. 1997, 2000). Response, survival and time spent in the hospital of the 33 MDS pts treated at a single center (University of Freiburg) are given. Patients and Methods: Between July 1996 and November 2000, 33 pts (21 males, 12 females) received i.v. DAC (provided by Pharmachemie and SuperGen) 45 mg/m^2/d over 72 hours, repeated every 6 weeks. Their median age was 73 years (range, 52-80). 32/33 pts had had no prior chemotherapy. 5 pts had RA, 9 RAEB, 13 RAEB-T and 6 CMML. By IPSS-Score 10 (30%) were „int-1“, 5 (15%) „int-2“ and 18 (54%) were in the „high“ risk group. Survival was from start of treatment to death (from AML or other causes, 27/33 pts) or last follow-up (6 pts) until 1.4.2001. All nights in the hospital (scheduled and all unscheduled admissions, e.g. for neutropenic fever, problems unrelated to MDS, terminal phase of MDS or AML) were captured from start of treatment to death, last follow-up or allografting (3 pts). Results: 33 patients received a median of 4 courses of DAC (range, 1-6) as initial treatment. A total of 111 courses were administered. 8 pts (24%) achieved a complete remission, 3 pts (9%) had partial remission (PR), 7 pts (21%) hematological improvement (HI), 4 pts (12%) stable disease (SD) and 9 pts (27%) had progressive disease (PD). 2 pts (6%) had toxic death. Median survival for all pts was 15 months (mths). Median survival for pts with „int-1“ was 18 mths, for „int-2“ 13 mths and for „high“15 mths. After initial response and subsequent relapse, 8 pts received a median of 3 additional courses as retreatment. 4 of them responded again with HI, 1 pt had SD, 3 pts had PD. Overall, pts spent a median of 72 nights in the hospital (range, 3-210 , mean 76 nights), resulting in 84 % of the median survival time from treatment start as outpatients. Conclusion: Intravenous DAC treatment resulted in 54% responses and had a moderate treatment-related mortality. Retreatment of initial responders at relapse showed activity in 50 % of cases. Overall survival from start of treatment was 15 months, and > 80 % of that time was spent on an outpatient basis. A randomized phase III study comparing survival and quality of life in high-risk elderly MDS treated with DAC vs. supportive care is in preparation.

677

Intensive chemotherapy is not recommended in patients with myelodysplastic syndromes aged over 60 years and karyotype anomalies

U. Germing, C. Strupp, B. Hildebrandt, T. Emde, G. Kobbe, A. Giagounidis, C. Aul, N. Gattermann (Düsseldorf, D)

Intensive chemotherapy in high-risk MDS aged over 60 years is still discussed controversial. Between 1991 and 2000 38 patients aged over 60 years (median age 66 years) with MDS were treated with induction chemotherapy (Idarubicin/ARA-C). Ini-tial diagnoses were RA in 5 cases, RARS in 1 case, RAEB in 10 cases, RAEB-T in 19 cases and CMML in 3 cases. At the start of treatment, in 29 patients transformation to AML was documented. In 34 patients cytogenetic data before the start of chemotherapy were available. According to the IPSS cytogenetic risk catego-ries 19 patients (56%) belonged to the low-risk group (A), 3 (9%) to the intermediate (B) and 12 (35%) to the high-risk group (C). 21 patients (55%) entered complete remission, 14 patients failed to respond or only achieved partial remission (37%) and 3 (8%) patients succumbed to treatment-related complications in aplasia. CR rate in cytogenetic group A was 77%, in group B 67% and in group C 25% (p=0.02). 6 of the patients with CR in group A relapsed (40%), all CR patients of groups B and C re-lapsed. Median survival of the entire group was 10 months with a cumulative survival rate of 25% after 2 years and 10% after 5 years. Patients who received CR with normal karyotype had a disease-free survival of 25 months. Patients with normal karyo-type have a benefit in survival as compared to age and me-dullary blasts matched patients that did not receive intensive chemotherapy. Only a very small group of patients achieve long-term remission. Patients with unfavourable karyotype have no benefit from intensive chemotherapy even if they enter CR, because of very early relapse. Intensive chemotherapy for high-risk MDS patients aged over 60 years with an unfavourable karyotype is not recommended.

Plenarsitzung: Quo vadis Onkologie?

678

Zukunft der Hämatologie und Onkologie aus der Perspektive des Gesundheitsreformgesetzes

W. Pföhler (Mannheim, D)

The German health care system has recently faced a far reaching reform of its hospi-tal sector. A new financing and remuneration system has to be implemented till the year 2003. It has to be based on a DRG-system. Other regulations of the new law are concerning the system of quality assurance and quality management in German hos-pitals and a new form of service organisation, the so called „integrated care“. This was introduced to overcome the fragmented system of ambulatory and stationary medical care in Germany.

679

The instrumentalization of ethics for private purposes

A.W. Bauer (Heidelberg, D)

Ethics in medicine and bioethics is a topic of large public and scientific interest at the beginning of the 21st century. Many biological and medical issues are discussed under moral aspects. Bioethics is expected to develop acceptable solutions for a range of difficult moral problems in the field of medicine and health care such as Preimplantation Genetic Diagnosis (PGD), research on embryonic stem cells, withholding and withdrawing treatment in intensive care medicine, euthanasia or physician-assisted suicide, and health care rationing. In this context, ethics is looked upon as a neutral and effective remedy for real dilemmata in medicine that might be left to philosophical experts. Nevertheless, it should be taken into account that ethics - unlike science - cannot discover objective „brute facts“ in the external reality. Moral facts, however, are a special sort of „institutional facts“ or social constructions with an intersubjective validity that may change during history. Like institutional facts, moral facts are not independent from social and political conditions. Therefore, moral standards are not the result of scientific discovery, which physicians and patients simply would have to accept without discussion. As each moral advice necessarily depends on specific values and is based on specific axiomatic foundations, modern ethics itself has to face critical arguments and has to deal frankly with contradictions. Otherwise ethics could be misused for authoritative purposes of the ethicist. Not many problems in medicine will be solved easily just because they are taken as „moral“ problems. Sometimes the term „moral“ may even address a problem as unsolvable. There can be no doubt about the fact that modern medicine needs a serious ethical supervision, but even bioethics itself should be subjected to scientific and social feedback mechanisms.

680

Present development and future trends in medical oncology

H. Ludwig (Wien, A)

Progress in understanding the malignant transformation and in the treatment of cancer: Medicine and in particular oncology has reached a revolutionary phase. Our understanding of the molecular events responsible for increased susceptibility and/or for development of cancer increases continuously. Gene arrays, proteonomics and other new technologies will allow precise definition of the genetic and molecular characteristics of an individual tumor. This knowledge will facilitate precise prediction of prognosis and set the targets for therapeutic interventions which no longer will rely on simple, rather primitive cytostatic substances but on specific and less toxic molecules, such as STI 571 which inhibit defined intracellular pathways important for cell growth. Progress with 3-dimensional computer modeling of target structures will allow significant advances in drug engineering with the ultimate outcome of receptor or target-structure tailored drugs. Chemoprevention will soon be extended to prostate and colorectal cancer and probably in due time to several other malignant diseases. Changing role of patients and care givers: The dramatic changes in our society with empowerment of the individual over the collective leveled the way for the empowerment of the patient. Contrary to this, the physician's role has been transformed to that of an 'expert moderator' playing still an important but less influential role than before, and acting as a member within a team of health care givers of various specialties. Restructuring of the healthcare system: The increasing use of oral medications, telemedicine, and self-medication will improve patient comfort and will reduce the need for treatment in large healthcare institutions which will be redimensioned to smaller, flexible health care facilities. The commercialization of scientific research: Academia has already lost it's primate for drug development which has almost completely been taken over by commercial industry. The same will apply to clinical research that is increasingly driven by the pharmaceutical industry. This tendency will transform the role of the clinical scientist from an independent investigator to an collaborator in an cooperative effort between industry and health care professionals. The beginning of the end of social health care: The proportion of public versus private provision of health services will be changed towards increasing use of private health care. This development is enhanced by the introduction of increasing numbers of expensive new drugs for which the public health care system will be no longer able or willing to pay for. As a consequence, after closely one hundred years of social health care offering free access to everyone in need in Europe, a two-class health care system with restricted care to those unable to pay for will reemerge. Increase of the 'healthy lifetime': Taken together, a plethora of new developments is on the way that will revolutionize prevention, diagnosis and treatment of cancer. They will also significantly change the situation of the patients and their healthcare givers as well as that of the entire healthcare system. The medical achievements that will be made will be more costly to the individual but should lead to a significant expansion of the time spent in excellent quality of health in.

681

Interdisciplinary surgical oncology at the inselspital: A patient directed model?

M.W. Büchler (Bern, CH)

Surgical oncology in Switzerland is defined by its interdisciplinary approach. The situation at the Inselspital is identical compared with other Swiss medical centres and chemotherapy and radiotherapy are planned and administered by the respective specialist. The implementation of neoadjuvant and adjuvant treatment concepts requires specific organizational tools which include:

1. dedicated interdisciplinary meetings with participation of radiation oncologists, radiologists and pathologists.
2. the permanent availability of an oncologist for in hospital surgical patients
3. facilitated availability of a consultant surgeon for oncological outpatients and hospitalised patients from the medical oncology unit.
4. participation of surgeons and medical oncologists in common study protocols
5. a coordinator that informs all concerned specialists with adequate notice

The advantages of such an interdisciplinary system are that a weekly forum exists where complex cancer patients can be discussed and treatment strategies can be decided upon. Interdisciplinarity often guarantees a rationale and optimal concept for the individual patient and should improve treatment quality. Regular exchange of information on outcome and complications are facilitated and formalized. Inclusion of patients from the surgical and medical oncology department in study protocols should increase patient accrual. The disadvantages of such a system are that considerable resources have to be dedicated from both sides on a regular basis and that the participation of several disciplines may reduce the efficiency of such a system. In conclusion, interdisciplinary surgical oncology is a valuable concept if the necessary organisational structures are maintained and focus on the treatment quality. The interdisciplinary concept has to be patient directed and must therefore guarantee easy accessibility for patients and referring doctors and institutions.

682

Economic evaluation of cancer treatment

V. Ulrich (Greifswald, D)

Increasing demands on health care resources have led to a situation where economic evaluation and cost analysis become more and more important. This is also true for cancer, despite the aura of sanctity often surrounding this disease. The paper looks at health system performance in Germany and Switzerland with respect to cancer. Many questions about health system performance have no clear answer, because outcomes are hard to measure, and it is hard to disentangle the health system's contribution from other factors. Nevertheless, the Swiss reforms since 1996 show that more competitive solutions may be a promising way for future reform steps also in Germany. The paper provides a survey of economic evaluations and cost analysis in cancer. The establishment of listings of different treatments according to the costs per unit of outcome is referred to as league table and has attracted considerable interest among researchers and policy makers. It can be shown that the costs per quality adjusted life year vary substantially between different interventions in cancer therapy. From the economists' viewpoint, these evaluations should raise the decision makers' awareness on interventions that greatly benefit patients and on ones that do patients little or no good at all.

Wissenschaftliches Symposium: Akute myeloische Leukämie II

683

FLT3-mutations in AML: Biological and clinical significance

H. Serve, B. Sargin, M. Mizuki, J. Schwäble, C. Müller-Tidow, W. E. Berdel (Münster, D; Osaka, J)

AML blasts show a characteristic block in myeloid differentiation and altered regulation of growth and survival. Growing evidence suggests that mutational activation of growth factor signaling pathways is an important event in leukemic transformation. Somatic mutations of Flt3 occur in more than 30% of AML patients. Most of these mutations are associated with a poor prognosis. We characterized the oncogenic potential and signaling properties of these mutations. They induced factor independent growth and apoptosis resistance in 32D cells and rapid development of a leukemia-type disease in syngeneic mice. Analysis of the involved signal transduction pathways revealed mutation-specific constitutive activation of STAT3 and STAT5 in addition to the activation of ras-dependent pathways shared with ligand-activated wild type Flt3. By microarray analysis of the target genes of the Flt3-isoforms, we could identify genes specifically repressed or induced by the Flt3 mutations. Hybridizations were performed on 18 Mu11K SubA oligonucleotide arrays containing 6800 genes. The statistical analysis of the expression data involved multiple mathematical algorithms. More than 200 differentially expressed genes could be identified with a predicted false positive rate of less than 5%. More genes were repressed than induced by Flt3 mutations. Approximately 50% of the genes that were repressed and 25% of the genes that were induced were not coregulated by IL-3. Flt3 mutations down-regulate the expression of several known inihibitors of cell cycle progression, inducers of apoptosis and, surprisingly, of transcriptional regulators of myeloid differentiation. Many of the induced genes are known STAT target genes or are known transforming genes. The functional relevance of some differentially expressed genes has been confirmed by cotransfection experiments. In addition, we analyzed the effect of tyrosine kinase inhibitors on Flt3-mediated proliferation and survival in vitro. Several compounds could be found that specifically inhibit the effects of wildtype and mutated Flt3 in 32D cells. Important issues of future research are how growth factor signals in AML are integrated into the cell cycle and apoptosis machinery, what is the effect of the mutations on the function of transcription factors involved in myeloid differentiation and finally the clinical evaluation of Flt3-specific tyrosine kinase inhibitors for AML therapy.

684

Analysis of FLT3-activating mutations in 712 patients with acute myelogenous leukemia: High incidence in FAB-subtype M5 and identification of subgroups with poor prognosis

C. Thiede, C. Steudel, B. Mohr, M. Schaich, U. Schäkel, M. Bornhäuser, M. Wermke, M. Ritter, A. Neubauer, G. Ehninger, T. Illmer and the Süddeutsche Hämoblastose Gruppe (Dresden, Marburg, D)

FLT3, a class III receptor tyrosine kinase, is expressed on early hematopoietic progenitor cells and plays an important role in stem cell differentiation. Mutational activation of the FLT3 gene, i.e. internal tandem duplications of the juxtamembrane domain or point mutations in the A-loop of the tyrosine kinase domain (TKD; codons D835/I836), has been described in patients with AML. We analyzed the prevalence, the association with cytogenetic and clinical features and the potential prognostic impact of FLT3-mutations in patients with AML. Methods: DNA was extracted from bone marrow or peripheral blood samples obtained at diagnosis from 712 patients with AML and 18 cases with RAEB-t treated in a prospective multicenter trial (AML-SHG 96). PCR screening was performed for the ITD cluster between exons 11 and 12 of the FLT3 gene and for point mutations in the 2.TKD, codons 835 and 836. Positive samples were sequenced directly or after subcloning to confirm the presence of a mutation. Results: FLT3-ITDs were detected in 143/712 AML-patients (20.1%) and 1/18 patients with RAEB-t (5.6%), mutations of the TKD were present in 57/712 AML patients (8%) and 1/18 RAEB-t cases (5.6%), twelve patients had both (ITD and TKD mutation). The presence of a FLT3 mutation (either ITD or TKD) was strongly associated with absolute leukocyte counts and with the percentage of blasts in the bone marrow ($p < .001$). No association was seen with age. FLT3 mutations were more frequent in AML FAB M5 (40.2%, $p < .001$) and less frequent in M2 (20.6%; $p = .024$) and M6 (0%, $p < .001$) cases. FLT3-ITD and FLT3-TKD mutations were significantly more prevalent in the group of patients with normal karyotype $p < .0001$). Although rarely (<5%) seen in patients with the aberrations t(8;21), inv(16), -7/7q-, -5/5q-, the number of FLT3-ITDs was significantly increased in patients with t(15;17) (34.4%; n =32) and with t(6;9) (88.9%; n = 9). Both FLT3 aberrations were associated with a significantly lower overall survival (OS) and disease free survival (DFS). In patients with ITD-mutations, this difference was especially pronounced in patients with a high allelic ratio of mutant to wt FLT3 (i.e.>0.8), as determined using a quantitative assay based on Genescan analysis (OS p = .0124). This was also found in the intermediate risk-group (OS: p= .037; DFS: p = .0094). Conclusions: Activating mutations of the FLT3 gene are amongst the most common changes in adult AML, affecting about 26% of all patients. They are especially prevalent in patients lacking other cytogenetic or molecular lesions. Mutational activation of FLT3 may favor differentiation into monocytic lineages. A high allelic ratio of the mutant/wt FLT3-ITD alleles appears to be associated with an unfavorable clinical outcome.

685

Biologically stratified therapy in adult acute myeloid leukemia: Experience of the SHG AML96 Study Group

G. Ehninger for the SHG AML Study Group (Dresden, D)

1015 patients with de novo (n = 829) and secondary leukemia (n = 149 with a history of MDS, n = 37 with tAML) were included in the ongoing prospective multicenter trial of the Süddeutsche Hämoblastosegruppe (SHG) AML96 with 34 participating centers. Patients with acute promyelocytic leukemia were entered into a different trial. Patients <= 60 ys (n = 593) were treated with double induction therapy and with priority-based post-remission therapy adapted to cytogenetic risk. Patients with t(8;21) (good risk group) received conventional chemotherapy. In first priority the other patients were transplanted with related allogeneic stem cells (n = 71), in second priority autologous (n = 88) after one course of conventional post-remission therapy and without donor two courses of conventional chemotherapy were applicated. Patients in the high risk group (-5/del(5q), -7/del/7q), hyplodiploid karyotypes other than -5,-7, -X, -Y, abnl3q, abnl11q, abnl12p, +11, +13 ,+21, +22, t(6;9), t(9;22), t(9;11), complex aberrations and tAML) without familiar donor were optionally transplanted with unrelated stem cells (n = 14). Patients > 60 ys received two induction chemotherapies.

67 % of patients <= 60 ys with de novo AML, 39 % with a history of MDS and 58% of tAML reached complete remission after induction therapy, 34 % of patients > 60 ys with de novo AML.

Survival was significantly worse in the older patients with de novo AML, in patients <= 60 ys with a history of MDS, tAML, with –5/del(5q), -7 and complex aberrations. Karyotype or disease status had no significant influence on survival in patients > 60 ys. No difference in overal and disease free survival and probability of relapse was seen in patients <= 60 ys with de novo AML comparing autologous with allogeneic PBSCT. In conclusion cytogenetic risk adapted therapy, including unrelated allogeneic PBSCT is feasible in AML patients <= 60 ys. Further the role of autologous and allogeneic PBSCT should be evaluated compared to standard strategies in a prospective fashion.

686

Risk adapted induction and consolidation therapy including autologous peripheral blood stem cell transplantation (PBSCT) in adults with de-novo AML aged ≤60 years

G. Heil, J. Krauter, H. Kirchner, A. Raghavachar, L. Bergmann, D. Hoelzer, W. Fiedler M. Lübbert, L. Noens, G. Schlimok, R. Arnold, M. Port, A. Ganser (Hannover, Ulm, Frankfurt, Hamburg, Freiburg, Augsburg, Berlin, München, D; Gent, B)

AML is a heterogenous disease and the prognosis of the patients is determined by a variety of clinical and biological risk factors such as age of the patients, the karyotype and chemosensitivity of the blasts. To improve the outcome of de novo AML patients aged up to 60 years a risk adapted induction and postremission therapy including autologous peripheral blood stem cell transplantation (PBSCT) was followed in a prospective multicenter trial initiated in 02/1995. A total of 291 AML patients (median age 44yearss) were treated according to the karyotype and the response to induction course I. Patients with CBF leukemias [t(8;21) or inv(16)] or a normal karyotype and a good response to induction I (< 5% blasts in bone marrow on day 15) were considered as standard risk (SR), all others as high risk (HR) cases. Patients with a t(15;17) were excluded. Good responders to induction I (idarubicin, Ara-C, etoposide, IVA I) received IVA II starting on day 21 („double induction“). Patients with >5% blasts after IVA I received ID-Ara-C (1 g/m^2, 8 doses) and m-AMSA (90 mg/m^2, 3 doses). Early consolidation included ID-Ara-C/daunorubicin or m-AMSA. For late consolidation SR patients received HD-Ara-C/DNR, while patients with a normal karyotype and a HLA-matched sibling were to be allotransplanted. HR patients were either allo- or autotransplanted. PBSC were mobilized after early consolidation by filgrastim. The overall CR rate was 76% (89% in the SR vs. 61% in the HR group) and 7% died during induction. After a median follow-up of 29 months, the CCR rate at 60 months was 50% for the SR group and 40% (median = 14 months) for the HR patients. The corresponding figures for overall survival were 50% for the SR group and 37% (median = 17 months) for the high risk patients. If the outcome of the standard risk patients is broken down by the karyotype the CCR rate at 60 months was 60% for the t(8;21) patients, 47% for the normal karyotype patients and 37% for the inv(16) group. In conclusion, the outcome of the patients confirm the prognostic relevance of our stratification criteria i.e. early response to induction course I and the karyotype. With respect to the CBF leukemias larger series of patients are needed to identify the precise prognostic relevance of these chromosomal lesions. The high risk patients are candidates for an intensified therapy including matched unrelated SCT early in CR and novel therapeutic approaches to improve their outcome.

687

Acute promyelocytic leukemia in children results of the AML-BFM 93/98 studies

D. Reinhardt, C. Lanvers, J. Ritter, U. Creutzig (Münster, D)

Background/patients: Acute promyelocytic leukemia (APL) was diagnosed in 32 (4.8%) of 668 children with acute myeloid leukemia (AML) enrolled in the AML-BFM 93/98 studies between 1/1994 and 4/2001. Morphological diagnosis was confirmed by detection of PML/RARa rearrangement or t(15;17) in 29 children. Patients (m: 18; f: 14) characteristics did not differ significantly from other AML patients. The median age was 10.8 years (0.9 to 15.5 yrs), median leukocyte count was 3690/µl (0.4 to 64.100/µl) and bone marrow blasts at diagnosis in median 88% (42% to 98%). All-trans retinoic acid (ATRA) treatment with a reduced dose of 25 mg/m^2/day started 3 days before induction therapy with cytosine-arabinoside (ARA-C), daunorubicin/idarubicin and etoposide. Children with elevated white cell counts received simultaneously ATRA and ARA-C. Intensive cytostatic treatment according to the AML-BFM protocol included AIE-induction (ARA-C, idarubicin, etoposide), consolidation (ARA-, prednisolon, vincristin, idarubicin, thioguanine, cyclophosphamide), intensification (HD-ARA-C/VP-16) and maintenance treatment (ARA-C, thioguanine). Results: Pharmacokinetics of low-dose ATRA were determined in 14 children (aged 0.9 to 18.4 years). ATRA plasma concentrations were characterized by large interpatient variability, with coefficients of variation of 84% for peak plasma concentrations, and by rapid ATRA clearance, which resulted in decreased plasma concentrations and reduced endogenous ATRA concentrations during treatment (P=0.03). Peak plasma concentrations 8.58–513.0 nM were in the lower range of those considered effective in vitro. Complete remission was achieved by 31/32 (97%) of the children, only one child died from fatal bleeding within the first days. However, in 18/31 children side effects associated with ATRA treatment like severe headache, pseudotumor cerebri or pulmonary infiltrations. The symptoms could be managed by administration of steroids, temporary cessation of ATRA therapy or drug reduction. Four children relapsed 1 to 2 years following diagnosis, two of them achieved second remission. The estimated 5 years event free survival was 80%±8% and the 5 years overall survival 92%±5%, significantly higher compared to non-M3 AML subgroups in this period (OS 62%±2). Conclusion: Children with AML M3 had a favorable prognosis if early fatal bleeding or infection could be prevented by ATRA-application. Low-dose ATRA proved effective for remission induction treatment in children with APL. However, ATRA treatment causes considerable but manageable toxic side effects. Intensive cytostatic treatment is required to achieve long term remission.

688

The leukemia-specific human fusion gene NUP98-HOXD13 causes myeloproliferation and collaborates with Meis1 in inducing acute myeloid leukemia (AML) in mice

C. Buske, N. Pineault, M. Feuring-Buske, C. Abramovich, P. Aplan, R.K. Humphries (München, D; Vancouver, CDN; Gaithersburg, USA)

Leukemia-specific chromosomal translocations involving the NUP98 gene, a member of the nucleoporin gene family, form a rapidly growing family of genetic alterations which are associated with therapy-related AML and CML in blast crisis. The NUP98 fusion genes involve multiple partner genes, mostly homeobox genes such as HOXA9 or HOXD13. We used retroviral gene transfer and the ES in vitro differentiation and bone marrow transplantation models to test directly the effect of the NUP98-HOXD13 (ND13) t(2;11) fusion gene on hematopoiesis and to identify potential collaborating genes in ND13-mediated leukemogenesis. Constitutive expression of ND13 and/or Meis1 cDNA was achieved by retroviral gene transfer using a MSCV-based vector with NeoR or GFP/YFP as a selectable marker. ND13 increased the proliferative potential of BM progenitors in liquid culture, increased the numbers of myeloid clonogenic progenitors (p<0.02) and induced a 43-fold increase in day 12 CFU-S after 7d in vitro culture (p<0.0001). The fusion gene induced a myeloproliferation in vitro in the ES cell system as well as in the BM system with a 50% increase in the absolute number of myeloid progenitors in differentiating ES cells and an increase of circulating myeloid cells (4.1 x 10^6/ml in ND13 mice versus 2.6 x 10^6/ml in GFP control mice) in transplanted mice at the early time point of 4 weeks post transplant (p<0.004). ND13 expressing BM cells transplanted into secondary irradiated recipients caused a lethal myeloproliferative syndrome with accumulation of granulocytes in the peripheral blood and splenomegaly with effacement of the normal follicular architecture by massive myeloid infiltration. Transplantation of BM cells expressing both ND13 and Meis1 (GFP+/YFP+) changed myeloproliferation to frank aggressive transplantable lethal AML with cachexia, 32-83 % blasts in the BM and leukemic infiltration of multiple organs in recipient mice. The leukemia showed myelomonocytic features and allowed the establishment of IL-3 dependent myelomonocytic cell lines. Deletion of the NUP98 portion of the fusion gene resulted in loss of hematopoietic activity. However, VP16 fused to the truncated HOXD13 showed, although reduced, hematopoietic activity in the CFU-S assay, suggesting that the NUP98 portion of the fusion gene acts as a transactivation domain. Our data provide direct evidence that ND13 is necessary and sufficient to induce myeloproliferation and identify Meis1 as a oncogenic collaborating gene in ND13 dependent leukemogenesis.

Symposium der Süddeutschen Hämoblastosegruppe (SHG): Multiples Myelom

690

Prognostic implications of cytogenetic aberrations in multiple myeloma

J. Drach, H. Kaufmann, E. Urbauer, J. Ackermann, T. Nösslinger, E. Krömer, H. Gisslinger, U. Jäger, N. Zojer, H. Ludwig, H. Huber (Wien, A)

Recent studies indicate that chromosomal abnormalities are among the most important prognostic parameters for patients with multiple myeloma (MM). Using metaphase analysis, Tricot and colleagues reported that partial or complete deletions of chromosome 13 (by metaphase cytogenetic analysis present in 15% of patients with MM) are independently associated with short event-free and overall survival (OS) following high-dose therapy and autotransplantation. Since MM cells are difficult to grow in vitro, molecular cytogenetic techniques like interphase fluorescence in situ hybridization (FISH) are required for the reliable detection of specific chromosomal aberrations in MM. FISH studies (mostly performed with two probes for distinct loci at band 13q14) have shown a much higher frequency of chromosome 13q deletions in MM, occurring in 39% to 52 % of newly diagnosed cases. Using a panel of 11 probes along the entire 13q arm in 50 patients (some were heavily pretreated), Shaughnessy et al. demonstrated 13q deletions in 86 % of these patients. Of particular note, current data indicate that deletions of 13q as detected by FISH remain an independent adverse prognostic factor after conventional-dose and high-dose therapy. We have recently observed that information about chromosome 22q adds prognostically relevant information to that already provided by chromosome 13q: Patients with a deletion of 22q had inferior response rates (55% vs. 77%; P = .02) and shorter median OS (29 vs. 53 months; P = .0008) compared to patients with normal chromosome 22q. Multivariate analysis in 120 patients with MM indicated that unfavorable cytogenetic features by FISH (deletion of 13q14 and/or 22q), beta-2-microglobulin > 4, and immature plasma cell morphology are independent adverse parameters for survival. According to the number of these risk factors present in individual patients, distinct prognostic groups of patients with MM can be discriminated (0, 1, 2, and 3 factors; median survival [months] 60+, 49, 23, and 10, respectively; P < .0001). Collectively, data indicate that interphase FISH should become a standard procedure in MM to estimate a patient's risk at diagnosis. Clinical trials employing a risk-oriented therapeutic approach are under way since current treatment strategies fail to improve the outcome of MM patients with high-risk cytogenetics.

691

New standards in the treatment of multiple myeloma

H. Einsele (Tübingen, D)

Most patients who fulfill the criteria for the diagnosis of multiple myeloma have symptomatic disease (bone pain, osteolytic lesions, anemia, infections, hypercalcemia or renal insufficiency) and require therapy. In recent years, treament of multiple myeloma has been increasingly become age- and stage-related. Whereas all patients in stage II/III require systemic therapy, chemotherapy of patients in earlier stages of the disease has not been shown to be beneficial. Current studies (DSMM3) are investigating the role of bisphophonates in slowing disease-progression in patients with MM stage I. Three comparative studies have shown that high dose chemo(radio)therapy followed by autologous stem cell transplantation is clearly superior to conventional chemotherapy in patients less than 60 years old, provding not only a higher response rate and rate of CR, but also a significantly improved progression-free and overall survival. In addition quality of response was shown to have a major impact on survival following high dose chemotherapy. Thus, current protocols for patients aged less than 60 years with advanced multiple myeloma try to increase the intensity of the conditioning regimen either by using combinations of non-cross-reacting agents or using radiochemotherapy or tandem high dose chemoprotocols, all to increase response rate and quality following high dose therapy.
Palumbo and co-workers have shown that also patients between 60 and 70 years old can be safely offered high dose chemotherapy (melphalan 100 –140 mg/m^2) followed by autologous stem cell support. High dose chemotherapy was also shown to be superior to standard chemotherapy with melphalan and prednislone in this age group. Standard therapy for patients not eligible for high dose chemotherapy is still treatment with melphalan/prednisolone. Chemotherapy should be continued until the patient is in a plateau phase. Interferon maintenance therapy has been reported by Wheatley to provide a survival benefit which has been confirmed by the EBMT for patients undergoing high dose chemotherapy and autologous stem cell transplantation.. In older patients with advanced stage multiple myeloma, randomized studies are currently investigating alternative combinations, e.g bendamustin/prednisolone and thalidomide/dexamethasone in comparison to standard chemotherapy with melphalan/prednisolone. Thus, new treatments target not only the MM cell, but also the bone marrow milieu. Thalidomide, an anti-antiogenic agent, has been shown to effect responses in one-third of patients with advanced and refractory MM including post-transplant relapses. The recognition of a dismal prognosis with chromosome 13 deletion even following high dose chemotherapy and autologous stem cell transplantation prompts trials of allogeneic stem cell transplantation for patients with MM and chromosome 13 deletions. Preliminary data from several centers offering an allogeneic transplant with a reduced intensity conditioning regimen after a preceding autotransplant have shown a high efficacy of this new treatment regimen even in patients with a chromosome 13 deletion.

692

Angiogenesis and new treatment options for multiple myeloma

H. Goldschmidt, T. Möhler, K. Neben, A. D. Ho (Heidelberg, D)

Angiogenesis plays an important role in the progression of solid tumours and increased angiogenesis has been demonstrated in most haematological neoplasms such as leukemia, lymphoma, myelodysplasia and multiple myeloma (MM). In MM a higher vascular density in bone marrow correlates with a poor outcome and can be confirmed by dynamic magnetic resonance imaging studies. Increased concentration of proangiogenic cytokines are detected in peripheral blood and bone marrow of MM patients. Data from preclinical studies by Folkman initially showed that thalidomide (Thal) has antiangiogenic properties. Based on a case report from 1965 Barlogie treated a cohort of end stage MM patients in 1997. We have initiated a phase II study evaluating the effects of Thal in MM, either as monotherapy or in combination with chemotherapy. Thal alone achieved average response rates of 30% in chemotherapy refractory and 50% in newly diagnosed MM patients (Abstracts containing data on Thal presented at the ASH- and ASCO-meeting, 2000). The combinations of Thal with dexamethasone with or without chemotherapy increased the response rates in poor risk MM to approximately 50% or 80%, respectively. The most frequent side effects observed were sedation, polyneuropathy, constipation and thrombosis. Recently, two classes of Thal analogous have been identified with selected cytokine inhibitory (SelCIDs) and immunomodulatory (IMiDs) effects. These drugs demonstrated promising anti-myeloma effects in vitro. Clinical trials with IMiDs have been initiated. Other new drugs tested in clinical trials are proteasome inhibitors and selective angiogenic inhibitors. Furthermore first results of a clinical trial evaluating a humanized monoclonal antibody specific for plasma cells have been promising. These new approaches targeting specifically the myeloma cell or the bone marrow microenvironment offer promising strategies to improve the outcome for patients with MM.

693

Bendamustine/Prednisone versus Melphalane/Prednisone in the primary treatment of multiple myeloma: a randomized trial of the East German Study Group for Hematology/Oncology (OSHO)

W. Pönisch, P.S. Mitrou, K. Merkle, M. Herold, M. Assmann, G. Wilhelm, K. Dachselt, P. Richter, V. Schirmer, M. Schulze, R. Subert, B. Harksel, N. Grobe, E. Stelzer, A. Schulze, R. Pasold, T. Friedrich, W. Helbig, D. Niederwieser (Leipzig, Frankfurt/Main, München, Erfurt, Riesa, Wernigerode, Nordhausen, Zella-Mehlis, Plauen, Zittau, Schwerin, Torgau, Neubrandenburg, Gera, Potsdam, D)

The 94BP01 protocol was designed to compare melphalane/prednisone (MP) and bendamustine/prednisone (BP) in multiple myeloma (MM). Between May 1994 and July 1998, 136 MM patients (stage II with tumor progression, n=11; stage III, n=125) were recruited by 31 hospitals in Germany. Five patients were not evaluable for response (inadequate modification of therapy: n=2; loss of follow up: n=3). Sixty eight patients received BP (bendamustine 150 mg/m^2 day 1+2, prednisone 60 mg/m^2 day 1-4) and 63 patients MP (melphalane 15 mg/m^2 day 1, prednisone 60 mg/m^2 day 1-4). Response to treatment was evaluated by SWOG criteria. Patients characteristics of both groups were similar and no significant differences were found with regard to age, stage, Ig isotypes, beta 2 microglobulin, and bone destruction. Results were analyzed after a median follow up of 45 months. Response rates were similar in both treatment groups (BP: 75 %, MP: 68 %). Complete remissions were significantly higher in the BP arm (32 % versus 11 %, p<0.003). In responding patients remission was achieved after 6,7±3,3 cycles in the BP arm compared to 8,5±4,7 cycles in the MP arm. The duration of response was significantly longer in the BP treated patients with a median of 14 months compared with 10 months in the MP group (p<0.03). The 30 months probability of progression free survival was 23 % (BP) versus 8 % (MP) The 60 months post-diagnosis probability of survival was 28 % in the BP arm versus 23 % in the MP arm (p=0.72). Minimally increased toxicity (leukopenia, gastrointestinal complaints) was observed with BP therapy. We conclude that the BP regimen shows important advantages over standard MP as primary treatment for MM. For the BP treated patients a higher rate of complete remissions with a shorter duration of therapy is achieved and for most patients a longer therapy free interval until the first progress was noted. The latter represents an important improvement in the quality of life of patients. Currently a cooperative clinical study group of the German Multiple Myeloma Study Group (DSMM) and the OSHO is ongoing to further evaluate this regimen in comparison to standard therapy.

694

DSMM pilot study: Age-adapted tandem melphalan with PBSCT in patients 60-70 years with multiple myeloma

C. Straka, H. Wandt, M. Freund, T. Kubin, J. Fischer, B. Metzner, M. Kropff, N. Frickhofen, N. Brack, H. Döhner, H. Einsele, B. Emmerich (München, Nürnberg, Rostock, Karlsruhe, Oldenburg, Münster, Wiesbaden, Ulm, Tübingen, D)

Older patients with multiple myeloma (MM) appear to benefit from high-dose melphalan followed by peripheral blood stem cell (PBSC) transplantation as do younger patients. Clinical studies specifically designed for older MM patients, however, are lacking. The DSMM carried out a pilot study on patients with multiple myeloma stage II and III with an age between 60 - 70 years who received tandem melphalan at a dose of 100 mg/m^2 or 140 mg/m^2 and PBSCT within two months. Sixteen centers reported 42 patients and a documentation after high-dose treatment is available from 22 patients at the moment. The median age of the patients was 65 years (range 60 - 68). Patients with only 0 - 1 cycle of chemotherapy (n=12; group 1) before mobilization with IEV-chemotherapy (ifosfamide, epirubicin, etoposide) and G-CSF could receive higher PBSC doses which were a median of 5.00 for the first autograft and a median of 5.50 x 10^6 CD34+ cells/kg for the second autograft, whereas patients with 2 or more cycles of pretreatment (n=10; group 2) received lower PBSC doses with a median of 3.13 for the first and a median of 2.67 x 10^6 CD34+ cells/kg for the second autograft. In group 1 the recovery of neutrophils (ANC>500/µl) took a median of 9 days and the platelet recovery to 20.000/µl a median of 10 days. In group 2 the hematopoietic recovery took one day longer. The tolerability of both melphalan dosages with PBSCT was excellent. Severe adverse events were reported for mobilization, first and second autograft in less than 10 %. The treatment-related mortality was 0 %. Neutropenic fever occurred in group 1 after the first and second autograft in 33 % and 29 %, respectively, in group 2 in 44 % and 67 %, respectively. The achieved remissons for all patients were PR in 63 %, CR in 21 % and no change in 16 %. In conclusion, age-adapted tandem high-dose melphalan with PBSCT in patients 60 - 70 years is very well tolerated and has a high anti-tumor potential. The impact of induction chemotherapy on the subsequent clinical course is now investigated in a phase III trial of the DSMM.

Wissenschaftliches Symposium: Transplantationsimmunologie

695

Innovative T cell manouvers to improve the outcome of clinical stem cell transplantation

Ch. Huber (Mainz, D)

A caveat of current strategies in allogeneic hemopoietic stem cell transplantation is due to the significant morbidity and mortality of graft versus host disease (GvH-D) following transfer of bone marrow or peripheral blood stem cell grafts across phenotypically matched major histocompatibility complex (MHC) and/or minor histocompatibility antigen (mHA) barriers. In spite of these disadvantages allogeneic transplantation is remarkably effective to prevent leukemia relapse by poorly understood mechanisms collectively designated as graft versus leukemia (GvL) effects.
This contribution summarizes three innovative approaches to either improve our understanding of allograft immunity or to even bypass the requirement for allografting. I will report on studies performed by C.Wölfel, D.Depres and G.Derigs to define new mHA or leukemia specific target antigens by means of T cell directed expression cloning or SEREX, on U. Hartwigs animal experiments to reduce the risk for development of GvH-D by selective elimination of alloreactive T cells from stem cell grafts by means of activation induced cell death and on M.Theobalds adoptive immunotherapy approach which aims to increase anti-leukaemic efficacy and to avoid allograft toxicity by transduction of autologous T cells with leukemia or pathogen-specific T cell receptors. Early results of these three projects are presented and will be discussed with respect to their clinical relevance to expand the therapeutic window between GvH-D and GvL.

696

Peripheral blood dendritic cells in patients during allogeneic hematopoietic cell transplantation

M. Pihusch, S. Böck, T. Heller, C. Schoch, T. Haferlach, B. Rolf, H. Diem, R. Pihusch, E. Hiller, W. Hiddemann (München, D)

Introduction: Although dendritic cells (DC) which originate from a bone marrow precursor and circulate via the blood to most body tissues fulfill a very important role in antigen uptake, processing and surveillance, little is known concerning peripheral blood DC after allogeneic hematopoietic cell transplantation (AHCT). We describe for the first time the analysis of the course and the chimerism of peripheral blood DC in patients during AHCT. Materials and Methods: During AHCT citrated samples were taken daily from 109 (69 m / 40 f; 61 PBSCT / 48 BMT) patients with different he-matological disorders (AML, CML, ALL, NHL, MM, CLL, Hodgkin's D., SAA, OMF, MDS, SCA). During engraftment (51 patients) and 6 months after AHCT (4 patients) we studied the DC-chimerism by using the FISH- or the STR-PCR-technique. The percentage of DC of periphe-ral blood NC's in our patients was between 0.01 – 3,00 % (CD123) and 0.01 – 6,00 % (CD11c). We identified the unstimulated DC in a flow-cytometer (FACSCalibur) using specific fluorescence-labelled antibodies and then isolated the DC in vivo by fluorescent activated cellsorting (FACSVantage SE) with > 95 % purity. The sorted DC were morpholo-gically identified by microscopy. Results: Peripheral host-DC decrease more quickly than peripheral leucocytes due to the conditioning before an AHCT. Peripheral graft-DC increase up to 10-fold of the normal range during engraftment. Patients with an early GvHD showed a decrease in peripheral graft-DC during engraftment already one day before the clinical diagnosis. Patients who were treated with prednisolone for clinical GvHD had no peripheral DC. 8 / 51 patients had an incomplete chimerism of DC during engraftment (32 FISH / 19 STR-PCR; 5 patients: complete Leukocyte-chimerism; 4 patients: GvHD; 2 patients: dead; 1 patient: relapse of MM). All 4 pati-ents analysed 6 months after

AHCT were found to be well and exhibited a complete peripheral DC-chimerism. Conclusion: Our study demonstrates that peripheral blood DC are affec-ted by AHCT. Further studies will show whether peripheral blood DC might be suitable as a laboratory marker for the clinical diagnosis of a GvHD and whether a relationship between the peripheral DC-chimerism and the development of a GvHD or a relapse in the patients may exist.

697

Tumor specific T cell activation by recombinant immunoreceptors: Integration of CD3-zeta signalling and CD28-mediated costimulation in a single receptor molecule

A. Wieczarkowiecz, T. Marquardt, C. Heuser, C. Pohl, H. Abken, A. Hombach (Köln, D)

Recombinant immunoreceptors with predefined antigen specificity can redirect grafted T cells to an efficient, MHC-independent anti-tumor response. We have recentely demonstrated that B7-costimulation did not affect target cell lysis but impacts substantially cytokine secretion, and effector cell proliferation (Cancer Res. 61, 1976-1982, 2001). To deliver both signals to receptor grafted T cells we have generated a recombinant anti-CEA immunoreceptor (BW431/26-scFv-Fc-CD28/CD3zeta) harboring intracellularily the signalling moities of both CD28 and CD3-zeta and compared its function to anti-CEA receptors that harbor intracellularily either the CD3-zeta (BW431/26-scFv-Fc-CD3zeta) or CD28 (BW431/26-scFv-Fc-CD28) signalling domain. T cells that were retrovirally engrafted with the BW431/26-scFv-Fc-CD28/CD3zeta receptor lyse CEA+ target cells with similar efficiency than T cells equipped with the BW431/26-scFv-Fc-CD3zeta receptor whereas T cells expressing with the BW431/26-scFv-Fc-CD28 receptor did not lyse CEA+ target cells. In contrast, only BW431/26-scFv-Fc-CD28/CD3zeta grafted T cells secreted high amounts of IL-2 upon specific receptor crosslinking whereas both BW431/26-scFv-Fc-CD3zeta and BW431/26-scFv-Fc-CD28 grafted T cells did not. Morover, signalling through the recombinant BW431/26-scFv-Fc-CD28/CD3zeta receptor resulted also in enhanced IFN-gamma secretion and was similar to simultaneous T cell stimulation by anti-CD3 and anti-CD28 mAbs, respectively. Our data demonstrate that the recombinant anti-CEA receptor BW431/26-scFv-Fc-CD28/CD3zeta integrates the signalling pathways of CD3-zeta and CD28 and combines highly efficient target cell lysis with efficient secretion of IL-2 and other cytokines that are highly important to sustain a long lasting anti-tumor response.

698

Non-invasive monitoring of anti-tumor activity of expanded CD8+ NK-T cells after syngeneic or allogeneic transplantation using real time in vivo imaging

M. Edinger, Y. Cao, M. Verneris, C. Schimmelpfennig, M. Bachmann, C. Contag, R. Negrin (Stanford, USA)

Since light is transmitted through mammalian tissues, cells expressing the bioluminescent reporter gene luciferase can be detected within living mice using low light imaging cameras. To establish animal models of leukemia and lymphoma, the B cell lymphoma BCL1 was retroviraly transduced with a luciferase-gfp fusion gene. Transduced tumor cells were sorted for gfp expression and tumor engraftment in Balb/c mice was visualized. As few as 7000 cells were detected early after iv injection. Measuring light emission over time enabled us to quantify tumor growth within internal organs. Reisolation of transduced tumor cells from animals in different disease stages showed a linear correlation of tumor load with signal intensity measured externally. Using this tumor model we investigated the therapeutic application of a cytotoxic cell population, previously described as cytokine induced killer cells (CIK). These cells co-express T cell and NK cell markers and are generated from splenocytes by stimulation with IFN-g on day 0, followed by anti-CD3-Ab and IL-2 stimulation and expansion over 2-3 wks. We examined the anti-tumor activity of these cells in syngeneic tumor models as well as after allogeneic bone marrow transplantation (BMT). After allogeneic BMT (C57BL/6 into Balb/c) these cells show anti-tumor activity with a significantly attenuated capacity to cause GvHD. Animals treated with $2.5\text{-}10\times10^6$ CIK cells show a long term survival of 75% and 30%, respectively ,while animals treated with as few as $1\text{-}2.5\times10^6$ splenocytes all succumbed to acute GvHD by day 50 and 14, respectively. Using the imaging technique we examined the anti-tumor activity of these cell populations in vivo and demonstrate the kinetics of tumor cell clearance. Labeling of the effector cells with the same retroviral transduction system allows us to examine the migration and proliferation of these cells in real time after allogeneic and syngeneic transplantation.

699

Cure of Burkitt's lymphoma in severe combined immunodeficiency mice by T cells, tetravalent CD3 x CD19 tandem diabody, and CD28 costimulation

B. Cochlovius, S. Kipriyanov, M. Stassar, J. Schuhmacher, G. Moldenhauer, M. Little (Heidelberg, Ladenburg, D)

To increase the valency, stability, and therapeutic potential of bispecific antibodies, we have constructed a tetravalent tandem diabody (Tandab) that is specific to both human CD3 (T-cell antigen) and CD19 (B-cell marker; Kipriyanov et al., J. Mol. Biol., 293: 41-56, 1999). It was generated by the functional dimerization of a single chain molecule that contained four antibody variable domains (V(H) and V(L)) in an orientation that prevented intramolecular pairing. Compared with a previously constructed heterodimeric CD3 x CD19 diabody (Cochlovius et al., J. Immunol., 165: 888-95, 2000), the Tandab exhibited a higher apparent affinity to both CD3+ and CD19+ cells and longer blood retention when injected into mice. Biodistribution studies in mice bearing Burkitt's lymphoma xenografts demonstrated specific accumulation of the radioiodinated Tandab in a tumor site with tumor-to-blood ratios of 1.5, 8.1, and 13.3 at 3, 18, and 24 h, respectively. Treatment of severe combined immunodeficiency mice bearing established Burkitt's lymphoma (5 mm in diameter) with human peripheral blood lymphocytes, Tandab, and anti-CD28 MAbs resulted in the complete elimination of tumors in all of the animals within 10 days. In contrast, mice receiving human peripheral blood lymphocytes in combination with either the diabody alone or the diabody plus anti-CD28 MAbs showed only partial tumor regression. These data demonstrate that the CD3 x CD19 Tandab may be a promising tool for the immunotherapy of human B-cell leukemias and lymphomas.

700

Adoptive T cell therapy following allogeneic stem cell transplantation. Different strategies to generate and expand antigen-specific T cells ex vivo

H. Einsele, L. Bissinger, G. Rauser, H. Hebart, C. Sinzger, L. Kanz, G. Jahn (Tübingen, D)

Adoptive transfer of donor-derived antigen-specific T cell clones or lines can be used to restore protective immunity to various infectious pathogens and to reduce residual tumor cells after allogeneic stem cell transplantation. To make T cell therapy with donor-derived T cells safe and effective, alloreactive T cells have to be depleted and antigen-specific T cells to be enriched for. We compared enrichment of antigen specific T cells either by repetitive stimulation in vitro or by selection based on antigen-specific cytokine secretion. These strategies were applied for ex vivo generation and expansion of CMV-specific T cells. For repetitive stimulation CMV protein or CMV pp65 peptide pulsed autologous monocyte-derived DCs and peripheral blood monocytes were applied. Following 4 specific stimulations alloreactivity was undetectable in proliferation as well as cytotoxicity assays in all T cell lines tested. Whereas pulsing of APCs with protein resulted in T cell lines which contained predominantly (median 76%) CD4+ T cells (> 7% CMV-specific T cells already after 2 weekly stimulations), following peptide-pulsing predominantly CD8+ T cells (> 30% CMV pp65 peptide-specific CTLs after 2 weekly stimulations) were obtained. During repetitive stimulation with APCs pulsed with the relevant peptide, CMV-specific T cells could be successful expanded by about 100-fold after 4 weeks. T cell lines showed effective killing of peptide pulsed as well as CMV-infected autologous fibroblasts (50% lysis, E: T ratio 10:1). Selection by antigen-specific cytokine secretion from PBMNCs (Cytokine Secretion Assay Miltenyi) and expansion with IL 2 resulted in T cell lines containing > 95% CMV pp65 -specific CD8+ CTLs as demonstrated by staining with tetrameric HLA/peptide-complexes and intracellular cytokine staining. After 3 and 6 weeks of ex vivo culture CMV-specific T cells could be expanded up to 1000 and 10 000 fold, respectively. The expanded T cells showed > 77% perforin expression and specific cytotoxicity of > 50% against peptide-pulsed T2 cells (E:T ratio 20:1) as well specific killing of CMV-infected autologous or HLA-matched skin fibroblasts. Thus, both strategies are highly effective in selecting CMV-specific T cells and their expansion in vitro without reducing their function in spite of prolonged ex vivo- culture.

Both approaches to generate CMV-specific T cells thus seem to be applicable for clinical use after allogeneic stem cell transplantation.

701

Frequency of dihydropyrimidine dehydrogenase mutation DPYD2*A (c.1905+1G/A)

M. Seule, M. Steiner, N. Laser, M. Freund, P. Schuff-Werner, C. Köhne (Rostock, Dresden, D)

Background: Dihydropyrimidine dehydrogenase (DPYD) is the principal enzyme involved in the detoxification of 5-fluorouracil (5FU), one of the most widely prescribed chemotherapeutic drugs in cancer treatment. DPYD gene mutations result in either complete (homozygosity) or partial (heterozygosity) loss of enzyme activity. Previous reports suggested that a G/A transition at the 5?-splicing donor consensus sequence in intron 14 leading to exon 14 skipping (c.1905+1G/A, DPYD2*A) might be responsible for 5FU-related toxicity in mutation carriers. However, the allele frequency in the general population has not been estimated. Objective: The aim of the present study was to determine the frequency of the gene mutation DPYD2*A. Study design: A population of 718 randomly chosen patients of Caucasian origin was investigated at the time of admission to the hospital. Genomic DNA isolated from peripheral blood samples was used for genotyping the DPYD2*A mutation by a PCR-mediated site-directed mutagenesis followed by restriction digestion. Positive results were confirmed by an independent PCR-RFLP analysis. Results: The frequency of the heterozygous DPYD*2A mutation was found to be 0.97% (7/718) in randomly selected patients of Caucasian origin. Identical results were obtained for both PCR-RFLP methods employed. As expected, no case of a homozygous mutation was identified as this would indicate a rare case of inherited thymine uraciluria. Implementing routine DPYD2*A mutation screening for patients before starting 5FU-containing cytotoxic cancer treatment (n= 29), we identified a heterozygous mutation in a 44-year-old female colon cancer patient. She suffered WHO grade IV gastrointestinal and haematological toxicity while undergoing 5FU/CPT11 treatment according to EORTC protocol 40986. Despite adequate supportive treatment, her status rapidly deteriorated leading to lethal outcome due to sepsis and mutiorgan failure. Conclusion: Heterozygosity for the DPYD2*A mutation represents a common pharmacogenetic syndrome. The clinical importance of the heterozygous mutation as related to 5FU-associated toxicity needs to be etablished in order to precisely define the feasibility of pretreatment mutation screening before application of 5FU-containing chemotherapeutic regimens.

702

The role of p53 in the response of germ cell tumors to chemotherapy

F. Mayer, A.-M. Kersemaekers, J.W. Oosterhuis, C. Bokemeyer, L. Looijenga (Tübingen, D; Rotterdam, NL)

BACKGROUND: Of all solid tumors of the adult, testicular germ cell tumors of adolescents and adults (TGCTs) are the most sensitive to systemic treatment. Even in metastatic stages, high cure rates can be achieved by multiagent, cisplatin-based chemotherapy. The exquisite chemosensitivity of TGCTs has been attributed to a high level of wild-type P53 protein as demonstrated by immunohistochemistry and mutation analysis. Accordingly an increased rate of p53 mutations in resistant cases has been suggested in one study. METHODS: To investigate these aspects independently, we studied a series of primary unselected TGCTs using immunohistochemistry and Western blotting for P53 and MDM2. Treatment resistant TGCTs were compared with sensitive cases by immunohistochemistry and mutation analysis of the p53 gene. RESULTS: Focal P53-positivity ($\geq$1% of tumor cells positive) was detected by immunohistochemistry in 7 out of 39 unselected TGCTs, while the remaining cases were negative. Western blotting confirmed the low level of P53 protein in TGCTs compared to well characterised cell lines and solid cancers. MDM2 was specifically identified in embryonal carcinoma, while a moderate to low level was detected in the other histological components. Analysis of a series of 18 treatment resistant TGCTs showed a similar staining pattern. In addition p53 mutations were found in one of the 20 responding patients and in one of the 18 refractory patients. CONCLUSION: In contrast to previous reports this study demonstrates that TGCTs have a relatively low level of wild type P53 protein. Furthermore, no significant difference in p53 status was found between refractory and responding TGCTs. We conclude therefore, that the level of wild-type P53 by itself does not explain the exquisite chemosensitivity of TGCTs, and that disruption of the P53 pathway may contribute to the resistant phenotype in only a small percentage of treatment resistant TGCTs.

703

Lack of overexpression of the EGF receptor family in refractory germ cell cancer (GCT)

C. Kollmannsberger, F. Mayer, W. Oosterhuis, L. Looijenga, L. Kanz, C. Bokemeyer (Tübingen, D; Rotterdam, NL)

Background: Metastatic GCT are highly sensitive to cisplatinum-based chemotherapy, however, only less than 5% of patients (pts) with cisplatin-refractory disease can be cured. A relationship between Erb-B2 overexpression and cisplatin-resistance has been suggested in-vitro. The recent development of EGF receptor interactive agents, particularly for EGF-R and Erb-B2 (Her2-neu), may offer a potential therapeutic option for these pts. The present study investigates the expression of the EGF receptors EGF-R, Erb-B2, Erb-B3 and Erb-B4 in pts with cisplatin-refractory GCT. Materials and methods: Paraffin-embedded specimens from 22 cisplatin-refractory pts, obtained either from the primary tumor (n=10) or from resection of postchemotherapy masses (n=12), were included in the analysis. Pts were heavily pretreated with a median of 9 cisplatin-containing cycles. Expression of EGF-1, Erb-B2, Erb-B3 and Erb-B4 was examined by immunohistochemistry (DAKO, Hamurg, Germany/Neomarkers, Freemont, CA). The presence of Erb-B2 gene (Her2-neu) amplification was additionally evaluated by fluorescent in-situ hybridization (FISH). EGF-R/Erb-B3 and Erb-B4 were classified according to the EGF receptor score and Erb-B2 based on the DAKO score. Staining was scored from 0 to 3+ with scores of 2+ or 3+ considered positive. All specimens were reviewed by a reference pathologist and EGF receptor expression was correlated to the different histological components within the tumor specimen (mature teratoma (MT), Yolk sac (YS), embryonal carcinoma (EC), seminoma (S))

Table 1: Overexpression (+2/+3) of EGF-R, Erb-B2, Erb-B3 and Erb-B4 in pts with cisplatin-refractory GCT (No- of pts):

	MT (n=8)	S (n=2)	EC (n=6)	YS (n=15)	NGCM (n=1)
EGF-R	1*	-	-	-	1**
Erb-B2	2	-	-	-	1**
Erb-B3	1*	-	-	-	-
Erb-B4	-	-	-	1	-

*/** same pt

Results: Expression of the EGF receptor family was generally weak and as follows (only membrane bound expression): EGF-R: 1 pt: +3 (MT), 1 pt: +2 (transformed non germ cell malignancy (NGCM)); Erb-B2: 3 pts: +1 (3 EC), +2: 3 pts (2 MT, 1 NGCM); Erb-B3: 1 pt: +1 (MT), 1 pt: +3 (MT); Erb-B4: 1 pt: +1 (MT), 1 pt: +2 (YS). No gene amplification of ErbB2 was found and no further EGF receptor expression was observed. Conclusions: Members of the EGF receptor family are only rarely overexpressed in GCT. Thus, it would be difficult to select GCT pts who would benefit from new therapeutic agents including monoclonal antibodies and tyrosine kinase inhibitors acting on the EGF receptor family. Erb-B2 overexpression seems not to play a role for cisplatin resistance in pts with refractory GCT.

704

Extrazellular domain of BMP receptor IA blocks BMP-2 dependent upregulation of ID proteins in breast cancer cell lines

P. Hortschansky, J. Clement, M. Lichtenberg, G. Maubach, S. Wölfl (Jena, D)

Bone morphogenetic proteins (BMPs) are important regulators of cell growth, differentiation and apoptosis. BMPs are involved in organogenesis depending on mesenchymal-epithelial interactions. In the mammary gland communication between epithelial and mesenchymal cell layers is also essential. BMPs act via membrane-bound serine/threonine receptors which transduce the signal after binding of BMPs to intracellular signalling molecules, e.g. SMADS. We previously show that BMP-2 is able to regulate the expression of the ID proteins, which are involved in cell cycle control. cDNA representing the extracellular domain of the BMPRIA was expressed as a fusion protein with the maltose-binding protein in E. coli. The fusion protein was extracted, refolded and purified by chromatography. Finally, the fusion protein was cleaved and the extracellular domain of BMPRIA (eBMPRIA) was further analyzed for functionality. Surface plasmone resonance studies of BMP-2 binding on eBMPRIA coated BIACOREâprobe sensor tips confirmed the 2:1 stochiometry of the eBMPRIA/BMP-2 complex in solution after preincubation of BMP-2 with eBMPRIA within different molar ratios. We incubated MCF-7 cells with recombinant human BMP-2 (rhBMP-2) and eBMPRIA with a ratio of 1:2 (25 ng/ml, 50 ng/ml and 100 ng/ml BMP-2). After an incubation period of 4 h or 24 h under serum-containing conditions RNA was isolated and reverse transcribed. Levels of gene expression of IDs and SKY were estimated by PCR. We produced a recombinant protein representing the extracellular domain of BMP receptor IA which is biologically active. Incubation of the breast cancer cell line MCF-7 revealed that eBMPRIA is able to repress the action of BMP-2 on the expression of target genes. Addition of BMP-2 alone induced the expression of the target genes after 4 h and even stronger after 24 h. Addition of eBMPRIA alone does not change the expression levels. Simultanous application of BMP-2 and eBMPRIA led to complete repression of BMP-2 specific induction of expression. In contrast, the addition of eBMPRIA 15 min after the application of BMP-2 was not able to block BMP-2 action. In conclusion, we could show, that the extracellular domain of BMP receptor IA act as an antagonist of BMP-2 with regard to the expression of target genes. This work was supported by the Deutsche Krebshilfe e.V. (grant 10-1245-Hö2) and the TMWFK.

705

Pharmacological telomerase inhibition leads to telomere dysfunction and consecutive growth arrest of human cancer cells

U. Martens, M. Pantic, A. Schnapp, K. Damm (Freiburg, Biberach, D)

Telomerase, the ribonucleoprotein enzyme maintaining the telomeres of eukaryotic chromosomes, is active in most human cancers and in germline cells but, with few exceptions, not in normal human somatic tissues. Telomere maintenance has been attributed to unlimited growth potential of tumor cells and inhibition of telomerase may lead to telomere shortening and cessation of unrestrained proliferation. To date, no potent selective telomerase inhibitors have been identified which might improve cancer therapy. Here, we describe novel chemical compounds which are capable of selectively inhibiting the enzyme telomerase. To investigate the cellular consequences of long-term treatment with a telomerase inhibitor, we cultivated exponentially growing NCI-H460 lung carcinoma cells in the presence of BIBR1532 or BIBR1591, respectively. Although the compounds had no effect on short-term viability or proliferation, inhibitor treated cells slowed their growth after more than 100 population doublings and stopped proliferation after 140 PD associated with hallmarks of replicative senescence. Additionally, a progressive decline of mean telomere length (TRF) was found from 4 kb to 1.5 kb at population doubling 140 in contrast to stable telomere length in control cells treated with solvent alone. A similar telomere erosion and growth pattern was observed in HT1080 fibrosarcoma cells, MDA-MB231 breast carcinoma and DU145 prostate carcinoma cells which are deficient of p53. Significant loss of telomeric sequences was also detected using the Q-FISH technique (Nat Genet 18:76, 1998). Furthermore, we found an increased number of chromosomes with no telomere signal (3.75/metaphase) compared to control cells (1.1/metaphase) as well as the presence of dicentric chromosomes and chromosome-end-fusions. In summary, these findings indicate that potent and selective, non-nucleosidic telomerase inhibitors can be designed as novel cancer treatment modalities. Since the antiproliferative effect is dependent on the telomere length of a given tumor cell, clinical response will be delayed until chromosomes run out of their telomeres. In a clinical setting, the most likely use of telomerase inhibitors would be as an adjuvant treatment in combination with other cytoreductive therapy approaches. Another potential application would be postremission therapy in order to eliminate minimal residual disease. Clinical studies will have to reveal whether this promising approach will result in an improvement of cancer therapy in the future.

706

Analysis of the Rb cell cycle pathway identifies patients with good prognosis in esophageal squamous cell carcinoma

P.G. Hemmati, D. Stuckenberger, S. Hermann, H. Petrowsky, S. Hauptmann, M. Lorenz, V. Budach, B. Dörken, P.T. Daniel (Berlin, Frankfurt am Main, D)

Deregulation of the Rb-pathway and G1-restriction point control is a frequent event in cancer. A variety of studies identified single components of the Rb pathway as prognostic factors. We previously observed that intact p16ink4/CDCN2 gene expression is associated with good outcome in patients with esophageal squamous cell carcinoma (J Clin Oncol 2001, 19:2272-2281). In the present study, we addressed the role of additional regulators of cell cycle and apoptosis. A cohort of 53 patients was investigated retrospectively. All patients underwent R0-resection of squamous cell carcinoma of the esophagus with a curative intent to treat. Protein expression was quantified by immunohistochemistry. In addition, tumor DNA was screened for BAX frameshift mutations by fragment length analysis and for p53 mutations by SSCP-PCR. Overall median survival was 13.7 months. High p16ink4a/CDKN2 protein expression was associated with a median survival of 23.8 months versus 9.7 months with low p16ink4a/CDKN2 (P=0.011). Patients Cyclin D1 overexpression had a median survival of 5,7 months versus 15,0 months with low Cyclin D1 expression (p=0.0034). The best survival (median 27,9 months) with a significantly better overal survival as compared with p16ink4a/CDKN2 status alone was seen in a subgroup of patients whose tumors beared the combination of both favorite phenotypes, i.e. low Cyclin D1 and high p16ink4a/CDKN2 protein expression (p=0.0009). Furthermore, loss of pRb was associated with an impaired prognosis and the inclusion of pRb in this multigene analysis further improves the prognostic model. Nevertheless, the deregulated expression of p53 or p21Cip/Waf protein expression or p53 mutation had no influence on the statistical model for identification of patients with good or poor prognosis. We conclude that such an analysis of the Rb pathway of cell cycle regulation could be useful for individualisation of risk-adapted therapeutic strategies in the future, and suggest prospective studies to confirm these results.

Wissenschaftliches Symposium: Diagnostik bei akuten und chronischen Leukämien

707

Quantification of PRV-1 expression, a molecular marker for the diagnosis of polycythemia vera

S. Klippel, E. Strunck, S. Temerinac, R. Kralovics, R. Kralovics, J. T. Prchal, M. Griesshammer, H. Heimpel, H.L. Pahl (Freiburg, Ulm, D; Houston, USA)

To date, diagnosis of the myeloproliferative disorders Polycythemia vera (PV) and Essential Thrombocythemia (ET), relies on clinical criteria. We have recently described the overexpression of a novel hematopoietic cell surface marker, PRV-1 (polycythemia rubra vera-1), in granulocytes from patients with PV. Here we describe the development of a quantitative TaqMan® RT-PCR assay for the measurement of PRV-1 expression from 5 ng of granulocyte RNA. In each sample, both PRV-1- and GAPDH-expression were quantified by determining the number of PCR cycles required to reach a preset threshold fluorescence value. From averages of triplicate measurements, the ratio of PRV-1 cycles to GAPDH-cycles was determined (PRV-1/GAPDH). Using this assay, 48 patients with PV, 4 patients with secondary erythrocytosis (SE) and 33 healthy controls were analyzed. PV patients express significantly higher amounts of PRV-1 than either healthy controls or patients with secondary erythrocytosis (Table 1, p< 0.0001 and p < 0.001 respectively). Moreover, PRV-1 expression is not detected in patients with CML, primary familial and congenital polycythemia (PFCP) or a patient with Chuvash polycythemia. Because there is no overlap between the PRV-1/GAPDH-ratio values in PV patients and the healthy controls or the SE, CML, PFCP and Chuvash patients, the assay has a sensitivity and a specificity of 100% for the diagnosis of PV in our population. In order to determine PRV-1 expression in the closely related disorder ET, 29 ET-patients were also analyzed by RT-PCR. In addition, peripheral blood mononuclear cells from these patients as well as the SE patients, 31 PV patients and 13 healthy controls were tested for the presence of „endogenous erythroid colonies" (EECs). The growth of these EPO-independent colonies has previously been proposed as a diagnostic assay for PV. All of the PV patients tested displayed EEC growth, while none of the healthy controls or the SE patients did. As previously reported, a substantial proportion of the ET patients also display EEC growth. Among our ET patients, only those displaying EEC growth showed elevated PRV-1 expression, EEC-negative ET patients expressed normal PRV-1 levels (Table 1). Since it has been reported that most EEC-positive ET patients develop PV during long term follow-up, EEC positive, PRV-1 overexpressing ET patients may comprise a distinct patient population, which might represent „preclinical PV". The quantitative RT-PCR assay described here provides a rapid, highly specific and sensitive tool for the diagnosis of PV.

708

Sensitivity, quality and interlaboratory comparability of BCR-ABL quantification by RT-PCR - results of an interlaboratory test

A. Weißer, P. La Rosée, K. Merx, S. Kreil, C. Kuhn, M.C. Müller, S. Saußele, P. Paschka, T. Burmeister, T. Raff, M. Kneba, R. Hehlmann, A. Hochhaus (Mannheim, Berlin, Kiel, D)

Quantification of residual BCR-ABL transcripts after therapy provides important information for clinical decision making in CML. Several real-time RT-PCR methods have been established and offer sensitive, fast and reliable quantitative determination of mRNA transcripts. To determine the quality and reliability of qualitative and quantitative RT-PCR methods for BCR-ABL detection an interlaboratory test was organized. 23 centers participated; 13 employed the hybridization probe approach using the LightCycler system (Roche Diagnostics) and 10 employed the hydrolysis probe approach (TaqMan) using the ABI PRISM system (Applied Biosystems). Nine participants performed qualitative nested PCR reactions in addition. To analyze sensitivity and linearity of cDNA synthesis and RT-PCR independendly, cDNA and RNA samples were prepared from dilution series of BCR-ABL positive cell lines (b3a2 BCR-ABL, K562; b2a2 BCR-ABL, BV173; and e1a2 BCR-ABL, SD1) in BCR-ABL negative HL60 cells (logarithmic dilution series of K562 and BV173: 10E-2 to 10E-6; of SD1: 10E-1 to 10E-5) and negative controls. Results from 479 of 484 RNA samples (99.0%) and 471 of 484 cDNA samples (97.3%) results were provided by the participants. 61/65 results from RNA samples (93.8%) and 71/78 results from cDNA samples (91.0 %) were determined correctly negative. Positive samples were analyzed in three sensitivity levels (level 1: dilution of 10E-4 for K562 and BV173, 10E-3 for SD1; level 2: dilution of 10E-5 for K562 and BV173, 10E-4 for SD1; level 3: dilution of 10E-6 for K562 and BV173, 10E-4 for SD1). At the first level 272/276 results from RNA (98.6%) and 256/262 results from cDNA (97.7%), at the second level 49/73 results from RNA (67.1%) and 56/66 results from cDNA (84.8%), and at the third level 14/65 results from RNA (21.5%) and 28/65 results from cDNA (43.1%) were correctly determined BCR-ABL positive. In total 335/414 (80.9%) of positive RNA and 340/393 (86.5%) of positive cDNA samples were correctly determined positive. Despite using various individual housekeeping genes for standardization, analysis of quantitative PCR data revealed a linearity of the analysis for most participating labs. Results (ratio BCR-ABL/housekeeping gene) of three 10fold dilutions of K562, BV173, and SD1 cells were compared for the individual center. The median ratio for each dilution step ranged between 4.3 and 13.3. We conclude that (i) differences in the proportion of false negative results between RNA and cDNA samples reveal that the quality and standardization of reverse transcription is crucial for RT-PCR based tests, (ii) various methods for quantification lead to a comparable linearity of the results. However, the actual results are lab specific. The inter-laboratory comparability of results needs the agreement on a common housekeeping gene for standardization of BCR-ABL transcript levels.

709

Prospective quantification of minimal residual disease (MRD) in adult acute lymphoblastic leukemia (ALL)

M. Brüggemann, J. Droese, T. Raff, C. Pott, U. Scheuring, R. Reutzel, N. Gökbuget, D. Hoelzer, C. Bartram, M. Kneba (Kiel, Frankfurt, Heidelberg, D)

The prognostic relevance of MRD detection has been proven for childhood ALL. However, only few data exist concerning the importance of MRD in adult ALL. We quantified MRD at different time-points during the first year of therapy using rearranged immunoglobulin (Ig) and T-cell-receptor (TCR) genes as molecular targets. 64 adult patients with ALL who attained morphological complete remission during induction and had at least one detectable clonal immune gene rearrangement at diagnosis were included into the study. All patients were treated according to the German Multicenter ALL Study (GMALL) 04/89 or 05/93. Follow-up material was taken at up to 6 time-points during the first year of therapy. After sequencing of the clonotypic immune gene segments allele specific oligonucleotides (ASO) matching the hypervariable N-(D)-N region were designed. Quantification of follow-up material was performed by real-time quantitative (RQ) PCR using the ASO primers together with a consensus primer/probe combination. For 59 patients specific assays with a sensitivity of at least 10^{-4} could be established. Three of the five (60%) patients with MRD levels $>10^{-2}$ after induction phase I (week 5) relapsed, but only 3 of 21 (14%) with MRD levels $\leq 10^{-2}$. From week 13 (before consolidation I) onwards a MRD level of $>10^{-4}$ was associated with a very poor outcome: 11 of 15 (73%) patients with MRD $>10^{-4}$ relapsed but only 4 of 31 (13%) of those with MRD levels of 10^{-4} or less at any time point beyond induction phase II. Our study shows that monitoring adult patients with ALL at consecutive time points gives clinically relevant insight into the effectiveness of treatment. Based on these results MRD quantification is used to improve risk assessment and treatment selection for standard risk patients that are enrolled into the present GMALL 06/99 study. Actual data from MRD evaluation of these patients will be presented.

710

Association of specific cytogenetic aberrations with mdr1 gene expression in adult myeloid leukemia

M. Schaich, T. Illmer, U. Schäkel, B. Mohr, S. Soucek, G. Ehninger (Dresden, D)

Objective. Cytogenetics and mdr1 expression are established prognostic factors for treatment outcome in adult acute myeloid leukemia (AML). The association, however, between specific cytogenetic aberrations and mdr1 expression has not yet been examined in a large patient cohort.
Patients and Methods. We therefore looked for mdr1 gene expression at diagnosis within specific cytogenetic aberrations in 331 previously untreated adult de novo or secondary AML patients (not included t(15;17)) entered into the German SHG AML96 treatment trial. Results. 25% of all examined patients were positive for mdr1, whereas this proportion was significantly higher in patients with aberrant karyotypes compared to normal karyotypes (39% vs. 15%; p<0.001). Looking at specific cytogenetic aberrations the proportions of mdr1 positive AML patients were significantly higher within t(8;21), +8, +21, del(7q), del(5q), -7, abn(3q) and multiple aberrations. In contrast, no patient with inv(16) was positive for mdr1. Induction treatment outcome was influenced by mdr1 expression. Only 26% of mdr1 positive patients with aberrant karyotypes achieved CR criteria compared to 54% of their mdr1 negative counterparts (p=0.002). Furthermore, within abn(11q), +21, +22, -5, abn(3q) or the group of other monosomies no mdr1 positive patient reached CR, whereas their negative counterparts had CR rates often comparable to the CR rate of patients with normal karyotype. This was most impressive in mdr1 negative patients with multiple aberrations achieving a CR rate of 63% (p=0.019).
In the multivariate analysis age, disease status and mdr1 expression were the strongest independent predictors for induction treatment failure, whereas high risk cytogenetics lost significance. Interestingly, four of five relapses of t(8;21) AML occurred in mdr1 positive patients. Conclusion. Specific cytogenetic aberrations correlated with mdr1 gene expression in adult acute myeloid leukemia. Within patients with adverse or intermediate cytogenetic aberrations mdr1 expression predicts induction treatment failure. This might explain the divergence of the prognostic value of some of these aberrations in different treatment trials due to the amount of mdr-drugs used within the protocols.

711

Cytogenetic profiles in acute myeloid leukemia (AML) with FAB subtypes M0, M1 and M2

T. Haferlach, M. Klaus, S. Schnittger, W. Kern, S. Bursch, W. Hiddemann, C. Schoch (München, D)

Close correlations between cytomorphology and cytogenetics have been established for the FAB subtypes AML M4eo showing an inv(16) (p13q22)/t(16;16) and AML M3/M3v characterized by t(15;17)(q22;q21). For all other FAB-subtypes no specific cytogenetic abnormality has been identified. We asked the question, whether a pattern of distinct karyotype abnormalities can be found within the more immature myeloid FAB subtypes M0, M1 and M2. Therefore we evaluated 48 AML M0, 179 AML M1 and 425 AML M2 that had been sent to our laboratory for cytomorphological and cytogenetic analysis and compared these subgroups to the total cohort of 1714 AML cases. The incidence of abnormal karyotypes was significantly higher in AML M0 compared to M1, M2 and the total AML group (71% vs. 49% vs. 53% vs. 56%, p<0.02). Complex aberrant karyotypes were observed in 20%, 10%, 16% and 16%, respectively. The only balanced rearrangement that is postulated to be specific for one FAB subtype was t(8;21), which was observed in 12.5% of all AML M2 but only in 1.7% of all AML M1 and just one single case with t(8;21) was classified as AML M4. While the most common numerical abnormality trisomy 8 was observed in all FAB subtypes, rare single trisomies +13 and +14 were only observed in AML M0, M1 and M2. The incidence of trisomy 13 was 4.2%, 1% and 0.1% and of trisomy 14 4.2%, 4.2% and 0.5%, respectively. Trisomy 11 as the sole abnormality is also most common in AML M0, M1 and M2 (2.1% vs. 3.4% vs. 0.7%) and was only observed in a single case of AML M4 and M5b each. We conclude that FAB subtypes M0, M1 and M2 show some specific cytogenetic characteristics in comparison to subtypes AML M3-M7: t(8;21) and AML M2 are closely correlated. Trisomies 13, 14 and 11 alone seem also to be a marker for immature myeloid leukemias. However, the cytogenetic pattern is not homogeneous and further molecular techniques such as gene expression analyses using microarrays will hopefully elucidate the underlying pathways that lead to a comparable cytomophological pictures despite different cytogenetic abnormalties.

712

Different mechanisms lead to rearrangements of the MLL gene in cases with acute myeloid leukemia and t(10;11)

M. Klaus, S. Schnittger, T. Haferlach, M. Dreyling, W. Hiddemann, C. Schoch (München, D)

Abnormalities affecting the MLL-gene on chromosome 11 band q23 occur in 2-5% of acute myeloid leukemia (AML). Today up to 40 different translocation partner genes are known to rearrange with the MLL-gene. In AML, predominantly in AML M5a, a rare but recurrent chromosome aberration involves 11q23 and the short arm of chromosome 10. Breakpoints in 10p were described to be heterogenous. However, molecular studies have shown that the AF10 gene at p12 is a common translocation partner gene in cases with t (10;11) generating a 5´MLL-AF10-3´fusion product. Because of the opposite orientation relative to the centromere/telomere of MLL and AF10, this rearrangement can not result from a simple reciprocal translocation. To further define the kind of rearrangements we selected 10 patients with cytogenetically identified t(10;11) and one patient who was picked up by fluorescence in situ hybridization (FISH) screening with a MLL probe. The involvement of the MLL-gene was confirmed by FISH with a probe that spans the MLL-gene in all cases. Whole chromosome painting probes (WCP) for chromosome 10 and 11 were additionally used on the same metaphases that demonstrated the MLL rearrangement to prove the involvement of chromosomes 10 and 11. Four patients showed a simple balanced translocation between chromosomes 10 and 11 with breakpoints in 10p12 and 11q23. The rearrangement of the MLL-gene in three other patients was caused by insertion of 11q-material into 10p. One of those cases was not visible using G-banding or WCP FISH. In a further case the insertion of 10p-material into 11q led to the MLL-rearrangement. Finally, three patients showed a very complex rearrangement including a t(10;11) as well as an inversion of the long arm of chromosome 11, which led to a disruption of the MLL-gene. In conclusion, in patients with t(10;11) three different mechanisms of MLL gene rearrangement including a) balanced translocations (n=4), b) insertions of either 10p into 11q or 11q into 10p (n=4) and c) very complex rearrangements including a t(10;11) as well as an inversion of the long arm of chromosome 11 (n=3) have been observed. The opposite orientation of MLL and AF10 may partially explain the necessity of such complex rearrangements.

Wissenschaftliches Symposium: Myelodysplasien

713

Vaccination strategies in patients with MDS and AML

C. Scheibenbogen, U. Keilholz, A. Schmittel, E. Thiel (Berlin, D)

Recent advances in understanding the mechanisms of recognition and destruction of tumor cells by T lymphocytes have provided the basis for development of antigen-specific vaccines. The two most interesting candidate antigens for use in vaccination strategies in MDS and AML are the Wilms` tumor antigen WT1 and proteinase 3. Both antigens are overexpressed in myeloid blasts and contribute to leukemogenesis. In patients with MDS increased WT1 expression levels are a sensitive predictor of the evolution to AML. HLA-A2.1-binding epitopes have been identified, which can induce specific T cells with the ability to lyse leukemia cell lines. Evidence that these peptide epitopes are immunogenic in patients was recently obtained: We could demonstrate CD8+ T cells specific for WT1 epitopes in AML patients in complete remission. Proteinase 3-specific T cells were identified by tetramer analysis in CML patients in complete cytogenetic remission. Thus, these data provide the basis for the development of peptide vaccine strategies. The use of dendritic cell stimulating adjuvants like GM-CSF in combination with helper proteins is the most promising approach, which in melanoma patients was recently shown to rapidly induce specific T cell responses.

714

Pathophysiology and immunomodulatory therapy of hypoplastic MDS

W. Verbeek, M. Stadler, M. Faulhaber, A. Ganser, E.M. Weissinger, M. Aivado, U. Germing, S. Koschmieder, K.O. Kliche, H. Schrezenmeier, N. Frickhofen, K.M. Josten, C. Aul (Hannover, Düsseldorf, Frankfurt, Jena, Berlin, Wiesbaden, Duisburg, D)

Recently, immune mechanisms have been implicated in the pathogenesis of marrow failure in the myelodysplastic syndromes (MDS) suggesting similarities between aplastic anemia and subgroups of MDS, especially hypoplastic MDS. In both diseases autoreactive T-cell clones were found to suppress hematopoiesis. In MDS this T-cell response is thought to be triggered by the preneoplastic clone and is most likely directed against both MDS and residual normal hematopoiesis. Growth suppression of myeloid and erythroid progenitors can be mediated by peripheral blood CD8-positive MHC class I restricted T-cells. Skewed V beta T-cell receptor (TCR) spectra in MDS suggest an oligoclonal expansion of selected T-cell clones. Interestingly, immunosuppressive therapy with ATG can reverse TCR V beta skewing and the growth suppressive effect of T cells in some MDS patients. Successful immunosuppressive therapy with ATG in hypoplastic MDS has been reported first by Tichelli in 1988. In a German randomized phase II trial, 37 patients (RA n=23, RAEB, n =9, CMML n=3) have been treated with horse (ALG, Merieux) or rabbit ATG. Overall, 8 patients had a clinically meaningful response, 2 complete responses, 6 good responses with significant hematological improvement. All responding patients belonged to the RA group (35% CR and GR (loss of a preexisting transfusion requirement or increase in hemoglobin level > 1.5g/dl and/or increase of platelets to > 100000/µl and/or neutrophils to > 1500/µl). In this study clonality assays revealed a polyclonal pattern in 3 female patients responding to immunosuppressive therapy as opposed to a monoclonal pattern in 6 nonresonders, suggesting that a residual polyclonal hematopiesis is present in the responding patients. In Europe, a phase III trial comparing ATG and cyclosporine A versus supportive care which can include EPO treatment is currently being conducted by the Swiss Group for Clinical Cancer Research (SAKK) and the German MDS Study Group. The study enrolls transfusion-dependent patients with low and intermediate risk MDS. This study should significantly contribute to defining the role of immunosuppressive therapy in respect to hematological response and long term benefit in MDS.

715

The 5q-minus syndrome

A. Gianounidis, U. Germing, S. Haase, H. Willems, C. Aul (Duisburg, D)

Deletions of the long arm of chromosome 5 (5q-) have been described in a wide variety of hematological and oncological disorders including acute myeloid leukemia (AML), myelodysplastic syndromes (MDS), acute lymphoblastic leukemia (ALL), chronic myeloproliferative disorders, non-Hodgkin's lymphomas, multiple myeloma, and benign and malignant solid tumors. The 5q-syndrome, however, is a distinct entity among MDS with specific clinical, morphologic, cytogenetic, and prognostic features. This disease usually occurs in patients with low-risk MDS, i.e. refractory anemia (RA)(65%) or refractory anemia with ringed sideroblasts (RARS) (16%), but may also be observed in patients with refractory anemia with excess blast (RAEB) (16%) or refractory anemia with excess blasts in transformation (RAEB/t) (2%). Striking clinical features are the combination of pronounced anemia with macrocytosis and a normal to elevated platelet count that may reach up to 1 G/l, while the leukocyte count is usually normal or slightly reduced. The morphologic hallmark of the 5q-syndrome is the presence of middle-sized megakaryocytes with hypolobulated nuclei, Regularly, only one spherical nucleus may be identifiable. One third of the patients shows marked erythroblastic hypocellularity. Dysplastic features are uncommon findings in the non-megakaryocytic lineage. Considerable efforts have been made to identify a critical deleted segment within 5q. To date, two distal breakpoints at 5q31 and 5q32 are under vigourous investigation, spanning from the IL-9 gene (centromeric) to the IL-12B gene (telomeric). This region comprises several genes of great importance in normal hematopoietic growth and differentiation. It includes several interleukin genes, the macrophage colony stimulating factor 1 receptor (CSF1R), as well as genes being tightly regulated during myeloid differentiation: the monocyte antigene CD 14, the early growth response-1 gene EGR-1, CD 74 (the invariant chain associated with the major histocompatibility complex class II molecules) and others. However, the breakpoints identified by different investigators do not necessarily overlap, and additional research to isolate the critical genetic lesion is needed. Survival of patients with 5q-syndrome is usually more favorable than for other MDS patients. In a series of 43 cases, we found a median survival of 84 months for the entire group. Factors indicating a bad prognosis were increased blast count, thrombocytes below 0.15 G/l and increase in LDH and – in our series- a distal breakpoint other than 5q31. The Düsseldorf score as well as the IPSS were good predictors of survival and are useful tools for the estimation of prognosis.

716

Stem cell transplantation (SCT) in children and adolescents with myelodysplastic syndromes (MDS): Impact of prior chemotherapy and karyotype

C.M. Niemeyer, U. Kontny, M. Zecca, F. Zintl, U. Göbel, W. Ebell, D. Niethammer, J. Stary, Chr. Bender-Götze, J. Ortega, M. Egeler, G. Dini, G. Souillet, G. Giorgiani, C. Messina, E. vanWering, B. Strahn, P. Noellke, F. Locatelli (Freiburg, Jena, Düsseldorf, Berlin, Tübingen, München, D; Pavia, Genova, Padova, I; Prague, CZ; Barcelona, E; Leiden, The Hague, NL; Lyon, F)

MDS in children and adolescents is a rare clonal hematopoietic disorder characterized by multilineage dysplasia and defined genetic abnormalities. It may develop in patients (pts) with congenital or acquired bone marrow failure disorders, previous chemo- or radiation therapy, or recognized syndromes (secondary MDS). About 70% of the cases arise in children without known predisposing conditions (primary MDS). Here we report on 74 children (48 male, 26 female) with high grade MDS transplanted in a prospective multi-center study EWOG-MDS. 55 pts had primary, 19 secondary MDS. For classification of MDS the FAB system was applied with the exception that Auer rods did not qualify for the diagnosis of RAEB-t. Data were analyzed according to the most advanced FAB-type prior to SCT: 26 pts were classified as RAEB, 25 pts as RAEB-t and 11 pts as myelodysplasia related AML (MDR-AML). Cytogenetics revealed monosomy 7 in 21 pts, complex abnormalities in 9 and other abnormalities in 11. In 31 pts the karyotype was normal and in 2 unknown. Median age at transplantation was 10.0 years and median time from diagnosis to SCT 5.4 months. Pts were transplanted from an HLA identical or 1 antigen mismatch family (MFD) (n=24) or unrelated (MUD) (n=50) donor with (n=33) or without (n=41) prior intensive therapy. Myeloablative therapy consisted of busulfan 16 mg/kg, cyclophosphamide 120mg/kg and melphalan 140 mg/m^2. 20 pts suffered a transplant related death, 13 pts relapsed. The probability of event-free survival (EFS) at 5 years was 0.45 ± 0.18 for the whole group, 0.58 ± 0.30 for MFD and 0.39 ± 0.18 for MUD transplants. There was no difference in EFS among pts transplanted for primary or secondary MDS. Neither FAB type nor blast percentage at the time of SCT were predictive for relapse or EFS. Pts who had received prior AML-like induction therapy had a similar outcome (EFS 0.49 ± 0.19) compared to pts who had not received intensive therapy (0.41 ± 0.27). Analyzing pts with RAEB-t/MDR-AML separately, EFS was 0.51 ± 0.20 in the intensive therapy group (n=30) compared to 0.56 ± 0.25 among pts without intensive therapy (n=17). Pts with a complex karyotype experienced an inferior outcome (EFS 0.11 ± 0.20) compared to pts with monosomy 7 (EFS 0.58 ± 0.23), other abnormalities (EFS 0.48 ± 0.43) or a normal karyotype (EFS 0.50 ± 0.25) ($p<0.001$). We conclude that there is no evidence that intensive therapy prior to transplantation improves survival and that a complex karyotype may be a poor prognostic variable.

717

Gene specific demethylation following treatment with 5-aza-2'-deoxycytidine in MDS

M. Daskalakis, T.T. Nguyen, P. Guldberg, C. Nguyen, G. Köhler, P. Wijermans, P.A. Jones, M. Lübbert, Universitätsklinik Freiburg, Norris Cancer Center, USC, Institute of Cancer Biology, Pathologisches Institut Freiburg, Leyenburg Hospital (Freiburg, D; Los Angeles, USA; Copenhagen, DK; The Hague, NL)

Background: 5-Azacytidine and 5-Aza-2'-deoxycytidine are potent DNA methylation inhibitors. MDS patients (pts) treated with low doses of either one showed 47-64 % overall response rates (Silverman et al., ASCO Proc. 17:14a, 1998; Wijermans et al., J. Clin. Onc.18:956, 2000). In MDS and acute leukemia the p15/INK4B gene is frequently hypermethylated. In MDS pts treated with Decitabine (DAC), we examined the methylation status and expression of the p15/INK4B gene in order to better understand the in vivo mechanisms of DNA methylation inhibitors. Methods: We assayed the methylation status of the p15 promoter/exon 1 region in bone marrow mononuclear cells (MNC) of MDS pts before and during DAC treatment by analysis of bisulfite-treated and PCR-amplified DNA using methylation-sensitive single nucleotide primer extension (Ms-SNuPE), denaturing gradient gel electrophoresis (DGGE), and sequencing of subcloned alleles. The protein expression of the p15/INK4B gene was detected by immunohistochemical staining of bone marrow biopsies using a polyclonal antibody. Results: p15 was methylated in 15/23 pts (65%) and unmethylated in 9/23 pts (39%). 17 pts were sequentially analyzed following initiation of DAC treatment. In 9 of 11 pts, initial p15 hypermethylation was reduced by > 25%. Both, the DGGE analyses and DNA sequencing were indicative of induction of hypomethylation at individual alleles. p15 protein levels in MDS bone marrow MNC were similar to normal bone marrow in 9/20 pts (45 %), and very low or absent in the remaining 11 pts (55%). Sequential results of p15 methylation and expression levels were available in 10 pts. Normal p15 expression was observed in 2 of the 3 pts with unmethylated p15, and lack of p15 expression in 4 of the 7 pts with hypermethylated p15. During DAC treatment, these 4 pts showed both p15 re-expression and a decrease in p15 methylation. In the remaining pts, pre-treatment expression levels were either stable (4 pts) or increased (2 pts). Conclusions: In a subgroup of MDS pts hypermethylation and silencing of p15 expression were reversed after DAC treatment. In other pts, either enhanced expression in the absence of initial hypermethylation, or lack of p15 expression despite unmethylated p15 were noted. The emergence of partially demethylated epigenotypes, and re-establishment of normal p15 protein expression following the initial DAC courses implicate pharmacologic demethylation as a possible mechanism resulting in hematologic response in MDS. To investigate the pharmacologic activity of DNA methylation inhibitors further, analysis of a larger series of MDS pts is necessary.

Wissenschaftliches Symposium:
Präklinische und klinische Erfahrungen mit STI571

718

Phase II study of Glivec (STI571) in adult patients with Philadelphia chromosome positive acute leukemias

O. Ottmann, C. Sawyers, B. Druker, J. Reiffers, S. O'Brien, J.M. Goldman, S. Reese, R. Capdeville (Frankfurt, D)

Glivec is an Abl specific tyrosine kinase inhibitor with promising single agent activity in advanced Ph+ acute leukemia patients in a phase I study (Druker et al, NEJM, 2001). We conducted a phase II study in 18 centers in France, Germany, Italy, Switzerland, UK and USA to confirm and expand these results. 48 patients with Ph+ ALL, 8 with CML in lymphoid blast crisis (CML-LBC) and 2 with Ph+ AML were enrolled from September 99 to May 2000. Eligibility criteria included relapse following standard chemotherapy, or stem cell transplant, or refractoriness to such treatment. 65% of the Ph+ ALL patients had relapsed after prior therapy, while 35% were treatment-refractory. 21% of Ph+ ALL patients and 1 patient with Ph+ AML had previously received an allogeneic transplantation. The median patient age was 51 years, with a male to female ratio of 1:1. Glivec was administered orally at doses of 400 or 600 mg daily until disease progression or toxicity occurred. Patients were monitored for hematologic response, survival and toxicity. The hematologic response rate in patients with Ph+ ALL was 58%, with 15% achieving a complete remission including complete recovery of PB counts. At the time of analysis, median follow-up based on time on treatment was 2 months (range 0.5-8.7), with 3 of 58 continuing therapy. 34/48 ALL patients, 4/8 CML-LBC, and 1/2 AML had discontinued therapy due to disease progression, 5 were transplanted and 3 died during therapy. Median survival was 5 months. Frequent side effects attributed to STI571 were mild to moderate nausea. Vomiting, muscle cramps, edema, diarrhea, headache and skin rash, generally mild to moderate have also been commonly reported. Severe neutropenia and thrombocytopenia were observed in 62% and 48% of patients, respectively. In all but one case, the deaths were attributed to progression of disease or its complications. One patient died severely neutropenic and a possible contribution of STI571 could not be excluded. Glivec has significant activity as a single agent against Ph+ acute leukemias, with a favourable toxicity profile. Clinical trials examining the earlier use of Glivec and combinations with other treatment modalities are needed to improve duration of leukemic remission, some of which are ongoing.

719

Molecular and chromosomal mechanisms of resistance in CML patients after STI571 (Glivec) monotherapy

S. Kreil, T. Lahaye, M.C. Müller, A. Weißer, C. Schoch, U. Berger, W. Linkesch, P. La Rosée, A. Corbin, B. Druker, N.C.P. Cross, H. Gschaidmeier, R. Hehlmann, A. Hochhaus (Mannheim, München, Nürnberg, D; Graz, A; Portland, USA; London, UK)

Specific inhibition of the tyrosine kinase BCR-ABL by STI571 (Glivec) is a promising new therapeutic strategy in patients with chronic myelogenous leukemia (CML) in chronic phase (CP) refractory or intolerant to interferon alpha or in advanced disease. Despite significant hematologic and cytogenetic responses, refractoriness or resistance has been observed in a proportion of patients after variable periods of STI571 monotherapy. We investigated four molecular and chromosomal mechanisms of resistance in 33 CML patients (16 m., 17 f.; median age 60, range 29-70 years) refractory to STI571 monotherapy (400-800 mg p.o./day). Prior to STI571 therapy, one patient was in CP, nine were in accelerated phase (AP), 21 in myeloid and one in lymphoid blast crisis (BC). Patients were treated in three multicenter phase II studies or an expanded access program. The median duration of therapy was 79 (14-219) days. Prior to STI571 and at the time of resistance we determined (i) the expression of BCR-ABL transcripts by quantitative RT-PCR with glucose-6-phosphate dehydrogenase transcripts as control, (ii) the number of genomic BCR-ABL copies by interphase fluorescence in situ hybridization (IP-FISH), and (iii) clonal evolution by metaphase cytogenetics. (iv) The STI571 binding site of the BCR-ABL tyrosine kinase domain was sequenced from cDNA derived from resistant blasts. The median expression level of BCR-ABL transcripts, expressed as the ratio BCR-ABL/G6PD was 8.6 (1.2-59) prior to STI571 therapy and 7.2 (0.2-87) at the time of resistance (n.s.). In seven cases indications for a selection of resistant clones have been detected: (i) Overexpression of BCR-ABL was found in two patients with an >10fold increase of the ratio BCR-ABL/G6PD in peripheral blood leukocytes; (ii) genomic amplification of BCR-ABL was found in two patients by IP-FISH; (iii) additional chromosomal aberrations resulting in clonal evolution with aneuploidy were observed in two patients; and (iv) a point mutation of the ATP binding site of the ABL tyrosine kinase domain resulting in an E255V amino acid exchange was observed in a CP patient with myeloid BC after 142 days on STI571. The mutation has been confirmed by DNA restriction digest analysis and excluded in DNA and cDNA from CP prior to STI571 therapy. The CRKL phosphorylation assay demonstrated a significant reduction of BCR-ABL inhibition at the time of resistance. The mutation induced an increase of the IC50 for STI571 from 0.025 µM to >5 µM. We conclude that various molecular and cytogenetic mechanisms of resistance exist which may lead to a clonal selection of resistant cells and ultimately result in hematologic resistance to STI571 monotherapy. The observations might contribute to the rationale of trials investigating the combination of STI571 and chemotherapy.

720

Telomere length in peripheral blood granulocytes correlates with response to treatment with STI 571

T. Brümmendorf, I. Ersöz, U. Hartmann, A. Wahl, P. Paschka, S. Kreil, T. Lahaye, U. Berger, H. Gschaidmeier, C. Bokemeyer, P. Lansdorp, R. Hehlmann, L. Kanz, A. Hochhaus (Tübingen, Mannheim, Nürnberg, D; Vancouver, CDN)

Telomeres in vertebrates are composed of TTAGGG repeats and associated proteins. In most somatic cells telomere repeats are lost with each cell division. Studies on telomere length dynamics in peripheral blood lymphocytes and granulocytes derived from > 300 samples normal human donors using fluorescence in situ hybridization and flow cytometry (Flow FISH) were described previously (J. Exp. Med. (1999)190:157-167). In follow up studies of patients with CML, we observed substantial telomere shortening in peripheral blood leukocytes compared to healthy donors and to Ph- CD3+ T-cells isolated from the same patients (Blood (2000) 95:1883-1890). In view of these findings we were interested to investigate whether telomere length in peripheral blood granulocytes from patients with CML is correlated with response to treatment with the tyrosine kinase inhibitor STI571. Flow FISH results were expressed in molecular equivalents of soluble fluorochrome units (kMESF) either as absolute telomere length (TEL) or as age-adjusted granulocyte telomere length based on the control population mentioned above (deltaTELgran). A total of 286 samples from 123 patients ranging from 0 to 560 days of treatment were included in the analysis. Age-adjusted telomere length and level of significance when compared to controls varied substantially between samples derived from patients in chronic phase (CP, mean + s.e.: -0.2 + 0.4 kMESF; n.s., n=157) to samples from patients in accelerated phase (AP, -1.1 + 0.5 kMESF; p=0.03; n=50) and blast crisis (BC, -1.4 + 0.4 kMESF; p=0.0008, n=71). Telomere length in 33 samples that were acquired in complete cytogenetic remission did not differ from controls (0.8 + 1.0 kMESF; n.s.; n=28), whereas the shortest telomere length was found in samples from patients lacking any cytogenetic response (-1.5 + 0.4 kMESF; p=0.001; n=59) or from patients that eventually became resistant to STI571 in the course of their disease (-1.5 + 0.6 kMESF; p=0.009; n=45). Furthermore, we found a positive linear correlation between duration of STI571 treatment and age adjusted telomere length, probably reflecting a steadily increasing fraction of Ph- cells (with normal or less reduced telomere length) contributing to the peripheral blood cell pool in a substantial number of these patients.

721

Synergistic activity of STI571 with chemotherapeutic drugs

S. Fruehauf, J. Topaly, A. Hochhaus, W.J. Zeller, A.D. Ho (Heidelberg, Mannheim, D)

The tyrosine kinase activity of the BCR-ABL oncoprotein results in reduced apoptosis and thus prolonged survival of chronic myelogenous leukemia (CML) cells. The ABL-specific tyrosine kinase inhibitor STI571 was reported to selectively suppress the proliferation of the BCR-ABL-positive cells; we were interested whether combinations of STI571 and chemotherapeutic agents with different mechanisms of action would display synergistic activity in these cells. A tetrazolium based MTT assay was used to quantify growth inhibition of BV173 CML cells after 48 hrs of exposure to cytotoxic agents alone and in simultaneous combination with STI571. The drug interactions were analyzed using the median-effect method of Chou and Talalay. The combination index (CI) was calculated according to the classic isobologram equation; CI values of 1, <1 or >1 indicate additivity, synergism or antagonistic effects, respectively. At growth inhibition levels of over 50%, the strongest synergims (CI at 75% inhibition, p<0.05, n>3, each) were observed for STI571 +cytarabine (CI 0.68+0.06), +etoposide (CI 0.7+0.07), +mitoxantrone (CI 0.74+0.17), followed by treosulfan, mafosfamide, ACNU, carboplatin, gemcitabine, busulfan, thiotepa, hydrodyurea, taxotere, methotrexate, cladribine and topotecan which showed antagonistic activity with STI571 (CI 1.264+0.461). Further compounds that play a role in the treatment of advanced phase CML such as fludarabin, idarubicin or corticosteroids for lymphoid blast crisis are being tested. A potentiation of toxicity in BCR-ABL-negative leukemic cell lines or normal CD34+ progenitor cells by addition of STI571 to chemotherapeutic drugs was not observed. Our data provide a rationale to develop and test clinical protocols with synergistic combinations. On the basis of our favorable experimental results we have now initiated a clinical study for patients in myeloid blast crisis of CML in which mitoxantrone/etoposide (NOVE) and cytarabine are combined with STI571.

722

STI571-therapy in patients with chronic myeloid leukemia causes regression of bone marrow fibrosis

Ch. Beham-Schmid, U. Apfelbeck, G. Höfler, O. Tsybrovsky, H. Sill, O. Haas, W. Linkesch (Graz, Wien, A)

Clinical trials starting in 1998 revealed encouraging results in the therapy of CML- patients with STI571 (GlivecR, Novartis, Switzerland). Many data have been obtained about the efficacy and safety of STI571 treatment in CML patients, but no information exists about morphologic bone marrow (bm) changes. Therefore, a clinicopathological study on sequential bm biopsies from 23 patients (12 females, 11 males, median age 60.5 years) with Ph1+-CML, resistant to interferon therapy, was performed to elucidate the effect of STI571 treatment. The median duration of CML in the patients, treated at the University Hospital in Graz, was 3.4 years. 9 patients were in chronic (group I), 8 in accelerated phase (group II) and 6 in blast crisis (group III). STI571 monotherapy included 400 mg/day in group I, and 600 mg/day in groups II and III. Complete haematological remission after 3 months STI571 therapy was seen in 9/9 patients of group I, 6/8 of group II and 2/6 of group III. Bm biopsies were performed before STI571 treatment, as well as 6 weeks and 3 months after starting STI571 therapy and examined by means of morphology, immunohistochemistry and morphometry.
After 3 months STI571 treatment, corresponding to the haematological response, a statistically significant decrease in bone marrow cellularity was seen (p<0.0000001), which was more pronounced in group I- compared to group II- and III-patients (p=0.0021). This was due to quantitative normalisation of erythropoiesis, and enormous reduction of granulopoiesis and megakaryopoiesis. Before treatment, no significant differences in the amount of CD61+ megakaryocytes was found between the 3 patient groups.

Morphometric and statistical analysis during 3 months therapy revealed a highly statistical decrease of initially rather small megakaryocytes (p<0.0000001) with reappearance of normal-sized forms. This reduction of megakaryocytes was accompagnied by significant regression of bm fibrosis during therapy in specimens of all patients (p=0.000000001). At time of the first biopsy, bm fibrosis was detected in 18/23 patients with increasing density from group I to groups II and III. At the time of the second biopsy, fibrosis was detectable in 9/23 patients, and at the time of the third biopsy in only 4/23 patients. However, the amount of fibrous matrix was markedly decreased compared to the first biopsy. STI571, which was originally developed as a specific platelet derived growth factor-receptor inhibitor, was found to selectively inhibit the proliferation of the Bcr/Abl tyrosine kinase and to cause apoptosis of Bcr/Abl-positive leukemic cells. In this study we were able to demonstrate regression of bm fibrosis in CML patients in a remarkably short time. Our findings concerning regression of bm fibrosis contrast exertion of fibrosis caused by interferon and busulfan. In this view, STI571 may offer hope and new possibilities in the treatment of disease with bm fibrosis.

723

Inhibitory effect of STI571 on autonomous CFU-GM growth in patients with CMML

K. Geissler, L. Öhler, A. Berer, E. Kabrna, M. Kollars, S. Stengg, O. Haas, P. Valent, K. Lechner (Wien, A)

Chronic myelomonocytic leukemia (CMML) is a hematopoietic malignancy of the elderly that is characterized by leukocytosis with monocytes and granulocytic cells in all stages of development, marked dysmyelopoiesis, and unresponsiveness to aggressive chemotherapy. We have previously described that in vitro formation of myeloid colonies (CFU-GM) in the absence of exogenous growth factors (autonomous CFU-GM growth) is a typical feature of the CMML clone. We have also found that high autonomous colony growth in CMML correlates with increased disease activity. STI 571 an inhibitor of the BCR-ABL protein is going to have a major impact on treatment strategies for CML. Although CMML cells lack the BCR-ABL translocation other molecular alterations have been found in some patients which could be potential targets for STI 571 (including t(5;12) with the PDGFRß/tel fusion gene). Therefore, we investigated the in vitro effect of STI 571 on spontaneous CFU-GM formation in 7 patients with CMML. STI 571 at a concentration of 10 µmol did not significantly suppress stimulated CFU-GM formation in normal bone marrow, but significantly inhibited autonomous myeloid colony growth in 6 of 7 patients with CMML. As compared to control cultures, in vitro CFU-GM colony formation in the presence of STI 571 ranged from 14% to 94% with a median value of 46%. This inhibitory effect was unexpected since a t(5;12) translocation was not found in any of these patients. Although the mechanism of the suppressive effects of STI 571 in patients with CMML remains to be established, our data clearly suggest therapeutic evaluation of this drug in CMML, in which current treatment strategies are unsatisfactory.

724

In vivo effects of the ABL tyrosine kinase inhibitor STI571 on bone marrow engraftment in lethally irradiated mice

J. Hoepfl, C. Miething, R. Grundler, F. Bassermann, Ch. Peschel, J. Duyster (München, D)

STI571 selectively inhibits the BCR-ABL tyrosine kinase, the causative molecule in the pathogenesis of chronic myelogenous leukemia (CML) and other Philadelphia chromosome-positive (Ph+) leukemias. The application of STI571 results in elevated apoptosis of leukemic cells. In addition to the blockade of the ABL and BCR-ABL tyrosine kinase activity STI571 is capable of inhibiting the PDGF-receptor and the c-kit receptor tyrosine kinases, both playing a critical role in hematopoiesis. C-kit kinases act at multiple levels to promote cell survival, proliferation, differentiation and adhesion and are of particular importance in multipotential stem- and progenitor cells. Current standard protocols discontinue the application of STI571 during stem cell transplantation as it is believed to interfere with bone marrow engraftment. In order to examine the effects of STI571 on bone marrow engraftment, we used a syngenic mouse transplantation model. So far we transplanted eight Balb/c-mice with different amounts of bone marrow cells, ranging from 5*104 to 2,5*105. After lethal irradiation (800 rad) mice were treated with 0.5 mg STI571 p.o. bid, whereas the control group received 0.1 ml H2O p.o. bid. The peripheral blood count of all animals was examined at least every two days. When compared to the control group the transplanted mice treated with STI571 revealed delayed recovery of the peripheral blood count. No significant difference was observed with regard to the red lineage. However, STI571-treatment was associated with a modest delay of platelet and leukocyte regeneration of about 2-3 days. While platelets in the control group reached the normal level at day 11, STI571-treated mice showed full regeneration at day 14. Although the regeneration of the normal blood count was delayed by STI571 treatment, all animals regained normal cell numbers after 14 days. Thus, in this experimental model the application of STI571 in the course of bone marrow transplantation seems to be a therapeutic regimen with tolerable suppressive influence on the engraftment. This result could have implications for the treatment of CML patients. We are currently investigating the effect of STI571 on transplantations performed with sorted CD34+ cells. This will help to further identify the specific STI571-mediated effects on stem cells involving c-kit kinases and to provide useful information for the establishment of new protocols enabling most beneficial STI571-treatment.

Wissenschaftliches Symposium: Krebsprävention

725

Genetic epidemiology of cancer

K. Hemminki (Huddinge, S, Heidelberg, D)

It is estimated that some 1 % of cancer is caused by the currently known cancer syndromes and up to 5 % by highly penetrant single-gene mutations. These data apply to dominant Mendelian conditions, which can be assessed in family studies covering two or more generations. However, such studies provide no data on recessive mendelian conditions and have a limited resolving power on polygenic conditions. Consequently, apart from highly penetrant single-gene mutations, the estimation of the total hereditary contribution is extremely difficult. The risks posed by low-penetrance single-gene mutations, polygenes and recessive genes are poorly understood. A cancer study was carried out by pooling data from the Swedish, Finnish and Danish twin registries for a joint analysis. The aim of this study was to provide reliable estimates of genetic and environmental effects for the most common cancer sites and to assess the modification of such estimates by age at diagnosis. Data from 90000 twins were combined to assess the cancer risks at 28 sites for co-twins of twins with cancer. A structural equation model, MX was applied in estimating the proportions of variation in cancers due to environmental and inherited causes. The nonshared random environmental effect was the largest factor for all cancers, accounting for 58-82% of the total variation. Statistically significant heritability estimates were detected for cancers of the colorectum (35%), breast (27%), and prostate (42%). The estimates for the shared environmental effects ranged from 0 to 20% but none were statistically significant. There were no significant differences between sexes at any of the sites. The Swedish Family-Cancer Database contains now 10 million people, organized in families and their 1 million cancers, retrieved from the Swedish Cancer Registry. The Family-Cancer Database is the largest population-based dataset ever used for studies on familial cancer. The Database has been used in some 80 studies characterizing familial risks of various cancers. This Database has been used in modeling for cancer causation, using the same MX program that was used in the above twin study. Because of the overwhelming size of the data, most estimates for environmental and heritable causes were statistically significant. However, the results were not different from the twin studies. Environmental causes explained most of the total variation for all neoplasms except for thyroid cancer, for which heritable causes were largest. For example, the estimate for heritable causes of 25% in breast cancer was in line with the estimate from twins. A third source of data on familial causes of cancer comes from multiple cancers in the same individuals. The risks for second cancer tend to be much higher than those for the first cancer. Very high risks for second cancer were noted for nose, skin (squamous cell carcinoma), connective tissue and leukemia. We have analyzed the effect of family history on some cancers, such as breast cancer, and it is an important factor but affects a relatively small proportion of patients with second breast cancer. The data suggest that patients with second cancer include a subgroup with a strong genetic predisposition to cancer, which cannot often be predicted by a family history. Such risks would be typical of polygenic diseases, and our tenet in the recent work on second cancers has been that they may serve as a unique population model for polygenic cancers.

726

The primary prevention of breast cancer

R. Morant (St. Gallen, CH)

Prevention of breast cancer includes the prevention of relapse or metastases (tertiary prevention) as well as early detection, primarily by screening mammography (secondary prevention) . These established interventions have significantly reduced mortality from breast carcinoma. Primary prevention, however, aims at preventing the development of breast carcinoma by eliminating or reducing known risk factors, and by using various chemical or biological drugs known to interfere with carcinogenesis. Chemical primary prevention in women at high risk of developing breast carcinoma is an attractive option in comparison to the surgical alternative of bilateral mastectomy. One of the first substances tested in an Italian trial was fenretinide, which reduced in women with operable breast cancer the incidence of contralateral breast cancer. Fentretinide failed to significantly prolong survival. Four large randomized studies have examined the value of tamoxifen in the primary prevention of breast cancer. The women in those studies had at least a twofold increased risk of developing breast . Tamoxifen was given at a dose of 20 mg daily for five years in all studies. Tamoxifen blocks the effect of estrogen at its receptors in breast tissue. Tamoxifen has, however, also estrogenic effects. It blocks the progression of osteoporis in postmenopausal women and has a beneficial effect on blood lipids. The first trial at Royal Marsden Cancer Hospital studied 2494 women beginning in 1986. The NSABP study (National Surgical Adjuvant Breast and Bowel Project Group) recruited 13388 women within 3 years from 1992 onward. After four years of follow-up a highly significant reduction of breast cancer incidence of 49% was reported in 1998. This significant reduction in breast cancer incidence could not be seen in the Royal Marsden Cancer Hospital study nor in an Italian study, which started in 1992 and recruited 5000 women after hysterectomy. The IBIS study (International Breast Cancer Intervention Study) was started in 1993 and is still recruiting. Currently more than 6000 women mainly from England and Australia, but also from continental Europe are followed on the study. Tamoxifen has been registered in the US in 1998 for the prevention of breast cancer, based on the results of the NSABP trial. Detrimental side effects of tamoxifen are an increase in thromboembolic complications and an increased incidence of endometrial carcinoma . Currently new attractive substances are considered for further testing in the prophylactic setting, such as aromatase inhibitors, inhibitors of signal transduction and blockers of neoangiogenesis. The NSABP has started a study comparing tamoxifen to the SERM (Selective Estrogen Receptor Modulator) raloxifen, which had shown a reduction of breast cancer incidence in a study examining osteoporosis and seems to be associated with fewer side effects than tamoxifen.

727

Decrease in prostate cancer mortality following introduction of prostate specific antigen (PSA) screening in the Federal State of Tyrol, Austria, 1993-99

W. Horninger, G. Bartsch, H. Klocker, W. Oberaigner, G. Severi, C. Robertson, P. Boyle (Innsbruck, A)

INTRODUCTION AND OBJECTIVES: Several factors support the use of screening tests for early detection of prostate cancer. First, patients do not experience symptoms in the early stages, and are unlikely to seek medical attention until the disease has progressed. Second, improvements in detection methods have increased the prospect for identifying the disease at an early stage when the lesion is still organ-confined, more easily treated, and often curable. METHODS: In 1993 PSA testing was made freely available to males aged 45-75 years in the federal state of Tyrol, Austria, and a mass screening project was launched. The Tyrol is an Alpine region in the western part of Austria with 331.410 inhibitants (65.123 males between 45 and 75). Previously (1990-1993) both PSA and digital rectal examination had been used as screening tools but not within an organised programme. The screening project was performed in collaboration with general practitioners, medical examiners, urologists, and medical laboratories in the Tyrol Blood Bank of the Red Cross. Informed consent was obtained from all volunteers participating in the program. In case of elevated PSA levels the volunteers were advised to undergo further urological evaluation, while men with normal PSA levels were invited to have a repeat PSA test twelve months later. The mass screening program was provided free of charge by the health and social security services of the Federal State of Tyrol and the University Hospital of Innsbruck. RESULTS: In the first year (1993) 32.3% of the 65,123 males aged 45-75 years underwent screening; 68% of all men in this age range were tested at least once in the first five years of the study. The incidence of prostate cancer in the Tyrol reached a peak in 1994 and has declined since. Significant migration to lower stages and an increase in the number of organ-confined, potentially curable prostate cancers have been observed since the introduction of this screening program. However, the percentage of clinically insignificant lesions in the screened group did not increase. Mortality from prostate cancer among Tyrolean men aged 40-79, which had remained constant from 1970 to 1993, has now declined, whereas in other parts of Austria the mortality rates for males in this age range have changed much less. Based on the age specific prostate cancer mortality rates in Tyrol between 1986 and 1990 there were 17 fewer prostate cancer deaths in 1997 (32% decrease), 22 deaths fewer in 1998 (42% decrease) and 18 deaths fewer in 1999 (33% decrease) than were expected. This decrease in mortality is statistically significant ($p<0.05$). CONCLUSION: The present study shows that the policy of making PSA screening freely available to the Tyrolean population and participation of a high percentage of males aged 45-75 have led to an increase in the number of organ-confined, potentially curable cancers detected as well as to a reduction in prostate cancer mortality.

728

Strategies to improve inclusion of patients in a clinical cancer registry

R. Brunner, N. Frickhofen (Wiesbaden, D)

Incomplete or incorrect patient data is a notorious problem for clinical cancer registries (CCR). Given the fact that clinical information systems (KIS) are typically designed for administrative purposes, the quality of medical data is often poor and essential data such as histology, TNM formula, and treatment are usually absent. Many CCR thus use medical records or medical reports as a source of data to be entered into the registry. The CCR of the Dr.-Horst-Schmidt-Kliniken, a tertiary care hospital and „Onkologischer Schwerpunkt“ in a town of about 250.000, evaluated strategies to improve access to data from its cancer patients. Methods: Initially, secretaries writing medical reports were asked to routinely send a copy of any report of cancer patients to the CCR. After written consent of all department chairs and approval by the data protection commissioner, data managers of the CCR were granted electronic access to the central server with medical reports as well as to the SAP-based KIS. Data were entered into the database of the „Giessener Tumordokumentationssystem“, developed by the Institute for Medical Informatics of the University of Giessen. Results: Medical reports, electronically retrieved by the data managers of the CCR, proved to be the most reliable source of cancer patient data. Electronic access eliminated reliance on the cooperation of physicians and secretaries, and the overall quality of the information was good, even though there was significant variability among the departments. However, data were incomplete, since physicians did not write reports every time the patient was seen in the hospital. ICD-coded diagnoses taken from the KIS were confirmed to be incomplete and unreliable. From 710 patients treated in the Oncology Department in the year 2000, a random sample of 300 patients was evaluated with regard to the accuracy of KIS-diagnoses. 15% of the cancer diagnoses were miscoded; 5% of patients with a primary diagnosis of cancer were missing; 94% of concurrent cancers were not reported. This pattern changed dramatically with the introduction of data entry in preparation for reimbursement by the system of diagnosis related groupings (DRGs). In a sample taken from February to April 2001, essentially all patients with a primary diagnosis of cancer were documented in the KIS; all cancer diagnoses were correctly coded; and all concurrent cancers were reported. Such a high quality of data required inclusion of secondary diagnosis, since cancer or concurrent cancer in it self may not be the primary diagnosis entered in a DRG-driven documentation. In addition, the DRG-driven KIS for the first time contained almost complete data on the administration of chemotherapy and radiotherapy. Conclusion: Direct electronic access to medical reports may be the most reliable way to assure high quality of the data. Access to data entered under the pressure of DRG documentation may be the best way to assure completeness of the data.

729

Pharmacological Aspects of anticancer drug dosing

K. Mross (Freiburg, D)

Purpose: To review the current procedure of dose calculation practice for the treatment of cancer patients and propose a pragmatic procedure without normalization to the body-surface area (BSA) or body weight (BW). Methods: Results supporting a normalization to BSA (introduced in the late 1950s) were critically reviewed. Variables like pharmackokinetic parameters, pharmacodynamics, pharmacogenomics, drug-drug interactions, influence of age and altered organ (hepatic and renal) function were examined. Alternative methods of dose calculation as therapeutic drug monitoring (TDM), use of estimates of function, flat dosing and adaptive control were reviewed too.
Results: The rationale for BSA-based drug dosing is only weakly evaluated. A scientific adequate evaluation of this common used procedure has not been performed. A reduction of the marked interpatient variation of PK parameters of anticancer drugs by BSA-based drug dosing cannot be observed. PK parameters cannot explain the variation in drug effect (at the tumor or normal tissue) in general. PK parameters correlate (weak to strong) with toxicity and in some (but by far not all) tumors with the response. The BSA-based drug dosing is used to administer a drug near at the maximum tolerated dose (MTD) because most of the anticancer activity of anticancer drugs were observed there. Another approach would be to determine the biologically effective dose (BED) of a drug and this is not always the dose at the MTD. The TDM system is highly impractical, costly, in general not available und cannot be recommended in general. Two distinct treatment situation must be separated: adjuvant/curative therapy and palliative therapy. For the first treatment situation drug dosing should be always performed according the published data (evidence or at least reference based medicine). For the drug dosing of palliative treatments a flat dose with adaption to the observed toxicity is an appropiate method. The flat dose can be derived from the re-analysis of phase I and II data and from daily clinical practice. Patients who are predicted to have abnormal drug handling (by elevated laboratory findings, excessive toxicity observed during prior therapies, age-related reduction of organ function) should receive a modified flat dose. Conclusions: The routine use of BSA for dose calculation needs a reevaluation procedure within studies with GCP quality. It is in fact not a fully validated procedure. The FMT dosing (flat dose, modified dose, toxicity-adjusted dose) ensures in the palliative setting that each patients condition is considered. Such a simple dose calculation system should be evaluated prospectively in comparative studies.

731

Slow intravenous release of doxorubicin from pegylated liposomes (Caelyx): Pharmacokinetics of encapsulated as well as free doxorubicin

R.A. Hilger, C. Oberhoff, H. Richly, D. Strumberg, M.E. Scheulen, S. Seeber (Essen, D)

Caelyx (Doxil) is a doxorubicin encapsulated formulation of polyethylene glycol-coated liposomes with a long terminal half life (t^2) and a minimal volume of distribution in steady state (Vss). It has demonstrated activity against solid tumors with mild myelosuppression, minimal hair loss and a low risk of cardiotoxicity. For this new form of doxorubicin, a better pharmacokinetic profile than for the administration of free doxorubicin was obtained. It was the aim of this study, to analyze the plasma levels of encapsulated and free doxorubicin. For this purpose, Caelyx was administered i.v. to patients with various refractory malignancies up to doses of 60 mg/m^2 (doxorubicin HCl equivalent) within one hour, once every 28 days, for as long as patients respond to and tolerate treatment. Pharmacokinetic analysis of caelyx as well as for the released doxorubicin in plasma and urine was performed on day 1, 2, 3, 4, 5, 8, 15, 22 and 28 of each cycle. For the quantification of caelyx, doxorubicin, doxorubicinol and 7-deoxy-doxorubicinolon, a RP-HPLC method was established, using a gradient solvent system consisting of 28% acetonitrile and 72% aqua (20 mM di-potassium hydrogen phosphate, 20 mM 1-heptanesulfonic acid sodium salt, pH = 7.50 with ortho-phosphoric acid) at 0 min to 40% acetonitrile and 60% aqua within 30 minutes, with fluorometric detection (excitation wavelength: 470 nm, emission wavelength: 550 nm). Separation of encapsulated doxorubicin from free doxorubicin levels was performed by the use of a silica absorption method. Quantification of caelyx was done as free doxorubicin after destruction of the liposomes using HCl/methanol. The limit of detection for doxorubicin and its metabolites was 0.1 ng/ml. The intra- and inter-day accuracy for this assay varied from 5% to 15% (from 10 µg/ml to 0.1 ng/ml, respectively), the precision was in the range of 7%. Up to now, blood and urine samples (n = 360 plasma- and n = 300 urine-samples) of 19 patients were enrolled into this study. The obtained accumulation of the liposomal as well as the free doxorubicin in plasma was only marginal in this setting. The mean t^2 for caelyx was calculated to 2.5 days, total clearance (CLtot) was determined to less than 1 ml/min. There was no renal excretion (Clren) of the intact liposomes detectable. The mean t^2 for released doxorubicin was calculated to about 3 days, Cltot to 180 ml/min, with a renal excretion of about 1% within the first 48 h. The mean area under concentration vs. time curve (AUC) was calculated to 260 ng/ml x d, which is as least 3fold higher than after conventional application of the equivalent dose (60 mg/m^2) of free doxorubicin. The mean peak plasma concentration (Cmax) of free doxorubicin after a 1 h infusion of 60 mg/m^2 Caelyx was calculated to 50 ng/ml, which is only 1/20 compared to the equivalent conventional dosing.

732

Selective cytotoxicity of a diphtheria toxin-interleukin-3 fusion protein on acute myeloid leukemia stem cells

M. Feuring-Buske, A. Frankel, R. Alexander, B. Gerhard, D. Hogge (München, D; Winston-Salem, USA; Vancouver, CDN)

Acute myeloid leukemia (AML) arises from the clonal expansion of a malignant transformed progenitor cell. Recent work has demonstrated that a hierarchy of malignant progenitors exists in AML similar to that found in normal hematopoiesis. The leukemic cells which initiate longterm hematopoiesis in stromal co-cultures (AML LTC-IC) and suspension cultures (AML SC-IC) and engraft in NOD/SCID mice (NOD/SCID leukemia-initiating cells) are rare progenitors that express a cell surface phenotype similar to that of normal multipotential longterm culture-initiating cells (LTC-IC) and lympho-myeloid NOD/SCID mouse competitive repopulating units (CRU). Thus, these assays identify candidate leukemic „stem cells" that may be responsible for maintaining the malignant clone. In this study we show that a diphtheria toxin (DT) human IL3 fusion protein (DTLIL3), is capable of eliminating selectively leukemic stem cells while it spares their normal counterparts. AML cells from 7 newly-diagnosed patients were exposed for 24h to 50ng/ml DTLIL3 prior to culture AML-CFC, AML LTC-IC and SC-IC assays and injection into NOD/SCID mice. The mean (range) % kill of AML-CFC, LTC-IC and SC-IC was 82 (47-100); 55 (28-91) and 76 (59-92), respectively, with most surviving progenitors being cytogenetically normal. AML cell engraftment in the NOD/SL-IC assay at wks 4 and 8 was reduced by a mean (range) % of 97 (93-100) and 53 (0-100), respectively. DTLIL3-treated cells from 2 samples showed no evidence of leukemic engraftment in mice up to wk 16. In contrast, 3 normal bone marrow (BM) showed a mean % CFC kill of 49 and 64 with 50 or 250ng/ml DTLIL3, respectively, while no significant kill was observed in the LTC-IC and SC-IC assays. The CRU frequency of normal CD34+ lin- BM cells was also not reduced by DTLIL3. In subsequent experiments NOD/SCID mice which received AML blasts i.v. followed in 24h by 0.045µg/g DTLIL3 daily i.p. x 5 showed a mean % (range) reduction in AML engraftment of 83 (14-100), 63 (0-100) and 57 (0-98) after 4, 8 and 12 wks, respectively (n=6). Complete eradication of leukemia was observed with 2 of 6 AML samples following one 5d course of DTLIL3. Repeating the DTLIL3 treatment every 4 wks enhanced its effectiveness against 2 of 3 samples. Thus, DTLIL3 selectively kills primitive leukemic progenitors from most AML patients sparing their normal counterparts and is a very promising candidate for leukemia treatment as well as an agent for purging leukemia cells before autologous BMT or PBSCT.

733

Exon-14 skipping mutation of the DPD-gene and severe 5-fluorouracil related toxicity

W. Schwabe, M. Raida, J. Clement, P. Hausler, K. Ott, K.O. Kliche, K. Höffken, D. Behnke (Jena, D)

Deficiency of DPD the rate limiting enzyme in 5-FU catabolism has been linked to toxic side effects of 5-FU. The most prominent mutation of the DPD gene resulting in severe DPD deficiency is a G to A mutation in the GT 5'-splice recognition site of intron 14 (exon 14 skipping mutation). The corresponding mRNA lacks exon 14 and the enzymatic activity of the translated DPD protein is virtually absent. We developed an RT-PCR based assay suitable for routine identification of the exon 14 skipping mutation and screened a control cohort of 851 individuals as well as a cohort of 33 cancer patients reported by their physicians to have suffered from WHO grade 4 toxicity upon 5-FU containing chemotherapy. Within the control cohort 8 heterozygotes for exon 14 skipping mutation were detected (0.94%). Among the 33 patients with severe 5-FU related toxicity 7 individuals (21%) were heterozygous and one individual (3%) was homozygous for the exon 14 skipping mutation. All 8 patients had experienced WHO grade 4 myelosuppression. Lethal outcome was reported in the homozygous patient and two of the heterozygous cases. We conclude that carriers of the DPD exon 14 skipping mutation are at significantly increased risk to experience life threatening myelosuppression upon 5-FU treatment even when the allelic status is heterozygous. These findings are presently under scrutiny by a prospective clinical trial. Nevertheless, these data lead us to suggest routine testing for the exon 14 skipping mutation prior to 5-FU treatment.

734

Preclinical and clinical properties of methotrexate-albumin (MTX-HSA)

G. Hartung, C. Schmidt, K. Burk, H.H. Fiebig (Mannheim, Rostock, München, Freiburg, D)

MTX-HSA is covalently bound MTX to human serum albumin (67000 D) in a 1:1 molar ratio. With respect to conventional MTX, the prodrug MTX-HSA reveals improved pharmacokinetic properties. Human serum albumin is a promising carrier system providing a long half-life of about two to three weeks and assuring an increased tumor uptake as it is the main nitrogen and energy source for tumor growth. Preclinical trials demonstrated an increased uptake of MTX-HSA in solid tumors and enhanced cytostatic effects of MTX-HSA as compared to. The distribution pattern of native albumin and the MTX conjugate in healthy and tumor bearing rats is similar. Thus, the tumor exposition time of this new drug is decisively prolonged. Preclinical experiments revealed enhanced cytostatic effects of MTX-HSA when compared to conventional MTX. In vitro tests revealed cytostatic activity of both MTX and MTX-HSA on Walker-256 carcinoma (rat tumor) cells and on the human leukemia cell line NALM-6. In human tumor xenografts MTX-HSA showed activity in 18 from 24 tested models, in 7 it was superior to MTX. Three phase-I studies have been performed with repeated weekly and biweekly administrations and every 4 weeks. Dose limiting toxicities were stomatitis, thrombopenia, transient transaminitis and bilirubin increase. No clinical signs of allergy were observed. For phase-II studies an initial weekly schedule of 50 mg/m^2 was recommended, with prolonged administration intervals (every 2 to 3 weeks) for long term treatment in case of toxicity during the weekly administrations. The half life of MTX-HSA in serum ranges between 10 to 19 days, being compareable to native human albumin. Currently phase-II studies for mesothelioma, prostate cancer, renal cell cancer, glioblastoma and head neck tumors are activated. MTX-HSA has shown activity in single patients with mesothelioma renal cell carcinoma and glioblastoma. One patients with mesothelioma and one with renal cell carcinoma have show response for more than 5 years meanwhile.

Wissenschaftliches Symposium: Kleinzelliges Bronchialkarzinom

735

Multimodality treatment of localized small cell lung cancer

M. Thomas, P. Broermann, C. Droege, U. Schäfer, M. Semik, W. Berdel (Münster, D)

Chemotherapy produces high response rates in small cell lung cancer (SCLC) and significantly improves median survival time. In patients with limited disease (LD) thoracic radiotherapy following chemotherapy results in an additional 3-year survival benefit of 5 %, compared to chemotherapy alone. Treatment strategies with thoracic radiation early concurrent to chemotherapy seem to provide further improvement. Three out of 4 randomized trials comparing early concurrent radiation with the sequential approach reported on an 8 – 10 % superiority in 3-year survival, being 1 clearly significant and 2 with borderline significance. In addition, it has been demonstrated that early concurrent hyperfractionated radiotherapy could contribute to superior outcome compared to early concurrent radiotherapy with conventional fractionation. Nevertheless, even with early concurrent hyperfractionated radiotherapy local relapse appears in 1/3 of patients. Thus, beyond more effective treatment of systemic disease better control of local failure is warranted to improve prognosis. Whether surgery might contribute to this improvement is an issue open to discussion. Clearly, it has been shown that in patients with complete response (CR) prophylactic cranial irradiation reduces the cumulative incidence of brain metastases substantially (59 % vs 33 %; 3 years) and impacts significantly on survival (21 % vs 16 %; 3 years). Thus, multimodality treatment is the standard of care for LD-SCLC patients, with sequential thoracic irradiation and additional prophylactic cranial irradiation in CR-patients being the mainstay of therapy. Further improvement of survival rates is facilitated by radiation early concurrent to chemotherapy. The impact of surgery on local control and improvement of survival needs further clarification.

737

Cell adhesion and constitutive Akt activation correlate with chemo- and radioresistance in small cell lung cancer cells

M. Schuermann, A. Kraus, S.-O. Bachmann, I. Ferber, M. Gross, J. Schlegel, G. Suske, A. Neubauer (Marburg, D)

Most small cell lung cancer (SCLC) patients relapse within 12 months of combination radio-chemotherapy, due to the development of secondary chemo- and radioresistance. This phenomenon relates to the induction of tumour differentiation resulting in apoptosis-resistant, morphologically variant cells, which often lack the neuroendocrine expression of classic SCLC cells. Adherent SCLC cells in vitro have been shown to mimic this phenomenon, with characteristic epithelial transdifferentiation, loss of neuroendocrine markers and gain of NSCLC markers such as CD44 and cyclin D1. In this study we examined whether the process of cell adhesion per se can trigger secondary chemo- and radioresistance. Using MTT chemo- and radiosensitivity assays we found that the response to etoposide, cyclophosphamide and gamma radiation was significantly diminished in adherent SCLC cells in comparison to the parent suspension cell lines (27- to 70-fold based on comparison of IC50 values). Western blot analysis using phosphorylation-specific antibodies to AKT and MAP kinase revealed markedly elevated activation in adherent SCLC sublines, where the levels of phosphorylated BAD protein and activated NFkB were likewise increased. Subcultivation of the adherent sublines on uncoated petri dishes reversed their adherent phenotype immediately and under these conditions AKT activity reverted to low levels. These results suggest that cellular adhesion triggers AKT-dependent inhibition of apoptosis in SCLC cells, leading to secondary chemo- and radioresistance.

738

Topotecan (T) / cisplatin (DDP) 3 days versus topotecan / cisplatin 5 days in patients with metastatic small cell lung cancer (SCLC)

U. Seifart, K. Jensen, J. Ukena, J. Lorenz, S. Petrasch, H.G. Fuhr, U. Keppler, A. Neubauer, H. Staab, M. Wolf (Marburg, D)

Objective: Topotecan is a new and active drug in SCLC. In first line therapy and in sensitive relaps, response rate of 40% have been reported. A combination with a DNA cross linking agent was suggested to be highly effective, in particular as Topotecan also inhibit DNA repair. Therefore, a phase II trial was initiated to 1) assess the activity of Topotecan / Cisplatin in patients (pts) suffering from extensive disease SCLC and 2) to compare the conventional 5 day regime with an experimental 3 day schedule. Material and methods: Chemotherapy naive pts. with histologically confirmed ED SCLC, Karnofsky 70% or better, age 18-72 years, adequate bone marrow, renal and liver function and written informed consent were eligible for the study. A total of 84 pts. were planned to enter the study. Pts. were randomized to receive either T 1,0 mg/sqm d 1-5 and DDP 75 mg/sqm d5 (Arm A) or T 1,5 mg/sqm d 1-3 and DDP 75 mg/sqm d3 (Arm B). Six cycles were given at a 3 week interval. RESULTS: Up to date 86 pts. are available for survival and 73 for toxicity rates. Pts. characteristics were 68 male, 18 female, median age 61 years, ED II 94%, Karnofsky 70-80%: 56%, Karnofsky 90-100%: 44%of the pts. 56% showed an elevated LDH. The toxicity was as followed:

WHO III/IV	Arm A	Arm B
Anemia	36,5%	25%
Leukopenia	65,7%	43,2%
Thrombopenia	46,9%	37,8%
Vomiting	6,5%	25%
Alopecia	38,7%	47,2%

Hycamtin*

Topotecan: Neue Chancen nutzen

Hycamtin
Wirkstoff: Topotecan. **Zusammensetzung:** 1 mg Durchstechfl. enth. 1 mg Topotecan als Topotecanhydrochlorid mit einer 10-% gen Überfüllung; 4 mg Durchstechfl. enth. 4 mg Topotecan als Topotecanhydrochlorid Sonst. Best.: Weinsäure, Mannitol, Salzsäure und Natriumhydroxid. **Anwendungsgebiete:** Zur Behandlung von Patientinnen mit metastasierendem Ovarialkarzinom nach Versagen einer Primär- oder Folgetherapie. **Gegenanzeigen:** Patienten mit schweren Überempfindlichkeitsreaktionen auf Topotecan und/oder einen der Hilfsstoffe in der Vorgeschichte. Patienten mit bereits vor Beginn des ersten Zyklus bestehender schwerer Knochenmarkdepression, Neutrophilenzahl von $< 1{,}5 \times 10^9$/l und/oder Thrombozytenzahl von $\leq 100 \times 10^9$/l sollten nicht behandelt werden. Hycamtin ist in der Schwangerschaft und Stillzeit kontraindiziert. Besondere Patientengruppen (z.B. mit Leber- oder Niereninsuffizienz): s. Fachinformation. **Nebenwirkungen:** *Hämatologisch:* Knochenmarksuppression, schwere Neutropenie (Neutrophilenzahl $< 0{,}5 \times 10^9$/l), ggf. mit Fieber und Infektionen (z. T. mit Sepsis), Thrombozytopenie (z. T. schwer und transfusionsbedürftig), Anämie. *Nicht-hämatologisch:* Gastrointestinale Nebenwirkungen wie Übelkeit, Erbrechen, Durchfall, Verstopfung, leichte Bauchschmerzen, Mukositis. Haarausfall (z.T. vollständig oder stark ausgeprägt), Müdigkeit, Asthenie, Hautausschlag, Pruritus. Selten Anorexie, Unwohlsein, Hyperbilirubinämie, schwere Überempfindlichkeitsreaktion. Keine Anzeichen für Kardiotoxizität, Neurotoxizität oder andere Organtoxizitäten. Regelmäßige Überwachung des Blutbildes einschließlich der Thrombozytenzahl! **Dosierung, Art und Dauer der Anwendung:** Weitere Hinweise zur Dosierung und Wechselwirkungen: siehe Gebrauchs- und Fachinformation. **Darreichungsform und Packungsgrößen:** Pulver zur Herstellung einer Infusionslösung. 1 Durchstechflasche 1 mg (N1), 5 Durchstechflaschen 1 mg (N1), 5 Durchstechflaschen 4 mg (N1). Verschreibungspflichtig. Stand: Januar 2001.
SmithKline Beecham Pharma GmbH & Co. KG, 80791 München

SmithKline Beecham Pharma GmbH & Co. KG
Ein Unternehmen der GlaxoSmithKline-Gruppe
www.gsk-med.de *Hycamtin ist eine eingetragene Marke.

gsk GlaxoSmithKline

Primary tumor progression was seen in 18%. Arm A shows a median survival of 8,6 months and arm B of 7,6 months. Conclusion: Combination of both, T/DDP, is active in ED SCLC with a low primary PD rate. The hematological and non hematological toxicity was tolerable and comparable to standard T single agent treatment. Toxicity and median survival seems comparable in both arms.

739

Gene expression profiling of chemo- and radioresistant variant small cell lung cancer

A. Kraus, H. Specht, A. Wimmel, M. Schürmann (Marburg, D)

Acquired resistance to chemo- and radiotherapy occurs commonly during treatment of small cell lung cancer (SCLC) and correlates with the appearance in the tumour of variant SCLC cells, exhibiting an epithelial 'adherent' phenotype. This appears to relate to the induction of differentiation, resulting in apoptosis-resistant, metastasis-prone, morphologically variant (v-SCLC) cells, in which the expression of neuroendocrine markers seen in classic (c-) SCLC is lost. We observed that the suspension c-SCLC cell lines NCI-H69 and NIH-N592 form adherent clones, which morphologically resemble undifferentiated epithelioid cells of non-small cell origin. The hypothesis that adherent SCLC clones represent the clinical variant phenotype has been investigated by differential gene expression analysis. RT-PCR analysis was performed on RNA from suspension and adherent states of the c-SCLC cell lines. The adherent sublines down-regulated expression of various neuroendocrine markers, including chromogranin A, L-dopa-decarboxylase and bombesin, relative to the parent cell lines. Expression of these markers was also diminished or absent in the variant (v-) SCLC cell line NCI-H841 and in the two NSCLC cell lines investigated in parallel. Conversely, expression of the NSCLC marker CD44, cyclin D1, thrombospondin and endothelin, as well as that of the adhesion molecules integrin a2 and b4, but not b1, subunits, was upregulated in the adherent sublines and the NSCLC cell lines. Differential expression in NCI-H69 and its adherent subline of CD44, chromogranin A and integrin a2 was verified at the protein level by immunocytochemistry. These results suggest that c-SCLC cells can differentiate spontaneously into v-SCLC, allowing these sublines to serve as an in vitro model of variant SCLC differentiation.

740

Differences in tropism of rAAV-2 into bronchial carcinomas

U.-P. Rohr, Stahn, S., Wulf, M.A., Haas, R., Kronenwett, R. (Düsseldorf, D)

Despite recent advances in the treatment of lung cancer, long term survival remains rare. New strategies such as viral transfer of therapeutic genes might improve conventional therapies. One promising candidate is the recombinant adeno-associated virus type 2 vector (rAAV-2). To examine tropism of rAAV-2 into bronchial carcinomas we used 7 established lung tumor cell lines in an attempt to reflect the heterogenicity of this tumor entity according to their different histological subtypes.

The lung cancer cell lines H23, H322 (adenocarcinoma), H157 (squamous cell carcinoma), H1299, H460 (large cell carcinoma), H82, H69 (small cell carcinoma) were transduced using a rAAV-2-EGFP-vector. After transduction using a MOI of 50, cells were applied to FACS analysis for EGFP marker gene expression on single cell level. All cell lines derived from non-small cell lung cancer (NSCLC) such as adenocarcinoma (H23, H322), squamous cell carcinoma (H157) and large cell carcinoma (H1299, H460) were highly susceptible for rAAV-2 showing transduction rates between 60% and 98.9% with a mean transduction rate of 84.1% (+19.9) 5 days after transduction. In contrast, cell lines (H82, H69) of small cell carcinomas (SCLC) were resistant to rAAV-2 showing mean transduction rates of 1.3% (+0.4) and 1.1%(+1.2) 5 days after transduction, respectively. In an attempt to explain the differences of tropism the expression of receptors possibly influencing rAAV-2 uptake such as epidermal growth factor receptor (EGFR) and alfaV integrin was assessed by flow cytometry. All highly rAAV-2 susceptible NSCLC expressed EGFR and aV integrin while low transducable SCLCs did not express these receptors. In conclusion, rAAV-2 has a strong tropism for NSCLC but not for SCLCs which might be explained by receptor status of the different cell types. For NSCLCs genetic aberrations of tumor supressor genes such as p53, loss of p16 and FHIT expression are described at high frequency. Therefore, rAAV-2 might be a useful vector for the replacement of a defect tumor supressor gene in a gene therapeutic approach of lung cancer.

Postersitzung:
Akute myeloische Leukämien - Prognosefaktoren

741

Transcriptional profiling of the AML cell cycle identifies the c-myc - p14ARF pathway as an important predictor for relapse free survival

C. Müller-Tidow, S.K. Metzelder, K. Kügler, J. Schwäble, M. Mizuki, B. Steffen, W.-D. Ludwig, A. Heinecke, T. Büchner, W. E. Berdel, H. Serve (Münster, Berlin, D)

Modern technologies allow parallel processing of large sample numbers and bioinformatics has evolved into a reliable tool for biomedical sciences. In the current study we systematically analyzed gene expression profiles of more than 20 cell cycle regulated genes in bone marrow samples from 78 AML-patients and 6 controls using high-throughput quantitative RT-PCR. In total, more than 4500 independent quantitative RT-PCR reactions were performed. Hierarchical clustering of the expression data clearly separated patient samples from controls. Also, the cluster analysis revealed AML patient groups with different risks of induction failure (P=0.03). In addition, the genes clustered according to two established functional modules, separating genes of the cyclinD-RB-E2F and of the myc-arf-p53 pathways. Two genes of the myc-arf-p53 pathway were of prognostic importance: Low p14ARF (p=0.01) and high c-myc (p=0.001) levels independently predicted AML relapse. Blasts from all long-term survivors contained high p14ARF and low c-myc levels. High expression levels of c-myc were found to be associated with FLT3 receptor mutations in patients´ blasts. In subsequent in vitro experiments, stable transfected FLT3 mutations derived from patient samples induced c-myc expression in hematopoietic progenitor cells. Thus, the poor prognosis conferred by FLT3 mutations might be associated with c-myc induction. Taken together, our systematic analyses of the cell cycle in AML patients led to the identification of c-myc and p14ARF as independent prognostic parameters in AML. These data provide evidence that a high-throughput quantitative RT-PCR approach is feasible for transcriptional profiling and might help to identify prognostic subgroups.

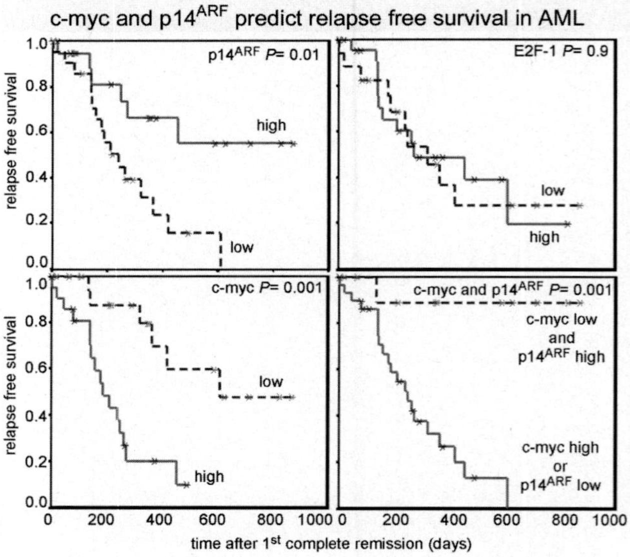

742

Clinical significance of cytogenetics in elderly patients with acute myeloid leukemia: analysis of 190 intensively treated patients entered into the multicenter trial AML HD98-B

S. Fröhling, C. Mayer, R. Schlenk, B. Addamo, C. Wieland, K. Döhner, H. Döhner (Ulm, D)

Karyotype is one of the most important prognostic factors in younger adults with acute myeloid leukemia (AML). To determine the prognostic value of pretreatment cytogenetics in elderly patients (pts.) with AML, we studied 190 cases with de novo or secondary AML other than acute promyelocytic leukemia entered into the multicenter treatment trial AML HD98-B. In this randomized trial, the impact of all-trans retinoic acid given during intensive induction and consolidation therapy in pts. above age 60 is being evaluated. Four cytogenetic categories were defined: normal, favorable [t(8;21), inv(16)/t(16;16)], unfavorable [-5/5q-, -7/7q-, 3 or more clonal aberrations],

Mittwoch, 3. 10. 2001

other [all other aberrations]. The prevalence of normal, favorable, unfavorable and other karyotypes was 51%, 5%, 26% and 18%, respectively. A total of 159 pts. had sufficient data to confirm their responses to induction therapy. The complete remission (CR) rate varied significantly among the groups, ranging from 52% for normal to 46% for other to 15% for unfavorable. This heterogeneity was largely due to the lower CR rate in the unfavorable group, compared with the other 2 combined (P<0.0001) Further analysis of the unfavorable group showed that pts. with aberrations of chromosome 5 and/or 7 had a particularly low CR rate (6%), whereas 4 of the 7 pts. with complex karyotypes but without -5/5q- and/or -7/7q- achieved CR. Unfavorable cytogenetics was associated with significantly inferior overall survival, compared with normal karyotype (P=0.0001) and other aberrations (P=0.007); the difference between normal and other groups was not significant (P=0.8). AML in the elderly is characterized by a very low frequency of favorable cytogenetic abnormalities. Pts. with unfavorable karyotypes do not seem to benefit from intensive remission induction therapy. However, further prognostic subdivision of this group may be possible.

743

Early allogeneic transplantation in patients with high risk acute myeloid leukemia defined by karyotype and response to first induction therapy: first results of the AML HD98A trial

R.F. Schlenk, S. Fröhling, B. Füllgarf, P. Dreger, J. Fischer, A. Glasmacher, K. Götze, W. Grimminger, U. Germing, F. Hartmann, E. Koller, H. Mergenthaler, H. Salwender, C. Waterhouse, F. del Valle, W. Weber, D. Bunjes, K. Döhner, H. Döhner, for the AMLSG Ulm (Ulm, Kiel, Karlsruhe, Bonn, München, Stuttgart, Düsseldorf, Homburg, D; Wien, A; Hamburg, Oldenburg, Trier, D)

The karyotype and the response to first induction therapy are one of the most important predicting factors for disease free and overall in acute myeloid leukemia (AML). In our ongoing multicenter treatment trial AML HD98A from the AML Study Group Ulm (AMLSG ULM) intensive consolidation therapy is stratified according to the karyotype and response to first induction therapy. High risk patients are defined either by karyotype (abn(3q), -5/5q-, -7/7q-, abn(12p), abn(17p), complex karyotype) or persistent AML after first induction therapy, which consists of ICE. Patients responding to double induction with ICE are assigned to HAM as first consolidation therapy. Patients with persistent leukemia after ICE-I are assigned to second induction therapy with A-HAM (cytarabine 3000mg/m^2/12h i.v. d. 1-3; mitoxantrone 12mg/m^2/day i.v. d. 2 and 3; ATRA orally 45mg/m^2/day d. 3-5 and 15mg/m^2/day d. 6-28). All high risk patients are assigned to allogeneic transplantation (Tpl) with grafts either from matched related donors, matched unrelated donors or haploidentical family donors irrespective of the response to A-HAM. At the interim analysis from March 2001, eighty-seven (23%) of 381 eligible patients were high risk patients. 22 high risk patients were defined by karyotype (group 1). All of them responded to ICE-I and were transplanted in complete remission (CR) after first consolidation with HAM, 7 received peripheral blood stem cells (PBSC) from matched related donors (MRD), 6 received PBSC from matched unrelated donors (MUD) and 2 received PBSC from haploidentical related donors (HRD). 65 high risk patients were defined by persistent leukemia after ICE-I (group 2). 59 patients received A-HAM, HAM or ICE as second induction therapy and 25 responded with partial remission (PR) or CR. 35 patients received an allogeneic Tpl, 16 with PBSC form MRD, 18 PBSC from MUD and one PBSC from HRD. The disease state at Tpl was 9 CR, 11 PR and 15 patients with persistent leukemia. In total 50 patients were transplanted after the following conditioning regimens: 36 patients with standard conditioning with total-body-irradiation/cyclophosphamid or busulfan/cyclophosphamid, in some patients with additional thiotepa or etoposide, 11 patients with antibody-targeted radionuclide therapy (188Re CD66 n=7) or bisphosphonate targeted radionuclide therapy (186Re HEDP n=4) in combination with busulfan/cyclophosphamid, 3 patients with melphalan, fludarabine and thiotepa. For rejection prophylaxis patients receiving grafts from matched unrelated donors or haploidentical related donors were treated with ATG. Tpl related mortality and relapse probability after Tpl are 46% (7/15) versus 29% (10/35) and 13% (2/15) versus 40% (14/35) in group 1 and group 2, respectively. The median follow-up is 11 months with 17 patients in CR after allogeneic Tpl. In conclusion our approach of early allogeneic Tpl in high risk AML is feasible. Further conclusions warrants longer follow-up.

744

Incidence and prognostic value of partial tandem duplications of the MLL gene in the elderly patient with acute myeloid leukemia: Prospective analysis within the multicenter treatment trial AML HD98-B

K. Tobis, S. Fröhling, R.F. Schlenk, H. Döhner, K. Döhner for the AML Study Group Ulm (Ulm, D)

Recently, in younger AML patients (< 60 years) molecular markers have been identified that are associated with an adverse prognosis. Partial tandem duplication (PTD) of the MLL gene represents such a molecular marker that has been identified with an incidence of 5-10%. Nevertheless, the role of MLL PTD in elderly AML patients (> 60 years) is not well defined. Aim of this study is to evaluate the incidence of MLL PTD and to determine its prognostic significance in the elderly AML patients (pts) treated within the ongoing AML HD98-B trial. So far, pretreatment samples of 174 pts were screened by Southern blot analysis. In total, 11/174 cases (6%) showed MLL rearrangements that were confirmed as PTD by RT-PCR and sequencing of the PCR product. In accordance to younger AML pts, PTD of the MLL gene was not associated with favorable cytogenetics [(t(8;21), inv(16), t(15;17)]. In pts presenting with normal cytogenetics MLL PTD was identified in 7/79 cases (9%) and in the group defined by other chromosomal aberrations in 4/69 cases (6%). A first correlation of these molecular findings with clinical outcome suggests a negative prognostic impact of this gene rearrangement in elderly pts with normal karyotype In 6/7 MLL PTD positive cases clinical information was available: 2 pts achieved CR and relapsed after 9 months. 4 pts presented with resistent disease and the overall survival was 3, 4 and 6 months, respectively, one pt is still alive with a follow up of 4 months. Our preliminary data show that the incidence and prognostic significance of MLL PTD in elderly AML are comparable to younger AML pts. In contrast, only a few pts exhibiting this gene rearrangement achieve a CR.

745

Partial tandem duplications of the MLL gene in acute myeloid leukemia:Prospective evaluation within the multicenter treatment trial AML HD98-A

K. Tobis, S. Fröhling, R.F. Schlenk, H. Döhner, K. Döhner, for the AML Study Group Ulm (Ulm, D)

40-50 % of adult patients (pts) with AML present with normal cytogenetics but have a heterogeneous clinical outcome. Thus molecular markers are warrented to identify subgroups of pts with a higher risk of relapse. The partial tandem duplication (PTD) of the MLL gene was the first molecular marker in this setting. This molecular rearrangement has been identified in 5-10% of AML with normal cytogenetics and has been associated with unfavorable prognosis . Our previous results from the multicenter treatment trial AML HD93 demonstrated the negative prognostic impact of PTD of the MLL gene in homogeneously treated pts. In this study the incidence of PTD of MLL was 8% in the group with normal cytogenetics. MLL PTD was not detected in the low risk group, whereas 5% of the high risk group had a PTD.Based on these data we initiated a prospective analysis for the identification of MLL PTD within the ongoing AML HD98-A trial for adult pts (16-60 years) with AML. Aim of this study is to prospectively evaluate the incidence and clinical significance of MLL PTD in a larger number of homogeneously treated pts. In the current treatment protocol pts receive double induction therapy consisting of idarubicin, cytarabine and etoposide (ICE), followed by early consolidation with high-dose cytarabine (18g/m^2) and mitoxantrone (HAM). For postremission therapy, pts are stratified according to their karyotype: low-risk [t(8;21),t(15;17)], intermediate-risk [normal cytogenetics, inv/t(16;16), t(11q23), all other karyotypes with £ 2 chromosomal aberrations] and high-risk [abn(3q), -5/5q-, -7/7q-, abn(12p), abn(17p), complex karyotypes].To date we screened pretreatment samples of 282 pts enrolled in the multicenter treatment trial by Southern blot analysis. In accordance with our previous study PTD of the MLL gene was not associated with the AML specific gene fusions [t(8;21), inv(16) and t(15;17); (n=64)]. The incidence of MLL PTD in pts with normal cytogenetics was 10/117 (9%) and in the group with other chromosomal aberrations 4/69 (6%). In the cases with normal cytogenetics correlations of the molecular findings with clinical outcome were done: 9 of the 10 pts achieved CR following induction therapy. 5 pts relapsed with a short disease free survival (range 2 to 16 months); 3 pts are still in CR with a disease free survival less than 6 months and one patient was lost to follow up.The results from our multicenter treatment trial AML HD98-A further demonstrate the negative prognostic impact of MLL PTD in AML with normal cytogenetics. A larger number of MLL PTD positive cases will allow statistical analyses that provide the basis using MLL PTD as a molecular marker for the stratification of postremission therapy.

746

Analysis of NRAS-mutations in AML according to cytogenetics and cytomorphology: high frequency in M4 and M4eo

S. Schnittger, C. Schoch, G. Mellert, S. Koneberg, W. Hiddemann, T. Haferlach (München, D)

Using cytogenetics and cytomorphology it is possible to classify acute myeloid leukemia (AML) into certain subgroups. Further subclassification of AML should be helpful also to develop more individualized therapies based on the underlying genetic defects. As cytogenetics has its limit in the resolution of the light microscope, we started out to investigate AML for mutations at the molecular level. It has already been shown that submicroscopic lesions like duplications in the MLL-gene and length mutations of the FLT3-gene play a mayor role in AML and help for further subclassification. In the present study we have analyzed NRAS-mutations that have been frequently described in AML, but have not been correlated to cytogenetics, so far. We performed a LightCycler screening for point mutations in the mutational hot spots of NRAS in codons 12/13 and 61. PCR products exhibiting an abnormal melting characteristic were directly sequenced. In an initial screening of 143 unselected AML the total frequency of NRAS-mutation was 24 (16.8%). Seventeen cases revealed a codon 12/13 and 7 a codon 61 mutation. It was obvious that the mutation was not distributed equally between the single cytogenetic subgroups. As a high frequency of NRAS mutation was detected in patients with inv(16) we analyzed patients with core binding factor leukemia in more detail and therefore we investigated a total of 60 patients with t(8;21) and 48 patients with inv(16). Three of 60 patients with t(8;21) (5%) and 16/48 (33.3%) of patients with M4eo were shown to be positive for a NRAS mutation. Thus, compared to the unselected group NRAS mutations are underrepresented in AML with t(8;21) and are significantly clustered in AML with inv(16)(Tab.1). Of 219 cases a cytomorphological classification was available. The NRAS mutations were significantly more common in M4 and M4eo than in all other subtypes (p<0.0001) (Tab2). This analysis confirmes that NRAS-mutations are common in AML but are not distributed equally in the different biological subtypes.

Tab. 1

Karyotype	analyzed n=238	NRAS mutation	
t(8;21)	60	3	5%
inv(16)	48	16	33.30%
t(15;17)	7	-	-
t(11)(q23)	6	-	-
-7/7q-	6	-	-
8	4	3	75%
complex	20	1	5%
other	20	3	15%
normal	67	9	13.40%

Tab. 2

FAB	analyzed n=219	NRAS mutation	
M0	4	1	20.00%
M1	16	1	5.90%
M2 /t(8;21)	60	3	5.00%
M2	42	2	4.80%
M3/M3v	7	-	-
M4	21	6	28.60%
M4eo	48	16	33.30%
M5	11	-	-
M6	10	1	10.00%

747

Single-Center Retrospective Analysis Of Outcome And Prognostic Factors In 139 Patients Aged > 55 Years With Acute Myeloid Leukemia (AML)

B. Behringer, R. Kunzmann, C. Schmoor, D. Behringer, M. Lübbert (Freiburg, D)

The majority of AML patients (pts) are over 60 years (yrs) of age but frequently underrepresented in multicenter trials. Since few studies focus on prognostic factors in unselected elderly AML pts, a retrospective analysis of 139 consecutive pts aged >55 yrs (median age 66, range 56-89) with AML diagnosed at a single center over an 8-yr period was performed (median follow-up 4 yrs). 68% of pts had de-novo AML, 32% secondary (s)AML. 74 pts (53%, median age 62 yrs) were treated with standard induction therapy protocols using full-dose cytarabine, anthracyclines -/+ VP-16, followed by cytarabine-based consolidation (1 course in 47%, 2 or 3 courses in 21 % of the remitters, no consolidation in 32%). 65 pts (47%, median age 73 yrs) received only palliative treatment. Univariate analyses of overall survival (OS) were performed for sex, age >/< 65 yr, de-novo vs. sAML, serum (s)LDH >/< 400 U/l, leukocytes >/< 50000/ml, induction therapy and karyotype. Cytogenetics (available in 68%) were normal in 41 pts, favorable (t[15;17]) in 1, adverse (3q abnormalities, 5q-, -5, -7, complex) in 23 and intermediate (others) in 13 cases. In pts receiving induction treatment, median OS was 0.4 yrs (range 0.1 - 8.3+), compared to 0.13 yrs (range 0-2) in patients not receiving induction. Overall CR rate was 47% (53% in pts with de-novo AML, 21 % in pts with sAML). Median survival from CR was 1.8 yrs. In a univariate analysis, significant factors for longer OS were induction, age <65 years (both p<0.001), sLDH <400U/l (p=0.002) and de-novo AML (p=0.03). In a multivariate analysis, significant factors for longer OS were sLDH < 400 U/l (p=0.0012) and induction therapy (p=0.002; difference possibly explained also by selection criteria not captured, such as performance status). Pts with normal karyotype had a lower risk of death than patients with any chromosomal abnormality (difference not statistically significant). Thus selected AML patients aged > 55 yrs and receiving standard-dose induction treatment had significantly longer OS than patients not receiving induction. This has previously been shown in a randomized trial (Löwenberg et al., J. Clin. Onc. 7:1268, 1989). In a multivariate analysis, sLDH emerged as an independent prognostic factor for OS. The effect of treatment in this patient group needs to be investigated in prospective trials in which clinical parameters of high relevance in elderly patients (such as performance status and comorbidity) are controlled for.

748

Role of immunophenotype as independent prognostic factor for risk stratification in AML

R. Repp, U. Schäkel, S. Soucek, U. Pascheberg, G. Helm, H. Wandt, W. Aulitzky, H. Bodenstein, R. Kuse, H. Link, G. Ehninger, M. Gramatzki (Erlangen, Dresden, Dortmund, Nürnberg, Stuttgart, Minden, Hamburg, Kaiserslautern, D)

Cytogenetic analysis has significantly contributed to assign AML patients to different treatment arms according to prognosis. However, only a limited number of patients has such informative chromosomal abnormalities. Here, we present data about the immunophenotype of 716 newly diagnosed AML patients treated in the German SHG-AML trials '91 and '96. A correlation with cytogenetic data could be done in 655 patients and 715 patients were evaluable for prognosis. Immunophenotyping allowed to define for each FAB subtype a characteristic set of informative antibodies. Some markers were closely associated with a certain karyotype. With a median follow-up of 2.5 years patients with AML blasts negative for CD9, CD13, CD11b, CD10 and CD34 had superior overall survival. In addition, the expression of antigen CD38, CD56 and CD64 and coexpression of 'T-marker' CD7 and TC-12 was associated with a significant better prognosis. Cox-regression analysis showed CD9, CD13, CD34, CD38, and CD56 as well as CD7 plus TC-12 as independent factors. With a step up approach to include cytogenetic, morphologic, and immunophenotypic parameter in patients <60 years (excluding M3) only CD13 and CD11b remained highly significant regarding overall survival. A prognostic score with these two markers defines three groups of 42, 179 and 156 patients, respectively, with a 3-year overall survival of 62.0%, 39.4%, and 28.1%. The highly significant differences between all groups remain after inclusion of patients without obvious karyotype abnormalities with overall 3-years survival rates of 74.6%, 44.2%, and 33.4%. In patients > 60 years (excluding M3) CD11b, CD19, and CD64 were identified. Thus, immunophenotyping is not only helpful for diagnosis but is of independent significance for prognosis and may be helpful for risk stratification in AML patients.

749

Is the inversion 16 positive acute myeloid leukemia an immunogenic form of the AML?

K. Ihlow, N. Usluoglu, S. Hoppmann, O. Christ, M. Hoeck, H.B. Pralle, A. Banat (Giessen, D)

As already described it is possible to generate leukemic dendritic cells from acute myelogenous leukemia (AML) cells by addition of distinct cytokines in culture to induce leukemia specific T cell response. In this study we report the ability of AML blasts with inv(16) karyotyp to stimulate allogeneic and autologous T cells without additional cytokines. AML blast cells were harvested from peripheral blood or bone marrow of 12 patients with AML. Theses cells with a variety of chromosomal abnormalities were studied for their ability to induce a significant T cells response. In 10 of the 12 cases the blasts changed their surface marker expression pattern after two weeks in proliferating culture while restored their typical morphology. They lost the expression of the progenitor markers like CD33, CD13 and CD34 and became expression of the costimulatory molecules B7.1 and B7.2, but failed to express the dendritic marker CD1a as well as the monocyte marker CD14.. In all of four cases the fusion protein CBFb/MYH11 for inv(16) could still be determined by western blotting in the following culture intervals. Allogeneic as well as autologous T cells could be activated by this cells followed by a significant proliferation with an classical expression of the activation molecules CD25 and CD69. On the other hand control AML blast with different karyotypes induced only a minimal initial activation without a significant proliferation of T cells. Beyond initial activated T cells could be expanded in a polyclonal way and were able to lyse inv(16) AML blasts significantly and specific in a MHC class I and B7.1 dependent manner. In this study we have shown that AML blasts cells from inv(16) positive patients have a distinct position in the valuation of immunogeneity. Furthermore we could illustrate that the inv(16) positive AML cells were able to differentiate away from the in vivo selection pressure into a immunogenic cell with significant stimulation capacity.

750

Individual prognosis in adult AML is well predicted by in-vitro chemosensitivity testing

P. Staib, T. Schinköthe, T. Siebert, R. Henrichs, T. Dimski, C. Schoch, M. Reiser, D. Voliotis, B. Lathan, V. Diehl (Köln, München, D)

In-vitro chemosensitivity tests like the differential staining cytotoxicity (DiSC) assay may be a useful tool for the prediction of the individual response to chemotherapy and also survival in patients (pts) with acute myeloid leukemia (AML). We prospectively investigated the predictive value of the DiSC assay in adult pts with AML by using the chemosensitivity index C_i, a newer evaluation system we described before. Patients and Methods: Pts with de-novo AML received TAD/HAM and pts with relapsed or secondary AML Ida-FLAG / sHAI±fludarabin for induction therapy. The DISC assay was performed as described previously (Weisenthal 1983) using minor modifications. The chemosensitivity index C_i was calculated according to dose-response curves for each clinically used drug and the area under the curve (AUC). The cut off point between resistance and sensitivity was adjusted at 0.5 for each drug by the AUC data of a subgroup of clinically resistant patients: C_i £ 0.5 predicts resistance, $C_i > 0.5$ sensitivity to the drug. Results: 154 pts are evaluable; 123 received TAD/HAM and 31 Ida-FLAG or sHAI ± fludarabin for induction therapy. 100 pts (65%) achieved CR and 38 (24.7%) reduction of BM blasts <10%, in both groups the maximum C_i (C_i-max) was >0.5 for at least one drug clinically used (TP = true positive correlation). 14 pts (9%) were non-responders and identified with C_i-max £0.5 of all drugs given for therapy (TN = true negative). 2 pts (1,3%) with C_i-max >0.5 did not reach CR (FP = false positive) (c^2 p<0.001). There were no false negative correlations. Overall predictive accuracy was 98.7%. Pts with C_i-max > 0.5 lived significantly longer with 26 month (median) than pts with C_i-max £0.5 who lived only two months (logrank, p<0.0001). Karyotypes were available in 123 pts. A cox regression multivariate analysis identified the chemosensitivity index C_i as the strongest independent prognostic factor for overall survival (p<0.0001) compared to the karyotype and age £/> 60 years (p<0.001 for both). Conclusions: In adult AML the DISC assay evaluated by the chemosensitivity index C_i provides a strong prognostic factor and, also, assessment of the pts´ individual prognoses. Especially, primary resistant disease was accurately predicted at the earliest time possible. It is justified that the C_i may be used for risk-adapted therapy strategies.

751

Gene expression profiles in good vs. high risk acute myeloid leukemias (AML)

M. Ritter, A. Neubauer (Marburg, D)

The AML is a heterogenous disease, about 30% of AMLs show primary resistance to chemotherapy, age and karyotype have been identified as the most important prognostic factors. Humans suffering from AML carrying aberrations of core binding factor (CBF)-complex show a good response to chemotherapy and a high cure rate using high dose ara-C containing regimens. On the other hand, AMLs with complex karyotypes or deletions of chromosomes 5 and 7 are poor responders to standard chemotherapy and have a very low cure rate. In cooperation with PD Dr. Wölfl, Jena, we have thus far analysed several AMLs, ALLs and samples from healtyh blood donors. We first used the „human gene expression array" (Clontech). Due to the high costs only a small number of samples could be analysed with the comercial arrays.so far. Therefore in cooperation with Prof. Dr. Eilers, IMT, Marburg we are now using an array-unit (MWG) to print our own slides. On one slide 3.000 genes can be printed. Using two flourescence colours on one slide two different samples can be analysed at once. In order to obtain comparable gene expression data each slide is hybridized with a sample and an internal expression standard. We therefore looked to the expression profiles of different leukemia cell lines. Kasumi and NB-4 showed reproducable and mainly balanced expression profiles and were thus used as internal standards. We are now hybridising AML-samples to our new slides. The obtained data will be correlated with clinical data (response; remission duration; overall survival) using our bioinformation facility. In summary, we hope that from these data, we will deduce clinically relevant gene expression data. For instance, we may find signal transduction pathways previously not implicated in leukemogenesis.

752

Methylation pattern of the E-cadherin promoter in two different subgroups of acute myeloid leukemia (AML)

K. Hölzle, M. Kersting, A. Wild, B. Alpen, A. Burchert, M. Ritter, C. Thiede, G. Ehninger, A. Neubauer, M. Schuermann (Marburg, Dresden, D)

Gene silencing by aberrant promoter methylation has been reported to occur frequently in patients with acute myeloid leukemia (Melki et al., Cancer Res. 59, 3730), affecting the genes coding for E-cadherin and HIC1 in 70% and 80% of all cases, respectively. To understand the process of aberrant hypermethylation in more detail, we have studied the DNA methylation pattern for both genes in two important cytogenetically defined subgroups of AML patients: in patients with t(8;21) translocation (favourable prognosis) and patients with monosomic deletions (del) of either chromosome 5 or 7 (poor prognosis). DNA from 9 patients with t(8;21) translocations, 2 patients with del 5, 5 patients with del 7 and 1 patient with both del 5 and del 7 were obtained from a sample collection of marrow blasts taken at the time of diagnosis during recruitment to the SHG-AML96 study trial. As control, bone marrow of patients undergoing bone surgery was analysed. The methylation status was investigated using the genomic bisulfite sequencing method, which was performed on 10 individual clones per patient. The number of CpG islands methylated in each individual was recorded. While methylation of individual CpG islands was observed only occasionally in the control group (mean 3.6%), we found methylation of the E-cadherin gene at multiple sites in most of the AML patients. Surprisingly, blasts with t(8:21) were more frequently methylated in the E-cadherin locus (mean 34.1%) as compared to those with monosomy 5 or 7 (mean 14.7%). The extent of promoter methylation seems to vary significantly in different subgroups of AML patients. Since hypermethylation directly affects gene expression, we conclude that the methylation pattern of blasts may provide an important parameter for assessing prognostic development and potential therapeutic response in the future.

753

Prognosis of acute myloid leukemia patients with an MLL/AF6 fusion gene detected by reverse transcriptase polymerase chain reaction (RT-PCR)

A. Kohlenberg, J. Krauter, S. Wittwer, H. Kirchner, A. Raghavachar, L. Bergmann, D. Hoelzer, W. Fiedler, M. Lübbert, L. Noens, G. Schlimok, R. Arnold, A. Ganser, G. Heil (Hannover, Ulm, Frankfurt, Hamburg, Freiburg, Augsburg, Berlin, D; Gent, B)

11q23 translocations involving the MLL gene fused to different partner genes occur in about 5-15% of patients with acute myeloid leukemia (AML) and are associated with an unfavourable prognosis. We wished to analyse the frequency and prognosis of MLL/AF6 [t(6;11)] positive cases among 293 de novo AML patients who were enrolled in two consecutive multicenter trials, the AML 02/95 and 01/99 study. For the detection of the MLL/AF6 fusion transcripts a modified RT-PCR protocol according to Pallisgaard et al. (Blood, 1998; 92: 574-588) was used. In eight (2.7%) of the 293 patients the leukemic blasts displayed MLL/AF6 fusion transcripts. Cytogenetic analysis showed a translocation t(6;11) in three of these patients and a deletion of 11q23 in another four patients. In one patient, no karyotypic abnormalities could be detected. The median age of the 8 patients was 47 years (range 36 to 59 years) versus 44 years of the whole patient cohort. All patients received a combination of Idarubicin, Etoposide and Cytarabine (IVA) for the first induction course. Four patients had a good response to IVA I (<5 % blasts in the day 15 bone marrow) while in another four patients a less effective cytoreduction could be achieved. In all eight patients a complete remission of the disease could be achieved by a second induction course (double induction). Consolidation treatment consisted of a HD-Ara-C based chemotherapy in two patients, three patients were auto- and another two patients were allografted. One patient received no consolidation therapy. Six of the eight patients relapsed with a median disease free survival of four months. One patient is in ongoing first CR two months after allo-PBSCT and one patient was lost to follow up. In none of the four patients receiving salvage therapy with Mitoxantrone, Topotecan and Cytarabine or Cytarabine in combination with Etoposide, a second CR could be induced. Only one of the six relapsed patients is still alive in second CR after a salvage protocol with Mitoxantrone, Topotecan and Cytarabine followed by a second autologous and a subsequent allogeneic PBSCT. Median overall survival of the seven evaluable patients is 12 months (range 5 – 18 months). In conclusion MLL/AF6 RT-PCR identifies a group of high risk AML cases with a high initial chemosensitivity but a poor long-term outcome. It remains to be shown whether allografting early in remission is a suitable approach to improve the outcome of these patients.

754

Tryptase as a novel disease-related marker of acute myeloid leukemia (AML)

W.R. Sperr, J.-H. Jordan, M. Baghestanian, A.W. Hauswirth, S. Natter, P. Samorapoompichit, G.H. Schernthaner, R. Valenta, K. Geissler, K. Lechner, P. Valent (Wien, A)

Alpha- and ß-tryptase genes encode serine proteases that are abundantly expressed by human mast cells. Under physiologic conditions other myeloid cells are virtually tryptase-negative. However, mast cell tryptases are also expressed in several myeloid leukemia cell lines. In this study, serum tryptase levels (a- plus ß-tryptases) were determined in 150 patients with acute leukemias (de novo AML, n=108; secondary AML, n=25; ALL, n=17) by fluoroenzymeimmunoassay (FIA). In healthy subjects, tryptase levels range between <1 and 15 ng/ml. Elevated tryptase levels (>15) were detected in 42/108 patients (39%) with de novo AML and 11/25 patients (44%) with secondary AML. No elevated tryptase levels were found in patients with ALL. In de novo AML, elevated serum tryptase levels (>15 ng/ml) were frequently detected in patients with FAB M0 (6/9), M2 (9/14), M3 (4/6), and M4eo (7/7), and less frequently in those with FAB M1 (7/20), M4 (6/26), M5 (2/18), M6 (0/5) or M7 (1/3). The highest serum tryptase levels (up to 881 ng/ml) were found in AML M4eo. Immunohistochemical staining of bone marrow sections with an anti-tryptase antibody as well as immunoelectron microscopy revealed tryptase expression in the cytoplasmic compartment of myeloblasts. A spontaneous increase of tryptase protein levels in cultured AML cells was found suggesting de novo enzyme-production. As assessed by Northern blotting and RT-PCR, AML cells expressed a-tryptase mRNA, but little or no ß-tryptase mRNA. In AML patients with elevated serum tryptase before therapy, who then entered complete remission in response to chemotherapy, tryptase levels returned to normal or near normal values. Blast cell persistence or regrowth was associated with a persistently elevated level or rapid increase of tryptase. In summary, „mast cell tryptase“ is expressed in myeloblasts in a group of patients with AML. Due to specificity for myeloid cells and high sensitivity in AML, tryptase measurements may serve as a useful tool to diagnose and monitor AML patients.

755

Genetic instability in hematopoietic progenitor cells of patients with AML in complete remission is associated with higher risk of relapse

B. Panzer, W. Dietmaier, M. Zaiss, A. Rehm, R. Andreesen, A. Reichle (Regensburg, D)

We have recently shown that abnormal growth pattern of colony forming cells and minimal residual disease (MRD) in leukapheresis products of patients with acute myeloid leukemia (AML) in complete remission (CR) are important predictors of relapse. In the present study we searched for genetic instability in CD34+ hematopoietic progenitor cells derived from leukapheresis products of AML patients in 1. CR as a possible marker for aberrant cell growth and therefore as prognostic factor for relapse. We tested CD34+ cells from 20 patients and 5 healthy control subjects. The CD34+ cells were purified from leukapheresis products collected after three cycles of induction chemotherapy (TAD - HAM - TAD) or after G-CSF treatment for 5 days (healthy controls). All patients received parts of the leukapheresis products after myeloablative chemotherapy as stem cell rescue. The median purity of the CD34+ cells in our samples was 96 % (18% - 99,8%). As negative controls we used CD34 negative cells from the same leukapheresis products, mainly composed of lymphocytes. After DNA purification 10 loci were tested for microsatellite instability (MSI) or loss of heterozygosity (LOH) using PCR. No signs of genetic instability were detectable in the cells from healthy subjects. In 8 of 20 patients (40 %) we found at least one locus with MSI or LOH in hematopoietic progenitor cells, in 3 patients (15 %) even more than one locus with MSI or LOH. 20 % of patients had MSI (± LOH), 20 % of patients had only LOH and 60 % of patients showed no signs of genetic instability. All patients with MSI relapsed (4 of 4 patients), while only 50 % of patients with LOH (2 of 4 patients) had a relapse. However, also 50% of patients without genetic instability experienced a relapse of AML. Altered loci in which MSI or LOH occurred in more than one patients were D7S486, D8S559, APC, D11S1356, Mfd15 and TP53ALS. In all leukapheresis products MRD tested by 4-parameter FACS analysis was lower than 0.6 %. Thus the possible presence of leukemic cells in our purified CD34+ samples would be too low to explain the occurrence of genetic instability by microsatellite analysis. This suggests that MSI or LOH can be found in progenitor cells of remission hematopoesis. Genetic instability in CD34+ hematopoietic progenitor cells of AML patients in 1. CR might therefore play a role in leukemogenesis and might be associated with an increased relapse rate.

756

Membrane expression of heat shock protein 70 (Hsp70) in acute myeloid leukemia: prognostic significance and target structure for NK cells

G. Multhoff, M. Gehrmann, C. Gross, M. Hantschel, R. Andreesen, K. Pfister, T. Haferlach, S. Schnittger, C. Schoch, W. Hiddemann (Regensburg, München, D)

An unusual heat shock protein 70 (Hsp70) localization has been found on a variety of human tumor cell lines, but not on normal cells. The tumor-selective membrane expression of Hsp70 correlates with an increased sensitivity against human NK cells. Here, we demonstrate by multiparameter flow cytometry using lineage specific markers combined with an Hsp70 specific antibody an Hsp70 plasma membrane expression on freshly isolated bone-marrow specimen and peripheral blood mononuclear cells (PBMNC) of patients suffering from primary, and secondary acute myeloid leukemia (AML), plasmocytoma, myelodysplastic syndrom acute lymphatic lymphoma and non-hodgkin lymphoma (NHL). In total 71 out of 95 (75%) interpretable bone marrow aspirates of untreated primary and secondary AML patients were positive for Hsp70. Bone marrow cells and CD34 enriched cells of healthy human individuals and of AML patients in complete remission did not express Hsp70 on the plasma membrane. With the exception of FAB M3, Hsp70 was frequently found on bone marrow of AML FAB M0, M1, M2, and M5 subtypes. In 75% of all AML cases (46 out of 65) Hsp70 expression was associated with an intermediate, in 78% (18 out of 23) with an unfavorable and in 63% (7 out of 11) with a good prognosis. Presently, we investigate as to whether the leukemia specific Hsp70 membrane expression provides an innovative leukemia specific target structure for autologous NK cells.

757

Minimal residual disease in acute myeloid leukemia in children: MRD-AML-BFM-98 study group

C. Langebrake, D. Reinhardt, F. Griesinger, M. Dworzak, O. Hrusak, S. Viehmann, J. Vormoor, U. Creutzig, J. Harbott (Münster, Göttingen, Giessen, D; Wien, A; Prague, CZ)

Background: Monitoring of minimal residual disease (MRD) is supposed to be useful in the treatment stratification of acute myeloid leukemia. Methods: Two methods are suitable to identify residual leukemic blasts in bone marrow specimens during therapy: first the polymerase chain reaction (PCR)-technique for the detection of specific gene-rearrangements (AML1/ETO, PML/RARα, CBFβ/MYH11 and MLL/AF9) and second the multiparameter flow cytometry (MFC) for the monitoring of leukemia-associated-immunophenotypes (LAIP). By PCR, a sensitivity of 10-4-10-5 can be achieved, however, gene-rearrangements are only available in 30 percent of the AML patients. MFC allows the determination of a LAIP in 80-90 percent of the patients, but specifity and sensitivity have not yet been defined. Heterogeneity of acute myeloid leukemias and the lack of a standardisation impede MRD-monitoring. Therefore, we established a three-colour-consensus-panel in the four participating laboratories. Results: To determine the specifity of the different immunophenotypes, we analyzed bone marrow samples of 17 children (ALL n=7, no malignant disease n=7, Ewing-sarcoma n=1; CML in chronic phase n=2) at diagnosis or in morphological remission. Furthermore, we tested sensitivity of our panel by quantifying descending dilutions of defined leukemic blasts in regenerating bone marrow specimens. Specifity of the different Immunophenotypes was variable. Coexpression of stem cell antigen CD34 and myeloid antigens CD13/CD33 could be found in normal bone marrow in a concentration of 0.5% (0-1,6%). By contrast, immunophenotypes with an aberrant coexpression of lymphatic antigens (i.e. CD2, CD7, CD19, CD56) or the antigen 7.1 with CD33/CD34 occurred in less than 0.1 % in normal or regenerating bone marrow. Actually the consensus panel was applied to 57 children with AML (median follow-up 5-6 months). During treatment in most AML patients 0.2 to 2.5% residual myeloid blasts were detectable until consolidation although several changes or loss of antigen expression hampered sensitivity. AML-specific rearrangements were analysed in 47 children at least at two time points during therapy. AML1/ETO-rearrangement persisted in 20 percent of the children (n=25) during the whole intensive treatment. By contrast, MLL/AF9 was negative in all patients (n=11) after induction. CBFb/MYH11 (n=11) became undetectable during consolidation. Conclusion: In conclusion, immunophenotyping and molecular genetics may enable the monitoring of residual blasts in a high percentage of AML patients. The prognostic impact of residual blasts or the kinetic of blast clearance awaits a longer follow-up and increasing patient numbers. Supported by the „Kompetenznetz Pädiatrische Onkologie und Hämatologie (BMBF)“ and „Elternverein für krebskranke Kinder Holzminden“

758

MDR-1 single nucleotide polymorphisms (SNP) in AML

T. Illmer, M. Schaich, U. Schuler, C. Thiede, G. Ehninger (Dresden, D)

Gene polymorphisms may influence both expression and function of the affected gene product. Recently different SNPs were described through the MDR-1 gene. These genetic changes are suspected to alter the quantity of functional P-glycoprotein both in leukemic blasts as well as in membranes of the genitourinary and gastrointestinal tract. We therefore investigated blast samples from 155 AML patients who were homogenously treated in the current SHG-96 trial. The three most frequent SNPs (position 1236 - Exon 12, position 2677 - Exon 21 and position 3435 - Exon 26) that were reported by large epidemiological studies were screened by the PCR-SSCP method. Representative allelic combinations were sequenced. The expression of the MDR-1 gene product was determined by a semiquantitative RT-PCR approach. The heterozygous variant „2“ (G/T) at position 2677 (47% of all patients), variant „2“ (C/T) at position 3435 (43% of all cases) and the homozygous variant „1“ (C/C) at position 1236 (42% of all cases) were the most frequent allelic variants detected. Thus distribution of allelic variants were not significantly different as compared to previously published Caucasian control persons (Cascorbi et al., Clin. Pharmacol. Ther., 2001). We detected a significant lower MDR-1 expression rate in AML patients with the homozygous variant 3(T/T) at position 2677(T/T vs. G/T - 10.7 vs. 30.9%; p=0.04) and a tendency towards lower MDR-1 expression in patients with the homozygous variant 1 (T/T) at position 3435 (T/T vs. C/T - 10%vs. 26%; p=0.057). Complete remission (CR) rate and overall survival were unaffected when investigated for the impact of SNPs in the MDR-1 gene. However, we observed clusters of combined SNPs. The most frequent association of SNPs was a complete heterozygous genotype 2-2-2 (C/T at position 1236 - G/T at position 2677 - C/T at position 3435)with 27% of all cases investigated. An overview of the observed SNPs, MDR-expression and CR-rate is given in table1. Moreover, the combined SNPs group 1-1-3 defined an AML subgroup with low OS (OS 15%, p=0.057) and a high probability of relapse (DFS 15% - p=0.058). These results may indicate a potential prognostic impact of MDR-1 SNPs and have to be substantiated in larger patient cohorts.

Table1:

	2-2-2	1-1-3	3-3-1	1-1-2	1-2-2	2-2-1	1-1-1
MDR-rate	28%	23%	12%	9%	36%	12%	0%
CR-rate	48%	49%	56%	63%	36%	38%	83%

759

No influence of additional chromosomal aberrations on prognosis in adult patients with translocation t(8;21)

B. Mohr for the SHG AML96 Study Group (Dresden, D)

The AML1/ETO(t(8;21))-positive acute myeloid leukemia (AML) is a distinct category with a generally better prognosis compared to AML harboring other chromosomal aberrations or a normal karyotype. Nevertheless there are some relapses within this group of AML patients. Thus, it is of great interest to discover patients at high risk of relapse for more intensified treatment strategies, like allogeneic blood stem cell transplantation (SCT). Secondary chromosome aberrations are frequent in AML with t(8;21) and their prognostic impact is controversially. One survey documented the loss of parts of 9q as an unfavorable prognostic factor (Schoch Leukemia 1996; 10: 1288-95) whereas another did not (Grimwade Blood 1998; 92: 2322-33). Within the AML96 trial of the Sueddeutsche Haemoblastose Gruppe 41 patients with t(8;21) were included so far. The characteristics of these patients were as follows: 1. age groups: n= 14 (19 to 40 years), n=17 (>40 to 60 years), n=10 (older than 60 years); 2. additional chromosome aberrations: n=24 (58%); 3. frequent additional chromosome aberrations: loss of a sex chromosome n= 18 (44%), partial deletion of 9q n=7 (17%, constant deleted band 9q22), partial deletion of 7q n=3 (7%), 4. complex aberrant karyotypes: n=13 (32%, 3 to 4 karyotype anomalies). 90% of patients with t(8;21) younger than 60 years achieved complete remission after double induction therapy. Five patient were transplanted (autologous SCT n=2, allogeneic SCT n=3). Two of these 5 patients had a deletion in 9q.Ten patients with t(8;21) relapsed, with six patients being younger than 60 years (one with del(9q), one with complex aberrations, two with loss of a sex chromosome). No significant influence (p>0.05) of deletion in 9q, complex aberrant karyotype, loss of a sex chromosome or additional aberrations could be found on the incidence of relapse. Up to now median disease-free survival for the t(8;21) positive patients <=60 years with an additional 9q-deletion (n=5) is 32.4 months (range 0.6-53.4), for the patients with complex aberrations with or without del(9q) (n=8) is 19.2 months (range 0.6-55.0), and for the patients without del(9q) or without complex abnormalities (n=19) is 21.8 months (range 2.0-57.6). In conclusion, we generally suggest the treatment of AML patients <=60 years with t(8;21) within the good risk group, independent of the existence of additional aberrations.

Postersitzung:
Anämien

760

Screening of iron deficiency in blood donors by soluble transferrin receptor

V. Adelberger, A. Willer, T. Bertsch, M. Kerowgan, J. Spindler, G. Metzgeroth, A. Dorn-Beineke, H. Wisser, R. Hehlmann, J. Hastka (Mannheim, D)

The level of soluble transferrin receptor in serum (sTFR) is a sensitive indicator of iron deficiency. The sTFR concentration is, however, dependent on the erythropoietic activity as well. We investigated, whether sTFR is a reliable parameter for screening of iron deficiency in blood donors. The sTFR concentration was determined using the Quantikine-ELISA (normal value: 1.54 ± 0.43 µg/ml, normal range: 0.85 – 3.05). 62 blood donors (48 males, 14 females) were included to the study. Iron deficiency was excluded in all these persons by the standard iron tests: hemoglobin 14,0 + 0.8 g/dl; MCV

91 ± 4 fl; ferritin 96 ± 65 µg/l, range 23-429; zinkprotoporphyrin 26 ± 7,4 µmol/mol heme, range 11-40. In 23 donors (37 %), the sTFR concentration was above the „Quantikine-upper value" of 3.05 µg/ml. Compared to our control group, which consisted of normal persons without iron deficiency (sTFR 1,72 ± 0.17 µg/ml), the mean sTFR measured in the donors was significantly elevated (2.82 ± 0.99 µg/ml, p < 0.001). We conclude that sTFR is not a reliable tool to detect iron deficiency in blood donors, because the sTFR values are elevated due to the activated erythropoiesis.

761

S-transferrin-receptor (sTfR), ferritin (F) and C-reactive protein (CRP) as markers in differential diagnosis of anemias of chronic inflammatory processes

P. Lehmann, M. Volkmann, A. Baldauf, J. Lotz, R. Röddiger (Heidelberg, Mainz, Mannheim, D)

Iron-deficient erythropoiesis results from a defect in iron-deficiency or recycling. In patients with anemia of chronic inflammation, there appears to be a defect in the release of iron from RES (e.g. macrophages), the loading of iron onto plasma transferrin, or both. Ferritin and CRP are well established parameters in the routine laboratories compared to sTfR whose importance as a routine parameter has to be evaluated in clinical practice. To differentiate in the daily routine between IDA and ACD sTfR, Ferritin and CRP was examined in 150 patients suffering from chronic diseases. The fact that ferritin is a known acute-phase reactant complicates its use as a marker in acute or chronic inflammatory diseases. The combination of sTfR, ferritin and CRP-measurements seems to give a clearer picture of functional iron status than ferritin alone. It may be rational to subdivide the results in different groups.

Ferritin [µg/L]	sTfR [mg/L]	CRP [mg/L]	
< 15	> 5	< 5	Depletion of functional iron
15 – 30	< 5	> 5	Anemia of inflammation
15 – 30	> 5	< 5	Iron deficient erythropoiesis (IDE)
15 – 300	< 5	< 5	Normal iron status
15 – 300	< 5	> 5	Anemia of chronic diseases
> 300	< 5	< 5	Iron overload

In patients with ferritin-levels < 30 µg/L and sTfR concentrations < 5 mg/L an additional erythrocyte or reticulocyte count should be recommended. The probability of disturbances of iron distribution (e.g. anemia of chronic diseases or chronic inflammation) in patients having ferritin-levels in the normal range and sTfR-levels in the normal range from < 5 mg/L and increased CRP-levels has to be in consideration.

762

Epoetin alfa increases hemoglobin levels and improves quality of life in anaemic cancer patients. Results of the German Epoetin alfa trial

U. Reinhardt (Bayreuth, D)

This German multicenter study investigates the impact of epoetin alfa to improve hemoglobin levels (Hb) and quality of life (QoL) in tumor patients. From Nov. 1998 until December 2000, 701 patients were enrolled to receive epoetin alfa 10.000 IU t.i.w. for 8-18 wks. during chemotherapy and/or radiotherapy. At this time data from 578 patients are evaluable for intent-to-treat analysis, 405 (70.1%) females and 173 (29.9%) males with a median age of 61 (23-85) yrs. 535 patients (92.6 %) were treated because of solid tumors and 43 (7.4%) because of hematological malignancies. The most frequent tumors were breast cancer (174 pts, 30.1%), ovarian cancer (100 pts, 17.3%) lung cancer (81 pts, 14.0%) and colorectal cancer (57 pts, 9.9%). All pts. were treated with chemotherapy, 94 received additional radiotherapy. Platinum was administered in 193 pts. (33.4 %). QoL was measured by a standardized questionnaire recording scores for separate aspects like social activities, mood, interest and exhaustion. QoL data were inquired from the physician, the patient, and a nurse. Results: The mean number of epoetin alfa applications was 31 ± 14, the mean treatment duration with epoetin alfa was 12 ± 6 weeks. Complete tumor response (CR) was found in 97 pts. (17%), partial response (PR) in 122 (21%), no change (NC) in 73 (13%), and progressive disease (PD) in 99 (17%). 75 pts. (13%) were not evaluable. 148 pts. received blood transfusions with a median of 3 units (1-14). Hemoglobin level increased from 10.1 ± 1 g/dl to 11.9 ± 2 g/dl during epoetin alfa treatment. This improvement was statistically significant (p<0.0001) in the total sample and in all tumor response groups. Hb increase was higher in pts. with CR (2.5 g/dl) as compared to pts. with PR (2.2 g/dl), pts. with NC (1.6 g/dl) and pts. with PD (0.9 g/dl). Changes in QoL were reflected above all in the reduction of exhaustion; according to the physician, exhaustion decreased significantly from 7.3 ± 1.2 to 4.8 ± 2.9 points on an 11-point rating scale (0-10) in the total sample. In the subgroups the reductions were as follows: CR 4.8, PR 3.7, NC 2.6 (p<0.0001) and PD 0.6 (p<0.05). The improvement of the exhaustion was strongly correlated with the increase of Hb (0.39, p<0.0001). The evaluation of QoL questionnaires answered by physicians, study nurses and patients yielded comparable results.
Conclusion: In patients with solid tumors as well as various haematological malignancies the treatment with epoetin alfa is accompanied by a significant increase of Hb resulting in a decline of exhaustion. The definition of parameters predictive for epoetin alfa response is under exploration.

763

The significance of anaemia in an outpatient clinic for oncology

T. Steinmetz, S. Schmitz (Köln, D)

Oncologists become more aware of the need to treat anemia of cancer patients (pat), due to the increasing importance for quality of life and prognosis. On the one hand red blood cell (RBC) transfusions still are the most common treatment burdened with complications as infections. On the other hand erythropoietin seems to be a very costly alternative. In order to focus on the requirements of outpatient clinics in future studies on management of anemia, it was the aim to analyse the incidence and clinical importance of anemia and the frequency of different therapies in an outpatient clinic for haematology and oncology. Data from all pat. in a 3 month period were analysed. The diagnosis of anemia according to the ICD-10 code was documented in 113 of 752 pat. (15.1%). However, thorough studies of each medical record revealed that in fact 301 pat. (40%) had an anaemia defined by Hb < 12 g/dl. 21 pat. (2.8%) with anemia were transferred for anemia diagnostics, 280 pat. (37,2%) had anemia caused by cancer and chemotherapy. 451 pat. (60%) had no anemia. Median age of the 280 anaemic cancer patients was 63y (range15-94), 153 (54.6%) were women and 127 (45.4%) men. 180/280 pat. (64%) had a Hb between 10-12 g/dl, 67 (24%) between 8-10 g/dl and 33 (12%) < 8 g/dl. Symptoms and therapies are listed in the table. For instance there was no specification of symptoms in 30-40% of all anemic pat. Conclusions: Anemia is more frequent (40%) than documented according to ICD-10 (15%). This may indicate a reduced perception of anemia, which is also reflected by the lean documentation of anemia related symptoms. Even mild anemia (10-12 g/dl) may cause relevant symptoms as fatigue and dyspnea in at least 25% pat. indicating the need of treatment.

764

Human mesenchymal stem cells (hMSC) are not of donor origin in patients with severe aplastic anemia (SAA) who underwent sex-mismatched allogeneic BMT

N. Stute, B. Fehse, A. Badbaran, J. Schröder, P. Adamietz, S. Arps, K.R. Held, R.A. Zander (Hamburg, D)

Objective: Since stroma defects are part of SAA etiology and hematopoietic engraftment is poor in unrelated and mismatched transplant we wanted to find out whether hMSC are in part of donor origin in patients with SAA years after successful BMT. Methods: Human mesenchymal stem cells (hMSC) were grown from patients' bone marrow aspirates after informed consent according to the method of Caplan AI et al. using a carefully selected FCS (Haynesworth SE, 1992 and Lennon DP, 1996). Primary cultures of adherent hMSC were further analyzed by interphase fluorescence in situ hybridization (FISH) analysis using centromer specific DNA probes for X and Y chromosome (DXZ1 and DYZ3, Vysis, IL, USA). For comparison we also analyzed bone marrow MNC after a Ficoll gradient. As evidenced by morphology, immunocytochemistry (APAAP) and flow-cytometry (SH2, SH3 and SH4) a homogeneous adherent population of hMSC was obtained after 2 weeks with < 0.1 % hematopoietic cells. Unlike stroma, hMSC obtained by this method showed multipotency in adipogenic, osteogenic and chondrogenic differentiation assays in vitro (Pittenger MF, 1999).
Results: Two male SAA patients 24 and 46 years old, 5 and 3 years after BMT (Cy, ATG) with filtered marrow from an HLA-identical female sibling donor were investigated. Bone marrow aspirates were taken and analyzed for donor contribution. The ability of the bone marrow to form hMSC was impaired, since the adherent cell number was significantly lower than in normal controls. FISH analysis did not show any evidence of female donor cells in the adherent hMSC : 0.0% and 0.2% out of 500 analyzed cells were XX type which is well within assay limits. In contrast, Ficoll separated

MNC (hematopoietic cells) from the same BM aspirate were 100% and 97.5% of donor origin as expected (FISH). - Also data from a female SAA patient 35 years old, 8 years after BMT (Cy, ATG) will be presented whose bone marrow apart from FISH is being analyzed for the Y chromosome of her brother by a quantitative TaqMan PCR established in our lab. FISH analysis of hMSC revealed 0.0% donor contribution in primary culture and 0.4% in 1st passage and TaqMan PCR of hMSC 0.005% donor chimerism in 1st passage. Hematopoietic donor chimerism was 92.9% at the same time (FISH).
Conclusion : In SAA, like previously found in patients with lysosomal and peroxisomal storage disease (Koc ON, 1999), bone marrow-derived hMSC remain host-derived despite successful hematopoietic engraftment after allogeneic BMT.

765

Clinical relevance of GBV-C/HGV viremia in aplastic anemia and blood donation

C. Seifried, H. Schrezenmeier, A. Raghavachar, H. Bialleck, D. Waschk, M. Weber, W. Roth (Frankfurt am Main, D)

Purpose: The aim of the study was to evaluate the clinical relevance of the GBV-C/HGV among blood donors and recipients and their familiy members and the consequences for the blood donor screening. The incidence of GBV-C/HGV in patients with aplastic anemia (AA) was measured to study whether the virus could be implicated in the etiology of AA. Method: Plasma samples were drawn from AA patients as well as from repeat donors and their family members and the recipients of GBV-C/HGV-positive blood components. All samples were tested by RT-PCR. To identify viruses, sequencing was done on an Applied Biosystems 373 or 310 sequencer. Results: 90/6,179 blood donors with ALT ≥45U/L were NAT-positive (1.46 %). 98 familiy members of these 90 positives could be tested and 14.3 % were viremic, only 1 % of the relatives of PCR-negative donors were viremic, suggesting intra-familial virus transmission. In 2 partner pairs and in 6 mother-child pairs virus transmission could be shown by a sequencing of isolates indicating horizontal or vertical virus transmission respectively. 13 recipients of a packed red cell unit, 2 recipients of a platelet concentrate and 3 recipients of fresh frozen plasma of GBV-C/HGV NAT-positive blood donors were infected. Interestingly, in two cases the virus could be followed from the blood donor to the recipient and to family members. Out of 53 patients with AA, 36 were GBV-C/HGV NAT-positive. All viremic patients were transfused with blood components during the course of their disease. No single patient without a history of transfusion of a blood component was NAT-positive. None of the patients with hepatitis-associated AA was NAT-positive at the time of diagnosis. No PCR-positive donor or recipient of blood products had signs of hepatitis. Conclusion: The prevalence of about 1.4 % of GBV-C/HGV NAT-positive blood donors confirms other studies in Europe. The high prevalance of 14.3 % among familiy members of NAT-positives indicates intra-familial transmission of the virus. 58% NAT-positive recipients of blood products demonstrate a high transmission efficacy of GBV-C/HGV by blood products. In accordance with other data, the virus does not play a causative role in AA or in the induction of hepatitis.

766

Mycophenolate mofetil (MMF) in the treatment of steroid-refractory autoimmune hemolytic anemia in patients with lymphoproliferative disorders

W.U. Knauf, B. Zimmer, E. Thiel (Berlin, D)

Autoimmune hemolytic anemia (AIHA) is a complication of chronic lymphocytic leukemia (CLL) and other types of non-Hodgkin's lymphoma (NHL). IgG-type antibodies against erythrocytes are detected in the vast majority of pts, while IgM-type antibodies indicate a worse prognosis. Despite remission induction of the underlying disease immunosuppressive drugs like steroids as well as splenectomy may control AIHA. However, a significant number of pts are non-responders to steroids, and splenectomy could be hazardeous particularly in elderly pts suffering from concomitant diseases. We therefore initiated a pilot study on mycophenolate mofetil (MMF), an inhibitor of lymphocyte proliferation, which is currently used to prevent and treat graft rejection in solid organ transplantation. Five consecutive pts (four males, one female) with steroid-refractory Coombs-positive AIHA were included. AIHA has been diagnosed within a median of 5 years (range 0 - 13 yrs) after diagnosis of CLL (2 pts), PLL (one pt), centroblastic NHL in CR (one pt), and another pt with combined immunodeficiency syndrome (cIDS), respectively. All pts had severe transfusion-depending hemolysis refractory against steroids (1-2 mg / kg body weight), and in three of them the steroids had induced diabetes mellitus demanding insulin. MMF was started with 1 g p.o. daily for one week followed by 2 g p.o. daily while tapering the steroids. Within a median of 4 weeks (range 3 - 6 weeks) hemolytic activity was diminished in 5 / 5 pts as shown by increasing hemoglobin, decrease of reticulocyte counts as well as of LDH levels, and by reduction of red blood transfusions. One of the 5 pts became Coombs negative. One pt with PLL discontinued MMF treatment and relapsed with AIHA shortly thereafter. However, hemolysis was stopped again by re-treatment with MMF. Interestingly, the pt with cIDS showed clinical signs of ITP in addition to AIHA (Evans syndrome). Platelet counts rose in parallel to hemoglobin while under MMF. This particular pt developed Guillian-Barre-syndrome 5 months and Burkitt-type NHL 8 months after onset of MMF. To what extent immunosuppression with MMF after his history of 13 years of cIDS has contributed to this remains unclear. In conclusion, MMF was well tolerated and showed effectiveness in recompensating severe steroid-refractory AIHA. The German CLL Study Group, in consequence, has initiated recently a multicenter phase II trial on MMF in the treatment of CLL-associated AIHA.

767

A phase 3, randomized, double-blind, placebo-controlled, study of novel erythropoiesis stimulating protein (NESP) in patients with lung cancer receiving chemotherapy

R. Pirker, J. Vansteenkiste, J. Gateley, P. Yates, A. Colowick, J. Musil, and the ARANESP 980297 Study Group (Wien, A; Leuven, B; Thousand Oaks, USA; Prague, CZ)

Purpose: NESP binds the erythropoietin (EPO) receptor and stimulates erythropoiesis by the same mechanism as recombinant human EPO (rHuEPO), but has a longer serum half-life. NESP has been shown to be safe and clinically effective in cancer pts when administered every 1, 2, and 3 wks in phase 1/2 studies. This phase 3 study compared the efficacy of NESP with placebo in lung cancer pts receiving platinum-containing chemotherapy (ctx). Methods: 320 anemic pts (hemoglobin £ 11 g/dL) receiving platinum-containing ctx, with ECOG 0-2, not iron deficient, and no rHuEPO therapy within 8 wks or < 2 RBC transfusions (tfn) within 4 wks, were randomized in a 1:1 ratio to NESP 2.25 mg/kg weekly or placebo. Study drug was administered SC once weekly for a maximum of 12 wks (tx phase). Results: NESP significantly ($p < 0.001$) reduced the Kaplan-Meier proportion (95% CI) of pts transfused during wks 5 to 12: NESP 21% (15, 28), placebo 51% (43, 60) and during the entire tx phase (wks 1-12): NESP 26% (20, 33), placebo 60% (52, 68). NESP pts received fewer standard units (mean [SE]) of RBC than placebo pts during wks 5 to 12: NESP 0.67 (0.14), placebo 1.92 (0.27) and during the tx phase: NESP 2.64 (4.32), placebo 1.14 (2.38). NESP pts were hospitalized fewer mean (SE) days than placebo pts: NESP 10.3 (1.1) days, placebo 13.0 (1.4) days. Patients were followed an average of 1 year after enrollment in the study. The median time (95% CI) to disease progression or death was 26 weeks (19, 31) for NESP pts and 20 weeks (17, 24) for placebo pts. The safety profile of NESP was similar to placebo and as expected for this population. Conclusions: NESP significantly reduced the proportion of subjects with RBC tfns and was well tolerated. The reduction in transfusions was observed when the entire treatment phase was included (wks 1-12), suggesting a rapid onset of action of NESP. Analysis of secondary endpoints in this study also indicate potential benefits with respect to a reduction in hospitalization and an increase in time to disease progression when treated with NESP that should be evaluated in future studies.

768

Prevalence of anemia in hematological cancer patients receiving cyclic non-platinum chemotherapy

M. Steurer, G. Gastl, H. Wagner (Innsbruck, Linz, A)

Background: Anemia is a frequent yet often overlooked morbidity in hematological cancer (HC) patients (pts). Negative sequelae include fatigue, exhaustion and overall impaired quality of life.
Aim: To determine prevalence of anemia and the need for its treatment, a prospective chart survey of 273 HC pts receiving cyclic, nonplatinum (NP) standard chemotherapy (CT) was conducted from 1 Jan 2000 through 30 Sept 2000 at 35 centers in Austria. Methods: Four different hematologic malignancies were studied: Chronic Lymphocytic Leukemia (CLL), Non Hodgkin Lymphoma (NHL), Multiple Myeloma (MM) and Hodgkin Disease (HD). All 273 HC pts were followed over four consecutive cycles of CT. Treatment of anemia, including either blood transfusion and/or recombinant human erythropoietin (rHu-EPO), was at physician's discretion. Results: All four hematologic malignancies studied revealed different hemo-

globin (Hb)-values at the beginning as well as further on during the observation period: At baseline, already 51.9% pts were anemic (Hb less than 12 g/dL) in the CLL group, compared to 35.1% in the NHL group, 77.4% in the MM group and 21.9% in the HD group. In every subgroup the percentage of pts with anemia increased with each successive cycle of CT, reaching 63% in the CLL group by the fourth cycle. In comparison, in the NHL group the prevalance of anemia doubled until the fourth CT cycle (74%). Concerning the MM patients, anemia occured in 92.5% and among HD pts in 54.5% by the fourth cylce. Out of all pts with final Hb levels below 12 g/dL, especially HD pts (37.5%) showed Hb decreases by more than 2 g/dL from baseline values. In contrast 22.2% of all anemic CLL pts, 31.2% of all anemic NHL pts and 28.9% of all anemic MM pts dropped by more than 2 g/dL in their Hb-values from baseline. Over the course of this study, 48% of all anemic patients received treatment with either blood transfusion and/or r-HuEPO. Conclusion: Anemia is prevalent in HC pts receiving NP standard CT, a trend that is likely to increase during CT, depending on the type of hematologic malignancy as well as on the chemoregimens used. This trend underscores the need to identify prognostic factors for development of anemia and to treat anemia to ameliorate accompanying morbidities. Further analyses regarding prognostic factors are underway and will be presented.

769

Allogeneic blood stem cell transplantation with dose reduced conditioning without ATG in patients with severe aplastic anemia

G. Kobbe, U. Germing, T. Emde, A. Hünerlitürkoglu, R. Fenk, U. Rohr, G. Meckenstock, M. Aivado, R. Haas, P. Schneider (Düsseldorf, D)

Graft failure is one of the major problems in allogeneic stem cell transplantation for patients with aplastic anemia. High dose cytoxan (200mg/m^2) combined with ATG has been used to lower the risk of rejection. Experience with new, dose reduced conditioning regimens is very limited in this disease.

Three patients with aplastic anemia (AA) in relapse after immunosuppressive therapy and 1 patient in pancytopenia after treatment for secondary AML following AA were treated with fludarabine (Flu) containing conditioning regimens and allogeneic blood stem cell transplantation. All patients had failed at least one prior therapy and had received >10 transfusions. Two female patients received a combination of Flu (3x 30mg/m^2), cytoxan (2x 60mg/kg) and basiliximab (a humanized mouse anti-IL-2 receptor antibody, 2x 20mg) followed by the infusion of unmodified mobilized blood stem cells from HLA-identical sibling donors (8.4 and 5.8x10^6 CD34+cells/kg, respectively). One male patient received a combination of Flu (3x 30mg/m^2), cytoxan (2x 50mg/kg) and low dose TBI (2Gy) before transplantation with unmodified bone marrow (4x10^8 MNC/kg) from a matched unrelated donor (MUD). Another patient was treated with Flu (3x 30mg/m^2) alone and 9.8x10^6 CD34+cells/kg from a MUD while on mechanical ventilation because of respiratory failure caused by bilateral pulmonary mycosis. After transplantation patients received prophylaxis for acute graft versus host disease (aGvHD) with mycophenolatemofetil and cyclosporine (3) or tacrolimus (1). Non hematologic toxicity of the regimens described above was minimal (<II°). Three patients did not need intravenous antibiotic treatment. The patient who was allografted while on mechanical ventilation recovered sufficient numbers of granulocytes to improve pulmonary function and was successfully weaned from the ventilator. All patients engrafted with a WBC>1000/µl at a median of 14 days (range 9-16) and a platelet count of >50000/µl at a median of 14 days (range 11-17). At present all patients are alive with stable blood counts. No acute graft versus host disease developed so far. We conclude that dose reduced conditioning with fludarabine containing regimens reliably ensures donor cell engraftment without producing major toxicities in heavily pre-treated patients with severe aplastic anemia.

770

Differential gene expression of EGR-1 and TAXREB107 in paroxysmal nocturnal hemoglobinuria

A. Lyakisheva, O. Felda, R.E. Schmidt, A. Ganser, J. Schubert (Hannover, D)

Paroxysmal nocturnal hemoglobinuria is an acquired clonal stem cell defect characterized by deficiency in GPI-anchored surface proteins. It is not yet known, how the GPI-deficient stem cells are able to expand within the bone marrow and contribute considerably to the patients hematopoiesis. In PNH as well as in AA and MDS, a genetic instability and increased mutation frequency has been detected. Therefore, a second event such as additional mutations is very likely leading to clonal expansion of GPI-deficient bone marrow stem cell in PNH. In order to elucidate the molecular basis of clonal expansion in PNH we identified several genes differentially expressed in normal and GPI-deficient cells of PNH patients by combination of RNA fingerprinting and cDNA array hybridization. Expression of two of these genes, EGR-1 and TAXREB107 has been further investigated. EGR-1 is upregulated in granulocytes of all PNH patients analyzed so far. In contrast, significant upregulation of TAXREB107 is present only in some of our PNH patients. Further analysis confirmed their overexpression in PNH and excluded possible secondary event character of observed overexpression. Moreover, similar level of expression in cases of other clonal diseases such as MPS and MDS has been identified. Our data suggest that additional genetic alterations apart from PIG-A mutations are present in PNH granulocytes. In addition, these genetic changes might contribute to clonal expansion of GPI-deficient cells in PNH.

771

Relapsing Evans' syndrome responsive to splenectomy: a case report

M. Starck, K. Rettler, F. Woitinas, C. Nerl (München, D)

We report the case of a 21-year-old woman with Evans' syndrome. The patient presented with severe hemolysis and thrombopenia. The lab analysis showed the Coombs test positive, cold agglutinins without specificity, HLA-antibodies and an inverse CD4/CD8-correlation. All other blood examinations were normal. Bone marrow biopsy excluded lymphoma, myelodysplastic disorder or hemoblastosis. Technical examinations, including ultrasound and computed tomography of the chest and abdomen showed a slightly enlarged liver and some small unspecific lymph nodes paraaortal. The spleen was normal in size. First line treatment with 150 mg prednisolone daily and intermittent application of high dose intravenous immunoglobulins could not improve hemolysis. Vincristine led to stabilized serum hemoglobin level at 9 g/dl, the platelets were normal. Two weeks later, the disease relapsed. The serum hemoglobin level reached 3 g/dl at this time, and the platelet count was at 0.1/nl. Treatment consisted of high dose dexamethasone, immunoglobulins, vincristine, cyclophosphamide and plasmaphereses. After twenty plasmaphereses and an immunosuppressive therapy with azathioprine orally severe hemolysis continued despite of increased platelet count. Regarding to the failure of efficient treatment for five months, we decided to perform a splenectomy which stopped the hemolysis immediately. The patient showed a slow reconstitution of her serum hemoglobin level and a reactive thrombocytosis. Blood transfusions were given repeatedly. Severe bleedings never existed. As complication, the patient developed a deep vein thrombosis after the sudden increase of the platelets and a painful microangiopathy that made a therapy with morphine necessary. Almost one year after the primary diagnosis and six months after splenectomy, the patient´s blood count is normal. The hemolysis parameters are all in range. The patient is in a good general condition without any specific therapy. We conclude that an endogenous immunomodulation therapy by splenectomy can be of benefit in cases of relapsing or refractory cases of Evans' syndrome.

Postersitzung:
Andere solide Tumoren

772

Has the FUS-CHOP fusion transcript detection a prognostic impact in soft tissue liposarcoma?

F. Willeke, M. Schwarzbach, R. Ridder, N. Attigah, G. Mechtersheimer, J. Weitz, U. Hinz, L. Bernd, T. Lehnert, C. Herfarth, M. von Knebel Doeberitz (Mannheim, Heidelberg, D)

Aim of the study: A distinct subgroup of liposarcomas express FUS-CHOP fusion transcripts encoded by gene fragments fused through translocation t[(12;16)(q13;p11)]. In-vitro and in-vivo assays suggest FUS-CHOP fusion transcripts harbour an oncogenic potential. We hypothesised that FUS-CHOP is associated with an unfavorable prognosis in patients with liposarcoma. Methods: Included in this trial were 33 patients with liposarcomas whose tumours had been snap frozen for consecutive RNA extraction. A nested RT-PCR was established for fusion transcript detection. After RNA extraction from snap frozen tissue, reverse transcription was carried out. We amplified resulting cDNA by nested PCR protocol and analysed the product on agarose gels. 30 tissue samples of normal tissues (skin, muscle) and various non-lipomatous tumours served as controls. Results: Fourteen of 33 liposarcomas investigated revealed a FUS-CHOP fusion transcript, but none of the control tissues. The presence of FUS-CHOP transcripts was neither associated with overall prognosis (p=0.81) nor the rate of local recurrence and / or distant metastases (p=0.39). Conclusions: FUS-CHOP fusion transcript in liposarcomas may help to establish a definitive diagnosis in soft tissue sarcoma. The mere existence of FUS-CHOP is not associated with a worse prognosis as compared to soft tissue liposarcomas not affected by this genetic event. The oncogenic potential demonstrated by in-vivo and invitro experiments does not translate into a simple correlation with prognostic parameters.

773

Double high dose chemotherapy (HDCT) with autologous hematopoetic stem cell transplantation (PBSCT) for metastatic soft tissue sarcoma (MSTS) after standard chemotherapy

A. Zumsprekel, H.J. Weh, K. Holstein, K. Höffken, J. Kienast, B. Metzner, A. Zander (Bielefeld, Hamburg, Jena, Münster, Oldenburg, D)

Purpose: To evaluate toxicity, response rate and progression free survival after double HDCT in patients with MSTS. Patients and Methods: Between 1995 and 2000 14 patients with MSTS who responded to a standard chemotherapy (partial response, minimal response or no change) were included in a prospective phase II study of double HDCT comprising DTIC (3 g/sqm)/cyclofosfamide (2,5 g/sqm) as first regimen and epirubicin (180 mg/sqm)/melphalan (140 mg/sqm) as second regimen. Results: Patients received a median number of 3.5 platelet transfusions (range, 1 to 10) and a median number of 3 RBC transfusions (range, 0 to 6) after the first HDCT. The median duration of fever days was 2.3 (range, 0 to 8). 10 patients needed systemic antibiotics after the first HDCT during aplasia, 4 patients did not. The median duration of parenteral infusion because of stomatitis was 2.5 days (range, 0 to 11). 10 of the 14 patients received a second course of HDCT. After this patients needed a median number of 6 platelet transfusions (range, 1 to 26) and a median number of 2.5 RBC transfusions (range, 0 to 6). The median duration of fever days was 2.4 (range, 0 to 4) and parenteral infusion was necessary during 7.3 days (range, 0 to 17) in median. 6 patients needed systemic antibiotics. Out of the 14 patients in partial remission (50%) or stable disease (50%) after conventional chemotherapy, complete remission (CR), partial remission (PR) and stable disease (NC) were achieved in 2 (14%), 6 (43%) and 4 (29%) patients, respectively, after the first HDCT, while one patient (7%) progressed and one patient (7%) was not evaluable. In 10 (71%) patients a second HDCT was performed. CR, PR and NC could be documented in 3 (30%), 3 (30%) and 2 (20%) patients, respectively, after the second course of HDCT. Two patients (20%) were lost to follow up. The median progression free survival (12/14 patients) after conventional chemotherapy was 10 months (range, 2 to 21+). Conclusion: The toxicity of the double HDCT/PBSCT was moderate notably, no therapy related mortality was observed. The response rate was encouraging because response was improved in 8/14 patients after the first HDCT and in 6/10 after the second HDCT. Median progression free survival of 10 months is notable. With a median follow up of 12.4 months (range, 2 to 25) 6/14 patients (43%) are alive and 6/14 patients (43%) have died. Two patients (14%) were lost to follow up. A phase III randomized trial is required to establish the role of HDCT in MSTS.

774

Phase II study with a weekly 24h-infusion of 5-fluorouracil (FU) and folinic acid (FA) combined with cisplatin (PLF) in patients with carcinoma of unknown primary site (CUP)

K. Sauer, M. Stahl, H. Wilke (Essen, D)

Treatment results in patients with CUP are still disappointing with median survival times of about 6-9 months. Inclusion of new drugs in chemotherapy regimens increased costs without the proof of superiority. As most of the patients with CUP present with adenocarcinoma, we investigated an optimised cisplatin/5-FU-regimen (PLF), which is effective in gastric adenocarcinoma. Patients with a performance status 0-3 (WHO) up to 75 years of age were included. Thirty-three patients with CUP after careful clinical and histologic workup were treated with cisplatin 50 mg/m^2 (1h-infusion) week 1,3,5 combined with folinic acid 500 mg/m^2 (2h-infusion) and 5-FU 2g/m^2 (24h-infusion) week 1-6, qd week 8 for up to 3 cycles. Major histologies were adenocarcinoma (AC, 63 percent) and undifferenciated carcinoma (30 percent). Patient characteristics: female/male 20/13; age 61 years (29-72), PS 1 (0-3); number of metastatic sites 1/2/>2 (9/17/7 pts.); 11 pts. had worse prognosis according to the van der Gaast score. Results: After a median number of 1.6 cycles (range 0.5-5.0) best response was CR/PR in 8 (24 percent), NC in 12 (36 percent) and PD in 10 pts., respectively (unknown in 3 pts.). Toxicity of grade 3 and 4 was rare: no leucopenia, thrombopenia 6 percent, anemia 3 percent, vomiting 18 percent, diarrhea 3 percent, stomatitis 3 percent. Most of the patients died due to tumor progression, 1 patient each died from suicide, cardiac disease, and hepatitis C, respectively. Median progression free survival was 4.1 months. Overall survival reached 6 months in pts. with AC and 7 months in all pts.. Survival rate at 1 year was 15 percent. Conclusions: PLF is a well-tolerated outpatient regimen with efficacy in patients with mostly bad prognosis CUP. Cost-effective comparisons with newer drug combinations are worthwile.

775

Good partial remission in primary metastatic adenoid cystic carcinoma of the parotid gland with 5-FU-containing chemotherapy

M. Wiermann, K. Jentsch-Ullrich, M. Koenigsmann, A. Franke (Magdeburg, D)

Introduction: Adenoid cystic carcinoma is a relatively rare tumor which arises in the parotid and submandibular salivary glands. Worldwide incidence rates of parotid gland tumors are 0.5-0.8/100,000/year. The overall 5-year survival rate is 45% to 70 %. Patients in stage I have a better survival than patients in stages II, III and IV. Initial management is surgical, often with post-operative radiotherapy, but local relapse is common and distant metastasis not infrequent. Chemotherapy is generally reserved for cases where symptoms are not controlled by other means. However, since the tumor is slowly growing, the response rates are usually disappointing. Patient: A 37-year-old woman had an adenoid cystic carcinoma of the right parotid gland. The tumor involved the surrounding tissues and metastasized to the lungs and right cervical lymph nodes. The initial stage was IV (T4N2M1). The tumor was incompletely resected. Therefore, the woman received local radiation therapy to the base of the skull (total dose 50.4 Gy), followed by combination chemotherapy with cisplatin 50 mg/m^2, adriamycin 35 mg/m^2 and 5-fluorouracil (5-FU) 500 mg/m^2 (PAF). After six cycles (8/2000 to 2/2001) of this regimen, there was total regression of her primary tumor and partial regression of her pulmonary metastases. The toxicities were mild (nausea). Subsequently the woman was treated with an oral fluoropyrimidine: Capecitabine (Xeloda 300 mg/m^2 daily for 4 weeks, repetition after 7 days). At the last follow-up at four months after PAF and at two months after beginning of Capecitabine, she had no signs and symptoms of local recurrence, and radiotherapy showed further regression of her pulmonary metastasis. Conclusion: Local therapy of malignant parotid gland tumor is the treatment of choice (surgery alone, in combination with postoperative radiotherapy or radiotherapy alone). Most responses to chemotherapy were achieved with 5-fluorouracil and cisplatin, which are the preferred drugs for initial treatment of metastatic disease or otherwise uncontrolled. Stable disease with chemotherapy is reported for a period of five to seven months. In our case with a primary metastatic adenoid cystic carcinoma of the parotid gland we could achieve a good partial response with a 5-FU containing chemotherapy (PAF). The result could be improved with oral 5-FU (further regression in pulmonary metastases).

776

Angiosarcomas of the heart-2 cases and new therapy-approaches

G. Dingeldein, M. Hoffmann, B. Weiss, G. Geibel-Werle, M. Uppenkamp (Ludwigshafen, D)

We present two cases of angiosarcomas of the heart. Since these are very rare solid tumors, so there is no conventional therapy known. Two young men (age 37+47y.) were transferred to our clinic of heart surgery. The patients were operated under the initial diagnosis of myxoma of the heart, but the intraoperative and histopathological findings revealed angiosarcomas of the right atrial and ventricle. In both cases the tumor could not be operated radically (R2) because of size and infiltration. Both tumors were highly mitotic (G3). The staging revealed no further manifestations of the angiosarcomas. A therapy with adriamycin 75mg/m^2 (d1) and ifosfamide 1500mg/m^2 (D1-4) was applied. G-CSF was given afterwards for 5-7 days. After 4 cycles of therapy a stem cell-harvest for high-dose chemotherapy was initialzed. A amount of 6,8 mill/kg KG CD 34-pos- cells were harvested.

We are planning to give both patients an autologous stem cell-transplantation after ICE-high-dose chemotherapy (Ifosfamide/Carboplatin/Etoposid), following an allogenic heart transplantation to prolong the overall survival. Angiosarcomas of the heart are very rare and are aggressive tumors with short overall survival. The literature indicates, that allogenic heart-transplantation alone is not sufficient to prolong survival, since there is a major risk for distant metastasis, especially in the brain. Therefore a maximum of therapy should be given to young patients with angiosarcomas of the heart.

777

Efficacy of temozolomide (TMZ) in heavily pretreated patients with various solid tumors and brain metastases, and patients with metastatic soft tissue sarcomas

A. Giagounidis, M. Schroeder, M. Heinsch, C. Aul (Duisburg, D)

Temozolomide is an imidazole tetrazinone, an orally applicable methylating agent with moderate toxicity and proven CNS activity due to its ability to cross the blood brain barrier. Pts with NSCLC and sarcomas (STS) with progressive disease under conventional chemotherapy have a poor prognosis. We investigated the influence of TMZ in heavily pretreated pts with various solid tumors and brain metastases and metastasized sarcomas and relapsed brain tumors. Schedule: TMZ was given at a dosage of 150 mg/m^2 d1-5 p. o. q 4 weeks, combined with cranial irradiation in pts with multiple brain metastases. The number of treatment courses ranged from 1 - 19 (median 4). Diagnosis and patients characteristics:

Sarcoma	N = 5	age 37–72 years	Melanoma	N = 2	age 63 years
NSCLC	N = 7	age 37–84 years	Colorectal	N = 1	age 65 years
Brain tumors	N = 4	age 31–49 years	Breast cancer	N = 1	age 64 years

Pretreatment: Pts with STS were pretreated with various combinations of ADM, IFO, DTIC including 1 pt receiving high dose chemotherapy with PBSC support. Pts with NSCLC were pretreated with different schedules including DDP, 5-FU, NVB, taxanes. Pts with brain tumors were pretreated with surgery and radiotherapy. Indication for TMZ in patients with solid tumors was the development of cerebral metastases. Toxicity: Hematotoxicity: Hb III° 5/20, Leuco III° 5/20, PTL IV° 2/20; GIT mild.
Conclusions: TMZ was well tolerated and showed efficacy especially in pts with STS. In patients with of cerebral metastases of NSCLC there was a clinical benefit and objective responses in 3/7 pts. (1 CR in STS, NSCLC and brain tumors each). Further investigations are needed to confirm the activity in pts with brain metastases and STS.

778

Paclitaxel and carboplatinum in recurrent heavily pretreated head and neck cancer

M. Schroeder, B. Alkemper, I. Blum, C. Aul (Duisburg, D)

An optimal salvage treatment for patients (pts) with recurrent head and neck cancer has not yet been defined. 20 pts (18 males, 2 females) with a median age of 54 years (range, 44 - 70) and a median performance score of 2 (range, 1 - 3) presenting with recurrent squamous cell carcinomas of the head and neck were treated with paclitaxel 175 mg/m^2 d1 (3-hour- infusion) and carboplatin AUC 6 d1 every 3 weeks. All pts had previously been treated with neoadjuvant and adjuvant simultaneous chemotherapy (4 cycles DDP/5-FU) and radiotherapy (ad 70 Gy) as well as radical surgery. Patients characteristics: 6 pts had a local recurrence, 12 pts local relapse + distant metastases and 2 pts had distant metastases. All pts had evidence of tumor progression at the start of chemotherapy. The median number of cycles administered was 3 (range, 1 - 6). Median response duration was 3 months for the 7/20 pts. who showed a partial remission. No drug related death occurred. Main toxicity was hair loss in all pts, mild gastrointestinal toxicity and no haematoxicity greater than grade II. We conclude that this schedule is well tolerated with moderate toxicity and moderate activity in heavily pretreated SCC pts with head and neck cancer.

779

German multicenter trial in adenocarcinoma of unknown primary (ACUP): Paclitaxel and carboplatin vs. gemcitabine and vinorelbine

G. Hübner, A. Eimermacher, J. Preiß, E. Laack, H.-J. Schmoll, H. Link (Kaiserslautern, Mainz, Saarbrücken, Hamburg, Halle/Saale, D)

With an incidence of 2-4% of all malignancies carcinoma with unknown primary (CUP-syndrome) is a relatively common problem, well known to all oncologists. Interestingly, the incidence has not changed during the past two decades in spite of improved diagnostic and histologic methods. The prognosis is dismal with a median survival of 5-8 months and a 5-year-survival-rate of less than 5% in patients with ACUP. Internationally accepted treatment strategies still do not exist. Scientifically, ACUP-syndrome remains an orphan disease with only comparatively few active research. The body of publications are dealing with phase-II-studies; the heterogeneity of patient cohorts being a cardinal problem. Phase-III studies are rare and suffer from very low numbers of patients per treatment group.In this situation, we have launched a multicenter phase-II-study in ACUP-syndrome. Although it is not a comparative trial, multicentricity and randomization is mandatory in order to obtain reproducible 'realistic' patient cohorts.Two chemotherapy regimens - paclitaxel / carboplatin and gemcitabine / vinorelbine, both well established in other tumor entities - will be given. With the first combination (combined with additional etoposide) three optimistic phase-II-studies have been done in selected patients with CUP-syndrome with a median survival of 10-13 months. The second regimen has been successfully applied in NSCLC and to a lesser extent in breast cancer, but controlled data in ACUP are missing. The rationale and details of the study protocol will be presented.

780

Grade I meningeoma with pulmonary metastasis - case report

R. Hebecker, S. Sola, M. Knoop, H.G. Bauer, W. Schmidt (Rostock, D)

Extracranial metastases of meningeomas have been described in the literature in a few cases. Most refer to atypical meningeoma with papillary differentiation. We present the history of a seaman admitted to hospital with dysphasia and mild psychotic behaviour due to a large frontal mass. On chest x-ray a focal lung lesion was found too. The latter was clinically asymptomatic. Both tumours have been surgically removed successively with uneventful postoperative courses. Histological examination showed meningothelial differentiation of the frontal meningeoma as well as in the distant metastasis to the lung (WHO grade I). Metastatic spread of tumours is generally an accepted sign of malignancy. One should consider that even „benign" meningeomas may cause pulmonary metastasis, important for the differential diagnosis of lung lesions in patients with proven meningeomas. The mechanisms of spread remain unclear.

781

Positron emission tomography in follicular dendritic cell sarcoma

U. Apfelbeck, C. Tinchon, W. Anderhuber, R. Aigner, W. Linkesch, C. Beham-Schmid (Graz, A)

Follicular dendritic cell sarcoma (FDCS) is a rare recently described uncommon neoplasm occurring primarily in lymphnodes but also in extranodal sites. It showes differentiation similar to that of follicular dendrtitic cells and must be differentiated from lymphomas, metastatic neoplasms and other rare, nonlymphoid lesions. Information about optimal treatment, staging procedures and follow up is scarce. 2-(18)Fluoro-2-deoxy-D-glucose (FDG) positron emission tomography (PET) is increasingly being used in the evaluation of patients with malignancies. PET with FDG is an effective technique to visualize soft tissue sarcoma. But no information exists regarding PET in patients with FDCS.
We report the case of a FDCS presenting as a parapharyngeal tumor in a 20-year-old female. CT-scans of neck, chest, abdomen and pelvis revealed the parapharyngeal mass and the FDG-PET whole body scan also showed a pathologic FDG uptake in the parapharyngeal region. The patient underwent surgery and radiation. Therapy monitoring with CT and PET is planned. In conclusion: the sarcoma lesion was clearly imaged by FDG-PET. PET may be a useful diagnostic method and valuable tool for disease monitoring in patients with FDCS.

782

Liposomes and hyperthermia: Phase I/II-study of Peg-liposomal Doxorubicin, Etoposide, Ifosfamide and regional hyperthermia (EILA + RHT) in the treatment of soft tissue sarcomas (STS)

L.H. Lindner, M. Dreyling, C.M. Wendtner, H. Sauer, W. Hiddemann, R.D. Issels (München, D)

Feasibility and activity of PEG-liposomal doxorubicin (Caelyx), etoposide and ifosfamide combined with regional hyperthermia (RHT) in the treatment of adult patients with primary, recurrent or inadequately resected non metastatic high-risk soft tissue sarcomas will be assessed. Doxorubicin and ifosfamide are the most active drugs in the treatment of STS. PEG-liposomal doxorubicin (Caelyx) has a significantly different pharmacologic profile from the native drug doxorubicin resulting in prolonged circulation time and intratumoral accumulation. In the treatment of STS Caelyx demonstrated equivalent activity to doxorubicin with an improved toxic profile. Although Caelyx is composed of phospholipids with high-phase transition temperature and cholesterin, it has been shown that Caelyx has properties of thermosensitive liposomes. At 42°C a 50% release of the encapsulated drug within 30 minutes in contrast to 2% at 37°C was demonstrated in vitro. A 15-fold higher intratumoral accumulation compared to free doxorubicin was reported for hyperthermia treatment (42°C) in animal models. Maximum effect was achieved applaying heat 1 hour (increased extravasation of intact liposomes) and 24 hours (increased drug release from intratumoral accumulated tumors) after i.v.-injection of liposomes. Recently it could been shown that RHT significantly enhances drug delivery and efficacy of Caelyx in metastatic breast cancer of the chest wall even if Caelyx was applied 1 hour after hyperthermia treatment (Park et al., Proc. ASCO 2001, 20: 184a). Patients will be treated with Caelyx at starting dose of 25 mg/m^2 on day 1, etoposide 125 mg/m^2 on days 1 and 4, and ifosfamide 1500mg/m^2 by 60 minutes on days 1 to 4 (EILA). On day 1 and 4, this EILA regimen will be combined with RHT (60 min, T > 42°C). Thermochemotherapy will be repeated every 3 weeks, a total of four courses will be given before assessment of tumor response. Dose escalation of Caelyx will be performed until DLT is reached. Serumpharmacokinetic of encapsulated and released doxorubicin will be measured following hyperthermia. Future development will focus on the development of new thermosensitive liposomes for tumor targeted drug delivery in combination with regional hyperthermia (RHT).

783

Benign metastasizing leiomyoma. A rare differential diagnosis of multiple pulmonary nodules

T. Yilmaz, Th. Südhoff, H. Stein, U. Graeven, W. Schmiegel (Bochum, D)

We report about a 51-year-old woman who was referred to our hospital to evaluate multiple asymptomatic pulmonary nodules in July 2000. The nodules were detected coincidentally four weeks before admission when a benign leiomyomatous tumor of the right gastrocnemicus muscle was resected. In 1983 she underwent hysterectomy for an uterus myomatosus and she had a right parotidectomy for a parotid adenoma in 1996. The patient did not have any complains especially no pulmonary symptoms, no weight loss, fever or night sweats. The comparison of chest-x-rays from 2000 and 1996 revealed a significant growth of the pulmonary lesions, therefore the abnormalities were suspected to represent metastatic disease and search for a primary tumor was initiated. Neither physical examination, endoscopical examinations nor laboratory tests showed any pathological findings. The only notable abnormality was a soft tissue tumor of the left chest wall detected by CT-scan. This tumor and a lung nodule were resected both revealing benign smooth muscle tissue. To confirm a diagnosis an immunhistochemical analysis of steroid hormone receptor content was performed. The stainings for estrogen and progesterone receptors showed a moderate to strong positive pattern for both receptors in each lesion (muscle, soft tissue, lung). On the basis of the morphological picture and the positive evidence of hormone receptors the diagnosis of a benign metastasizing leiomyoma (BML) was established. According to our knowledge this is the first description of metastatic manifestations in more than two organs (lung, sceletal muscle and soft tissue) in one individuum. There is no standard therapy for BML, but antihormonal therapy strategies seem to be effective interfering with the hormonal dependence of the lesions. We started treatment with an antiestrogen (tamoxifen 20 mg p.o. daily) in combination with a luteinizing hormone-releasing hormone agonist (goserilin 3.6 mg depot s.c. monthly). The serial determination of serum hormone concentrations confirmed a nearly total suppression of estrogen and progesterone within three weeks after the treatment was started and the levels kept being suppressed in the follow-up period. The first control CT-examination of the thorax after six months of therapy showed a decrease of the lung nodules in size by nearly 50 percent. The number of nodules was also decreased. In conclusion dealing with pulmonary nodules in asymptomatic middle aged women the rare possibility of BML should be kept in mind since after correct diagnosis antihormonal therapy may offer very satisfactory remissions like in the case described. This is a typical case concerning the delay of primary uterine myoma and diagnosis of metastatic lesions.

784

Temozolomide chemotherapy offers effective palliation for patients with symptomatic brain tumor relapse after initial surgery and irradiation

U. Keilholz, M. Brock, M. Sternemann, E. Thiel (Berlin, D)

Temozolamide has been licensed for relapsed anaplastic gliomas since one year. We here report on a series of patients with brain tumor relapse after initial surgery and irradiation, who presented with neurologic symptoms and a Karnofsky performance status exceeding 50% at time of initiation of treatment. In case of grade IV tumors, the classical tomozolomide dose of 5 x 200 mg/m^2 was administered over 5 days every four weeks for a maximum of 12 cycles. In case of grade III tumors, this regimen was modified to 75-100 mg/m^2/day for two weeks every 4 weeks. Twelve consecutive patients were enrolled between 1999 and April 2001. Grade IV glioblastoma was present in 6 patients, grade III glioma or astrocytoma in 5, and PNET in one. Two of 12 patients progressed despite therapy, five had stable disease ranging from 3+ to 14 months, and 5 had partial remissions for 3+ to 8 months. Remarkably, all 10 patients with SD or PR experienced improvement of initial neurologic symptoms for a period of 3 to 14 months. Toxicity was generally mild and primarily hematological, although one patient developed opportunistic pneumocystis pneumonia after 4 cycles of treatment. In conclusion, oral temozolomide offers valuable palliation in patients with recurrent brain tumors after initial surgery and irradiation, who present with neurologic symptoms.

Postersitzung:
Autologe Stammzelltransplantation

785

Salivary levels of tumor necrosis factor alpha (TNFalpha) during autologous stem cell transplantation

C. Keulertz, O. Manzke, H. Loevenich, D. Soehngen, D. Waldschmidt, O. Cornely (Köln, D)

Objective: High dose chemotherapy causes the release of inflammatory cytokines such as TNFalpha. We examined the changes in TNFalpha salivary levels and the severity of mucositis during autologous SCT. Methods: From 5-9/00 the TNFalpha levels in saliva of 7 patients (age 26-56; median 39) undergoing PBSCT for relapse of Hodgkin´s Disease or high-grade Non Hodgkin´s Lymphoma were measured. Days (d) of sampling: before myeloablative BEAM chemotherapy (d-6), the day of peripheral blood stem cell transplantation (d0), d+6, d+14 and d+100. To induce salivation patients chewed on a paraffin block. Saliva was taken by the investigators. TNFalpha was quantified by enzyme immunometric assay (EIA). Severity of mucositis was determined by the Oral Assessment Guide and oral pain was self-assessed using a 0-100 visual analogue scale by the patient. Results: Mean values were as follows for TNFalpha [pg/ml] d-6: 256; d0: 309; d+6: 1638; d+14: 251). Simultaneously the severity of mucositis and oral pain increased. The mean patients´ self assessment for „severest pain during the last 24 hours" was on days d-6: 1/100; d0: 5/100; d+6: 65/100; d+14: 6/100. The mean mucositis grade was on d-6: 8; d0: 13; d+6: 16; d+14: 9, respectively. Conclusions: In high-dose chemotherapy oral mucositis grade and oral pain appeared to be associated with an increase of TNFalpha.

786

Autologous stem cell transplantation for acute myeloid leucemia in complete remission in Austria

S. Sormann, W. Linkesch, H. Greinix, D. Niederwieser (Graz, Wien, Innsbruck, A)

Between 1/1991 and 30.04.2001 95 evaluable patients over 16 years of age with acute myeloid leucemia were reported to the Austrian stem cell transplantation registry for autologous stem cell transplantation. In this patient group 44 were male and 51 female, median age 38 years (range 16 to 69).13 belong to FAB M1,12 to M2,5 to M3,23 to M4,10 to M5,2 to M6,1 to M7. Out of this group 66 pts. (29 male and 37 female)have been transplanted in first complete remission (CR1), 21 in CR2 and 8 pts. in subsequent CR. They received bone marrow(32 pts) or peripheral blood progenitor stem cells (36 pts.) or both (19 pts) after a median of 7 months (range 3 to 115) between diagnosis and transplant. Cytogenetics were available in 55 (52%) of pts. Eleven (10% of all) belonged to the favourable, 10 pts. (10% of all) to the unfavourable and 34 (32% of all) to the standard risk group. In the group transplanted in CR1 29 were male and 37 female, median age 38 (range 17 to 69), the distribution of FAB types M1 13, M2 12,M3 5, M4 23, M5 10, M6 2, M7 1; median time to transplant 6 month.In the group transplanted in CR2 10 were male and 11 female, median age 39 (range 18 to 59), distribution of FAB types M1 4, M2 6, M3 2, M4 6, M5 3; median time to transplant 20 months. The 5 year probability (Kaplan Meier analyses) for leucemia free survival of pts. transplanted in CR1+CR2 was 48%in CR1 56%, in CR2 38% (median observation time 33 months, range 6 to 198). The 5 year probability (Kaplan Meier analyses) for event free survival of pts. transplanted in CR1+CR2 was 43 % in CR1 48 %, in CR2 32%. The 5 year probability (Kaplan Meier analyses) for event free survival of pts. transplanted in CR1 with peripheral blood stem cells was 62% and for pts. transplanted with bone marrow 30% respectively. This results show that peripheral autologous stem cell transplantation in first remission is an effective treatment for acute myeloid leucemia.

787

Autologous blood stem cell transplantation (PBSCT) for malignant lymphoma after high dose chemotherapy (HDT) with BEAC: an analysis on efficacy and toxicity

M. Sandherr, C. von Schilling, R. Peter, C. Peschel (München, D)

The therapeutic potential of high dose chemotherapy followed by PBSCT in refractory or relapsed malignant lymphoma has been demonstrated. The role of the preparative regimen used before transplantation still has to be determined. Between 09/97 and 05/01, 57 autotransplants have been performed at our institution for refractory or relapsed lymphoma. In 26 transplant procedures, BEAC was used for HDT: BCNU 600 mg/m^2, Etoposide 400 mg/m^2 x 4, Cytarabine 2x3 g/m^2 x 4 and Cyclophosphamide 90 mg/kg. After a median follow up of 20 mo.(range 1-44), 6 pts. have died. Five pts. died due to progressing lymphoma and one pt. died after cardiac failure during HDT. Histology included 11 follicular lymphoma (FL), 12 aggressive B-cell lymphoma (DLBCL), 1 mantle cell lymphoma, 1 peripheral T cell lymphoma and 1 AILD. Indication to transplant was sensitive relapse in 16 pts. (FL n=9), primary refractory disease in 3 pts. and response to second line chemotherapy after induction failure in 7 pts. (FL n=2). The overall response rate was 85% while in 3 pts. the lymphoma progressed. All FL pts. and 75% of DLBCL pts reached a CR or PR. After a median f/u of 20 mo., 64% of FL pts. are still in remission, while 36% have relapsed without any death. In DLBCL, 58% are in ongoing remission, while 4 pts. died due to relapsing or progressive disease. Overall disease free survival (DFS) is 54% after a median f/u of 20 mo. Main adverse events are related to BCNU: 65% of pts. experienced serious allergic reactions with significant cardiovascular symptoms in 50%. One pt. died due to cardiac failure one day after BCNU. In 2 pts., we observed a BCNU related pneumonitis to be treated with sytemic steroids. Other toxicity comprised transient elevation of liver enzymes in 12% and neutropenic fever in 92% of pts. The transplant related mortality is 4%. BEAC is an efficacous preparative regimen for stem cell autografting in malignant lymphoma. However, there is significant toxicity which is related mainly to the use of BCNU. After the short f/u the relapse rate for FL pts. is high and the DFS is inferior to FL pts. which have been treated with TBI/Cyclophosphamide. We conclude, that chemotherapy alone as preparative regimen is insufficient for this population of FL pts. and should be supplemented by for example radioimmunotherapy. The use of BCNU in the dose of 600 mg/m^2 in this setting should be reconsidered.

788

Appearance of a cerebral tumor after autologous blood stem cell transplantation of a patient with a follicular lymphoma

J. Koenig, H. Kasparu, M. Girschikofsky, O. Krieger, S. Machherndl-Spandl, R. Koplmüller, D. Lutz (Linz, A)

A patient (60 y) in the second CR of a follicular germinal center lymphoma grade I was transplanted with an autologous blood stem cell graft. Conditioning therapy was performed according to the BEAM-protocol (BCNU, VP16, ARA-C, Melphalan); Rituximab was added on days -9 and -1. There were no unexpected or major complications until day +9 (emesis grade II, mucositis grade II, enteritis grade II, subfebrility and moderate elevation of inflammatory parameters). On day +10 the patient complained for the first time about double vision and headache. The clinical symptoms at this time were ptosis of the right upper eyelid and paresis of oculomotor n. and abducent n. Nuclear magnetic resonance imaging revealed a 2.8 x 1.7 x 2.3 cm mass located at the base of the skull, displacement of the pituitary body to the left, and infiltration of the cavernous s. on both sides. Differential diagnosis implied intracranial inflammations of bacterial, viral or mycotic origin or intracerebral bleeding or tumors of the pituitary gland, as well as cerebral manifestation of the already existing lymphoma. But cerebral appearance of the lymphoma only a few days after high-dose chemotherapy seemed very unlikely.
After transsphenoidal puncture histological findings indicated a prolactinoma. Therapy with cabergolin 0.5 mg twice weekly was started. After five months, the clinical course showed partial disappearance of the neurological symptoms. The size of the tumor was roughly the same. Prolactinomas are the most frequent endocrine-active tumors, with a prevalence of 1:1600. They appear in all age groups; the prevalence among older patients is divided equally between the sexes, whereas among younger patients, it is more common in women. The main symptoms are galactorrhea, amenorrhea, infertility, hypogonadism and impotence. Other symptoms include visual disturbances, in particular bitemporal hemianopia, eye muscle paralysis, hormonal disturbances due to tumor invasion of the pituitary gland and the hypothalamus, and cerebral convulsions. A retrospective inquiry of the patient revealed that he had been impotent for many years. Why the symptoms caused by local pressure first occurred 10 days after transplantation remains unclear. We searched the medical literature (Medline) and found no evidence for increased incidence for prolactinomas in patients trated for Non-Hodgkin's lymphoma. Therefore we conclude that the coincidence of lymphoma and prolactinoma in our patient was a random occurrence. A computerized tomography of the cerebrum was not performed before transplantation.

789

Phenotypic characterization of a rare T cell subset expanded after autologous peripheral stem cell transplantation (PBSCT)

V. Kunzmann, H. Berghammer, P. Reimer, F. Weißinger, M. Wilhelm (Würzburg, D)

Posttransplant infections and immunotherapeutic strategies for minimal residual disease are both dependent on cellular immune reconstitution after high-dose chemotherapy followed by autologous PBSCT. The CD8+CD28-CD57+ T cell subset has been reported to significantly expand in the peripheral blood of patients receiving allogeneic and autolgous stem cell grafts. Similar oligoclonal expansions of CD8+CD28-CD57+ T cells have been observed in several clinical settings associated with impaired thymic function such as HIV-infection or elderly people and are therefore attributed to thymic independent pathways of CD8+ T cell regeneration. The aim of this study was to further characterize the phenotype and of this T cell subset using three-color flow cytometric analysis of multiple cells surface markers. This analysis revealed an expansion of CD8+CD28-CD57+ T cells in all 17 enrolled patients after PBSCT ranging from 12% to 78% of all lymphocytes (mean absolute count: 271/microl; range 50-737/microl). The increase of CD8+CD28-CD57+ T cells was associated with up-regulation of activation markers (HLA-DR, CD29, CD38), although early activation markers such CD69 and CD25 were absent. In addition, we observed a coincidence of increased CD95 and CD11a with decreased CD62L expression that is highly suggestive for an antigen-experienced, non-naive T cell subset. Interestingly, CD8+CD28-CD57+ T cell counts correlated with serologic CMV immune status (CMV negative: 97/microl ± 21 vs. CMV positive: 366/microl ± 61; $p = 0.0078$) and age (< 50 years: 116/microl ± 55 vs. > 50 years 319/microl ± 59; $p = 0.0476$). In conclusion, our results indicate that the expanded CD8+CD28-CD57+ T cell subset in immunosuppressed patients after PBSCT represents an in vivo activated memory cell population derived from a thymic independent maturation pathway. The association with CMV status suggests that CMV antigens might be involved in this chronic stimulation.

790

Ex vivo B cell purging in multiple myeloma: A prospective comparative trial

C. Straka, M. Mitterer, F. Oduncu, E. Drexler, B. Schnabel, I. Suttmann, R. Bleicher, I. Langenmayer, H. Buhmann, K. Auchter, R. Bartl, R. Lamerz, H. Dietzfelbinger, M. Sandherr, M. Starck, N. Fischer, W. Gesierich, N. Brack, C. Nerl, P. Coser, B. Emmerich (München, Herrsching, D; Bozen, I)

In multiple myeloma, clonotypic B cells are present in the circulation and form a reservoir of drug resistant tumor cells which might contribute to relapse when reinfused into the patient with an autologous stem cell autograft. Since 7/95, 110 patients with multiple myeloma received sequential chemotherapy according to the „total therapy" concept of Barlogie which consisted of VAD, IEV (ifosfamide 2500 mg/m^2 iv day 1-3, etoposide 150 mg/m^2 iv day 1-3, epirubicin 100 mg/m^2 iv day 1), EDAP and then tandem high-dose therapy with melphalan 200 mg/m^2 within 3-6 months followed by autotransplantation of peripheral blood stem cells (PBSCs) which were collected after IEV-chemotherapy and G-CSF (Filgrastim, Amgen). Immunomagnetic ex vivo B cell purging of PBSCs (CD10, CD19, CD20, CD22, CD37; MaxSep, Baxter/Nexell) was applied in approximately half of the patients in order to remove clonotypic B cells which may contribute to relapse after high-dose therapy. Whether the PBSC autograft of a patient was purged or not depended on the availability at the treatment center and logistical reasons but not on individual patient characteristics. The treatment-related mortality for all patients was 5%. The median follow-up of surviving patients after the first autograft was 29 months. For B cell purged (n = 53) versus unpurged (n = 57) autografting, the median event-free survival (EFS) was 40 versus 22 months, respectively (p = 0.038). The median overall survival (OS) for unpurged autografting was 48 months. For B cell purged autotransplantation the median OS was not reached; the OS was 72% at 4 years after the first autograft (p = 0.048). In conclusion, immunomagnetic ex vivo B cell purging of PBSC autografts in multiple myeloma improved the clinical outcome after high-dose therapy and autograft. This result needs confirmation."Curative" autograft strategies possibly should include a B cell depletion of PBSC autografts.

791

Long-term outcome of 130 patients with acute leukemia autografted with mafosfamide-purged marrow in a single institution

G. Egerer, A. Abdallah, H. Goldschmidt, M. Koerbling, R. Haas, A.D. Ho (Heidelberg, Düsseldorf, D; Houston, USA)

There are few reports on the long-term outcome for acute leukemia patients 10 years after autologous bone marrow transplantation (ABMT). We retrospectively analyzed data from 130 patients with acute leucemia (32 with ALL and 98 with AML) transplanted at our institution in Heidelberg from July, 1983 to November, 1994, most of whom were referred from other institutions. As conditioning regimen all patients have received high dose cyclophosphamide (200 mg/kg) preceded by total body irradiation with a median dose of 14.4 Gy (range, 12.1-16.7) followed by ABMT. Bone marrow was purged with mafosfamide which was added at a median dose of 70 µg/2 x 107 WBC according to the individual sensitivity of granulocyte-macrophage colony-forming units. A retrospective analysis in March 2000 showed that 39 out of 130 ALL/AML patients (12 ALL, 27 AML) were alive and in continuous complete remission for a median of 141 months (range 0-180) after ABMT with overall survival and disease-free survival of 30 % and could be considered as „cured". The analysis of the characteristics of such long-term survivors showed that the 53 patients transplanted at first complete remission (CR) had significantly higher chance of survival (47 %) than those transplanted at second or subsequent remissions (p=0.001). There was no significant difference in age, gender, time from diagnosis to transplantation or time from harvest to transplantation between long-term responders and patients who decreased. Our retrospective analysis shows that a significant number of adult patients with ALL and AML in first CR derived long-term benefit from HDCT and ABMT with mafosfamide purged preparation. However, we observed 4 patients with secondary myelodysplasia (MDS) 35, 46, 134 and 142 months after transplantation. The latter patient, in addition to MDS, developed secondary AML 145 months after transplantation. The cumulative risk of therapy related MDS and secondary AML increased 32 months after bone marrow transplantation for leukemia and reached 3.8% after 145 months.

792

High dose melphalan and autologous stem cell support after VAD chemotherapy for light chain (AL) amyloidosis

J. Perz, S. Schönland, R. Singer, U. Hegenbart, G. Egerer, A. Ho, H. Goldschmidt (Heidelberg, Ladenburg, Leipzig, D)

AIMS: The median survival of patients (pts) with AL amyloidosis treated with standard chemotherapy (melphalan/prednisone) is 17 months. High dose melphalan (HD-M) chemotherapy followed by autologous stem cell transplantation (ASCT) has recently been shown to be effective in pts with AL amyloidosis due to the reduction of clonal light-chain producing plasma cells. In order to evaluate the efficacy and the toxicity of conventional VAD (vincristine/ adriamycin/ dexamethason chemotherapy) and high dose chemotherapy, we treated 10 pts with AL amyloidosis. METHODS: Since May 1998 to May 2001, 10 pts (7 male, 3 female) with biopsy-proven AL amyloidosis of heart (n=7, n=2 main manifestation), kidney (n=7, n=6 main manifestation) and liver or intestinum (n=7, n=2 main manifestation) were treated with 2-4 cycles of VAD, followed by stem cell mobilisation with cyclophosphamide or ifosfamide and G-CSF. Finally, 9 of 10 pts received HD-M and ASCT. In 5 pts at least three organs were involved. RESULTS: Survival: One patient (pt) with main kidney manifestation
died during therapy due to a cerebral hemorrhage 9 days after ASCT. Two pts succumbed to progression of amyloidosis (in 1 pt with heart manifestation progress before HD-M, in 1 pt with heart and kidney manifestation progress 6 months after HD-M). The 1 month survival after HD-M and ASCT is 8 of 9 (88 %). The median follow-up is 12 months. Six pts survived longer than 6 months after HD-M. Response: 6 months after HD-M 4 of 6 pts achieved a complete remission of their plasma cell disorder, 2 of 6 pts a minimal response, in all 6 pts a clinical improvement in organ function can be observed. CONCLUSIONS: HD-M followed by ASCT after conventional VAD is feasible in pts with AL-amyloidosis. By reduction of the clonal plasma cells an improvement of amyloid-related organ dysfunction can be achieved.

793

Rituximab in vivo purging is safe and effective in combination with CD34-positive selected autologous stem cell transplantation for salvage therapy in B-NHL

T. Flohr, G. Heß, K. Kolbe, H. Gamm, T. Stanislawski, C. Huber, H.G. Derigs (Mainz, D)

Purpose To evaluate feasibility and efficacy of rituximab included into a sequential salvage protocol for CD20+ B-NHL in relapse or induction failure. Patients and methods Twenty seven patients with CD20+ B-NHL in relapse or induction failure received Rituximab combined with DexaBEAM (R-DexaBEAM) for stem cell mobilization. Ex vivo selection of CD34 positive cells was performed using the CliniMacs® positive-selection method in patients who mobilized more than 5x10^6/kg CD34+ cells (n = 19). Two doses of Rituximab were included into the high dose therapy regimen (HDT). Results R-DexaBEAM was well tolerated and 26 of 27 patients mobilized sufficient numbers of CD34+ blood stem cells to receive HDT. Application of R-DexaBEAM resulted in significant depletion of peripheral B-cells. However, in peripheral blood of three patients, lymphoma cells were still detected by flowcytometry or PCR analysis at the day of graft collection. The additional ex vivo selection cleared leukapheresis products in all subjects. No treatment related death occurred after HDT and all patients showed stable engraftment of hematopoesis. A prolonged combined immunodeficiency was observed post HDT, with half of patients demonstrating a T4-cell count below 200/µl at day +180 and subnormal serum immunoglobuline levels for up to 2 years in the majority of the population. Eight patients developed CMV antigenemia, but with preemptive ganciclovir medication no CMV disease occured. Remission rate post HDT was 96% (CR: 24/26; PR: 1/26). Overall and progression-free survival (PFS) at sixteen months post HDT (range: 6 to 27) is 95% and 77%, respectively. With regard to histology PFS was 71% in aggressive lymphoma (n = 11), 74% in indolent FCL (n = 10) and 100% in MCL (n = 5). Conclusion The treatment protocol has proven feasible with high purging efficiency and encouraging remission rates. Prolonged impairment of the immune system post transplant needs special attention.

794

Immune reconstitution after autologous peripheral blood stem cell transplantation (PBSCT) combined with rituximab for refractory B-cell non-Hodgkin's lymphoma

T. Flohr, G. Heß, S. Kreiter, R. Meyer, C. Huber, H.G. Derigs (Mainz, D)

Combining ex vivo CD34 positive selection and use of Rituximab for in vivo purging leads to effective tumor cell depletion in autologous grafts of pts with B-NHL. This strategy involves the risk of prolonged immunodeficiency post transplantation by intensive B- and T-cell depletion. For this reason we intensively studied lymphoid reconstitution in this setting with special regard to CMV serostatus and occurence of CMV reactivation post PBSCT. Pts with relapsed or refractory CD20+ B-NHL received Rituximab combined with DexaBEAM for stem cell mobilization. In pts collecting >5x10^6 CD34+ cells, stem cell products were purged using the Clinimacs device. Two doses of Rituximab were administered as part of conditioning. Reconstitution of lymphocyte subsets were characterized by flowcytometric analysis. Immunoglobuline levels and clinical complications were prospectively documented. In HLA-A2 positive pts the frequency of CMV-pp65495-503 reactive CTL is currently assesed using interferon gamma ELISPOT assay. Immune reconstitution was evaluated in 26 pts with a minimum follow up of 180 days post PBSCT. 18 pts had a CMV positive serostatus pretransplant and 16/18 pts showed CMV-DNAemia at least once post PBSCT. B-cells did not reappear in peripheral blood before day +180. Serum levels of IgA and IgM remained below normal values for the whole study period in the majority of pts, while IgG restored to at least pretransplant levels at about 1 year post PBSCT in most pts. CD4+ T-cell count slowly increased and approx. half of pts presented with a T-Helper-cell count below 200/µl at 6 mths post PBSCT. NK cells showed a fast regeneration kinetic with maximum values reached by approx. two months post PBSCT. Reconstitution of CD8+ T-cells differed significantly between pts with a positive CMV serostatus pretransplant and those being negative. In the CMV positive group (n=18) CD8+ T-cells represented the predominant lymphocyte-subset throughout the whole studied time post day +30. This led to an emphasized inverse T4/T8 ratio and a higher absolute T-lymphocyte count, when compared to the CMV negative group at every timepoint studied. This T-cell-subpopulation explicites an activated phenotype coexpressing the markers CD38, CD57, HLADR. Surprisingly, statistical analysis comparing reconstitution of CD3+ T-cells and CD19+ cells in patients receiving selected vs unselected CD34+ cells did not reveal any significant difference. The present study confirms the severe combined immunodeficiency status of pts post autologous transplantation of CD34 selected peripheral blood stem cells.The lymphoproliferation seen in pts under CMV infection remarkebly alters the composition of the cellular immunesystem and this proliferating T-cell-subset is currently under investigation.

795

High dose chemo/radiochemotherapy can be administered safely as consolidation therapy for patients with indolent NHL treated with CHOP and rituximab

G. R. Heß, T. Flohr, W. Hiddemann, Ch. Huber, H. G. Derigs, T. Fischer (Mainz, München, D)

Optimal treatment of patients with newly diagnosed indolent NHL is still of debate and cure is almost never achieved by conventional chemotherapy. Rituximab is a new promising treatment modality for indolent NHL. As more potent combination therapies including this antibody might by able to reduce tumor burden to a higher extent, high dose chemotherapy could eradicate remaining disease. Therefore we initiated a treatment protocol for the evaluation of the feasibility and safety of this treament approach. Untreated pts with proven CD20+ stage IIB to IV indolent NHL (FCC I-II or MCL) were enrolled in an ongoing study, in which treatment consisted of 2 cycles of CHOP to prove chemosensitivity and 4 additional cycles of CHOP plus rituximab (375mg/m^2 day 1). Pts with at least stable disease received mobilization chemotherapy with Dexa-BEAM/G-CSF. HDT (TBI/Cy or BEAC/M) was scheduled within 8 weeks after mobilization. Monitoring for MRD was performed by standard PCR-techniques. 15 pts (median age 48y, 9 pts FCC, 6 pt MCL) have been enrolled and 9 pts completed HDT sofar. 1 pt had stage IIB, 5 pts had stage III and 9 pts had stage IV disease. 9/15 pts had BM-involvement. Evaluation upon the first 2 cycles of CHOP (n=13) showed 7 pts achieving MR or SD, 5 pt achieving PR and 1 pt achieving CR. Restaging before mobilization showed 6 pts achieving PR and 4 pt with CR. Sufficient numbers of PBSC could be collected in all pts mobilized (10/10). Side effects of mobilization were pneumonia in one pt and instable angina in 1 pt. 9 pts received HDT (7 pts TBI/Cy and 2 pt BEAC or BEAM) and are at least at day +60. Neutropenic fever and mucositis represented the major toxicities of HDT. Median time to hemopoetic recovery was 10 days (8-13). With a follow up of up to 480 days following adverse events are documented: gastric bleeding, pneumonia and venous thrombosis, each in 1 pt.. One patient with stage IV disease achieveing PR after HDT relapsed at day 180. Restaging at day +60 to 90 (9 pts) revealed 6 CR and 3 PR. Monitoring of MRD by PCR revealed negativity at day 60 post PBSCT in 3/3 pts tested positive initially. Immune reconstitution was monitored closely during the treatment course and post HDT. Rituximab treatment transiently reduced levels of PB B-lymphocytes. Conclusions: Taken together, the results of this study show that high dose chemotherapy can be performed safely following treatment with CHOP plus Rituximab. Preliminary analysis suggests an additional therapeutic effect of rituximab combination chemotherapy and of HDT. Randomized trials are warranted to evaluate potential superiority of this treatment protocol in newly diagnosed patients.

796

Ultrasonographic process of a hepatic veno-occlussive disease after autologous peripheral blood stem cell transplantation – A case report

S. Wilhelm, A. Holle, M. Leithäuser, J. Casper, M. Freund (Rostock, D)

Hepatic veno-occlussive disease is one of several manifestations of regimen-related toxicity that can occur after high dose cytoreductive therapy. We report about ultrasonographic criteria in the process of the therapy of a hepatic veno-occlussive disease in a case report. We have recently performed an autologous peripheral blood stem cell transplantation on a 60 year-old man, with a CML in acceleration phase. The conditioning regime consisted of busulfan 4 mg/kg d1-4. The patient had an uneventful neutropenic phase. On day +48 he began to experience right upper quadrant abdominal pain and tenderness which progressed. An abdominal color flow Doppler US showed a hepatomegaly and a marked reduction in portal vein blood flow. Within 24 hours we showed a hepatofugal flow with ascites and visualization of the paraumbilical vein. Serial color flow Doppler US were performed to follow the course of the disease. Treatment with heparin and prostacyclin E1 (Minprog® 500µg) resulted in a progressive disease. The patient was then given defibrotide (DF). DF was started at an initial dose of 20 mg/kg/d (increased to 40 mg/kg/d). No toxicities attributable to its use were observed during treatment. At 3 days after onset of DF therapy we could detect an improvement of objective ultrasonographic criteria. This was followed by marked improvement of the clinical condition. On day +14 after the onset of VOD his liver function tests and color flow Doppler US are normal. This case demonstrates that color flow Doppler US can be used to confirm and a process of the VOD diagnosis. Furthermore this technique provides a way off easy and reliable evaluation of the therapeutic effect.

797

CBV as preparative regimen for ASCT in relapsed or refractory aggressive non-Hodgkin's lymphoma

M. Eibl, W. Zinke-Cerwenka, P. Neumeister, U. Apfelbeck, W. Linkesch (Graz, A)

Introduction: Patients with aggressive Non Hodgkin´s lymphoma, who fail induction or relaps therapy, have a very poor prognosis. The addition of etoposide to the standard preparative regimen for autologous stem cell transplantation (ASCT) has been associated with lower relaps rates in recent reports. Therefore we evaluated the effect of augmented CBV as preparative regimen in this high risk patient subgroup. Patients and methods: From December 1999 to April 2001 seven patients (male:female = 5:2; median age: 37 years, ranged from 30 to 54) suffering from diffuse large B-cell lymphoma (n=5) and anaplastic large T-cell lymphoma (n=2) respectively, underwent ASCT after failing induction therapy. Prior to conditioning two patients were in partial remission (PR), and and five suffered from progressive disease (PD) or chemoresistant relaps. The conditioning regimen consisted of: BCNU 15 mg/kg (based on ideal body weight) on day –6, Etoposid 60 mg/kg (based on actual body weight) on day –4, Cyclophosphamide 100 mg/kg (based on ideal body weight) on day –2. Premedication and dose adjustment for obese patients was performed according to the protocol of Stiff et al (JCO 1998). Results: Response to ASCT was achieved in 6/7 patients (85.7%): complete remission (CR) in 5 patients and PR in 1 patient. So far, after a median observation period of 7 months, 3 patients (42.8%) are in continuous complete remission (CCR), 2 patients relapsed after 6 and 9 months respectively. 5/7 patients (71.4 percent) are alive, 3 of them disease free. Cause of death was tumor progression in both cases. Side effects possibly related to the preparative regimen were: reversible tumor lysis syndrome in one patient, hypotension and bradycardia in 1 patient.
Conclusion: Patients undergoing ASCT for resistant disease appear to have an improved outcome, if an augmented preparative regimen is used, compared to reports with standard conditioning regimens.

798

Stem cell mobilisation after two cycles high dose methotrexate

T. Kiefer, F. Schüler, M. Wimmer, M. Montemurro, M. Stockschläder, M. Schwenke, G. Dölken (Greifswald, D)

The most effective regimens for treating primary central nervous system lymphoma contain cytotoxic drugs like methotrexate (MTX), which penetrate the blood–brain barrier easily in addition to radiotherapy. However, chemotherapy combined with radiotherapy may result in delayed severe neurotoxic effects. To avoid this toxicity we have initiated a study protocol using two cycles of MTX (8000 mg/m^2, given at a ten days interval) followed by high dose chemotherapy (Busulfan: 16 mg/kg BW, thiotepa 10 mg/kg BW) and peripheral blood stem cell transplantation (PBSCT). In our protocol radiotherapy is restricted to patients that do not achieve a CR after chemotherapy alone. Combining the therapeutic effects of MTX with the mobilisation of peripheral blood stem cells would be an ideal task. Since nothing is known about the mobilising effect of high dose MTX we determined the number of CD34 positive cells in the peripheral blood after each high dose MTX combined with G-CSF s.c.. These are the results of the first three patients:

No.	CD34 (PB after MTX I)	CD34 (PB after MTXII)	No.of Leuka-pheresis	collected CD34 /kg BW x10E6	Recovery (days after BSCT)	
					WBC > 1000/µl	PLT > 50000/µl
1	0,63/µl	20,76/µl	2	6,71	8	10
2	13,89/µl	45,60/µl	2	6,19	9	11
3	16,5/µl	68,64/µl	1	8,69	9	9

PB: peripheral blood, MTX: methotrexate, kg: kilogram, BW: body weight, WBC: white blood count, PLT: platelets

Compared to the first MTX dose there is a clear increase in circulating CD34 positive cells after the second dose. A successful mobilisation of more than 5 x 10^6 stem cells/kg BW was achieved in in all three patiens after the second cycle. The hematopoietic recovery after the high dose chemotherapy protocol was fast (WBC > 1000/µl day 8-9 and PLT > 50000/µl day 9-11 after transplantation).

Summary: We conclude that two cycles of high dose MTX (8000 mg/m^2, given at an interval of ten days) are an excellent mobilisation regimen in patients with primary central nervous system lymphoma.

Postersitzung:
Chronische myeloische Leukämie und andere CMPE

799

Lack of prognostic significance of ABL-BCR mRNA expression in chronic myeloid leukemia patients treated with interferon alpha

K. Merx, M. C. Müller, A. Weißer, O. Maywald, S. Kreil, U. Berger, R. Hehlmann, A. Hochhaus and the German CML Study Group (Mannheim, D)

Recently it was proposed that extensive submicroscopical deletions 5' of ABL and 3' of BCR on the derivative chromosome 9q+ may accompany BCR-ABL rearrangement and that disease associated with a deletion was more refractory to treatment (Sinclair et al., Blood 2000; 95:738). Patients with such a deletion lack the reciprocal ABL-BCR fusion on the genomic level. We therefore sought to establish a relationship between ABL-BCR mRNA expression and prognosis in chronic phase CML patients treated with interferon alpha. Single-step reverse transcriptase polymerase chain reaction (RT-PCR) for ABL-BCR transcripts has been performed in 344 BCR-ABL+ CML patients initially treated with interferon alpha combined with hydroxyurea and/or cytarabine. In order to detect all possible fusion transcripts, primers mapping to ABL exon Ia and Ib and BCR exon b5 have been used. Only samples within the first year after diagnosis with high expression of BCR-ABL transcripts b3a2 or b2a2 were tested. ABL-BCR expression was detected in 188 patients (55%). 137 (73%) of them presented ABL(Ib)-BCR(b4) either alone or as main transcript in combination with Iab3 or Iab4. 47 patients (25%) showed expression of ABL(Ib)-BCR(b3) or combinations of Ibb3/b4, Iab3/b4. In one patient only ABL(Ia)-BCR(b3/b4) was found. Six patients expressed aberrant transcripts smaller than Ibb4, in three cases in addition to regular transcripts. Median follow up of all 344 patients was 2.1 years (0-6.2 years). 54 patients had died during this period. Correlation of ABL-BCR expression and survival showed no significant differences between ABL-BCR positive and negative patients. There was no significant survival difference between the ABL-BCR fusion subgroups. While patients stratification according to Hasford was associated with a significant survival difference (p=0.0062), ABL-BCR expression was not relevant for prognosis within the three strata of the score. This data suggests that in contrast to genomic 9q+ deletions ABL-BCR mRNA expression is not associated with prognosis in CML patients treated with interferon alpha.

800

Lack of ABL-BCR transcription indicate structural variants of standard t(9;22) in BCR-ABL positive CML

J. Römer, C. Bleck, U. Claussen, A. Heller, H. Sayer, H. Starke, M. Ziegler, T. Liehr, I.F. Loncarevic (Jena, D)

ABL-BCR expression is only found in about 60-70% in BCR-ABL positive chronic myeloid leukemia (CML). ABL-BCR transcription is suggested not to correlate with the response to interferon alpha therapy, although, a deletion in der(9) chromosome, where the ABL-BCR gene resides, seems to be associated with a worth prognosis. In order to investigate whether ABL-BCR RNA negativity results from cytogenetically detectable deviations from standard der(9)t(9;22)(q34;q11) we have analyzed 65 patients by RT-PCR and subsequently by FISH. We found 17 CML patients without an ABL-BCR-transcript among the 65 with b2a2 and/or b3a2 BCR-ABL transcription. Out of the 17 ABL-BCR negatives, 5 showed a deletion of both the 5'ABL and the 3' BCR gene sequences. Two patients showed a an apparently normal karyotype after GTG-banding because BCR-ABL gene fusion was generated by gene insertion (5'ABL insertion into chromosome 22q11 in one patient, and 5' BCR insertion into chromosome 9q34 in the other patients). Three patients showed a complex translocation with a BCR-ABL but not the ABL-BCR fusion gene. A weak 5'ABL- and BCR FISH signal was obtained in 2 patients, potentially resulting from a partial deletion of ABL-BCR. Three patients showed no deletion with the 5' ABL or BCR specific probes. Two samples could not be analyzed by FISH. None of the 48 patients with an ABL-BCR transcript exhibit a variant t(9;22). We conclude, that ABL-BCR-RNA negativity in BCR-ABL positive CML indicates: 1. a deletion in der(9)t(9;22)(q34;q11)in 33% (5/15), 2. a variant of the standard t(9;22)in 33%(5/15), and 3. In a minority 33% (5/15), cytogenetics could not explain the lack of the ABL-BCR transcript. We conclude, that in most cases, ABL-BCR RNA negativity is explainable by structural deviations from standard translocation t(9;22) in CML.

801

Detection of BCR-ABL m-RNA 18 months before the development of Ph positive chronic myeloid leukemia

M. Bornhäuser, B. Mohr, P. Bornhäuser, C. Felber, G. Ehninger, C. Thiede (Dresden, Metzingen, D)

There are some reports on spontaneous remission of Philadelphia-chromosome positive chronic myeloid leukemia (CML). Other investigators have found BCR-ABL m-RNA in healthy individuals using very sensitive RT-PCR. None of these individuals has been reported to have acquired Ph-positive CML during the follow-up. Here we describe a 63 year old patient who had received a coronary stent after myocardial infarction in March 1999. He initially presented with unspecific symptoms of fatigue without pathological findings in a physical examination. The initial leukocyte count was normal but the differential showed 1 myeloblast, 3 myelocytes and 10 bands. Surprisingly, positivity for b3a2 and b2a2 BCR-ABL m-RNA was detected in a peripheral blood sample using nested RT-PCR. Cytogenetic analysis of a bone marrow aspirate performed two months later showed a normal male karyotype. During the next 12 months the blood count remained unchanged but interestingly the message for b3a2 decreased whereas quantitative PCR revealed an increasing amount of b2a2 m-RNA in two subsequent analyses. In January 2000 an increase of CD34+ cells in the peripheral blood up to 65/µl was detected indicative for the development of active leukemia. Another bone marrow aspirate was dry but histological analysis of the bone marrow showed trilineage hyperplasia with moderate fibrosis. At the same time plasma lactate dehydrogenase increased up to two times above the upper limit and splenomegaly developed. Hypermetaphase FISH and conventional karyotyping now showed BCR-ABL positive metaphases in the peripheral blood for the first time. Since blood counts remained stable, no treatment has been started until now. This is one of the first reports on the clinical and molecular monitoring of the development of Ph-positive myeloproliferative disease in a patient. The long latency between detection of BCR-ABL and the first clinical signs is surprising. It may be speculated that immunological surveillance has kept the b3a2 positive clone from proliferating favouring the expansion of b2a2 positive cells in this HLA-B8, DR0901 positive subject.

802

Various nonrandom whole arm translocations in addition to t(9;22)(q34;q11) in chronic myeloid leukemia (CML) are leading to a loss of 17p

C. Scherer, S. Schnittger, A. Hochhaus, T. Haferlach, U. Berger, R. Hehlmann, W. Hiddemann, C. Schoch (München, Mannheim, D)

CML is characterized by the occurrence of the fusion gene BCR-ABL in the leukemic cells. This fusion gene originates through a balanced translocation between chromosome 9 and 22, t(9;22)(q34;q11). In this event the ABL-onkogene on chromosome 9 fuses with the BCR-gene on chromosome 22. The fusion gene can be detected by fluorescence in situ hybridization (FISH) with loci specific probes and polymerase chain reaction (PCR). In 90% of all CML cases the leukemic cells show this specific genetic abnormality cytogenetically as a shortened rearranged chromosome 22, the Philadelphia chromosome (Ph). The Ph marker originates through other rearrangements than the classic t(9;22) in 5-10% of CML cases. In simple variant translocations, the segment lost from chromosome 22 is translocated onto another chromosome than 9, whereas three or more chromosomes are involved in complex variant translocations. In 10% of CML cases the translocation occures on the microscopic level and cytogenetically normal chromosomes 9 and 22 are detected (Ph-negative CML). In addition to this typical CML aberration, several frequent chromosomal abnormalities were observed in CML patients, above all in the accelerated phase and blast crisis. Trisomy 8, i(17)(q10), trisomy 19 and an extra Ph are the most common additional aberrations. These secondary aberrations have prognostic relevance. We observed several new additional chromosomal abnormalities exclusively in chronic myeloid leukemia, showing whole arm translocations involving 17p. der(7;17)(p10;q10) was observed in three patients, two of them were in blast crisis and one in an accelerated phase of the disease. der(17;18)(q10;q10) was observed in four patients, two of them were in blast crisis and one in accelerated phase. One patient showed both chromosomal abnormalities. Further additional aberrations observed were der(16;17)(q10;q10) in one patient being in the chronic phase of the disease and der(1;17)(q10;q10) in a patient in blast crisis. In all cases whole chromosome painting (WCP) was performed to prove the fusion of the two chromosomes involved. In some cases FISH with probes for the centromers revealed that the fusion product was dicentric. All aberrations led to loss of 17p, like i(17)(q10), the most frequent abnormality observed during the disease progression of Ph-positive CML. Several genes located on 17p are mapped: the genes for TOP3A (topoisomerase III alpha), HIC1 (hypermethylated in cancer) and p53 at 17p12-p11.2, 17p13.3 and 17p13.1, respectively. The p53 gene product, TP53, plays an important role in tumor suppression. Deletion or mutations of p53 are the most common genetic abnormalities observed in human cancer especially during tumor progression. A deletion of p53 was also proven by FISH with a loci specific probe in all eight patients described above. Further molecular studies will reveal if these known genes in 17p or as yet unidentified genes are keyplayers for disease progression.

803

Acute vision loss due to interferon related retinopathy in a patient with CML

A. Berndt, R. Naumann, G. Ehninger (Dresden, D)

A 70 years old male patient with confirmed diagnosis of CML presented to our institution with painless visual deterioration of his left eye. The patient was initially treated with hydroxyurea for 2 weeks, followed by subcutaneous interferon alpha (INF). INF had been started at a dose of $1.5*10^6$ IU/m^2 and was later escalated up to $3.5*10^6$ IU/m^2 daily. Visual deterioration developed 2 weeks after initiation of interferon therapy. At the time of consultation white blood cell count was in normal range ($5.0*10^6$/l). Medical evaluation before CML revealed coronary artery disease. Conditions predisposing to capillary nonperfusion, i.e. diabetes and hypertension were not found. On examination in the Eye Clinic corrected visual acuity was 0.1 in the left eye and 1.0 in the right eye. Both anterior segments were normal without significant lens opacities. Fundoscopy revealed cotton-wool-spots, intraretinal haemorrhages and marked macula deem in the left eye. 4 weeks later also his right eye developed cotton-wool-spots and intraretinal haemorrhages. Therefore, INF was discontinued. During an examination 6 weeks later the patient noticed a subjective improvement in his visual symptoms. Ophtalmoscopy revealed pronounced improvement in the macular deem in both eyes with substantial resolutions of the cotton wool spots and the intraretinal hemorrhages and considerable improved

It is well known that visual deterioration may occur in patients with CML, usually as a result of hyperleukocytosis, predisposing to intravascular thrombus formation. Manifestations of retinal ischaemia, however, are rarely reported complications of interferon alpha treatment. In our patient the resolution of much of the retinopathy after discontinuation of the interferon therapy with no change in patients overall condition argues strongly for the patients vision loss representing a case of interferon related retinopathy. It is concluded that interferon related retinopathy should be considered in the differential diagnosis of acute visual loss in CML patients, even if hyperleucocytosis could be excluded.

804

Whole blood histamine level: A proposed novel marker of chronic myeloid leukemia (CML)

A.W. Hauswirth, H. Semper, M. Ghannadan, H. Agis, K. Geissler, K. Lechner, W.R. Sperr, P. Valent (Wien, A)

Leukocytosis with left shift and basophilia are common findings in patients with CML. We have recently shown that basophils in CML produce huge amounts of histamine and lower levels of mast cell tryptase. In the present study, we have determined the levels of whole blood histamine (WBH) and whole blood tryptase (WBT) in patients (pts) with CML (n=17; chronic phase, n=15; accelerated phase, n=2), other myeloproliferative disorders (IMF, n=14; ET, n=9; PV, n=18) as well as healthy volunteers (n=30). WBH levels were measured by RIA and WBT by fluoroimmunoenzyme-assay (FIA). The median levels of WBH were 5,687 ng/ml (range: 936-28,570) in CML, 145 ng/ml (range: 10-735) in IMF, 162 ng/ml (range: 35-1,004) in ET, 170 ng/ml (range: 16-241) in PV, and 60 ng/ml (range: 27-108) in healthy controls. WBH levels in CML pts were significantly higher compared to other myeloproliferative disorders or healthy controls ($p<0.005$). The median levels of WBT were 294 ng/ml (range: 36-3,150) in CML, 9 ng/ml (range: 0-153) in IMF, 10 ng/ml (range: 7-17) in ET, 11 ng/ml (6-78) in PV, and 6 ng/ml (1-17) in healthy controls. The highest levels of WBT were measured in a case of accelerated phase CML. In the CML group, significant correlations were found between the WBH levels and absolute basophil counts (r=0.91) as well as between WBH levels and leukocyte counts (r=0.63) ($p<0.05$). No significant correlations were found when WBH levels were compared with LDH levels (r=0.31) or WBT (r=0.32). Follow-up during initial cytoreductive therapy with hydroxyurea p.o. (n=3) or hydroxyurea p.o. plus IFNalpha2a s.c. (n=1) showed a decrease in WBH and WBT in all pts. In one patient, a normal WBH and WBT was recorded after 3 months of hydroxyurea treatment, whereas in the other patients no biochemical remission could be achieved by hydroxyurea (2 and 4 months) or hydroxyurea plus IFNalpha2a (24 months). In summary, our data show that histamine is a novel additional marker useful for the monitoring of CML. Studies are in progress to compare WBH and WBT levels with quantitative BCR-ABL and Ph-chromosome determinations in pts with CML during treatment.

805

p190 BCR-ABL-positive myeloproliferative disorder: a heterogenous disease resembling features of ALL, CML, and CMML

M.C. Müller, A. Reiter, S. Kreil, K. Merx, A. Weißer, U. Berger, R. Hehlmann, A. Hochhaus (Mannheim, D)

In BCR-ABL-positive chronic myeloid leukemia (CML) BCR exons e13 (b2) or e14 (b3) are fused to exon a2 of the ABL gene in the vast majority (>99%) of cases. However, atypical chimeric transcripts resulting from aberrant genomic breakpoints have been described. The e1a2 BCR-ABL fusion transcript encodes for the chimeric p190 protein which is found almost exclusively in Philadelphia-chromosome (Ph) positive acute lymphoblastic leukemia (ALL). But very rarely it has also been found in patients displaying features of a myeloproliferative/myelodysplastic disorder with prominent monocytosis. We describe three p190 BCR-ABL-positive cases demonstrating the heterogeneity of this disease. All patients were Ph-positive at diagnosis and multiplex RT-PCR revealed the e1a2 fusion transcript. (1) A 81 year old woman presented with anemia (Hb 6,0 g/dL), thrombocytopenia (45/nl) and leukocytosis (24/nl) with a differential displaying typical features of a myeloproliferative syndrome with 24% blasts, 1% promyelocytes, 16% myelocytes, 1% metamyelocytes, 4 % bands and a monocytosis of 28%, but without eosinophilia or basophilia. Bone marrow aspiration cytology showed 50% myeloid blasts. No organomegaly was found. Four weeks after diagnosis the patient died from fulminant pneumonia. (2) A 55 year old woman presented with leukocytosis (549/nl; 14% blasts, 3% promyelocytes, 5% myelocytes, 7% metamyelocytes, 9 % bands, and 6 % monocytes, no eosinophilia or basophilia), thrombocytopenia (31/nl) and hepatosplenomegaly. Bone marrow aspiration and immunophenotyping (CD19+, TdT+) suggested the diagnosis of a pre-pre-B-ALL. After high-

Mittwoch, 3. 10. 2001

dose chemotherapy a typical CML-like chronic phase (100% Ph+) was seen at several occasions. 21 months after diagnosis the patient died from refractory monoblastic blast crisis (CD11C+, CD14+).(3) A 66 year old man presented with leukocytosis (126/nl) and splenomegaly. The differential showed 3% blasts, 14% promyelocytes, 1% metamyelocytes, 8% monocytes, no eosinophilia, no basophilia. Bone marrow revealed CML in chronic phase. Complete hematologic and complete cytogenetic response were achieved on interferon alpha although persistent monocytosis (up to 50%) was observed. RT-PCR for e1a2 BCR-ABL transcripts was repeatedly positive on very low levels (Ratio BCR-ABL/ABL 0.061%). The patient died five years after diagnosis in complete cytogenetic remission due to cerebral hemorrhage. We conclude that the p190 BCR-ABL-postive myeloproliferative disorder resembles phenotypical features of CML, CMML and ALL and demonstrates various clinical courses. Monocytosis, lack of eosinophilia and basophilia, and splenomegaly are common features of this entity. However, response to therapy and prognosis is heterogeneous ranging from complete cytogenetic remission after IFN-therapy to rapidly progressing disease. Multiplex RT-PCR is therefore suggested in all typical or atypical cases with a phenotype of CML, CMML, or ALL.

806

Coexistence of chronic myeloid leukemia and chronic lymphocytic leukemia - Evidence of two independent clonal disorders

S. Kreil, T. Lahaye, S. Mende, U. Berger, J. Hastka, E. Lengfelder, M. Kneba, S. Stilgenbauer, C. Schoch, C. Fonatsch, R. Hehlmann, A. Hochhaus (Mannheim, Ravensburg, Kiel, Ulm, München, D; Wien, A)

Over a period of 12 years we studied two patients who developed chronic myelogenous leukemia (CML) and chronic lymphocytic leukemia (CLL) independently, i.e. two different clones could be differentiated by hematological, cytogenetical, and molecular methods:

I) A 68 year old winegrower developed Ph+ CML in 5/1990 27 months after the onset of B-cell (CD5+, CD19+, CD/22+) CLL (Rai stage III) in 2/1988. When CML was diagnosed, both malignant cell populations were detected in bone marrow (BM) and peripheral blood (PB). Cytogenetic analyses were performed on mononuclear cells and granulocyte enriched populations. Myeloid BM metaphases were 100% Ph+, whilst lymphoid metaphases were Ph negative. Southern blot analysis revealed a BCR rearrangement in DNA from the granulocyte-enriched fraction, b2a2 BCR-ABL mRNA transcripts were detected by RT-PCR. CLL has been treated with chlorambucil for two years. After diagnosis of CML therapy was changed to busulfan, later to hydroxyurea and low dose ara-C. Myeloid blast crisis occurred in 6/1991. The patient died from blast crisis four years after diagnosis of CLL and two years after diagnosis of CML in 1/1992.

II) A 60 year old man was diagnosed with Ph+/b3a2 BCR-ABL+ CML in 12/1997. An initial interferon-alpha based therapy was not tolerated. Therapy was continued with hydroxyurea. Due to lack of hematologic control STI571 has been started in 1/2000. Cytogenetics demonstrated the typical Ph translocation in 7 metaphases, 9 metaphases showed an additional Ph chromosome. STI571 failed to achieve a complete hematologic remission. Proportion and the absolute count of lymphocytes in PB increased to 64% and 26,000/µl, respectively. In 10/2000, clonal lymphocytosis of B-cell phenotype (CD5+, CD19+, CD20+) was detected. CLL (Rai stage 0) was diagnosed. Fluorescence in situ hybridization revealed a del13q14 but excluded the BCR-ABL fusion in B cells. High resolution PCR fragment analysis for identification of clonally rearranged immunoglobulin heavy chain genes demonstrated rearranged IgH in a PB sample from 11/2000 as well as in stored DNA from 1/2000 prior to STI571 therapy excluding the association of STI571 therapy and onset of CLL. STI571 therapy was continued; progressive BM fibrosis was accompanied by modest anemia. BM histology excluded the suppression of hematopoiesis by clonal lymphocytes.

We conclude that the coexistence of CML and CLL is a rare phenomenon. The two diseases are probably associated by chance. However, various lines of evidence suggest that acquisition of the BCR-ABL gene may not be the first or only molecular lesion causing CML. Therefore, unknown coexisting mechanisms could be responsible for the manifestation of both CML and CLL in the same individual.

807

Improved disease free-survival after transplantation of peripheral blood stem cells as compared to bone marrow from HLA-identical unrelated donors in patients with first chronic phase chronic myeloid leukemia

A. Elmaagacli, S. Basolglu, R. Peceny, R. Trenschel, H. Ottinger, A. Lollert, V. Runde, D. Beelen, U. Schaefer (Essen, D)

Peripheral blood stem cell (PBSC) transplants from HLA-A, -B, -DR compatible unrelated donors (n=34) were compared to bone marrow (BM) (n=52). The recovery of both neutrophils and platelets was faster with PBSCs than with BM (p<0.05). PBSCT had an improved immune reconstitution compared to BMT with elevated peripheral blood counts of naïve (CD4+CD45RA+) and memory (CD4+CD45RO+) helper T cells at 3 months and 12 months post transplant (p<0.03) The cumulative incidence of acute (grade II-IV) and chronic graft-versus-host disease (GVHD) were similar. But patients after BMT had a slightly higher incidence of cumulative severe acute GVHD (grade III+IV) with 27% compared to 13% (p<0.09). A molecular relapse, as defined by two consecutive positive BCR-ABL –PCR assays within a 4-week interval was detectable in 11 of 44 evaluable patients (25%) after BMT and 3 of 34 (9%) after PBSCT (p=0.06). Cytogenetic relapse developed in 5 of 52 patients after BMT (10%), but in none of the 34 patients after PBSCT (not significant). Median follow-up time was 15 months after PBSCT and 29 months after BMT. Sixteen of the 52 patients died after BMT (31%) compared to two of 34 patients after PBSCT. The predominant causes of death in the BMT-group were mainly associated with fatal infections and severe acute GVHD. The estimated probability of transplantation-related death (TRM) and disease free survival (DFS) at 1000 days after transplant were 31% and 66% in the BMT group and 7% and 93% in the PBSCT group (p<0.03). Overall survival 1000 days after transplant was 68% after BMT and 93% after PBSCT (p<0.03). In the multivariate analysis only acute GVHD had a significant influence on TRM (p<0.01).

808

Pegylated interferon alpha 2b (PegIntron) in the treatment of CML patients

C. Guenther, J. Lärmer, T. Graf, C. Schoch, A. Hochhaus, R. Hehlmann, B. Stade, C. Waterhouse, C. Nerl (München, Mannheim, D)

Patient population: Between 7/2000 and 5/2001 7 pts with CML in chronic phase have been treated with PegIntron (5 male, 2 female, median age 51, range 34 -65 yrs). 3 pts were in late chronic phase with prior treatment of conventional IFN (two with hematologic IFN-resistance after initial response, one with IFN intolerance), 2 pts had PegIntron therapy after diagnosis and 2 pts received peg. IFN after autologous blood stem cell transplantation. Pts were treated according to the German CML study protocol (CMLIIIA). Median treatment duration was 6.7 months (range 4 wks to 10 mths). Median dose was 180 µg (range 1-5 µg/kg/wk corresponding to 100-400 µg/ wk). Hydroxy-Urea was not given concomitantly. Results: Tolerance: Five pts tolerated the weekly injection well. Two female patients complained of psychic alterations (depression, aggressiveness), one of them stopped treatment after 5 months. Other acute side effects reported consisted of fever, myalgias and headache and lasted for 24 h after injection. Prolonged side effects were asthenia and dry skin. 6 pts used paracetamol. Local irritation was not reported. The compliance to the treatment is excellent. Hematological response (HR): 5/6 evaluable pts achieved a HR after 4 wks to 4 mths of treatment, 4/5 pts had a complete HR. 1 pt turned to a partial HR after 5 mths of treatment. 1 pt in late chronic phase did not respond hematologically. 1 pt after autologous BMT suffered from prolonged thrombopenia and leukopenia after administration of PegIntron and had to be withdrawn from treatment after 4 weeks. The PegIntron treatment was stopped for 8 mths. The second pt after auto. BMT had transient thrombocytopenia leading to a temporary stop of PegIntron. Cytogenetic response (CR): 6 pts were evaluable for cytogenetic response (4-10 mths of treatment). 4/6 pts achieved a minor cytogenetic response after 3 –6 mths of treatment. One of the responding pts was resistent to conventional IFN. 2/6 pts did not respond. Summary: Pegylated interferon alpha 2b in a dose of 1-3 µg/kg/ week is tolerated well. The side effects are comparable to conventional interferon and short-lasting (24 h after injection). Two female pts complained of psychic alteration. The hematological response was good (5/6 pts). 2 pts after aut. BMT exerted thrombopenia. A minor cytogenetic response was achieved in 4/6 pts after 4-6 mths of treatment. One of the responding pts was resistent to conventional IFN-treatment. Further data are necessary to evaluate the potency of pegylated IFN in the treatment of CML.

809

Effect of interferon-alpha on myelofibrosis during the course of chronic myeloid leukemia - a randomized comparison with chemotherapy

G. Büsche, A. Schmeil, H. Hecker, B. Heinze, M. Kappler, M. Pfirrmann, R. Hehlmann, H.H. Kreipe, A. Georgii and the German CML Study Group (Hannover, Ulm, München, Mannheim, D)

Interferon-alpha has become a standard of non-myeloablative therapy of Ph+ CML. Evaluations of morphologic bone marrow changes under this medication are exceptional and controversial, especially considering the risk of myelofibrosis, and a comparison with chemotherapy on a randomized basis does not exist so far. 854 bone marrow biopsies from 402 patients with Ph+ CML recruited in the German CML therapy trial were histostereologically examined for the marrow volumes with myelofibrosis before and under therapy. Developing therapy resistance significantly correlated with developing myelofibrosis, a complication affecting more than 70 % of patients during course of CML. Interferon-alpha and hydroxyurea medication prevented myelofibrosis from progression for 1.5 to 2 years. The impact of therapy on marrow fibrosis significantly influenced the survival times of patients independently from risk score and superior to all the other follow up parameters of the study, especially the hematologic and cytogenetic remission, detection of accelerated phase as well as the therapy applied. Conclusion: The development of marrow volume with fibrosis appears to be one of the most important prognostic parameters during interferon-alpha therapy of CML allowing for an early prediction of developing therapy resistance.

810

Allogeneic stem cell transplantation in a patient with chronic neutrophilic leukemia and multiple myeloma: a case report

I. Genvresse, V. Krenn, G. Reich, G. Massenkeil, C. Saati, R. Pasold, R. Arnold (Berlin, D)

Because chronic neutrophilic leukemia (CNL) is rare and occurs predominantly in elderly patients, only 2 cases of patients treated with allogeneic BMT have been reported. We report the case of a 28 year old male patient with the coexistence of CNL and IgA multiple myeloma who underwent allogeneic stem cell transplantation (SCT). The patient was admitted because of acute pain in the left humerus. The blood count showed hemoglobin 11.4 g/dl, leucocytes 39,900/µl with 95% neutrophils, and platelets 146,000/µl. The serum ß2-microglobulin and C-reactive protein were normal. The serum vitamin B 12 was >2,100 pg/ml and LAP score 321 (normal 20-100). Total serum protein was 89.9 g/l. An IgA l paraprotein was detected in the serum at a level of 19.7 g/l and Bence Jones protein was present. MRI revealed no osteolytic lesions. Sonography showed a moderate hepatosplenomegaly. The bone marrow was hypercellular with a plasma cell infiltration (40%) and granulocytic hyperplasia without dysplastic features. The karyotype was normal, BCR/ABL was negative by FISH and by molecular biology methods. An allogeneic SCT from HLA-identical brother was performed. Conditioning consisted of TBI and cyclophosphamide. Cyclosporin and methotrexate were given for GvHD prophylaxis. SCT was uncomplicated. No GvHD occurred. At day 28 after SCT, a bone marrow examination revealed persisting plasma cell infiltration and granulocytic expansion. Discontinuation of immunsuppression and one infusion of donor lymphocytes caused severe acute GvHD III° of the gut. At last follow-up (day 110), blood count was normal. Bone marrow analysis is pending. One can speculate that the 2 different committed progenitors contain the same gene defect and immunomodulation will be helpful if chemotherapy has a low effectivity in cells with low proliferation rate (i.e. granulocytes and plasma cells).

811

Criteria and classification of mastocytosis: Presentation and discussion of a novel WHO consensus proposal with special reference to clinical and hematopathological aspects

P. Valent, W.R. Sperr, H.-P. Horny (Wien, A; Lübeck, D)

The term mastocytosis denotes a heterogeneous group of disorders characterized by abnormal growth and accumulation of mast cells (MC) in one or more organ systems. A number of different classification systems for this disease have been presented in the past. Today, mastocytosis is considered as a myeloproliferative disorder. Recently, consensus criteria and a consensus classification for mast cell proliferative disorders have been worked out by participants of the Year 2000 Working Conference on Mastocytosis (Wien 2000): Peter Valent, Hans-P. Horny, Luis Escribano, B. Jack Longley, Chin Y. Li, Lawrence B. Schwartz, Gianni Marone, Rosa Nuñez, Cem Akin, Karl Sotlar, Wolfgang R. Sperr, Klaus Wolff, Richard D. Brunning, Reza M. Parwaresch, K. Frank Austen, Karl Lennert, Dean D. Metcalfe, James W. Vardiman, John M. Bennett. The final concept has been approved by the WHO. Seven major types of mastocytosis are recognized by the working group: cutaneous mastocytosis (CM), indolent systemic mastocytosis (ISM), SM with an associated clonal hematologic non-mast cell lineage disease (AHNMD), aggressive systemic mastocytosis (ASM), mast cell leukemia (MCL), mast cell sarcoma, and extra-cutaneous mastocytoma. Most patients with CM are children - they have a good prognosis and high rate of spontaneous regression. Patients with SM have persistent disease and are characterized by typical histological lesions in visceral organs (bone marrow in most cases) together with other (cytological or biochemical) signs of systemic disease (SM-criteria) with or without skin lesions. Whereas patients with ISM have an excellent prognosis, patients with SM-AHNMD, ASM, and MCL and those with mast cell sarcoma, have a grave prognosis. Depending on the clinical course of mastocytosis and development of an AHNMD, patients can shift from one category of MC disease into another. In all categories, mediator-related symptoms may occur and may represent a serious clinical problem. All categories of mastocytosis should distinctively be separated from reactive MC hyperplasia, MC activation syndromes, and a more or less pronounced increase in MC in myelogenous malignancies other than mastocytosis. Novel criteria to diagnose and classify the disease should be helpful in this regard.

812

Correction of hyperhomocysteinemia in polycythemia vera and essential thrombocythaemia by folate and cyanocobalamin substitution

G. Dimou, I. Pabinger-Fasching, A. Kaider, C. Bieglmayer, C. Cerni, H. Gisslinger (Wien, A)

In the cohort of 76 patients with polycythaemia vera (PV) or essential thrombocythaemia (ET) who have been repeatedly tested for fasting plasma homocysteine levels we observed hyperhomocysteinaemia (HHC) in 22 patients. HHC was defined by plasma levels above 13,6 mmol/l (female) or 16.3 mmol/l (male). Since HHC occurs more frequently in PV and ET compared to healthy controls and there is some evidence that homocysteine metabolism is altered in malignant transformed cells, we investigated if elevated plasma homocysteine levels can also be normalized by folate and cyanocobalamin substitution in PV and ET. The patients with HHC (16 PV, 5 ET; 12 female, 9 male; median age 70 years, range 32-88) were treated with a daily dose of 5mg folate and 500µg cyanocobalamin for three months. One patient refused to participate in the trial. Two patients had side effects due to the folate and cyanocobalamin substitution with an itching exanthema and had to be withdrawn from the study. Fasting homocysteine levels were determined at least twice before therapy and monthly during folate and cyanocobalamin substitution. In 16 of 20 patients plasma homocysteine levels reached normal values within the first month of treatment. After two months all patients under folate and cyanocobalamin substitution showed normalization of plasma homocysteine levels. Five of 7 patients who were withdrawn from the substitution for three months according to the study protocol showed reappearance of HHC. Our data demonstrate that folate and cyanocobalamin substitution can effectively correct HHC also in PV and ET patients. However, the increase of plasma homocysteine levels after withdrawal of substitution seems to occur earlier in these patients compared to healthy individuals as observed in previous studies.

813

Unrelated peripheral blood stem cell (PBSC) transplantation for myelofibrosis with myeloid metaplasia

M. Koldehoff, N. Basara, S. Günzelmann, G. Lentini, M.G. Kiehl, A.A. Fauser (Idar-Oberstein, D)

Myelofibrosis with myeloid metaplasia is a clonal stem-cell disorder that leads to ineffective erythropoiesis, dysplastic-megakaryocyte hyperplasia, and an increase in the ratio of immature granulocytes to total granulocytes. This clonal disease is accompanied by reactive myelofibrosis and by extramedullary hematopoiesis in the spleen or in multiple organs. The use of stem cell transplantation using related HLA-identical donor offers the potential to cure the patients with myelofibrosis. However, only few patients have been treated with unrelated donor PBSC. We describe our patients with myelofibrosis with myeloid metaplasia successfully treated with PBSC from an unrelated donor. A 52 years old man with a history of disease duration of 144 months, his bone marrow histology revealed severe myelofibrosis without signs of osteosclerosis. The patient received a PBSC with 4.2 x 10^6/kg b.w. CD34+ cells from HLA-identical unrelated donor. Neutophil engraftment occurred on day +22, platelets engraftment on day +42 after PBSCT. Myelofibrosis in bone marrow at day +100 and day +220 after PBSCT was decreased. The patient is doing well 2 years after the unrelated PBSCT. The second patient is a 46-years old woman with a history of disease duration of 33 months. Bone marrow histology has shown myelofibrosis with osteosclerosis. The patients received PBSC with 3.3 x 10^6/kg b.w. CD34+ cells from HLA-non identical unrelated donor. Neutrophil engraftment occurred at day +18 and platelets engraftment at day +29 after PBSCT. Myelofibrosis in bone marrow at day +100 was slightly decreased in comparison to the one prior to PBSCT. The patient is doing well also one year after PBSCT. In conclusion, allogeneic PBSCT from unrelated donor is a feasible and effective treatment in selected patients with myelofibrosis with myeloid metaplasia.

814

Development of a new quantitative PCR-based assay for the polycythemia rubra vera-1 (PRV-1) gene: Diagnostic and therapeutic implications

S. Fruehauf, J. Topaly, M. Villalobos, M.R. Veldwijk, B. Gentner, S. Laufs, A.D. Ho (Heidelberg, D)

Chronic myeloproliferative syndromes (MPS) are characterized by defects in hematopoietic stem cells resulting in variable degrees of expansion of myeloid elements, thrombocytes and erythrocytes. In one group of MPS, the recognition of BCR-ABL as molecular marker of the disease has greatly facilitated diagnostic evaluation and the development of therapeutic strategies. BCR-ABL negative MPS have remained elusive in this respect. Recently, a novel gene subsequently named PRV-1 was isolated and selective expression in peripheral blood (PB) granulocytes of polycythemia vera (PV) patients was shown by Northern blot. To validate this interesting new marker we developed a real-time polymerase chain reaction assay for PRV-1 cDNA using TaqMan technology. Expression of the PRV-1 gene relative to the GAPDH housekeeping gene (PRV-1/GAPDH quotient) was quantified by comparison with a standard curve obtained with a control plasmid containing fragments of both genes. Granulocytes and mononuclear cells of 65 patients and 5 normal control subjects were analyzed. As learning sample we defined a group of 12 PV patients who fulfilled the diagnostic criteria of the German PV study group, including growth of endogenous erythroid colonies (EEC). These patients had been pretreated with phlebotomy (PT) or hydroxyurea (HU) (Table). Interestingly, a second group of 10 PV patients who met the same diagnostic criteria, including EEC but who had received interferon-alpha (IFN) for 5 or more months had a significantly lower PRV-1 expression ($p<0.05$), comparable to normal controls. PRV-1 expression was in the range of normal controls in chronic myelogenous leukemia (n=2) and all but one essential thrombocythemia patients (n=8). As expected, PRV-1 expression was hardly detectable in normal PB lymphocytes (PRV-1/GAPDH <0.006). Of 33 patients with suspected PV not meeting the required diagnostic criteria of the PV study group 16 were PRV-1 positive (PRV-1/GAPDH quotient >0.09), rendering the diagnosis likely. All 7 patients who were later found to have secondary erythrocytosis had normal PRV-1 values. Apart from the obvious implications of PRV-1 analysis for the diagnosis of PV and differential diagnosis of MPS our data allow to hypothesize that PRV-1 expression is a surrogate marker of therapeutic response to IFN in PV.

815

Comparison between polycythemia vera study group (PVSG) and European criteria for the diagnosis of myeloproliferative syndromes (MPS)

C. Annaloro, F. Radaelli, P. Massaro, E. Pozzoli, G. Lambertenghi Deliliers (Milano, I)

The criteria proposed by the PVSG, which are widely accepted for the diagnosis of BCR-ABL- MPS, see the diagnosis of essential thrombocythemia (ET) as one of exclusion, since it relies on the absence of findings suggesting alternative diagnoses. Furthermore, they do not take into account histological findings. Conversely, MPS classifications relying on histology alone have been proposed by German groups, but have not gained such wide acceptance. The more recent European Criteria for the diagnosis of MPS take into account both histological and clinical data. In order to evaluate the feasibility of this new approach, we started reviewing bone marrow biopsies (BMB) taken at the time of diagnosis of polycythemia vera (PV) and ET. We began by considering our first 94 ET and 87 PV, all of whom had been diagnosed on the basis of the PVSG proposals. The diagnosis was confirmed in 78 of the 87 PV patients, but could not be confirmed in six cases because of the absence of full-blown panmyelosis; in the remaining four patients, the absence of clearcut erythroid hyperplasia coexisted with granulocytic, megakaryocytic hyperplasia and some degree of maturational defect, thus suggesting (according to the European criteria) an otherwise inconsistent diagnosis of idiopathic mielofibrosis (IMF), which was prefibrotic in one and early in three. The ET diagnosis was confirmed in 80/94 patients, but could not be confirmed in four cases because of small megakaryocyte proliferation and in three because of the absence of megakaryocyte hyperplasia; in the remaining seven patients, a diagnosis of IMF seemed more appropriate (prefibrotic in three and early in four cases). Although they are widely used, the lack of positive elements for the diagnosis of ET makes the PVSG criteria somewhat unsatisfactory. On the other hand, a purely histological classification of myeloproliferative disorders raises problems of reproducibility outside Reference Centres. The recent European Classification is an interesting attempt to combine clinical and histological findings and allows a substantial proportion of conventional diagnoses to be confirmed using positive criteria. At the same time, our findings support the view that a proportion of ET patients, with some degree of maturational defect could be better diagnosed as having early IMF, an opinion that can be less easily held in the cases of PV. Furthermore, a minority of patients do not fit in any diagnostic category because of the absence of a single positive criterion. It is not clear whether these patients should be considered as independent subgroups or whether additional elements should be sought in order to fit them into one of the positive MPS forms. The review of our entire MPS series may indicate whether they have some particular clinical characteristics.

816

Platelet activation in patients with essential thrombocythemia

H.-H. Wolf, M. Zejewski, A. Frühauf, H.-J. Schmoll (Halle, D)

In patients with myeloproliferative (MPS) disorders bleeding complications may occur despite elevated platelet counts. In order to characterize platelet activation in MPS patients we measured glycoprotein IIb/IIIa expression both in the resting and activated state by flow cytometry in patients with essential thrombocythemia (ET), chronic myelogenous leukemia (CML), osteomyelofibrosis (OMF) and polycythemia vera (PCV) as well as in healthy volunteers. METHODS: After standardized venous punctures we analysed blood samples of 17 patients with myeloproliferative diseases (ET n=5, other myeloproliferative disorders n=12) and 20 healthy volunteers. Platelet activation has been analysed by PAC 1 antibody (Becton Dickinson pure FITC) expression of glycoprotein (Gp) IIb/IIIa complex using FACS Calibur (Becton Dickinson). Tests have been done twice: immediately after blood sampling and after a 30 minutes RGDS incubation according to the manufacturers recommendation in order to achieve platelet stabilization.

RESULTS:

		#	mean platelet count Gpt/l	mean PAC 1 expression without RGDS % platelets
myeloproliferative disorders	Essential Thrombocythemia	5	709,75	34,91
	Other Myeloproliferative Disorders	12	253,42	54,9
healthy volunteers		20	260,50	54,31

After RGDS incubation „resting“ platelets did not present any significant Gp IIb/IIIa expression. Without previous RGDS stabilization, however, we found lower activation in ET patients compared to patients suffering from other myeloproliferative disorders or healthy volunteers. CONCLUSIONS: Our preliminary data indicate lower „spontaneous“ platelet activation in patients with ET compared to other myeloproliferative disorders despite elevated platelet counts. These findings correlate with spontaneous hemorrhagic complications in ET patients and should be proved by further examination.

817

Efficacy of cytoreduction by anagrelide and safety of this drug in patients with primary thrombocythemia and other CMPDs: an overview of nine prospective non-randomized trials

P. Petrides, M.K. Beykirch (München, D)

Background: In 1988 anagrelide was first introduced as a platelet reducing agent and in 1997 approved in the United States. Aim: To evaluate clinical studies published in peer reviewed journals between 1988 and 2001 for efficacy of platelet reduction, prevention of thromboembolic complications and safety in terms of short term and long term adverse effects. Methods: A metaanalyis was performed on 9 non-controlled, non-comparative clinical trials carried out in the US as well as Italy, Germany, Switzerland, Australia and Argentina. Results: The largest patient cohort was reported in the US (942), followed by Germany (60) and Italy (20). Alltogether clinical data were reported on 1069 patients (F/M = 631/438) with a median age of 52 years. Of these 773 had a complete hematological response (defined differently in various studies), 92 a partial response and 203 were treatment failures. Treatment failures may be due to a lack of the production of an active anagrelide metabolite or heterogeneity of the underlying disease. The maintenance daily dosage ranged from 2 to 2,5 mg. Minor side effects were headache, palpitations, nausea and diarrhea and probably due to adaptation to the positive inotropic/vasodilatatory effects of the drug. In the majority of the patients they resolved within four weeks. Serious adverse events included congestive heart failure, myocardial infarction, pulmonary fibrosis or gastrointestinal ulcerations. Although it was impossible to prove or disprove any relationship with anagrelide therapy, cardiac side effects, most importantly congestive heart failure, have to be considered as a risk factor in patients taking anagrelide. Hence the drug should either not be offered to patients with known cardiac risk factors. Long term treatment results were only reported by the US and German groups: Of 35 young patients in the US 82 % remained on anagrelide therapy (median 10.8 years). In these patients anemia was the only new side effect. In Germany patients have been treated for up to ten years (median 5 years): this was accompanied by a reduction in thromboembolic complications in most but not all patients which indicates that normal platelet values do not completely prevent complications. Conclusions: Anagrelide effectively reduces platelet counts and causes minor side effects who are reversible in the majority of patients and do not lead to therapy discontinuation. Several patients have been treated for up to ten years. There is no evidence for increased leukemogenic transformation which may become a major advantage over alternate treatment options. At present reduced platelet levels have been accepted as a surrogate marker for reduced thromboembolic complications. Future comparative studies, however, have to reveal which cytoreductive agent does not only reduce platelet levels but also thrombocythemia associated symptoms and prevents thromboembolic complications and with this morbidity and mortality in individuals with MPDs.

Postersitzung:
Diagnostische Verfahren

818

Telomerase activity in pericardial effusions: diagnostic significance

K. Wittstock, S. Meyer-Neumühlen, M. Kersting, S. Pankuweit, B. Maisch, A. Neubauer, M. Schuermann (Marburg, D)

The differential diagnosis between malignancy-related and non-malignant pericardial effusions has remained an essential problem in clinical practice. Our purpose was to determine the diagnostic value of telomerase activity in discriminating between these two categories compared with that of cytological analysis. Fluid samples obtained from 31 patients undergoing pericardiocentesis were examined by two different PCR-based telomeric repeat amplification protocols (TRAP), by conventional TRAP and by GeneScan-analysis of fluorescent products detected in ultrathin gels. The samples were blinded before analysis. Both methods were validated by an externally determined telomerase standard based on limited dilution of lung tumor cells and checked against a background of peripheral mononuclear blood cells. Positivity for telomerase activity was detected in 9 cases using the conventional TRAP assay and in 8 cases using the GeneScan-analysis which allowed the determination of a threshold limit. In contrast, cancer cells were detected by cytopathology in 7 cases (reviewed by two pathologists). 14 pericardiac effusions showed signs of inflammatory or chronic disease, 10 remained cytologically normal. Compared to cytopathology as the gold standard, the telomerase activity detected by the conventional TRAP assay yielded a 100% sensitivity for malignant cells providing a 91% specificity. Semi-quantitative detection by GeneScan-analysis can even improve the specificity to 96% at 100% sensitivity. Thus, the use of cell extract-based TRAP assay in addition to cytologic examination may make the diagnosis of malignancy-related pericardial effusions more reliable.

819

Procalcitonin - a useful discriminator between febrile conditions of different origin in hemato-oncological patients?

S. Schüttrumpf, L. Binder, L. Trümper, C. Binder (Göttingen, D)

Plasma concentrations of procalcitonin (PCT) have been shown to be elevated in bacterial and fungal infections, probably induced by circulating endotoxins and cytokines. In contrast to the C-reactive protein (CRP), PCT is not elevated in inflammations of non-infectious origin. Febrile inflammatory conditions are frequent in patients (pts) with hemato-oncological diseases. A reliable marker to discriminate inflammations caused by microbial infections from drug-induced and tumor-related fever, is still lacking. To evaluate the usefulness of PCT for this purpose, PCT and CRP were prospectively measured in 85 pts with leukemias, lymphomas and various solid tumors. All pts had fever > 38.5°C, 43/85 (50.6%) were neutropenic (< 1000/µl). In 36/85 pts (42.3%), an infectious origin could be identified: 26/36 suffered from clinically defined infections (pneumonia, abscess, etc.), 1/36 from varicella zoster infection, 9/36 had documented bacteraemia. 49/85 (57,6%) had fever of unknown origin (FUO). In 9 of the FUO-pts, temperature rose during infusion of high-dose cytarabin, suggesting drug-related fever. 5/49 were suspected to have „tumor-fever" because of long-lasting temperature elevation, high lymphoma load and lacking clinical signs of infection. As expected, there was no significant difference in median PCT values between pts with infectious fever or FUO (p=0.06), the latter group containing a mixture of infectious and non-infectious cases. However, in pts with suspected drug-related or tumor fever, PCT levels were significantly lower than in pts with infections (p<0.001) or FUO (p=0.009). Differences were still significant, comparing pts with suspected drug-related fever alone with the infection (p=0.002) or the FUO cohort (p=0.022). All 5 pts with putative „tumor-fever" had PCT levels within the normal range (< 0.5 µg/l). Despite the small number of these pts, differences to the infectious (p=0.01) or the FUO group (p=0.046) were statistically significant. Regardless of the fever cause, PCT values were generally higher in non-neutropenic than in neutropenic pts (p=0.027). CRP values did only partially allow to discriminate between the various subgroups. Differences were significant between pts with „drug-fever" and the infectious (p<0.001) or FUO group (p=0.003). However, as CRP levels were far above the normal range also in the „drug-fever" pts, the significance of individual values was rather limited. Taken together, PCT may provide useful additional information to assess the clinical significance of febrile conditions, especially in non-neutropenic pts. PCT may facilitate the decision on when to initiate antimicrobial or cytotoxic therapy. Further studies are needed to confirm these data and to show whether PCT might help to avoid unnecessary and potentially toxic antibiotic treatment.

820

Characterization of hematopoietic cells collected from sternum or vertebra

S.W. Reif, U. Neumann, P. Neuhaus (Berlin, D)

Purpose: Hematopoietic cells collected from the sternum in comparison to cells collected from the vertebra were characterized in order to find the appropriate cell source for in-vitro bone marrow studies and in-vivo cellular therapy. Material and Methods: Hematopoietic cells from sternum (n=63) and vertebra (n=4) were collected and analyzed for a) cell content, b) vitality, c) morphology and d) clonogenic ability of freshly isolated aspirates as well as during different cultivation procedures and time periods. Results: Fresh hematopoietic cells from vertebra were characterized by a 6.32-fold higher proportion of mononuclear cells, a higher viability, a 2.95-fold respectively 1.89-fold higher clonogenic ability (CFU-e resp. CFU-GM) but only 84.8% amount of CD34-positive stem cells in comparison to cells collected from the sternum (vertebra 0.89% versus 1.05% sternum). In vertebra less immuncompetent cells (CD3: 1.72% vs 18.86%, CD19: 0.46% vs 2.68%) as well as less monocytes (CD14: 1.90% vs 12.35%) in comparison to sternum-aspirates could be detected. In contrast to sternum derived cells, less stroma cells could be grown from vertebra aspirates.
After 14 days of cultivation the relative amount of CD34-positive stem cells as well as lymphocytes (CD3, CD19) increased, whereas the clonogenic

potential decreased in vertebra-aspirates in comparison to sternum-aspirates. When cells were cultured in a medium containing GM-CSF the amount of monocytes (CD14) as well as of granulocytes (CD15) increased during the first 7 days, irrespective of the cell source. Conclusion: Cells aspirated from the vertebra are characterized by a higher content of mononuclear cells, a higher clonogenic ability and vitality and a lower lymphocyte content, offering studies on specific hematopoietic cells. Sternum derived aspirates on the other hand are characterized by a higher amount of CD34-positive stem cells and a higher content of stroma cells, offering the possibility for co-cultures of stem cells with mesenchymal and stroma cells.

821

Immunomagnetic enrichment and detection of isolated tumor cells in bone marrow of patients with epithelial malignancies

M. R. Weihrauch, E. Skibowski, A. Draube, A. Geller, H. Tesch, V. Diehl, H. Bohlen (Köln, Villingen-Schwenningen, D)

[Introduction] The detection of isolated tumor cells (ITC) in the bone marrow of patients with epithelial cancers has been shown to be an independant significant prognostic factor for several entities as breast cancer, colorectal cancer or non-small lung cancer. However, with conventional immunocytology using FICOLL density gradient and APAAP staining only a small proportion of the bone marrow samples can be scanned for cytokeratin expressing (CK+) cells. To improve detection rates, we evaluated the enrichment of ITC by magnetic activated cell sorting (MACS) compared to regularly stained cytospins. [Patients and Methods] Recovery experiments with a CK+ breast cancer cell line (SKBR3) were performed to calculate the MACS enrichment rate. Bone marrow was obtained by aspiration from 20 patients with carcinomas of epithelial origin (10 breast cancers, 5 non-small lung cancers, 3 colon cancers, 1 bladder cancer, 1 carcinoma of unknown origin) and of different stages. Bone marrow from 7 controls with non-epithelial malignancies served as negative controls. ITC were enriched and stained with magnetically labeled CAM 5.2 antibodies directed to cytokeratin 7 and 8.
[Results] MACS of SKBR3 seeded in peripheral blood revealed average recovery rates of 62% and 48% and average enrichment factors of 104-fold and 8139-fold of the CK+ cells after one and after two separations, respectively. After immunomagnetic enrichment, CK+ cells were detected in 16 of 20 (80%) cancer patients, whereas only 7 (35%) patients showed CK+ cells without magnetic enrichment. Ten of 12 (83%) patients with metastatic disease (stage M1) and 6 of 8 (75%) patients without any overt metastases (M0) had CK+ cells in their bone marrow. None of the negative controls showed any CK+ cells. [Conclusion] Enrichment with magnetically labeled anti cytokeratin antibodies increases the detection rate of epithelial cells in bone marrow of cancer patients compared to immunocytology.

822

A fast, simple and cost effective method to circumvent the limitations of bone marrow dry tabs

J.R. Novotny, C. Rosenthal, H. Nückel, S. Mahlmann, U. Schmücker, B. Staats, U. Dührsen (Essen, Dortmund, D)

To allow the application of cytomorphology, flow cytometry and even cytogenetical analysis in cases of dry tab marrow aspiration we developed a fast, simple and cost effective method using a simple chopping device similar to a small scale kitchen food processor. The diagnosis, subclassification and risk assessment of haematological neoplasms largely depends on morphological, immunological and cytogenetical analysis of aspirated bone marrow. However, occassionly it is impossible to aspirate sufficient cells from bone marrow even with optimal technique and equipment. This so called dry tab can be due to bone marrow fibrosis or marrow densely packed with leukemia, lymphoma or solid tumor cells. In these cases the diagnosis of the underlying disease solely depends on bone marrow histology or touch preparations of native core biopsy cylinder with certain disadvantages such as longer time to diagnosis in the former und often unsufficient material in the latter. Cytological studies can not be done. In our study native core biopsy cylinders were obtained from patients with suspected leukemia and at least one unsuccessful attempt to aspirate marrow. The marrow cylinders were cut into shorter segments and put into the Medimachine (DAKO) for 30 seconds at fixed speed. The obtained cells were diluted, counted and tested for viability with dye exclusion. We could obtain bone marrow cell suspensions suitable for cytology, immunophenotyping and even cytogenetical analysis in the majority of 16 cases. In 11/16 cases cytomorphology of cytospins, in 8/16 immunophenotyping by FACS and in 2/5 cases mitoses for cytogenetical analysis were obtained. Especially in newly diagnosed cases of ALL with packed marrow this technique gave good cytomorphological and immunophenytipical results (5/5) leading to more accurate and faster diagnosis. In summary, we could obtained valuable additional information in more than half of the investigated cases, allowing clinical decisions before the result of bone marrow histology was available.

823

Real-Time single molecule imaging of adeno-associated virus uptake in living cells

H. Büning, M. U. Ried, G. Seisenberger, T. Endreß, C. Bräuchle, M. Hallek (München, D)

Recombinant adeno-associated virus (rAAV) is a promising vector for gene therapy. However, its broad host range might prevent the delivery of transgenes into a specific organ or cell lineage. In an attempt to control the tropism of rAAV, we have generated several capsid mutants and achieved some receptor specific retargeting of rAAV (Girod et al., Nat Med 1995(9):1052). However, some of the capsid mutants showed changes in their efficiency to infect specific target cells. This observation led us to develop methods to study the receptor binding, uptake, endosomal processing and nuclear transport of AAV capsid mutants in more detail. Based on Single Molecule Detection technologies, in particular Single Dye Tracing (SDT), we established a method for the detection of single AAV particles during infection. Single Cy5-labeled wild-type AAV particles were visualized and their cellular processing characterized by a real time observation of virus trajectories in living cells. The technology was used to study the viral diffusion in aqueous solution, as well cellular docking, surface movement, receptor mediated uptake, kinetics and characterization of endosomal processing, nuclear uptake and nuclear transport of the virions. More than 1000 wild type AAV trajectories were analyzed. Only in 5 %, entry of AAV into the cell was seen. In 32 %, AAV made a membrane contact and disappeared in solution thereafter. 60 % of AAV particles did not show any interaction with the cell. After binding to the cell membrane, AAV2 entered the cell within 80 ms. On average, AAV needed 1.2 s from the first contact to the cell membrane to the final entry. After 2 minutes, AAV was observed in the cytoplasm. 15 min later, AAV was detected in the nucleus of half of the infected cells. 102 trajectories of viral particles in the nucleus were analyzed. Surprisingly, when arrived in the nucleus, AAV underwent a directed motion along apparently pre-defined ways, suggesting that AAV moved along tubular structures within the cell nucleus. The nature of these nuclear structures remains to be determined. Taken together, fluorescent SDT allows the exact and real-time determination of virus movement from the first membrane contact to the nucleus. This new technology might shed some light on the poorly understood mechanisms and pathways of AAV infection and rAAV transduction.

824

Fiberoptic bronchoscopy in immunocompromised hematological patients with pulmonary infiltrates

A. Mulabecirovic, H. W. Auner, H. Popper, R. Krause, M. Trinker, W. Linkesch, H. Sill (Graz, A)

We retrospectively analyzed the results of fiberoptic bronchoscopies (FOB) in immunocompromised hematological patients with pulmonary infiltrates. Seventy FOB consisting of bronchoalveolar lavage (BAL) and protected specimen brushing (PSB) with or without transbronchial lung biopsy (TBLB) were performed in 67 patients suffering from various hematological malignancies. The male:female ratio was 35:31 and the median age 60 years (range 26-79). FOB was performed in according to standardized guidelines in cases with nonresolving diffuse (n=61) or solid (n=9) pulmonary infiltrates as shown on conventional chest radiographs and/or CT scans. Thirty six patients had received standard chemotherapy for their malignancy, 11 autologous or allogeneic stem cell transplantation; 23 patients had not received any antineoplastic therapy before FOB. Previous antimicrobial treatment had been applied to 48 cases; in 41 of them as empirical intravenous strategy. Microbiological investigation disclosed an infectious agent in 18 cases (26%) which was regarded causal and lead to appropriate antimicrobial therapy in 11 patients. The following microorganisms were detected: bacterial, n=6; fungal, n=3; viral, n=6; protozoal, n=1; chlamydia, n=2 and mycoplasma, n=2. Chemotoxic pneumonitis could be identified in 11 (16%) cases requesting therapeutic intervention in two. Furthermore, a malignant pulmonary infiltration was diagnosed in nine individuals (13%) necessitating antineoplastic therapy in four of them. In one case FOB revealed hemorrhage as the cause of a lung infitrate. Including miscellaneous reasons, FOB was able to clarify pulmonary infiltrates in 35 (50%) investigations which was followed by therapeutic intervention in 18 (26%). No severe complications of FOB like pneumothorax or bleeding were observed in this cohort. With respect to the total sensitivity of FOB with and without TBLB a difference (55% vs.43%) could be demonstrated in this patients cohort which was statistically significant for toxic pneumonitis (p=0.0186) and neoplastic infiltration (p= 0.008). In conclusion, FOB provides a valuable tool in the evaluation of pulmonary infiltrates in immunocompromised patients with hematological malignancies. Outweighing the individual risk, TBLB should be taken into account as it cleary increases the sensitivity of this procedure.

825

A standardised method for detection of circulating tumor cells: the MAINTRAC analysis for direct and timely monitoring of minimal residual disease in solid tumors

K. Pachmann, U. Pachmann, J. Clement, K. Höffken (Jena, Bayreuth, D)

Adjuvant chemotherapy in patients with solid tumors has shown to offer survival advantage in a fraction of patients but it is still impossible to control the efficacy of therapy in the individual patient. Treatment results are available only after several years. We here present a method to directly and timely monitor the response to therapeutic regimen using magnetic bead enrichment and Laser Scanning Cytometry to determine the number of circulating tumor cells from peripheral blood before and after chemotherapy. Epithelial antibody positive cells were quantified per volume and per 10^6 white blood cells in a control population of 10 breast cancer patients, in whom chemotherapy was delayed or who denied treatment and 44 treated patients in response to chemotherapy. Cell numbers increased in all untreated patients, whereas under chemotherapy there was an up to 100 fold decrease in 23 patients, almost no change in 6 patients and an increase in circulating tumor cells in 9 patients during 4 cycles of chemotherapy. In 6 patients an initial decrease was followed by a reincrease. This method allows monitoring of the response of tumor cells to therapy in vivo and will enable individual tailoring of chemotherapy in breast cancer patients before development of metastases.

826

Frequency and etiology of bi- and pancytopenia in a section of hematology and oncology - validity of the bone marrow puncture: A retrospective study on 342 consecutive patients

C. Boewer, T. Politz (Berlin, D)

Introduction: Frequency and etiology of bi- and pancytopenias in a section of hematology/oncology are investigated with special concern to the validity of bone marrow cytology in the diagnostic concept.
Methods: Blood pictures and the corresponding cytological bone marrow preparations from 2737 consecutive patients over a period of 4 years were available. Pancytopenia was defined as anemia (Hb <7.4 mmol/l for females and <8.6 mmol/l for males) plus leukopenia (WBC < 5.0 G/l) plus thrombocytopenia (PLT < 150 G/l). Bicytopenia was defined as any combination of two of these disturbances. Results: The frequency of bi- or pancytopenias amounted to 13.4 % yearly on an average. Totally, 342 patients with bi- or pancytopenia were registered. Etiologically, 10 different groups could be distinguished: malignant hematologic diseases with involvement of the bone marrow (27.5 %, n=98/342), hypersplenism (14.9 %, n=51/342), myelodysplastic syndrome (14.6 %, n=50/342), iron deficiency anemia (11.4 %, n=39/342), megaloblastic anemias (10.5 %, n=36/342), toxic bone marrow disturbances (5.6 %, n=19/342), carcinosis of the bone marrow (2.6 %, n=9/342), chronic myeloproliferative disorders (2.3 %, n=8/342), aplastic anemia (0.6 %, n=2/342) and miscellaneous diseases (8.8 %, n=30/342). In 87.1 % (298/342) the bone marrow cytology was conclusive as the first diagnostic measure. In 44 out of 342 patients (12.9 %) a further diagnostic procedure was necessary i. e. second bone marrow puncture (5.6 %, n=19), bone marrow trepanation (3.8 %, n=13) and autopsy (1.2 %, n=4). In the remaining 8 cases (2.3 %) no conclusive clarification could be reached because of discharge or exitus. Conclusions: Bi- and pancytopenias are caused by similar and often serious diseases. In this connection the cytological investigation of the bone marrow is the first and most important diagnostic aid.

Postersitzung:
Hochmaligne Non-Hodgkin-Lymphome

827

AIDS-associated primary central nervous system lymphoma: New perspectives

C. Hoffmann, E. Wolf, A. Stoehr, A. Plettenberg, S. Tabrizian, C. Eggers, H.-J. Stellbrink, T. Buhk, H. Jaeger, H.-A. Horst, T. Rosenkranz (München, Hamburg, Kiel, D)

Background: Primary Central Nervous System Lymphoma (PCNSL) represents a late complication of HIV infection. AIDS patients with PCNSL usually have a dismal prognosis with a median survival of less than three months. Cranial radiation can prolong median survival for up to six months. In this setting, only few data exist concerning the role of the immune recovery achieved by highly active antiretroviral therapy (HAART). Objective: To evaluate the impact of immune recovery induced by HAART on the survival of AIDS patients with PCNSL. Methods: In a multicentric retrospective analysis, 29 HIV-infected patients were identified in whom PCNSL had been histologically confirmed between 1990-2000. Last follow up was May 2001. To evaluate median survival, Kaplan-Meier statistics were used. To explore the effects of different variables on survival, a Weibull accelerated failure time regression analysis was performed. Results: Median age at manifestation of PCNSL was 39.1 years and median CD4 cell count was 11 cells/µl. Seventy percent of the patients had had a prior AIDS-defining illness. The most common subtypes were diffuse, large B-cell-lymphoma (15 cases) and high-grade B-cell lymphoma including B-lymphoblastic and Burkitt's lymphoma (9 cases). Cranial radiation (CR) was given to 12/29 patients. Six patients were treated with HAART, five of them with additional CR. Survival time of these patients and of the patients treated with CR alone differed significantly from those receiving neither CR nor HAART (median Kaplan-Meier survival estimate: 1093, 132, and 33 days, respectively). In the multivariate regression model, HAART and CR were identified as the only variables independently associated with prolonged survival. HAART versus no HAART and CR versus no CR increased the time to event by a factor of 6.5 (95 % CI 2.5-17.0, p=0.0001) and 3.1 (1.5-6.4, p=0.002), respectively. Four of six patients on HAART showed a marked immune recovery. All four patients survived for more than two years, with two patients still being alive. Two patients died of a progressive neurological syndrome which was at least in one case due to radiation induced encephalopathy. Conclusion: Data from this cohort indicate that immune recovery induced by HAART leads to dramatic improvement in survival of patients with AIDS-associated PCNSL. These findings may have important implications for future treatment strategies. Accurate prediction of the effectiveness of new therapeutic approaches requires detailed information on the patient's immune recovery and antiretroviral therapy.

828

Combined systemic and intraventricular chemotherapy in primary CNS lymphoma

A. Glasmacher, H. Pels, R. Kleinschmidt, C. Helmstaedter, M. Deckert, D. van Roost, A. Zellner, F. Kroschinsky, G. Schackert, K. Neben, A. Ho, A. Engert, V. Diehl, U. Bode, T. Klockgether, U. Schlegel, I. Schmidt-Wolf (Bonn, Dresden, Heidelberg, Köln, D)

OBJECTIVES: To evaluate the response rate, response duration and toxicity after systemic and intraventricular chemotherapy in primary central nervous system lymphoma (PCNSL). PATIENTS AND METHODS: Since 1995, 46 consecutive patients (pts) with PCNSL (mean age 61 years, range 27-72 years) were enrolled in a study evaluating chemotherapy without radiotherapy. High dose methotrexate (5 g/m^2, courses 1, 2, 4, 5) and cytarabine (3 $g/m^2/d$, d1-2, courses 3, 6) based systemic therapy (including dexamethasone (courses 3-6 only), vinca-alcaloids (VCR/VDS), ifosfamide and cyclophosphamide) was combined with intraventricular MTX, prednisolone and cytarabine (via an Ommaya/Rickham reservoir). RESULTS: In 44 evaluable pts, complete response (CR) was achieved in 27 (61%) and partial response (PR) in 5 (11%) pts. Seven pts (16%) had progressive disease (PD) and five (11%) died from treatment-related complications. Two pts treatment discontinued
because of their refusal. In 20 pts. < 60 yrs. a CR rate of 80% (PR 5%) was obtained. Observation time is zero to 67 months (median 18). Kaplan Meier estimate for median time to treatment failure (TTF) in all patients is 32 months (95%CI 11-53 months), for median survival 55 months (95%CI 21-87 months). In patients under 60 years during a median follow-up of 16

months only 6/20 had a treatment failure and only 5/20 died. Systemic toxicity was mainly hematologic. Ommaya reservoir infections occurred in seven pts and acute (subacute) transient MTX induced encephalopathy in two (one) pts. Permanent cognitive dysfunction due to treatment was seen in only one patient after 12 courses (six at relapse). CONCLUSIONS: Chemotherapy without radiotherapy based on high-dose methotrexate and cytarabine is highly efficient in primary CNS lymphoma. In patients younger than 60 years the results reported here are superior to previously reported therapy regimes. Treatment-related hematological toxicity is considerable, particularly in elderly patients. Permanent treatment-associated neurotoxicity, however, is infrequent.

829

Successful treatment of aggressive, recurrent subcutaneous panniculitis-like T-cell non-Hodgkin's lymphoma (SPTLC) with high-dose chemotherapy followed by allogeneic peripheral blood stem cell transplantation

E. Leo, H. Bertz, G. Köhler, J. Finke (Heidelberg, Freiburg, D)

Aggressive subcutaneous panniculitis-like T-cell non-Hodgkin's-lymphoma (SPTLC) is a rare, but distinct subtype of peripheral postthymic non-Hodgkin's-lymphoma (NHL). It is often misdiagnosed as panniculitis, which may precede, accompany or mimic this peculiar form of NHL. The main clinical features include lymphocytic infiltration of the subcutaneous tissue, hemorrhage, necrosis and erythrophagocytosis. The nature of the disease is usually very aggressive. Treatment options are limited and 5-year survival is below 30%. Especially patients with presence of hemophagocytosis have a very poor prognosis. Reported treatment strategies consist in the use of immunosuppressants and conventional chemotherapy. Here we report a case of subcutaneous lymphoma with an aggressive clinical behaviour reaching complete remission after high-dose chemotherapy with consecutive allogeneic peripheral stem cell transplantation. The patient remains in complete remission 70 months after transplantation. High-Dose chemotherapy in combination with Allogeneic Stemcell Transplantation may provide a new and successful treatment approach for this disease entity by combining the cytoreductive impact of high-dose chemotherapy with the Graft-versus-Lymphoma effect of allogeneic transplantation.

830

BMPD chemotherapy with/without radiotherapy for primary CNS lymphoma: First randomized multicenter study

A. Korfel, M.R. Nowrousian, D.K. Hossfeld, W. Weber, J. Brücher, W. Berdel, J. Birkmann, C. Peschel, R. Pasold, B. Gleißner, L. Fischer, E. Thiel (Berlin, Essen, Hamburg, Trier, Aschaffenburg, Münster, Nürnberg, München, Potsdam, D)

The prognosis of nonimmunocompromised patients with primary CNS lymphoma (PCNSL) can significantly be improved by treatment with high-dose methotrexate (HD MTX)-based chemotherapy as compared with whole-brain irradiation (WBI) alone. Late neurotoxicity is the major treatment-related problem especially after both chemotherapy and WBI. Eliminating the use of WBI may significantly reduce the risk of long-term neurotoxicity, but may result in an increased relapse rate. This is the first randomized multicenter trial worldwide to address the role of adjuvant WBI after chemotherapy. The primary chemotherapy regimen (BMPD) incorporated BCNU 80g/m^2 i.v.d1, procarbacine 100 mg/m^2 orally d1-8, a 24-hour infusion of 1.5 g/m^2 MTX d1 with leucovorin rescue, and dexamethasone 3x8 mg orally d1-14 and intrathecal MTX d1 in course 1 only (repeated in patients with meningeal involvement). A salvage chemotherapy with idarubicine 12 mg/m^2 d1-2 and ifosfamide 3g/m^2 d1 or cytarabine 2x3g/m^2 d1-2 was given to patients who failed to achieve CR after 3 BMPD courses. Patients in CR were randomly asigned to WBI with 50 Gy or to one additional chemotherapy course. Assessment of meningeal involvement was performed using immunocytochemistry on poly-l-lysin coated slides and PCR analysis of Ig-gene rearrangement in addition to morphologic and biochemic evaluation. Interim treatment results are presented. As the randomisation rate was low, the study was closed after inclusion of 63 patients. The treatment concept of randomized adjuvant WBI after chemotherapy has been transferred recently in a nationwide interdisciplinary study applying HD MTX alone.

831

Detection of meningeal involvement of primary central nervous system lymphoma (PCNSL) by immunocytochemistry using poly-L-lysine coated slides

L. Fischer, A. Korfel, S. Schwartz, B. Gleißner, E. Thiel (Berlin, D)

The incidence of meningeal involvement in PCNSL is controversial due to limitations of conventional investigation methods. Morphologic distinction between reactive and neoplastic lymphatic cells may be difficult whereas immunocytology in cell suspensions is often hampered by the tiny amount of cells evaluable. By using optimized cell attachement on poly-L-lysine-coated multispot slides followed by immunocytochemistry cell loss can be minimized and the sensitivity for detection of meningeal involvement can be enhanced. We investigated 21 cerebrospinal fluid (CSF) samples from patients with newly diagnosed PCNSL. In all cases the diagnosis had been confirmed by stereotactic tumor biopsy. All lymphomas were of B cell origin. Nineteen patients had been pretreated with corticosteroids at the time of CSF sampling and one patient was also pretreated with methotrexate intrathecally. CSF sample volumes ranged between 0.7 and 1.5ml. The number of cells measured in the samples ranged from 1.3 to 87/µl (median 4/µl) . In all but one cases no meningeal involvement was diagnosed morphologically. Monoclonal antibodies against CD3, lambda and kappa light chains, CD19, CD20 and CD22 were applied. Antibody reactivity was visualized by the alkaline phosphatase anti alkaline phosphatase (APAAP) method. 50 to 500 cells could be evaluated for each surface marker. Of the 21 samples investigated 2 samples showed marked relative increase of B lymphocytes. In both samples also a predominant expression of one immunoglobulin light chain could be detected (kappa/lamda ratio 52:6% and 6:29% respectively). The number of CSF cells in both samples was within the normal range (3/µl and 4.3/µl resp.). Morphologic analysis revealed malignant cells in one of both cases. In 2 further cases with elevated CSF cell count (32 and 31/µl) and morphologically normally appearing cells a normal B-cell propertion was found but a kappa/lamda ratio of 3:7% and 0:5% was suspicious. The technique described above allows the detection of meningeal involvement in PCNSL otherwise missed by routine cytology especially in minute CSF cells specimens. The low rate of positive results in our series as compared to the literature is most probably due to corticosteroid pretreatment. The value of this method needs to be further investigated, especially in comparison to other sensitive methods, i.e. detection of immunoglobuline gene rearrangement by polymerase chain reaction.

832

Molecular genetic analysis of cerebrospinal fluid (CSF) in primary CNS lymphoma (PCNSL) patients and its clinical value in the context of actual sampling procedures

B. Gleissner, J. Siehl, A. Korfel, E. Thiel (Berlin, D)

Background: Cytological evaluation of CSF reveals malignant cells at a varying rate in patients with primary CNS lymphoma (PCNSL) and is hampered by the lack of a clear distinction between neoplastic and reactively transformed lymphoid cells. Although sensitive, the molecular assessment of monoclonality has rarely been reported in CSF samples. Patients and Methods: We studied nested PCR of the complementary determining region III (CDRIII) on 76 CSF specimens of PCNSL patients. Seventeen patients with systemic disseminated B-cell non-Hodgkin's lymphoma and 17 patients with no history of lymphoma were analysed in comparison. PCR products were evaluated by automated fluorescent fragment analysis (ALF). Patients were treated within one multicenter PCNSL trial. Results: Despite 60 patients were under treatment with corticosteroids, we analysed in 68 PCNSL patients the first obtained CSF sample. CDRIII PCR consistently revealed monoclonal PCR products in eight PCNSL samples and polyclonal results in 52 PCNSL specimens. In 16 PCNSL samples, amplifiable DNA was not yielded. CSF cytology is available in 74 PCNSL patients and correlate with the CDRIII PCR data in 64 patients. In ten PCNSL patients, the CDRIII PCR result and the CSF cytology were discordant.

Table 1: Correlation of CDR III PCR result and cytology in PCNSL patients

CDRIII PCR result	Monoclonal	Polyclonal	No DNA
Cytology reactive	6	47	15
Suspicious cells	2	4	0
Cytology not done	0	1	1

Concerning clinical outcome in the context of actual clinical sampling procedures, leptomeningeal tumor spread did not predict tumor response in the patients analysed. Conclusions: This study confirms CDRIII PCR as a routine diagnostic technique applicable even on CSF samples with low cell counts. In the context of actual clinical practise, our data present a low incidence of leptomeningeal involvement in PCNSL patients. Since the CSF evaluation did not predictive outcome in our patients, further analysis in PCNSL patients should focus on CSF samples that are obtained very early after diagnosis.

833

High-dose chemotherapy followed by autologous stem cell transplantation in aggressive NHL first-line therapy is superior to conventional chemotherapy only in selected patients: An analysis of the published data

J.W. Strehl, U. Mey, T. Sauerbruch, I.G.H. Schmidt-Wolf, A. Glasmacher (Bonn, D)

BACKGROUND: In intermediate to high-grade non-Hodgkin-lymphoma (NHL), conventional induction chemotherapy regimens (e.g. CHOP) achieve an unsatisfactory overall survival (OS). Thus, intensive efforts have been made to evaluate therapeutical improvements with high-dose chemotherapy (HDT) followed by autologous stem cell transplantation. In the published data, large differences in overall survival of patients after HDT can be found. We analyzed the published data to identify subcategories of patients benefiting from HDT. METHODS: Published, randomized, controlled studies with HDT in first-line therapy were identified by Medline search and reference lists. The histological subtypes included were the intermediate- to high-grade lymphomas (Kiel-classification) or aggressive lymphoma (WHO-classification) excluding follicular lymphoma. The statistical analysis used the odds ratio (OR, confidence intervals 95%) and a fixed effects model. RESULTS: In eight studies with 886 patients, a statistically significant benefit regarding OS and disease free survival (DFS) of HDT compared with conventional therapy was found (OS: OR=0.61 [95% CI: 0.46-0.81], p=0.0007; DFS: OR=0.46 [0.35-0.61], p<00001). But, there was also a significant heterogeneity between the compared studies (OS: p=0.0002, DFS: p<0.00001). Therefore, we formed categories of patients according to the international NHL prognostic index (IPI), induction therapy, response to induction therapy, and control therapy. Compared with conventional chemotherapy, patients with high-intermediate and high risk IPI benefit from HDT (OR=0.40 [0.26-0.63], p=0.00007) in three studies where this analysis was possible, while heterogeneity is not significant in this group (p=0.057). Data is also screened for differences in the following categories: induction therapy; response to induction therapy, without additional information. CONCLUSION: This preliminary meta-analysis demonstrates a significantly better disease-free and overall survival in patients with intermediate to high-grade lymphomas and high-intermediate / high risk IPI.

834

Phase I/II study of pegylated liposomal doxorubicin in combination with etoposid and dexamethasone (ELID) for the treatment of patients ≤60 years with relapsed aggressive NHL

N. Gaartz, C. Pott, B. Glass, M. Brüggemann, T. Raff, M. Kneba (Kiel, Göttingen, D)

Antineoplastic activity of antracyclines even in relapsed aggressive NHL is proven, however treatment options for the elderly are limited because of cumulative cardial toxicity. Pegylated liposomal doxorubicine (Caelyx) has an increased half-time compared to the free drug (40-55 h vs. 16min) and a low cardial toxicity. It has proven to be highly effective in combination therapy of solid tumors and multiple myeloma. We initiated a phaseI/II study in patients ≤60 years with relapsed aggressive NHL to prove applicability of CaelyxÒ in a combination protocol with etoposid and dexamethasone and to evaluate remission rate, tumor control, toxicity and disease free survival after treatment. 10 pts. with stage IV disease (7 diffuse large cell, 1 transformed follicular, 1ALC and 1 MALT NHL) were enrolled in the study with a median age of 71 years. All pts. were extensively pretreated with a median of 3,5 therapies (Range 1-7) The study treatment consisted of 6 cycles of CaelyxÒ 20 mg/sqm i.v. day 1, Etoposid 100 mg/sqm i.v.day 1-3 and dexamethasone 8 mg orally twice daily day 1-5. Pts. received a median of 4,2 cycles (2-6). Myelosuppression occured with grade 3/4 neutropenia in 4 of 10 patients, in 3 of them only once after the first cycle. Grd. 3 mucositis occured in 1 and palmar-plantar erythrodysthesia (PPE) in 2 patients, respectively. No other grade 3/4 toxicity has been observed. CR has been documented in 1 patient with unmaintained remission now 10 months after start of treatment, 1 patient achieved a PR and further 5 a minor response. 2 pts. progressed under treatment with ELID , one did not show a response. The median time to treatment failure of 9 patients was 2,5 months, one patient was lost to follow-up. Conclusion: ELID is a feasable treatment option for elderly pts. with aggressive, relapsed NHL and is able to induce even CR in this heavily pretreated patients.

835

Enhanced detection of malignant lymphoma in cerebrospinal fluid by multiparamter flow cytometry

H. Nückel, J. Dürig, U. Schmücker, J.R. Novotny, R. Noppeney (Essen, D)

Background: Cerebrospinal fluid (CSF) is routinely analysed in patients with certain Non-Hodgkin-Lymphomas, and the demonstration of central nervous system involvement is often an important determinant of clinical management. Currently, the morphologic evaluation of CSF samples by routine light microscopy is the method of choice for analysis in patients with lymphoma. However, morphologic examination can be hindered by the presence of normal or reactive lymphocytes or by paucity of cells in the sample. While immunphenotyping by flow cytometry has been applied to fine-needle aspiration biopsy specimens, it has not been widely applied to clinical CSF samples. The potential advantages of flow cytometry include the rapidity of analysis compared with immunhistochemistry, the ability to enrich for possible malignant cells via gating techniques in samples with mixed cell populations, and the simultaneous evaluation of multiple surface markers on each cell. In the past, limited sample volume, low cell counts and the presence of cellular debris or reactive lymphocyte populations could lower the sensitivity of flow cytometry in recovering clonal lymphocyte disorders. Materials and methods: A retrospective comparison of 10 consecutive CSF flow cytometry results with the corresponding cytologic diagnosis was undertaken. One sample could not be analysed by flow cytometry because of insufficient number of cells. Flow cytometric immunophenotyping was performed by 3-color immunufluorescent staining. Detection of lymphoid clones was optimized by careful selection of antibody combinations for use in limited cell samples. Results: A total of 8 lymphomas was detected. 3 by both flow cytometry and cytology; 3 by cytology alone and 2 by flow cytometry alone. This represents a 33,3% increase in the detection of lymphoproliferative disorders in CSF by a combination of flow cytometry and cytology versus cytology alone. Conclusion: Multicolor flow cytometry is a rapid and sensitive technique that enhances detection of lymphoma in paucicellular CSF samples. We conclude that flow cytometry analysis markedly improves sensitivity when used in combination with cytology in the evaluation of lymphoid cells in CSF.

836

Cologne high-dose sequential chemotherapy in relapsed and refractory Hodgkin's and aggressive non-Hodgkin's lymphoma - Results of a multicenter phase-II study

A. Josting, C. Rudolph, M. Mapara, Y. Ko, D. Söhngen, B. Dörken, V. Diehl, A. Engert (Köln, Cottbus, Berlin, Bonn, D)

Combination chemotherapy can cure patients (pts) with Hodgkin`s disease (HD) or Non-Hodgkin`s lymphoma, but those with treatment failure or relapse still have a poor prognosis. High-dose chemotherapy (HDCT) with autologous stem cell support (ASCT) can improve the outcome of these pts as shown in the HR-R1 study of the GHSG/EBMT and the PARMA trial. Chemosensitivity and achievement of a minimal disease status prior to HDCT are important prognostic factors in HD and NHL pts treated with HDCT. We thus designed an intensified salvage program with a final myeloablative course. Eligibility criteria include age 18-60 years, histologically proven primary progressive or relapsed HD and NHL. Our treatment program consists of two cycles DHAP; pts with PR or CR receive cyclophosphamide 4g/m^2, followed by PBSC harvest; methotrexate 8g/m^2 plus vincristine 1,4mg/m^2; and etoposide 2g/m^2. The final myeloblative course is BEAM followed by ASCT.

161 pts (median age 36 years, range 18-65) with HD (n=102) or NHL (n=59) have been enrolled. So far 146 pts are available for the final evaluation (97 pts with HD, 49 pts with NHL). At 18 months of median follow-up (range 3-31 months) results are as follows: Response rate (RR) after DHAP: pts with HD: 87% (23% CR, 64% PR); pts with NHL: 69% (10% CR, 59%PR). RR at the final evaluation: pts with HD: 77% (68% CR, 9% PR), pts with NHL 48% (32% CR, 16% PR). PBSC harvest was succesful in 96% of pts. Toxicity was tolerable. FFTF/OS for pts with HD are: early relapse 64%/87%; late relapse 68%/81%; PD: 30%/58%; multiple relapse 55%/88%. FFTF/OS for pts with NHL: relapse: 42%/54%; PD 8%/22%. After 2 cycles of DHAP the rapid sequential administration of high doses of cyclophosphamide, methotrexate and etoposide is feasible and did not affect the tolerability of the final myeloablative BEAM. The Cologne high-dose

sequential chemotherapy regimen allows effectice mobilization of stem cells, is well tolerated and very effective in pts with relapsed HD and aggressive NHL. In pts with aggressive NHL relapse a multicenter trial combining HDSCT with Mabthera® is planned. In pts with relapsed HD the GHSG/EBMT/EORTC is comparing this regimen in a prospective randomized study (activated 01/2001) with 2 cycles DHAP followed by BEAM (HD-R2 protocol).

837

The monoclonal anti-CD20 antibody rituximab in the treatment of patients with aggressive B-cell non-Hodgkin's lymphoma

M. Reiser, H. Schulz, T. Elter, T. Klein, U. Winkler, M. Jensen, A. Engert (Köln, D)

Rationale/Objectives: While response rates of 50% have been observed after treatment of patients (pts) with relapsed low-grade follicular lymphoma (FL) using the monoclonal antibody rituximab (anti-CD20), only few data on therapeutic results in pts with aggressive CD20-positive NHL are currently available. Here we report 23 pts with aggressive B-NHL treated with rituximab at a dosage of 375mg/m^2 once weekly for 4 weeks. Methods: Pts either had DLBCL (n=14), Burkitt-like lymphoma (n=1), follicular lymphoma (n=7) having transformed into diffuse large B-cell NHL and one patient with initial HD and DLBCL at the time of relapse. Median age was 54 years (range 33-64 Years), and most pts (n=16) had stage III-IV disease. At the time of treatment 21 pts were in 2nd or higher relapse with a median of 2,5 prior chemotherapeutic regimens. Median time after first diagnosis was 19,2 month for all pts with DLBCL and 64,5 month for transformed secondary high-grade lymphoma. Results: So far 23 pts are evaluable for response: 23,5% (no CR and 4 PR) of 17 pts with primary DLBCL and 43% (0 CR and 3 PRs) of pts with secondary aggressive NHL responded to therapy. Conclusion: While toxicity is very low in pts with aggressive NHL during treatment with rituximab, efficacy seems to be dependent on the lymphoma subtype. Further clinical studies are currently performed to investigate the activity of rituximab in aggressive NHL in combination with chemotherapeutic regimens.

838

Malignant tumors in patients with primary intestinal lymphoma (PIL)

G. Egerer, C. Harter, J. Rudi, M. Treiber, M. Wannenmacher, C. Herfarth, A.D. Ho, T. Lehnert (Heidelberg, Mannheim, D)

The risk of additional malignant tumors in patients with lymphoma is well recognized. Since primary lymphoma of the intestinal tract is rare, only little information is available on the incidence and prognosis of associated malignancy in this condition. Methods: A retrospective, single-center analysis was performed upon 51 patients (pts.) with lymphoma involving the small or large bowel treated between 1981 and 3/2001. Patients with gastric lymphoma were excluded. Results: The median age at the time of diagnosis of PIL was 55 years (range 22-84). 25 pts. were male, 26 pts. female. In 42 pts. a B-cell lymphoma, in eight pts. a T-cell lymphoma and in one pt. a Hodgkin's lymphoma was diagnosed. Five patients (10 percent) had an additional tumor diagnosed a median of 8 years (range 5 to 12) years before diagnosis of PIL (Hodgkin's lymphoma, non-Hodgkin's lymphoma, colon cancer, breast cancer and corpus carcinoma, one each). Another six patients (12 percent) had a metachronous malignoma at a median of 8 years (range 3-17 years) after the diagnosis of PIL (two breast cancer, 1 rectum cancer, 1 MDS, 1 urethral carcinoma, 1 gastric cancer). The overall survival of all 51 patients was 33 % at 5 years. Surprisingly the 5 year overall survival rate of the 11 pts. with additional malignant tumor was 64 %. Patients received either surgical resection (n = 6) followed by either chemotherapy (n = 23), radiotherapy or both of these modalities (n = 12) in combinations or chemotherapy alone (n = 7) or in combination with radiotherapy (n = 3). In one patient an intestinal lymphoma was diagnosed after heart transplantation and 3 patients with coeliac disease developed a non Hodgkin's lymphoma of the small intestinum. 32/51 pts. are still alive (22 complete remission, 9 partial remission, 3 progressive disease), while 19/51 pts. died. Conclusion: The incidence of second malignant tumor in pts. with PIL is 3,4-fold higher compared to a comparable population described in the cancer registry of the state Saarland. Close surveillance to detect metachronous malignancies in pts. with PIL is warrented, as excellent long-term survival can be achieved.

839

A case of a primary diffuse large B-cell non-Hodgkin's lymphoma, T-cell-rich variant, of the uterine cervix

S. Mewes, K. Jentsch-Ullrich, M. Koenigsmann, A. Franke (Magdeburg, D)

Introduction: Non-Hodgkin lymphomas are frequent tumors. However, extraglandulary forms are very unusual, and the location in the uterine cervix is also extraordinary. A case of a 28 years old woman with malignant non-Hodgkin's lymphoma of the cervix uteri is presented. At the time of diagnosis no systemic involvement could be established. Patient: A 28-year-old woman presented for a routine check in ambulant gynecology. A suspicious focus was observed. Transvaginal biopsy showed a diffuse large B-cell lymphoma, T-cell rich variant. The CT scan showed no abnormality of the uterine body, other organs or lymph nodes. Initial Ann Arbor stage was IE. Serum levels of lactate dehydrogenase (LDH) was normal. Three weeks later cervical lymphadenopathy and a modest splenomegaly were observed. At the same time we found a positive serology of chlamydia. Antibiotic therapy and combination chemotherapy was started, consisting of cyclophosphamide, vincristine, adriamycin, etoposide and prednisolone (CHOEP). After 2 courses of this therapy the patient achieved a complete remission. Conclusion: Primary malignant lymphoma of the uterine body is extremely rare. In this localisation is a diffuse diffuse large B-cell lymphoma, T-cell rich variant, a very rare finding. Solitary malignant lymphomas of the cervix must be seen as systemic diseases. They should therefore be treated by local surgical intervention and consecutive systemical chemotherapy in an interdisciplinary conception.

840

Taxane, cyclophosphamide plus rituximab in patients with relapsed and refractory aggressive non-Hodgkin's lymphoma

M. Zoz, A. Krämer, F. Krasniqi, M. Görner, E. Leo, A.D. Ho (Heidelberg, D)

Based on the single-agent activity of docetaxel, cyclophosphamide, and rituximab and the promising results of paclitaxel plus high-dose cyclophosphamide in the treatment of non-Hodgkin's lymphoma (NHL), we conducted a phase II study to evaluate the efficacy of a combination of paclitaxel/docetaxel, cyclophosphamide, and rituximab in patients with refractory and relapsed aggressive NHL. Patients received 375 mg/m^2 rituximab intravenously on day 1, 900 mg/m^2 bolus of cyclophosphamide from day 2 to day 4 with a concurrent infusion of 150 mg/m^2 of paclitaxel over 72 h or 700 mg/m^2 bolus of cyclophosphamide from day 2 to day 4 and 75 mg/m^2 bolus of docetaxel on day 2. 6 days after completion of chemotherapy, patients received subcutaneus injections of 5 µg/kg of granulocyte colony stimulating factor (G-CSF) daily until white blood cell recovery. Cycles were repeated every 3 weeks. Patients who had at least a partial response (PR) after two courses continued to receive a maximum of four courses, one patient received 6 courses. Patients with responding disease were allowed to undergo high-dose chemotherapy followed by stem cell/bone marrow transplantation if they were eligible. A total of 22 patients were treated, including 15 with diffuse large cell lymphoma, 2 with mantle cell lymphoma, 1 with follicle center cell lymphoma grade III, and 3 with follicle center cell lymphoma grade III transforming into diffuse large cell lymphoma. 11 patients received docetaxel. Another 11 patients were treated with paclitaxel. There were 15 male and 7 female with a median age of 54.2 years (range 32-71 years). The patients were pretreated with radiotherapy (n= 9) and a median number of 6.6 chemotherapy cycles (range 3-12) including one cycle of high-dose chemotherapy plus autologous stem cell transplantation (n=6). The overall response rate was 50%, 36% in the paclitaxel group and 64% in the docetaxel group. Eleven patients who received treatment after their disease relapsed from a complete response had a 73% response rate (36% CRs), whereas eleven patients with refractory disease had only a 27% response rate (0% CRs). Thus, the combination of a taxane and cyclophosphamide plus rituximab is an effective regimen in the treatment of relapsed and refractory B-cell NHL.

841

Rapid engraftment after T-cell depleted allogeneic hematopoietic stem cell transplantation using a non-myeloablative conditioning regimen

M. Görner, M. Thalheimer, S. Pfeiffer, T. Luft, M. Moos, H. Goldschmidt, A. Ho (Heidelberg, D)

Background: GvHD remains a significant complication after allogeneic stem cell transplantation using both myeloablative and reduced intensity conditioning (RIC). In myeloablative transplantations T-cell depletion (TCD) results in a significant reduction of aGVHD, but is associated with increased rate of graft failure. Preliminary studies suggest that non-myeloablative transplantations - which are usually performed in older patients - might be associated with even higher rates of aGVHD, probably due to the well-known association of age and GVHD. Therefore TCD might be a useful tool to reduce transplant-related mortality of non-myeloablative transplantation. We present 17 patients transplanted after RIC. 5/17 received ex-vivo TCD grafts. Patients and Methods: From 9/99 to 1/01 17 patients with hematologic diseases (MM=13, AML=2, ALL=1, NHL=1) have received related (n=10) or unrelated (n=7) allogeneic bone marrow (n=2) or blood stem cell transplantations (n=15). In 5 related transplantations a mismatch was accepted (2 in HLA class I and 3 in HLA class II). For conditioning 3 patients received 2 Gy TBI alone, in the other patients fludarabin was combined with TBI (n=14). Post transplant immunosuppression consisted of cyclosporine and mofetil mycophenolat. 5 patients received T-cell depleted and 12 patients received unmanipulated grafts. Mean age of patients was 43 years. At time of transplant 5 patients were in CR, 4 were in PR, 3 had SD and 5 had progressive disease. Results: Treatment was well tolerated with no transplant-related deaths. All patients engrafted with 6 patients never dropping ANC $<0.5x10^9/l$. All but 2 evaluable patients had mixed chimerism >50% by day 50 and >75% by day 100 with the majority achieving >95% donor chimerism documented by molecular techniques. The two graft failures occured in non-TCD transplants. Of the 5 TCD transplants no graft failure occured despite mismatches in HLA I or II in 2 patients. 2/5 patients of the TCD group developed a medullary relapse and consecutively lost their grafts. 5/17 patients developed acute GVHD >grade II, which responded well to therapy. In 4 TCD patients no aGVHD developed, one patient of this group developed Grad IV GVHD of the liver 6 weeks post DLI given for evidence of relapse. With a median follow up of 202 days 12 patients are alive, while 5 patients died due to progressive disease. 9 patients are either in stable disease (n=1) or CR (n=8) and 8 patients progressed after transplant. Conclusion: Allogeneic transplantation after non-myeloablative conditioning can be safely performed in heavily pretreated patients. Transplant-related mortality and morbidity is low, but GVHD still remains a significant risk. TCD as GVHD prophylaxis can be performed without increased risk for graft failure and was in our series associated with no aGVHD so far. For a more definite interpretation of aGVHD, response and survival rates longer follow-up is needed.

842

Allogeneic hematopoietic stem cell transplants (HSCT) from related and unrelated donors in patients with AML following minimal conditioning

U. Hegenbart, T. Lange, W. Pönisch, B. Sandmaier, M. Maris, D. Maloney, T. Chauncey, L. Uharek, S. Leiblein, J. Edelmann, H. Greinix, D. Niederwieser, R. Storb (Leipzig, D; Seattle, USA; Wien, A)

Allogeneic HSCT for patients with acute myeloid leukemia (AML) are usually limited to patients below 55 years in good general condition because of high transplant-related mortality (TRM) in elderly patients. We applied allografting from related and unrelated donors using 200 cGy total body irradiation (TBI) before and a combination of mycophenolate mofetil (MMF) and cyclosporine (CSP) after transplantation patients with AML not eligible for a conventional HSCT. In order to avoid rejection, Fludarabine (FLU) 30 mg/m^2 per day on days -4, -3, and –2 was added to the regimen in patients with unrelated donors. Here, we report on 38 patients with AML given HSCT from related (n=21; CR1=11, CR2=5, >CR2=1, PR=1, progr.=3) and unrelated donors (n=17, 2 pts. with 1-Ag-mm; CR1=4, CR2=3, >CR2=5, PR=2, progr=3). Patients were 36–71 (median 59 years) years old and have a follow-up of 224 days (range 8 to 994 days). Median donor engraftment on day 28 was 75 (range 20-100)% for T cells and 94 (range 32-100)% for unseparated blood cells. Three patients rejected and had autologous recovery. Acute GVHD occurred in 33% patients after related and in 53% patients after unrelated HSCT and was fatal in 8%. Chronic GvHD developed in 43% of patients after related and in 53% after unrelated HSCT. TRM was 9,5% after related and 25% after unrelated HSCT. Of the 21 patients given related HSCT, 11 are alive in complete remission (52%), 8 (38%) died of relapse and 3 (14%) died of transplant causes. Of the 17 patients given unrelated HSCT, 14 (82%) are alive in CR, 2 (12%) died of relapse, and 1 (6%) died of transplant causes. We conclude that allogeneic HSCT after minimal conditioning is a feasible and safe strategy for elderly patients with AML. Hematopoietic chimerism could be induced in the majority of patients. Unrelated HSCT was highly effective in AML patients despite advanced disease at the time of HSCT in the majority of patients. A phase II study for AML patients > 60 years to compare consolidation with chemotherapy or allogeneic related HSCT after minimal conditioning was initiated by the OSHO/FHCRC/HOVON/SAKK and FRENCH Study Groups.

843

Allogeneic bone marrow transplantation for CML with reduced intensity conditioning is associated with increased detection of the BCR-ABL fusion transcript

M. Weisser, S. Schnittger, O. Stoetzer, M. Schleuning, H. Kolb (München, D)

Background: Despite major progress in the therapy of CML, allogenic bone marrow transplantation (BMT) remains the only potential cure for this hematologic malignancy. To date overall survival is approximately 75%, but the data for the eldery patients is less encouraging. In order to reduce toxicity in elderly CML-patients we established a conditioning protocol consisting of 8 Gy TBI, fludarabin, cyclophosphamid and ATG. Since November 1999 twenty three patients 45 years of age and older with BCR/ABL+ CML were transplanted at our center. In this evaluation we compare the outcome of our reduced intensity (RIC) protocol with a historic control group of 67 elderly CML patients treated with a standard intensity conditioning (SIC) regimen either 12 Gy TBI/Cy/ATG or Busulfan/Cy/ATG.
Patients and Methods: RIC consisted of 8 Gy TBI, 4x30 mg/m^2 fludarabin, 120 mg/kg cyclophosphamid, 40 mg/kg ATG. Patients with blast crisis were excluded. GvHD prophylaxis consisted of cyclosporin and methotrexate. Median age was 51 years, 11 male, 12 female. There were 9 MUD and 14 HLA-identical sibling-donors. Patient characteristics between the RIC and the SIC group were well balanced. Using nested RT-PCR we analyzed bone marrow and peripheral blood at the following time intervals: 0-90, 0-180, 181-360 and 361-720 days post BMT. BCR-ABL was detected via qualitative nested RT-PCR in 2x35 cycles. The percentage of BCR-ABL positive PCR in the reduced and standard intensitiy conditioning group was compared at each time interval.
Results: At 180 days after BMT 87 % of RIC- patients were BCR-ABL positive versus 46% of the SIC-patients ($p<0.05$). After 360 days BCR-ABL-detection decreased to 40% (RIC) versus 35% (SIC). So far there is no significant difference concerning overall survival. Engraftment was 100%, transplant related mortality was 21%. Relapse rates were 1/23 (RIC) versus 9/67 (SIC) . Acute GvHD occurred in 9/23 RIC, 26/67 SIC patients. Chronic GvHD was diagnosed in 9/23 RIC, 18/67 SIC patients.
Conclusions: Reduced intensity conditioning for CML patients above 45 years of age is well tolerated. It is associated with increased detection of the BCR-ABL fusion transcript in the first 180 days after bmt. In the early phase after bmt SIC seems to clear the BCR-ABL positive clone more effectively. After 360 days BCR-ABL detection via RT-PCR decreases in the RIC-group, probably due to a potent GvL-effect. RIC was associated with a lower relapse incidence at the expense of a slightly higher incidence of chronic GvHD.

844

In vitro and in vivo characterization of graft-versus-tumor responses in melanoma patients after allogeneic peripheral blood stem cell transplantation with reduced intensity conditioning

T. Kurokawa, K. Fischer, H. Bertz, J. Finke, A. Mackensen (Freiburg, Regensburg, D)

It has been shown that after allogeneic peripheral blood stem cell transplantation (PBSCT) donor T cells can induce potent graft-versus-tumor (GvT) effects in hematological malignancies and possibly solid tumors such as renal cell carcinoma. Two patients (27 and 30 years old) with metastatic melanoma received allogeneic PBSCT from an HLA-identical sibling donor after reduced intensity conditioning with fludarabine, carmustine, and melphalan. One patient showed a delayed mixed response with complete regression of lymph node metastases but progressive liver metastasis at day +120 consistent with a GvT response. In order to generate tumor-reactive cytotoxic T lymphocytes (CTL) from the donor peripheral blood mononuclear cells were stimulated with donor monocyte-derived dendritic cells (DC) loaded either with tumor lysate or apoptotic tumor cells from the patient. We have shown previously, that two weekly stimulation cycles with tumor-loaded DC followed by restimulation with irradiated tumor cells alone were optimal for induction of tumor-specific CTL responses in vitro (Int.J.Cancer 91:749, 2001). Using these culture conditions a marked increase of CD8+ CTL was observed exhibiting a strong MHC class I-restricted cytotoxic activity against the patient's tumor without cross-reactivity against EBV-transformed B cells from the donor. Length pattern analysis of the complementary determining region 3 (CDR3) of the T-cell receptor (TCR) Vß chain revealed expansion of oligoclonal CTL populations. Functional (Elispot) and phenotypic (CDR3 spectratyping) analysis of patients' T cells at different time points after transplantation revealed an expansion of alloreactive T cells with a polyclonal TCR Vß pattern. Altogether, our results suggest that (A) clinical GvT responses against melanoma can be induced after allogeneic PBSCT and (B) oligoclonal tumor-reactive donor CTL can be generated and expanded in vitro that may be used for adoptive T cell transfer after allogeneic transplantation.

845

Decreased transfusion requirements for patients receiving non-myeloablative compared to conventional peripheral blood stem transplants from HLA-identical siblings

F. Weissinger, B.M. Sandmaier, D.G. Maloney, W.I. Bensinger, T. Gooley, R. Storb (Würzburg, D; Seattle, USA)

Red blood cell (RBC) and platelet transfusion requirements in patients given nonmyeloablative peripheral blood stem cell (PBSC) transplants from HLA-matched siblings were compared to those after conventional transplants. Between December 1997 and March 2000, 40 patients, aged 21 to 67 (median 51) years, with hematologic malignancies underwent nonmyeloablative allografts after either 2 Gy total body irradiation (TBI) alone (n=30) or 2 Gy TBI preceded by fludarabine 30 mg/m^2/day on days -4, -3, -2 (n=10). All received postgrafting mycophenolate mofetil (MMF) and cyclosporine (CSP). Controls included 67 concurrent patients, aged 23 to 66 (median 46) years, given conventional PBSC transplants following high-dose conditioning and postgrafting methotrexate (MTX) and CSP. Among patients given nonmyeloablative transplants, 23% required platelet tranfusions compared to 100% among patients given conventional grafts (p<.0001). Further, the number of platelet units given to nonmyeloablative recipients was reduced with a median of 0 (range 0 to 214) compared to a median of 24 (range 4 to 358) after conventional transplantion (p<.0001). Sixty-three percent of nonmyeloablative recipients required RBC transfusions compared to 96% of those with conventional grafts (p=.0001). The number of RBC units transfused was also reduced with a median of 2 (range 0 to 50) compared to 6 (range 0 to 34) after conventional transplants (p=.0001). High transfusion requirements before transplant and donor-recipient ABO-incompatibility increased transfusion requirements in both patient groups, though neither significantly influenced the outcome of the analysis. Neither patient age, splenomegaly at transplant, development of GVHD nor posttransplant cytomegalovirus (CMV)-antigenemia or CMV-disease had statistically significant influences on posttransplant transfusions.

846

Kinetic of lymphohematopoietic chimerism in dose reduced allogeneic stem cell transplantation

F. Keil, K. Moser, E. Prinz, P. Kalhs, C. Mannhalter, H. Loidolt, I. Schwarzinger, P. Höcker, K. Lechner, H. T. Greinix (Wien, A)

Dose- reduced allogeneic stem cell transplantation (SCT) allows establishment of donor hematopoiesis without eradication of recipient stem cells by chemoradiotherapy. This procedure has a decreased conditioning related toxicity but graft rejection and GVHD remain clinically challenging. In the case of mixed chimerism monitoring of chimerism of T cells and hematopoietic precursor cells is of high importance. It allows to define origin of hematopoietic recovery and to manipulating favorably the graft by infusion of donor leukocytes or increasing or tapering of immunosuppressive therapy. We performed dose reduced SCT with conditioning the recipient with fludarabine 90 mg/m^2 followed by TBI of 200 rad and infusion of unmanipulated PBSC. Cyclosporine and mycophenolate mofetil were given to prevent GVHD and rejection. Chimerism was evaluated in unseparated blood and marrow and various cell subsets including myeloid (CD33+, CD34+)and lymphatic lineage (CD3+, CD4+ CD8+) after cell sorting. Complete donor T or myeloid chimerism was defined if 90 to 100% of cells were from donor origin and mixed chimerism if at least 5% of cells but less than 90% were of recipient origin. In case of sex mismatch between recipient and donor monitoring was performed by FISH, otherwise by quantitative variable number of tandem repeats (VNTR)-PCR. So far, 19 patients 7 with AML, 2 with MDS, 1 with CLL, 5 with NHL, 1 with CML, 2 with MM and one with hypernephroma were analysed. Median age was 55 with a range from 33 to 70 years. 16 of 19 were evaluable (one early death and two patients are too early to asses). After 28 days 11 (69%) patients showed complete and 4 (24%) mixed myeloid donor chimerism. For T cells a complete chimerism was seen in 8 (50%) patients and in another 7 (43%) patients a mixed chimerism. One patient rejected the allograft. All patients who had less than 50% of donor T cells after SCT relapsed whereas all other patients are in remission betweeen one and 8 months after SCT. These results emphasize the importance of sequencial analyse of lineage specific chimerism to predict outcome after allogeneic nonmyeloablative SCT with dose reduced conditioning.

847

Allogeneic transplantation after dose-modified conditioning in acute myeloid leukemia - a retrospective multicentre analysis

H.G. Sayer, M. Bornhäuser, R. Schwerdtfeger, J. Kienast, H. Martin, H. Wandt, J. Beyer, W. Siegert, V. Runde, M. Kiehl, G. Ehrlinger (Jena, Dresden, Wiesbaden, Münster, Frankfurt, Nürnberg, Marburg, Berlin, Essen, Idar-Oberstein, D)

Between 02/98 and 12/00 117 patients [pt.] (70 male/47 female) from 10 German transplant centres with acute myeloid leukemias (FAB: M0-M7, 56 [47,9%] with cytogenetic abnormalities), not eligible for conventional allogeneic hematopoetic stem cell transplantation were grafted after a fludarabine / busulfan-based modified conditioning regimen. Ninty-seven pt. received fludarabine (90 to 180 mg/m^2 iv.) and busulfan (6.4 mg-16 mg orally or iv.). Twenty pt. received 4 to 8 Gy total body irradiation instead of busulfan. The median age for the entire group was 51 years (range 16-67), median time to transplant 8 months (range 0.5-111). Thirty-seven (31.6%) pt. were in first or second CR, whereas 80 (68.4%) pt. were in 3rd CR, PR, refractory or in relapse. The reason for modifying the transplant regimen were age (n =36/ 30.8%), organ dysfunction (32/27.4%), fungal infection (18/15.4%), previous auto- or allogeneic transplant (14 /11.9%) or other (17/14.5%). Graft source were blood stem cells (106/90.6%), bone marrow (10/8.5%) or both in one patient. Fifty-one (43.6%) donors were family HLA matched, 53 (45.3%) unrelated fully matched and 13 (11.1%) mismatched family or unrelated. One-hundred-eleven (94.9%) showed neutrophil engraftment and platelet recovery after a median time of 14 (range 0-35) days and 13.5 (range 0-105) days, respectively. The probability of overall survival [OS] of the whole group was 48.7% at a median observation time of 305 days (range 9-930). Death occurred due to relapse in 29 pt. (47.5%), infection in 13 (21.3%), bleeding and organ-failure in 7 (11.5%), and acute GvHD in 8 pt. (13.1%). The probability of OS in complete [CR] and partial remission [PR] at day 930 was 50 % versus 18 % in the non-CR/PR patients at day 731 (p = 0.000). Acute GvHD occurred in 71 (60.7%) pt. [40 with grade I-II; 31 with grade III-IV]. The probability of transplant-related mortality [TRM] at day 930 was 49% (34,3% for the CR/PR group and 68,2% for the non-CR/PR group). In summary, this analysis of a cohort of AML-patients after allogeneic stem cell transplantation using a dose-modified risk-adapted conditioning regimen revealed a high incidence of delayed TRM, a high relapse rate according to disease-status and a favourable outcome in CR patients. Risk-adapted conditioning regimen in AML-patients otherwise not eligible for allografting are promising in controlled disease, rather than in advanced disease and need to be confirmed in controlled prospective trials.

848

EBV reactivation in mycosis fungoides (MF) patients referred for mini-BMT

C. Annaloro, A. Ibatici, M. Corbellino, D. Soligo, A. Della Volpe, C. Parravicini, G. Lambertenghi Deliliers (Milano, I)

Oportunistic virus reactivations are described with increasing frequency in patients with different hematological diseases whether or not they undergo BMT. Apart from CMV, many other species have emerged including most DNA viruses. We have observed EBV reactivations in all three MF patients referred to our institution for BMT. These male patients were aged 53, 48 and 37 years, and had received more than three lines of conventional therapy. The first patient had last been treated with two monthly cycles of fludarabine, when he developed hyperpyrexia, huge generalized lymphoadenopathies and respiratory insufficiency. A diagnosis of EBV-related lymphoproliferative syndrome (LPS) with concomitant EBV interstitial pneumonia was established on lymph-node biopsy and broncho-alveolar lavage. The patient achieved remission with weekly administrations of rituximab and soon after underwent allogeneic mini-BMT with a conditioning regimen including fludarabine and 200 cGy TBI. On day +25 after mini-BMT, the second patient developed a clinical picture of meningo-encephalitis, rapidly followed by respiratory failure. The PCR-revealed presence of 30,000 copies of EBV genome/μg DNA and the demonstration of clonally restricted B-lymphocytes at cytofluorimetry allowed a diagnosis of EBV-related meningoencephalitis to be established; the patient was admitted to an intensive care unit until the resolution of the acute disease. The third patient had EBV copies in four consecutive blood samples taken from day +12 after mini-BMT for the purpose of periodic monitoring; no clinical counterpart could be identified and spontaneous normalisation occured after two weeks. No case of EBV positivity was observed in eight patients undergoing mini-BMT for other hematological diseases. The three patients are still alive and in CR (as shown by cutaneous immuno-histochemistry) 2-8 months after BMT. The above case of EBV-related LPS is one the first to be reported in a patient not having received BMT. Its occurrence after the start of fludarabine therapy may add EBV to the list of uncommon infectious complications ascribed to the drug's immunosuppressive activity, an opinion that is supported by the observation that the other patients had also received fludarabine as a component of their conditioning regimens. On the other hand, as EBV reactivation is very unusual after fludarabine-including non-myeloablative regimens in other diseases, it could be inferred that both MF and fludarabine may be simultaneous risk factors for this complication. Our preliminary data show that mini-BMT is possibly an effective therapeutic choice for MF. The need for periodic monitoring of EBV genome is an obvious conclusion to be drawn from this limited experience; alternatively, other drugs (particularly pentostatin) could be included in non-myeloablative regimens for MF patients.

849

Treosulfan and fludarabine as toxicity reduced conditioning regimen in allogeneic blood stem cell transplantation

J. Casper, W. Knauf, B. Steiner, D. Wolf, A. Lueck, H. Kleine, U. Hammer, R. Wegener, M. Freund (Rostock, Berlin, D)

Toxicity reduced conditioning regimens allow allogeneic transplantation of patients otherwise ineligible. Moreover, transplant related mortality /morbidity might be reduced in general. However reduction of dose intensity may increase risk of transplant failure or relapse. We have, therefore, combined fludarabine with the alkylator treosulfan for conditioning of patients (pts.) with hematologic diseases otherwise not suitable for allogeneic transplantation. 25 pts. with a median age of 50 years (20-60) have been transplanted with CML (6), MDS (4), AML (6), multiple myeloma (4), CLL (2) and NHL cbcc (3). 6 pts. had been transplanted in autologous programs and 1 pat. had received two allogeneic transplantations previously. 10 matched (MRD), 1 mis-matched related and 14 matched unrelated donors (MUD) donated bone marrow (10) or peripheral blood stem cells (15) (median 2,3x10^6 CD34+ cells/kg BW, range 0.3 to 15.9x10^6). Treosulfan 10g/m^2 i.v. was given from d-6 to d-4, fludarabine 30mg/m^2 i.v. d-6 to d-2. Pts. with unrelated donors received ATG (rabbit) 10mg/kg i.v. d-4 to d-2 in addition. Leukocyte recovery was fast reaching 0.5 Gpt/l by median d+9 (range 0-18) and 1 Gpt/l by median d+10 (range 0-19). Thrombocyte recovery (>50 Gpt/l) occurred by median d+15 (range 8- not reached). Conditioning related toxicity was mild (< CTC °3 except for 1 pat. with bilirubine and 5 pats. with ASAT/ALAT elevations CTC°3). After a median observation time of 233 d (range 1-590) aGvHD > °2 occurred in the pat. with the mis-matched related donor (°IV) and in 1 pat. with MRD (°III). Chronic limited and extensive GvHD occurred in 3/16 resp. 4/16 pts. evaluable so far. 4/25 (16%) of the pts. died due to treatment related causes (repeated virus infections and septic shock). One pat. with an accompanying severe cardiac amyloidosis died on d+61 of a sudden cardiac death/septic shock. Complete chimerism was reached in 20 of 23 pts. by d+28 (2 pts. too early). One pat had a primary graft failure and recovered with his own hematopoiesis. The donor chimerism of the other two pts. is still increasing. Complete donor chimerism has been maintained by all pts. except for 2 pat. with CML. Both became Philadelphia chromosome positive again, but could be cured by reduction of the immune suppression resp. repeated DLI. Conditioning with treosulfan and fludarabine was very well tolerated and the results so far compare favourably with other toxicity reduced regimens.

850

Non-myeloablative allograft to induce graft versus myeloma effect after cytoreductive autotransplant for multiple myeloma

N. Kröger, R. Schwerdtfeger, H. Renges, W. Krüger, N. Stute, T. Zabelina, J. Schrum, F. Tögel, G. Wittkowsky, R. Kuse, S. Hegewisch-Becker (Hamburg, Wiesbaden, D)

To reduce transplant-related complications of allogeneic stem cell cell transplantation in multiple myeloma but retaining its curative potential, we conduct a study starting with 200 mg/m^2 melphalan supported with autologous PBSC followed by a non-myeloablative regimen consisting of 100 mg/m^2 melphalan, 180 mg/m^2 fludarabine and allogeneic stem cell transplantation. GvHD prophylaxis consisted of ATG (30mg/kg), short course of methotrexat and cyclosporine. So far, 12 patients (pts) had completed autologous and allogeneic transplantation with HLA-identical sibling (n=5), HLA-mismatched sibling (n=2) or unrelated donor (n=5). The interval between autologous and allogeneic transplanation was 110 days (range 60-210). After allogeneic transplantation all patients achieved a leukocyte count >1.0x10^9/l at a median of 15 days and a platelet count >20x10^9/l at a median of 20 days. All pts had complete donor engraftment. One patients died on day +23 of pulmonary infection (day 100 mortality 8%) Only one pts developed GvHD III. After a median follow-up of 12 months (range 6-16) 11 pts (92%) are alive without disease progression. 8 (73%) are in continous CR with negative immunofixation and 3 pts achieved PR with still decreasing monoclonal bands.
This preliminary results show that autologous transplantation followed by non-myeloablative allogeneic transplantation from related or unrelated donor is a feasible approach in patients with multiple myeloma. Further accrual and follow-up is necessary to determine its curative efficacy.

851

Allogeneic blood stem cell transplantation after minimal conditioning in high risk patients with hematologic malignancies

G. Kobbe, P. Schneider, M. Aivado, F. Zohren, D. Schubert, R. Fenk, U. Rohr, F. Neumann, R. Kronenwett, H. Pape, U. Germing, A. Heyll, R. Haas (Düsseldorf, D)

Twenty three patients with hematologic malignancies (AML 3, MDS 4, CML 4, NHL 4, Hodgkin 1, MM 7, median age 47 years, range 39-64) were treated with Fludarabin 3x 30mg/m^2 and low dose TBI (2GY) followed by transplantation of G-CSF mobilized peripheral blood stem cells (median 7.6x10E6 CD34+cells/kg, range 3.5-13.1) from HLA-identical sibling (19) or a matched unrelated donors (3). Mycofenolatemofetil (22) and Cyclosporine (22) or FK506 (1) was given for GvHD prophylaxis. Sequential chimerism analysis was performed with STR-PCR and FISH on unfractionated bone marrow and peripheral blood as well as on selected CD34+cells, CD3+cells and granulocytes. Patients who maintained mixed chimerism and/or had disease recurrence received donor lymphocyte infusions. Twenty of 22 evaluable patients (90%) engrafted. Eight of 14 evaluable patients who had active disease at the time of transplantation had a response. While 3 patients showed only transient disease regression (1 AML, 1 CML, 1 MDS) stable remission was achieved in 5 patients (2 MM in PR, 2 CML in molecular CR and 1 NHL in CR). Four of 5 patients who were transplanted in remission are disease free (2 MDS, 1 HD, 1 NHL). Eight patients (35%) died at a median of 106 days after transplantation (range 50-129). Seven (2 AML, 2 NHL, 1 CML, 1 MM, 1 MDS) died of progressive disease and 1 (AML) of treatment related complications (TRM <5% at day 100). Fifteen patients are alive with a median follow up of 287 days (range 16-626). Twelve of 21 evaluable patients (55%) have developed acute GvHD (6x II°, 6x III°-IV°) and 5 of 13 patients (31%) alive beyond day 100 have chronic GvHD (extensive 1, limited 4). Allogeneic blood stem cell transplantation after minimal intensive conditioning is associated with low regimen related toxicity. Remissions can be achieved or maintained in a variety of hematologic malignancies but GvHD is common and related with considerable morbidity. This transplantation modality seems to be a promising new treatment option for patients with CML, myeloma or lymphoma as well as for MDS/AML in remission.

852

Treatment of secondary myelodysplastic syndrome (MDS) after heart transplantation with chemotherapy and non-myeloablative blood stem cell transplantation

G. Kobbe, U. Germing, M. Aivado, D. Schubert, F. Zohren, G. Meckenstock, R. Haas, A. Heyll, P. Schneider (Düsseldorf, D)

Secondary cancer is a relevant complication of long term immunosuppression after solid organ transplantation. Acute leukemia and myelodysplastic symdrome have been described after heart and kidney transplantation. Treatment of these malignancies is difficult as most patients are unable to tolerate intensive chemotherapy and conventional myeloablative blood stem cell transplantation due to coexisting medical conditions, advanced age or reduced organ function. Allogeneic stem cell transplantation using minimal intensive conditioning has so far not been reported after solid organ transplantation. A 60-year-old man developed secondary (treatment related) MDS (RAEB-T) ten years after orthotopic heart transplantation. Remission was induced with two courses of chemotherapy (Idarubicine/Cytarabine, 2xHAM [1g Cytarabine]). The patient did not qualify for a conventional myeloablative allotransplant because of advanced age, reduced performance status and chronic renal failure. In first complete remission he received minimal conditioning with fludarabine (3x 30mg/m^2) and low dose TBI (2Gy) followed by mobilised stem cells (4.7x10^6 CD34+cells/kg) from his HLA-identical sister. Following transplantation mycophenolatemofetil (MMF) was added to cyclosporine for the first 28 days. The patient experienced low toxicity from the conditioning regimen and achieved complete donor chimerism in blood and marrow within 100 days. Following a period of steroid therapy for grade II acute graft-versus-host-disease (aGvHD) he had to be treated for clostridium difficile infection and CMV reactivation. The patient then developed pure red cell aplasia caused by persisting host derived anti-A isoagglutinins which was refractory to therapy with high-dose immunoglobulin (HD-ivIg), low-dose donor lymphocyte infusions and erythropoietin. He finally recovered after combined treatment with MMF, HD-Epo and HD-ivIg. The patient is now alive in remission with 100% donor hematopoiesis 612 days after transplantation. We conclude that allogeneic blood stem cell transplantation using minimal conditioning may be a new promising treatment option for patients with hematologic cancers after solid organ transplantation who are unable to tolerate intensive myeloablative conditioning regimens due to advanced age or comorbid medical conditions.

853

Different chimerism kinetics in myeloid versus lymphoid malignancies after non-myeloablative blood stem cell transplantation

F. Zohren, D. Schubert, M. Aivado, P. Schneider, U. Germing, R. Fenk, U. Rohr, R. Haas, G. Kobbe (Düsseldorf, D)

Twenty-two patients (median age 47 years, range 39–64) with advanced myeloid (3 AML, 4 CML, 4 MDS, median number of transplanted cells 8.3x10^6 CD34+/kg, range 3.5-13.1) and lymphoid (6 MM, 4 NHL, 1 HD, median number of transplanted cells 6.1x 10^6 CD34+/kg, range 3.6-11.6) malignancies received mobilised blood stem cells from HLA–identical related (20) or unrelated (2) donors after treatment with Fludarabine (3x30 mg/m^2) and low dose TBI (2 GY). After transplantation sequential analyses of chimerism were performed on total nucleated peripheral blood cells, selected CD3+ cells and granulocytes, as well as on unfractionated bone marrow and selected CD34$\pm$ cells. Chimerism studies were done using both STR–PCR (HUMARA, HUMTH01, TPOX, CSF1PO, D8SI179) and FISH with probes for the X and the Y chromosome (in sex-mismatched transplantation). In addition flow cytometric studies were performed to monitor levels of circulating T, B and NK cell subpopulations.
In the early post transplant period (day 28-65) patients with lymphoid malignancies developed a significantly higher degree of donor chimerism than patients with myeloid malignancies (p<0.05). This observation was made throughout all analysed cell compartments but was most evident among the CD3+ cells and least in circulating granulocytes. The lymphoid group achieved 100% donor chimerism among peripheral blood cells early between day 28 and 42 after transplantation, whereas the myeloid group maintained mixed chimerism of 54-71% donor cells up to day 100 after transplantation when either complete donor chimerism was established or disease progression occurred in line with a falling percentage of donor cells. Early high levels of donor chimerism among patients with lymphoid malignancies were associated with a trend towards higher levels of CD3/ CD8/HLA-DR positive cytotoxic T-Cells which was not significant due to large variations among a small number of patients.
The differences observed may originate from a more intensive, immunosuppressive pre-treatment in the group of patients with lymphoid malignancies, allowing an early immunologic elimination of host haematopoietic cells. However, the involvement of the myeloid- and the CD34+ compartment in myeloid malignancies may also account for differences in homing of the transferred stem and progenitor cells which could be another reason for different chimerism kinetics. If confirmed in a larger number of patients, these results argue for further reduction of the conditioning regimen (omission of TBI) in patients with lymphoid malignancies.

Postersitzung:
Qualitätssicherung / Gesundheitsökonomie

854

Interactive teaching methods for medical students. Cooperative pilot project between a medical school and a oncology group practice

R. Weide, J. Heymanns, H. Köppler (Koblenz, D)

Delivering high quality medical education to medical students is one of the primary tasks of medical schools in Germany. However often lecturers and students are not satisfied with the current practice. This is due to the relatively high numbers of medical students, limited resources and a system where clinical research, patient care and the education of medical students are not separated and often have to be accomplished by one person. With the aim to improve the education we have started a pilot project where medical students get additional academic lectures in haematology and oncology in a community based group practice. Several requirements are necessary to make these lectures attractive for students:
1. Small groups (a maximum of 4 students per lecturer)
2. The subjects of the lectures are chosen by the students
3. The education must be interactive rather than monodirectional
4. Case reports help delivering the educational message
5. Modern picture visualisation systems help to get insights in haematological morphology

Between 4/98-5/01 we have performed 12 weekend workshops for students of the Philipps-University. 122 medical students have participated so far. Students arrive in our practice on Friday afternoon and get their lectures on Friday evening and Saturday. The didactics and the effectivity of the education has been judged by the students as being optimal, especially in explaining difficult interrelations.

855

Methodology of patient oriented research

H. Lövenich, U. Bethe, G. Fätkenheuer, V. Diehl, O.A. Cornely (Köln, D)

Objective: To describe the structure and the working methods in patient oriented clinical research in the field of infections in the immunocompromised host. Most of the patients undergoing chemotherapy for the treatment of hemato-oncological malignancies are at some point in need of antimicrobial therapy. We conduct up to 15 studies concurrently in order to assess the efficacy of antimicrobial substances in the prevention and treatment of infections in neutropenic patients. Even though the number of haemato-oncological patients is high, recruitment figures for infectious diseases (ID) studies often fall below expectations. In part, this can be due to organisational circumstances, that can be overcome by establishing a study center within a hospital or department. We describe the work of our ID study center, which has become the backbone of the institution´s patient oriented ID research. After the patient is informed about his or her diagnosis and recommended antineoplastic therapy we provide background information about medical studies in general and ID complications of neutropenia. In order to provide all patients with this information we conduct a screening for eligible patients every other day. After identifying a patient, who meets the most essential eligibility criteria for one of the current studies, a sheet with a signalizing color is attached to his/her medical record. The eligible patient is then informed about a suitable clinical study. This is done by an investigator, either the responsible ward physician or the ID consultant, before the expected onset of fever. Thus the patient can give informed consent in full awareness and without any time pressure. If the patient agrees, the sheet in his/her file gets marked; if he refuses to take part, the sheet gets withdrawn promptly. When a patient meets the final inclusion criterion, e.g. becoming febrile, he can immediately be included 24h a day without any delay. Randomisation is conducted by the study coordinating doctor on call, practically before the second set of blood cultures is taken. The ward physician marks the allocated treatment arm on the colorized sheet. Trial documentation and sampling and laboratory work is done exclusively by the study team. A continuous documentation of all neutropenic episodes in an anonymous data base helps to check the screening log and to detect systematic errors in the screening procedure. Moreover, during the planning phase of new protocols the patient numbers expected can most accurately be estimated. In order to assure high compliance and data quality, all doctors, pharmacists, nurses and other ward staff are provided with detailed information about the study protocols. We use a flow chart that gives the appropriate indications of all study protocols along with the state of the art treatment for non-eligible patients. Additionally, protocols, clinical record files and other essential documents can be downloaded from our homepage.

856

Qualitative research in psycho-oncology

L. de Lange, M. Koenigsmann, J. Frommer, A. Franke (Magdeburg, D)

Coping with disease and attitudes towards therapeutic treatments in hematological patients is currently being examined in our clinic using interviews for qualitative research. In psychological interventions with 71 patients (acute lymphocytic and myeloid leukemia: 23, including 9 with recurrence; chronic lymphocytic and myelogenous leukemia: 9; non-Hodgkin's lymphoma and Hodgkin's disease: 34, including 8 with recurrence and multiple myeloma and other plasma cell neoplasms: 5) it was found that articulating and expressing their feelings over the situation were important elements in order to be able to cope with the disease and to build a trusting relationship to the members of the treatment team. Furthermore, stimulating the patients to show a „fighting spirit" and to think positively was, at least primarily, not helpful in all cases, but on the contrary sometimes hindering in coping with the disease. Finally, statements of patients as: „I do live now more consciously as in the past", or: „I am gratefull for every day that I live", show that a serious illness can be understood as a chance, in fact a developmental chance.

857

ChemoDoQ, a software for planning, documentation and evaluation of chemotherapy regimen

I. Schneider (Ludwigsburg, D)

The basic features of the programme are twice:First,the database of the different chemotherapy schemes and secondly a databank to store patient individual chemotherapeutic data including calculated and administered dose per cycle. Ad 1:The scheme-database includes the definition and a precise dose regimen of the drugs with their timely administration.Most schemes used in clinical routine are available in the data base.They can be easily adopted according to specific clinic situation,state of art and/or new schemes or drugs may be added.These adoptions can be performed by the user without external updates.Concurrent medication and antiemetics may be chosen according to the specific clinic setting and once defined can be used with each new cycle planning.For safety reasons changes in the scheme database can be restricted to authorized persons. Ad 2:In the patient databank patient data,tumor diagnosis and the specific chemotherapeutic scheme are collected.With the chosen scheme further planning will be based upon the modalities defined in the scheme-database.Dose calculation of the different drugs automatically is performed after input of necessary patient parameters (body weight,serum-creatinine/creatinine clearance, AUC).Despite this, the physician's responsibility to decide the percentage of the dose of each drug administered fully remains by confirmation of the planning physician.Future treatment cycles are recalculated on the basis of possibly changed patient parameters.All further data are stored in the patient databank to be used for new cycle-planning.With every new planning an information is given of the total dose of a specific drug so far,and whenever defined,of the maximal total dose allowed.Additionally with short descriptions of the clinical course during each chemotherapeutic cycle an epicrisis accumulates with ongoing therapy. After having completed to plan a new cycle a therapy protocol is printed,which clearly shows the different drugs and administration routes in their chronological listing offering a reasonable reduction of workload for doctors and nurses.Automated family doctor information is offered.The program offers quality assurance of chemotherapeutic procedures at the following points:The definition/adoption of a chemotherapeutic scheme by an authorized physician guarantees unchanged scheme parameters as the basis for any calculation.Updating of the scheme database easily can be performed.Only the dose of the drug that has been approved by the planning physician will be printed in the therapy protocol.With use of a relational databank an overview of the therapy of an individual patient is visualized,as well as the total chemotherapeutic activities of a department of oncology may be statistically evaluated with different criteria.

858

Quality assurance and documentation of chemotherapy by use of a central control system

W. Bernhardt, M. Herling, M. Behrens, E. Jüttner, H. Henß, D. Behringer (Freiburg, D)

Application of chemotherapy represents one of the key expertises of practicing hematologists /oncologists. While most recommendations for polychemotherapy specify information about the chemotherapeutic agents, the application protocol for the individual patient must contain details with respect to the antiemetic regimen, intravenous hydration, duration, sequence and application mode as well as specific dilution directions for the chemotherapeutics. In order to standardize and control these parameters in our department, a panel of established protocols for oncologic and hematologic diagnoses has been specified, categorized (GCP and non GCP protocols) and made available as a „blue book" hard copy since 1994 and electronically via intranet since 2000. This protocol panel is reviewed on a regular basis. As a standard procedure, chemotherapies are ordered on forms addressing patient data, diagnosis and protocol identification according to the „Blue Book" and supplying the patient consent form. At the GCP-station technicians in cooperation with a physician perform an independent data review and categorize the protocol (e.g. standard, nonstandard). The data are then transferred to an individual patient protocol, which serves as the documentation sheet on the ward. Linking this information to a clinical documentation system, this data base can be exploited i) to review all chemotherapeutic agents, including cumulative dose of individual patients, ii) to quantify diagnosis related use of chemotherapy protocols according to internal standards iii) as a source for improving patient care (e.g. to improve treatment time for the patient) and finally iiii) to electronically transfer OPS related procedures. In 1999, >7500 patient chemotherapy days/year have been administered according to these guidelines and transferred to the database. Statistical evaluations retrieved from the day clinic with 400 patient chemotherapy days/month will be presented.

859

Computer aided diagnosis of leukemic diseases using digital image processing techniques

M. Pandit, H. Link, F.G. Hagmann, H. Hengen (Kaiserslautern, D)

The computer aided diagnosis of leukemic diseases can be carried out using May/Gruenwald/Giemsa stained smears of peripheral blood and bone marrow. Images of these smears are acquired using a combination of optical microscope and CCD-camera. The image data is normalized during the acquisition process. This is achieved using colour space transforms (RGB to HSI). In order to be processed in detail, the white blood cells have to be segmented from the images first. Secondly the segmented cells are decomposed and characteristic features are extracted using digital image processing algorithms. One problem is - especially concerning images of bone marrow smears - that the cells are clustered and that in order to perform a suitable cell analysis, these clusters have to be decomposed. This task has been completed sucessfully by invention of (mathematically) morphologic algorithms for the cluster decomposition. Besides the morphologic decomposition, active contour models can be used for the determination of the cells' true boundaries. Although this step is not necessarily required for the decomposition of blood smear images, it will be essential in the further analysis of bone marrow smears. The overall aim of the project is however to obtain a histogram of the different development stages of white blood cells encountered in the smears of peripheral blood and bone marrow in order to detect:
- precursors of ripe white blood cells which must not be encountered in peripheral blood - find atypic cells which have obvious aberrations (e.g. hairy cells). In order to fulfill this task, manifold algorithms have to be invented and their results experimentally verified:

1. Normalized image acquisition (standardization)
2. Presegmentation (colour-model based)
3. Segmentation of cell clusters (using nonlinear image processing methods and optimal contours)
4. Analysis of features of different cell types:
   - Experimental decomposition of the cells' cytoplasm and nucleus
   - Determination of specific outstanding features of specific cell types and identification of these using mathematical models
   - Wavelet decompositions and multirate/multiresolution processing

- Fractal stochastic processes
- Metrics and Fuzzy classification
5. Composition of the components 1-4 using a suitable classificator (Bayesian classifier)
6. Analysis of a large number of cells
7. Classification and validation of AML data of study patients
8. Prepare the system for a wide usage in hematology

The present research is focused mainly on 4, whereas the aspects 1,2 and 3 and 5 have been completed sucessfully. Some of the cell's features can be already sucessfully determined, such as e.g. number of segments of the nucleus, nucleus to plasma ratio etc., further results are expected soon.

860

Effect of networking and patient monitoring on patient accrual in the German multicenter trials for adult acute lymphoblastic leukemia

N. Gökbuget, R. Reutzel, S. Hug, D. Hoelzer (Frankfurt, D)

Introduction: The question whether leukemia patients are included in clinical trials or treated „off" study according to ongoing or closed treatment protocols is discussed intensively and it is a major aim of the Competence Network for Leukemias in Germany (founded in 1/2000) to increase the patient accrual of clinical trials. Intensified cooperation with laboratories for central diagnosis and increased visibility of clinical studies within the Network may contribute to this goal. In addition the GMALL study group started in 1/2000 a programme for monitoring of adult ALL patients. All patients with diagnosis of ALL according to central immunophenotyping (available since 1980) were followed in order to identify additional study patients of the ongoing GMALL studies. Methods: For this report patients with central diagnosis of ALL in months 1-3/1999 are compared to 1-3/2000. In both periods a study for elderly ALL patients >65yrs (ALL-ELD) and a study for mature B-ALL (B-NHL90) were open for accrual. For patients with B-precursor and T-ALL < 65 yrs in 1-3/1999 the GMALL Study 05/93 and in 1-3/2000 GMALL 06/99 were active. Results: In 1-3/1999 central diagnosis of de novo ALL was made in 73 patients. This included 60 patients with ALL <65 yrs (82%), 7 with ALL >65 yrs (10%) and 6 with mature B-ALL (8%). Out of 60 patients with ALL <65 yrs 40 (67%) were reported to GMALL 05/93 (37 incl./3 excl.). 2 of 7 ALL patients > 65 yrs and 5 of 6 patients with mature B-ALL were reported to ALL-ELD and B-NHL90 respectively. Overall 26 newly diagnosed ALL pts (36%) were not reported to the GMALL study center. In 1-3/2000 87 de novo ALL patients were diagnosed. This included 75 patients with ALL <65 yrs (86%), 10 with ALL >65 yrs (11%) and 2 with mature B-ALL (2%). 92% (N=72) of 75 patients with ALL < 65 yrs were reported to GMALL 06/99 (69 incl./3 excl.). 5 of 10 ALL patients >65 yrs (4 incl./1 excl.) and 2 of 2 patients with mature B-ALL were reported to ALL-ELD and B-NHL90. Overall 8 patients (9%) were not reported. Conclusions: The accrual of newly diagnosed ALL patients to GMALL trials increased from 64% to 91% in two comparable time periods. This may be related to several factors: (1) Monitoring of all centrally diagnosed patients based on data exchange with the central immunologic laboratory since 1/2000 (2) Improved knowledge on aims and legal requirements of clinical trials within participating centers. Furthermore according to our experience specialised staff responsible for documentation of study patients is now available in more participating centers. Results may be limited by the fact that additional patients were probably diagnosed in specialised laboratories or at local institutions. The GMALL monitoring programme will be continued and extended to other specialised laboratories. In addition monthly updates on course of treatment and toxicities in individual patients are performed by telephone. Overall this approach may contribute to improved care for ALL patients and to realistic results in a nearly population based study. Supported by the BMBF, Competence Network Leukemias MRD in ALL, Grant No 01GI 9971

861

Counseling of participating hospitals in the German Multicenter Study Group for Adult Acute Lymphoblastic Leukemia (GMALL)

N. Gökbuget, R. Reutzel, S. Hug, D. Hoelzer (Frankfurt, D)

The GMALL conducts clinical trials in adult ALL and B-ALL/Burkitt's lymphoma with more than 100 participating centers and offers the opportunity for counseling by telephone to the study hospitals. Specific questions are recorded in a data base. General questions related to active protocols, entry criteria etc. are excluded from this report. Between 2/2000 and 4/2001 284 questions were recorded, 50% from university hospitals and 50% from regional hospitals. 187 (66%) questions were related to ALL and 97 (34%) to lymphoma. ALL: 75% (N=141) of the questions were related to details of treatment protocols in de novo ALL. The most frequent issues were protocol steering in 32% (N=45) with BMT (N=12) or irradiation (N=9) procedures and treatment of CNS involvement (N=6) being the most frequent issues. 47% (N=66) were related to toxicities, most frequently liver toxicity (N=14) and prolonged cytopenias (N=18). 11% (N=15) referred to treatment failure and 11% (N=15) to complications at timepoint of diagnosis. In 14% (N=27) of all cases recommendations for relapse (REL) treatment were requested (Specific REL location: 7; REL on treatment: 6; REL after BMT: 6). Lymphoma: 62 questions (64%) were related to B-ALL/NHL and 35 to T-lymphoblastic lymphoma (T-LBL). In B-ALL/NHL a treatment suggestion was requested in 15 cases (24%). 50% (N=31) referred to the B-NHL90 protocol and were mostly related to protocol steering (N=13) and toxicities (N=14). 24% were related to options for REL treatment (REL after CHOP/CHOEP: 6; REL during B-NHL90: 5; secondary B-ALL/NHL: 3). In T-LBL 80% (N=28) of questions were related to treatment according to the GMALL 06/99 protocol. The most frequent issues were irradiation procedures (N=8), pts with residual TU (N=5) and toxicities (N=10). Counseling included the explanation of protocol experience in 45 cases whereas detailed recommendations were discussed in 237 cases. In the latter case options for experimental treatment were mentioned in 34 cases (antibody therapy: 14; STI571:9; „mini" transplantation:7 and CLAEG, a cladribine/cytarabine based regimen for T-ALL/LBL: 16). ALL, B-ALL and T-LBL are rare diseases and treatment protocols are complicated. Responsible physicians often change in participating hospitals. Although treatment protocols comprise detailed description of procedures e.g. in case of toxicities this is apparently not sufficient to cover individual problems in the course of disease. Therefore a need for counseling appears to be present in university as well as in regional hospitals. It provides additional information for final decision on treatment procedures in participating hospitals For the GMALL counseling proved to be helpful to identify frequent problems with protocol steering, toxicities and requirements for additional treatment studies. Based on this experience the GMALL plans to extend the counseling by definition of additional standard procedures, written recommendations and initiation of new studies e.g. for first line treatment of T-LBL or phase II studies for new treatment elements such as CLAEG. Furthermore the results of recommendations will be recorded for quality control.

862

Experience in establishing a cleanroom laboratory for preparation of autologous and allogeneic stem cells

M. Ritter, J. Beyer, R. Mutters, A. Neubauer (Marburg, D)

The preparation of stem cells and other cells for either autologous or allogeneic transplantation is a standard procedure necessary in each transplant unit. Since the law for production of drugs in Germany demands highest cleanroom conditions for the handling with stemcell products, we have built a laboratory with cleanroom conditions A under background conditions B. A company was engaged to construct the laboratory including air conditioning technology and clean room walls. The „CliniMACS" cell selection system was placed under a specially developed laminar-airflow workbench. The centrifuge which might cause cleanroom contamination in operation was housed in a big board with air backflow-system inside. The qualification procedures of the laboratory and the „drug production" was divided in 1. installation qualification IQ; 2. operational qualification OQ; 3. process qualification PQ. Methodes used in proving the aseptic environment: electronic particle counter „Metone, Typ 2408", microbial contamination was tested with air sampler „Biotest RCS plus", fingerprints of gloves and prints of benchs on „Rodac"-plates, „Sabouraud" sedimentation plates. Quality of the product was controlled by counting MNCs, CD34-positive cell number, colony forming units, vitality and microbiological testing. The results of our „at rest" testing were all within the limits given by European Union. The process qualification PQ had to be done for each different blood preparation procedure. This meant not only single steps of preparation had to be qualified, but the whole process of production. Our „in operation" testing of the steril product, the particle count, the air sampler count and sedimentation plates showed only result within the limits given by European Union. However single „rodac" plates showed slightly elevated numbers of bacterial colonies in fingerprints of cleanroom area B. The mean of fingerprints however was within the limits of European Union. Therefore we conclude, that the concept of our laboratory is valid to prepare stem cells under international standard conditions.

863

Leukemia information system for classification and documentation of microscopic description, based on an object oriented data base – multimedia electronic patient record

I. Schestag, J. Dengel, F.G. Hagmann, H. Link in cooperation with the AML96 Study Group of the SHG; (Darmstadt, Kaiserslautern, D)

Leukemia Information System for classification and documentation of microscopic description, based on an object oriented data base – multimedia electronic patient record. The mainstay of leukemia diagnosis is based on light microscopy. Description of smears varies between laboratories as well as physicians. A clear standardization and quantification of descriptive terms is not available. Therefore, comparison of results is difficult, electronic and quantitative documentation of these descriptive terms for an electronic patient record is not in use. The aim of this deveopment was, 1) to generate an object oriented data base, which uses the descriptive terms of leukemia diagnosis and includes the microscopic knowlege of experts. 2) to standardize and quantify descriptive terms of leukemia diagnosis. The descriptive and quantitative terms which are necessary for mircoscopic diagnosis of leukemia were used for the database. Electronic images of leukemia cells were included as well. A prototype of an information sytem which supports leukemia diagnosis was extended to a service platform which supports the following services: Definition of material and its preparation, registration of the associated laboratory data, diagnostic algorithm for descriptive data for standard and cytochemical stains, quantitative or semiquantitative description of cell properties, definition of cells etc., final diagnosis, inquiry for information according to different criteria, storage of associated digitalized microscopic pictures; definition of significant parameters concerning the digitalized pictures, based on a parameter directory which was defined in order to characterize digitalized cell material, maintenance of the parameter directory as a superuser functionality, statistical information, clinical data. Upon validation of the system with the data of the AML96 SHG study group (Dresden), the database will be extended for documetation of all bone marrow diseases. The data base can be extended for immunocytochemistry, cytogenetics, molecular genetics and research projects.

In addition to the client/server-application which provides the services mentioned above, a web application supports inquiry for information via intranet/internet, based on the central object oriented data base server. The web application fits for interchange of digitalized pictures inside a clinical intranet as well as it can be considered as the basis for a reference database within an alliance of hospitals: findings can be compared and material is available to support teaching of students and training of physicians. The integration into existing hospital information systems is feasible. The standardized interfaces HL7 and DICOM are supported.

864

Chemotherapy-planning-system for medical documentation in hematology and oncology (OnkoDAT): Release of a new version

K. Huber, P. Köberl, A.T. Mackensen, S.W. Krause, R. Andreesen, C.M. Reng (Regensburg, D)

Correct application of chemotherapy is critical to patients with malignancies. Mistakes in planning of the regimen, the dosing of the generics and the duration of the application can be fatal. In order to obtain standardisation of these procedures, the computer program OnkoDAT was set up and subsequently refined. Based on user-defined therapy regimes and patient-specific parameters like body weight or body surface OnkoDAT is able to calculate individual therapies automatically. The software has functionality to generate a summary sheet, a detailed application schedule, a prescription for the hospital pharmacy and an accompaning sheet with information about the chemotherapy including important precautions and side effects. The program allows modification of dosage, scheduling of therapy-interruptions or stops of therapies according to clinical requirements. The definition of new therapy regimens and scheduling of supportive therapy can be easily done by physicians without the need of any special programming skills. Exceeding of daily, maximum and cumulative dosage of each individual drug is reported by the built in dosage control. The system runs on computers with 32-bit MS-Windows-systems either as single user application with an independent administration of patients and diagnoses or integrated into the hospital information system (his) SAP IS-H and IS-H*Med as a network solution with all relevant interfaces to this his including all DRG-relevant administrative documentation. In summary, a high amount of standardisation is achieved for end-users with low demand on computer skills. For development and support there is a cooperation with the MedicDat-Project. This project is sponsored by a grant of the German Ministery of Science and Education. Further information is available via Internet at http://www.onkodat.de.

865

Outpatient treatment in patients with metastatic colorectal cancer, breast cancer and non small cell lung cancer. Assessment of treatment quality in a community based oncology group practice

R. Weide, J. Heymanns, K. Balzer, J. Grass, S. Haase, H. Köppler (Koblenz, D)

Objective: Evaluation of feasibility, effectivity, toxicity and outcome in patients with metastatic colorectal cancer (CRC), metastatic breast cancer (BC) and advanced or metastatic non small cell lung cancer (NSCLC), who were treated completely on an outpatient basis. Methods: Retrospective analysis of 260 unselected, consecutive patients (73 CRC, 90 BC, 97 NSCLC) treated between 6/95-6/2000 in a community based oncology group practice. Results: CRC: Median age was 60 years (range 29-82). In 60/73 pts 1st line treatment was 5-FU/FA (Mayo-Clinic) and high dose 5-FU/FA in 13/73. 47 pts received 2nd line therapy with high dose 5-FU/FA, oxaliplatin or irinotecan. 20/73 pts received 3rd line treatment (oxaliplatin or irinotecan). Median survival is 16 months (range 3-44) from time of start of palliative chemotherapy. The 1, 2, 3 and 5 year survival is 57%, 28%, 14% and 14% respectively. Only 3 pts were hospitalised due to treatment related causes. BC: Median age was 49.5 years (30-85) with a performance score of 0-2. A total of 855 chemotherapy applications were performed during the study period. A reversible grade 3 or 4 hematotoxicity was seen after 27 applications (3.2%). Therapy associated hospitalisation occurred in 1 patient. The median survival of the whole cohort was 29 months (2-256). The 1, 2 and 3 year survival is 83%, 57% and 34% resp. NSCLC: The median age of the pts was 62 years (37-83) and their performance status ranged from 1-3. Clinical stage distribution was as follows: 5 stage I (functional inoperability), 3 stage II, 41 stage III (8 IIIA, 33IIIB), 48 stage IV. 82 pts received cytotoxic treatment (7 radiotherapy only, 44 radiotherapy and chemotherapy, 31 chemotherapy only). 15 pts received best supportive care only. A total of 1008 chemotherapy applications were performed during the study period. A reversible grade 3 or 4 hematotoxicity was seen after 48 applications (4.8%). Therapy associated hospitalisation was seen in 4 pts. The median survival of the whole cohort form diagnosis is 21 months (range 1-118), survival of stage IIIB is 22 months (range 5-66) and stage IV is 11 month (range 1-118). The 1, 2, 3 and 5 year survival is 60.7%, 43.3%, 36.8% and 17.9% respectively. Conclusion: Treatment of patients with metastatic solid tumors can be performed with minimal toxicity and a high patient acceptance in the outpatient setting. The median survival in this unselected cohort is high and hospitalisation could be avoided in the majority of patients.

866

A new approach to evidence-based oncology: The Cochrane Haematological Malignancies Group (CHMG)

A. Engert, T. Kober (Köln, D)

The Cochrane Haematological Malignancies Group (CHMG) was registered in October 2000
to bridge a gap in the knowledge transfer between scientists, clinicians and patients involved with the management of Hodgkin's lymphoma and other haematological disorders, and to promote the concept of evidence-based oncology. This concept embraces the conscientious, explicit and judicious use of current evidence in making decisions about the care of individual cancer patients. The Cochrane Collaboration is an international organisation aiming to assist to make well informed decisions in the delivery of health care by producing high quality, regularly updated systematic reviews of the effectiveness of interventions used for the prevention, treatment or control of health problems, including cancer. The CHMG is the only Cochrane review group worldwide with a clear focus on haematological malignancies. The editorial base of the CHMG is located in Cologne, Germany, and a subproject of the Kompetenznetz Maligne Lymphome, funded by the German Ministry of Education and Research (BMBF). Since late 2000, the newly established CHMG has published one systematic review and five review protocols (preliminary reviews) on the Cochrane Library. Twelve review titles have been registered with the Editorial Base and are in various stages of the editorial process. Half of the CHMG's output to date deals with treatment outcomes of Hodgkin's disease. The review process is transparent and all information on the Cochrane Library is accessible to interested clinicians, other health care workers and health consumers (i.e. those affected by haematological malignancies) either by subscription or free via the Internet. Each review and protocol is based on a extensive literature search of primarily randomised controlled trials (RCTs) for Hodgkin's-lymphoma, non-Hodgkin's- lymphoma acute lymphoblastic leukemia, acute myeloid leukemia, chronic myeloid leukemia, chronic lymphatic leukemia, multiple myeloma, myelodysplastic syndrome and aplastic anaemia. CHMG contrib-

utors are international and multidisciplinary. All review activities are co-ordinated by the Editorial Base (Cologne) and through a panel of nine international and national oncology experts from five countries. The review process follows stringent Cochrane Collaboration guidelines, and puts particular emphasis on methodological aspects. The CHMG has currently 39 active international and multidisciplinary contributors in Canada, Denmark, England, Germany, Netherlands, USA and Spain. The work of the CHMG is considered a valuable adjunct to other innovative approaches in the control of Hodgkin's disease and the advancement of evidence-based oncology at international level.

Postersitzung:
Signaltransduktion

867

SMIF is a crucial novel smad4-interacting co-activator in TGF-ß signalling

R.Y Bai, T. Ouyang, J. Sänger, S.A. Hahn, C. Peschel, J. Duyster (München, D)

The TGF-ß receptor family regulates diverse cellular responses including cell growth and differentiation. Hereby Smad proteins play a central role in transducing signals from receptors to the nucleus. The receptor-associated Smads (R-Smads) are phosphorylated by the receptor enabling them to form a complex with the common mediator Smad4/DPC4. The complex then translocates to the nucleus where it is able to specifically activate transcription. For the plethora of cellular responses induced by TGF-ß, additional levels of transcriptional regulation have been proposed. Recently we have described a novel, ubiquitously expressed specific Smad4-interacting transcriptional coactivator, SMIF (Smad4-interacting transcription factor). SMIF forms a complex with Smad4 but not with R-Smads or inhibitory Smad6 and translocates to the nucleus in a TGFß/BMP4-inducible and Smad4-dependent manner, which is demonstrated by Immunofluorescence and cellular fraction. We now demonstrate TGFß/BMP4-inducible complex formation of both overexpressed and endogenous SMIF with Smad4. Without significant homology to any known protein, SMIF possesses strong intrinsic TGFß-inducible transcriptional activity, which depends on Smad4 in mammalian cells. The interacting sequence of Smad4 on SMIF has been mapped to the C-terminus of the linker region of Smad4. A point mutant of Smad4 lacking the affinity to SMIF was generated, which failed to bind to SMIF in vitro and in vivo, and lost the transcriptional activity in Gal4 transcription assay and partially lost the ability in activating ARE-lux luciferase construct. Binding experiment of Smad4 and the mutant excluded the possibility that this functional reduction of the mutant is due to the loss of its affinity to p300/CBP, a known transcriptional coactivator involved in TGF-ß signaling pathway. Overexpression of the wildtype SMIF enhances, while the dominant negative SMIF construct suppresses the induction of a TGF-ß regulated target gene and is able to block TGFß-induced growth inhibition in mammalian cells. Together these data suggests a crucial role for SMIF in the TGF-ß signalling pathway.

868

Ablation of cancer cells with a telomerase promoter driven pseudomonas exotoxin construct

U. Martens, F. Osterroth, F. Rosenthal, W. Lange (Freiburg, D)

Cancer-specific regulatory elements provide an excellent opportunity to express cytotoxic genes for a potential antineoplastic approach. hTERT, coding for the catalytic subunit of the telomerase holoenzyme, has been shown to be a good candidate gene, because it is expressed in about 90 % of human cancers, whereas in normal somatic cells, hTERT expression is down-regulated. Here, we developed a novel recombinant DNA molecule, pAB, in which the hTERT-promoter was cloned upstream of Pseudomonas exotoxin (PE). As a control, pAB was frameshifted to generate pABfs in which the PE-gene is non-functional. Both constructs were transfected into several human cancer cell lines, murine embryonic and primary human fibroblasts. To monitor the effects of pAB and pABfs on survival of transfected cells quantitatively, a co-transfection assay was performed using pEGFP-N1, which produces the enhanced green fluorescent protein (EGFP) as reporter gene. We can demonstrate that the cytotoxic construct pAB is capable of inhibiting preferentially growth of cancer cell lines in vitro in the range of 80-94%.

869

Bcl10 is a crucial mediator of antigen receptor incuded NF-kappaB activation

J. Ruland, G. S. Duncan, A. Elia, L. Nguyen, D. G. Millar, P. S. Ohashi, T. W. Mak (Toronto, CDN)

The most common type of lymphoma arising in extranodal sites are B cell lymphomas of mucosa-associated lymphoid tissue (MALT lymphomas). MALT lymphomas typically develop in the context of prolonged reactive lymphoid proliferation at sites of chronic infections or in autoimmune disorders. Bcl10, a caspase recruitment domain (CARD)- containing protein identified from the recurrent t(1;14)(p22;q32) breakpoint in MALT lymphomas, has been shown to induce apoptosis and activate NF-kappaB in vitro. However, it's physiological role and the molecular mechanism through which Bcl10 translocation contributes to lymphomagenesis has remained elusive. We generated bcl10 deficient mice by gene targeting in embryonic stem (ES) cells. About one-third of bcl10 -/- embryos died during embryogenesis as a consequence of a neural tube closure defect leading to exencephaly. But surprisingly bcl10 -/- ES cells, primary embryonic fibroblasts, thymocytes and mature lymphocytes retained normal susceptibility to various death receptor or mitochondrial mediated apoptotic stimuli. Surviving bcl10 -/- mice had a severe immune deficiency characterized by reduced levels of serum immunoglobulins and defective B and T cell responses to virus infections in vivo. We show that bcl10 is essential for antigen receptor or PMA mediated activation of peripheral lymphocytes but dispensable for LPS induced proliferation of B cells. TCR-induced early tyrosine phosphorylation as well as Ca2+- and MAP kinase signaling after BCR- or TCR-ligation were normal in bcl10 -/- lymphocytes. In contrast antigen receptor or PMA initiated NF-kappaB activation was absent in bcl10 -/- B and T cells. In addition, neither IKK activation nor IkappaBalpha phosphorylation or degradation occurred in PMA stimulated bcl10-/- T cells. However, activation of NF-kappaB by TNF-alpha, IL-1 or LPS was unaffected in mutant mice. Together, these results indicate that bcl10 is not required to execute apoptosis but functions as a positive regulator of lymphocyte proliferation. The data provide a mechanistic insight into bcl10 function upstream of the IKK complex to specifically connect the antigen receptor signaling machinery in B and T cells to NF-kappaB activation. Thus, we identified a novel pathway with therapeutic potential in lymphoid malignancies, autoimmune diseases or transplant rejections and offer a molecular explanation for the upregulation of Bcl10 in MALT tumors.

870

The proteasome-specific inhibitor lactacystin induces cleavage of caspase-3 and Bax in resting and activated B-CLL cells

C. Bogner, F. Schneller, S. Hipp, C. Peschel, T. Decker (München, D)

In B-cell chronic lymphocytic leukemia (B-CLL), accumulation of mature CD5 B cells is caused by defective apoptosis and cell cycle deregulation. Although peripheral blood B-CLL cells are arrested in G0/early G1phase of the cell cycle, a proliferating pool of cells in proliferation centers in lymph nodes and the bone marrow might well contribute to the expansion of the malignant clone. We have previously described an in vitro model of this proliferating pool of cells using B-CLL cells stimulated with immunostimulatory oligonucleotides (CpG-ODN) and Interleukin-2. Lactacystin is a microbial metabolite that inhibits protease activity only in the proteasome and is a potent apoptosis inductor in resting peripheral B-CLL cells. In our study we show that lactacystin inhibits cell cycle progression in activated B-CLL cells and inhibits degradation of p27. Upregulation of cyclin D2 and D3 in activated B-CLL cells was also inhibited but cdk4, cdk6 and cyclin E remained unchanged. Apoptosis induction was even more effective in activated as compared to resting B-CLL cells and accompanied by a complete loss of the mitochondrial membrane potential. Caspase 3 cleavage was observed but a broad spectrum caspase inhibitor (zVAD-fmk) only partially inhibited apoptosis. Interestingly, a 18kd cleavage product of bax occurred in lactacystin treated cells that has been described to have high proapoptotic potential. Thus proteasome inhibition is highly effectiv in activated B-CLL cells and induces apoptosis using a caspase dependent and independent pathway.

871

Activation of protein kinase A (PKA) as an antileukemic therapy for acute or chronic leukemias

E.M. Weissinger, K. Öttrich, K. Aitken, W. Kolch, H. Mischak, B. Hertenstein, H.J. Kolb, A. Ganser (Hannover, München, D; Glasgow, UK)

Chronic myelogenous leukemia (CML) is a disease of the pluripotent stem cell, hallmarked by the philadelphia chromosome (Ph1; t:9/22), which leads the expression of an atypical fusion protein, p210bcr/abl, which in turn activats the Ras-signalling cascade. In about 50% of acute myeloid leukemia the Ras-pathway is also activated either following the same mechanism or by different stimuli. Here we have studied the effect inhibiting the Ras-signalling cascade by activation of PKA in bone marrow or peripheral blood cells obtained from 8 patients with CML and 7 with acute leukemias. From the CML patients 3 were in chronic phase after cytoreductive treatment, 2 at diagnosis and 3 in accelerated phase of CML. When bone marrow obtained in chronic phase was treated for 24h with 100 μM 8-Cl-cAMP and cultured for 4 weeks in long term culture, no Philadelphia Chromosome (Ph1) positive colonies could be detected (n=3; 0/15; 0/15 and 0/37). Only Ph1 negative cells gave rise to colonies and no toxic side effects on the normal cells were observed. When bone marrow was obtained from patients in more advanced stages of the disease, fewer colonies grew and some were Ph1 positive (Ph1 positve colonies per total colonies 7/15; 5/6; 1/3; 2/4; 5/8). Treatment of bone marrow from healthy donors only showed a slight, reversible growth inhibition, but no apoptosis was observed. Peripheral blood cells from 7 patients with therapy-refractory AML were also treated for 24h ex vivo with 100μM 8-Cl-cAMP. A massive cytoreduction and apoptosis was seen in 3 of the 7 samples, as compared to the untreated control cells. Cells of 2 patients did not grow ex vivo and 2 samples did only show slower growth patterns as compared to the controls. In the last case only few apoptotic events were seen. We have initiated a compassionate use trial for 8-Cl-cAMP in patients suffering from therapy-refractory AML.

872

Pharmacological activation of G protein-coupled receptor signalling in hematopoietic progenitor cells upregulates CXCR4 function: Potential improvement of bone marrow homing by the immunomodulator FTY720

T. Kimura, V. Brinkmann, L. Kanz, R. Möhle (Tübingen, D; Basel, CH)

Chemokine receptors, particularly CXCR4, have been shown to play key roles in the homing of hematopoietic progenitor cells (HPC) to the bone marrow. Enhancement of CXCR4 expression or function has, therefore, been proposed as a measure to improve the homing capacity of HPC. However, ex vivo culturing of HPC to upregulate CXCR4 expression is laborious and may be associated with loss of stem cell function. Here we describe activation of CXCR4 function by a short-term incubation (2 hours) with the immunomodulator FTY720, which modulates chemokine receptor signalling most likely by increasing PI3-kinase activity and cytoskeleton assembly. FTY720 at 10-100 nM increased SDF-1-induced in vitro transendothelial migration of mobilized peripheral blood CD45low/CD34+ cells nearly 3-fold, including also more primitive CD34+/CD38low cells, and particularly CD34+/CXCR4+ cells. FTY720 (100 nM) did not affect CXCR4 expression, but upregulated both rapid intracellular calcium fluxes and actin polymerisation in response to SDF-1 (100 ng/ml), indicating activation of CXCR4 signalling pathways. Interestingly, FTY720 at 100 nM was not toxic in hematopoietic long-term cultures and even increased in vitro growth of primitive HPC (week 5 CAFC) 2.4-fold. Only at high concentrations (>1 μM), colony formation was significantly suppressed in a dose-dependent manner. We conclude that activation of CXCR4 signalling by FTY720 at moderate doses increases CXCR4 function in HPC without affecting the proliferative capacity of primitive HPC. Since FTY720 is well tolerated in humans as an immunosuppressant, it may be a useful agent to improve bone marrow homing also in the clinical setting of stem cell transplantation.

873

Activated Notch1 signaling promotes tumor cell proliferation and survival in Hodgkin and anaplastic large cell lymphoma

F. Jundt, I. Anagnostopoulos, R. Förster, S. Mathas, H. Stein, B. Dörken (Berlin, D)

Notch signaling controls cell fate decisions of hematopoietic progenitors by inhibiting certain steps of differentiation and inducing either self-renewal or differentiation toward lymphoid or myeloid lineages. In addition, truncated Notch1 alleles could be associated with 10% of all cases of human T lymphoblastic leukemia and, when introduced into mouse bone marrow stem cells, cause T-cell neoplasms. However, functional links between abundant expression of intact Notch1 and oncogenesis are still lacking. Here we show that Notch1 is strongly overexpressed in B- and T-cell derived tumor cells of Hodgkin and anaplastic large cell lymphoma. We demonstrate a novel mechanism for the oncogenic capacity of Notch1 by showing that interaction of intact Notch1 on tumor cells with its ligand Jagged1 on bystander cells induces both proliferation and inhibition of apoptosis in vitro. Given the co-localization of Jagged1-expressing cells and Notch1-positive tumor cells in malignant lymphoid tissues it seems likely that this interaction also contributes to lymphomagenesis in vivo. Thus, our data suggest that activated Notch1 signaling plays a fundamental role in the pathobiology of Hodgkin and anaplastic large cell lymphoma and might be a potential new therapeutic target for the treatment.

874

Cloning and characterization of IAN-5: a mitochondrial GTP-binding protein that constitutes the first human member of the IAN-family conserved in plants, mice and humans

T. Zenz, L. Daheron, B. Calabretta (Ulm, D; Poitiers, F; Philadelphia, USA)

We recently identified a murine mitochondrial GTP-binding protein – IAN-4 – (Immune Associated Nucleotide-binding protein) that is specifically induced by BCR/ABL. Here we present the cloning and characterization of IAN-5, the first human member of the IAN family. IAN-5 has a predicted molecular weight of 34.8 kDa. It shows significant homology with a number of GTP-binding proteins. IAN-5 contains the typical G-motifs of GTP-binding proteins (G-1 to G-4) in the N-terminus, N-glycosylation sites, tyrosine phosphorylation sites and a C-terminal transmembrane domain. IAN-5 was cloned by RT-PCR from the leukemic cell line U937 and standard subcloning techniques. After expressing IAN-5 in different cell lines, IAN-5 was shown to be localized strictly in the mitochondria as shown by subcellular fractionation and immunofluerescent microscopy. Expression profiling of IAN-5 mRNA by dot blot analysis showed expression of IAN-5 mRNA in all human tissues outside the CNS. IAN-5 is highly expressed in lymph node and spleen but was not detectable in lymphoblastic or other cancer cell lines. The IAN-5 gene is localized on chromosome 7q36.1 in a critical region that is commonly deleted in AML del 7q35/36. Interestingly, two other human hypothetical proteins that show significant homology to IAN-4 and IAN-5 are also located on chromosome 7q35/36. Overexpression of IAN-5 in the leukemic cell line HL-60 did not alter resistance to apoptosis or the production of reactive oxygen species. The physiological function of IAN-5 remains to be determined. IAN-5 extends the recently discovered IAN-family of GTP-binding proteins conserved in arabidopsis thaliana, mice and humans.

875

The granulocytic inducer C/EBPalpha inactivates the myeloid master regulator PU.1: possible role in lineage commitment decisions

V. Reddy, A. Iwama, M. Schulz, A. Elsässer, R. Vangala, T. Khare, S. Singh, D. Tenen, W. Hiddemann, G. Behre (München, D; Tsukuba, J; Boston, USA)

Several transcription factors have been implicated to play a role in myelopoiesis. PU.1, an ets-family transcription factor, is required for the development of both myeloid and lymphoid lineages while the transcription factor CCAAT/enhancer binding protein family member C/EBPa is essential for granulocytic development. We present here the first evidence that C/EBPa blocks the function of PU.1. PU.1 and C/EBPa interact physically and co-localize in myeloid cells. As a consequence of this interaction C/EBPa can inhibit the function of PU.1 to activate a minimal promoter containing only PU.1 DNA binding sites. We further demonstrate that the

DNA binding domain of C/EBPa interacts with the b3/b4 region in the DNA binding domain of PU.1, and as a result displaces the PU.1 co-activator c-Jun. Finally, C/EBPa blocks PU.1 induced dendritic cell development from CD34+ human cord blood cells. The functional blocking of PU.1 by C/EBPa could be the mechanism by which C/EBPa inhibits the cell fates specified by PU.1, and directs cell development to the granulocytic lineage.

876

Ras signaling enhances the transcriptional activity of C/EBPalpha by phosphorylation of serine 248

S. Singh, L. Bortolin, M. Christopeit, W. Hiddemann, A. Friedman, D. Tenen, G. Behre (München, D; Boston, Baltimore, USA)

The transcription factor C/EBPa regulates early steps of normal granulocyte differentiation since mice with a disruption of the C/EBPa gene do not express detectable levels of the G-CSF receptor and produce no neutrophils. We have recently shown that C/EBPa function is also impaired in acute myeloid leukemias. However, how the transcriptional activity of C/EBPa is regulated both in myelopoiesis and leukemogenesis, is not fully understood. The current study demonstrates that activated Ras enhances the ability of C/EBPa to transactivate the G-CSF receptor promoter and a minimal promoter containing only C/EBP DNA binding sites. Ras signaling activates C/EBPa via the transactivation domain, because it enhances the transactivation function of a fusion protein containing a Gal4 DNA binding domain and the C/EBPa transactivation domain, and does not change C/EBPa DNA binding. Ras acts on serine 248 of the C/EBPa transactivation domain, as it does not enhance the transactivation function of a C/EBPa serine 248 to alanine point mutant. Interestingly, serine 248 of C/EBPa is a PKC consensus site, and a PKC inhibitor blocks the activation of C/EBPa by Ras. Finally, Ras signaling phosphorylates C/EBPa on serine 248 in vivo. These data suggest a model where Ras signaling enhances the transcriptional activity of C/EBPa by phosphorylation of serine 248.

877

Activation of p21ras in MDS, AML, JMML and other haematopoietic malignancies measured by a quantitative and non-radioactive method

J. Scheele, D. Raepple, F. v. Lintig, T. Zenojtel, M. Duchniewicz, P. Shayan, C. Niemeyer, M. Lübbert, G. Boss (Freiburg, D; San Diego, USA)

There is evidence of mutational and non-mutational activation of the small G-binding protein p21ras in various tumors and haematopoetic malignancies. We applied a highly sensitive, non-radioactive method to determine absolute amounts of GTP and GDP bound to p21ras. The malignancies assayed were derived from patients' peripheral blood (PB), bone marrow (BM) or from spleen culture (SC). The samples comprised 10 determinations from 7 patients with MDS and AML (6 from BM and 4 from PB), 4 determinations from 4 patients with JMML (all SC), 2 determinations of one patient with CML (all BM), 3 determinations of one patient with CLL (all PB) and one determination of a patient with Burkitt lymphoma (BM).The cells were separated using a common density gradient. After purification of the cells we performed immunoprecipitations with a commercial p21ras-specific antibody. Both nucleotides, GTP and GDP were eluted from the immunoprecipitate and determined quantitatively in a luciferase/luciferin system. Therefore coupled reactions were used in order to convert both nucleotides to ATP as substrate for the luciferase/luciferin reaction. GTP was converted to ATP by NDP kinase, GDP was converted to GTP by pyruvate kinase and could be used in the latter reaction for determination of the total amount of G-nucleotides bound to p21ras (GTP+GDP). From both results we calculated the proportion of GTP-bound p21ras to the total amount of p21ras in percent. For standardisation of the GDP and GTP values we used the amount of DNA in the lysate used for immunoprecipitation as a measure for the amount of cells. For validation of the results we measured several human and murine cell lines either with a known mutation in p21ras or transfected with an mutational activated p21ras-construct. HL-60 cells (AML M2, N-Ras/61 mutation) used for control showed an activation of 28.6%, Kasumi-1 cells (AML M2, K-Ras/12 mutation) an activation of 21.9%. Three different clones of murine NIH3T3-cells transfected with mutational activated p21ras (V12 and HA-tagged V12) showed an average activation level of 22.0%. Untransfected wildtype NIH3T3 cells showed an activation of 7%. In MDS and AML (n=10) we found an average activation level of 3.1% (0.1%-18.4%). One sample showed an activation of 18.4%. The JMML samples showed an average level of activation of 17.0% (1.1%-38.7%). 50% of the values were higher than 15%. In CML we found an average level of activation of 1.6% (1.1% and 2.1%), in CLL 3.2% (1.2%-5.6%). The Burkitt lymphoma showed an activation of 16.3%. This method is a sensitive tool to measure p21ras activation. Our results are consistent with the levels of activation described in literature. Since both mutational and non-mutational activation of p21ras occurs, our goal in future is to sequence n-ras and k-ras exons in the samples measured above and to have a closer look at the pathways signalling via p21ras especially in the non-mutational activated samples.

878

Constitutive activation of STAT3 by Hodgkin and Reed-Sternberg-cells in Hodgkin's lymphoma

T. Ahmadi, C. Wickenhauser, U. Holtick, D. Kube, M. Vockerodt, H. Tesch, J. Wolf (Köln, Tübingen, Villingen-Schwenningen, D)

We previously demonstrated constitutive STAT3 activation in HD-cell lines (Blood 2001). In contrast to multiple myeloma where constitutive STAT3 activation has been reported previously our results indicated a IL-6 and IL6 type cytokine independent mechanism of STAT3 activation. (Blood 2001) As STAT activation is marked by tyrosine phosphorylation and nuclear translokation we analysed 50 primary tissues of Hodgkin's and non-Hodgkin's lymphoma for nuclear phosphorylation of STAT3 proteins. Immunohistochemical staining revealed STAT3 activation in all 20 primary cases of classical Hodgkin's disease and 5 cases of multiple myeloma but in no other case of non-Hodgkin's lymphoma including CLL, follicular lymphoma, mantle cell lymphoma, diffuse large cell and anaplastic large cell lymphoma. Our findings now provide evidence that STAT3 is constitutively activated in the malignant cells of Hodgkin's disease. Moreover constitutive STAT3 activation seems to be highly specific for classical Hodgkin's disease when compared to a broad spectrum of other B-cell malignancies. These results contribute to our hypothesis that constitutive STAT3 activation is involved in the pathogenesis of Hodgkin's disease.

879

The BCR-ABL oncogene utilizes the GTP-exchange factor Vav for efficient transformation of bone-marrow cells in vitro, but Vav seems not to be required for leukemogenesis in mice

C. Miething, C. Mugler, F. Bassermann, T. Jahn, J. Hoepfl, R. Grundler, Ch. Peschel, J. Duyster (München, D)

The protooncogene p95 Vav is expressed predominantly in hematopoetic cells. It has been shown that tyrosine-phosphorylated Vav catalyses GDP/GTP exchange on Rac-1, a Rho-member of small GTP-binding proteins implicated in cell proliferation and cytoskeletal organization. The oncogenic tyrosine kinase BCR-ABL (BA) of chronic myelogenous leukemia (CML) has been shown to activate Rac-1, which is important for BA induced leukemogenesis. We have shown recently, that BA is able to bind and phosphorylate Vav, leading to Rac-1 activation. In order to examine the pathophysiologic significance of this interaction in vivo, we have now retrovirally infected bone-marrow (BM) cells from Vav knock-out mice and performed liquid culture bone marrow assays. We have also recently obtained data on disease development in a mouse model of CML utilizing a retroviral transduction/transplantation assay. Bone marrow from 5-FU treated C57BL6 Vav-/- mice was infected with p210 BA in a MSCV-based retroviral bicistronic vector coexpressing the enhanced green fluorescent protein (EGFP). Two days after transfection, the BM cells were analyzed by FACS for EGFP expression, showing similar transduction efficiencies for the Vav-/- BM and control BM (between 1-20% in different experiments). The BM was plated at a density of 1*105 cells per well in 12-well plates without growth factors. After 9 days, the BM cells were counted and analysed by FACS. The Vav+/+ BM cells grew to up to five-fold higher cell numbers than the Vav-/- BM in three independent experiments, whereas control-infected cell number did not increase, implicating that BA needs Vav to efficiently transform primary BM cells. FACS-analysis of the transformed cells demonstrated an outgrowth of EGFP-positive cells staining positively for the murine myeloid marker Mac-1 in the Vav+/+ as well as in the Vav-/- BM. In ongoing experiments, though, there was no difference in disease latency between control (n=3, 38±11 days) and Vav-/- mice (n=3, 40±9 days) after transplantation of Vav deficient and control BM infected with BA into lethally irradiated Balb/C mice. We are currently analyzing the mice for differences in disease phenotype. In conclusion the results suggest that BA utilizes Vav as a guanine nucleotide exchange factor (GEF) in order to fully transform murine BM cells in vitro. If Vav also plays a role in leukemogenesis in vivo will be determined in ongoing experiments. These findings also highlight the importance of in vivo models for the characterization of the contribution of single molecules to the development of neoplasia.

880

Galpha13 induces interleukin-3-dependent chemokinesis and change of cell-shape in myeloid progenitor cells

J. Schwäble, M. Lopes, W. Grüning, A. Rudat, C. Müller-Tidow, M. Mizuki, B. Sargin, B. Surmann, W. E. Berdel, H. Serve (Münster, D; Osaka, J)

Chemotaxis and proliferation of myeloid progenitors is a highly integrated process regulated by multiple extracellular stimuli. The molecular mechanism of this interplay is largely unknown, but it involves cytokine- and heterotrimeric G-protein coupled (chemokine) receptors. Previous reports indicated that heterotrimeric G proteins of the Galpha12 family (G12 and G13) are important mediators of cell growth, apoptosis and organization of the cytoskeleton in non-hematopoietic cells. Therefore, we were interested in the expression and function of these proteins in hematopoiesis. As judged by real time RT-PCR and Western blot analyses, human CD34+ cells showed high G13 mRNA and protein levels. To determine the function of G13 in this cellular context, we overexpressed the wildtype form (WT) and a constitutively active mutant Q226L (QL) in the interleukin-3 (IL-3)-dependent murine myeloid progenitor cell line 32Dcl3. The expression of these proteins had no effect on proliferation or clonal growth. In contrast, cells expressing either form of G13 showed a dramatic change in cell shape and actin reorganization, with the formation of cellular processes and membrane ruffles. This change in morphology was reversible upon IL-3 starvation and reappeared within less than 30 minutes after readdition of IL-3. In the presence of IL-3, cells transfected with G13QL, but not with G13WT maintained this phenotype during serum-starvation. This argues for a factor present in the serum stimulating G13-dependent receptors, a signal constitutively provided by the QL-mutant. To investigate the effect of G13 on chemotaxis and chemokinesis, we measured undirected and directed migration of the transfectants towards IL-3 and SDF-1alpha, a chemoattractant that mediates its chemotactic effects by binding to the heptahelical receptor CXCR-4. In the presence of IL-3, we found a greatly enhanced mobility of G13-expressing cells. Migrating across a membrane with 5µm pores from an upper to a lower chamber, 4±2.5% of mock-transfected cells were found in the lower chamber compared to 33±2% and 35±7% of cells transfected with G13WT and G13QL, respectively. This was even more enhanced if cells migrated towards an SDF-1alpha gradient: 11,6±8,3% (control), 56±4,5 % (WT) and 67,1±13,8% (QL). Interestingly, the increased migration of the G13 transfectants was completely dependent on the presence of IL-3, regardless of the presence or absence of SDF1-alpha. In summary, this is the first report linking the chemotactic response of hematopoietic cells to the Galpha12 family of heterotrimeric G-proteins. Furthermore, the data provide an unexpected signaling mechanism of cytokine receptors by demonstrating dependency of IL-3 induced chemokinesis on the activity of heterotrimeric G-proteins.

881

Constitutive activation of STAT3 and STAT5 is induced by leukemic fusion proteins with protein tyrosine kinase activity and is sufficient for transformation of hematopoietic precursor cells

K. Spiekermann, M. Pau, R. Schwab, K. Schmieja, S. Franzrahe, W. Hiddemann (München, D)

Signal transducers and activators of transcription (STAT) factors are critical mediators in the signal transduction of cytokine receptors. In hematopoietic and epithelial cells activation of STAT 1 induces apoptosis and growth arrest whereas activation of STAT3 and STAT5 transduces growth-promoting signals. We and others have previously described a high expression and constitutive activation of STAT1, 3 and 5 in AML blasts. In this report we focussed on the mechanisms and also the biologic relevance of STAT-activation in AML. We report here that a constitutive STAT-activation can be detected in up to 95% of primary AML blasts. In vitro, neither induction of the leukemic fusion protein PML-RARa in U937 cells nor expression of transforming ras-mutants, but several leukemic PTK strongly induced activation of STAT3 and 5. Stable transfection of BA/F3 cells with TEL-JAK2, TEL-ABL, and BCR-ABL resulted in IL-3 independent growth and strong activation of STAT3 and STAT5 by TEL-JAK2 and TEL-ABL, but not by BCR-ABL. In addition, expression of constitutive active mutants of STAT3 and STAT5 alone were sufficient to transform BA/F3 cells. These results show that STAT3 and STAT5 are activated in the majority of primary AML blasts and are major targets of leukemic fusion proteins with PTK activity. However, the STAT-activation pattern by leukemic PTKs differed signficantly and might reflect their transforming potential in acute (TEL-JAK2 and TEL-ABL) and chronic leukemias ($p210^{BCR\text{-}ABL}$). The transforming capacity of constitutively activated STAT3 and STAT5 mutants strongly supports their fundamental role in the process of malignant transformation in hematopoietic cells.

882

Deregulation of chondroitin-4-O-sulfotransferase 1 (C4ST1) gene in a B-cell chronic lymphocytic leukemia with a t(12;14)(q23;q32)

H. Schmidt, V. Dyomin, N. Pallanisamy, G. Nanjangud, H. Pirc-Danoewinata, O. Haas, R. Chaganti (Graz, Wien, A; New York, USA)

The t(12;14)(q23;q32) breakpoints in a case of B-cell chronic lymphocytic leukemia (B-CLL) were mapped by fluorescence in situ hybridization (FISH) and Southern blot analysis and cloned using IGHG switch (S-gamma) probe. The translocation affected a productively rearranged IGH allele and the chondroitin-4-O-sulfotransferase (C4ST1) locus at 12q23, with a reciprocal break in intron I of the C4ST1 gene. C4ST1 belongs to the HNK1 family of Golgi-associated sulfotransferases, a group of glycosaminoglycan-modifying enzymes and is expressed mainly in the hematopoietic lineage. Northern Blot analysis of tumor RNA using C4ST1-specific probes showed expression of two C4ST1 forms of abnormal size. 5'- and 3'- Rapid Amplification of cDNA Ends (RACE) revealed IGH/C4ST1 as well as C4ST1/IGH fusion RNAs expressed from the der(14) and der(12) chromosomes. Both fusion species contained open reading frames making possible translation of two truncated forms of C4ST1 protein. IGH/C4ST1 fusion RNA predicted a protein lacking the transmembrane domain but retaining the catalytic domain, while the C4ST1/IGH fusion RNA predicted a protein comprising the C4ST1 transmembrane domain alone. Possible functional implications of the expression of the aberrant C4ST1 isoforms are discussed. Southern blot analysis of a panel of non-Hodgkin's lymphoma (NHL) cases with 12q23-24 cytogenetic abnormalities did not identify additional rearrangements using junction probes from the index case. The C4ST1 locus comprises approximately 300 kb of genomic DNA; we are currently establishing a FISH assay to screen a panel of CLL cases for additional rearrangements.

883

Effects of PPARg- und RAR-ligands on glioblastoma cells

E. Elstner, M. Wächter, C. Zang, M.G. Posch, J. Reichardt, J. Fritz, A. Klügel, H. Liu, A. von Deimling, K. Possinger, H.P. Koeffler (Berlin, D; Los Angeles, USA)

High-grade gliomas are the most common primary tumor of the brain. These highly aggressive tumors are characterized by a rapid proliferative rate, invasiveness, and angiogenesis. The absence of any curative treatment promotes the development of novel therapeutic approaches. The primary roles that apoptosis plays in neural development, oncogenesis, and therapeutic response suggests that alterations in apoptotic pathways may be one of the molecular determinants of tumorigenesis in the brain. Thus, the induction of apoptosis should be an important goal of glioblastoma therapy. Our previously data showed that the combinations of ligands for peroxisome proliferator-activated receptor gamma (PPARg) and retinoic acid receptor (RAR) induced apoptosis of breast cancer cells in vitro and in mice model (Elstner et al., PNAS, 1998). Therefore, the aim of this study was to test the effect of the combinations of these ligands on the proliferation, apoptosis and invasiveness of 7 glioblastoma cell lines in vitro. The PPARg-ligand, pioglitazone (PGZ), was combined with ligand for RAR, all-trans retinoic acid (ATRA). All studied glioblastoma cell lines and tissues from glioblastoma patients expressed high level of PPARg as measured by RT-PCR and/or Western blot. No mutation or polymorphism in the PPARg gene were observed in all studied glioblastoma cell lines by using SSCP. Dose-response studies with PGZ showed that PGZ alone is a weak inhibitor of clonal proliferation of glioblastoma cell lines. ATRA inhibited the clonal proliferation of 5/7 glioblastoma cell lines (ED50 10-9–10-6 M). The combination of PGZ (10-5 M) and ATRA (10-6 M) reduced the clonal proliferation of 5/7 cell lines up to 59% of control and induced apoptosis in 5/7 glioblastoma cell lines as measured by TUNEL. Experiments using caspase inhibitor clearly indicated that such an apoptosis induction was caspase-dependent. Furthermore, migration and invasion of these treated glioblastoma cells is decreased compared to untreated cells. Taken together, this combination may be a new approach in the therapy of patients with glioblastoma.

884

Population based documentation and follow-up of patients with newly diagnosed colorectal cancer in the region of Bonn/Rhein-Sieg

C. Lamberti, K. Di Blasi, D. Archut, J Mezger, I. Schmidt-Wolf, T. Sauerbruch (Bonn, Karlsruhe, D)

Introduction: In Germany, incomplete epidemiological data exist about colorectal cancer patients. Most of the data are based on results of the cancer registries of the Saarland and the former GDR.
Objective: To evaluate the treatment and 5-year-prognosis of non-selected and population based patients with colorectal cancer. Method: Between 6-11/1994 tumor stage, surgical and other primary treatment regimens were documented of all patients with newly diagnosed colorectal cancer in the region Bonn/Rhein-Sieg. To avoid underestimation, data were ascertained independently according to the reports of the medical practitioners and pathologists. Each patient was followed up by a standardised questionnaire sent to the practitioners and/or the patients themselves. Results: 98 % (n=349) of all newly diagnosed patients of the region could be documented. The median age at diagnosis was 69 years for male patients (n=161) and 72 years for female patients (n=188). The distribution according to the UICC-Criteria I-IV was 18.3%, 25.8 %, 22.9 % and 26.1 %, respectively; in 6.9% of the patients the initial tumor stage was unclear. 316 patients (90.5%) were treated by surgery, the treatment intention was curative in 72.8%. A stage adjusted adjuvant (radio)chemotherapy was conducted in only about 50 % of the respective patients. During follow-up 42.0% of patients with initial curative treatment died due to recurrent disease. The 5-year survival rate of all patients was 45 % (n=157). Conclusion: Most of the newly diagnosed patients with colorectal cancer were treated by surgery. However, stage adjusted adjuvant radio/chemotherapy was conducted only by half of respective patients. Although the 5-year survival was comparable to data from other countries, the proportion of patients with advanced colorectal cancer and poorer outcome was alarming. Thus, the improvement of prevention and diagnosis of premalignant or early malignant lesions is mandatory.

885

Tumor M2-PK: a promising tumor metabolic marker for the diagnosis of gastrointestinal cancer

P. Hardt, H. Schnell-Kretschmer, B. Ngoumou, J. Rupp, H. Klör (Giessen, D)

Tumor M2-PK is the dimeric isoform of the tetrameric glycolytic enzyme pyruvate kinase M2. It is strongly overrepresented in a metabolic state specific for a variety of tumor cells. The marker can be quantified in EDTA-plasma with an ELISA-test kit (ScheBo *Biotech AG, Giessen, Germany). The diagnostic value of Tumor M2-PK has been investigated in different tumors, and showed interesting results in cases of breast cancer, renal cancer, pancreatic cancer, lung cancer and some cases of gastric cancer. The following study was designed to evaluate the use of Tumor M2-PK as a tumor metabolic marker for the diagnosis of various gastrointestinal cancers. Methods: In this study we investigated the Tumor M2-PK concentrations in EDTA plasma of 68 patients with gastrointestinal cancer, 22 patients with inflammatory bowel disease (IBD) and 60 healthy controls. The group of GI-cancer patients comprised the following malignancies: gastric or oesophageal cancer: n=20; colorectal cancer: n=34, pancreatic cancer or hepatocellular carcinoma: n=14. Results: see Table 1

**	RR(CR+PR)	RR(PD)	RR(adj.)	DR(CR+PR)	DR(PD)	DR(adj.)
L25	33.3%	33.1%	15.3%	27.3	37.0	30.0
L05	15.0%	6.7%	8.2%	28.4	All censored	20.6
MA	10.4%	15.4%	19.7%	18.3	8.6	18.7
p-value	0.0297	0.0484	0.1461	0.02	-	0.4416

** RR(X) resp. (DR(X) = RR resp. median duration of objective response (months) to L05, L25 or MA of patients with response X to prior antiestrogen therapy. Adj. = patients with only adjuvant prior antiestrogen treatment.

Sensitivity of Tumor M2-PK was 70.6 % for all GI-Cancer, compared to 55,4 % for CA19-9 and 53,3 % for CEA. In pancreatic cancer CA19-9 showed a better sensitivity than Tumor M2-PK. In oesophageal/gastric cancer Tumor M2-PK was most sensitive and in colorectal cancer CEA and Tumor M2-PK showed the best results. Specificity of Tumor M2-PK was 90-96,7 %. In IBD, some patients showed elevation of Tumor M2-PK, but there was no correlation to CRP or inflammatory activity according to the clinical activity score. The mean Tumor M2-PK concentrations allowed a significant discrimination between colcorectal cancer vs. IBD (p=0.002) or controls (p<0.0001), oesophageal and gastric cancer vs. IBD (p=0.056) or controls (p<0.001) and pancreatic cancer vs. IBD (p<0.01) or controls (p<0.001). Conclusions: The data obtained so far indicate that Tumor M2-PK is a promising tumor metabolic marker for the diagnosis of gastrointestinal cancer and might become a routine diagnostic parameter.

886

Serum levels and tumor expression of HER-2 in colorectal cancer (CRC) - a prospective analysis

M. Benecke, A. Frühauf, H.J. Holzhausen, H.J. Schmoll, A. Grothey (Halle, D)

RATIONALE: The oncogene HER-2 is overexpressed in 20-25% of breast and ovarian cancers and is the target for novel therapeutic strategies with monoclonal antibodies or small molecules. Reports on the expression levels of HER-2 in CRC have been inconsistent, potentially due to differening methods of detection. Recently, soluble HER-2 has been characterized in the serum of patients with various carcinomas, however the correaltion between tumor expression of HER-2 and respectibe HER-2 serum levels is still unknown. Therefore, we prospectively analyzed sera and tumor samples of patients with metastatic CRC for the expression of HER-2 using standard ELISA and immunohistochemistry (IHC) techniques and tried to correlate it with the clinical course of the disease.
PATIENTS AND METHODS: Serum levels of HER-2 were determined in 88 pts (52 m, 36 f) with metastatic CRC and 20 healthy controls using a commercially available ELISA system (Dako). In 46 pts, tumor expression of HER-2 was analyzed by IHC (Hercept-Test®, Dako). Routine lab parameters and tumor markers CEA and CA19-9 were recorded in all patients. RESULTS: HER-2 serum levels were significantly higher in pts with CRC compared with the control group (median 2511 vs. 2184 U/ml; p=0.038). 21 of 88 pts (23.9%) showed elevated serum levels (>3000 U/ml), but in only 6 of 46 pts (13%) weak expression of HER-2 could be detected in tumor samples (2x 1+, 4x 2+, 0x 3+ Dako-score!). However, a positive correlation could be found between serum levels and tissue expression of HER-.2 (p=0.035). HER-2 serum levels were further significantly correlated with CEA (p=0.031), bilirubin (0.04) and with ongoing anti-tumor therapy (p=0.0083), but not with CA19-9, CRP, LDH, creatinine, hematological parameters (Hb, leukocytes, platelets) and age. Patients with liver metastases were more likely to demonstrate elevated serum levels of HER-2 (p=0.0033), but did not show HER-2 overexpression in the respective primary tumors. In preliminary analysis, HER-2 serum levels did not correlate with overall survival. CONCLUSIONS: Very few colorectal cancers overexpress HER-2 when assessed by standard IHC. Therefore, molecular approaches targeting this oncogene in CRC are not necessarily warranted. The pathophysiological role of elevated levels of soluble HER-2 in the serum of a substantial number of patients with CRC has yet to be determined. Supported by a research grant from Hoffmann-La Roche.

887

Transduction efficiency of adenoviral vectors in colorectal cancer cells is determined by the presence of the coxsackie adenovirus receptor

Th. Heinicke, S. Hemmi, T. Sauerbruch, W. Caselmann (Zürich, CH; Bonn, D)

In the present study we investigated the factors that influence the gene transfer efficiency of adenoviral vectors (AdV) in colorectal cancer (CRC) cells. Adenoviruses have been shown to use a common receptor (CAR) with Coxsackie B viruses for the binding to target cells. Virus entry is then mediated by interaction with alphaV-integrins. In addition, it has been suggested that AdV may also use HLA class I molecules as receptors. In order to assess the relative contribution of these virus host cell interactions for the susceptibility of CRC cells to AdV we determined the expression levels of CAR as well as those of alphaVbeta5-, alphaVbeta3-integrins and HLA class I molecules on various CRC cell lines and primary CRC cell cultures by flow cytometry and compared them to the respective transduction efficiencies. CRC cells were transduced using an AdV encoding the green fluorescent protein (GFP). High transduction efficiencies were found in those CRC cell lines which express large amounts of CAR. In contrast, when CAR was low or absent, the transduction efficiencies were 20 to 50-fold decreased. However, the presence of HLA class I molecules was not correlated with a high transduction efficiency. The primary cell cultures obtained from freshly resected

CRC which have been tested so far showed no detectable CAR expression and high doses of AdV were necessary for efficient transduction of these cells. We conclude, that CAR is an important factor for the transducibility of CRC cells by AdVs. Results from CRC cell lines may not reflect the situation in primary tissue. Determination of CAR expression levels should allow to predict the quantity of AdV necessary to achieve efficient gene transfer to CRC cells.

888

Weekly irinotecan/folinic acid/fluorouracil as first-line chemotherapy in patients with metastatic colorectal cancer: Prospective multicenter phase IV study

M. Möhler, C. Zanke, H. Hohl, P. Ehscheid, P. Schwindt, K. Kurfeß, E. Gracien, G. Gröger, P.R. Galle, M. Heike (Mainz, Bad Ems, Neuwied, Simmern, Andernach, Bad Soden, D)

The weekly combination of irinotecan (CPT-11), folinic acid (FA) and 5-fluorouracil (FU) has been approved as a new first-line standard regimen for metastatic colorectal cancer in the U.S. (Saltz et al. NEJM, 2000; 343: 905-914). In this study, the combination (CPT-11 125 mg/m^2, FA 20 mg/m^2; FU 500 mg/m^2 days 1, 8, 15, 22, q 6 weeks) was superior to both arms, CPT-11 and FA/FU (Mayo-Clinic-regimen) regarding response rates, progression-free and overall survival. Recently, unexpectedly high early death rates between 2,2 % and 4,8 % presumably due to toxicity have been observed in 2 subsequent trials (Sargent et al. NEJM 2001, July 19). In this phase IV the combination CPT-11/FA/FU according to Saltz, 2000 for first-line treatment of patients with metastatic colorectal cancer was analysed for response rates, quality of life and side effects in regional outpatient clinics. Patients and Methods: 40 patients (64 (42-73) years) were included in 12 centers of Rhineland-Palatinate from June 2000 to May 2001. So far, 24 patients were evaluable for toxicity. Their tumor manifestations were 19/24 liver metastases, 4/24 lung metastases, 2/24 peritoneal carcinosis and 8/24 local or lymph node recurrence. Overall, 210 weekly doses were given in 59 cycles. So far, 19 of 24 patients were evaluable for response rates. Results: 8/19 (42%) patients had objective responses (1 CR; 7 PR), 5/19 (26%) stable disease. The mean time to progression was 5,5 months (2-11 months). Treatment delays of 1 week were reported 14 times. 6 hospital admissions occurred during treatment: 2x neutropenia plus diarrhoea grade IV for rehydratation; 1x psoriasis, 2x pulmonary embolism, 1x bowl obstruction due to local recurrence. In one patient, treatment had to be stopped due to recurrent diarrhoea despite dose reduction. There were no drug-related deaths observed. Conclusions: Tumor control was achieved in 68% of patients with metastatic colorectal cancer by CPT-11/FA/FU with acceptable toxicity, concordant with the above cited U.S. phase III study results. Diarrhoea grade III/IV occurred in about 20 percent and can be treated with early high dose loperamide or budesonide, early rehydratation plus additional dose reduction. Up-dated results will be presented at the meeting.

889

Phase III study of adjuvant immunotherapy with edrecolomab (Panorex) of stage II colon cancer - interim analysis after one and half year median observation time

Y. Dencausse, G. Hartung, R. Hehlmann, W. Queißer for the Studiengruppe Gastrointestinale Tumoren (Mannheim, D)

Background: Adjuvant therapy with the murine monoclonal antibody 17-1A (Edrecolomab) has been shown by Riethmüller et al. (Lancet 1994;343:1177-83) to be an adequate alternative to chemotherapy in stage III colorectal cancer presenting similar efficacy and being less toxic. Thus, it seems conceivable that also patients with stage II colon cancer might benefit from this treatment. Therefore we started in January 1997 a prospective randomized multicentre study to compare Edrecolomab with observation. Patients and Methods: Patients with curative (R0) resection of stage II colon cancer are randomized to either arm A (treatment with Edrecolomab) or arm B (observation). The arm A patients got once a 500 mg infusion intravenously from Edrecolomab 4 weeks postoperatively, than every 4 weeks a 100 mg infusion for four months. All patients have follow-up examinations every 6 months. Results: Since January 1997, 377 patients have been randomized; 51 were excluded, 326 are at present evaluable. The median follow up is 16,8 months; 163 patients were treated and 163 only observed. The overall relapse rate is 9,3% / 12,6% (Arm A / Arm B, p>0.5). No difference for overall survival is observed between both arms. A total of 13 patients died, 4 in arm A and 9 in arm B. Six deaths were not cancer related, 2 in arm A and 4 in arm B. We can confirm the moderate and rare side effects of Edrecolomab. Adverse events are detected by only 119 (15,9%) of the 746 courses, only 1 case toxicity WHO 3-4 (severe vasculitis). The most common side effects are transitory diarrhoea, nausea and vomiting. Conclusions: Finally, treatment is well tolerated. The median observation of 16,8 months is to short to demonstrate efficacy of Edrecolomab in stage II colon cancer.

890

Pilot study of oxaliplatin or irinotecan in combination with capecitabine in patients with untreated and pretreated metastatic colorectal cancer

K. Jordan, O. Kellner, L. Müller, B. Schädlich, T. Kegel, A. Grothey, H. Schmoll (Halle, D)

BACKGROUND: The addition of oxaliplatin (OXA) or irinotecan (CPT-11) to 5-FU/FA based chemotherapy has been shown to improve the efficacy in patients with metastatic colorectal cancer (MCRC) treated in both first- and second-line setting. Capecitabine (CAP), an oral 5-FU prodrug, has shown superiority to Mayo clinic protocol in two phase III studies and may replace intravenous 5-FU as first line treatment for MCRC. Therefore we initiated a pilot study with CAP plus either CPT-11 or OXA to determine the efficacy and toxicity of a particular dosage and schedule in patients with untreated and pretreated MCRC. DOSAGE AND SCHEDULE: CAP 2000 mg/m^2/d p.o. days 1-14 qd 21 and OXA 70 mg/m^2 day 1, 8 or CPT-11 100 mg/m^2 day 1, 8 were given in an outpatient setting.
RESULTS: 22 pts. are evaluable for toxicity (CAP/CPT-11: 14; CAP/OXA: 8) and 16 for efficacy (CAP/CPT-11: 10; CAP/OXA: 6). Median age was 68 (range 53-73), colon: rectum (14:8). Except diarrhoea grade 3 (NCI-CTC) in 5 pts. (CAP/CPT-11: 3/14; CAP/OXA: 2/8) and 1 hand-foot-syndrome grade 3 in the CAP/CPT-11 pts. group, both regimen were well tolerated. There were no grade 3/4 peripheral neuropathy or myelosuppression in both protocols. Efficacy in untreated pts.: CR/PR: CAP/OXA: 2/2 pts. evaluable for response. CAP/CPT-11: 0/2 pts. evaluable for response. SD: CAP/CPT-11: 2/2 pts. evaluable for response. Efficacy in pretreated pts. (2-line-5-line chemotherapy): CR/PR: CAP/OXA: 1/4 pts. evaluable for response. CAP/CPT-11: 2/8 pts. evaluable for response. SD: CAP/OXA: 2/4 pts.. CAP/CPT-11: 2/8 pts.. PD: CAP/OXA: 1/4 pts.. CAP/CPT-11: 4/8 pts..
CONCLUSION: The combination of CAP/OXA and CAP/CPT-11 has shown an excellent safety profile and clinical activity in MCRC. The main toxicity was managable diarrhea without a fatal event. As no central venous port access is required for treatment administration potential medical complications associated with port systems can be avoided. To more precisly identify complications, toxicity and efficacy this pilot study served as basis for an ongoing randomised phase II study comparing CAP/OXA and CAP/CPT-11.

891

Multifluorescent labelling of immunmagnetic enriched circulating colon tumour cells and cellclusters

B. Molnar, A. Ladanyi, J. Bocsi, L. Floro, L. Sreter, Z. Tulassay (Budapest, H)

Background: Detection and characterisation of circulating tumour cells may contribute to the better understanding of metastatic process in human. With a recently developed intracellular cytokeratin immunmagnetic microbead cell separation method intact cells and mixed cell clusters were detected in patients with different stage and duration of colorectal cancer. Aims: Development and evaluation of a fluorescent cytokeratin/hemopoetic labelling technique with immunmagnetic microbead cell separation method for the characterisation of enriched tumour cell and cell clusters in patients with colorectal cancer. Methods: 16 colon cancer patient (Stage Dukes B, C, and D) and 20 healthy donor samples were evaluated. The colon cancer samples were evaluated before adjuvant chemotherapy. Immunmagnetic cell separation was performed from whole blood/ ficoll separated cell blood fractions by the surface epithelial markers (HEA 125, Ber-EP4). The enriched cell fraction (400ul from 5ml blood) was immuncytochemically labelled using a FITC labelled cytokeratin antibody (CAM 5.2) and a Texasred labelled CD45. Cell death was detected by Propidium Iodide nuclear staining. Spiking experiments were performed using the HT29 colon cancer cell line. The lowest detectable concentration was 1 cell/1 ml blood. A fluorescent inverse microscope was used to evaluate and manipulate the labelled cell fraction. Results: Two of the twenty healthy donor samples contained CK+CD45+ cells (3 and 4 such cells in 5 ml blood). In cancer patients the lowest cell number was 9 CK+CD45- cells/ml. Cell clusters containing CK+CD45- and CK-CD45+ cells were detected only in carcinomatous patients. Conclusions: Immunmagnetic cell separation with immuncytochemical labelling is a useful method for detection and characterisation of circulating colon cancer cells and clusters. CD45+CK+ cells appear both in healthy and carcinomatous patients in low number, as well. Their further characterisation and evaluation needs additional fluorescent markers and studies.

892

Detection of circulating tumor cells in the peripheral blood of colorectal cancer patients using cytokeratin 20 RT-PCR and immuncytochemistry

A. Ladanyi, B. Molnar, L. Floro, L. Sreter, Z. Tulassay (Budapest, H)

Background: The detection of circulating tumour cells and micrometastases may have important therapeutic and prognostic implications. Aims: Our objective was to compare the two gold standard techniques, such as RT-PCR and immuncytochemistry (ICC), of circulating tumor cell detection in clinical samples with advanced disease. Patients and methods: 30mL of peripheral blood was obtained from 11 Dukes D colorectal cancer patients at two separate times within a month. The first 5 mL after puncturing the skin were discarded. 5 mL of the sample was used for mRNA isolation and subsequent RT-PCR with cytokeratin 20 primers. 20 mL was subjected to immunmagnetic tumour cell enrichment with positive selection of cytokeratin 7/8 expressing cells on high-gradient magnetic columns (Carcinoma Cell Enrichment Kit, Miltenyi Biotech, Germany). The enriched cell fraction was cytocentrifuged on poly-L lysin coated slides. The cytospins were immuncytochemically labelled using a pancytokeratin antibody (Dako, MNF 116) and biotin-streptavidin-peroxidase technique (Dako, LSAB2). In spiking experiments the sensitivity of both assays were in the 1 cell/mL range. Results: Of the 22 samples in this study, 19 (86.36%) showed concordance between the two methodologies. The three disparity cases were due to RT-PCR positivity, but ICC negativity. With using RT-PCR 9/22 (40.90%) samples were positive, while with ICC 6/22 (27.27%) samples showed positive results. In five cases (45.45%) with RT-PCR and in four (36.36%) with ICC the result of the second sample conflicted with the first one. Conclusions: Our results indicate that both RT-PCR and immunmagnetic enrichment with subsequent ICC are sensitive and reliable methods for circulating tumour cell detection. The concordance rate in our study is high, but because both techniques have significant limitations it may be advisable to use them complementary in further investigations. Even in advanced staged patients with a high probability of circulating tumour cells multiple sampling can increase the risk of positive finding.

893

Thymidine phosphorylase (TP), thymidylate synthase (TS) and dihydropyrimidine dehydrogenase (DPD) gene expression levels in tumor tissue and peripheral blood mononuclear cells of solid cancers

S. Saußele, C. Kuhn, A. Weißer, M. Hartel, R. Sparla, A. Willer, R. Hofheinz, T. Landmann, A. Schreiber, R. Hehlmann, A. Hochhaus (Mannheim, D)

Combination therapy containing the antimetabolite 5-fluorouracil (5-FU) is considered standard treatment of several human epithelial malignancies. The in vivo and in vitro response to 5-FU shows a great variation in individual patients despite of identical cancer type. TP is of crucial importance in the 5-FU drug activation pathway, TS is the main target for 5-FU, and DPD is the rate-limiting enzyme in the catabolism of 5-FU. Patients with DPD-deficiency or with mutations in the DPD gene are at risk of developing severe 5-FU associated toxicity. In several studies it has been shown that increased protein levels or gene expression of TS, TP, and DPD in pretreatment tumor biopsies are associated with insensitivity to 5-FU as independent variables. Decrease of TS has been observed after bolus injection of 5-FU. For further prospective studies and molecular analysis of patients not responsive to chemotherapy, easier methods to determine the gene expression levels of TS, TP, and DPD are required. Therefore, mRNA levels in peripheral blood mononuclear cells (PBMNCs) and in primary tumor cells of 9 patients at diagnosis with colon (n=6), pancreatic (n=1), and rectal cancer (n=2) were investigated by a novel relative quantification real-time PCR assay using LightCycler technology. PBMNCs were isolated by ficoll gradient centrifugation, total RNA of PBMNCs and tumor cells was extracted by ultracentrifugation after cell lysis. Reverse transcription was performed using random primers. The amount of TP, TS or DPD transcripts of the samples was calculated as a ratio of target gene copies (T) to glucose-6-phosphate dehydrogenase transcripts (R) as reference gene and divided by the T/R ratio of a calibrator RNA that was run with each sample in parallel. The ratios of gene expression in PBMNCs and tumor cells were 0.65-6.18 versus 0.53-6.96 (TP), 0.001-0.04 versus 0.36-2.18 (TS) and 17.52-48.88 versus 3.92-99.12 (DPD). Statistical analysis of the ratios showed significant lower levels of TS in PBMNCs as compared to tumor tissue ($p<0.01$), whereas TP and DPD levels did not differ significantly. No correlation of the mRNA expression in PBMNCs and tumor cells could be demonstrated for all three genes. We conclude that the expression levels of TP, TS, and DPD in PBMNCs do not parallel the level in tumor tissues. However, PBMNCs are a useful source to screen for inborn DPD deficiency in order to avoid exaggerated 5-FU toxicity. Correlation of the molecular data with response to chemotherapy and clinical outcome is ongoing.

894

Smad7 is a prognostic marker in patients with colorectal cancer

C. Rochlitz, G. Mild, A. Lowy, J. Reuter, M. Lagrange, L. Terracciano, U. Laffer, R. Herrmann, J. Boulay (Basel, CH)

Deletion of chromosomal region 18q21 is frequently observed in colorectal cancer (CRC) and is associated with poor prognosis. This region includes genes encoding the TGFß signaling mediators SMAD2, SMAD4 and SMAD7, and the adhesion molecule DCC. In order to evaluate the clinical relevance of the deletion of each of these single genes, we took advantage of 325 archived biopsies available from a randomized study on benefit of adjuvant chemotherapy in patients with CRC. Gene copy dosage revealed that SMAD7 had a lower deletion frequency (43%) than SMAD4, SMAD2, or DCC (67%, 64%, and 54%, respectively). Multivariate analysis of these genetic data with patient survival were undertaken on a subset of patients with clinical data (n between 178 and 206 depending on the marker considered) and showed a favorable outcome for SMAD7 deletion [overall survival (OS): hazard ratio (HR) = 0.43, p = 0.001; disease free survival (DFS): HR = 0.50. p = 0.003]. In contrast, SMAD7 duplication had a hazardous effect on patient survival [OS: HR = 1.97, p = 0.049; DFS: HR = 1.99, p = 0.047]. Since SMAD7 blocks TGFß signaling, these data are consistent with the hypothesis that loss of SMAD7 renders carcinoma cells more sensitive to apoptosis induction through the TGFß pathway, whereas gain of SMAD7 function might result in TGFß-resistance. No significant association was observed between deletion of SMAD2, SMAD4 or DCC and patient outcome. This suggests that the previously described poor prognosis associated with loss of the 18q21 region is not related to on any of these genes. Interestingly, patients with SMAD4 deletion had less benefit from 5FU-based perioperative chemotherapy than patients with normal SMAD4 copy number [OS: HR=3.23, p=0.06; DFS: HR=2.89, p=0.05], suggesting that 5FU may act through components of the TGFß pathway. In addition, our obvservation raises the hypothesis that - if confirmed in larger, independent series - SMAD4 deletion might be useful as a predictive marker in the adjuvant, 5-FU-based chemotherapy of CRC. In conclusion, our findings strengthen the notion that loss of response to the apoptotic function of TGFß is a key factor in cancer progression and might be clinically useful in the treatment of colorectal carcinomas.

Postersitzung:
Zytostatika

895

A rapid new generation chemosensitivity assay: High prediction of ex-vivo responses in cancer cells

H.R. Thinneberg, R. Loewe (Bielefeld, D)

This retrospective study was assessed on 30 patients with late stage carcinoma, in the majority mamma carcinoma, to consider the different outcomes for drug sensitivity. The assay called ChemoSelect® was designed on a sensorchip based technology and is used to evaluate changes in metabolic activity of cells during the challenge of antineoblastic drugs. The clinical correlation of the assay prediction compared with the clinical outcome in all 30 patients was > 80%, a high success rate, not known with other assays in the past. All tumor entities could be tested. The results obtained with this new assay are very promising and offer the therapeutical benefit to avoid ineffective chemotherapy. Another advantage is speed (20 hours) and the absence of culturing the tumorcells. Tested drugs are all currently approved cytostatics and their combinations.

896

Pharmacokinetics of MTX-HSA and MTX in cancer patients receiving intravenous injections of MTX-HSA every two weeks

G. Hartung, K.B. Mross, M.E. Scheulen, B. Hermann, U. Heese, A. Bucholski, G. Lang, A. von Niecięcki, A. Möller (Mannheim, Rostock, Essen, München, D)

Background: Methotrexate-albumin (MTX-HSA) is a prodrug conjugate showing promising properties in preclinical and Phase-I trials. A low MTX-loading provided optimal tumor targeting and similar pharmacokinetic properties as native albumin. As part of a multicenter phase-I study a population of 27 cancer patients was investigated for basic pharmacokinetic parameters of the methotrexate-human serum albumin conjugate (MTX-HSA) administered as intravenous injection every two weeks.
Material and methods: The following assay methods were used to analyze the serum samples (pharmacokinetic profile and trough level samples): specific ELISA for MTX-HSA, LC-MS/MS for MTX, and clinical routine immunoassays for MTX cross-reacting with MTX-HSA: EMIT®, TDx®.
Results: The concentrations determined with the immunoassays match with good linear correlation (r^2>0.8) but with different slopes of the regression line leading to mean conversion factors of 1.5 (ELISA/EMIT®) and 0.5 (ELISA/TDx®). Main pharmacokinetic parameters calculated by non-compartmental analysis (WinNonlin 3.0) are: The parameters AUC and Cmax are dependent on the administered dose, $t_{1/2}$, CL and Vd do not change (dose range 50 to 90 mg/m^2). Conclusion: MTX-HSA is eliminated with a prolonged half-life of 10 days, free MTX was found in low concentrations leading to a slow elimination half-life of 7 days. The long half-life of free MTX may indicate a prolonged cleavage of MTX from MTX-HSA.

897

Therapeutic drug level monitoring of MTX-HSA administered to cancer patients as an intravenous injection every week

A. Möller, B. Hermann, U. Heese, A. Bucholski, G. Lang, A. von Niecięcki, G. Hartung (München, Mannheim, Rostock, D)

Background: Methotrexate-albumin (MTX-HSA) is a prodrug conjugate showing promising properties in preclinical and Phase-I trials. A low MTX-loading provided optimal tumor targeting and similar pharmacokinetic properties as native albumin. As part of 3 multicenter phase-II studies an interim population of 33 cancer patients (kidney, prostate, mesothelioma) was evaluated for therapeutic drug level concentrations during weekly i.v. administrations. Material and methods: The following assay methods were used to analyze the trough level serum samples: specific ELISA for MTX-HSA, LC-MS/MS for MTX, and TDx® (a clinical routine immunoassay for MTX cross-reacting with MTX-HSA).
Results: No center effect is found by comparison of trough level profiles. On the basis of the trough level concentration a half-life time ($t_{1/2}$ omega) can be calculated (fit to the equation f=a*(1-exp(-b*x))/(1-exp(-b*7): Steady state concentrations are reached after 4 to 5 injections of MTX-HSA. The mean plateau concentrations are 18 µM (ELISA), 8 µM (TDx®), 15 nM (LC-MS/MS). Conclusion: MTX-HSA is eliminated with a prolonged half-life of about 10 days, which may correspond to albumin half life in tumor patients. For free MTX a comparable slow elimination half-life is found suggesting prolonged cleavage of MTX from MTX-HSA.

898

Multi-targeted antifolate-phenylalanine: a novel prodrug for antibody-directed enzyme prodrug therapy

P.M. Deckert, S. Welt, L.J. Old, F.M. Huennekens, J.R. Bertino, W.G. Bornmann (Berlin, D; New York, La Jolla, USA)

In metastasized colon carcinoma, chemotherapy is the mainstay of therapy, but its efficacy is low. Antibody-Directed Enzyme-Prodrug Therapy (ADEPT) utilizes the dual specificity of antibody-enzyme conjugates for the targeted activation of an inert prodrug, generating higher local drug concentrations with lower toxicity. Multi-targeted antifolate (MTA) is a new, non-conventional antifolate drug that has shown considerable potential in the treatment of advanced colon carcinoma, making it an attractive candidate for ADEPT. Here we investigate whether dual specificity can be achieved by conjugates of the colon cancer-targeting antibody huA33 with carboxypeptidase A (A33-CPA) and a phenylalanine derivative of MTA (MTA-Phe). MTA-Phe was synthesized de novo. The complete and cleaved derivatives were physically identified by HPLC in a time-course incubation of MTA-Phe with CPA. To establish enzyme specificity of the MTA-based prodrug system, A33-positive LIM1215 and SW1222 cells and, as controls, A33-negative HT29 cells were incubated with the prodrug either in the presence of absence of CPA. Growth inhibition and cell kill were assessed by an MTT-uptake assay and a clonogenic assay, respectively. In all cell lines, the presence of CPA increased the toxicity of MTA 100 to 1000-fold. The complete ADEPT system was tested in vitro using the same cell lines as above. Cells were incubated for 1 hr with a chemical conjugate of A33 and CPA. After repeated washing and media exchange, the prodrug was added and the incubation continued for 72 hrs. At this point, cell viability was assessed by an MTT uptake assay, or cells were stained and counted in the clonogenic assay. Compared to controls without the A33-CPA conjugate, we found increased toxicity specifically in the A33-positive cell lines, but not in the A33-negative controls. Pre-incubation with unconjugated A33 blocked this effect, confirming the dual specificity of an ADEPT system based on MTA-Phe and A33-CPA. We conclude that such a system has a potential for future targeted chemotherapy that warrants further exploration.

899

Intradermal skin testing, a useful tool to confirm hypersensitivity reactions to oxaliplatin

L. Meyer, T. Zuberbier, H. Riess (Berlin, D)

Background: Hypersensitivity reactions (HSRs) following the systemic administration of platinum compounds are well known complications during chemotherapeutic treatment of different types of malignancies. A few case reports describing anaphylactic reactions that may be due to the third generation platinum agent oxaliplatin (L-OHP) have been published so far, but methods confirming this clinical suspicion are not accepted. Patients and methods: We assessed the use of intradermal allergy tests on 8 patients with metastatic colorectal adenocarcinomas and clinical signs, consistent with HSRs while treated with L-OHP-containing regimens. Different concentrations of the platinum compounds carboplatin, cisplatin, and L-OHP were used for intradermal skin testing. Results: 6/8 patients showed positive results to L-OHP and carboplatin. Intradermal skin tests to cisplatin were negative in all patients. The clinical severity of the observed HSRs was considered mild with regard to kind and intensity in every patient. Conclusion: HSRs may start unsuggestive. Intradermal skin testing is an effective tool in distinguishing between HSRs and other adverse events in L-OHP-containing antitumor therapy. Crossreactivity to carboplatin but not to cisplatin was observed.

900

Increased risk of lung toxicity under high-dose BCNU in elder female patients

R. Noppeney, U. Dührsen (Essen, D)

BEAM followed by transplantation of PBSC is a generally accepted high-dose-chemotherapy-protocol in non-Hodgkin`s lymphomas (NHL) and Hodgkin`s disease. Lung toxicity, described as „non-infective pulmonary complications“ or „idiopathic pulmonary syndrome“ is a well-known complication of protocols containing high-dose BCNU. Its reported frequency and mortality varies and data about risk factors are inconsistant. Several publications suggest that female gender and BCNU-dose
of > 450 mg/sqm could be independent risk factors for the development of lung toxicitiy. Between 1997 and 2001, 21 patients were treated in our department with high-dose-chemotherapy according to BEAM-protocol (BCNU 300mg/sqm , etoposide 900mg/sqm, cytarabin 1200 mg/sqm, melphalan 140mg/sqm) followed by transplantation of autologous PBSC for high-grade NHL (1pat.), relapsed or refractory high-grade NHL (11pat.), low-grade NHL (4pat.), Hodgkin's lymphoma (3pat.) or mantle-cell lymphoma. Median age of 13 male patients was 43 years (19-65) and 51.5 years (18-60) in 8 female patients. Two of the male patients (64 resp. 65 y) received reduced doses of BCNU (200mg/sqm) and melphalan (100mg/sqm) in consideration of their age.Transplant-related-mortality (3 of 21 patients, 14%) was limited to female patients whose ages ranged from 55 to 57. Patient 1 died of lung failure at d24 after HDC. Ten days after transplantation she had developed fever and interstitial infiltrates concomitant with leukocyte regeneration. There was no evidence of infectious cause. Patient 2 also showed lung toxicity with need of artificial ventilation after hematological reconstitution. She died of an ischemic stroke which superveined 10 days after the beginning of artificial ventilation. Patient 3 died of respiratory failure due to pneumonia aquired during bone-marrow aplasia. Blood cultures showed enterobacteriaceae. Respiratory failure in patients 1, 2 and possibly 3 may have been related to BCNU-induced lung toxicity. We hypothesize that elder female patients may run an increased risk for BCNU-related lung toxicity even at doses < 450 mg/sqm. Alternative conditioning regimens may be more appropiate for this group of patients

901

Nitrate sensitive contraction of the brachial artery following 5-FU containing chemotherapy

T. Südhoff, T. Yilmaz, C. Petz, U. Graeven, M.-D. Enderle, W.-H. Schmiegel (Bochum, D)

5-Fluorouracil (5-FU) belongs to the most often used cytostatic drugs and 7-10% of the patients treated will develop cardiovascular complications. Nevertheless, the underlying pathophysiology is unknown and noninvasive diagnostic procedures which allow a risk stratification are not established. Methods: We investigated prospectively the influence of 5-FU infusions on the diameter of the brachial artery using high resolution ultrasound (Esoate/Biosound, 13.0 MHz tranducer) in 13 patients (12x GI cancer, 1x head and neck cancer, 7x bolus, 6x continuous infusions). 6 patients with cyclophosphamide infusions were used as control group, the investigators were blinded concerning the applicated cystostatic drug. Additionally, plasma levels of big endothelin were serially assessed using a commercially available ELISA assay. Results: 6 of 13 patients showed 5-FU associated contraction of the A. brachial artery (median 10%, range 4-15) after the end of 5-FU applications, whereas after cyclophosphamide infusion all patients demonstrated a dilatation of the vascular diameter (median 10%, range 1-28). The big endothelin plasma levels were significantly increased 2 hours after start of 5-FU infusions when compared with baseline levels (median 2.34 fmol/ml (0.19-4.89) vs 0.71 fmol/ml (0.09-6.69), U-Test: p=0.02). Increases of big endothelin concentrations were similar in 5-FU treated patients independant of contraction of the A. brachialis. In contrary, there was no increase in big endothelin plasma levels after cyclophosphamide infusions (median 0.84 fmol/ml (0.34-2.96) vs median 1.25 fmol/ml (0.23-2.68) p=0.86). Inhalation of glyceroltrinitrate resulted in a stronger dilatation of the brachial artery in patients following 5-FU as compared to cyclophosphamide (median 20%, range 12-47 vs 7%, range 2-19, p=0.01). Discussion: Big endothelin may be one mediator of 5-FU induced cardiovascular complications as indicated by increased big endothelin plasma levels following 5-FU but not cyclophosphamide applications. This finding was accompanied by a contraction of the brachial artery in a substantial proportion of the patients which have been exposed to 5-FU. 5-FU induced vessel contraction was mostly reversible by nitrate inhalation. Future investigations will clarify whether high resolution ultrasound may serve as an easy to use means to identify tumor patients at risk of 5-FU induced cardiovascular complications.

902

Development of high-throughput in vitro and in vivo testing strategies for the discovery of novel anticancer agents of natural origin

T. Greiner, G. Maier, A. Maier, A.M. Burger, H.H. Fiebig (Freiburg, D)

Natural products have proven to be a rich source of novel drugs. They provide more structural diversity than standard combinatorial chemistry and thus, offer excellent opportunities for finding low molecular weight compounds that are active in a wide range of diseases. The majority of anticancer drugs which successfully completed clinical trials in the past decade were of natural origin e.g. paclitaxel, CPT-11, or recently ET 743. With the continuing need for novel lead structures against increasing numbers of molecular targets; the diversity derived from natural products will remain important to the future of anticancer drug discovery. In view of this, we have opted to test a collection of over 5.000 chemically defined, pure natural products from the Hans-Knöll-Institute and have developed high-throughput in vitro and in vivo testing strategies. The major obstacle in using isolated versus synthetic material is its limited availability - often only a few milligrams can be provided. To obtain maximal information with least amounts of drug, we employed a cellular screen using 10 xenograft-derived human cell lines comprised of 7 slow or intermediate growing and sensitive as well as chemoresistant solid tumor types. The cell lines resemble the original patient tumor and xenograft histology after s.c. injection into nude mice. Each drug was tested in 2 concentrations (0.3, 3 µg/ml) and triplicates. Of 10 registered anticancer agents within the natural product pool, 9 were found active applying our screening criteria. One % of the total number of novel compounds screened, possessed potent and differential in vitro cytotoxicity profiles. Eleven of these hits were forwarded for in vivo testing. To save precious compound, we developed a model system, which would give a hint of in vivo antitumor activity and enable estimation of the maximal tolerated dose. Two or 4 different xenografts with similar growth rates and doubling times were implanted s.c. between the fore and hind flanks of nude mice. They were validated for stable growth and response to clinical agents as well as maintaining the morphology. The xenograft doublet was composed of mammary (MAXF 401) and lung cancer models (LXFL 529) which had also been used for in vitro screening. Against these tumors, leads were tested at 10, 3 and 1 mg/kg given i.p. on days 1,5, and 9 in a single mouse. If toxicity occurred, doses were decreased and the experiments were repeated. Hence 4/11 drugs were shown to possess potential antitumor activity within a therapeutic window. Further screening will involve testing in xenograft tumor quartets composed of e.g. fast growing chemoresistant cancers (RXF 944, MX-1, OVXF 899, XF 575), or slower growing chemoresponsive tumors (LXFL 529, MAXF 401, MEXF 989, CXF 280). Compounds with confirmed in vivo efficacy against several tumor types will be preclinically developed. By using this approach, we hope to discover clinical candidate agents on an economical and fast track.

903

Comparison of the cytostatic activity of three nitrosoureas in O6-AGT proficient and -depleted tumor cells

T. R. Büch, W. J. Zeller (Heidelberg, D)

Background: Previous experimental results indicated that BCNU had an about 6 to 8 fold less cytotoxic potency than ACNU in cell lines naturally deficient of DNA repair protein O6-alkylguanine-DNA alkyltransferase (O6-AGT). Objective: To investigate which of the three nitrosoureas, BCNU, ACNU or HeCNU, had the highest cytotoxic potency against tumor cells with different natural O6-AGT activity and against tumor cells with O6-AGT, depleted by co-treatment with O6-benzylguanine (O6-BG), an inactivator of this DNA repair protein. Material and Methods: The cytotoxicity of BCNU, ACNU and HeCNU against seven tumor cell lines was investigated by the MTT assay. O6-AGT activity was determined by measuring the O6-demethylating rate of a [3H]-methylated DNA substrate.
Results: Without O6-BG treatment, BCNU was superior to both other drugs in tumor cells with high O6-AGT activity. O6-BG treatment resulted in effective sensitization of tumor cell lines with middle or high O6-AGT activity towards BCNU, ACNU and HeCNU. After O6-AGT depletion, the cytotoxic potency (comparison of IC50 values) of ACNU was higher than that of BCNU (p=0.016) or that of HeCNU (p=0.016) in all tumor cell lines. Conclusion: We would expect that without O6-AGT depletion BCNU is the drug of choice, especially in tumors with high transferase activity. The higher cytotoxic potency of ACNU after O6-AGT depletion is compensated by the higher bone marrow toxicity of ACNU necessitating clinical dose reduction of ACNU as compared to BCNU. Thus, we would not expect a superiority of ACNU + O6-BG over BCNU + O6-BG after systemic administration.

904

The effect of aminopterin-human serum albumin conjugates (AP-HSA) on three human melanoma cell lines

I. Kiprianová, G. Hartung, D. Schadendorf, H. Sinn, M. Wesseler, F. Neuwirth, E. Frei (Mannheim, Rostock, Heidelberg, D)

Aminopterin (AP), an antifolate, was first synthesised in 1947 and was found to be effective in the treatment of haematological cancers, but has since been displaced by methotrexate (MTX) due to its toxic side effects and narrow therapeutic window. Conjugation of this compound to human serum albumin (HSA) might result in greater selectivity for the tumour tissue and reduce toxicity as was seen for MTX-HSA, which is currently in phase II clinical trial. Albumin is an ideal shuttle of compounds to the tumour, because plasma proteins accumulate in tumours. The plasma half-life of MTX-HSA was found to be in the range of native albumin, if the loading ratio with MTX is below 3 moles /mole albumin. Three cell lines isolated from human melanoma were tested for their sensitivity against AP-HSA. The three lines had been found to have differing sensitivities against MTX-HSA, if grown as xenografts on athymic mice. The toxicity of AP-HSA was tested in cell cultures of the three lines using the sulforhodamin assay and compared to that of MTX-HSA. The cells had been adapted to medium with half the standard amino acid content, 2.5% human serum and HSA to mimic conditions in the tumour, and force the cells to take up albumin as a source of amino acids. Both conjugates were purified of contaminating free antifolate to below 0.03% to exclude effects elicited by the unbound agents. Since AP-HSA has to be cleaved and AP liberated intracellularly to exert its effect, the time dependence of AP-HSA toxicity was investigated. AP-HSA with a loading ratio of 1.4:1 was not toxic to any line at concentrations up to 30 µM. We therefore synthesised AP-HSA with 3.5 moles AP/mole HSA and MTX-HSA with a loading ratio of 5.8. AP-HSA at 10 µM was toxic to all three cell lines after 24h and nearly complete cell loss was observed after 48h. MTX-HSA at the same concentration elicited the same level of toxicity only after 96h incubation. The effects of low doses over long incubation times is currently under investigation and shall be presented. The selectivity of the albumin conjugates observed in vivo for the different lines was not observed in vitro, probably because it is very difficult to mimic the tumour conditions adequately, but the cell lines nevertheless offer a rapid method to screen new albumin conjugates for their cytotoxicity. From our data we expect AP-HSA to be even more effective in vivo against these xenografts than MTX-HSA.

Nachträglich eingereichte Abstracts

Wissenschaftliches Symposium: Multimodale Therapieansätze in der Onkologie

36

Multimodal therapy in oesophageal cancer

M. Stahl (Essen, D)

About two third of the patients with oesophageal cancer present with advanced disease at first diagnosis in Germany and the western countries. Therefore, even today more than 80% of the patients will die from their disease. Combination treatment with chemotherapy, radiotherapy and surgery offers the chance for improving the rate of long term survivors in patients with localized disease. Numerous controlled trials have been performed within the last years, but many questions remain unsolved. In *squamous cell carcinomas* there is no single phase III trial to prove the superiority of pre- and/or postoperative treatment compared with surgery alone. Chemoradiation is superior to radiation alone, but we do not know, which patients may take profit from surgery after chemoradiation. Most investigators beliefe that responders to chemo(radio)therapy should undergo surgery, whereas nonresponders can at best be treated with palliative intent. If this is true methods are needed to early identify nonresponders to prevent them from overtreatment. On the other hand we know that about 25% of the patients present with pathologic complete response after preoperative chemoradiation, which makes surgery unnecessary in these patients. But methods to identify these patients prior to surgery are still lacking. In *adenocarcinomas* phase III trials demonstrated that preoperative chemoradiation improves long term survival in localized resectable tumours compared with surgery alone. Nevertheless, these trials have been criticized due to insufficient staging methods and possible bias by the type of surgery. Contradictory results have been reported with preoperative chemotherapy in these tumours. In locally advanced stages preoperative chemotherapy or chemoradiation clearly improves the rate of R0-resection and by this way long term survival, but we do not know which intensity and duration of the preoperative therapy may be best in terms of tumour control and postoperative mortality. The high incidence of distant recurrences strongly asks for an improved efficacy of systemic treatment. The high number of involved intraabdominal lymph nodes particularly in locally advanced tumours favours chemoradiation with surgery over chemoradiation alone in adenocarcinomas, but it remains controversial which surgical technique may be most suitable after intensive preoperative treatment. Current data of controlled clinical trials clearly indicates that multimodal treatment is indicated in high risk patients with locally advanced tumours and may also be useful in resectable adenocarcinomas. As many questions remain unsolved this treatment should not be performed outside of clinical trials and it should be restricted to centers with experience in multidisciplinary approaches.

Postersitzung: Immuntherapie

263a

The receptor for hyaluronan mediated motility (RHAMM) is expressed in a high percentage of renal cell carcinoma and is able to elicit humoral immune responses

M. Ringhoffer, J. Greiner, J. Gschwend, H. Kirsche, A. Schenk, S. Huebsch, A. Szmaragowska, R. Hautmann, H. Döhner, M. Schmitt (Ulm, D)

In recent years several tumor associated antigens in renal cell carcinoma (RCC) have been identified. Their potential therapeutic value is often vague since they show a low percentage of expression in a cohort of RCC, an unfavourable tissue distribution, or fail to elicit immune responses. We generated an expression library from the cell line K562, which is derived from a patient, who suffered from chronic myeloid leukemia in the state of blast crisis. Serological screening of this library (SEREX) with allogeneic sera from patients with acute and chronic myeloid leukemia yielded a reactive clone, which was then sequenced and identified as the receptor for hyaluronan mediated motility (RHAMM). RHAMM is known as a gene with an impact on cell growth and motility. Overexpression of RHAMM leads to transformation of cells and the gene is essential for ras mediated transformation. There are several tumor antigens, which are expressed in hematological milignancies as well as in solid tumors like for example genes belonging to the MAGE gene family. Subsequently we examined the immunogenicity of RHAMM in RCC. To evaluate the expression of RHAMM in RCC we investigated 43 renal cell carcinoma samples and 20 corresponding normal kidney tissue samples, which were cryopreserved during nephrectomy, by means of RT-PCR. RHAMM mRNA expression was detected in 34/43 of the tumor samples but in none of the normal tissue samples. Assessment of cDNA from a panel of normal tissues showed mRNA in placenta, testis and thymus, thus very similar to a CT (cancer/testis) antigen. Next, we screened the RHAMM expressing clone with 25 preadsorbed sera of RCC patients, collected during nephrectomy and with 12 preadsorbed sera from healthy volunteers. Antibody reactivity was detected in 13/25 RCC patients but not at all in healthy volunteers. Antibody detection correlated in all but two cases with the PCR result (PCR positive, antibody negative). We conclude that due to the high expression in RCC, the missing or very limited expression in normal tissues and the ability to induce humoral responses in renal cell carcinoma patients, RHAMM might be obviously an interesting target for immunointervention.

Postersitzung: Allogene Stammzelltransplantation: Klinik

474a

A reduced intensity conditioning regimen introducing radioimmunotherapy with a ^{188}Re–labeled anti-CD66 monoclonal antibody for patients at high risk of transplant related mortality: results of a pilot study of 12 patients

M. Ringhoffer, I. Buchmann, H. Martin, B. Hertenstein, M. Wiesneth, S. von Harsdorf, G. Glatting, J. Kotzerke, L. Bergmann, S. Reske, H. Döhner, D. Bunjes (Ulm, Frankfurt, Hannover, D)

In recent years several investigators have initiated trials using reduced – intensity conditioning for patients (pts) with high risk of transplant – related mortality (TRM). In most of these studies total body irradiation (TBI) is avoided, reduced doses of cytoreductive chemotherapy are given and immunoablative treatment is intensified. However, many of the pts entering these trials suffer from graft failure or the occurence of graft versus host disease (GVHD). We have attempted to reduce the organ toxicity but still provide potent immunosuppression by the introduction of targeted radiotherapy with a radiolabeled monoclonal antibody instead of TBI. An anti-CD66 mouse antibody was applied, which targets normal granulopoeisis and rapidly localizes to the bone marrow thus producing a powerful crossfire effect. We chose ^{188}Re as radionuclide because it combines the properties of a high-energy beta-emitter for treatment with just enough gamma radiation for dosimetry. During the last 26 months, we have recruited 12 pts with high risk haematological malignancies (8 acute leukemias > 1.CR, 3 chronic myeloid leukemias > 1.c.P, 1 multiple myeloma). 6 pts were more than 55 years old, and 6 pts were younger than 55 but had received a prior stem cell transplantation (SCT)(4 allogeneic, 2 multiple autologous). All pts showed a favourable dosimetry with a higher dose to the marrow or spleen than to any other organ (especially liver and kidney). The therapeutic dose in the marrow was 15 Gy (median). In all pts, the regimen was completed by the application of fludarabine 180 mg/ m^2, ATG 40 mg/kg plus/minus melphalan. The graft consisted of G-CSF mobilized peripheral blood progenitor cells from a HLA-compatible family donor (n=11) or a MUD (n=1). We depleted the T cells by either using the campath-1H antibody or using positive selection techniques. No acute toxicity of the radiolabeled antibody and low organ toxicity after transplantation was observed, but one pt developed idiopathic pneumonitis (lung dose: 0,07 Gy). All pts engrafted and became full donor chimeras. No primary acute GVHD was observed, but one pt developed grade III acute GVHD after a donor lymphocyte infusion and two pts subsequently developed chronic GVHD. Three pts with advanced leukemia, who received prior SCT have relapsed. Four pts have died of transplant – related complications (1 idiopathic and 1 P. carinii pneumonia, 2 systemic infections) and 6 pts are currently alive. (5 disease free, 1 in relapse). These preliminary data demonstrate the feasibility of the approach, however, a longer follow – up is required for the evaluation of the therapeutic efficacy.

474b

Fatal outcome in a patient developing Epstein-Barr virus - associated lymphoproliferative disorder (EBV-LPD) without measurable disease

M. Ringhoffer, K. Döhner, S. Scheil, H. Wilts, T. Mertens, W. Grimminger, H. Döhner, D. Bunjes (Ulm, D)

EBV-LPD is a serious complication after allogeneic PBSCT or BMT and occurs in up to 25% of the patients. The grade of severity ranges from post-transplant infectious mononucleosis to monoclonal B-cell-lymphoma with a lethal outcome in most of the cases. Risk factors for the development of EBV-LPD include haploidentical transplantation or the use of a marrow from a matched unrelated donor (MUD), the grade of graft-versus-host disease (GVHD) and T cell depletion of the transplant. Two strategies have been used to treat severe cases of EBV-LPD: first, the infusion of donor lymphocytes, which has been proven to be effective in several clinical trials and second, the application of a monoclonal CD-20 antibody (rituximab), if donor lymphocytes are not available or the interruption of immunosuppression is not feasible. Successful treatment seems to be dependent upon early initiation of therapy and therefore early detection of the disease is mandatory. Here, we report a very untypical case of EBV-LPD without any characteristical clinical findings and therefore with a particular difficult diagnosis. A 51-year-old female patient in the first chronic phase of CML received an allogeneic PBSCT from a matched unrelated donor. The conditioning regimen included fractionated total body irradiation in divided doses of 2 x 2 Gy on d -9 to d -7 (total dose 12 Gy); thiotepa 10 mg/kg once daily i.v. on d -5, cyclophosphamide 60 mg/kg once daily on d -3 and -2. Anti-thymocyte-globulin (Fresenius) 5 mg/kg i.v. once daily for 4 consecutive days starting from d -4 was administered as rejection prophylaxis. On day 0 the patient received an ex vivo CD34 positively selected transplant (7.69 x 10^6 /kg CD34+ cells).The transplant was manipulated by CD34+ selection. Hematological recovery was fast and all analysis during the posttransplant course showed complete donor chimerism. On day +193 after transplantation the patient was readmitted to the hospital with recurrent fever of unknown origin and cough. Clinical, radiographic and sonographic evaluation revealed no characteristical findings beside a mild splenomegaly. Screening for EBV, CMV, RSV and HSV did not indicate an active infection. On day +203 the patient developed generalized seizures, respiratory failure and died within 24 hours in multiorgan failure. The macroscopic postmortem was still not enlightening, the histological examination however, demonstrated diffuse organ infiltration by monoclonal lymphoblastoid cells due to EBV-LPD.

Postersitzung:
Urologische Tumoren

651a

2nd-line protocol for the treatment of metastatic renal cell carcinoma with a combination therapy containing interleukin-2, interferon-alpha, 13 cis-retinoic-acid (13CRA) and gemcitabine

M. Ringhoffer, J. Gschwend, M. Schmitt, J. Greiner, A. Schenk, R. Hautmann, H. Döhner (Ulm, D)

Chemoimmunotherapy with interleukin–2 (IL-2), interferon-alpha, (IFN-alpha) and 5-fluorouracil (5-FU) is currently used as a standard protocol for metastatic renal cell carcinoma (RCC). Phase III studies had been initiated and overall response rates (PR and CR) of 25 – 30% have been reported. Many patients, however, endure progressive disease, because they are refractory to this front line therapy. Synergistic effects between IFN–alpha and 13 CRA and further between IFN-alpha and gemcitabine have been demonstrated. Therefore we initiated a phase II trial for a 2nd-line therapy with IL-2, IFN-alpha, 13 CRA and Gemcitabine to treat metastatic RCC. The study was activated in January 2000 and in an intention-to-treat analysis eleven patients have been enrolled so far. The applicated dosage was 20 mg of 13CRA three times daily p.o., between 9 and 18 Mio IE IFN-alpha s.c., between 9 and 18 Mio IE IL-2 s.c., and 750 mg or 1000 mg Gemcitabine i.v. per dose, depending on the age of the patient and the side effects. All patients recruited in the study were suffering from progressive disease while receiving a cytokine containing (chemo)-immunotherapy. Up to May 2001, nine patients completed the regimen. One patient dropped out for violating the exclusion criteria. Beside a transient elevation of transaminases (isolated WHO grade III toxicity) in one case no severe toxicity occured and except for the starting week of the therapy no further hospitalization of the patients was necessary. The results show 1 PR for 10 months, 4 SD for at least 6 months, and 3 PD. 5 patients died from progressive disease with a median survival time of 261 days after initiating the therapy. The other patients are still alive for a median survival time of 201 days (May 2001). One patient is receiving a third-line vaccination protocol with Her-2/neu derived peptides. Taken together these preliminary results show that a 2nd line therapy for refractory metastatic RCC with the mentioned combination regimen is feasible. A larger cohort of patients, however, is required for the final evaluation of its therapeutic efficacy.

ONKOLOGIE

Vol. 24, Sonderheft 6, Oktober 2001

ONKOLOGIE

Onkologie 2001;24

Hinweise für Autoren

Die Zeitschrift ONKOLOGIE ist ein Forum für alle Themen der Onkologie. Veröffentlicht werden Original- und Übersichtsarbeiten aus der internistischen Hämato-Onkologie, der chirurgischen, gynäkologischen und urologischen Onkologie, der Strahlentherapie und der klinisch-experimentellen Onkologie. Auch Beiträge zur pädiatrischen, HNO- und dermatologischen Onkologie, zu Pathologie und Biometrie, Epidemiologie, Prävention und Frühdiagnose bis hin zur psychologischen Betreuung werden berücksichtigt. Alle Beiträge unterliegen einem Begutachtungsverfahren.

1. Allgemeines, Verlagsrecht

Bitte senden Sie Ihr Manuskript in vierfacher Ausfertigung (jedes Exemplar komplett mit Abbildungen und Tabellen) direkt an den S. Karger Verlag, z. H. Dr. Heike Endemann, Postfach, D-79095 Freiburg. Eine weitere Kopie des Manuskripts soll grundsätzlich beim Autor verbleiben.

Alle Autoren müssen in einem **Begleitschreiben** durch Unterschrift bestätigen, daß die eingereichte Arbeit bislang weder ganz noch teilweise (Ausnahme: Kongreßberichte) veröffentlicht wurde noch anderweitig zur Veröffentlichung eingereicht ist und daß alle Autoren das Manuskript gelesen haben und mit der Veröffentlichung in der vorliegenden Form einverstanden sind. Es ist Sache des Autors, die Nachdruckerlaubnis für Abbildungen, Tabellen usw. aus anderen Publikationen zu beschaffen. Über die Annahme wissenschaftlicher Beiträge wird aufgrund mehrerer Gutachten entschieden (peer-review). Mit der Annahme des Manuskripts zur Publikation gehen alle Rechte auf den Verlag über. Ohne schriftliche Genehmigung des Verlages dürfen diese Publikationen oder Teile daraus nicht in andere Sprachen übersetzt oder in irgendeiner Form mit mechanischen oder elektronischen Mitteln (einschließlich Fotokopie, Tonaufnahme und Mikrofilm) reproduziert oder auf Datenträger oder einem Computersystem gespeichert werden.

Ethische und rechtliche Voraussetzungen: Zur Publikation eingereichte Manuskripte, bei denen Untersuchungen an Probanden oder Patienten durchgeführt wurden, müssen die Erklärung enthalten, daß das Versuchsprotokoll von einer Ethikkommission genehmigt wurde und den ethischen Standards der Deklaration von Helsinki in ihrer erweiterten Form von 1975 und ihren Zusätzen von 1983, 1989 und 1996 [JAMA 1997;277:925–926] entspricht. Zugleich muß im Manuskripttext auf die Einwilligung der Versuchsperson nach Aufklärung hingewiesen werden. Angaben, die auf die Identität der Person schließen lassen, sind unzulässig.

Tierversuche bedürfen gleichfalls der Genehmigung und müssen den Forderungen des Tierschutzgesetzes in seiner jeweils gültigen Fassung entsprechen. Auf diese Genehmigung ist im Manuskript hinzuweisen. Arbeiten, die nicht aus Deutschland stammen, müssen einen vergleichbaren Hinweis tragen.

Beiträge erscheinen in einer der folgenden **Rubriken:** Übersichtsarbeit, Kurze Übersicht, Originalarbeit, Laufende klinische Studie, Kurzmitteilung, Kasuistik, Brief an die Herausgeber, Kommentar, Meinung, Onkologische Fortbildung, Klinische Information, Kongreßbericht, Wilsede-Schule.

Übersichtsarbeiten, Originalarbeiten, Laufende klinische Studien, Kurzmitteilungen, Kasuistiken, Briefe an die Herausgeber, Kommentare und Editorials sind in englischer Sprache abzufassen. Alle anderen Beiträge können wahlweise in deutscher oder englischer Sprache eingereicht werden.

2. Manuskripte

Die Manuskripte sollen in $1^1/_2$zeiligem Abstand mit beidseitig je 3 cm Rand eingereicht werden. Folgender Umfang einschließlich Zusammenfassung, Abbildungen, Tabellen und Literaturzitate sollte nicht überschritten werden:

	Druckseiten	Entspricht Manuskriptseiten (einschließlich Zusammenfassung, Abbildungen, Tabellen und Literaturzitate)
Übersichtsarbeit, Originalarbeit	5	14
Kurze Übersicht, Kasuistik	3	8
Laufende klinische Studie Onkologische Fortbildung, Klinische Information	4	11
Brief an die Herausgeber	1	3
Übrige Beiträge	2	6

Für Umfangüberschreitungen wird dem Autor ein Mehrkostenanteil in Rechnung gestellt (s. unten). Kontrollieren Sie bitte das Manuskript sorgfältig auf einheitliche Bezeichnungen und korrekte Schreibweise. Typographische Gestaltung und redaktionelle Bearbeitung werden vom Verlag durchgeführt. Generelle Hinweise zur Abfassung von Manuskripten können den «Uniform Requirements for Manuscripts Submitted to Biomedical Journals» des «International Committee of Medical Journal Editors», veröffentlicht in Ann Intern Med 1997;126:36–47 bzw. http://jama.asma-assn.org/info/auinst_req.html, entnommen werden.

Manuskripte auf Diskette: Jedem Manuskript ist eine Diskette beizulegen, deren Inhalt exakt mit dem Ausdruck übereinstimmen muß. Bitte beachten Sie auch die «Hinweise für Manuskripte auf Diskette».

Bitte numerieren Sie die Seiten des Manuskripts durchlaufend, beginnend mit der Titelseite. Die Arbeit soll gegliedert sein in **Schlüsselwörter und Zusammenfassung, Key Words und Summary, Einleitung, Material und Methoden** (oder **Patienten und Methoden**), **Ergebnisse, Diskussion,** gegebenenfalls **Dank, Literatur, Tabellen, Abbildungslegenden und Abbildungen.**

Die **Titelseite** muß enthalten: **Titel und Kurztitel** der Arbeit in Deutsch und Englisch, die vollständigen Autorennamen mit der jeweiligen Institutsangabe in der Landessprache der Autoren, die vollständige Postadresse des Korrespondenzautors, gegebenenfalls mit Telefon- und Faxnummer sowie E-mail-Adresse.

Originalarbeiten müssen eine englische und deutsche **strukturierte Zusammenfassung** enthalten. Für Informationen zur Gliederung der Zusammenfassung siehe Haynes RB, Mulrow CD, Huth EJ, Altman DG, Gardner MJ: Ann Intern Med 1990;113:69–76. Die Zusammenfassungen sollen 250 Wörter nicht überschreiten und sind zu strukturieren in **Hintergrund, Material und Methoden, Ergebnisse, Schlußfolgerungen.** Die Zusammenfassung darf keine Literaturzitate, Tabellen, Abbildungen, Abbildungs- und Tabellenhinweise enthalten. Falls Hinweise auf Literaturzitate unvermeidlich sind, müssen sie vollständig sein. Zu einer Arbeit gehören 3–5 **Schlüsselwörter** (Englisch und Deutsch), die möglichst der «Medical Subjects Heading List» des Index Medicus entnommen sein sollen (die Liste ist unter www.nlm.nih.gov erhältlich). Übersichtsarbeiten folgen nicht der Gliederung von Originalarbeiten; Titelseite, Zusammenfassung, Schlüsselwörter und Literaturverzeichnis sind jedoch erforderlich.

Fußnoten sollten nur in Tabellen verwendet werden. **Abkürzungen** im Text bitte möglichst sparsam verwenden. Bei nicht allgemein üblichen Abkürzungen jeweils beim ersten Auftreten das Wort ausschreiben und die später verwendete Abkürzung in Klammern folgen lassen: z. B. Adenosin-Monophosphat (AMP). Geschützte **Produkt- bzw. Markennamen** müssen beim ersten Auftreten im Text mit einem hochgestellten ® bzw. ™ bezeichnet werden. Die Verantwortung für die korrekte Befolgung dieser Vorschrift liegt ausschließlich beim Autor. **Danksagungen** sollten unmittelbar dem Text folgen und dem Literaturverzeichnis vorangestellt sein.

Das **Literaturverzeichnis** soll nur die im Text zitierten Publikationen enthalten. Bei Übersichtsarbeiten sollten nicht mehr als 60 Literaturstellen angegeben werden, bei Kurzen Übersichten nicht mehr als 40. Die Literatur ist entsprechend der Reihenfolge des Auftretens im Text zu numerieren. Die Referenzen werden als Zahlen in eckigen Klammern in den Text eingefügt. Bitte alle Autoren einer Arbeit angeben, die Abkürzung «et al.» ist zu vermeiden. Zitiert werden sollen nur Arbeiten, die entweder publiziert oder zur Publikation angenommen sind. Der Passus «in Vorbereitung» oder «persönliche Mitteilung» ist zu vermeiden. Die Abkürzung der Zeitschriftentitel erfolgt nach dem Index-Medicus-System (www.nlm.nih.gov).

Beispiele zur Abfassung von Literaturzitaten:

(a) Zeitschriftenbeiträge: Lelie PN, Cuypers HTM, van Drimmelen AAJ, Quint WGV: Quality assessment of hepatitis C virus nucleic acid amplification methods. An international proficiency study. Infusionsther Transfusionsmed 1998;25:102–110.

(b) Bücher mit Herausgeber: Hardy WD Jr, Essex M: FeLV-induced feline acquired immune deficiency syndrome: A model for human AIDS; in Klein E (ed): Acquired Immunodeficiency Syndrome. Prog Allergy. Basel, Karger, 1986 vol 37, pp 353–376.

(c) Monographien: Schaefer UW, Beelen DW: Bone Marrow Transplantation, ed 2, revised. Basel, Karger, 1996.

In **Abbildungen** und **Tabellen** genannte Informationen bedürfen keiner Wiederholung im Text. Hinweise auf Abbildungen und Tabellen erfolgen im Text mit (Abb. 1) bzw. (Tab. 1). Abbildungslegenden sind auf einer gesonderten Seite zusammenzustellen; jede Tabelle muß eine Legende haben. Die Legenden sind kurz und präzise zu halten. **Farbabbildungen** sind möglich, erfordern jedoch einen Zuschuß des Autors zu den Druckkosten (s. unten). **Abbildungsvorlagen** sind 4fach einzureichen (Original und 3 Kopien). Auf der Rückseite der Vorlage (nicht aufkleben oder anheften) bitte den Namen des Autors, ein Titelstichwort, die Angabe, wo «oben» ist, sowie die Bildnummer angeben. Die Qualität der gedruckten Abbildungen hängt von der Qualität der Vorlagen ab.

Randomisierte, kontrollierte klinische Studien sollen nach den Empfehlungen des CONSORT-Statements [Begg C, Cho M, Eastwood S, Horton S, Moher D, Olki I, Pitkin R, Rennie D, Schulz F, Simel D, Stroup D: Improving the quality of reporting of randomized trials: The CONSORT statement. JAMA 1996;275:637–639] abgefaßt werden. Das Verfahren, nach dem die Vergleichsgruppen ermittelt wurden, muß erkennbar sein, so daß der Leser die Methode der Randomisierung nachvollziehen kann. Im Titel und in der Zusammenfassung der Arbeit ist darauf hinzuweisen, daß es sich um eine randomisierte kontrollierte klinische Studie (RCT) handelt. Wird eine RCT eingereicht, sind die «Checklist for Authors Submitting Randomized Controlled Trials» und das CONSORT-Flußdiagramm zu beachten, welche im ersten Heft jedes Jahrgangs in der englischen Version dieser Hinweise abgedruckt sind. Die CONSORT-Checklist und das CONSORT-Flußdiagramm sind auch unter www.ama-assn.org/public/journals/jama zugänglich, eine deutsche Version mit weiteren Informationen kann unter www.med.uni-muenchen.de/gmds/pub/consort.html heruntergeladen werden. Die ausgefüllte Checkliste und das Flußdiagramm müssen der Arbeit beigelegt werden, da sie in das Begutachtungsverfahren einbezogen werden.

Druckfahnen: Das lektorierte Manuskript und die Druckfahnen werden dem Korrespondenzautor zusammen mit einem Formular zur Sonderdruckbestellung zugesandt. Die Fahnen sind sorgfältig mit dem Manuskript zu vergleichen und notwendige Korrekturen auf der Fahne einzutragen. Die Korrespondenz betreffend angenommene Arbeiten und Druckfahnen bitte direkt mit dem S. Karger Verlag in Freiburg führen.

3. Kosten und Sonderdrucke

a) Sollte die Arbeit den vorgegebenen Umfang überschreiten (siehe «2. Manuskripte» und vom Autor nicht gekürzt werden, wird der Mehrumfang mit DM 650,– pro Druckseite berechnet.
b) Für Farbabbildungen wird dem Autor ein Reproduktions- und Druckkostenanteil von DM 450,– pro Abbildung in Rechnung gestellt.
c) Übersteigen die Kosten der Autorkorrektur 10% der Satzkosten, werden die Zusatzkosten dem Autor in Rechnung gestellt.

Der Korrespondenzautor erhält 25 kostenlose Sonderdrucke seiner Arbeit.

Die vollständigen Hinweise erscheinen im ersten Heft eines Jahrgangs.

KARGER

Fax +49 761 4 52 07 14
E-mail Information@Karger.de
www.karger.com

© 2001 S. Karger GmbH, Freiburg

Accessible online at:
www.karger.com/journals/onk

Hinweise für Autoren

ONKOLOGIE

Onkologie 2001;24

ONKOLOGIE is a peer-reviewed journal intended to provide a forum for presentation of all aspects of oncology, including reviews and original articles from internal hematology and oncology, surgical, gynecological and urological oncology, radiation therapy as well as clinical-experimental oncology. Contributions to pediatric, ENT and dermatological oncology, pathology, biometry, epidemiology, prevention, early diagnosis and psychological care will also be considered.

1. General Information and Publishing Rights

Manuscripts for publication should be submitted in quadruplicate (each copy accompanied by a complete set of figures and tables) to S. Karger Publishers, attn.: Dr. Heike Endemann, Postfach, D-79095 Freiburg. Each author should always keep a copy of the manuscript.

In a **covering** letter all authors must confirm with their signature that the submitted paper or parts of it (with the exception of congress reports) were not previously published and are not submitted for publication elsewhere. All authors have to confirm that they have read the manuscript and agree to have it published in the submitted form. It is the authors' responsibility to obtain permission from other publishers to reproduce illustrations, tables, etc. Acceptance of a manuscript is based on the evaluation by several reviewers (peer review). Once the manuscript is accepted for publication, all rights are retained by the publisher. The publication, or parts of it, may neither be translated into other languages nor reproduced by any mechanical or electronic means (including photocopying, recording, and microfilm) nor stored in a retrieval system without the publisher's written permission.

Ethical and Legal Prerequisites: Manuscripts involving examinations of probands and patients must include a statement that the trial protocol has been approved by an ethical committee and thus meets the standards of the Declaration of Helsinki in its revised version of 1975 and its amendments made in 1983, 1989, and 1996 [JAMA 1997;277:925–926]. Accordingly, the proband's informed consent has to be included in the manuscript. Information suitable to reveal the proband's identity is to be avoided. Experiments involving animals demand consent as well and must meet the requirements of current German laws for the protection of animals. A reference to this approval must be included in the manuscript. Articles from countries other than Germany must include a comparable note.

Categories: Contributions are classified as Review Article, Short Review, Original Article, Ongoing Clinical Trial, Short Communication, Casuistic Contribution, Letter to the Editors, Commentary, Opinion, Oncological Education, Clinical Information or Congress Report, or Wilsede School. Review articles, original articles, ongoing clinical trials, short communications, casuistic contributions, letters to the editors, commentaries and editorials must be written in English (American or British spelling and grammar). All other contributions may be submitted either in German or in English.

2. Preparation of Manuscripts

Please print the manuscript $1^1/_2$-spaced with 3-cm margins (left and right). The text should be limited to the following numbers of pages:

	Printed pages	Equals approximately manuscript pages (including summary, illustrations, tables, references)
Review Article, Original Article	5	14
Short Review, Casuistic, Contribution	3	8
Ongoing Clinical Trial Oncological Education, Clinical Information	4	11
Letters to the Editors	1	3
Others	2	6

Each additional complete or partial page to print may be charged to the author (see below). Please check the final copy carefully for consistent notations and correct spelling. Accepted manuscripts are edited according to the publisher's style. General information for the preparation of manuscripts may be found in 'Uniform Requirements for Manuscripts Submitted to Biomedical Journals' by the International Committee of Medical Journal Editors as published in Ann Intern Med 1997;126:36–47 (available from: www.asma-assn.org/public/journals/jama/sc6336.htm).

Electronic Manuscript Submission: Each manuscript must be accompanied by a disk exactly matching the printed version. Please see 'Guidelines for Electronic Manuscript Submission' in the first issue of each volume for assistance in preparing manuscripts on disk.

All pages should be numbered consecutively, starting with the title page. Articles should be structured as follows: **Key Words and Summary, Schlüsselwörter** and **Zusammenfassung, Introduction, Material and Methods (**optionally **Patients and Methods), Results, Discussion,** optionally **Acknowledgements, References, tables, figure legends,** and **figures.**

The **title page** must include the following information: Title and short title in English and German, each author's full name, followed by hospital, institute, etc.; exact postal address of the corresponding author including phone and fax numbers as well as E-mail address.

Original articles must contain an English **and** a German (translation by the editor is possible) **structured summary** (for more information refer to Haynes RB, Mulrow CD, Huth EJ, Altman DG, Gardner MJ: Ann Intern Med 1990;113: 69–76) of not more than 250 words in four paragraphs: **Background, Material and Methods, Results, Conclusions.** Summaries should not include any references, tables, figures, or references to figures and tables. If it is inevitable to cite a reference, complete publication data must be given. Each article must have 3–5 **key words** (English and German). If possible, use terms from the Medical Subject Headings List (MeSH) of Index Medicus (available at www.nlm.nih.gov). Review articles do not adhere to the framework of original articles; however, the title page, summaries, key words, and a list of references are necessary.

Footnotes should be used only for tables. Please use **abbreviations** as economically as possible. Special abbreviations should be written in full when first mentioned, followed by the abbreviation in parentheses, e.g. adenosine monophosphate (AMP). When first mentioned, **registered trade names** must be marked with the superscript registration symbol ® or ™. It is the sole responsibility of the author to ensure that this rule is followed properly. **Acknowledgements** should be placed directly after the text and before the list of references.

The list of **references** should include only the publications cited in the text. References must be numbered sequentially in the order cited in the text, do not alphabetize. Insert textual references in square brackets. In each reference all authors should be mentioned; the abbreviation 'et al.' is to be avoided. It is recommended to quote only articles already published or accepted for publication. Please avoid notices like 'in preparation' or 'personal communication'. Authors are responsible for the accuracy and completeness of their references and for correct text citation. The journal titles should be abbreviated according to the Index Medicus system (www.nlm.nih.gov). For references, please follow the examples given:

(a) Papers published in periodicals: Lelie PN, Cuypers HTM, van Drimmelen AAJ, Quint WGV: Quality assessment of hepatitis C virus nucleic acid amplification methods. An international proficiency study. Infusionsther Transfusionsmed 1998;25:102–110.

(b) Edited books: Hardy WD Jr, Essex M: FeLV-induced feline acquired immune deficiency syndrome: A model for human AIDS; in Klein E (ed): Acquired Immunodeficiency Syndrome. Prog Allergy. Basel, Karger, 1986, vol 37, pp 353–376.

(c) Monographs: Schaefer UW, Beelen DW: Bone Marrow Transplantation, ed 2, revised. Basel, Karger, 1996.

The number of **figures** and **tables** should be kept low. They are listed in the text as (fig. 1) and (table 1). Data explained in tables need not be repeated in the text. Figure legends must be listed together on a separate page. Figure and table legends should be kept short and precise. **Color illustrations** are reproduced at the author's expense (see below).

Figures should be submitted in quadruplicate, 1 original and 3 copies. Please do not fasten them, and mark on the back the first author's name, a short form of the title, the 'top' instruction, and the number of the figure. The quality of the illustrations printed depends on the quality of the originals provided.

Reports on Randomized Controlled Trials (RCTs) should be prepared according to the CONSORT statement [Begg C, Cho M, Eastwood S, Horton S, Moher D, Olki I, Pitkin R, Rennie D, Schulz F, Simel D, Stroup D: Improving the quality of reporting of randomized trials: The CONSORT statement. JAMA 1996;276:637–639].The report must state explicitly how the comparison groups were assembled, allowing the reader to access the method of randomization. Please specify in the title and summary that the manuscript is a report on an RCT. When an RCT is submitted, the authors should refer to the checklist 'Checklist for Authors Submitting Randomized Controlled Trials' reprinted in the guidelines of the first issue of a volume. Checklist and flow diagram are also available at: www.ama-assn.org/public/journals/jama. Both should be submitted with the manuscript, as they will be included in the review process.

Galley Proofs: The edited manuscript and the galley proofs are forwarded together with a reprint order form to the corresponding author. The author is asked to verify the galleys against the manuscript and to mark appropriate corrections in the galley proofs. Correspondence regarding accepted manuscripts and proofs should be directed to the publisher.

3. Costs and Reprints

a) If a paper exceeds the allotted size (see «2. Preparation of Manuscripts» and will not be abridged by the author, each additional printed page is charged to the author at DEM 650.00.
b) Color illustrations are reproduced at the author's expense. Each one is charged to the author at DEM 450.00 as a contribution to the additional reproduction and color printing costs.
c) The author will be charged for corrections to the galley proofs entailing expenses in excess of 10% of the original typesetting costs.

The corresponding author will receive 25 reprints free of charge.

The complete guidelines are printed in the first issue of each volume.

KARGER

Fax +49 761 4 52 07 14
E-mail Information@Karger.de
www.karger.com

© 2001 S. Karger GmbH, Freiburg

Accessible online at:
www.karger.com/journals/onk

Karger journals in . . .

. . . Oncology

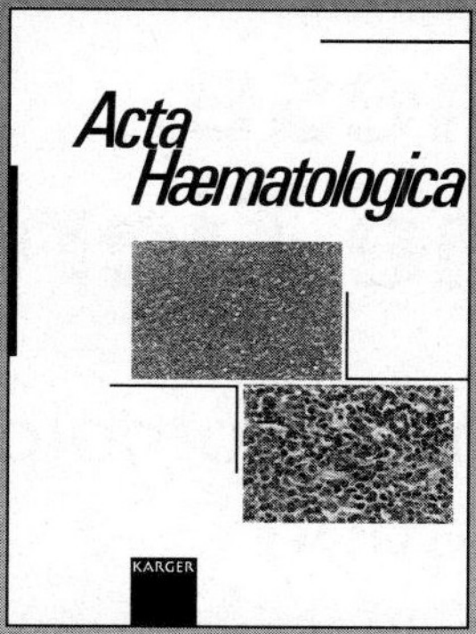

From hairy cells and sickle cells to heavy chains and neutrophils – uniting all lines of evidence relevant to blood disease

Up-to-date information on the efficacy, toxicology and interaction of new drugs in various applications

Assumes the invaluable function of accelerating the consolidation of experimental and clinical cancer research

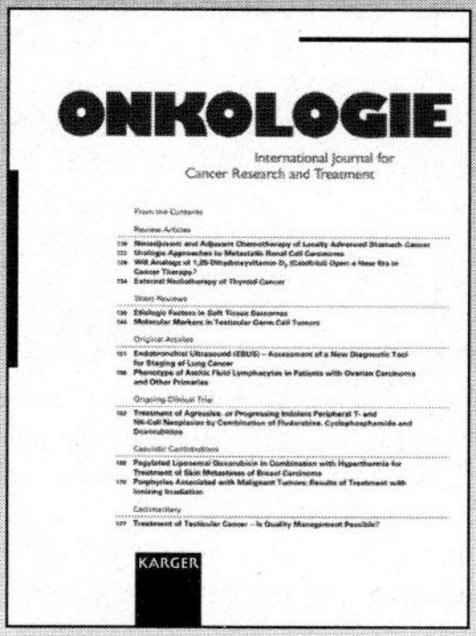

An international journal giving a complete overview of all aspects of basic and clinical cancer research and treatment

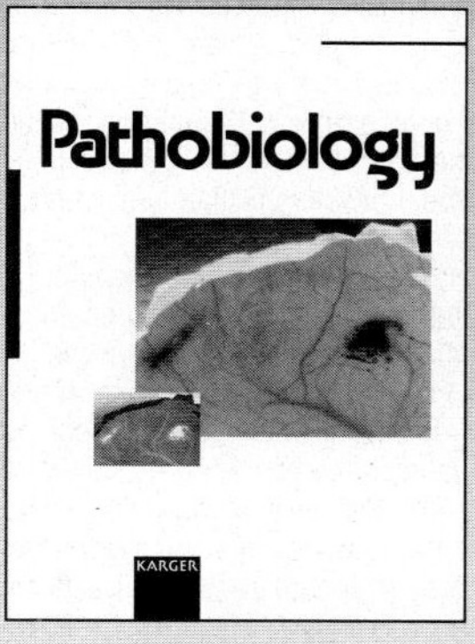

Original reports and interpretive reviews address timely topics in pathobiology, immunopathology, cellular and molecular biology, and medicine

An international forum publishing original data and reviews in experimental and clinical cancer research, with a focus on translational research and tumor markers

Free sample copies available!

KARGER

S. Karger AG
P.O. Box, CH–4009 Basel
(Switzerland)
Tel. +41 61 306 11 11
Fax +41 61 306 12 34
E-mail karger@.karger.ch

S. Karger Publishers, Inc.
26 West Avon Road
P.O. Box 529
Farmington, CT 06085 (USA)
Toll free: 1-800-828-5479
Fax (860) 675-7302

For easy ordering or further information on Karger publications log on to:
www.karger.com

KI 02209

Zur Einsicht in den geringsten Teil ist die Übersicht über das Ganze nötig *(J.W. v. Goethe)*

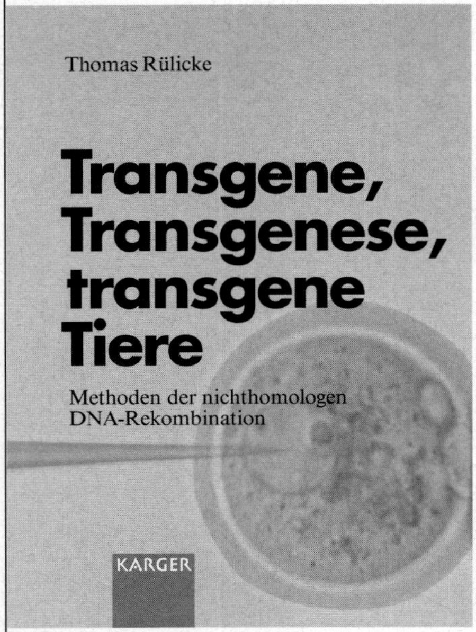

T. Rülicke

Transgene, Transgenese, transgene Tiere

Methoden der nichthomologen DNA-Rekombination

Genetik; Biotechnologie, Humangenetik, Molekularbiologie

Inhalt

1. Die Labormaus in der experimentellen Genetik
2. Zum Inhalt der Monographie
3. Die Entwicklung von Techniken zur Transgenese bei Labormäusen
4. Die Vorkerninjektion zur Transgenese bei Säugetieren
5. Transgenese via retrovirale Vektoren
6. Spermien als Vektoren für transgene DNA-Moleküle
7. Transgene Labormäuse durch nichthomologe Rekombination in ES-Zellen
8. Transchromosomale Labortiere
9. Ausblick
10. Literatur
11. Sachwortverzeichnis

Der Begriff *Transgenese* umfasst eine Reihe von Techniken des Gentransfers, deren Ziel die stabile Integration isolierter genetischer Elemente *(Transgene)* in das Genom eines Organismus ist. Auf diese Weise erzeugte, sogenannte *transgene Tiere* bieten ein weites Spektrum möglicher Anwendungen. Von besonderem Interesse sind sie in der biomedizinischen Forschung und Entwicklung, wo sie einen entscheidenden Beitrag zur Aufklärung der in vivo Regulation und Funktion von Genen innerhalb komplexer genetischer, biochemischer und/oder physiologischer Netzwerke leisten.

Gegenwärtig nutzt man zum Gentransfer (1) die Infektion von Embryonen in frühen Stadien mit viralen Vektoren, (2) die DNA-Mikroinjektion in die Vorkerne befruchteter Eizellen und (3) die Transformation embryonaler Stammzellen mit nachfolgender Entwicklung von Keimbahnchimären. Für verschiedene Tierarten ohne etablierte Stammzellkulturen gelang es unlängst, (4) mittels Kerntransfer aus transformierten Fibroblasten in Oozyten (Klonen) transgene Nachkommen zu erzeugen.

Die stabile Integration eines Transgens erfolgt durch Mechanismen der Rekombination mit der DNA des Wirtsgenoms. Prinzipiell unterscheidet man dabei die homologe Rekombination, d.h. Rekombination zwischen gleichen Sequenzen (legitim), und die nichthomologe Rekombination, d.h. die Insertion des Transgens in eine ungleiche DNA-Sequenz (illegitim).

Die vorliegende Monographie beschäftigt sich vorrangig mit der Biologie der Transgenese nach dem Prinzip der nichthomologen DNA-Rekombination. Dabei wird der Vorkerninjektion, der direktesten und am häufigsten genutzten Technik, besonderes Augenmerk gewidmet. Diskutiert werden nicht nur Strategien zur Entwicklung transgener Tiere, sondern auch deren genetische Besonderheiten. Für die kompetente Zucht und Beurteilung transgener Tiere ist das Verständnis dieser Strategien und Besonderheiten unerlässlich.

Das vorliegende Buch richtet sich an Wissenschaftler mit biologisch-medizinischer Ausrichtung, deren Forschung bzw. Interesse die Erzeugung und experimentelle Nutzung transgener Tiere einschliesst.

KI 01226

www.karger.com/genetics

Rülicke, T. (Zürich):
Transgene, Transgenese, transgene Tiere
Methoden der nichthomologen DNA-Rekombination
X + 166 S., 29 Abb., 1 in Farbe, 5 Tab., broschiert, 2001
CHF 52.– / DEM 68.– / USD 45.25
Preisänderungen vorbehalten
DEM-Preis nur für Deutschland, USD-Preis nur für USA
ISBN 3–8055–7163–1

KARGER

Bestellformular

Bitte senden Sie: ____ Expl.

Bei Vorauszahlung Porto und Verpackung zu Lasten des Verlags

Ich bezahle:
☐ Mit Scheck ☐ Nach Erhalt der Rechnung

Bitte belasten Sie diese Bestellung meiner Kreditkarte
☐ American Express ☐ Diners ☐ Eurocard
☐ MasterCard ☐ Visa

Kreditkarten.-Nr.: ____________

Gültig bis: ____________

Senden Sie Ihre Bestellung an Ihre Buchhandlung, Agentur, direkt an den Verlag oder an eine seiner Vertretungen.

Fax: +41 61 306 12 34

S. Karger AG, Postfach, CH–4009 Basel (Schweiz)
E-Mail karger@karger.ch, **www.karger.com**

Name/Adresse: ____________

Datum: ____________

Unterschrift: ____________